Bioinspired Materials Science and Engineering

Bioinspired Materials Science and Engineering

Edited by

Guang Yang, Lin Xiao, and Lallepak Lamboni

Huazhong University of Science and Technology, Wuhan, China

WILEY

This edition first published 2018

Registered Offices
John Wiley & Sons, Inc., 111 River Street, Hoboken, NJ 07030, USA

Editorial Office
111 River Street, Hoboken, NJ 07030, USA

For details of our global editorial offices, customer services, and more information about Wiley products visit us at www.wiley.com.

Wiley also publishes its books in a variety of electronic formats and by print-on-demand. Some content that appears in standard print versions of this book may not be available in other formats.

Library of Congress Cataloging-in-Publication Data

Names: Yang, Guang, 1968 April 18– editor. | Xiao, Lin, 1986– editor. |
Lamboni, Lallepak, 1988– editor.
Title: Bioinspired materials science and engineering / edited by Guang Yang,
Lin Xiao, Lallepak Lamboni.
Description: First edition. | Hoboken, NJ : John Wiley & Sons, 2018. |
Includes bibliographical references and index. |
Identifiers: LCCN 2018002997 (print) | LCCN 2018009868 (ebook) | ISBN
9781119390336 (pdf) | ISBN 9781119390343 (epub) | ISBN 9781119390329
(cloth)
Subjects: LCSH: Biomimetics. | Materials–Biotechnology. | Materials science.
| Engineering.
Classification: LCC QP517.B56 (ebook) | LCC QP517.B56 B4796 2018 (print) |
DDC 610.28–dc23
LC record available at https://lccn.loc.gov/2018002997

Cover Design: Wiley
Cover Images: © SergeOstroverhoff/Getty Images; © TonyBaggett/Getty Images; ©vitstudio/Shutterstock; © me4o/Getty Images; Shamrock logo courtesy of Guang Yang

Set in 10/12pt Warnock by SPi Global, Pondicherry, India

Printed in the United States of America

V10002143_072018

Contents

List of Contributors

Mitsuru Akashi
Osaka University
Osaka, Japan

Mosa Alsehli
Taibah University
Madina, Saudi Arabia

Georg Bold
Freiburg Institute for Interactive Materials and Bioinspired Technologies
Freiburg, Germany

Tao Du
Lanzhou Institute of Chemical Physics
Lanzhou, China

Mario Gauthier
Waterloo University
Ontario, Canada

Robert G. Gilbert
Queensland University
Brisbane, Australia

Eric Habib
Université de Montréal
Québec, Canada

Sanming Hu
Huazhong University of Science and Technology
Wuhan, China

Lixia Huang
Huazhong University of Science and Technology
Wuhan, China

Xiaoze Jiang
Donghua University
Shanghai, China

Xingyu Jiang
National Center for NanoScience and Technology
Beijing, China

Tetsuo Kondo
Kyushu University
Fukuoka, Japan

Lallepak Lamboni
Huazhong University of Science and Technology
Wuhan, China

Sixiang Li
Huazhong University of Science and Technology
Wuhan, China

Chun-Yen Liu
Osaka University
Osaka, Japan

Dong Liu
Peking University
Beijing, China

Fengwei Liu
Donghua University
Shanghai, China

Li Liu
Huazhong University of Science and Technology
Wuhan, China

Yuyun Liu
Fudan University
Shanghai, China

Jiu-an Lv
Fudan University
Shanghai, China

Sehrish Manan
Huazhong Agricultural University
Wuhan, China

Lin Mao
Huazhong University of Science and Technology
Wuhan, China

Tom Masselter
University of Freiburg
Freiburg, Germany

Michiya Matsusaki
Osaka University
Osaka, Japan

Frank A. Müller
Friedrich Schiller University Jena
Jena, Germany

Akihiro Nishiguchi
Osaka University
Osaka, Japan

Daniel Van Opdenbosch
Technische Universität München
Munchen, Germany

Zhenchao Qian
Beijing National Laboratory for Molecular Sciences
Beijing, China

Amarachi Rosemary Osi
Ningbo Institute of Material Technology and Engineering
Ningbo, China

Takahiro Sato
Osaka University
Osaka, Japan

Heng Shen
Beijing National Laboratory for Molecular Sciences
Beijing, China

Zhijun Shi
Huazhong University of Science and Technology
Wuhan, China

Tiatou Souho
University of Kara
Kara, Togo
Huazhong University of Science and Technology
Wuhan, China

Olga Speck
University of Freiburg
Freiburg, Germany

Thomas Speck
University of Freiburg
Freiburg, Germany

Lixue Tang
National Center for NanoScience and Technology
Beijing, China

Ken Terao
Osaka University
Osaka, Japan

Marc Thielen
Freiburg Materials Research Centre
Freiburg, Germany

Francisco Vilaplana
KTH Royal Institute of Technology
Stockholm, Sweden

Muhammad Wajid Ullah
Huazhong University of Science and Technology
Wuhan, China

Li Wang
Huazhong University of Science and Technology
Wuhan, China

Ruili Wang
Donghua University
Shanghai, China

Shutao Wang
Technical Institute of Physics and Chemistry
Beijing, China

Xiao-Wei Wang
Peking University
Beijing, China

Yazi Wang
Donghua University
Shanghai, China

Lin Xiao
Huazhong University of Science and Technology
Wuhan, China

Jian Xu
Beijing National Laboratory for Molecular Sciences
Beijing, China

Deyue Yan
Shanghai Jiaotong University
Shanghai, China

Guang Yang
Huazhong University of Science and Technology
Wuhan, China

Yuan Yao
East China University of Science and Technology
Shanghai, China

Guang-Zhong Yin
Peking University
Beijing, China

Yanlei Yu
Fudan University
Shanghai, China

Ning Zhao
Beijing National Laboratory for Molecular Sciences
Beijing, China

Wen-Bin Zhang
Peking University
Beijing, China

Wenfu Zheng
National Center for NanoScience and Technology
Beijing, China

Feng Zhou
Lanzhou Institute of Chemical Physics
Lanzhou, China

Yongfeng Zhou
Shanghai Jiaotong University
Shanghai, China

Meifang Zhu
Donghua University
Shanghai, China

X.X. Zhu
Université de Montréal
Québec, Canada

Cordt Zollfrank
Technische Universität München
Munchen, Germany

Wei Zou
Yangzhou University
Yangzhou, China

Foreword

Centered on materials, this book explores the full scope of products inspired by nature. The process of learning from biological structures and principles for the development of advanced and multifunctional materials as novel resources that revolutionize human life is discussed, presenting fundamental concepts and methods of biofabrication. Examples are offered that showcase currently trending compounds and macromolecules with their properties, their potential, and their contribution to the fabrication of bioinspired materials. Concrete applications are discussed as well with an accent on biomedically engineered materials, that will take the reader into the realm of such seductive biomaterials.

With currently captivating topics such as biotemplating, microfluidics, self-assembly, mussel-inspired surface modification, 3D biofabrication and more, this book represents a source of inspiration for the design of novel materials, and an important tool for updating active researchers. Additionally, its comprehensive approach will be of great interest to the beginner in the field who will discover the concept of bioinspiration from its fundamentals to its applications. Although the book emphasizes biomedical engineering, the multidisciplinary aspect of the subject will make it appeal to many research areas, such as biologists and engineers, while not leaving out chemists, physicists, and technicians.

Although an old concept, by proposing natural materials with superior features and low cost as models, bioinspiration has re-emerged as an essential tool for overcoming various limitations in current materials science and engineering, thereby solving many of mankind's substantial problems, such as the shortage of resources and the environmental concerns. Hence, this book deals with an important topic of the moment, which concerns numerous researchers across the world and should also be of interest to the general public. As illustrated by the authors, many different talents need to come together to make this approach a reality, and this book will inspire, instruct, and involve both current and the next generations in advancing the field.

April 13, 2017

Lei Jiang
Technical Institute of Physics and Chemistry
Chinese Academy of Sciences, China

Preface

Bioinspiration is an old concept which can be described as the process of learning from nature and its biological principles. Taking advantage of the properties and nanostructures of natural compounds, the science of bioinspired materials aims at developing new and formerly non-existent materials, which exhibit novel and multifunctional properties, in the attempt of meeting the current requirements of human well-being. The idea is to take inspiration from natural mechanisms and the problems they are set to solve, in order to design advanced materials which are solutions to problems encountered in human life. Indeed, the focus of materials science is being increasingly shifted towards the development of bioinspired materials, prompted by the shortage of resources, the low cost and superior characteristics of natural materials, and the environmental and climatic concerns. The first step to engineering bioinspired materials is understanding biological materials and the processes involved in their production, and thence, develop biofabrication or bioinspired fabrication approaches. This leads to the highly interdisciplinary character of this field, which brings together natural scientists (biologists, chemists, and physicists), engineers, and technicians. Thus, an active interaction across disciplines is the key to the real development of this old research area, which is now attracting many researchers worldwide. However, as underlined by several recent reviews on the subject, this condition is yet to be fully met, due to the rather limited understanding of the building principles of living entities which are numerous and complex, and because the definition of the scope and novel applications remain to be further clarified. Hence, approaches for conveying information in the field and storing the bioinspired solutions already uncovered are of real importance, and would contribute significantly in propelling this promising research area.

Biofabrication approaches are developed by studying and exploiting unique and basic biological aspects, including evolution, growth, and structure (formation and performance), which are non-existent in engineering materials. Based on the "growth and functional adaptation" concepts, the strategies adopted aim at creating hierarchical structures and self-assemblies (dynamic strategies), while those associated with the "damage repair and healing" principles design self-repair or self-healing materials.

The purpose of this book is to introduce a comprehensive view of the bioinspired materials science and engineering, discussing biofabrication approaches and applications of bioinspired materials as they are fed back to nature in the guise of biomaterials. Some biological compounds will also be brought up, as of what is learned from them, and how they can be useful in the engineering of bioinspired materials. Thus, this book will include 3 main sections: biofabrication, biomacromolecules, and biomaterials. Illustrating the bioinspiration process from materials design and conception to application of bioinspired materials, this book will present the multidisciplinary aspect of the concept, and represent a typical example of how knowledge is acquired from nature, and how in turn this information contributes to biological sciences, with an accent on biomedical applications. We anticipate that this book will be suitable for different classes of the scientific community including undergraduate, graduate, and senior researchers in all areas of bioinspired materials. We hope that it will stimulate new thoughts and research in this field.

We would like to acknowledge the National Natural Science Foundation of China for the financial support on this book. Then we would like to express our appreciation to China-Germany Center for Science, which supported the Sino-German Sympoisum on Bioinspired Materials and Engineering held from May 11-15, 2014 in Wuhan, China, co-chaired by Prof. Guang Yang (HUST) and Prof. Cordt Zollfrank (TUM). This memorable symposium brought together outstanding scientists working in the bioinspired material field from China, Germany, Japan and the rest world. It is this symposium that first inspired the conception of this book. Many of the symposium attendees then accepted our invitation to contribute to this book. We offer special thanks to them. We also would like to express our appreciation to all contributing authors and to staff at John Wiley & Sons, Inc., for their patience and never-failing support of this project. Finally, we would like to express our sincere gratitude to Prof. Lina Zhang at Wuhan University, Prof. Lei Jiang at Technical Institute of Physics and Chemistry CAS, and Prof. Deyue Yan at Shanghai Jiaotong University for their on-going support and guidance.

Guang Yang, Lin Xiao, Lallepak Lamboni
February 2018

Introduction to Science and Engineering Principles for the Development of Bioinspired Materials

Muhammad Wajid Ullah[1,2], *Zhijun Shi*[1,2], *Sehrish Manan*[3], *and Guang Yang*[1,2,*]

[1] *College of Life Science and Technology, Huazhong University of Science and Technology, Wuhan, China*
[2] *National Engineering Research Center for Nano-Medicine, Huazhong University of Science and Technology, Wuhan, China*
[3] *College of Plant Sciences and Technology, Huazhong Agricultural University, Wuhan, China*

I.1 Bioinspiration

Bioinspiration refers to the process of learning from nature and its biological principles. The science of bioinspired materials aims to develop novel functional materials with advanced and multi-functional properties by using the nano-, micro-, meso-, and macro-structures and features of natural materials with the aim to meet the requirements of human well-being. Natural mechanisms and biological materials can be exploited to design advanced materials to solve the problems encountered in human life. Indeed, the focus of materials science is being increasingly shifted toward the development of bioinspired materials, prompted by the shortage of resources, the low cost, and the superior characteristics of natural materials, and the environmental and climatic concerns. To this end, understanding the biological phenomenon, natural biological materials, and the processes involved in their natural production is essential, and hence, developing biofabrication or bioinspired fabrication approaches.

I.2 Bioinspired Materials

Bioinspired materials are synthetic products fabricated to mimic the structure and mechanical features of natural biological materials [1]. Biological materials are inherently multi-functional in nature but may have evolved to optimize a principal mechanical function such as the impact of fracture resistance, for armor and protection, for sharp and cutting components, for a light weight for flight, or special chemical and mechanical extremities for reversible adhesive purposes. These functions are regulated by the nano-, micro-, meso-, and macro-structures of the materials. Further, these structures determine the mechanism of the biological systems to adapt themselves to the external mechanical stimuli. These inherent functions and structural properties are inspiring scientists and engineers to design novel multi-functional synthetic materials with a wide range of structural features and a broad spectrum of potential applications. In the past few decades, several natural biological materials have been examined by researchers for various aspects to explore their potential in different fields. Studies reveal that the inherent multi-scale structures of natural biological materials possess several functions. Nature as a school for scientists and engineers has served as a great source of inspiration to fabricate new materials [2]. At present, biomimetic and bioinspired approaches have been adopted for the fabrication of several biological materials with multi-scale structures for function integration, as summarized in Table I.1. An interdisciplinary collaboration of materials science and engineering, chemistry, biology, physics, and bioinformatics, etc. may lead to the design and fabrication of novel multi-functional bioinspired materials.

To date, several biofabrication approaches have been developed by studying and exploiting unique and basic biological aspects, including evolution, growth, and structure (formation and performance) which are not found in engineering materials. Based on the "growth and functional adaptation" concepts, the strategies adopted mainly aim at creating hierarchical structures and self-assemblies (dynamic strategies) and those associated with the "damage repair and healing" principle designs, and self-repair or self-healing materials. To achieve these objectives, several models have been presented by the researchers to describe the design, fabrication, and optimization of properties of bioinspired materials. Modeling of biological materials helps in rational understanding of

* Email: yang_sunny@yahoo.com

Bioinspired Materials Science and Engineering, First Edition. Edited by Guang Yang, Lin Xiao, and Lallepak Lamboni.

Table I.1 Typical biological materials with function integration.

Biological materials	Functions	Ref.
Butterfly wing	Superhydrophobicity, directional adhesion, structural color, self-cleaning, chemical sensing capability, fluorescence emission functions	[3–7]
Brittlestar	Mechanical and optical functions	[8]
Cicada wing	Anti-reflection, superhydrophobicity	[9]
Fish scale	Drag reduction, superoleophilicity in air, superoleophobicity in water	[10]
Gecko foot	Reversible adhesive, superhydrophobicity, self-cleaning	[11]
Lotus leaf	Superhydrophobicity, low adhesion, self-cleaning	[12]
Mosquito compound eye	Superhydrophobicity, anti-reflection, anti-fogging	[13]
Nacre	Mechanical property, structural color	[14, 15]
Peacock feather	Structural color, superhydrophobicity	[16]
Polar bear fur	Optical property, thermal insulation	[17]
Rice leaf	Superhydrophobicity, anisotropic wettability	[12]
Rose petal	Superhydrophobicity, structural color, high adhesion	[18–20]
Shark skin	Drag reduction, anti-biofouling	[21]
Spicule	Mechanical and fiber-optical properties	[22–24]
Spider capture silk	Water collection ability, mechanical property, elasticity, stickiness	[25–27]
Spider dragline silk	Mechanical property, supercontraction, torsional shape memory	[28–35]
Water strider leg	Durable and robust superhydrophobicity	[36]

Source: Reproduced from [2] with permission from Elsevier.

the design principles which can lead to subsequent designing of bioinspired complements. For example, mechanical modeling of biological materials based on natural materials has attracted immense attention owing to their diverse applications in medicine and engineering. This can be attributed to the structurally hierarchical biomaterials which possess a highly desirable structure-properties relationship and can serve as templates for the fabrication of bioinspired materials. Several approaches, such as single- and multi-scale, micro-structural and phenomenological, and continuum and discrete, etc. have been developed for the mechanical modeling of biological and bioinspired materials [37]. However, further extensive research is required to fabricate bioinspired materials due to their greater flexibility in design variables, such as the selection of material components, the varying degree of constraints among the different available components, the variable boundary conditions, and the novel architectural conformations.

I.3 Biofabrication

Biofabrication is the combination of two words: "bio" means living and "fabrication" means to synthesize or design using templates etc., thus biofabrication refers to the synthesis of living structures using some standard templates or models. Precisely, biofabrication refers to the application principle of engineering and information science to produce an automated robotic assembly of living cells, tissues, and organs, etc. [38]. Further narrowing down the concept, biofabrication refers to the biomedical applications of rapid prototyping or computer-aided additive technologies. It is closely related to tissue engineering and is considered an integral part of it and uses engineering approaches in the assembly of complex tissues and organs. Despite extensive developments in the field of tissue engineering, the transformation of this labor-intensive technology into an automated industry still requires further innovative and creative strategies.

I.3.1 Summary of Part I Biofabrication

In Part I, "Biofabrication," we discuss various biotemplating principles and recent advances in the one-dimensional and two-dimensional biotemplated formation of inorganic functional materials using natural templates. The chapters in Part I (Chapters 2–6) also discuss microbial-mediated material manufacturing techniques for the fabrication of a variety of functional materials. Recently developed tubular structures are discussed, which serve as templates for *in vitro* recapitulating of highly complex tissues such as blood vessels, etc. and microfluidics-based cell manipulation and development of tubular tissues. This Part also illustrates the fabrication of three-dimensional (3D) tissues with capillary networks by controlling the cell microenvironment with emphasis on 3D-tumor invasion models with blood- and lymph-capillary networks. Furthermore, biofabrication of ordered cellulose scaffolds (nematic ordered) to mediate 3D cell culturing and biomineralization is discussed. As an example of bioinspiration, the preparation and application of biomimetic materials inspired by muscle adhesive proteins are overviewed in detail. Finally, the self-assembly of poly(lactic acid)-based amphiphilic diblock copolymers and their applications in biomedical field are presented.

I.4 Biofabrication Strategies

Biofabrication strategies mainly aim to improve the existing strategies and develop reliable biomaterials-based cell culturing strategies for advances in tissue engineering and regenerative medicines. To achieve such goals, scaffolds have been developed from various biocompatible materials. A scaffold refers to a temporary structure made of biocompatible material and provides support to the growing cells. A scaffold is declared biocompatible when it remains in direct contact with living host tissues without causing any toxic, allergic, or side effects. Scaffolds with well-defined 3D topologies and geometries have been fabricated to introduce various biological molecules with various shapes and sizes. Tissue engineering applications of scaffolds require high porosity, tunable pore sizes, and better mechanical features. For example, scaffolds with large pore sizes allow easy penetration of the impregnating materials, the diffusion of nutrients, the removal of wastes, and the exchange of gases, etc. Further, an ideal scaffold supports adhesion, proliferation, and migration of cells [39]. The following sections describe a few conventional and advanced biofabrication strategies.

I.4.1 Conventional Biofabrication Strategies

To date, a multitude of fabrication strategies have been devised to fabricate 3D scaffolds using various natural and synthetic materials, mainly polymers. These strategies aim to design scaffolds in such a way as to mimic the natural environment of a living cell. To achieve this goal, earlier scaffolds were fabricated followed by the seeding of viable cells. The following overviews some of these strategies.

I.4.1.1 Solvent Casting Strategy

In this strategy, a polymer solution prepared in an appropriate solvent with uniformly distributed salt particles (i.e. porogen) of known size is poured into a mold and the solvent is allowed to evaporate, leaving behind a composite with uniformly distributed salt particles [40]. Thereafter, the composite is immersed in water to allow the leaching out of the salt particles, leaving behind pores according to the size and shape of the salt particles. Thus, a highly porous uniform 3D scaffold is formed on which different types of cells can be seeded. It is worth mentioning here that the size and shape of the pores are directly related to the size and shape of the salt particles, respectively. The size and shape of the pore can be optimized according to the type of cells and specific application. Further, the solvent used should be non-toxic to the seeding cells.

I.4.1.2 Freeze-drying or the Lyophilization Method

In this strategy, the temperature of a polymer solution is lowered well below its freezing point which results in the solidification of the solvent molecules and leads to the aggregation of the polymer within the interstitial spaces of the scaffold matrix. Thereafter, the solvent molecules are allowed to evaporate via sublimation, leaving behind a highly porous polymeric structure containing well-distributed interconnected pores on the surface and within the matrix of scaffold [41]. Different types of cells can be seeded with the formed interconnected pores. It is worth mentioning here that the pore size of the scaffold depends upon the freezing regime, the concentration of the polymeric material, the size of the ice crystals formed, and the pH of the solution [42].

I.4.1.3 Gas Foaming

Gas foaming is another biofabrication strategy where a polymeric scaffold is first completely saturated using a foaming agent at high pressure, followed by the release of pressure, which results in the solubility of the gas in the polymer. The gas bubbles are formed which grow in the polymer due to the thermodynamic instability [43]. Different types of foaming agents such as CO_2 [44], N_2 [45], or H_2O [46] are used for such purposes, which results in highly porous structures with varying pore size in the range of 100–500 μm [47].

I.4.2 Advanced Biofabrication Strategies

Advanced biofabrication strategies are classified into bioprinting and photolithographic techniques.

I.4.2.1 Bioprinting

Bioprinting is one of the most advanced and innovative technology of this century which has received growing interest worldwide and revolutionized the medical technology and pharmaceutical industries [48]. It refers to the use of 3D printing technology to print various biomaterials with incorporated viable cells to engineer tissue construct applications in tissue engineering and regenerative medicines. Currently, this technology has received immense attention and is widely used for broad spectrum applications, such as regenerative medicines, tissue engineering and transplantation, screening of drugs, and cancer research, etc. It offers several advantages, such as the precise and controlled deposition of cells, hormones, drugs, and growth factors, etc., thus directing improved tissue formation. Further, it provides a base for the development of tissue constructs, organs and organoids, and organ-on-a-chip mimicking the natural ones [49]. Bioprinting is carried out using a 3D printer, which has the ability to print 3D structures such as tissues and organs, etc. using various bioink solutions. A bioink solution refers to a mixture of biomaterial and live cells. A general bioprinting process is shown in Figure I.1. A typical 3D bioprinter has the ability to simultaneously dispense various biomaterials to fabricate structures with high resolution and accuracy and maintain high degree of freedom motion and ensure sufficient motion speed. These 3D printers are user-friendly, fully automatic, easily sterilized, affordable, durable, versatile, and compact instruments [50]. Bioprinting technology is advancing rapidly, however, the technological modalities are based on three fundamental strategies including the inkjet or droplet, extrusion, and laser-based bioprinters, which are described as follows.

I.4.2.1.1 Droplet-based Bioprinting

The droplet-based bioprinting strategy is based on the thermal, piezo, or acoustic-driven mechanisms and uses heat energy, electrical energy, and sound energy, respectively for the generation of droplets of cell suspension in a high-throughput fashion. These bioprinters have received immense attention owing to their simplicity, versatility, agility, and high-throughput potential to dispense a variety of biologics, such as viable cells, growth factors, genes, and pharmaceutics, etc. [51]. These types of printers have a high speed of fabrication of scaffolds, however, this high speed make the strategy difficult to apply to most of the polymer systems as it requires the gelation time to be in accordance with the drop deposition time.

I.4.2.1.2 Extrusion-based Bioprinting

The extrusion-based bioprinting system is a hybrid of a fluid-dispensing system and an automated robotic system for extrusion and bioprinting, respectively. These bioprinting systems use mechanical or pneumatic-driven systems and deposit the viable cells in the form of a filament [52]. In this system, the bioink is dispensed using a

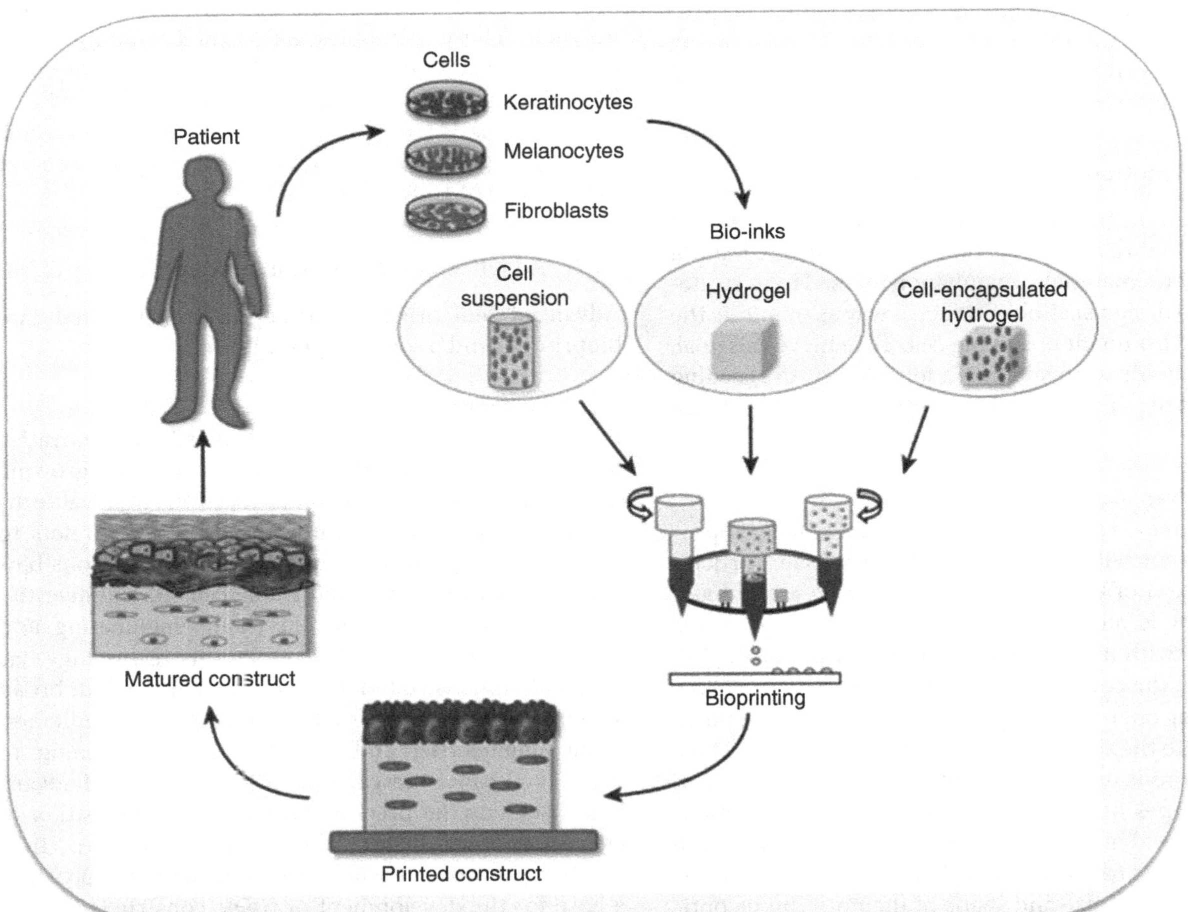

Figure I.1 Illustration of isolation of viable cells, bioprinting of tissue constructs, and implantation into the patient. *Source:* Reproduced from [148] with permission from "Cell-press."

deposition in a computer-aided design (CAD) system which ensures the precise dispensing of viable cells encapsulated in a cylindrical filament. During the printing process, a stage or a surface is moved in a directed pattern controlled by CAD, which ensures the spatial deposition of bioink from a nozzle to fabricate materials of specific structural conformations [53].

I.4.2.1.3 Laser-based Bioprinting

Compared to extrusion-based bioprinting, the viable cells from a donor-slide to a receiver-slide are dispensed without the assistance of a nozzle, using laser energy in a laser-based bioprinting system. This modality offers several advantages in dispensing a variety of biologics such as live cells, biomaterials, growth factors, genes and vectors, and drugs, etc. [54].

I.4.2.2 Photolithographic Strategies of Biofabrication

The photolithography bioprinters have been modified from laser-assisted printers. Similar to a laser-assisted bioprinter, the stereolithography modality uses light or photons for the selective solidification of bioink in a layer-by-layer pattern during the fabrication of a scaffold. Usually, these bioprinters use a digital projector that ensures the same printing time even for the complex planes in a structure and thus this system is more advantageous than conventional bioprinters. Further, such printers are simpler in operation, offer high resolution (100 μm) printing in a short time, and maintain high cell viability [55]. These strategies are used for the fabrication of 2D scaffolds for the growth of cells [56] or the encapsulation of cells in a 3D network of polymers [57]. Photolithographic strategies are further classified into mask-based photolithography, stereolithography, and multiphoton lithography. The mask-based photolithographic strategies of biofabrication use a patterned mask to illuminate selected regions of a polymer. For this purpose, the prepolymer solution is exposed to UV light, which results in the polymerization of the exposed regions of the polymer and thus prevents the formation of a network of 3D porous scaffolds. The unnecessary unpolymerized solution is washed out by immersing in a buffer [58]. Similarly, the stereolithography is a maskless photopatterning CAD strategy used for the fabrication of prototypes. In this strategy, the design of the desired scaffold is first developed using a 3D computer drawing software, which is then processed by software and sliced into a number of layers (25–100 μm thick). The information is then passed to the SL apparatus which prints one layer at a time using a UV laser. Similarly, the multiphoton lithography is also a maskless lithographic strategy which uses a focused laser or a confocal microscope [59–61]. This lithographic strategy offers high lateral (x–y) resolution but little to no control over the axial (z) direction. To solve this issue, several photochemistries have been expanded to multiphoton-based approaches with the potential to confine photochemical reactions in 3D orientation. A comparative analysis of various bioprinters in use is shown in Table I.2.

I.5 Part II Biomacromolecules

Part II of the book deals with biomacromolecules. The term "biomacromolecules" refers to the biological molecules with high relative molecular masses whose structure is essentially comprised of multiple repeated units derived from low molecular mass molecules. Generally, a biomacromolecule is synthesized through the polymerization of smaller subunits generally referred to as monomers. Compared to monomers, the macromolecules have exceptionally different physical properties. Similarly, these biomacromolecules are relatively insoluble in water and other common solvents compared to their smaller units and instead form colloids. In general, there are three classes of biomacromolecules discussed in this book: carbohydrates, proteins, and nucleic acids.

I.5.1 Summary of Part II Biomacromolecules

In Part II, "Biomacromolecules," details of the synthesis approaches and applications of electroactive bioartificials are provided. Further, chemical modification of starch and the conformational properties of various linear and cyclic polysaccharide derivatives are discussed. Thereafter, structure, basic properties, and fabrication strategies of silk-based materials with a special emphasis on biomimetic structures are described. Finally, recent developments in polypeptides synthesis by ring-opening polymerization, micro- and nano-structures through the self-assembly of polypeptides, and their applications are presented.

I.5.2 Carbohydrates

Carbohydrates are biological molecules consisting of three main components: carbon, hydrogen, and oxygen. These are represented by an empirical formula $C_m(H_2O)_n$, where m and n can have the same or different values. Chemically, carbohydrates are polyhydroxy aldehydes, ketones, alcohols, acids, their simple derivatives, or their polymers with linkages of the acetal-type. These are categorized according to their degree of polymerization into three main classes: sugars, oligosaccharides, and polysaccharides. Sugars include monosaccharides (e.g. glucose, fructose, galactose, and xylose, etc.), disaccharides

Table I.2 Comparison of four types of bioprinting techniques.

Parameters	Inkjet	Laser-assisted	Extrusion	Stereolithography	Reference
Cost	Low	High	Moderate	Low	[62–65]
Cell viability	>85%	>95%	40%–80%	>85%	[66, 67]
Print speed	Fast	Medium	Slow	Fast	[68–70]
Supported viscosities	3.5–12 mPa/s	1–300 mPa/s	30 mPa/s to above 6×10^7 mPa/s	No limitation	[70–72]
Resolution	High	High	Moderate	High	[63]
Quality of vertical structure	Poor	Fair	Good	Good	[73]
Cell density	Low <10^6 cells/mL	Medium < 10^8 cells/mL	High (cell spheroids)	Medium <10^8 cells/mL	[70]
Representative materials for bioinks	Alginate, PEGDMA, Collagen	Collagen, Matrigel	Alginate, GelMA, Collagen	GelMA, GelMA-PEGDA hybrid hydrogel	[65, 73–77]
Reported applications	Tissue engineering (blood vessel, bone, cartilage, and neuron)	Tissue engineering (blood vessel, bone, skin, and adipose)	Tissue engineering (blood vessel, bone, cartilage, neuron, muscle, tumor) Controlled release of biomacromolecules Organ-on-a-chip	Tissue engineering (blood vessel and cartilage) Organ-on-a-chip	[78–81]

Source: Reproduced from [147] with permission from Elsevier.

(e.g. sucrose, lactose, maltose, and trehalose, etc.), and polyols (e.g. sorbitol and mannitol, etc.). Similarly, oligosaccharides include malto-oligosaccharides (e.g. maltodextrins) and others (e.g. raffinose, stachyose, and fructo-oligosaccharides, etc.). Similarly, the polysaccharides include starches (e.g. amylose, amylopectin, and modified starches, etc.) and non-starches (e.g. cellulose, hemicellulose, pectins, and hydrocolloids, etc.). Common examples of carbohydrates used in biomedical research are briefly described here.

I.5.2.1 Starch

Starch, also known as amylum, is a polymeric carbohydrate comprised of repeated glucose units linked via glycosidic linkages. Generally, it is produced by green plants where it serves as an energy storage material. Chemically, it consists of two types of molecules: the linear and the helical-shaped amylose that accounts for 20–25% of the total starch content, and the branched amylopectin accounting for 75–80% of the total starch content. Amylose is a comparatively long linear chain of α-glucans containing about 99% (1 → 4)-α-linkages and around 1% (1 → 6)-α-linkages. Similarly, amylopectin contains about 95% (1 → 4)-α-linkages and 5% (1 → 6)-α-linkages [82]. Its insolubility in cooled water limits its various applications. This issue has been resolved to great extent by chemical modification, by introducing different functional groups onto its back bone [83]. It possesses several properties which are significant from a biomedical perspective, such as high biocompatibility, biodegradability, and non-toxicity, etc. It has been extensively used for various applications, such as tissue engineering, drug delivery, and enzyme immobilization, etc. [84]. Starch and its derivatives serve as promising materials for various tissue engineering applications, such as artificial skin, scaffolds, bone and cartilage, vascular regeneration, and teeth, etc. owing to its biocompatible, biodegradable, non-toxic, non-immunogenic, and porous structure. Starch has the potential to form nanofibers, highly porous scaffolds, and injectable hydrogels, etc.

I.5.2.2 Cellulose

Cellulose is the most abundant polysaccharide available on Earth and consists of chain of glucose monomers. The molecular formula of cellulose is $(C_6H_{10}O_5)_n$ and is an unbranched homopolysaccharide composed of α-D-glucopyranose units linked by β-(1 → 4) glycosidic bonds (Figure I.2). The linear glucan chain forms highly stable regular intra- and inter-molecular hydrogen bonds which stabilize its reticulate structure [85]. It is produced by various sources such as plants, microbial cells (*Acetobacter, Rhizobium, Agrobacterium, Aerobacter, Achromobacter, Azotobacter, Salmonella, Escherichia,* and *Sarcina*, etc.) [86], and enzymes (the cell-free system) [87]. Compared to microbial cellulose, also known as bacterial cellulose (BC), and bio-cellulose produced by cell-free enzymes system [85], which represent the purest forms, the plant cellulose contains several impurities in the form of lignin and hemicellulose, which necessitate its further treatment by various chemical methods. However, BC and bio-cellulose are directly used in various applications without further pretreatment. Further, these possess unique structural, physico-chemical, mechanical, and biological features, such as high water-holding capacity (WHC) (100–200 times its dry weight), slow water release rate (WRR), higher crystallinity (60–90%), high tensile features (with elastic modulus 1 to 15 MPa and elongation at break 10–30%), an ultrafine fiber network, and moldability into 3D structures. which bestow BC with high potential value [85, 88–90] and broaden its applications in different fields, such as biomedicine, opto-electronics, food technology, and separation processes, etc. [91]. BC is rarely soluble in common solvents such as water and organic and inorganic solvents, owing to its highly extended hydrogen bonding [92, 93]. However, several solvents systems have been developed such as LiCl/dimethylacetamide (DMAc) [94], N-methylmorpholine-N-oxide (NMMO) [40], ionic liquids (ILs) [95], and alkali/urea (or thiourea) aqueous, which can dissolve BC [96]. BC has mainly found applications in the biomedical field where it is used as wound dressing material, for burns, artificial skin, vascular grafts, scaffolds for tissue engineering, tissue regeneration, and artificial blood vessels, etc. [89, 97, 98]. Also, it has been used for the preparation of several commercial products such as tires, headphone membranes, high performance speaker diaphragms, high-grade paper, makeup pads, diet food, and textiles, etc. [98]. Furthermore, it is used as carrier in drug delivery systems, enzyme immobilization, and ion exchange membrane, and as biodegradable and biocompatible sensors and actuators [99, 100].

I.5.2.3 Chitosan

Chitosan is an abundant polysaccharide present in nature. Chemically, it is composed of randomly arranged β-(1 → 4)-linked D-glucosamine and N-acetyl-D-glucosamine, representing deacetylated and acetylated units, respectively. Its chemical structure is similar to that of cellulose except for the replacement of the $-NH_2$ group instead of –OH moieties in the glucose units of cellulose backbone. It is synthesized through deacetylation of chitin shells of crustaceans, such as shrimps, using an alkaline solvent such as sodium hydroxide (NaOH) in excess as a reagent and water as a solvent. The chemical reactions are completed in two steps under first-order kinetic control where activation energy (48.76 kJ/mol) of the first step is higher than the second and yields about 98% chitosan as the final product.

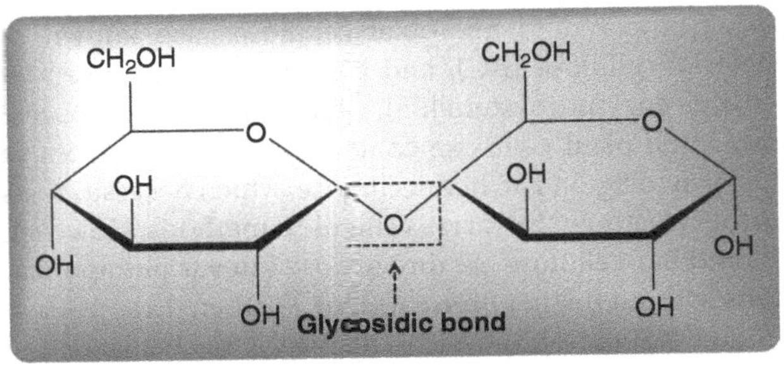

Figure I.2 Chemistry of glycosidic bond formation in carbohydrates.

The degree of deacetylation (% DD) is determined via nuclear magnetic resonance (NMR) spectroscopy, where it ranges from 60–100% for commercial chitosan. The molecular weight of commercial chitosan ranges between 3.8–20 KDa.

Chitosan is highly soluble in dilute acids, highly biocompatible, non-toxic, biodegradable, does not provoke the immune system, is anti-cancerous, and environmentally friendly [101]. It biodegrades slowly to harmless products of oligomers and is absorbed slowly in the body. The biodegraded chitosan accelerates wound healing [102]. Owing to these features, it has found immense use in biomedical applications. It is used for drug delivery and wound care, both alone and in the form of composites with other materials such as BC, owing to its antibacterial activity [103]. Also, it is heavily used in agriculture as a seed treatment and biopesticide against fungal invasion. In the wine industry, it is used as a fining agent and for the prevention of spoilage. Further, it is used as self-healing polyurethane paint coating.

I.5.2.4 Alginate and Other Seaweed-Derived Polysaccharides

Alginate is an anionic polysaccharide isolated from seaweed. Chemically, it is composed of mannuronic acid and guluronic acid units where mannuronic acid units form β (1→4) linkages while guluronic acid units are linked via α-(1→4) linkages. Structurally, mannuronic acid units are distributed linearly and exhibit a flexible conformation while guluronic acid units display a steric hindrance in the vicinity of carboxyl groups. Besides alginate, several other seaweed-derived polysaccharides, such as carrageenan, fucoidan, laminaran, and ulvan have also been used for various biomedical applications owing to their high biocompatibility, easy availability, and simple isolation strategies [104]. Such polysaccharides can form hydrogels through ionic interaction between carboxylic groups (–COOH) present on their surface with a cationic cross-linking agent [105]. Besides, divalent cations such as Ca^{2+}, Zn^{2+}, Ba^{2+}, or trivalent cations (e.g. Al^{3+}) may also exist in these hydrogels. Of the various seaweeds, polysaccharides-based hydrogels, alginate-based hydrogels are extensively used as biomaterials for various biomedical applications including as scaffolds in tissue engineering, as carriers in drug delivery, and as model ECMs for biological studies [106]. Similarly, the carrageenan-based hydrogels are extensively used for the encapsulation of cells, the transformation of growth factors, and the formation of bone and cartilage tissues [107, 108]. The fucoidan and ulvan-based hydrogels are used in the culturing of cells and improving their activity [109, 110].

I.5.2.5 Hyaluronic Acid

Hyaluronic acid (HA), a non-sulfated glycosamino-glycan is constituted of repeated units of D-glucuronic acid and D-N-acetylglucosamine linked together via alternating β-1,4 and β-1,3-glycosidic linkages [111]. In nature, it is present in the form of a long straight chain of anioinc polysaccharide. It is extensively found in the human body where it is dispersed in different tissues. Generally, it has the ability to cross-link by simple freeze-thawing in the absence of any cross-linker or organic solvent [112]. It can be potentially modified into different forms due to the presence of several functional groups in its chemical structure, such as –COOH, –OH, and N-acetyl functional group. It can form hydrogels through various cross-linking methods, such as chemical, enzymatic, or photo-cross-linking which find different applications in the biomedical and electronics fields. It has been extensively used for biomedical applications, such as cell motility, wound care and healing, cell signaling analysis, fabrication of different matrices, and angiogenesis, since it accounts for a major portion of ECM of skin, cartilage, and vitreous humors [91].

I.5.3 Proteins

Proteins are biomacromolecules comprised of one or more long chains of amino acids. Each amino acid is comprised of an α-carbon, amino group ($-NH_2$), carboxyl group (–COOH), and a variable side chain designated as R-group (–R). The amino acid residues are linked together through a peptide bond between the –COOH and $-NH_2$ groups of two consecutive amino acids (Figure I.3). The peptide bond has two resonance forms which contribute some double-bond character and inhibit rotation around its own axis which lead to the coplanar conformation of α-carbon. The protein synthesis is carried out inside the living cells in a two-stage process: transcription and translation, which take place in the nucleus and the cytoplasm of the cell, respectively. During transcription, the information from the segment of DNA known as the gene encodes the information for the formation of messenger RNA (mRNA). The mRNA

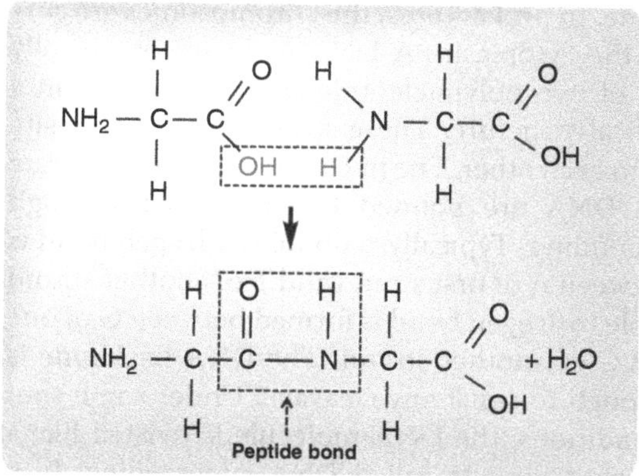

Figure I.3 Chemistry of peptide bond formation in proteins.

then forms the respective polypeptide through translation which involves the formation of an assembly of ribosome, ribosomal RNA (rRNA), and transfer RNA (tRNA). The sequence of amino acids in a protein is determined by the sequence of nucleotide in its respective gene designated in the form of a genetic code. A single genetic code is comprised of three nucleotides (triplet codon). These genetic codes encode 20 different amino acids designated as standard amino acids, however, some organisms contain additional non-standard amino acids, such as selenocysteine and pyrrolysine. Besides biological synthesis, proteins are also synthesized chemically by various methods collectively known as "peptide synthesis" which mainly use organic synthesis methods [113]. These methods include liquid-phase synthesis, solid-phases synthesis, and using solid supports, etc. Usually, the structure of protein is determined by X-ray crystallography and NMR spectroscopy.

A linear chain of amino acid is called a polypeptide, however, polypeptides containing less than 20–30 amino acid residues are generally not designated as proteins, rather they are called peptides or oligopeptides. Besides amino acids, certain proteins contain additional parts termed prothetic groups or cofactors, which impart specialized function to a protein. Proteins exist in four levels of structural organization: the linear sequence of amino acids (i.e. polypeptide) represent the primary structure of the protein. The secondary structure of proteins contains α-helices and β-sheets which are stabilized by hydrogen bonds. The spatial distribution of the secondary structure represents the tertiary level of protein organization. It is stabilized by non-local interaction such as hydrophobic interaction, salt bridges, sulfide bonds, and post-translational modifications. The quaternary structure of proteins is the complex form which is formed by several polypeptide chains designated as subunits. These subunits either perform individual functions or account for the cumulative function of complex protein molecule. Similar to other biomacromolecules, proteins are essential components of living organisms and perform different functions. For example, some proteins functions as enzymes where these catalyze biochemical reactions to perform various metabolic activities. Similarly, some proteins perform structural and mechanical functions, such as actin and mysoin proteins in muscles and the cytoskeleton and maintain the cell shape by forming a system of scaffolding. Other classes of proteins serve as cell-signaling and ligand-binding components, transcription factors, membrane receptors, antibodies, transport molecules, and vaccines, etc. Other proteins, for example, collagen and gelatin, are extensively used for biomedical application owing to their potential to form hydrogels. Collagen, the main component of ECM, possesses a right-handed triple helix and is comprised of three left-handed polypeptide helices that are twisted into a right-handed coil. It forms hydrogels with other materials through chemical, enzymatic, and photo-cross-linking [114–116]. Similarly, gelatin biopolymer derived from collagen through hydrolytic degradation is widely used in pharmaceutical, food, and biomedical applications [117]. However, it has limited thermal stability and possesses weak chemical cross-linking ability.

I.5.4 Nucleic Acids

Nucleic acids are biomacromolecules essential for all forms of life. Chemically, these are composed of nucleotides which are linked to each other through covalent bonds. Typically, sugar of one nucleotide forms a covalent bond with the phosphate of subsequent nucleotide. Each nucleotide is composed of 5-carbon sugar (pentose), phosphate (PO_4^{3-}) group, and a nitrogenous base. The sugar and phosphates in nucleic acids are linked together in alternate fashion through a phosphodiester bond which is formed through the reaction of –OH groups in phosphoric acid with –OH group of another molecule [118] (Figure I.4). There are two types of nitrogenous bases, purines and pyrimidines, based on their chemical structures (Figure I.5). Two types of nucleic acids are known, DNA and RNA. Strings of nucleotides bind together to form helical backbones. Typically, one and two strings form the backbones of RNA and DNA, respectively. DNA and RNA differ from each other in their chemical composition, DNA contain adenine (A), guanine (G), cytosine (C), and thymine (T), while RNA contains uracil (U) instead of thymine besides other three bases. Besides structural arrangement and chemical composition, DNA and RNA also differ from each other in the sequence of nucleotides. The sequence of nucleotides is of immense importance and contains the information for the synthesis of biological molecules

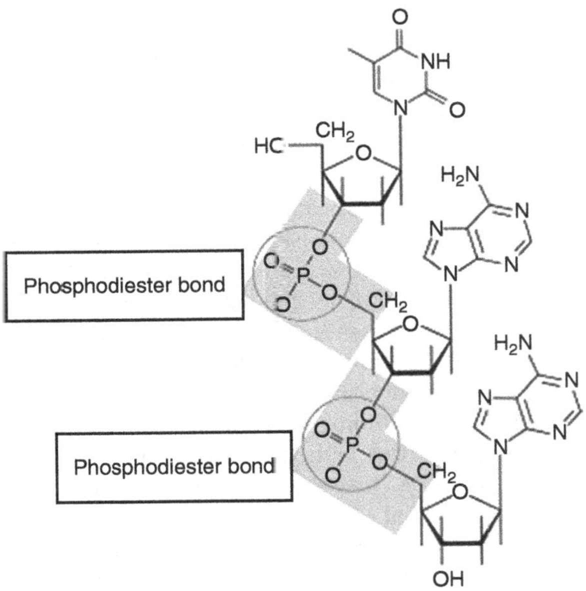

Figure I.4 Chemistry of phosphodiester bond formation in nucleic acids.

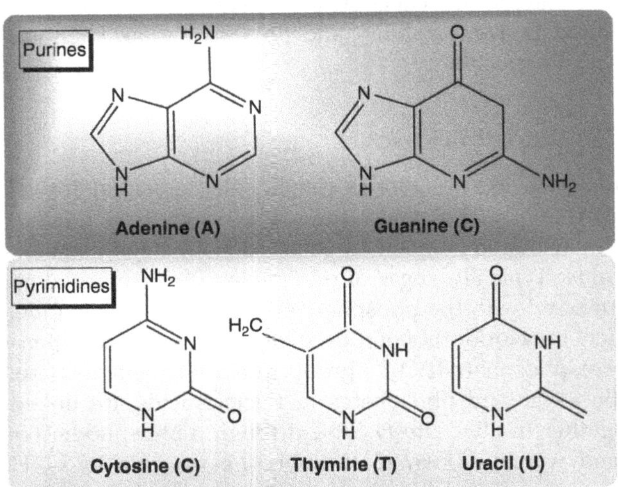

Figure I.5 Illustration of chemical structures of various nitrogenous bases in nucleic acids.

(i.e. protein), molecular assemblies, cellular and sub-cellular structures, organs, and organisms, etc.

I.5.4.1 DNA

Deoxyribonucleic acid (DNA) carries the genetic information of a cell and is referred to as a "genome." A genome is the structural and functional organization of a cell. It contains all the information for the growth, shape, size, pattern, development, and functionality, etc. of a cell, tissue, organ, and organism. Typically, the DNA molecule is present in the nucleus of eukaryotes and forms a complex with histone proteins, referred to as a chromosome. In prokaryotes, the chromosomes are suspended in the cytoplasm. A DNA molecule is typically comprised of two polynucleotide strings which form a double-helical structure. These strands run in opposite directions to each other. The nucleotides on two separate strands in DNA are bonded to each other through hydrogen bonding. Typically, a double hydrogen bond is formed between A of first strand and T of another strand while a triple hydrogen bond is formed between G of one strand and C of another strand. The DNA backbone is flexible enough to resist any cleavage. Under some specialized conditions, the DNA molecule is twisted like a rope through a process called DNA supercoiling. In a relaxed state, one strand in DNA molecule encircles the axis of the double helix once after every 10.4 base pairs (bp) on average, however, in a twisted state, the strands become tighter or are loosely arranged [119]. Supercoiling can be either positive (twisting in the direction of the strand) or negative (twisting in the opposite direction of the strand). Most DNA molecules in nature exist as negatively supercoiled, which is caused by an enzyme topoisomerase. DNA molecules can exist in different conformations such as A-DNA, B-DNA, and Z-DNA, however, only B-DNA and Z-DNA are known to be identified in functional organisms [120]. Both DNA strands contain the same biological information which is replaced when two strands separate from each other during mitotic cell division. The strand of DNA which is the same as the mRNA formed as a result of transcription is termed the sense-strand, while the complementary strand is called the anti-sense strand. However, both sense and anti-sense sequences can co-exist on the same DNA strand. To date, the function of the anti-sense sequences is unknown [121]. It has been suggested that anti-sense RNAs might be involved in regulating gene expression through RNA-RNA base pairing [122]. It is worth mentioning here that about 98% of the total DNA in humans does not code for any protein and is referred to as non-coding or junk DNA.

DNA forms temperature, pH, enzymes, and UV light-responsive hydrogels through physical entanglement or chemical cross-linking upon extraction from the nucleus (eukaryotes) or pooling out from cytoplasm (prokaryotes). For example, it forms hydrogels with cationic polyelectrolytes [123] or metal ions, such as silver [124] for their applications in cell culture and differentiation studies [125] and 3D bioprinting [126].

I.5.4.2 RNA

Ribonucleic acid (RNA) is typically a single-stranded molecule, however, it forms an intra-strand double helix through complementary base pairing. Similar to DNA, it is also comprised of nucleotides linked via phosphodiester bonds, except it contains uracil (U) nitrogenous base instead of thymine (T), which is linked to adenine (A) via a triple hydrogen bond. Unlike DNA, it contains ribose sugar at the C-2 position, hence it is less stable and prone to hydrolysis. The presence of this functional group leads to A-type conformation of RNA [127] which results in the formation of a deep and narrow major groove and a shallow and wide minor groove [128]. Further, due to the presence of the –OH group in the conformationally flexible regions of the RNA molecule, it can chemically attack the adjacent phosphodiester linkage and cleave the sugar-phosphate backbone [129]. It plays several vital roles in living systems, such as coding, decoding, gene expression and regulation, catalyzes biological reactions, and sense and communicates responses to cellular signals, etc.

Typically, there are three types of RNA molecules: messenger RNA (mRNA), ribosomal RNA (rRNA), and transfer RNA (tRNA). Collectively, these three types of RNA molecules direct the protein synthesis in living systems, an important event of life. mRNA represents the coding genetic material which is produced inside the cytoplasm through the transcription of the sense strand of the DNA (serving as a template) molecule using the enzyme RNA polymerase. tRNA delivers amino acids from the "amino acids pool" to the ribosome where tRNA links the amino acids together via peptide bond to form a polypeptide or protein after post-translational modification. Specifically, each tRNA molecule translocates a specific amino acid from the pool, thus, there are 20 different types of tRNA molecules to bring specific amino acids to the ribosome assembly. Recent research has shown that besides three typical RNAs, there also exist some regulatory RNAs in the cell, such as micro RNA (miRNA), small interfering RNA (siRNA), Piwi-interacting RNA (piRNA), CRISPR RNAs, antisense RNA, and Xist, etc. These RNA molecules are generally involved in the regulation and down-regulation of genes. For example, miRNA, comprised of 21–22 nucleotides, forms a complex with enzymes and cleaves the complementary mRNA and thus blocks its translation into a polypeptide [130]. The siRNAs, comprising 20–25 nucleotides, are produced through viral breakdown and interfere with the translation of mRNA using the same mechanism as miRNA [131]. The piRNA found in germline cells in animals is comprised of 29–30 nucleotides and serves as defense mechanism against the transposons and plays a vital role in gametogenesis [132]. CRISPR RNAs serves as a regulatory system similar to RNA interference [133]. Similarly, the antisense RNAs mostly downregulate the gene expression by binding to a mRNA and form a double-stranded RNA which is degraded enzymatically, however, a few are known to activate transcription [134, 135]. Besides the above-mentioned coding RNAs, a non-coding RNA also exists such as Xist which inactivates the X chromosome in females by coating it [136].

I.6 Part III Biomaterials

Biomaterials are substances that are engineered to interact with a living system for medical purposes which can be either therapeutic (tissue engineering and regenerative medicines) or diagnostic. These are non-viable materials and are implanted into the body to replace a missing part or repair a damaged tissue. The study of biomaterials is called "biomaterials science" or "biomaterials engineering," which is about five decades old. It is a multi-disciplinary field which encompasses different fields, such as tissue engineering, materials science and engineering, medicine, cell and molecular biology, and biochemistry, etc.

By nature, biomaterials are either derived from nature or synthesized using different strategies from various materials, such as organic and inorganic metals, polymers, ceramics, or clay, etc. The role of biomaterials can be passive, such as their use as a heart valve, or bioactive, where they function interactively, for example, hydroxyapatite-coated hip implants. Biomaterials are routinely used for broad spectrum applications, such as in dental applications, surgery, drug delivery, tissue engineering, wound healing, pharmaceutics, and several others. In general, different types of materials are used for various biomedical and tissue engineering applications which can be broadly classified into four classes [137]:

1) *Naturally derived polymer materials*, for example, collagen, gelatin, starch, alginate, and chitosan, etc. These materials are better due to their recognition by biological systems [138].
2) *Acellular tissue matrices*, for example, bladder submucosa and small intestinal submucosa. Similar to naturally derived polymeric materials, these are also recognized by biological systems and are preferred candidates for tissue engineering [138].
3) *Synthetic polymers*, for example, poly(glycolic acid) (PGA), poly(lactic acid) (PLA), poly(caprolactone) (PCL), and poly(lactide-co-glycolide) (PLGA), etc.
4) *Ceramics*, for example, hydroxyapatite (HA) and β-tricalcium phosphate (TCP).

I.6.1 Summary of Part III Biomaterials

In Part III, "Biomaterials," we discuss different classes of biomaterials, such as thermal- and photo-deformable liquid crystal polymers, intriguing bioinspired movements and their potential applications, dendritic polymer-based micelles and their function in drug delivery, bone-inspired biomaterials and their applications in bone tissue engineering, human dental caries restoration, protein-based hydrogels and their applications, and various fiber-reinforced composites as well as damping structures and self-healing materials, etc.

1.6.2 Features of Biomaterials

1.6.2.1 Biocompatibility

Biocompatibility of biomaterials refers to their ability to remain in direct contact with living tissues without causing a toxic effect or provoking an immunogenic response [139]. Usually, the biocompatibility of biomaterials is the function of their liquid-absorbing capabilities that support prolonged and sustainable growth of living cells. Biocompatibility is determined for individual materials and composites. Under certain circumstances, the biocompatibility of biomaterials is not up to the desired levels, therefore, their biocompatibility is enhanced by making their composites with more compatible polymers. For example, the biocompatibility of BC has been enhanced by making its composite with several polymers, such as gelatin, hydroxyapatite, and chitosan [103, 117]. Generally, the implantation of an acellular material triggers a series of events similar to a foreign body response, which leads to further acute inflammation response and in some cases developing a chronic response. However, a biomaterial should not provoke such a response, which is highly related to its biological properties, internal and external architecture properties, surface features, porosity, and the source of materials used for the fabrication of biomaterial implant [140]. Any deviation from the normal wound healing process determines the level of biocompatibility of biomaterials [141].

1.6.2.2 Biodegradability

Biomaterials when used as *in vivo* implants must degrade into non-toxic products at a controlled degradation rate that closely matches the regeneration rate of the host tissues. When biomaterials are used as scaffolds, these are degraded when cells are deposited, and they secrete their own ECM and eventually form a 3D structure that closely mimics the natural microenvironment of the host tissues. The *in vivo* degradation process of the scaffold is a function of certain variables, such as the material structure (e.g. the chemical composition, the molecular weight and the molecular weight distribution, crystallinity, and structural and morphological features, etc.), the macroscopic features (e.g. the shape and size of the implant, and porosity etc.), and the microenvironment (e.g. the pH and temperature of the culture medium, and the presence of enzymes, cells, and tissues, etc.). Optimization of the degradation process and minimizing its effects are crucial, for example, it may result in local drop of pH if the released degradation products contain acidic compounds, which may lead to necrosis or inflammation.

1.6.2.3 Porosity, Pore Size and Shape, and Micro- and Macro-Structure

Porosity of biomaterials is defined as the percentage of void spaces [142]. Tissue engineering applications of biomaterials require a high level of porosity (>90%) to offer a high surface area for the seeding of high cell density, migration and the proliferation of cells, and neovascularization [143]. The porous structure of the biomaterials enhances the biological and mechanical interlocking between the implanted scaffold and the host tissue at the site of implantation and ensures greater stability and avoids fibrous encapsulation. Similarly, appropriate pore size and shape of the pores in biomaterials ensure cell viability and tissue regeneration by accumulating a large number of cells and the absorption of high nutrient concentrations. However, a balance must be maintained between the pore size and the mechanical strength of the biomaterials since an increase in pore size generally leads to lower mechanical strength. Usually, pores of biomaterials, that are too large or too small, result in the failure of cell adhesion, proliferation, and migration into different parts of the biomaterial. Therefore, it is essential to define an appropriate size and shape of the pores of biomaterials for a specific cell type and desired application. In general, small pores favor greater cell adhesion and differentiation *in vitro* while larger size pores favor greater cell adhesion and proliferation *in vivo*. However, the pore size is not the only parameter that ensures the successful development of regenerated tissue, several other factors, such as pore interconnectivity and distribution are also essential which guarantee the transport of nutrients, diffusion of gases, and removal of wastes from cells.

1.6.2.4 Mechanical Features of Biomaterials

Biomaterials used for tissue engineering applications must demonstrate high mechanical properties to withstand *in vitro* manipulation and stressed condition in the *in vivo* microenvironment. High mechanical strength of biomaterials preserves the required space for cell attachment, proliferation, and migration as well as matrix formation. The biomaterial implant must demonstrate mechanical properties closely mimicking those of the host tissues to provide temporary support to the regenerating tissues.

1.6.2.5 Surface Topography

Surface features of the biomaterials play a vital role in their applications in tissue engineering and regenerative medicines. The surface roughness of biomaterials ensures efficient cell adhesion, proliferation, differentiation, and viability. Also, the scale of topography, such as micro-, nano-, and mezo-scale biomaterials, the shape of the biomaterials, such as ridges, steps, grooves, pillars, and pits, and their orientation have profound effects on the overall cell behavior and phenotypic expression [144]. In general, nano-scale (100–300 nm) biomaterials offer improved cell adhesion and proliferation for most cell types. It is generally believed that nano-scale materials modulate the interfacial forces between implants and cells which leads to the regulation of cytoskeleton formation and the organization of the cell membrane receptors and also triggers the complex intra-cellular molecular signaling [145]. Besides the scale of biomaterials, their surface roughness also determines the cell-to-cell interactions, adhesion, and proliferation. For example, highly rough surfaces of biomaterials usually fail to establish interconnections, which leads to compromised adhesion and the proliferation of cells. Thus, surface properties and interactions between cells and ECM are of immense consideration for cell signaling and subsequent cell attachment, proliferation, and formation of phenotypes [146].

I.6.2.6 Simple Manufacturing and Sterilization Strategy

Biomaterials used for various applications should be manufactured by simple, robust, reliable, and cost-effect strategies. Moreover, the manufactured materials must be easily sterilized by common sterilization strategies, such as heat, chemicals, autoclavation, and UV sterilization, etc.

I.6.3 Current Advances in Biomaterials Science

Some of the advanced applications of biomaterials are:

- *Smart polymers:* These are injectable hydrogels that are used for the delivery of growth factors, gene and gene segments, and viable cells to the target site, such as damaged tissues.
- *Biomimetic elastomers:* These are elastic polymers which mimic the mechanical features of the host tissue and play the vital functions of the proteins found in the host body.
- *Engineered skin:* This is fabricated from collagen and collagen-synthetic composites through electrospinning strategies to improve the collagen strength and elasticity for various applications.
- *Heart patches and blood vessels:* Researchers across the world have evaluated the potential of different biomaterials for the formation of synthetic heart patches and blood vessels. These biodevices have been fabricated by various strategies, such as electrospinning and several others.
- *Immuno-isolation systems:* Attempts have been made to provide protection to the cells against various factors by using different biomaterials to avoid immune attack. This practice has also resulted in the improved efficacy of cell transplantation.
- *Aligned fiber assays:* Using electrospinning strategies, highly aligned fibers have been fabricated for their application in diagnosis and treatment of several highly metastatic cancers.
- *Islet delivery:* Researchers have developed biomaterials for pancreas replacement to preserve the functionality of islets after their harvesting from the donor and to assist their delivery to the recipient. It can result in the long-term survival and health of the patient.
- *Tendon repair:* Using biomaterials, tissue-engineered constructs have been designed to shorten the recovery time for repair of tendons and to improve their function.

I.7 Scope of the Book

The importance of bioinspired materials leads to the highly interdisciplinary character of this field, which brings together natural scientists (biologists, chemists, and physicists) and engineers. Thus, an active interaction across disciplines is the key to the real development of this old research area, which is now attracting many researchers worldwide. However, as underlined by several recent reviews on the subject, this condition is yet to be fully met, due to the rather limited understanding of the building principles of living entities which are numerous and complex, and because the definition of the scope and novel applications remain to be further clarified and exploited. Hence, conveying information in the field and disseminating the bioinspired solutions already discovered are of real importance, and would contribute significantly in propelling this promising research area. Therefore, the aim of this book is to introduce a comprehensive overview of the bioinspired materials science and engineering, discussing biofabrication approaches, and applications of bioinspired materials as they are fed back to nature in the guise of biomaterials. Some biological compounds, such as carbohydrates, proteins, and nucleic acids, etc. have been mentioned with regard to what is learned from them, and how they are useful in the engineering of bioinspired materials.

The book is divided into three main Parts: biofabrication, biomacromolecules, and biomaterials. Illustrating the bioinspiration process from materials design and conception to the application of bioinspired materials, this book presents the multidisciplinary aspect of the concept and represents a typical example of how knowledge is acquired from nature, and how in turn this information contributes to biological sciences, with an accent on biomedical applications. This book is devised both for novice readers and experts in the field who will find it valuable as it presents recent advances in topics of current interest, such as biotemplating, microfluidics, self-assembly, mussel-inspired surface modification, and several others. Furthermore, its different Parts, each of which is a multidisciplinary research area, will appeal not only to chemists, but also to biologists, physicists, and engineers, etc.

Acknowledgments

This work was supported by the National Natural Science Foundation of China (21574050, 31270150, 51603079), the China Postdoctoral Science Foundation (2016M602291, 2015M572132), Fundamental Research Funds for the Central Universities, Open Research Fund of State Key Laboratory of Polymer Physics and Chemistry, Changchun Institute of Applied Chemistry, Chinese Academy of Sciences.

References

1 McKittrick, J., Chen, P.Y., Tombolato, L. et al. (2010) *Mat. Sci. Eng. C.*, **30**, 331–342.
2 Liu, K., Jiang. L. (2011) *Nano Today*, **6**, 155–175.
3 Zheng, Y.M., Gao, X.F., Jiang, L. (2007) *Soft Matter*, **3**, 178–182.
4 Vukusic, P., Sambles, J.R. (2003) *Nat.*, **424**, 852–855.
5 Potyrailo, R.A., Ghiradella, H. Vertiatchikh, A. et al. (2007) *Nat. Photon.*, **1**, 123–128.
6 Vukusic, P., Hooper, I. (2005) *Sci.*, **310**, 1151–1155.
7 Sato, O., Kubo, S., Gu, Z.Z. (2009) *Acc. Chem. Res.*, **42**, 1–10.
8 Aizenberg, J., Tkachenko, A., Weiner, S. et al. (2001) *Nat.*, **412**, 819–822.
9 Sun, T.L., Feng, L., Gao, X.F., Jiang, L. (2005) *Acc. Chem. Res.*, **38**, 644–652.
10 Liu, M.J., Wang, S.T., Wei, Z.X. et al. (2009) *Adv. Mater*,. **21**, 665–669.
11 Hansen, W.R., Autumn, K. (2005) *Proc. Natl. Acad. Sci. U.S.A.*, **102**, 385–389.
12 Feng, L., Li, S.H. Li, Y.S. et al. (2002) *Adv. Mater.*, **14**, 1857–1860.
13 Gao, X.F., Yan, X., Yao, X. et al. (2007) *Adv. Mater.*, **19**, 2213–2217.
14 Mayer, G. (2005) *Science*, **310**, 1144–1147.
15 Tan, T.L., Wong, D., Lee, P. (2004) *Opt. Express*, **12**, 4847–4854.
16 Zi, J., Yu, X.D., Li, Y.Z. et al. (2003) *Proc. Natl. Acad. Sci. U.S.A.*, **100**, 12576–12578.
17 Stegmaier, T., Linke, M., Planck, H. (2009) *Philos. Trans. R. Soc. A*, **367**, 1749–1758.
18 Feng, L., Zhang, Y.A., Xi, J.M. et al. (2008) *Langmuir*, **24**, 4114–4119.
19 L. Feng, Y.A. Zhang, M.Z. Li, et al. (2010) *Langmuir*, **26**, 14885–14888.
20 Bhushan, B., Her, E.K. (2010) *Langmuir* **26**, 8207–8217.
21 Ball, P. (1999) *Nature* **400**, 507–509.
22 Sundar, V.C., Yablon, A.D., Grazul, J.L. et al. (2003) *Nature*, **424**, 899–900.
23 Aizenberg, J., Sundar, V.C., Yablon, A.D. et al. (2004) *Proc. Natl. Acad. Sci. U.S.A.*, **101**, 3358–3363.
24 Aizenberg, J., Weaver, J.C., Thanawala, M.S. et al. (2005) *Science*, **309**, 275–278.
25 Zheng, Y.M., Bai, B., Huang, Z.B. et al. (2010) *Nature*, **463**, 640–643.
26 Becker, N., Oroudjev, E., Mutz, S. et al. (2003) *Nat. Mater.*, **2**, 278–283.
27 Swanson, B.O., Blackledge, T.A., Hayash, C.Y. (2007) *J. Exp. Zool.*, **307A**, 654–666.
28 Vollrath, F., Knight, D.P. (2001) *Nature*, **410**, 541–548.
29 Work, R.W. (1985) *J. Exp. Biol.*, **118**, 379–404.
30 Shao, Z.Z., Vollrath, F. (1999) *Polymer*, **40**, 1799–1806.
31 Shao, Z., Vollrath, F., Sirichaisit, J., Young, R.J. (1999) *Polymer*, **40**, 2493–2500.
32 Perez-Rigueiro, J., Elices, M., Guinea, G.V. (2003) *Polymer*, **44**, 3733–3736.
33 Liu, Y., Shao, Z.Z., Vollrath, F. (2005) *Nat. Mater.*, **4**, 901–905.
34 Bell, F.I., McEwen, I.J., Viney, C. (2002) *Nature*, **416**, 37–137.
35 Emile, O., Le Floch, A., Vollrath, F. (2006) *Nature*, **440**, 621–1621.
36 Gao, X.F., Jiang, L. (2004) *Nature*, **432**, 36–136.
37 Wu, M.S. (2011) *Mat. Sci. Eng. C.*, **31**, 1209–1220.
38 Mironov, V., Kasyanov, V., Markwald, R.R. (2011) *Curr. Opin. Biotechnol.*, **22**, 667–673.
39 Annabi, N., Nichol, J.W., Zhong, X. et al. (2010) *Tissue Eng. Part B Rev.*, **16**, 371–383.
40 Khan, S., Ul-Islam, M., Khattak, W.A. et al. (2015) *Cellulose*, **22**, 565–579.
41 Madihally, S.V., Matthew, H.W.T. (1999) *Biomat.*, **20**, 1133–1142.
42 Kang, H.W., Tabata, Y., Ikada, Y. (1999) *Biomat.*, **20**, 1339–1344.
43 Mooney, D.J., Baldwin, D.F., Suh, N.P. et al. (1996) *Biomat.*, **17**, 1417–1422.

44 Quirk, R.A., France, R.M., Shakesheff, K.M., Howdle, S.M. (2004) *Curr. Opin. Solid State Mater. Sci.*, **8**, 313–321.

45 Zellander, A., Gemeinhart, R., Djalilian, A. et al. (2013) *Mater. Sci. Eng. C.*, **33**, 3396–3403.

46 Haugen, H., Reid, V., Brunner, M. et al. (2004) *J. Mater. Sci.: Mater. Med.*, **15**, 343–346.

47 Sachlos, E., Czernuszka, J.T. (2003) *Eur. Cells Mater.*, **5**, 29–40.

48 Ozbolat, I.T. (2015) *Trends Biotechnol.* **33**, 395–400.

49 Peng, W., Unutmaz, D., Ozbolat, I.T. (2016) *Trends Biotechnol*, **34**, 722–732.

50 Dababneh, A., Ozbolat, I.T. (2014) *J. Manuf. Sci. Eng.*, **136**, 061016.

51 Saunders, R.E., Derby, B. (2014) *Int. Mater. Rev.*, **59**, 430–448.

52 Ozbolat, I.T., Hospodiuk, M. (2016) *Biomaterials*, **76**, 321–343.

53 Lewis, J.A. (2006) *Adv. Funct. Mater.*, **16**, 2193–2204.

54 Riggs, B.C., Dias, A.D., Schiele, N.R. et al. (2011) *MRS Bull.*, **37**, 1043–1050.

55 Gou, M., Qu, X., Zhu, W. et al. (2014) *Nat. Comm.*, **5**, 3774.

56 Sun, J., Tang, J., Ding, J. (2009) *Chin. Sci. Bull.*, **54**, 3154–3159.

57 Lee, S.H., Moon, J.J., West, J.L. (2008) *Biomaterials*, **29**, 2962–2968.

58 Ivanov, V.V., Decker, C. (2001) *Polym. Int.*, **50**, 113–118.

59 Luo, Y., Shoichet, M.S. (2004) *Nat. Mater.*, **3**, 249–253.

60 Hahn, M.S., Miller, J.S., West, J.L. (2006) *Adv. Mater.*, **18**, 2679–2684.

61 Musoke-Zawedde, P., Shoichet, M.S. (2006) *Biomed. Mater.*, **1**, 162–169.

62 Billiet, T., Vandenhaute, M., Schelfhout, J. et al. (2012) *Biomaterials*, **1**, 6020–6041.

63 Ozbolat, I.T., Yu, Y. (2013) *IEEE Trans. Biomed. Eng.*, **60**, 691–699.

64 Orloff, N.D., Truong, C., Cira, N.J. et al. (2014) *RSC Adv.*, **4**, 34721.

65 Ozbolat, I.T., Chen, H., Yu, Y. (2014) *Robot. Comput. Integr. Manuf.*, **30**, 295–304.

66 Xu, T., Jin, J., Gregory, C. et al. (2005) *Biomaterials*, **26**, 93–99.

67 Catros, S., Guillotin, B., Bačáková, M. et al. (2011) *Appl. Surf. Sci.*, **1**, 5142–5147.

68 Xu, T., Baicu, C., Aho, M. et al. (2009) *Biofabrication*, **1**, 035001.

69 Guillotin, B., Souquet, A., Catros, S. et al. *Biomaterials*, **31**, 7250–7256.

70 Murphy, S.V., Atala, A. *Nat. Biotechnol.*, **32**, 773–785.

71 Guillemot, F., Souquet, A., Catros, S. et al. (2010) *Acta Biomater.*, **1**, 2494–2500.

72 Chang, C.C., Boland, E.D., Williams, S.K., Hoying, J.B. (2011) *Biomed. Mater. Res. B Appl. Biomater.*, **1**, 160–170.

73 Wang, Z. Abdulla, R., Parker, B. et al. (2015) *Biofabrication*, **7**, 045009.

74 Nahmias, Y., Schwartz, R.E., Verfaillie, C.M., Odde, D.J. (2005) *Biotechnol. Bioeng.*, **92**, 129–136.

75 Cui, X., Breitenkamp, K., Finn, M.G. et al. (2012) *Tissue Eng. A*, **63**, 1304–1312.

76 Xu, C., Chai, W., Huang, Y., Markwald, R.R. (2012) *Biotechnol. Bioeng.*, **109**, 3152–3160.

77 Kolesky, D.B., Truby, R.L., Gladman, A.S. et al. (2014) *Adv. Mater.*, **26**, 3124–3130.

78 Chang, R., Emami, K., Wu, H., Sun, W. (2010) *Biofabrication*, **2**, 045004.

79 Ker, E.D.F., Nain, A.S., Weiss, L.E. et al. (2011) *Biomaterials*, **32**, 8097–8107.

80 Gou, M., Qu, X., Zhu, W. et al. (2014) *Nat. Comm.*, **5**, 3774.

81 Huang, T.Q., Qu, X., Liu, J., Chen, S. (2014) *Biomed. Microdevices*, **16**, 127–132.

82 Xie, F., Pollet, E., Halley, P.J., Avérous, L. (2013) *Prog. Polym. Sci.*, **38**, 1590.

83 Chen, Q., Yu, H., Wang, L. et al. (2015) *RSC Adv.*, **5**, 67459–67474.

84 Kenawy, E., Kamoun, E.A. Mohy Eldin, M.S., El-Meligy, M.A. (2014) *Arabian J. Chem.*, **7**, 372.

85 Ullah, M.W., Ul-Islam, M., Khan, S. et al. (2016) *RSC Adv.*, **6**, 22424–22435.

86 Seo, C., Lee, H.W., Suresh, A. et al. (2014) *Korean J. Chem. Eng.*, **31**, 1433–1437.

87 Ullah, M.W., Ul-Islam, M., Khan, S. et al. (2015) *Carbohyd. Polym.*, **132**, 286–294.

88 Fu, L. Zhou, P., Zhang, S., Yang, G. (2013) *Mat. Sci. Eng. C.*, **33**, 2995–3000.

89 Czaja, W., Krystynowicz, A., Bielecki, S., Brown, R.M. (2006) *Biomaterials*, **27**, 145–151.

90 Ul-Islam, M., Khan, T., Park, J.K. (2012) *Carbohyd. Polym.*, **88**, 596–603.

91 Shi, Z., Gao, X., Ullah, M.W. et al. (2016) *Biomaterials*, **111**, 40–54.

92 Ul-Islam, M., Khattak, W.A., Ullah, M.W. et al. (2014) *Cellulose*, **21**, 433–447.

93 Xu, D. Cai, J., Zhang, L. (2016) *Chinese J. Polyn. Sci.*, **34**, 1281–1289.

94 Striegel, A.M. (2003) *J. Chil. Chem. Soc.*, **48**, 73–77.

95 Li, L. Lin, Z., Yang, X. et al. (2009) *Chinese Sci Bull.*, **54** (9), 1622–1625.

96 Luo X., Zeng, J. Liu, S., Zhang, L. (2015) *Bioresource Technol.*, **194**, 403–406.

97 Cienchanska, D. (2004) *Fibres Text. East. Europe*, **12**, 69–72.

98 Evans, B.R., O'Neill, H.M., Malyvanh, V.P. et al. (2003) *Biosens. Bioelectr.*, **18**, 917–923.

99 Mahmoudi, K., Hosni, K., Hamdi, N., Srasra, E. (2015) *Korean J. Chem. Eng.*, **32**, 274–283.
100 Finkenstadt, V.L. (2005) *App. Microbiol. Biotechnol.*, **67**, 735–745.
101 Singh, K.D., Alok, R.R. (2000) *Macromol. Chem. Phys.*, **40**, 69.
102 Phisalaphong, M., Jatupaiboon, N. (2008) *Carbohydr. Polym.*, **74**, 482.
103 Ul-Islam, M., Shah, N., Ha, J.H., Park, J.K. (2011) *Korean J. Chem. Eng.*, **28**, 1736–1743.
104 Venkatesan, J., Lowe, B., Anil, S. et al. (2015) *Starch – Stärke*, **67**, 381–390.
105 Rowley, J.A., Madlambayan, G., Mooney, D.J. (1999) *Biomaterials*, **20**, 45–53.
106 Augst, A.D., Kong, H.J., Mooney, D.J. (2006) *Macromol. Biosci.*, **6**, 623–633.
107 Rocha, P.M., Santo, V.E., Gomes, M.E. et al. (2011) *J. Bioact. Compat. Pol.*, **26**, 493–507.
108 Santo, V.E., Frias, A.M. Carida, M. et al. (2009) *Biomacromolecules*, **10**, 1392–1401.
109 Thébaud, N., Pierron, D., Bareille, R. et al. (2007) *J. Mat. Sci. Mater. Med.*, **18**, 339–345.
110 Dash, M., Samal, S.K., Bartoli, C. et al. (2014) *ACS Appl. Mater. Inter.*, **6**, 3211–3218.
111 Burdick, J.A., Prestwich, G.D. (2011) *Adv. Mater.*, **23**, 41–56.
112 Zhang, H., Zhang, F., Wu, J. (2013) *React. Funct. Polym.*, **73**, 923–928.
113 Bruckdorfer, T., Marder, O., Albericio, F. (2004) *Curr. Pharma. Biotechnol.*, **5**, 29–43.
114 Yan, L.P., Wang, Y.J., Ren, L. et al. (2010) *J. Biomed. Mater. Res.*, A. **95**, 465–475.
115 Jus, S., Stachel, I., Schloegl, W. et al. (2011) *Mater. Sci. Eng. C.*, **31**, 1068–1077.
116 Mizuno, S., Glowacki, J. (1996) *Biomaterials*, **17**, 1819–1825.
117 Khan, S., Ul-Islam, M., Ikram, M. et al. (2016) *RSC Adv.*, **6** 110840–110849.
118 Berg, J.M., Tymoczko, J.L., Stryer, L. (2007) *Biochemistry*. San Francisco, W.H. Freeman.
119 Benham, C.J., Mielke, S.P. (2005) *Annu. Rev. Biomed. Engin.*, **7**, 21–53.
120 Ghosh, A., Bansal, M. (2003) *Acta Crystallographica. Section D, Biological Crystallography*, **59**, 620–626.
121 Hüttenhofer, A.,Schattner, P., Polacek, N. (2005) *Trend. Genet.*, **21**, 289–297.
122 Munroe, S.H. (2004) *J. Cell. Biochem.*, **93**, 664–671.
123 Morán, M.C., Miguel, M.G., Lindman, B. (2010) *Soft Matter*, **6**, 3143–3156.
124 Guo, W., Qi, X., Orbach, R. et al. (2014) *Chem. Commun.*, **50**, 4065–4068.
125 Wu, Y., Li, C., Boldt, F. et al. (2014) *Chem. Commun.*, **50**, 14620–14622.
126 Li, C., Jones, A.F., Dun, A.R. et al. (2015) *Angew. Chem. Int. Ed.*, **54**, 3957–3961.
127 Salazar, M., Fedoroff, O.Y., Miller, J.M. et al. (1992) *Biochem.*, **32**, 4207–4215.
128 Hermann, T., Patel, J. (2000) *Structure*, **8**, 47–54.
129 Mikkola, S., Stenman, E., Nurmi, K. et al. (1999) *Perk. Tans.*, **2**, 1619–1626.
130 Wu, L., Belasco, J.G. (2008) *Mol. Cell.*, **29**, 1–7.
131 Watanabe, T., Totoki, Y., Toyoda, A. et al. (2008) *Nature*, **453**, 539–543.
132 Horwich, M.D., Li, C., Matranga, C. et al. (2007) *Curr. Biol.*, **17** 1265–1272.
133 Horvath, P., Barrangou, R. (2010) *Science*, **327**, 167–170.
134 Gilbert, S.F. (2003) *Develop. Biol.* (7th edn.). Sinauer, pp. 101–103.
135 Wagner, E.G., Altuvia, S., Romby, P. (2002) *Adv. Genet.*, **46**, 361–398.
136 Heard, E., Mongelard, F., Arnaud, D. et al. (1999) *Proc. Nat. Acad. Sci. U.S.A.*, **96**, 6841–6846.
137 Bártolo, P.J., Domingos, M., Patrício, T. et al. (2011) in *Advances on Modeling in Tissue Engineering, Computational Methods in Applied Sciences*. P.R. Fernandes, P.J. Bártolo (eds.) Springer, Spain, pp. 137–176.
138 Chan, G., Mooney, D.J. (2008) *Trend. Biotechnol.*, **26**, 382–392.
139 Ul-Islam, M., Khan, S., Ullah, M.W., Park, J.K. (2015) *Biotechnol. J.*, **10**, 1847–1861.
140 Anderson, J.M. (1988) *Trans. Am. Soc. Intern. Organs.*, **24**, 101–107.
141 Anderson, J.M. (1998) In *Frontiers in Tissue Engineering*, C.W. Patrick, A.G. Mikos, L.V. McIntire (eds.), Elsevier, Oxford.
142 León, C. (1998) *Adv. Colloid Interface Sci.*, **76/77**, 341–372.
143 Kuboki, Y., Takita, H., Kobayashi, D. et al. (1998) *J. Biomed. Mater. Res.*, **39**, 190–199.
144 Stevens, M.M. (2005) *Science*, **310**, 1135–1138.
145 Curtis, A.S.G., Gadegaard, N., Dalby, M.J. et al. (2004) *IEEE Trans. Nanobioscience*, **3**, 61–65.
146 Quirk, R.A., Chan, W.C., Davies, M.C. et al. (2001) *Biomaterials*, **22**, 865–872.
147 Mandrycky, C., Wang, Z., Kim, K., Kim, D.H. (2016) *Biotechnol. Adv.*, **34**, 422–434.
148 Ng, W.L. Wang, S., Yeong, W.Y., Naing, M.W. (2016) *Skin*, **34**, 689–699.

Part I

Biofabrication

1

Biotemplating Principles

Cordt Zollfrank and Daniel Van Opdenbosch*

Technische Universität München, Munich, Germany

1.1 Introduction

Materials science has several frontiers such as the quest for bottom-up manufacturing in ever greater detail, alloying metals for improved mechanical or electrical properties, doping semiconductors for improved energy efficiency, nanoparticle synthesis for imaging and sensing applications or creating optical computing devices with improved information density and processing speeds. In human technology, this usually involves changing the material structure on a limited number of hierarchical levels, for example, by changing the chemical composition and grain structure. However, as we will explore in this chapter, this is not the only possible strategy to create functionalities.

Also, an observable trend has emerged, toward materials whose properties are prefixed "self-": Self-healing, self-cleaning or self-assembly are desirable properties in materials covering a wide range of intended applications [1–5]. They share a common property: Passivity, i.e. an action that does not require much or any external stimulus or energy input. We are moving toward integrated passive systems, such as passive houses, cooling systems, sensors, etc. They all have the same underlying motivation: costs of fossil fuel will rise, owing to increased demand, slowly depleting deposits, and the self-imposed restraint due to the consequences of man-made climate change. Further, after two actual, grave and several near-accidents, nuclear energy should be – and has been -re-assessed as highly risky [6, 7] and, due to mankind's failure to provide a working concept for the recycling or even storage of spent fuel, as unsustainable. Even accounting for a greater proportion of our energy needs being generated directly from primary sources such as solar radiation or wind, the cost of energy may rise by 32%, and the importance of energy efficiency with it [8]. Rising energy costs along with Earth's increasing human population will necessitate both the abandonment of our most energy-inefficient habits, and developing more energy-efficient product lifecycles to reach sustainability.

It is therefore sensible to look at manufacturing principles used in living nature where energy efficiency is vital. Before the rise of purely synthetic materials due to the abundance of energy from fossil fuels, the largest volume of materials employed by mankind were biologically manufactured and then refined [9–11] to make the final products: Cordage, horn, leather, wood, (cetacean and plant) oil, and wool have since been replaced by various types of duromers, elastomers, thermoplasts, and many other mineral oil-based products. Previously, materials that could not be directly sourced had to be obtained in the most energy-efficient manner possible. Open pit and hydraulic mining, forest glass manufacturing, or bloomery smelting relied on the on-site winning of raw materials and energy [12]. In light of a potential future energy scarcity, it may be the right time to return to such processing techniques and materials structures that emphasize energy efficiency, both in manufacturing and application. This is the underlying thought of the concept of retaining the intricate structuring of natural materials in engineering materials instead of restructuring additively or subtractively. It has been explored in the past, but experienced recent advances. This concept is called biotemplating.

Biotemplating is a set of techniques that take the refining of biologically manufactured materials one step further by entirely replacing the biological structure with a functional engineering material. In manufacturing terms, it is a casting technique using a lost mold, see Figure 1.1.

However, man-made casting molds are subject to manufacturing limitations in the first place, with attainable minimal sizes being on the n·100 nm scale [13]. Therefore, they do not attain the same degree of three-dimensional hierarchical complexity as natural templates. Biological

* Email: cordt.zollfrank@tum.de

Bioinspired Materials Science and Engineering, First Edition. Edited by Guang Yang, Lin Xiao, and Lallepak Lamboni.

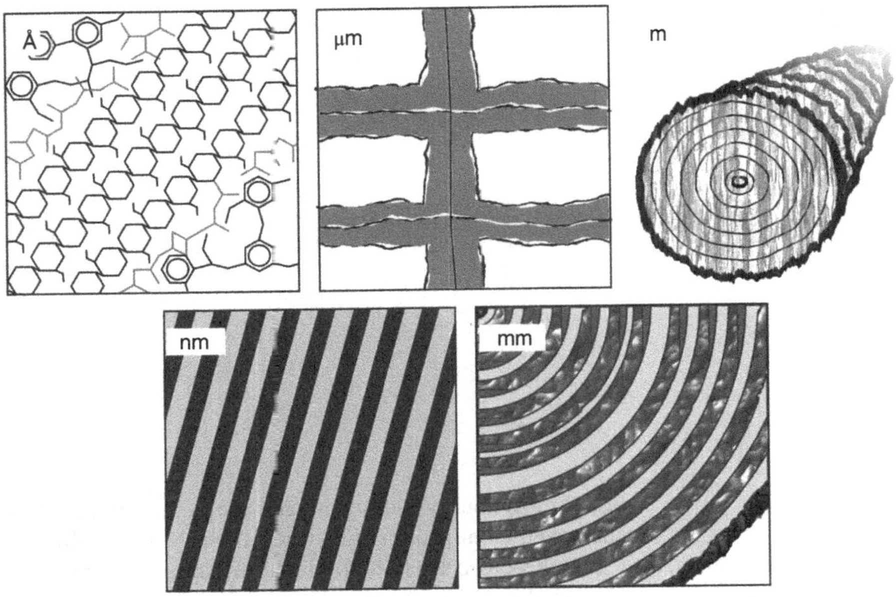

Figure 1.1 Scheme of the deposition of inorganic phases (red) in a biological material (black, adaptation of the wood structure) on several levels of hierarchy. The shown levels of hierarchy may themselves contain sublevels. For example, in the actual wood structure, the nm scale contains the hierarchical levels of the cellulose micro- and elementary fibrils.

structures, on the other hand, are created by molecule-assisted low-energy bottom-up assembly, resulting in complexly structured organic phases, often in composite with inorganic phases [14]. Another aspect of natural materials is their multifunctionality. Arthropod cuticle "produced by a single layer of epithelial cells, is called upon to provide a large number of functions, such as shape, structure, hinges, barrier, filter and similar functions" [15]. Another example is wood, which provides mechanical support, fluid transport [16] and even actuation [17–19]. Straddling the borders of biotemplating is the use of structures made from shaped biological materials, such as corrugated fiberboards [20, 21] or preceramic papers [22].

As further detailed in this chapter, the biotemplating approach has already enabled casting such materials as compact, fibrous ceramics with a non-catastrophic fracturing behavior, photonic reflectors with an internal continuous Schwartz surface, or quasicrystalline structuring on the $n{\cdot}100$ nm scale.[1]

In the following we will first outline the discovery and description of the naturally occurring paragon for biotemplating, namely, fossilization. This will be followed by its human adaptation and the transition from a purely academic interest toward its use for the manufacturing of novel functional materials. The characteristics and properties of biotemplated materials will be explored, highlighting prominent examples. Finally, we will discuss the challenges and limitations of this technique, while sketching directions of further development and instances of usage. Since research in this field is currently strong, and continues to produce exciting results, this chapter not only presents well-established general findings. It also endeavors to point out the most recent findings in order to show the complete history, the present, and the future of biotemplating.

1.2 Mineralization in Nature

1.2.1 Biomineralization

Biomineralization denotes the precipitation of mineral phases by living organisms, typically aided by biomacromolecules.[2] Examples include the silica deposition in various species of grass, the formation of nacre in aquatic animals, or tooth enamel in vertebrates. Due to the abundance of exhaustive and detailed literature on the subject [23–29], biomineralization, or its immediate [30, 31] or abstracted [32] application for materials synthesis will not be detailed here.

However, there are important lessons to be learned and adapted from biomineralization. Jerominidis wrote:

> an engineer is not very good at extracting maximum benefit from *structural shapes* with indifferent materials. In Nature it is the other way round, yet the properties of shapes manufactured from materials possessing poor attributes [are] truly staggering ... [33]

Natural mineralized materials are tailored by evolution to be adaptive and versatile, yet each utilizes a very specific structuring to achieve this goal. It is not surprising that there are no known cases where an organism relies on classical, undirected mineralization to incidentally give a useful crystal morphology. The main point of any templating process, such as the biotemplating discussed here, is to mold a material into a shape that it would not automatically assume. As a consequence, the following four aspects are worth noting and learning.

First, the generalized common principle in biomineralization: While the details are subject to discussion, it is assumed that all minerals are transported to their intended deposition site in solution or colloidal form. Since this would lead to a supersaturation and immediate uncontrolled amorphous precipitation, the dissolved or colloidal minerals are stabilized by molecules [34–38]. Once an appropriate amount of minerals has been transported, a different set of molecules is introduced in order to achieve a directed precipitation, usually with the incorporation of polymer phases in the inorganic precipitate [39]. In the following, specifically in Section 1.7.1, we will explore how this principle relates to artificial mineralization processes in the laboratory, and can be used for biotemplating purposes.

Second, the phases employed by nature reflect the materials available in the respective organism's environment, and are usually limited to a selection of light elements [40]. They are also indicative of which materials can be used, based on other chemical or physical properties. As we will discuss in the following, these include, among others, a low crystallization energy, low solubility in water and having amorphous polymorphs [41].

Third, the resulting biominerals themselves are typically, but with exceptions [42], only hierarchically structured on two levels, namely, their micrometer-scale outer shapes as defined by the tissue cells, and the nanometer-scale internal structuring as defined by the interwoven polymer phase [43]. However, for a functional material in a living organism to be of use, all the hierarchical levels have to be shaped to function in concert [44, 45]. Therefore, mechanisms of active structuring are employed by living nature to achieve structuring on larger length scales. This was also pointed out by Addadi et al. [46, 47], who noted that

> [t]he more sophisticated space-filling strategies used by organisms rely heavily on the intervention and abilities of the cells themselves over and above the presumably exquisite structural design properties of the associated matrix macromolecules. These are properties which are well-nigh impossible to mimic in synthetic systems.

Such multiscale hierarchical structuring will be aspired to in biotemplating technology, as we will discuss in Section 1.5.1.

Fourth, and finally, the range of functions that natural biominerals fulfill is wide, ranging from protection [48–50], actuation [51, 52], substance transport [53, 54], predation [55, 56], propulsion [57, 58], optical [59–63], acoustic [64], and substance-borne [65] signaling. This already indicates that the range of functionalities attainable by biotemplating will be equally wide, as we will see in Section 1.6.

Keeping these aspects in mind, we will now leave the realm of living nature and explore geological and man-induced mineralization processes.

1.2.2 Geological Mineralization

Fossilization of living organisms is the naturally-occurring equivalent of biotemplating. In the broadest meaning of the term, it is any imprint of a deceased plant, animal, or fungus with an inorganic phase through a geological process. The range of minerals in which fossils are found is impressive, and dependent on the locale in which the fossilization takes place. They may also contain unexpected structures and properties, which are exemplified in opals.

Geologically mineralized wood has attracted the interest and imagination of great philosophers and rulers from all the ages, from Pliny the Elder to Georgius Agricola, Kaiser Franz I and Johann Wolfgang von Goethe to Richard Willstätter. In the nineteenth century, William Nicol already recognized the presence of silica, carbonate of lime, oxides of iron, alumina and carbonaceous matter in petrified wood [66]. Nowak et al. applied X-ray diffraction and spectroscopy to find that in single pieces of naturally petrified wood, locally different deposited phases may be present [67]. In some specimens, certain phases such as pyrite were found in the lumen, while in others, the same phase was found in the cell walls. This points to a combination of geological factors and biological decay determining the location of the deposited phases. Equally, Buurman used X-ray diffraction and scanning electron microscopy to determine the phase of the deposited material and the retained structures [68]. They found that the structures of petrified wood can be so detailed as to be virtually indistinguishable from normal wood on the micrometer scale. Importantly, in the literature, there has been no indication that the nanoporous-scale structuring was retained in natural petrifaction [69–71].

John Martyn already noted in 1756 that fossils from every eukaryote kingdom had been found and described [72]. The range of fossilized organic materials further includes roots, trunk, branches, bark, foliage, flowers, fruit [73], and animal tissue [74], with varying degrees of conserved organic components [75–79]. A case of rapid mineralization was described by Renaut et al., who found that in hot springs, microbes were petrified rapidly, possibly even while alive [80]. Similarly rapid petrifactions can be observed in Yellowstone National Park, where upwards petrifaction of standing plants due to the wicking action of the stem can be observed [81]. This was attributed to a combination of the high dissolved mineral contents in the waters, supersaturation by evaporation, and pH reduction due to degassing of carbon dioxide [81], as well as to rapidly onsetting decay of the microbes, providing more reactive sites for the nucleation of silica [82].

The most common mechanism, however, begins with the rapid burial of the organism to avoid aerobic bacterial decay. It is followed by an impregnation with minerals from solutions and their precipitation. This mechanism usually leads to a complete filling of the organism, which in turn can either slowly decay, or become petrified itself, i.e. turn into the various grades of coal, mineral oil and gas.

The formation of coal, i.e. the autonomous petrifaction, is the alternative to a fossilization with a non-carbonaceous second phase. In this process, the wood is carbonized by pressure and temperature under anaerobic conditions. In a similar fashion, artificially carbonized wood for biotemplating purposes is obtained in the laboratory, as discussed in Section 1.6.1. The result is a porous material in which the organic molecules are replaced by an inorganic carbon material. The exact atomic structure depends on the grade of coal and contains, with lower grade, decreasing amounts of carbon and increasing amounts of impurities, typically sulphur, ash, or moisture. Since coal is formed from deposits of vegetation matter, many structural features of the original plant are retained (Figure 1.2).

With regard to fossilization including a second, inorganic phase, polymorphs and suboxides of silicon dioxide, collectively termed "silica," are the most widespread fossil minerals found in nature [84, 85]. The foremost reason for this is the abundance of silicon oxides in the Earth's crust, at 60% by mass [86–88]. The exact sequence of events during silica fossilization is a matter of academic debate, and dependent on the geological conditions, as outlined in the following.

Michael Landmesser identified three basic different mechanisms of petrifaction, each leading to a distinctly different product [89]. He uses the terms "*Einkieselung*," "*Verkieselung*," and "*Durchkieselung*," which can be translated as "in silicification," "per silicification," and "through silicification." These mechanisms are the same as encountered during artificial mineralization, and therefore worth analyzing. The first proceeds via the precipitation of solids from solution, resulting in the pore filling of a buried material. The second is the partial or full replacement of a preexisting mineral with silica by metasomatosis, leading to its imprint in the replacing phase. The third is the – often subsequent – filling of existing pores in an already petrified material. In case of wood petrifaction, this step is held to be accompanied by a simultaneous decay of the wood structure, in turn, opening up new pore spaces for

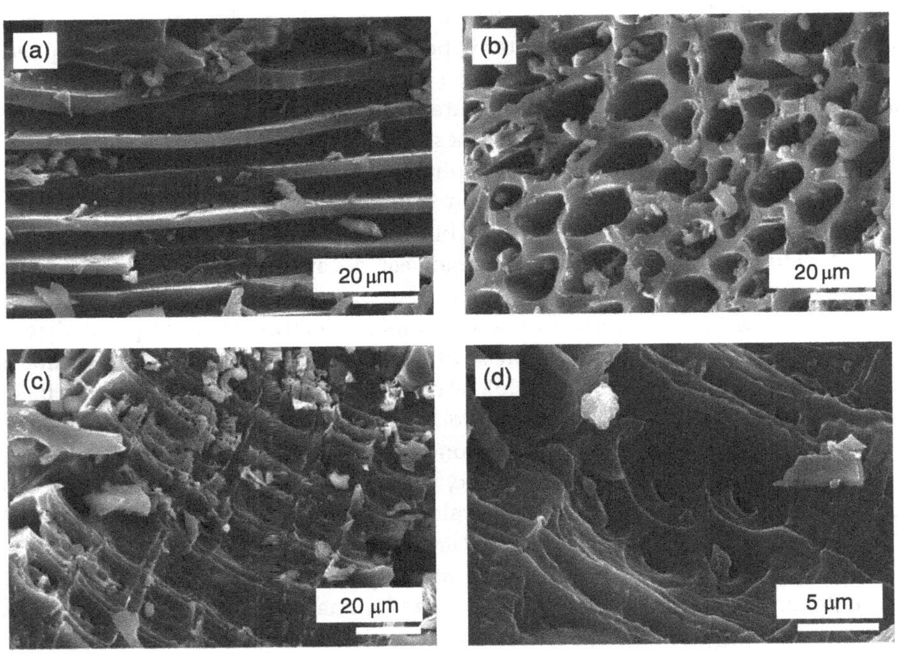

Figure 1.2 Micrometer-scale structures of coal, showing several well-replicated features such as (a, b) the wood cells, (c) ray fibril intersections, and (d) pit holes. *Source:* Modified from Zhang (2014) [83].

Durchkieselung [90]. The exact nature of the decay-petrifaction dynamic is the matter of a long and ongoing debate, and in detail varies from specimen to specimen [89, 91, 92].

In line with the mechanism of *Verkieselung*, petrified wood has been described as a mineral that has displaced the wood structure [93]. However, studies on geologically petrified [94] as well as artificially petrified wood [95–97] have shown that the mechanism is in detail a mixture of *Einkieselung* and *Durchkieselung*. However, the situation may be non-uniform within a single piece of petrified wood:

> [C]ell walls may or may not be demarcated from cell lumina; cell wall petrifaction may or may not be accompanied by partial or complete infilling of pit chambers and canals, intercellular spaces, tyloses and cell lumina; cells, even adjacent cells, within a specimen, may differ in appearance, and mineral-filled lumina appear non-uniform; and cell walls may appear to be swollen, mineral-filled intercellular spaces enlarged, and partly or completely separated cells rejoined by mineralization of the gaps between them. [98]

In geological petrifaction, the pH value influences the solubility of minerals in groundwater: The solubility of silica in water decreases to lower pH values, leading to the precipitation of silica from oversaturated solutions [99]. Studies on the degradation of wood under natural conditions have shown that the pH value of water passing through wood exposed to typical wood rot fungi can drop [100] to values of down to 3.2 [101]. This in turn means that solutions of aqueous silicic acid passing through decaying wood would preferably precipitate locally.

The shortest polymerization chain length of water-soluble silica is, in this context, of great importance: It has been determined in a laboratory study that silica oligomers larger than 4-membered rings, with particle sizes larger than 2 nm, are sterically unable to enter the finest pores of wood cell walls, which can be seen as a structural proxy for most plant cell walls [102]. Willstätter et al. determined that silicic acid monomers exist in aqueous solution, but thought that their lifespan was limited and therefore their relative concentration at any time was low [103]. Belton et al. confirmed that at close to neutral pH, monomeric silica is condensed to oligomeric species within 40 min., and that the condensation rate drops to nil at a pH of 3.4 [104].

This sterical limitation is the reason why the smallest details of geological petrifaction are typically accurate only down to the micrometer scale. One finds, however, that for archeologists and geologists, different aspects are more important than for material scientists, performing biological templating as a means of material synthesis. To identify the origins of the fossil, the geological horizon, and whether the fossil is autochtonous or allochtonous, i.e. originating from the finding site or not, are important. Further, any retained organic material from the original biological material is highly valued, since it can provide information on the chemical composition, physical structure, function, coloring, and so forth [105]. The nanometer length scale is rarely of interest to the geologist, simply because it is seldom replicated in geological fossils. As an exceptional counterexample, Buurman prepared a micrograph showing a fibrillar cell wall structure imprint [68]. The material scientist, on the other hand, will be interested in a consistent structuring of the biological templates and quality of the reagents that he employs. In the laboratory, it is possible, and often the aim of the study, to create replicas of biological structures retaining the smallest levels of structuring, as discussed in Sections 1.6 and 1.7.

However, the interest of the material scientific community in biotemplating approaches arose from the already-known fact that fossilization occurs in nature, and that (1) it is demonstrably possible to retain many levels of structuring; and (2) fossils can be used to learn in great detail about the structuring of their templates, with the advantage that minerals provide great contrast and thermal stability, making them suitable for electron microscopy. In order to trace the motivation for creating bio-replica, it is helpful to consider the history of artificially mineralized biological structures. In the following two sections, we will show how the initially isolated fields of applied engineering and structural analysis with petrified substances were combined to create biotemplating as an academic discipline.

1.3 Petrified Wood in Construction and Technology

Naturally petrified wood has been used in the construction of the agate house pueblo in the Petrified Forest National Park in Arizona. Here, sections of petrified wood provided material that only needed lengthwise fracturing to obtain building blocks. Similarly, but for its decorative properties, fossil wood is being used for construction in Texas, where deposits abound [106]. More common, however, is the use of native wood which then becomes petrified *in situ*. Examples are the foundations of Venice or the Roman bridges over the River Danube in Belgrade [107].

The first author since antiquity to systematically describe minerals by properties, and in greater detail than ever before, was Agricola [108]. He asserts that petrified wood was being widely used as whetstones in Bohemia. He also names many places where petrified wood was found, such as the Elbogen ("Elboganum") fig tree, several trees in Meissen ("Misena") near Chemnitz or the petrified oak of Hildesheim.

1.4 Structural Description and Emulation

1.4.1 Antiquity

From the observation of the locations in which petrified organic substances can be found, it is conclusive that minerals dissolved in waters are responsible for their formation. Theophrastus of Eresos (third and fourth centuries BC) is reported to have written a two-volume treaty on petrifactions [109].[3] Pliny the Elder (first century AD) wrote of various bodies of water with the ability to petrify embedded organic substances, or induce stone growth. A river near the Cicones (a tribe, i.e. in Thrace) and a river in Colchis (the southern Caucasus) named Sirius, lead to embedded organic substances being covered by a skin of stone. Further, he writes that the Sele River near Sorrento (and Salerno) bears potable, yet petrifying water, while the marshlands of the Rieti plain induce the growth of embedded stones.[4] Considering that these observations were only briefly noted in an encyclopedia, one can assume that they were preceded by more systematic research. Even without mining operations, petrified remains of archaeoplantae can be found. Prominent examples likely to have been known to writers of antiquity[5] are the Lybian and Lesbian petrified forests, readily presenting petrified wood above ground [111].

1.4.2 Modern Age: Advent of the Light Microscope

It first required the development of the light microscope to analyze microstructures and identify and name petrified biological materials, which in return led to staggering micrographs. The first analyses of the micrometer structures – and functions – of biological materials were made by Anton van Leeuwenhoek [112]. Robert Hooke, who expanded on Leeuwenhoek's work, imaged fossilized wood in side illumination geometry as well [113]. His engravings thereof show tracheae, tracheids, and an annual growth ring. Hooke also correctly hypothesized about the general pathway of petrifaction, namely,

> by precipitation, cohesion, or coagulation, abundance of stony particles from that permeating water; which stony particles, having by means of the fluid vehicle, conveyed themselves not only into the microscopical pores, and perfectly stoped [*sic*] up them, but also into the pores, which may perhaps be even in that part of the wood, which through the microscope appears most solid, do thereby so augment the weight of the wood… [114]

As discussed in Section 1.2.2, one only has to take "stony particles" to mean colloidal oligomers of silica, or any other mineral, to find that this is the correct description of the mechanism of both natural and artificial mineralization.

The first step toward imaging smaller levels of structural hierarchy in wood, both native and petrified, was Nicol's preparation of thin sections for transmitted light microscopy [115]. These showed the layered cell wall structures, as well as the pit holes (alternatively termed areolae or discs). In the latter, the typical birefringence patterns were observed, whose "beautiful appearance" and "minute assemblages of prismatic colors as seen in the microscope, cannot be represented either by the pencil or the tool of the engraver."[6] Since this birefringence is due to the tangential alignment of the cellulose crystals around the pit hole, the resulting images illustrate three hierarchical levels of hierarchy in wood. Another important finding by William Nicol was that siliceous fossil pit holes do not display birefringence. We will see later, that this indicates not only a loss of the cellulose crystallinity, but also of the nanometer structuring of the fibrillar arrangements of these crystals. Conversely, it was noted that crystallization of calcium carbonate during fossilization occurs

> in divergent prisms, and has thus given rise to a kind of cellular appearance, very readily distinguishable from the true, but which a novice may easily interpret into the regular structure of a cellular or agamic plant; while the siliceous or chalcedonic has arranged itself in parallel undulations or series of curves. [66]

We will discover that such crystallization is a major issue in synthetic mineralization as well.

Nowadays, the confocal light or scanning electron microscopes are arguably the most versatile microanalysis tools, because they afford good depth resolution and therefore allow one to image the often curved contours of biological materials or their fossil forms. An alternative, or additional, method of preparing samples for light microscopy involves increasing the surface structuring by etching with a suitable acid such as hydrochloric or hydrofluoric acid. Then, a thin film of cellulose acetate is placed, or a carbon film evaporated, on the sample [116]. Selmeier noted, among other practical notes on the preparation of microscopic sections, that this peel method only requires one-hundredth of the material required for the preparation of thin sections [117–119]. An advantage of this approach is that it allows evenly thick, microscopically analyzable structures, even of curved surfaces, to be rapidly created. Thereby, the limited depth of focus of the light microscope is overcome at the cost of the original overall outer sample shape and information on the internal structuring.

1.4.3 Aqueous Silicon Dioxide, Prime Mineralization Agent

If organic substances are to be encrusted or replaced by a non-organic phase, this phase has to be transported into the material either on its own, or in a carrier medium. In both cases, the phase or medium has to be either liquid or gaseous, with liquids having the obvious advantages of being denser, and therefore more efficient in transportation rate, as well as being easier to handle.

In this respect, polymorphs of silicon dioxide stand out, not only in nature, but in science and application as well. Most importantly, they have a low tendency to crystallize. This fact forms the basis of all glass-making techniques since it allows forming and then freezing a melt of glass. It also means that nanometer-scale silica structures can remain stable without distortion due to crystal growth [120, 121], even without the use of surfactants or capping agents. Due to its mostly amorphous structure and highly symmetrical building blocks, its mechanical properties are highly isotropic as well. Mann et al. noted that exactly this isotropicity of mechanical and crystallization behavior is what enables nature to mineralize complex shapes with silica [36]. Similarly, Perry et al. noted in their comprehensive review on biomineralized silica that it occurs in nature "in a variety of structural motifs, e.g. sheet-like, globular and fibrillar" [122]. We -will see that the same holds true for materials derived from artificial petrifaction.

The silica polymorph quartz has been put forward as the possible origin of life itself. Reversing the idea of templating silica by biological materials, Larry Hench noted that the bioactive interaction between quartz and amino acids could tightly bond them to specific crystal planes. By thus lowering the entropic energy barrier for photocatalytic reactions with other randomly formed amino acids, "the inorganic origin of biopolymers is irreversibly and immutably locked into the very beginning of the genetic code" [123].

With regard to commercial viability, silica is abundant, literally sand at the seashore. By contrast to silicon, which must be obtained by reduction, it can be used directly in its native form. However, to be used directly, it must still be molten and softened by an energy-intensive process. Instead, there is nowadays a wide variety of mixtures and compounds of silica as its silane derivatives that can be precipitated from solution for a wide variety of applications.

Of great importance is the ability of solutions of silica to glaze, i.e. deposit hard, clear films even on rough surfaces. For this reason, an aqueous solution of silica, often with added sodium to improve the solubility, has always been desirable. In modern times, many alchemists and chemists have researched the manufacture of water-glass, which has gone by the names *Spiesglas* [124, 125], *succo lapidescenti* [108], *liquor silicium/-is* [126, 127], finally settling on *Wasserglas*, water-glass. This has a wide range of applications [128–130], among them the preservation and restoration of cultural artifacts [131, 132]. To deposit either water-glass or any other solution of silicon dioxide in water, one merely needs – similar to the processes in nature – to change the solution's acidity, or remove the solvent by evaporation [133–135].

1.4.4 Artificial Petrifaction of Wood

Basilius Valentin, a Benedictine monk and a very eager alchemist from Erfurt [136], about whose life little is known,[7] allegedly not only created the first water-glass in 1520, but also used it to petrify wood [126, 127]. Another early description, also reported by Leo and Barghoorn [142], is passed down by Meyer, citing Kentmann (1518–1574) from Torgau. He presents a conclusive method of petrifying wood, namely, by burying wood soaked with beer mash, which is rich in dissolved silica, for "3 years in a fresh sand or gravel in a cellar" [143]. Kentmann attests to the qualities of the resulting *lapides igne liquescentes* as whetstone or, strangely, as flintstone.

More recent works on artificially replicating biological substances in the laboratory have yielded structures that bear close semblance to fossil structures found in nature. The range of artificial fossils that were prepared in this manner is large, even including bacteria [144]. However, by far the most commonly used structure for templating purposes is wood. The reason for this is that it provides a structuring that is well known, straightforward to assess, and also desirable for applications, as we will see in Sections 1.5 and 1.6. We will therefore often use wood as a model structure in the following.

The general processing steps of any biotemplating method, of which the artificial mineralization is a part, are outlined in Figure 1.3. The individual steps may vary greatly in actual implementation and may include harvesting fresh and preserving or stabilizing the structure, for example, by preventing it from drying [145]. It may also include removing components from the structure by organic extraction [146] or regioselective oxidation (bleaching), thereby opening up pore spaces into

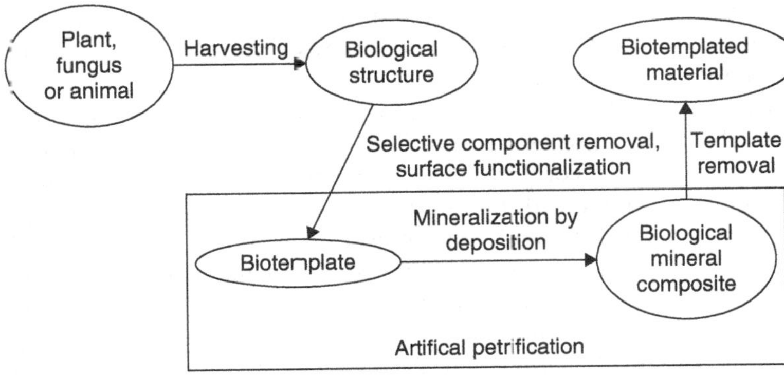

Figure 1.3 Scheme of the general pathway to creating biotemplated materials starting from a living object. The boxed steps are those followed during the artificial petrifaction of structures. The surface functionalization is an optional step; the deposition may be reactive, or purely sterical, as outlined in the text.

which minerals can be deposited [147]. Organic extraction is usually necessary due to the varied and complex nature of organic low molecular weight extractives, which would lead to unwanted reactions with the precursor fluid [148].

Functionalization of the surface of the resulting biotemplate may increase the polar affinity toward a subsequently infiltrated substance [149–151], or increase its reactivity [152–154]. A widely used and effective reactive functionalization agent for polysaccharide biotemplates is maleic acid anhydride [155, 156] When applied in a solution of dimethyl acetamide, the solvent swells the biological tissue [157] and the anhydride reacts with the surface hydroxyl groups with its carboxylic acid functionalities, which in turn have a higher reactivity toward impregnated phases [158], such as dissolved minerals [96].

A good non-reactive coupling agent combines a chain end that condenses with wood cell wall hydroxyl groups and another end that nonreactively couples with the infiltration agent. Aminopropyltriethoxysilane, 3-glycidoxypropyltriethoxysilane or 3-isocyanatopropyltriethoxysilane are suitable silica coupling agents since the amino-, epoxy- and isocyanate end groups can react with the hydroxyl groups, while the oxysilane groups couple sterically with deposited material [159–161].

A review of nanometer-scale coating processes showed how hydrolyzing and condensating silica precursors (mainly tetraethyl orthosilicate) by acids and bases leads to coatings of structures with silica or combinations of silica and other materials, such as titania, zirconia or gold nanoparticles with varying thicknesses [162]. Priming with gelatin was described as a method to allow the coating of particles that do not readily bond to silica, such as iron [163]. The fact that gelatin is a polypeptide and biological materials such as keratin and DNA are good templates for silica templating illustrate the intimate bonding that is possible between biopolymers and silica.

Any biological template, whether artificially functionalized or not, must be infiltrated with a precursor material in gaseous [164, 165] or, more commonly, liquid form. The former was demonstrated by Hyde et al., who used atomic layer deposition to evenly coat cotton fibers with increasing thickness per deposition cycle [166]. Gas phase templating of biological structures can also be carried out by autoclaving diatoms consisting of silica together with pieces of metal or metal fluoride gas and resulting in a magnesia or titania replica of the material [167, 168]. Following up on these processes, such titania replicas were used as templates for the preparation of perovskites by treatment with titanium fluoride, sublimated in a protective atmosphere, followed by barium or strontium hydroxide to give titania, barium titanate or strontium titanate replica [169]. Entire gold replica of diatoms with catalytic rate constant were attained by functionalizing its surface with 3-[2-(2-aminoethylamino)ethylamino]propyl-trimethoxysilane and then depositing gold from tetrachlorauric acid, followed by the removal of the diatom by hydrofluoric acid [170].

Liquid deposition phases can be melts, solutions of salts or condensable reactive precursors, or particulate slurries, although the latter method is rarely used [171]. Of the examples presented in Section 1.6, we will explore in detail which structures result from applying either of the named substances. Generally, it is advantageous to choose a solvent that has a high solubility for the salt or precursor, is able to wet the cell wall, does not inhibit the reactivity of the precursor, and optionally swells the biological template without dissolving or greatly structurally altering it. Alcohols are solvents which fulfill the aforementioned criteria. They are therefore widely used for the infiltration of alkoxide precursors. They are sufficiently hydrophilic to wet the cell walls [172], but do not prematurely hydrolyze the precursors [173]. The latter can then condense to oxide ceramic phases in the presence of hydrolyzing agents, such as water or a carboxylic acid [174–176]. In nature, diatoms mineralize by changing the pH within the silicon deposition vesicle [177]. Coccolithophores deposit calcium carbonate, in the presence of acidic polysaccharides [178, 179]. We propose that, similarly, the inherent acidity of wood aids in the spatially defined deposition of silica in the cell wall pores during natural or artificial mineralization [180].

Navrotsky said: "The living world is wet," [181] and therefore, water is present in the cell walls of all but the most elaborately dried biotemplates, which then rapidly rehydrate to equilibrium moisture from air [182–184]. In many sol-gel reactions, the precursor is first hydrolyzed under loss of alcohol, and then condensed under loss of water, in which case the initial presence of water can serve as a spatially confined catalyst. The aforementioned deposition of silica from an alkoxide precursor such as tetraethyl orthosilicate is an example of this.

Predicting the deposition rate of minerals from solution into organic materials in nature and in the laboratory must be based on the concentration of the dissolved salt or precursor in the solution, on the transport rates into and within the organic material, the enthalpy of reaction of the formation of the solid phase on the template by heterogeneous crystallization or a condensation reaction, and the specific surface of the material. Consequently, all other factors being predetermined by the type of template, we propose that the progression of the mineralization of hierarchically porous biological templates depends on their pore structures in the following way. If the offset to the bulk energy gained by nucleation is the surface energy spent by the surface tension, then the "more heterogeneous" the nucleation, the faster the nucleation rate. The rate in a closed thermodynamic system is proportional to the negative exponent of the Gibbs free energy of the critical nucleus $\exp(-\Delta G/(k_B T))$. The nucleation will be more heterogeneous, if the template surface is highly curved, and the pore radius correspondingly small. It was determined that nucleation in pores with radii no larger than two times the critical pore radius for homogeneous nucleation reduces the ratio of the free energy of heterogeneous to homogeneous nucleation $\Delta G_{het}/\Delta G_{hom}$ by a factor of 2 [185]. Radii of the same size as the critical pore radius for homogeneous nucleation reduce the ratio by a factor of 10. The nucleation rate in these examples is then greatly increased, namely, by a factor of $\exp(0.5\cdot\Delta G_{hom}/(k_B T)$ and $\exp(0.9\cdot\Delta G_{hom}/(k_B T)$, respectively. Similarly, another study determined, via Monte Carlo simulations, that pore sizes of within 20% of the size of the critical nuclei in homogeneous nucleation show a greatly increased nucleation rate, an effect that diminishes at larger pore sizes [186]. It was also found that the critical pore radius for the formation of silica oligomer colloid spheres is 3 nm, which is in turn similar to the smallest pore size determined for wood biotemplates in previous studies [96, 97]. As already mentioned in Section 1.1 and further discussed in Section 1.5.1, one can therefore state that a typical biotemplate consists of distinct pore size domains, the smallest of which greatly promotes nucleation of the mineral phases. The larger pore domains would still give rise to heterogeneous nucleation, albeit without the enhancing effect of the small radii of curvature.

The afore-mentioned pore curvature dependence and therefore size dependence imply that nucleation first takes place in the smallest of pores. Since the mineral phase is typically in solution and close to its maximum solubility concentration, the energy gain by grain growth will be comparatively small. Further, silicates are initially deposited in amorphous form, further reducing the bulk energy gain by expanding existing phases. Therefore, the high initial rate of deposition will decrease markedly once all smallest pores are filled. This may be combated by supersaturating the solution, either by adding more mineral phase, or by evaporating the solvent. The latter was routinely done for the mineralization of wood in the laboratory [96], but it is also widely found in nature. Here, periodic wetting and drying of wood have been identified as a factor leading to a good replication of the wood structure [93]. We found that applying dissolved ethanolic silica precursor in different portions, and drying the templates in the solution [96] yielded different weight percentage gains in the final composites (Table 1.1). Our interpretation of this data is that multiple infiltration cycles open up more pore spaces during each drying step, thereby opening up further spaces for the afore-mentioned nucleation in small pores.

The question whether a material that has been deposited within a porous biological structure has reacted to the surface of the material, i.e. formed covalent bonds, has been the subject of numerous studies [188–192]. In this context, chemical imaging via confocal Raman microscopy is able to provide valuable insights into the spatial distribution, and bonding states of wood modification agents [187]. For deposited inorganic phases, impedance spectroscopy has been shown to be a suitable method for the determination of the reaction of the template. Since the dielectric properties of wood are mainly determined by the functional groups or water contained therein [193–195], this technique can be used to analyze the reaction of wood cell wall groups with the

Table 1.1 Comparison of the mass increase due to multiple infiltrations with constant masses of dissolved ethanolic tetraethyl orthosilicate precursor as opposed to single infiltrations with larger amounts.

Number of infiltration cycles	Mass ratio of total silica precursor per wood (g/g)	Resulting weight percentage gain
1	0.1	0.5
1	1	6.9
1	3	7.1
1	10	8.9
2	2	14.4
3	3	17.0

Source: Van Opdenbosch (2011) [96].

deposited material [196, 197]. In particular, the relaxation times, dielectric constants, and dispersion parameters of the entire material allow one to assess whether the wood hydroxyl groups are still freely oscillating in the electric field, or bound by reactions. The incentives for creating covalent bonds between the wood structure and embedded phases for wood protection purposes are: (1) creating a stable link, thereby preventing the opening of micropores through mechanical actuation; and (2) preventing the hydroxyl ligands from forming hydrogen bonds with water molecules. For biotemplating purposes, the latter aspect is arguably less important, since the ultimate step of the processing is to remove the template.

The removal of the organic phase is complicated by the fact that it comprises a significant proportion of the mineralized structure, typically more than 50% [198–201]. This necessitates an approach, which does not induce large stresses on the mineral phase in order to avoid cracking. This could, as happens in nature, be achieved through biological removal [202, 203]. However, the mineralization of natural structural materials is known to lead to a decrease in its susceptibility to biological decrease [204–206]. One finds organic residuals, even in millennia-old petrified materials [207–210]. Therefore, and due to the long time frame required for a thorough organics removal, this approach has not been used in biotemplating approaches. Instead, Ryan Drum used chromic acid, a strong oxidant and highly toxic compound, to selectively remove the wood from his artificially mineralized composites [116]. The most common approach, however, is using thermal oxidation of the organic phase [96, 211]. This approach allows fine control of the degradation velocity of the organic phase by matching the temperature profile to the existing thermal analysis data, a typical example being shown in Figure 1.4. The regions of high mass loss should be traversed slowly, either by holding at a set temperature, or via a slow heating rate. For each mineral phase, there is a maximum temperature, after which it will undergo sintering, destroying the finer structural details of the template [96, 97, 212].

1.5 Characteristic Parameters

1.5.1 Hierarchical Structuring

All materials are structured on more than one length scale. Even monolithic and monocrystalline or amorphous materials possess an outer shape, with dimensions of, for example, several millimeters, as well as the crystal structure, or the structure of the amorphous building blocks, at the sub-nanometer level. Therefore, the *number* of hierarchical levels is what sets biological materials apart from their technological counterparts.

The number of hierarchical levels may be defined in two different ways. One is to look at the dimensions of all structural features present in the sample and sort them into categories, e.g. the decimal metric scale. We are using this approach in the following. The other way is to determine which composite *functional* units are present.

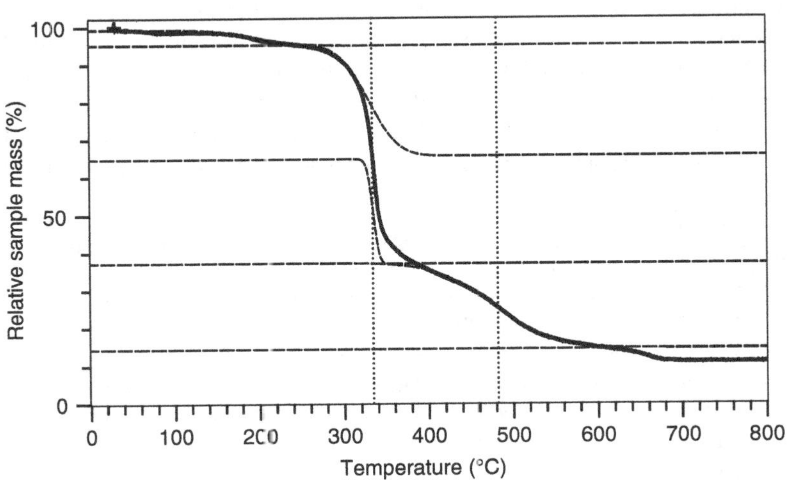

Figure 1.4 Deconvolution of a thermal analysis curve (in air, solid line) of an artificially mineralized biological template (phenylated silicon dioxide on wood) [213] by Gaussian cumulative distribution functions (dashed lines) to give a fit of the curve with a goodness $R^2 = 0.9999$. The mass losses associated with the proposed processes are (low temperature to high): 4.4 Ma% loss of absorbed moisture, 29.8 Ma% and 27.7 Ma% stepwise devolatilization of the wood structure,[1] 22.7 Ma% combustion of the carbonized material, 3.6 Ma% combustion of the temperature-resistant phenyl substituents 11.9 Ma% ash. The vertical dotted lines mark appropriate holding temperatures for a gentle template removal.
Source: Van Opdenbosch (2013) [213].
Note:[1] The widely accepted sequence is the carbonization of the hemicelluloses followed by the more crystalline cellulose, both of which have an initial endothermic and a subsequent exothermic phase [214–216].

An example of such a unit is the assembly of collagen molecules into fibrils, reinforced with embedded and extrafibrillar apatite platelets. Then, one determines the length scales of such assemblies. Since an assembly may contain smaller sub-units, one arrives at a lower figure for the hierarchical levels in the material. These do, however, have a direct functional relevance which can be assessed, first independently, and then comprehensively [217].

Engineering materials typically contain three distinct levels of hierarchy [218]. The atoms of which the material consists are arranged in molecules or a crystal lattice at the sub-nanometer level. The crystalline lattice is segmented into grains, each typically with a uniform phase on the micrometer level, and each material has an outer shape at the millimeter or meter level. Occasionally, an intermediate structuring in the shape of nanometer scale sheets, lamellae, fibers. or grains can be found. The same argument can also be made for materials that are doped, or microstructured or deposited in layers, since every such structuring method is at the expense of another structural level.

In biological materials, on the other hand, as few as six and as many as nine [219] levels of hierarchy have been found (Figure 1.5). Conclusively, by comparing problem-solving techniques in technology and nature, Vincent et al. found that biological materials rely heavily on structuring, and the information required to create such structuring with minimal energy effort, while engineering materials require a large amount of energy, as well as specific substances for their manufacture [218].

The importance of hierarchical structuring is shown by the range of functional materials from nature. The adhesive mechanism of Gecko feet, among many other examples [220], was found to be based on an intricate branching network of keratinous hair, setae, each branching several times and terminating into hundreds of so-called spatulae a few hundred nanometers in diameter [221]. As also compared in Table 1.2, bone [222, 223], wood [224–226], or lobster cuticle [227, 228] use hierarchical structuring to fulfill a load-bearing role. In both cases, a common structural motif, is used to build different structures on larger length scales, thus allowing the formation of materials with widely varying properties. Bone, for instance, can organize mineralized collagen fibrils into parallel bundles, woven fiber structures, plywood-like structures or radial arrays [229]. Similar to the properties of carbon-fiber

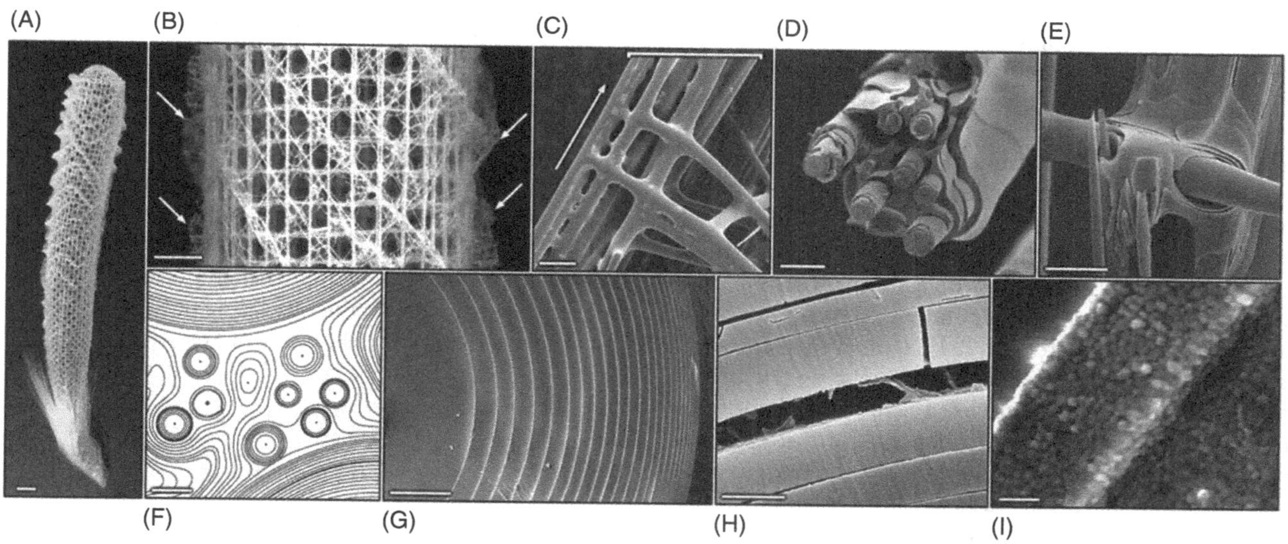

Figure 1.5 Structural analysis of the mineralized skeletal system of *Euplectella sp.* (A) Photograph of the entire skeleton, showing cylindrical glass cage. Scale bar, 1 cm. (B) Fragment of the cage structure showing the square-grid lattice of vertical and horizontal struts with diagonal elements arranged in a chessboard manner. Orthogonal ridges on the cylinder surface are indicated by arrows. Scale bar, 5 mm. (C) Scanning electron micrograph (SEM) showing that each strut (enclosed by a bracket) is composed of bundled multiple spicules (the arrow indicates the long axis of the skeletal lattice). Scale bar, 100 μm. (D) SEM of a fractured and partially HF-etched[1] single beam revealing its ceramic fiber-composite structure. Scale bar, 20 mm. (E) SEM of the HF-etched junction area showing that the lattice is cemented with laminated silica layers. Scale bar, 25 mm. (F) Contrast-enhanced SEM image of a cross-section through one of the spicular struts, revealing that they are composed of a wide range of different-sized spicules surrounded by a laminated silica matrix. Scale bar, 10 mm. (G) SEM of a cross-section through a typical spicule in a strut, showing its characteristic laminated architecture. Scale bar, 5 mm. (H) SEM of a fractured spicule, revealing an organic interlayer. Scale bar, 1 mm. (I) Bleaching of biosilica surface revealing its consolidated nanoparticulate nature (25). Scale bar, 500 nm. *Source:* [219]. Reprinted with permission from AAAS.

Note:[1] Hydrofluoric acid treatment and bleaching experiments: Small sections (2 cm by 2 cm) of the Euplectella sp. Skeletal lattice were soaked for 10 min in 5 M NH 4 F:2.5 M HF or 2.5% NaOCl solution, respectively. The remaining skeletal material was removed, rinsed with water and 95% ethanol, dried, and prepared for examination by scanning electron microscope.

Table 1.2 Comparison of levels of hierarchy found in selected natural materials [230]. Note that plant elementary fibrils are sometimes termed micro- [231] or nanofibrils, and subsequently microfibrils become macrofibrils [233], or just fibrils [232].

Hierarchical length scale	Wood	Tendon	Bone
Ångström	Glucose monomers	Amino acids	Hydroxylapatite unit cells, amino acids
Nanometer	Cellulose elementary fibrils, hemicellulose and lignin molecules	Collagen molecules	Collagen molecules, apatite platelets
10 Nanometer	Lignin and hemicellulose encased cellulose fibrils	Helical molecular arrangement	Extrafibrillar apatite platelets
100 Nanometer	Cell wall layers	Collagen fibrils	Mineralized collagen fibrils
Micrometer	Cell wall	Fibroblasts in proteoglycan matrix	Osteon sublamellae
10 Micrometer	Tracheid cells	Collagen fascicles	Osteon lamellae in rotated plywood structure
100 Micrometer	Trachea cells	Tendon fibers	Vessel pores
Millimeter	Growth rings	Outer shape bounded by epitenon, attachment points	Bone struts
10 Millimeter	Trunk sections		Anatomical sections of bone
100 Millimeter	Trunk sections		Outer shape of bone
Meter	Outer shape		

compound materials, these arrangements result in materials that are less and less anisotropic, parallel bundles being the most anisotropic.

The hierarchical structuring of a material is highly adapted to its functionality. Fratzl and Weinkamer found that the animal and plant kingdoms apply similar structural principles to create mechanical materials that are remarkable tough, yet sufficiently stiff and resistant to fracture [40]. Fracture resistance has been found to be strongly dependent on the angle between the structural orientations and the crack propagation direction. The latter in turn is commonly dependent on the direction of the strain on the material [234]. However, Gao et al. determined that there is a critical structuring size below which the fracture strength of a structured biological composite material becomes independent of stress concentration at interfaces, which limits the strength in brittle engineering materials [235]. With the example of the *Euplectella aspergillum*, it was hypothesized that nanometer scale structuring may influence the manner of crack deflection [236]. Deflection at the micron-scale organic interlayers was determined with certainty. If the first hypothesis were found to be true, this would indicate that multiscale strengthening mechanisms are implemented by nature.

The biopolymers incorporated in many biological materials serve not merely as a glue holding the load-bearing components together. Instead they play a more active role in determining the mechanical properties, namely, by providing complex and stepwise elastic resistance to stresses, thereby dissipating energy and preventing or delaying material failure [237, 238]. While composite engineering materials also combine elastic with strong phases, they lack the highly adapted molecular unraveling mechanisms that characterize the response of biopolymers in biological mechanical materials.

We have compiled a selection of hierarchical biological materials for comparison purposes (Table 1.2). The selection comprises materials from the animal and plant kingdoms as well as one example for an organic-inorganic, i.e. mineralized material. Using the concept of hierarchical order, Roderic Lakes reviewed that the ratio of strength to density increases with the increasing number of hierarchical levels distinguishable in a material and given by Eq. (1.1) [239].

$$\frac{\sigma_n}{\rho_n} = \frac{E_0}{\rho_0} a k^n \left(\frac{\rho_n}{\rho_0}\right)^{r-1+(q-r)/n} \tag{1.1}$$

With σ_n and ρ_n being the strength and density of a material with hierarchical order n, E_0 and ρ_0 being the bulk elastic modulus and density of the material, a and q the strength multiplier and exponent, k and r the stiffness parameter and exponent, respectively. The values of a, q, k and r depend on the structure as well as on the mode of deformation and are in the range between 0 and 3, a being an exception with possibly higher values [239].

Vincent et al. elaborated on the importance of hierarchy: in an engineering material with a small number of structural hierarchies, a problem can be solved by modifying the material on the same level of hierarchy on which it occurred [240]. As already pointed out, and further discussed in the following, the several levels of hierarchy found in natural materials interact to provide an overall behavior. The cause of a problem, for example, the occurrence of mechanical failure observed at a given level of hierarchy need not necessarily originate from the same level. Vice versa, the solution to a problem may lie in changing the material on an entirely different hierarchical level [241].

For biotemplating purposes, it is therefore not only important to know the species of plant or animal tissue that is being used, but the specific location where it was derived from. This is further illustrated by the example of wood, where two sample specimens from the same branch can show a considerably different structuring on the 10 nm scale, with accordingly greatly altered properties in the native wood templates [241].

Many structures that nature builds for protective purposes are composed of the brittle mineral calcium carbonate: chalk. It has been found that nature relies heavily on hierarchical structuring to greatly alter its properties. Seashells [242], for instance, achieve a remarkable fracture toughness of up to 2.26 $MPa{\cdot}m^{1/2}$. The similarly composed egg shells have been used to template materials as varied as zinc, cobalt and palladium oxide in a comprehensive study [243], silicon carbide [244], titania [245], or zirconia [246]. In the latter two cases, the calcium carbonate was removed by treatment with hydrochloric acid, washed with water, and dehydrated with alcohol. The resulting biotemplates were impregnated with metal alkoxides and dried. The membrane template was then removed by thermal oxidation. This process followed the generalized procedure outlined in Figure 1.3, and yielded intricate tubular ceramic networks with specific surface areas on the order of several 10 m^2/g [245, 246], By using the intricate structure responsible for the seashell's or egg shell's great strength, it was shown that one can create a material structuring not achievable with any other currently known approach. Further, Wang et al. demonstrated that mixed silver and carbon-doped titania structures obtained by egg shell templating showed improved photocatalytic activities over a standard commercial product [247].

By calcining the horsetail plant *Equisetum hyemale*, a grass with high silica content, Sapei et al. demonstrated that the complete retention of the interconnected, complexly structured silica phase within the native grass is possible [248]. Zampieri et al. templated *Luffa* sponges by depositing a zeolite phase which is itself porous [249]. The resulting specific surface areas were between 370 m^2/g and 410 m^2/g, providing a suitable scaffold for catalytic reactions. Valtchev et al. used a combination of these two processes by templating and retaining the silica structure of *Equisetum arvense* with a zeolite precursor material, resulting in a material with a specific surface area of several 100 m^2/g, more than 500 m^2/g, typical of zeolites [250]. The novelty of this process lies in the addition of a level of hierarchy to the already existing hierarchically structured silica material, increasing its already high specific surface area. This led to a multimodal pore distribution, which is atypical for zeolites. Another example of an added level of hierarchy was presented by Gonzales et al. [251]. By depositing bioactive glass on a previously wood-templated silicon carbide ceramic, they prepared a material that combines the high anisotropic strength of biotemplated silicon carbide with the functionality of the porous bioactive glass.

We have already stated that knowledge of the nanometer scale structure and its effect on the properties of the biotemplating product are of great significance. In this context, the structure of wood illustrates well the importance of also choosing the correct micrometer structuring, i.e. species, to begin with. Wood densities range from 0.1 g/cm^3 in Balsa wood to 1.3 g/cm^3 in Guaiacum wood (*lignum vitae*) [252]. It can be comparatively fragile, highly flammable, lubricating,[8] dark in color, with visible trachea, rot-resistant, or the opposite of all these afore-mentioned characteristics [16]. It is apparent that by merely changing the type of wood used for the templating process, one will receive a material with widely different properties and processabilities. The dense, oily wood of oak or *lignum vitae* would, for example, not be suited for biotemplating purposes without the application of extensive organic extraction procedures. Two examples of how the choice of wood changes the properties of the final product will be discussed in Section 1.6.1.

Not only natural mechanical materials derive their unique properties from hierarchical structuring. Optical materials, as prominently found in the carapaces of beetles, show combined optical effects derived from structuring on the nanometer, up to the millimeter scale. In the simplest case, natural reflectors are simple sheet or lattice reflectors. However, some beetles have carapace reflectors that include further levels of hierarchy which add further functionalities.

An example of this is circularly polarized iridescence in beetles [253, 254]. The polarization dependence of the structure is due to a chirally arranged cholesteric structuring within the cell walls of the reflecting structure. Another example is the moisture-dependent coloring in beetles. By contrast to most photonic reflectors [253, 255], some beetles' photonic reflectors consist of structures that are porous on a length scale below that required for optical reflection [256]. This allows the passage of moisture through the structure, thereby allowing passive changing color on moisture change. Similarly, there exists a beetle that can actively fill fluid channels to change its coloring when stressed [257].

Skipping the photonic lattice itself, i.e. two scales of hierarchy larger, the arrangement of multiple domains of uniform photonic lattices also determines the visual impression provided by natural reflectors. Mixing the colors reflected from domains in the direction of an observer leads to an observed range of wavelengths and a mixed optical impression [258, 259]. This also means that a uniform overall coloring over a large angular range is observed. Such a strategy to create a uniform coloring from anisotropic photonic reflecting domains is not only found in animal reflectors but also in plants [260]. There is also a combination of pixilated coloring and polarization dependence that might be described as ingenious if it were man-made. It is based on geometries of the reflecting multilayers that selectively reflect light of different polarization by means of trapezoid cavities [258]. The afore-mentioned examples provide not only inspiration for bioinspired materials but also for biotemplated materials obtained from the structures themselves [261, 262]. In Section 1.6, we will discuss how these properties can be transferred to engineering materials.

Nature not only uses hierarchical structuring to create functionalities, it also uses low-energy pathways to create these materials. The complex structure presented in Figure 1.5 is grown underwater in a cool and dark, low-energy environment by an immobile sponge. The body of *Euplectella sp.*, was found to have a layered structuring not unlike a man-made glass fiber, as seen in Figure 1.5. It was proposed that it fulfills a similar light-guiding function [219]. Man-made glass fibers on the other hand are drawn from a melt at about 1200 °C. With vitreous silica having a heat capacity of $700\,J{\cdot}kg^{-1}{\cdot}K^{-1}$, making an 'artificial Euplectella' that weighs the same as the original, around 18 g, from drawn glass fibers would require an energy of 15 kJ for the melting step alone, enough to bring 36 g of water to the boil.

1.5.2 Specific Surface Areas

A measure of the replication detail is the specific surface of the resulting materials. However, it is elusive for two reasons. First, materials with the same specific surface areas may have different pore hierarchies. An aerogel with uniform pore sizes may have the same specific area as a hierarchically structured material with a smaller minimal pore size. Second, biotemplated materials may have high specific surface areas due to retention of the smallest structural details. However, this is not a guarantee that these details have been retained, since the finest pores may become closed, or crystallites may precipitate that provide a large surface area but do not reflect the nanoscale structure of the template. Therefore, most studies combine a nitrogen sorption analysis with transmission electron microscopy or Small-angle X-ray scattering (SAXS) to determine all pore sizes that are present in the sample, and whether they relate to the structuring of the template [263–265].

1.5.3 Pore Structures

The actual pore structures are, therefore, a more suitable measure of the structure retention during biotemplating processes. There are ample descriptions and depictions of geologically petrified wood sections fossilized in their entirety, while retaining their original structuring from the meter to the micrometer level, i.e. the stem cross-section, bark, wood, and pith, as well as the pores [66]. Such a replication of hierarchical structures requires that there are available pores on all levels of hierarchy.

In plants, the smallest pore volumes are partially filled by extractives, such as terpenes, fats, and low molecular extractives [146]. Also, fresh plants may contain a significant amount of water. Its thermal removal would greatly alter the pore structure. Instead, as exemplified by work on plant leaf templating [266], chemical dehydration, i.e. the exchange of water for an organic solvent, is usually necessary. Further, the selective removal of cell wall components, as happens in nature by white- or brown rot, opens up volumes in plant cell walls into which minerals can be deposited [267]. The field of wood modification [267–269] can give important insight into the accessibility of wood cell walls, since wood cell wall mineralization for protective purposes is only one step – the template removal – away from biotemplating.

A property that can be achieved by replicating biological materials is a chiral porosity at the 10 nm or 100 nm scale. As comprehensively reviewed by Sotiropoulou et al., many biological structures, among them lipids and DNA, show chiral structuring by self-assembly [270]. It is possible to artificially create chiral structures by self-assembly of nematic cellulose with the same methods also employed in biotemplating [271, 272]. The resulting material was very similar to chiral protein stacks found in the cuticles of beetles, as discussed in Section 1.6.4.

As further discussed in Section 1.6.4, chiral structures on the 100 nm scale are responsible for the polarization-dependent optical response of photonic biological materials. These structures, containing rotation-translational symmetry axes describing corkscrewing planes running through the material, are thought to be assembled from cytoplasmic membranes in beetles [273]. However, the exact process – if there is a unifying route – of the structure formation is, to our best knowledge and due to lack of observations *in vivo*, unknown. It requires systematic analysis to correctly identify a chiral motif in such lattice works, since the helical pitch is not apparent when viewed from any fixed direction in the microscope [274]. Correspondingly, Stavenga et al. describe how the internal structuring of *Callophrys rubi* was successively identified as simple cubic, face-centered cubic, as inverted face-centered cubic opals and finally, the gyroid structure [273]. The gyroid is described as a net of points centered between and normal to adjacent surfaces with a space group of $I4_132$, indicating its rotation-translation axis [275]. Further, even in materials without a screw axis that is apparent by observing the pore channels, directions-containing screw axes may exist. One example is the diamond surface with space group Fd $\overline{3}$ m, d signifying the diamond (¼,¼,¼) translation [275].

Arguably, SAXS is the most practical, rapid, and efficient method to obtain statistical information about sample cross-sections up to several 100 μm [276]. This is particularly true in contrast to transmission electron microscopy (TEM), which requires more elaborate sample preparations, yet yields information about slices that are less than 100 nm in thickness. The nanometer-scale structural information obtained by TEM consists of images, which are, for the most part, intuitively interpretable. Scattering patterns are recorded in the frequency domain, and generally require knowledge of the underlying principles for interpretation. In the case of SAXS, the patterns are generally highly convoluted due to the scattering process itself, the non-uniform wavelength distribution, the beam intensity profile and divergence, as influenced by the pinhole geometry and atmospheric scattering [277, 278]. This convolution is an obstacle to their interpretation. This is well illustrated by patterns obtained from arrangements of fibrous structures, as often found in natural structures. In wood, 2 nm and 10 nm cellulose elementary and microfibrils are helically wound around the roughly cylindrical lumen [279, 280]. Due to this, parallel and coherent X-rays are scattered perpendicular to the fiber orientations [281]. Their angle relative to the lumen axis is called the microfibril angle. There is a scattering set-up-specific minimum threshold below which this angle cannot be clearly discerned except by detailed knowledge of the machine parameters and data fitting (Figure 1.6).

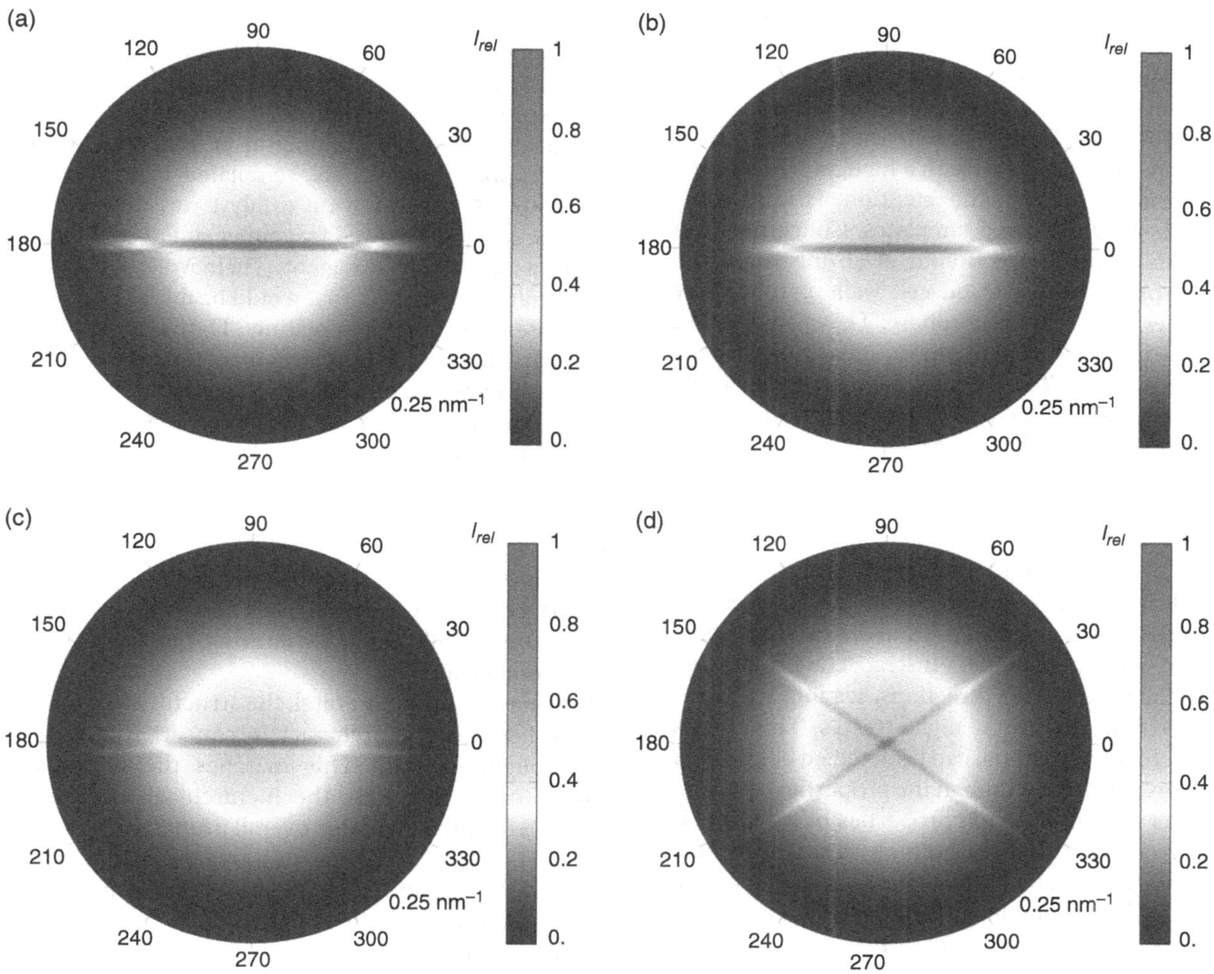

Figure 1.6 Simulated small-angle X-ray scattering patterns from closest-packed arrangements of each 500 fibrillar structures (with angular uniformity of σ_{ang}:0.2 $^{\circ-1}$), 2 nm in diameter, embedded in cylindrical cell walls, which were probed centrally, accounting for isotropic scattering from randomly oriented structures (10 nm) and for a Gaussian beam profile (σ_{beam}:8 nm^{-1}) [282]. The simulated microfibril angles (cylinder front to back) are (a) 0 °, (b) 2 °, (c) 3 ° and (d) 35 °. The separation of the directional scattering patterns originating from the fibrils can, under such practical conditions, be observed starting at 3 °. (*See insert for color representation of the figure.*)

1.6 Applications

1.6.1 Mechanoceramics

The concept of biotemplated ceramics for mechanical applications was pioneered by the group of Ota [283]. It was initially a study on the microstructure formation that can be achieved by infiltrating a wood template with tetraethyl orthosilicate or titanium tetraisopropoxide [284] as silica or titania precursors respectively, followed by firing under a protective atmosphere to give reaction-formed silicon carbide by carbothermal reduction. The results were materials composed of parallel cylindrical cells with diameters of several micrometers. Similarly, it could be shown that titanium carbide materials could be prepared by a titanium alkoxide infiltration and high temperature treatments [285].

Porous ceramics are commonly prepared by using polymer foams to produce isotropic material with spherical pores by evaporation [286, 287]. Byrne et al. and Greil et al. used silicon-, polymer melt or gas infiltration of previously carbonized wood to create cylindrically porous silicon/silicon carbide/molybdenum silicide composite materials [288–292]. These combine the inherently high strength of the ceramic phase with a high and directional fracture resistance, based on the retention of the micrometer-scale fibers of the wood templates. Concerning the smaller levels of hierarchy, Brandt et al. measured the mechanical properties of wood cell walls pyrolized by heating under protective atmosphere. They showed that they became isotropic after pyrolysis, indicating that the destruction of the smallest structural details already takes place during pyrolysis [293]. This was further shown by Paris et al. through combined heat treatment and structural analysis [294].

In monolithic, unstructured materials, cracks propagate freely, leading to an immediate catastrophic failure of the material. In wood-templated ceramics, on the other hand, it was demonstrated that the stress-strain behavior does not end with an immediate catastrophic failure of the material [295, 296]. Instead, crack deflection takes place along the fiber directions, leading to a lowered stress-strain gradient at high stresses. This means that the failure of the material can be detected before fatal fracture takes place. Even though the properties of silicon carbide differ greatly from the properties of wood, their common structuring leads in both cases to anisotropic mechanical properties that are atypical for their material class. Singh et al. showed how the choice of wood influences the bending strengths of wood-derived silicon carbide ceramics [297]. Notably, the material prepared from the wood with the lowest density (maple) showed not only increased flexural strength compared to silicon carbide derived from denser woods (African Bubinga and Zebra), but also compared to commercially obtained reaction bonded silicon carbide ("Cerastar RB-SiC").

Since machining of ceramic materials incurs great cost in energy and tool abrasion, the outer shape of wood-biotemplated materials is best machined before the reactive ceramic formation. Therefore, the final geometry, accounting for shrinkage, should be machined from the wood before carbonization. Since the final material will be near net shape, this aids in the formation of the final geometry by removing the need for a casting mold.

The previous examples had unique properties that could be traced to their micrometer-scale structuring. Returning to natural materials, one finds that "interpretations [of data] in terms of structure, mechanical behaviour and function, however, are difficult, because they incorporate the contributions of different hierarchical structural levels" [298]. Correspondingly, the mechanical properties of multiscale hierarchical biotemplated materials diverged from the behavior of bulk materials, even when subjected to nanoscale indentation [299, 300]. Wang et al. artificially mineralized bacteria with nickel to increase their elastic modulus 17-fold and their hardness 50-fold [301]. Similarly, Zhang et al. prepared striking hollow copper microspirals, templated from *Spirulina* algae [302]. A palladium-tin catalyst was deposited on the algae surface, followed by copper sulfate and formaldehyde as the reduction agent. Their work showed not only resulting structures that would be difficult to obtain otherwise than by biotemplating. They also attributed the low hardness and high elasticity of the material, compared to a template-less deposited copper, to the geometry of the hollow corkscrews. The explanation put forward to account for this behavior, namely, deformation of the thin copper sheet into the hollow lumen of the spiral, is an example of the structure-property relationships attainable through biotemplating. In a wider context, this example also shows how a lightweight elastic material could be made, which has not only elasticity due to its wall and hollow tube structuring, but also due to its spiral outer shape.

As reviewed by Peter Fratzl, the structures of bone on several levels of hierarchy act in concert to give it its large toughness [40]. This matches the findings of Vincent et al. that only the hierarchical interplay of forces leads to the favorable overall properties of natural materials [218]. In this context, it was noted that natural materials typically possess a stress -over-strain behavior with a low initial gradient [33]. The material initially does not transmit strains between compartments and therefore requires a high concentration of force to produce a fracture. In wood-biotemplated silica, stress-strain curves with similarly reduced gradients compared to bulk silica were found, indicating that the concept of a microstructure-induced alteration of the mechanical properties can be retained by biotemplating [97]. Götze et al. compared wood mineralized with nanoparticles as opposed to wood mineralized with tetraethyl orthosilicate. They found that the latter possessed a superior hardness and reduced water intake, illustrating that a continuous mineralization phase leads to superior properties in biotemplating and mineralization approaches [303].

In Section 1.5.1, we briefly discussed the necessity of choosing a material with a suitable structure on the micrometer scale, exemplarily using the properties of different species of wood. Wang et al. studied the wear behavior of aluminum-carbon and aluminum-silicon carbide + carbon composites where the latter was derived from the pyrolysis of different species of wood [304, 305]. By choosing wood from three deciduous trees with different pore structures, they could prepare three types of composite materials with different aluminum contents and bending strengths. They demonstrated that the composites with the highest carbon contents, made from elm and oak, significantly reduced the coefficient of friction and the wear rate of the composites when rotated against hardened steel: The long cylindrical channels of the wood-derived carbon preform allow for a simple preparation of the final material by a melt infiltration approach, while the interconnected carbon phase in the composite provides an uninterrupted lubrication action. This combines favorably with the species-specific structuring of the wood preforms, which allows the properties of the final material to be tailored. Similarly, carbonized wood fiber-derived silicon-silicon carbide with a residual carbon content of ~5% was tested as a potential material for ceramic brake pads [306]. It showed a high coefficient of friction, compared to metal and reaction bonded silicon carbide-based pads. Nevertheless, they showed good wear resistance, which could be related to the same lubricating effect by the biomorphous carbon.

The same aluminum-carbon composites also showed directional thermal conductivities that could be related to the directional wood structure [307]. In particular, the heat conductivity in the direction of the wood fiber axes was about twice as high as perpendicular to this axis, demonstrating the barrier function of the embedded interconnected wood-derived carbon phase. Arellano et al. described how wood-derived channeled silicon carbide was being researched as receptors for heliothermal power plants [308]. By focusing sunlight from arrays of mirrors on a directionally porous plate of silicon carbide, it is heated up to several 100 °C. Air passing through the pores is consequently heated, and used to heat a steam turbine, which in turn produces electrical energy.

1.6.2 Nanoparticle Substrates

Biological materials are not only of interest because of their bulk properties when structured on the micrometer scale. Also, their structuring on the nanometer scale provides many possibilities for usage. To access these structures, biological materials can be processed to break down or open up their structure, followed by tailored reassembly.

Plant cell walls can be mechanically pulped, i.e. fibrillated, biochemically digested, or chemically decomposed. Mechanical pulping of wood requires either simultaneous wetting and heating, resulting in thermomechanical pulp, or chemical degradation of the lignified cell walls. The latter may, for example, industrially use sodium hydroxide and sulfide in the Kraft process, or the sulfite process. For small batch laboratory purposes, more expensive or more toxic delignification agents may be used, such as hydrogen peroxide or sodium chlorite. If the aim is to regenerate cellulose in a non-native state, i.e. as cellulose nanocrystals, cellulose powder from bleached softwood or cotton is typically hydrolyzed with a sulfuric acid, then reprecipitated by washing with distilled water. The resulting fibrillar cellulose or cellulose solution can then be stabilized or reprecipitated in various solvents to give fibers of varying sizes [309]. These in turn can be functionalized [310, 311] or coated to provide different surface properties. They provide excellent substrates for metal nanoparticles [312, 313].

These substrates provide a large surface area for the deposition of nanoparticles, while prohibiting their agglomeration, which would reduce the accessible surface of the active metal component. Further, they can be prepared from simple processing steps and an abundant natural source. Metal nanoparticles deposited on such substrates can act as great enhancers of Raman signals in what is called surface-enhanced Raman spectroscopy [312]. Structures that can be created in this manner include whiskers [314], fibers and nanofibers [315–318], percolating nanofiber arrangements [319], transparent nanofiber sheets [319], cellulose fibers [320], and sheets [321, 322] regenerated from solution as well as nematic cellulose nanorod suspensions [271, 323–325].

1.6.3 Filter and Burner Assemblies

By combining a scaffold obtained by templating entire biological structures with a large surface area obtained by replicating the overall and micrometer scale structuring, down to even finer structural details, one can produce structural materials for non-loading applications. Instead, if such materials are made from a ceramic phase, they can be valuable for their thermal and chemical resistance. Suitable scaffolds for such materials are highly porous, spongy structures, such as wood, citrus peels, sea urchins [326], or sponges [327]. Suitable target phases include aluminum oxide [328], silicon nitride [327], or yttria-stabilized zirconia [329]. The processing includes the infiltration of organometallic alkoxide precursors. The resulting materials are characterized by large porosities and a large variation of pore sizes [330], providing a large surface area. All of these are intended and suitable for filters [327], sensors, catalyst carriers, thermal insulators or electrolyte materials for solid oxide fuel cells [329], or burner applications [331].

1.6.4 Photovoltaic and Sensing Materials

Photonic materials, by definition, guide the propagation of light. The guidance is based on the allowed electromagnetic resonant frequencies, as defined by the dimensions of the alternating optical densities of the constituent phases of the materials [332]. In the simplest case, the alternation is one-dimensional and leads to a strongly directional reflection of wavelengths that correspond to the lengths of the repeating units in the material. One-dimensional materials are very commonly found in the carapaces of beetles [253]. However, as already discussed in Section 1.5.1, such structures are not the only ones found in nature. On the contrary, structures in nature are often more complexly structured and may even contain areas of deliberate disorder [333–335]. Straddling the border between order and disorder is the crystallographic motif that was only recently discovered in man-made materials, the quasicrystalline lattice. It has been synthesized by nature for millennia in the shape of photonic crystals, for example, in *Eupholus magnificus* [336]. It has further been shown that such quasi-ordered structures can be transformed into engineering materials by the biotemplating approach, enabling the synthesis of structures whose existence has only just been shown [337].

Importantly, biotemplating is a processing method that is often required instead of alternative bottom-up manufacturing techniques [338]. While intricate structures mimicking natural photonic crystals can be manufactured, this is often only possible through elaborate combined bottom-up and top-down approaches [339]. These have the advantage that the obtained structure may be more freely tuned and altered. Biotemplating only allows for a certain degree of control over the dimensions of the final structure through controlling the shrinkage during processing [340]. On the other hand, it allows for a rapid replication even of complex three-dimensional biological structures in a comparatively simple and rapid manner.

The exact procedure to biotemplate photonic materials differs from the processing routes used to create mechanoceramics, filters, and substrates in several ways. The resulting material should be transparent to allow the propagation of light through the structure. This means that the material should have no energy transitions in the visible range. The filling factor, i.e. the fraction of solid material, is an important factor to determine the optical response of a material [341]. In natural photonic materials, it is carefully tuned to provide the maximum band gap width of the material, and should therefore be retained [342]. Further, the target material should have an index of refraction of at least 3 in order to achieve a complete band gap [343]. As discussed in Section 1.5.1, the second phase is most commonly air, but may also be actively or passively filled with a fluid, thereby reducing the dielectric contrast.

Suitable material precursors are metal alkoxide precursors that have been prehydrolized. They provide a strong pore volume filling as well as a large dielectric contrast when condensed to a metal oxide such as titanium oxide [337], or aluminum oxide [344]. Some structures require a double-infiltration approach during which the structure is first inverted by an intermediate phase [337]. Structures that require double infiltration are all structures whose inverts are not identical to their originals. As noted earlier, structures in nature are optimized for a maximum band gap width, while their inverts are typically not. Further, most structures, such as densely packed opals, cannot readily achieve complete band gaps [345]. Therefore, many structures found in nature require a double inversion process to yield a positive copy of the original material [346]. Templating routes using a two-step double inverse replication yielded materials with high dielectric contrasts such as alumina or titania [347, 348]. A prominent example of this is the gyroid structure, which produces a circularly polarized reflectance and transmittance, and is found in nature [349–351]. It is therefore a highly interesting optical material, with potential applications as circular polarization-dependent beamsplitters or dichroic mirrors [352]. However, manufacturing "full" gyroid structures [353], as found in nature, poses a manufacturing challenge due to the complex three-dimensional structuring, which is currently only achievable with bottom-up assembly using low-dielectric contrast materials [354, 355]. Therefore, several studies have undertaken to assess routes to either biotemplate, or template from self-assembled, polymer phase, gyroid-structured materials [356, 357]. Further suitable precursors are silicone melts, which provide a trade-off between a smooth resulting structure, albeit with a lower dielectric contrast to air [340]. Thermal template removal has been used in all known studies, which can be traced to chitin's relative resistance to chemical degradation [358, 359]. Photonic crystals can be designed and tailored to a variety of applications including anti-reflective [360, 361] and highly reflective coatings [362], waveguides [363–365], surface-enhanced Raman scattering [366], as well as gas, temperature, and pH sensors [367, 368].

Photonic structures for photoanode applications have been produced by replicating butterfly wings with a titanium sulfate precursor [369]. In this case, the target phase titanium oxide not only provides a structure that directs the propagation of light on the 100 nm scale and a large specific surface area due to its large porosity in the 10 nm range, but also has photocatalytic properties due to its atomic structuring on the Angstrom scale. Gutbrod et al. used soft rush as templates to obtain low-density titania networks, with high photo-catalytic activities [370]. These were in part due to their accessibility for streaming fluids and large surface areas, but also due to doping of the resulting titania with carbon derived from the template.

Li et al. have succeeded in replicating the thylakoid structures, the sites of the photosynthesis, of various green leaves with nanoscale precision [371]. The resulting material showed a high photocatalytic activity due to structuring of the titania replica, consisting of the same 15 nm stacks as found in the original leaf photosynthetic thylakoid stacks. This is one of the rare examples where a nanometer-scale precision templating of an entire hierarchical structure with a precision on the 10 nm scale was carried out with a crystallizing target phase. The photocatalytic activities of leaves from different plants replicated in titania by the same process were compared. While the measured photodegradation rate constants showed a relative variation of 0.25, they were on average four times as high as in a non-structured control sample, at less than twice the specific surface area.

Also, photonic structures, when applied to the surfaces of solar cells would selectively allow the passage of light wavelengths. This, in turn, would lead to the retention of the light-induced photovoltaic current, while reflecting wavelengths that would cause a temperature increase in the silicon structure, thereby reducing its effectivity [372]. Zhang et al. replicated the dark black photonic wing structures of *Papilio paris* and *Thaumantis diores* with titanium oxide [373]. They found that the absorbance of the structure was greatly increased by the biotemplated structuring, which included a latticework on the micrometer scale, and irregular hexagonal structures on the n·100 nm scale. Courchesne et al. demonstrated how biotemplated silica materials could be reduced to give complexly structured, interconnected nanoporous silicon networks [374]. The future challenge in these fields will be to transfer these separate bioinspired insights into technical solutions.

Biological photonic materials that were converted to engineering materials can be used as gas sensors, owing to the changed dielectric contrast on exchange of the gas phase from air against another gas or vapor [375, 376]. By determining the relative intensity change [376] of optical reflections, changes in gas concentrations could be detected. Zhu et al. successfully created butterfly wing-templated copper-doped tungsten oxide and showed its large sensitivity for detecting hydrogen sulfide and trimethylamine [377]. In this study, the large surface area of the butterfly wing, combined with its latticework structuring allowing for a free flow of gas provided for a template that could be replicated by a four-step process involving cleaning, infiltration by dipping followed by drying and calcination, in agreement with the general processing steps outlined in Figure 1.3. Liu et al. demonstrated that wood-templated zinc oxide is highly effective as a gas sensor for the detection of hydrogen sulfide [378]. In this example, however, the effectivity was not based on the hierarchical structuring, since the templated materials were ground before usage. Rather, the influence of the templating on the porosities on the 100 nm and micrometer scale was exploited. Further, they could be used as mechanical sensors, as has already been demonstrated on artificial photonic structures [379].

1.6.5 Wettability Control

A material that is not apparently hierarchically structured is the surface of the lotus leaf. Its surface consists of waxy knobs, 5 μm in diameter. However, Gao et al. demonstrated that the finer structuring of the knob surfaces at the length scale of 100 nm is responsible for a significant reduction of the dynamic wetting/dewetting hysteresis [380], a concept that could be confirmed for various biological and bioinspired materials [381, 382]. Fürstner et al. noted that pockets of air trapped in cavities of multiscale hierarchically structured surfaces underneath a passing drop are required for a lotus-effect, i.e. self-cleaning hydrophobicity [383]. Consequently, Sun et al. were able to biotemplate lotus leaves under retention of its hydrophobic properties [384]. By twice replicating the original wax structure with polydimethylsiloxane and deploying a trimethylchlorosilane antistick intermediate layer on the first (negative) siloxane replica, they obtained a positive replica. It was then shown that the wetting properties of the original wax surface and the second (positive) silicone replica were visually indistinguishable even though composed of a different material. Comparison of the first (negative) replica with the second showed greatly different wetting behavior via contact angle and wetting hysteresis, even though composed of the same material. This example is illustrative of the necessity to replicate all structural levels of hierarchy to obtain the desired effect through a biotemplating process. Saison et al. also succeeded in biotemplating the surface structure of lotus leaves [385]. By changing the annealing temperature of the final material, the wettability of the resulting surface could be tuned over a wide range, from contact angles of 120° to complete wetting.

1.6.6 Image Plates

Images made from phosphor plates suffer from a trade-off between signal strength and resolution. In granular phosphor materials, large grains and thick phosphor powder layers lead to a large signal to noise ratio while small grains and thin layers lead to a fine resolution, due to light scattering from particles during the laser read-out process [386, 387]. Therefore, it is advantageous to create phosphor grain structures that consist of columnar crystals that are pointing in the direction of the incoming X-rays [388, 389]. Typical lateral columns for such an application are between 1 μm and 10 μm in diameter, which are in turn grouped into columns of 50 μm in diameter, and separated by a void of 1 μm in diameter [390]. This was found to lead to a contrast transfer that is improved by more than 10% below spatial frequencies of 4 line pairs per millimeter [391]. In this way, a large absorbing cross-section in the direction of the incoming X-ray beam is retained while ensuring that the lateral resolution is retained. There are several ways of obtaining such materials, including gas phase deposition [389], epitaxial growth from liquid phases [392], or by biotemplating the cylindrically porous wood structure [393, 394]. The structuring of wood, with columns on the 10 μm scale separated by voids on the 1 μm scale, corresponds very closely to the structures needed for such phosphor plates. Consequently, storage phosphors, such as $BaFBr:Eu^{2+}$, when templated from wood, showed promising imaging potential [395].

1.7 Limitations and Challenges

1.7.1 Particle Growth

If the aim of the biotemplating approach is to achieve the most detailed replication possible, the main obstacle is particle growth of the deposited mineral phase. It is brought forth by the gain in volume energy once the surface tension for the formation of a new particle is overcome. The need to stabilize a normally metastable amorphous phase or restrict particle growth during crystallization is best illustrated by nature's use of cell compartments and surfactant polymers to slow the attachment of dissolved ions to the face of existing particles [396].

This is not only true for the most prominent biomineral, silica. The mineralization of cellulose structures with calcium carbonates both in nature and the laboratory typically proceeds via amorphous intermediates [397–401]. It applies even more when the mineralizing substance is deposited in prehydrolized or nanoparticle form. While there have been successes infiltrating nanometer-scale porous biological materials – as opposed to a surface deposition on biological molecules – with nanoparticles, these have also been singular successes [402]. By stabilizing calcium phosphate with the nature-derived dentin matrix protein 1, He et al. managed to create nanoparticles that would be able to mineralize the interstitials of collagen strands [403].On the other hand, a study of the mineralization of wood with calcium carbonate shows that precursors that are not stabilized by surfactants will form granular precipitates on the fiber surfaces, with the particle size depending on the precursor system used [404]. Interestingly, calcium hydroxide precursors produced the least visible calcium carbonate precipitates. In nature, calcium ions are also transferred into the deposition vesicles in hydrated form, giving a further example where materials scientists may look to nature for inspiration.

As discussed in Section 1.5, and in some of the examples in Section 1.6, retaining the entire structural hierarchy of a material is important to achieve true unique structure-derived material properties. A general trade-off is the amount of material deposited versus the achieved structural detail. The first is improved by using prehydrolized precursors, or nanoparticles that are already fully crystallized. The second is achieved by small molecular particles that can penetrate into the smallest of pores in the biological structure. These are typically no larger than 10 nm [405], setting the maximum size of precursor molecules.

Navrotsky reviewed the surface and transformation enthalpies of stable and metastable polymorphs of several common biological mineralization agents [181]. As discussed before, these are also the best agents for biotemplating. She found that the surface enthalpies for amorphous silica are one order of magnitude lower than those of titania, zirconia, or alumina polymorphs, lowering the nucleation free energy barrier and facilitating their nucleation. She also noted that the other phases would most likely follow the Ostwald step rule and nucleate as amorphous phases with a lower free energy barrier, which then undergo recrystallization and growth. Comparing the surface and volume terms for stable and metastable polymorphs, she determined that the free energy over particle radius for the reviewed materials follows a progression as shown schematically in Figure 1.7.

The commonly applied rationalization is "that the metastable phases are somehow structurally more similar to their precursors in solution, melt, or glass and are able to nucleate more readily" [181], which would not

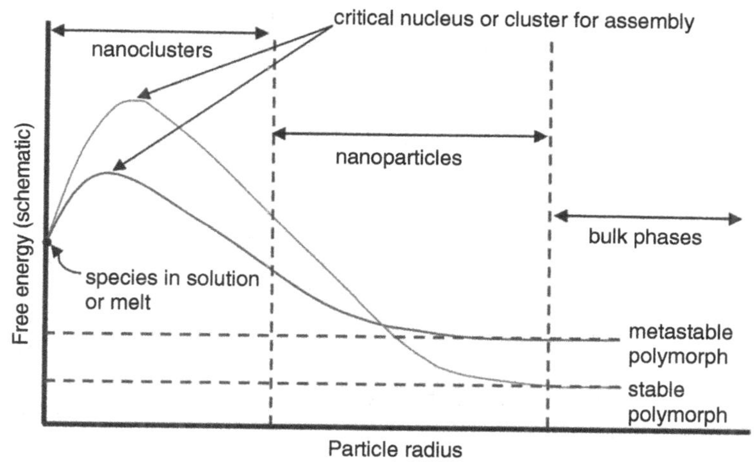

Figure 1.7 Scheme of the free energy over particle radius for alumina, titania, zirconia and silica and their metastable, often amorphous, polymorphs, showing that the free energy barrier for nucleation is lower for metastable (often amorphous) polymorphs [183]. *Source:* Navrotsky (2004) [181]. Copyright (2004) National Academy of Sciences, USA.

only explain the processes occurring in nature, but also show pathways for nanometer-detail replications, namely, by precipitating an amorphous polymorph first. Similarly, differences between the precursors were determined, with clear silica sols typically containing low-numbered oligomers together with monomers and polymeric clusters up to 4 nm in size [181]. This is due to the fact that the energy required for hydrolysis increases if the silicon atom already possesses hydrolyzed ligands. By contrast, "[a]luminum-containing solutions at near-neutral pH contain extensive clusters, often containing a central tetrahedral AlO_4 group bonded to 12 other aluminum ions in octahedral coordination, with water and hydroxyl species holding the cluster together" [181]. Using a rounded-off aluminium to oxygen bond length of 0.2 nm [406], such a cluster would be about 1.5 nm in diameter, making it just large enough to enter the pore channels of wood, provided there is no steric or electrostatic hindrance.

The challenge presented by crystal growth is immense and typically restricts the replicated detail to the 100 nm scale, or larger. For example, Liu et al. used pine wood to template nickel oxides with nickel nitrate as the precursor, following a route as outlined in Figure 1.3 [407]. In the final products, the wood cells were well replicated with a granular texture of the cell walls. They found that the materials had specific surface areas and NiO crystallite sizes which decreased and increased with calcination temperature, respectively. The smallest grains were estimated from X-ray diffraction data via the Scherrer equation to be larger than 20 nm, exceeding even the sizes of the microfibrils found in the wood cell wall. In this example, the smallest pores were most likely no longer related to the internal cell wall structuring, but to the crystal morphology of the nickel oxide grains.

Templating approaches applied to regenerated cellulose with various metal oxide target phases illustrate the crystallization behavior of different metal oxides deposited from sol-gel precursors: while silica produced a smooth replica of the nanometer-scale cellulose fibers [408], the granularity of the replica increased from titania [409, 410] over tin oxide [411] to indium tin oxide [412].

Similarly, Zhang et al. systematically replicated wood using each time the same fundamental approach, but with different material precursors. After boiling in 5% ammonia and washing to remove extractives, they prepared ethanol-based precursor solutions with nitrates of zinc [413], chromium [414], manganese [415], iron [416] and nickel [417], or chlorides of titanium [418]. After impregnation at a temperature of 60 °C between 1 d and 3 d and drying, the samples were calcined at 600 °C, 800 °C and 1000 °C each. These studies provide a systematic comparison of the behavior of the precursors. In each case, minimal pore diameters between 10 nm and 25 nm were found, save for ZnO, where they were 50 nm. The specific surfaces were for all samples on the order of magnitude of $n{\cdot}10\,m^2/g$. Most importantly, these values were taken from the presented calcination temperature of 600 °C. In each study the minimal pore diameter increased and the specific surface area of the replica diminished markedly with temperature. The structural change was also visible from electron micrographs, which showed a strong granular cell wall structure when calcined at high temperatures. The exception to these findings is the replication using titanium chloride, which used calcination temperatures from 450–600 °C, and correspondingly showed a large specific surface area of $55\,m^2/g$ and pore diameters of down to 8 nm. This, however, seems to be more indicative of the final calcination temperature than the choice of precursor material. In another study, Zhu et al. remarked a decrease of the specific surface of the material with increasing calcination temperature, "probably due to the collapse of the porous structures and the growth of the TiO_2 crystallites" [418]. These findings therefore show a certain "robustness" of the biotemplating approach toward the choice of material if micron-scale replication detail is to be achieved.

If even higher temperatures are involved, crystal growth may lead to fibrous particle growth inward from the cell walls, as shown by Luo et al. during the templating of wood with silicon nitride [419]. Even if the initially deposited phase is an amorphous intermediate, temperature treatment will result in a phase transformation and crystal growth to particle sizes on the size order of 100 nm [420]. A further processing detail that has consistently been found to aid the good replication of biological structures was drying the sample in the replication liquid [96, 211].

1.7.2 Comparison with Alternating Processing Principles

When comparing biotemplating to other processing principles, one can draw up a list of definitive differences. We will use two materials that are challenging to manufacture for our comparisons, namely, ceramic gradient foams and D-surface structured photonic crystals. These stand proxy for other sought-after structures and illustrate the relative ease, as well as the problemd, by which such materials can be made via biotemplating approaches.

Gradient foams in technology can be obtained by mixing and firing alumina with carbon powder [421]. Routes to obtain hierarchical silica materials include sintering of graded particles, pulsed electric current sintering, pressure filtration of mixed particles, infiltration of compression-molded sponge, multiple and differentiated impregnation, multiple tape casting and lamination, solid free form fabrication, centrifugation of a suspension and freeze drying, introduction of a graded biodegradable phase, electrospraying into a template, freeze casting, microspheres for scaffolds, templating of emulsions, polymer foams, colloidal spheres settled by sedimentation or ice and phase separation, as reviewed by Miao et al. and Triantafillidis et al. [422, 423]. For example, gradient ceramic foams find application in implantable biomaterials for tissue regeneration, strengthening, or drug delivery. Gradient foams in nature, on the other hand, are abundant and appear as cellular soft tissue with many outer dimensions and gradient pore structures, for example, in citrus peels, sponges or grasses [293, 424]. Due to their accessible pore structure, they provide suitable biotemplates, as has been shown with the examples of the pomelo peel [425], sponges [426, 427], and the *Luffa*, among many others [428–430].

There are also several non-biotemplating routes to obtain three-dimensional photonic materials with complete band gaps. Spherically projected thin film reflectors prepared by coating polyethylene terephtalate with dichromated gelatin and alternating refractive indices produced directionally uniform reflections [431]. Titania woodpile structures, fabricated by two-photon polymerization and then twice replicated by zinc oxide and then titania showed complete directional band gaps, a first in entirely man-made materials [432]. Photonic structures found in nature do not typically possess complete band gaps, owing to the low refractive index of chitin [273]. However, certain biological photonic structures attain a complete band gap when replicated with a material with a higher refractive index. Galusha et al. used scales of the beetle *Lamphrocyphus augustus* as templates for a double replication process, as outlined in Section 1.6.4 [337]. The result was a titania structure possessing a lattice with diamond symmetry and a complete band gap in the visible range. The routes utilized by Frölich et al. and Galusha et al. are very similar, the main difference being the starting template. It can be argued that, since structures from nature are highly adapted, artificially manufactured templates can at best achieve the same quality and "adaptability" of structuring. However, biotemplates do not provide the possibility for targeted tailoring, e.g. the inclusion of waveguiding defects.

Assessing the complexity of a manufacturing process would require an in-depth study of its economic viability, which has to our best knowledge not been performed for biotemplating processes. However, we have compiled in Table 1.3 a comparative list of two materials together with the steps to obtain them via a biotemplating and an alternative route.

1.7.3 Availability

The biotemplating process relies on templates found in nature. However, we have outlined many examples of structures that are highly complex and adapted. As pointed out by Philip Ball, geometries that were determined, forgotten, or overlooked and then rediscovered by man were ultimately found to already exist in nature [433]. To obtain the best structuring for a given task requires a good knowledge of structures that exist in the different classes of animals. Since similar living objects may contain similarly structured materials with minute, but important differences (Figure 1.8), it may also require a comprehensive analysis of specimens beforehand.

The best starting point is, however, the ever-growing body of literature, of which we could only present a minute fraction in this chapter. In the literature, it has been shown that biotemplating approaches are consistently able to replicate even the most delicate of structures, including rattan [434], pomelo peel [425], even leaves [426]. Bhushan presented an overview of natural structural materials with widely differing structures and properties, as well as derivatives created by bioinspired, i.e. abstracted approaches [435]. Fan reviewed the state of biotemplating in a 118-page review in 2009 [436]. He discussed many topics outlined

Table 1.3 Comparison of the processing steps of two biotemplated materials and each one alternative processing route. The columns termed 'Complexity' are assigned to the processes to their left as a measure of the work time required.

Step	Biotemplated gradient foam	Complexity	PU-based gradient foam	Complexity	Biotemplated photonic crystal	Complexity	2-P polymerized photonic crystal	Complexity
1	Selection and growth or purchase of fruit	Medium	Purchase of materials	Low	Selection and capture or purchase of insect	High	Mathematical desription of desired structure	Medium
2	Washing	Low	Synthesis of components	Medium	Detachment of photonic structure	Medium	Programming and calibration of machine	Medium
3	Ethanol extraction	Medium	Mixing	Low	Preparation of precursor liquid	Low	Mixing of two-photon-curable material	Low
4	Silica precursor preparation	Low	Shaping	Medium	Infiltration by capillary action	Low	Writing of structure	Low
5	Vacuum infiltration	Low	Drying	Low	Curing/drying	Low	Steps 1 to 8 of "Biotemplated photonic crystal"	High-low
6	Drying	Low	Calcination	Low	Preparation of second precursor liquid	Low		
7	Vacuum infiltration	Low			Infiltration by capillary action	Low		
8	Drying	Low			Calcination	Low		
9	Calcination	Medium						

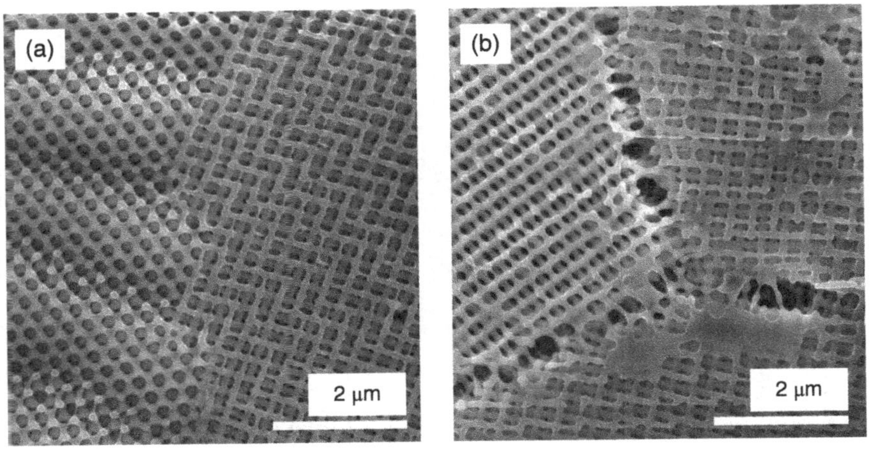

Figure 1.8 Comparison of the photonic structures found in *Entimus imperialis* and *granulatus*. While the scales from both animals show a similar internal D-surface structure, they differ in their domain boundary, and outer shell cover. This leads to a different appearance of the beetle, (a) *imperialis* being brilliant and (b) *granulatus* dull. *Source:* Courtesy of Dr. Helge-Otto Fabritius.

in this chapter, together with biotemplated materials for electronic and magnetic applications. His review includes numerous examples of microbial and viral templating, whereas we have focused on hierarchically structured materials.

1.8 Conclusion and Future Topics

Studies mimicking biomineralization approaches have been found to result in structures that are far from what is predicted from classical crystallization theory. To the best of our knowledge, there have only been a few studies of biomineralization in combination with self-assembled templates. One study showed that multiscale hierarchically structured silica materials can be prepared by using dicationic peptides and tetraethyl orthosilicate [437]. These materials showed a remarkable similarity to wood fibers. In particular, the SAXS patterns and polarized light micrographs are at first sight virtually indistinguishable from those obtained from silica replicated wood fibers. Another study showed how viruses, which can be used as biotemplates, self-assemble on a larger template [438]. Arguably, this concept could be adapted to a range of templates, even making hierarchical layers of templates thinkable.

Since biological materials with ever new unexpected functionalities such as photonic [439] or acoustically reflecting [440] plants are being discovered and described, it can be expected that the range of biotemplated functional materials to emerge will follow a similar trend.

Acknowledgments

We thank the German Science Foundation (DFG) for their generous support of our work through numerous grants. We thank our colleagues and students for their invaluable support in deed and thought.

Notes

1 We will use the metric length scale definitions for different levels of hierachical structuring. However, since structural dimensions will at best coincidentally adhere exactly to this scale, we mean any length within 0.1 to 100 times the said scale For instance, "micrometer scale" will denote any structure with a dimension between 100 nm and 100 μm.

2 Mineralization denotes any formation of an inorganic phase by a natural process. Therefore, we will use the term "artificial mineralization" to denote a man-made, but nature-inspired formation of an inorganic material in this chapter.

3 At the beginning of Chapter ii, Diogenes Laertius states that he obtained information about Theophrastus from Apollodorus: 28*V, 2, 42.

4 It reads:

> in Ciconum flumine et in Piceno lacu Velino lignum deiectum lapideo cortice obducitur et in Surio Colchidis flumine adeo, ut lapidem plerumque durans adhuc integat cortex. similiter in flumine Silero ultra Surrentum non virgulta modo inmersa, verum et folia lapidescunt, alias salubri potu eius aquae. in exitu paludis Reatinae saxum crescit ...[110]

5 Theophrastus was a native of Lesbos.
6 An illustration is nevertheless shown in Plate III. 5.
7 It has been pointed out that the attribution of various writings to him are barely cross-referenced [137]. The hereafter mentioned citation has been passed on [138–140] without us being able to identify the primary source. The fact that the description of the process is not found in the most complete collection of his works [141] casts some doubts on its reliability as the very first instance of the manufacture of aqueous sodium silicate.
8 *Lignum vitae* wood was used as railroad cross-ties or propeller shaft bearings on ships.

References

1 Cordier, P., Tournilhac, F., Soulié-Ziakovic, C., Leibler, L. (2008) *Nature*, **451**(7181), 977–980.
2 Kessler, M.R., Sottos, N.R., White, S.R. (2003) *Compos, Part A Appl. Sci. Manuf*, **34**(8), 743–753.
3 Liu, J., Kim, A.Y., Wang, L.Q. et al. (1996) *Adv. Colloid Interface Sci.*, **69**(1), 131–180.
4 Masuda, Y., Itoh, T., Koumoto, K. (2005) *Adv. Mater.*, **17**(7), 841–845.
5 Toohey, K.S., Sottos, N.R., Lewis, J.A. et al. (2007) *Nat. Mater.*, **6**(8), 581–585.
6 Fahlquist, J.N., Roeser. S. (2015) *Risk Res*, **18**(3), 333–346.
7 Bredberg, I., Hutter, J., Philippczyk, F., Dose, J. (2012) *State and Development of Nuclear Energy Utilization in the Federal Republic of Germany 2011*, Bundesamt für Strahlenschutz (BfS), Berlin, Germany.
8 Conti, J.J., Holtberg, P.D., Diefenderfer, J.R. et al. (2014) *Annual Energy Outlook 2015*, Energy Information Agency, Washington, DC, USA.
9 Blakiston, B. (1876) US Patent 176,269.
10 Preston, C.F. (1938) US Patent 2,111,494.
11 Torrey, W. (1873) US Patent 135,865.
12 Agricola, G. (1556) *De re metallica*, Hieronymus Froben and Nicolaus Episcopius, Basel, Switzerland, Books 8, 9, 12.
13 Kim, E., Xia, Y., Whitesides, G.M. (1996) *J. Am. Chem, Soc*, **118**(24), 5722–5731.
14 Lichtenegger, H.C., Schöberl, T., Bartl, M.H. et al. (2002) *Science*, **298**(5592), 389–392.
15 Vincent, J.F., Bogatyreva, O.A., Bogatyrev, N.R. et al. (2006) *J. Roy. Soc*, **3**, 471.
16 Fengel, D., Wegener, G. (1983) *Wood: Chemistry, Ultrastructure, Reactions*, Walter de Gruyter, Berlin, Germany, pp. 19–23.
17 Burgert, I., Eder, M., Gierlnger, N., Fratzl, P. (2007) *Planta*, **226**(4), 981–987.
18 Burgert, P Fratzl, (2009) *Philos. Trans. Roy. Soc., A*, **367**(1893), 1541–1557.
19 Burgert, I., Fratzl, P. (2009) *Integrative Comp. Biol.*, **49**(1), 69–79.
20 Kercher, K., Nagle, D.C. (2003) *Carbon N.Y.*, **41**, 3.
21 Hofenauer, A., Treusch, O., Tröger, F. et al. (2003) *Adv. Eng. Mater.*, **5**, 794.
22 Travitzky, N., Windsheimer, H., Fey, T., Greil, P. (2008) *J. Am. Ceram. Soc.*, **91**, 3477.
23 Mann, S. (1995) *J. Mater. Chem*, **5**, 935.
24 Nudelman, F. (2015) *Semin. Cell Dev. Biol.*, **46**, 2–10.
25 Knoll, A. (2003) *Rev, Mineral. Geochem.*, **2000**.
26 Mann, S., Behrens, P., Bäuerlein, E. (2009) *Handbook of Biomineralization: Biomimetic and Bioinspired Chemistry*, Wiley-VCH, Weinheim, Germany.
27 Weiner, S., De Yoreo, J.J., Dove, P.M. (2003) *Biomineralization*, Mineralogical Society of America, Washington, DC, USA.
28 Young, J.R. (2003) *Rev, Mineral. Geochem*, **54**(1), 189–215.
29 Voznesenskiy, S., Kul'chin, Y.N., Galkina, A.N. (2011) *Nanotechnologies Russ.*, **6**(1–2), 43–78.
30 Li, C., Kaplan, D.L. (2003) *Curr. Opin. Solid State Mater. Sci.*, **7**(4–5), 265–271.
31 Nudelman, F., Sommerdijk, N.A. (2012) *Angew. Chem.*, **124**(27), 6686–6700.
32 Davis, M.E., (1993) *Nature*, **364**(6436), 391–393.
33 Jeronimidis, G., Atkins, A.G. (1995) *J. Mech. Eng. Sci.*, **209**,4, 221–235.
34 Beniash, E., Aizenberg, J., Addadi, L., Weiner, S. (1997) *Proc. Roy. Soc. Biol. Sci.*, **264**, 461.
35 Addadi, L., Weiner, S. (1992) *Angew. Chem. Int. Ed., Engl.*, **31**, 153.
36 Mann (1983) In *Inorg Elem Biochem.*, Springer, Berlin, Germany, p. 125.
37 Hildebrand, (2008) *Chem. Rev.*, **108**, 4855.
38 Sumper, E. Brunner, (2008) *Chem Bio Chem*, **9**, 1187.
39 Kato, A. Sugawara, N. Hosoda, (2002) *Adv. Mater.*, **14**, 869.
40 Fratzl, P. (2007) *J. Roy. Soc. Interface*, **4**, 637.
41 Addadi, L., Weiner, S. (1992) *Angew. Chem. Int. Ed. Engl.*, **31**, 153.
42 Voznesenskiy, S., Kul'chin, Y.N., Galkina, A.N. (2011) *Nanotech. Russ.*, **6**(1-2), 43–78.
43 Ehrlich, H. (2010) *Int. Geol. Rev.*, **52** (7–8), 661–699.
44 Fratzl, P. & Weinkamer, R. (2007) *Prog. in Mat. Sci.*, **52**(8), 1263–1334.
45 Bouligand, Y. (2004) *Comptes Rendus Palevol*, **3**, 617–628.
46 Addadi, L., Weiner, S. (1992) *Angew. Chem.*, **104**, 159.

47 Addadi, L., Weiner, S. (1992) *Angew. Chem. Int. Ed. English*, **31**, 153.
48 Massey, F.P., Ennos, A.R., Hartley, S.E. (2006) *J. Anim. Ecol.*, **75**, 595.
49 Jackson, P., Vincent, J.F.V., Turner, R.M. (1988) *Proc. R. Soc. B*, **234**, 415–440.
50 Raabe, D., Romano, P., Sachs, C. et al. (2006) *Mat. Sci. Eng.: A*, **421**(1–2), 143–153.
51 Burgert, I. Fratzl, P. (2009) *Integr. Comp. Biol.*, **49**(1), 69–79.
52 Harrington, M.J., Razghandi, K., Ditsch, F. et al. (2011) *Nat. Comm.*, **2**, 337.
53 Siau, J. (1984) *Transport Processes in Wood*, Springer, Berlin, Germany, p. 245.
54 Van den Honert, T. (1948) *Disc. Faraday Soc.*, **3**(146), 146.
55 Gosline, J.M., DeMont, M.E., Denny, M.W. (1986) *Endeavour*, **10**(1), 37.
56 Lin, L.H., Edmonds, D.T., Vollrath, F. (1995) *Nature*, **373**(6510), 146–148.
57 Gao, X., Jiang, L., (2004) *Nature*, **432**(7013), 36.
58 Gao, H., Wang, X., Yao, H., Gorb, S., Arzt, E. (2005) *Mech. of Mat.*, **37**(2–3), 275–285.
59 Galusha, J.W., Richey, L.R., Gardner, J.S. et al. (2008) *Phys. Rev.*, **77**(5), 2.
60 Srinivasarao, M. (1999) *Chem. Rev*, **99**(7), 1935.
61 Vukusic, P., Sambles, J. R. (2004) *Nature*, **424**, 852.
62 Vignolini, S., Rudall, P.J., Rowland, A.V. et al. (2012) *Proc. Nat. Acad. Sci. U.S.A.*, 1–4.
63 Land, M.F. (1972) *Prog. Biophys. Mol. Bio.*, **24**, 75.
64 Simon, R., Holderied, M.W., Koch, C.U., Von Helversen, O. (2011) *Science*, **333**(6042), 631.
65 Murlis, J., Elkinton, J.S., Cardé, R.T., Carde, R. (1992) *Ann. Rev. Entomol.*, **37**(1), 505.
66 Witham, H. (1831) *Observations on Fossil Vegetables, Accompanied by Representations of Their Internal Structure as Seen Through the Microscope*, Blackwood, Edinburgh, pp. 28–30.
67 Nowak, J., Florek, M. Kwiatek, W. et al. (2005) *Mater. Sci. Eng. C*, **25**, 119.
68 Buurman, P. (1972) *Scr. Geol.*, **12**, 1.
69 Kuczumow, A., Pikus, S., Un-Ro, C. et al. (2001) *Spectrochim. Acta B*, **56**(4) 339.
70 Kuczumow, A., Vekemans, B., Schalm, O. et. al. (1999) *J. Anal. Atom. Spectrom.*, **14**(3), 435.
71 Kuczumow, A. et. al. (2000) *Atom. Spectrosc.*, **55**(10), 1623.
72 Martyn, J. (1747) *The Philosophical Transactions (From the Year 1743, to the Year 1750)*, Vol. 10 (Ed: J. Martyn), Royal Society: London, Chap. III, p. 590–671.
73 Agricola, G. (1546) *De Natura Eorum Quae Effluunt Ex Terra*, Hieronymus Froben and Nicolaus Episcopius, Basel, Switzerland, Book 2.
74 Stankiewicz, A., Briggs, D.D.E.G., Evershed, R.P. et al. (1997) *Science*, **276**(5318), 1541.
75 Mustoe, G.E. (2008) *Geol. Soc. Am. Spec. Paper*, **435**, 127.
76 St. John, R.N. (1927) *Econ. Geol.*, **22**(7), 729.
77 Sigleo, C. (1978) *Geochim. Cosmochim. Ac.*, **42**, 1397.
78 Petersen, D., Link, R., Carll, C., Highley, T. (1999) *J. Test. Eval.*, **27**(2), 150.
79 Kim, Y.S., Singh, A.P. (2000) *IAWA J.*, **21**(2), 135.
80 Renaut, R.W., Jones, B., Sciences, A., Tiercelin, J. (1998) *Sedimentology*, **45**, 1083.
81 Channing, A., Edwards, D. (2009) *Plant Ecol. Divers.*, **2**, 111.
82 Ferris, F.G., Fyfe, W.S., Beveridge, T.J. (1988) *Geology*, **16**(2), 149.
83 Zhang, F.-Q., Chen, H.-L., Batt, G.E. et al. (2014) *Palaios*, **29**, 325.
84 Dernbach, U., Glas, M., Hochleitner, R. (1994) *Versteinertes Holz: aus Holz wird Stein - die Mineralogie der Holzversteinerung*, Christian Weise, Munich, Germany, pp. 49–54.
85 Kuczumow, A., Pikus, S., Un-Ro, C. et al. (2001) *Spectrochim. Acta B*, **56**(4), 339.
86 Crook, T. (1922) *Nature*, **110**, 253.
87 Correns, C.W. (1940) *Die Naturwissenschaften*, **28**(24), 369.
88 Leo, E., Barghoorn, E.S. (1976) *Bot. Mus. Lealf. Harv. Univ.*, **25**, 1.
89 Dernbach, U., Glas, M., Hochleitner, R. (1994) *Versteinertes Holz: aus Holz wird Stein - die Mineralogie der Holzversteinerung*, Christian Weise, Munich, p. 50.
90 Scurfield, G., Segnit, E.R. (1984) *Sediment. Geol.*, **39**, 149.
91 Buurman, P. (1972) *Scr. Geol.*, **12**, 1.
92 Dernbach, U., Glas, M. & Hochleitner. R. (1994) *Versteinertes Holz: aus Holz wird Stein - die Mineralogie der Holzversteinerung*, Christian Weise Munich, Germany, p. 63.
93 Harder, H. (1993) *Der Aufschluss*, **44**, 23–31.
94 Siever, R. (1959) *Soc, Econ. Paleontol. Miner.*, **7**, 55.
95 Leo, R.E., Barghoorn, E.S. (1976) *Bot. Mus. Lealf. Harv. Univ.*, **25**, 1.
96 Van Opdenbosch, D., Fritz-Popovski, G., Paris, O., Zollfrank, C. (2011) *J. Mater. Res.*, **26**(10), 1193.
97 Fritz-Popovski, G., Van Opdenbosch, D., Zollfrank, C. et al. (2013) *Adv. Funct. Mater.*, **23**, 1265.
98 Scurfield, G., Segnit, E.R. (1984) *Sediment. Geol.*, **39**, 149.
99 Alexander, G.B., Heston, W., Iler, R.K. (1954) *J. Phys. Chem.*, **58**, 453.
100 Correns, W. (1940) *Die Naturwissenschaften*, **28**(24), 369.
101 Humar, M., Petri, M., Pohleven, F. (2001) *Holz als Roh- und Werkstoff*, **59**(4), 288.
102 Reinsch, S., Unger, B., Bu, M. et al., (2012) *Wood Sci. Technol.*, **47**, 83.

103 Willstätter, R. (1931) *Nat. Mus.*, **61**, 332.
104 Belton, D.J., Deschaume, O., Patwardhan, S.V., Perry, C.C. (2010) *J. Phys. Chem. B*, **114**(31), 9947.
105 Radtke, J. (2013) Eine Eidechse aus den Plattenkalken bei Ettling (Markt Pförring). *Archaeopteryx*, **14**.
106 Garett, C. (2012) *Texas Books in Review 9*, Rockstone Press, Spicewood, Texas, USA.
107 Quenstedt, F.A. (1856) *Sonst und jetzt: Populäre Vorträge über Geologie*, Verlag der H. Laupp'schen Buchhandlung, Tübingen, pp. 200–201.
108 Agricola, G. (1546) *De Natura Fossilium*, Hieronymus Froben and Nicolaus Episcopius, Basel, Book 10.
109 Laertius, D., Hicks, R.D. (1925) *Lives of Eminent Philosophers*, Harvard University Press, Cambridge, MA, Book 5, Chap. 2, Sec. 40.
110 Maior, GPS. *The Wonders of Waters, Fountains and Rivers*, Naturalis Historia, Rome, 79 AD, Book 2, Sec. 106, Par. 226.
111 Dernbach, U., Glas, M., Hochleitner, R. (1994) *Versteinertes Holz: aus Holz wird Stein - die Mineralogie der Holzversteinerung*, Christian Weise, Munich, Germany, pp. 24–31.
112 van Leeuwenhoek, A. (1700) *A Philos. Trans.*, **22**(260–276), 867.
113 Hooke, R. (1665) *Micrographia, or Some Physiological Descriptions of Minute Bodies Made by Magnifying Glasses with Observations and Inquiries Thereupon*, Martyn J And Allestry J. The Royal Society, London, pp. 11, 172.
114 Birch, T. (1756) *The History of the Royal Society of London*, Vol. 1, The Royal Society, London, p. 262.
115 Nicol, W. (1834) in *The Edinburgh New Philosophical Journal*, Vol. 16, (Ed: R. Jameson), Longman, Edinburgh, Scotland; p. 434.
116 Drum, R.W. (1968) *Science*, **161**(3837), 175.
117 Selmeier, A. (1990) *Holz als Roh- und Werkst.*, **48**, 111.
118 Juniper, B.E., Bradley, D.E. (1958) *J. Ultra. Mol. Struct. R.*, **2**(1), 16.
119 Thomassen, L., Williams, R.C., Wyckoff, R.W.G. (1945) *Rev. Sci. Instrum.*, **16**(6), 155.
120 Chan, S. (1989) *Geothermics*, **18**, 49.
121 van der Weijden, C. H. (2007) *Cahiers of Geochemistry Silica I: Silicon Analytical, Physical and Terrestrial Geochemistry*, Utrecht University, Utrecht, the Netherlands, p. 4.
122 Perry, C.C., Keeling-Tucker, T. (2000) *JBIC, J. Biol. Inorg. Chem.*, **5**, 537.
123 Hench, L.L. (1996) *Thermochim. Acta*, **1**, 280.
124 Valentin, B. (1677) *Chymische Schriften Theil Eins*, Naumans und Wolff, Hamburg, Germany, pp. 89–94.
125 Fuchs, G.F.C. (1786) Versuch einer natürlichen Geschichte des Spiesglases.
126 Glauber, J.R. (1658) *Bücher und Schrifften, so viel deren von ihme bißhero an Tag gegeben worden; Jetzo vom neuen übersehen und vermehret*, Thomas Matthias Gößen, Frankfurt am Main, Germany, p. 185.
127 Kohn, C. (1862) *Zeitschrift des Österreichischen Ingenieur-Vereines*, **168**(6), 394.
128 Zwick, (1877) *Das Wasserglas: Seine Natur und seine Bedeutung für die Industrie und Technik*, O. Füssli & Company, Zürich, Switzerland.
129 Vail, J.G. (1928) *Soluble Silicates in Industry*, Little and Ives, New York.
130 Hicks, H. (1986) US Patent 4,612,050.
131 Christ, A. (1994) *Zeitschrift für Kunsttechnologie und Konserv.*, **8**(1), 25.
132 Riederer. J. (1971) In: The treatment of stone-proceedings of the meeting of the joint committee for the conservation of stone, Bologna October 1–3, pp 105–138.
133 Storz, M. (1928) *Die sekundäre authigene Kieselsäure in ihrer petrogenetisch-geologischen Bedeutung.* Gebrüder Borntraeger, Berlin, Germany.
134 Hellmers, J.H. (1949) *Abhandlungen der Geologischen Landesanstalt Berlin*, **218**, 1.
135 Akahane, H. (1996) *Muki Materiaru*, **3**(264), 492.
136 Pierce, C.S. (1898) *Science*, **8**(189), 169.
137 von Lippmann, E.O. (1919) *Entstehung und Ausbreitung der Alchemie*, Vol. 1, Verlag von Julius Springer, Berlin, Germany.
138 Kohn, C. (1862) *Zeitschrift des Österreichischen Ingenieur-Vereines*, **168**(6), 394.
139 Vail, G., (1928) Little and Ives, New York, USA, p. 11–12.
140 Leo, R.F., Barghoorn, E.S. (1976) *Bot. Mus. Lealf. Harv. Univ.*, **25**(1), 1.
141 Valentin, B. (1677) *Chymische Schriften Theil Eins*, Naumans und Wolff, Hamburg, Germany.
142 Leo, R.F., Barghoorn, E.S. (1976) *Bot. Mus. Lealf. Harv. Univ.*, **25**(1), 1.
143 Meyer, A. (1791) *Briefe über einige mineralogische Gegenstände*, Johann Christian Dieterich, Göttingen, Germany, p. 226.
144 Francis, S., Margulis, L., Barghoorn, E.S. (1978) *Precambrian Res.*, **6**(1), 65.
145 Weatherwax, R.C. (1977) *J. Colloid Interf. Sci.*, **62**(3), 432.
146 Jungnikl, K., Paris, O., Fratzl, P., Burgert, I. (2008) *Cellulose*, **15**(3), 407.
147 Ahlgren, P.A., Yean, W.Q., Goring, D.A.I. (1971) *Tappi*, **54**(5), 737.
148 Shebani, A.N., van Reenen, A.J., Meincken, M. (2008) *Thermochim. Acta*, **471**(1–2), 43.
149 Valadez-Gonzalez, A., Cervantes-Uc, J. (1999) Compos. *Part B Eng.*, **30**, 321.
150 Abdelmouleh, M., Boufi, S., ben Salah, A. et al. (2002) *Langmuir*, **18**, 3203.

151 Ly, B., Belgacem, M.N.N., Bras, J., Brochier Salon, M.C.C. (2010) *Mater. Sci. Eng. C*, **30**, 343.
152 Rowell, R.M., Clemons, C.M. (1992) *Proc. Wash. State Univ. Int. Part, Mater. Symp. 26th*, 251.
153 Rowell, M. (2006) *Wood Mater. Sci. Eng.*, **1**, 29.
154 Heinze, T., Liebert, T. (2001) *Prog. Polym. Sci.*, **26**, 1689.
155 De Melo, J.C.P., Da Silva Filho, E.C., Santana, S.A.A., Airoldi, C. (2009) *Colloid. Surface, A*, **346**(1–3), 138.
156 Roussel, C., Marchetti, V. Lemor, A. (2005) *Holzforschung*, **55**(1), 57.
157 Timar, M.C., Maher, K., Irle, M., Mihai, M. D. (2000) *Holzforschung*, **54**(1), 77.
158 Sèbe, G., Brook, M.A. (2001) *Wood Sci. Technol.*, **35**, 269.
159 Coutts, R.S.P., Campbell, M.D. (1979) *Composites*, **10**(4), 228.
160 Elvy, S.S.B., Dennis, G.G.R., Ng, L.L.-T. (1995) *J. Mater. Process, Technol.*, **35**(20), 3417.
161 Tingaut, P., Weigenand, O., Mai, C., Militz, H., Sèbe, G. (2006) *Holzforschung*, **60**, 271.
162 Caruso, R.A., Antonietti, M. (2001) *Chem. Mater.*, **13**, 3272.
163 Wang, G., Harrison, A. (1999) *J. Colloid Interface Sci.*, **217**, 203.
164 Vogli, E., Mukerji, J., Hoffman, C. et al. (2001) *J. Am. Ceram. Soc.*, **40**, 1236.
165 Vogli, E., Sieber, H., Greil, P. (2002) *J. Eur. Ceram. Soc.*, **22**(14–15), 2663.
166 Hyde, G.K., Park, K.J., Stewart, S.M. et al. (2007) *Langmuir*, **23**, 9844.
167 Sandhage, K.H., Dickerson, M.B., Huseman, P.M. et al. (2002) *Adv. Mater.*, **14**, 429.
168 Unocic, R. R., Zalar, F.M., Sarosi, P.M. et al., (2004) *Chem. Commun.*, **1**(7), 79.
169 Dudley, S., Kalem, T., Akinc, M. (2006) *J. Am. Ceram. Soc.*, **89**(8), 2434.
170 Yu, Y., Addai-Mensah, J., Losic, D. (2010) *Langmuir*, **26**, 14068.
171 Rambo, C.R., Sieber, H., Genova, L.A. (2008) *J. Porous Mater.*, **15**(4), 419.
172 Mantanis, G.I., Young, R.A. (1997) *Wood Sci. Technol.*, **31**(5), 339.
173 Assink, R.A., Kay, B D. (1993) *Colloids Surfaces A Physicochem. Eng, Asp.*, **74**, 1.
174 Brinker, C.J., Scherer, G.W. (1990) *Sol-Gel Science*, Academic Press, London.
175 Buckley, A.M., Greenblatt, M. (1994) *J. Chem. Educ.*, **71**(7), 599.
176 Pope, E.J.A., Mackenzie, J.D. (1986) *J. Non. Cryst. Solids*, **87**(1–2), 185.
177 Vrieling, E.G., Gieskes, W.W.C., Beelen, T.P.M. (1999) *J. Phycol.*, **35**(3), 548.
178 Young, J.R., (2003) *Rev. Mineral. Geochemistry*, **54**, 189.
179 Marsh, M. (1994) *Protoplasma*, **177**, 108.
180 Heymann, E., Rabinov, G. (1941) *J. Phys. Chem.*, **45**(8), 1152.
181 Navrotsky, A. (2004) *Proc. Nat. Acad. Sci. U,S,A.*, **101**, 12096.
182 Tiemann, H. (1906) *USDA Forest Service Bulletin* 70.
183 Spalt, H. (1958) *Forest Prod. J.*, **8**(10) 288.
184 Jinzhen, C., Guangjie, Z., Zhao, G. et al. (2001) *Holzforschung*, **56**(3), 655.
185 Hörstermann, H., Kundin, J., Emmerich, H. (2014) *Eur. Phys. J. Spec. Top.*, **223**(3), 481.
186 Page, A.J., Sear, R.P. (2006) *Phys. Rev. Lett.*, **97**, 1.
187 Gierlinger, N., Schwanninger, M. (2006) *Plant Physiol.*, **140**, 1246.
188 Castellano, M., Gandini. A. (2004) *J. Colloid Interf. Sci.*, **273**, 505.
189 Sèbe, G., Brook, M.A. (2001) *Wood Sci. Technol.*, **35**(3), 269.
190 Tingaut, P., Weigenand, O., Mai, C., Militz, H., Sèbe, G. (2006) *Holzforschung*, **60**(3), 271.
191 Mai, C., Militz, H. (2004) *Wood Sci. Technol.*, **37**(5), 339.
192 Van Opdenbosch, D., Dörrstein, J., Klaithong, S. et al. (2013) *Holzforschung*, **67**(8), 1.
193 Norimoto, M., Yamada, T. (1972) *Wood Res. Bull. Wood Res. Inst. Kyoto Univ.*, **52**, 31.
194 Einfeldt, J., Meißner, D., Kwasniewski, A. (2001) *Prog. Polym. Sci.*, **26**, 1419.
195 Sugimoto, H., Takazawa, R., Norimoto, M. (2005) *J. Wood Sci.*, **51**(6), 549.
196 Fu, Y., Zhao, G. (2007) *Wood Sci. Technol.*, **41**(6), 511.
197 Van Opdenbosch, D., Dörrstein, J., Klaithong, S. et al. (2013) *Holzforschung*, **67**(8), 1.
198 Saka, S., Sasaki, M., Tanahashi, M. (1992) *Mokuzai Gakkaishi*, **38**(11), 1043.
199 Saka, S., Ueno, T. (1997) *Wood Sci. Technol.*, **31**(6), 457.
200 Donath, S., Militz, H., Mai, C. (2004) *Wood Sci. Technol.*, **38**(7), 555.
201 Persson, P.V., Hafrén, J., Fogden, A. et al. (2004) *Biomacromolecules*, **5**(3), 1097.
202 Blanchette, R., Obst, J., Timell, T. (1994) *Holzforschung*, **48**, 34.
203 Danninger, E., Messner, K., Rohr, M. (1980) *Holzforsch. Holzverw.*, **32**(5), 126.
204 Suttie, E.D., Hill, C., Jones, A.S. et al. (1998) *Mater. Organismen*, **32**(3), 159.
205 Tanno, B.F., Saka, S., Yamamoto, A., Takabe, K. (1998) *Holzforschung*, **52**(4), 365.
206 Mai, Militz, H. (2004) *Wood Sci. Technol.*, **37**(5), 339.
207 Sigleo, A.C. (1978) *Geochim. Cosmochim. Ac.*, **42**(9), 1397.
208 St. John, R.N. (1927) *Econ. Geol.*, **22**(7), 729.
209 Leo, R.F., Barghoorn, E.S.H. (1976) *Bot. Mus. Leafl. Harv. Univ.*, **25**(1), 1.

210 Mustoe G.E., (2008) *Geol. Soc. Am. Spec. Papers*, **435**, 127.
211 Deshpande, A.S., Burgert, I., Paris, O. (2006) *Small*, **2**(8–9), 994.
212 Shin, Y., Liu, J., Chang, J.H. et al. (2001) *Adv, Mat.*, **13**(10), 728.
213 Van Opdenbosch, D., Dörrstein, J., Klaithong, S. et al. (2013) *Holzforschung*, **67**(8), 1.
214 Browne. F. (1958) *Theories of the Combustion of Wood and its Control*; The University of Wisconsin, Madison, WI.
215 Branca, C., Di Blasi. C. (2004) *Fuel*, **83**(1), 81.
216 Bamford, C.H., Crank, J., Malan, D.H., Wilson, A.H. (2008) *Math. Proc. Cambridge Philos. Soc.*, **42**(02), 166.
217 Bechtle, S., Ang, S.F., Schneider, G.A. (2010) *Biomaterials*, **31**(25), 6378.
218 Vincent, J.F.V., Bogatyreva, O.A., Bogatyrev, N.R. et al. (2006) *J. Roy. Soc.*, **3**, 471.
219 Aizenberg, J., Weaver, J.C., Thanawala, M.S. et al. (2005) *Science*, **309**(5732), 275.
220 Arzt, E., Gorb, S., Spolenak, R. (2003) *Proc. Nat. Acad. Sci.*, **100**, 10603.
221 Gao, H., Wang, X., Yao, H. et al. (2005) *Mech. Mater.*, **37**, 275.
222 Rho, J.-Y., Kuhn-Spearing, L., Zioupos, P. (1998) *Med. Eng. Phys.*, **20**(2), 92.
223 Weiner, S., Wagner, H.D. (1998) *Annu. Rev. Mater. Sci.*, **28**(1), 271.
224 Hoffmann, B., Chabbert, B., Monties, B., Speck, T. (2003) *Planta*, **217**, 32.
225 Keckes, J., Burgert, I., Frühmann, K. et al. (2003) *Nat. Mater.*, **2**(12), 810.
226 Milwich, M., Speck, T., Speck, O. et al. (2006) *Am. J. Bot.*, **93**(10), 1455.
227 Raabe, D., Sachs, C., Romano, P. (2005) *Acta Mater.*, **53**(15), 4281.
228 Raabe, D., Romano, P., Sachs, C. et al. (2006) *Mater. Sci. Eng. A*, **421**(1), 143.
229 Weiner, S., Wagner, H. D. (1998) *Annu, Rev, Mater. Sci.*, **28**(1), 271.
230 Kastelin, J., Baer, E. (1980) In *The Mechanical Properties of Biological Materials*,. Vincent, J.F., Currey, J.D. (Eds.), Cambridge University Press, Cambridge, pp. 397–435.
231 Fratzl, P., Weinkamer, R., (2007) *Prog. Mater. Sci.*, **52**(8), 1263.
232 Paris, O., Fritz-Popovski, G., Van Opdenbosch, D., Zollfrank, C. (2013) *Adv. Funct. Mater.*, **23**(36), 4408.
233 Gibson, L.J. (2012) *J. R. Soc. Interf.*, **9**(76), 2749.-
234 Fratzl, P., Weinkamer, R. (2007) *Prog. Mater. Sci.*, **52**(8), 1263.
235 Gao, H., Ji, B., Jäger, I.L., Arzt, E., Fratzl, P. (2003) *Proc. Nat, Acad. Sci.*, **100**, 5597.
236 Woesz, A., Weaver, J.C., Kazanci, M. et al. (2006) *J. Mater. Res.*, **21**, 2068.
237 Fantner, G.E., Hassenkam, T., Kindt, J.H. et al. (2005) *Nat. Mater.*, **4**(8), 612–616.
238 Fantner, G.E., Oroudjev, E., Schitter, G. et al., (2006) *Biophys. J.*, **90**(4), 1411.
239 Lakes, R., (1993) *Nature*, **361**(6412), 511.
240 Vincent, J.F. (2005) *J. Theor. Biol.*, **236**, 73.
241 Booker, R.E., Sell, J., (1998) *Holz als Roh- und Werkst.*, **56**(1), 1.
242 Kamat, S., Su, X., Ballarini, R., Heuer, A.H. (2000) *Nature*, **405**(6790), 1036.
243 Dong, Q., Su, H., Xu, J. et al. (2007) *Mater. Lett.*, **61**, 2714.
244 Maddocks, A.R., Harris, A.T. (2009) *Mater. Lett.*, **63**(9), 748.
245 Yang, D., Qi, L., Ma, J. (2002) *Adv, Mater.*, **14**(21), 1543.
246 Yang, D., Qi, L., Ma, J. (2003) *J. Mater. Chem.*, **13**(5), 1119.
247 Wang, Q., Jiang, Z., Wang, Y. et al. (2009) *J. Nanoparticle Res.*, **11**(2), 375.
248 Sapei, L., Noske, R., Strauch, P. et al. (2008) *Chem. Mater.*, **20**, 2020.
249 Zampieri, A., Mabande, G.T.P., Selvam, T. et al. (2006) *Mater. Sci. Eng. C*, **26**, 130.
250 Valtchev, V., Smaihi, M., Faust, A.-C., Vidal L. (2004) Paper presented at 14th International Zeolite Conference on Recent Advances in the Science and Technology of Zeolites and Related Materials, Cape Town, South Africa.
251 Gonzalez, P., Serra, J., Liste, S. et al. (2003) *Biomaterials*, **24**, 4827.
252 Byrne, C.E., Nagle, D.C. (1997) *Mat. Res. Innov.*, **1**(3), 137.
253 Seago, A.E., Brady, P., Vigneron, J.-P., Schultz, T.D. (2008) *J. R. Soc. Interf.*, **6** Suppl. 2, 165.
254 Sharma, V., Crne, M., Park, J.O., Srinivasarao, M. (2009) *Science*, **325**, 449.
255 Vukusic, P., Sambles, J.R. (2003) *Nature*, **424**(6950), 852.
256 Rassart, M., Simonis, P., Bay, A. et al. (2009) *Soft Matter Phys.*, **80**, 31910.
257 Vigneron, J.P., Pasteels, J.M., Windsor, D.M. et al., (2007) *Phys. Rev. E*, **76**, 31907.
258 Vukusic, P., Sambles, J.P., Lawrence, C.R. (2000) *Nature*, **404**(6777), 457.
259 Bartl, M.H., Nagi, R.K. (2014) In *Bioinspiration, Biomimetics, and Bioreplication*, Vol. 9055, Lakhtakia, A (Ed.), SPIE, San Diego, California, USA.
260 Vignolini, S., Rudall, P.J., Rowland, A.V. et al., (2012) *Proc. Nat. Acad. Sci. U.S.A.*, **109**(39), 1.
261 Biro, L.P., Vigneron, J.-P. (2011) *Laser Photon. Rev.*, **5**(1), 27.

262 Barrows, F.P., Bartl, M.H. (2014) *Nanomater. Nanotechnol.*, **4**(1), 1.
263 Shin, Y., Liu, J., Chang, J.H. et al. (2001) *Adv,. Mat.*, **13**(10), 728.
264 Persson, P.V., Fogden, A., Daniel, G. et al. (2004) *Biomacromolecules*, **5**(3), 1097.
265 Deshpande A.S., Burgert I., Paris, O. (2006) *Small*, **2**(8–9), 994.
266 Zimmerman, A.B., Nelson, A.M., Gillan, E.G. (2012) *Chem. Mater.*, **24**, 4301.
267 Allan, G.G., Carroll, J.P., Negri, A.R. et al. (1992) *Tappi J.*, **75**, 175.
268 Donath, S., Militz, H., Mai, C. (2004) *Wood Sci. Technol.*, **38**, 555.
269 Xie, Y., Hill, C.A.S., Xiao, Z. et al. (2010) *Compos. A Appl. Sci. Manuf.*, **41**, 806.
270 Sotiropoulou, S., Sierra-Sastre, Y., Mark, S.S., Batt, C.A. (2008) *Chem. Mater.*, **20**(3), 821.
271 Dujardin, E., Blaseby, M., Mann, S. (2003) *J. Mater. Chem.*, **13**, 696.
272 Shopsowitz, K.E., Qi, H., Hamad, W.Y., MacLachlan, M.J. (2010) *Nature*, **468**, 422.
273 Michielsen, K., Stavenga, D.G. (2008) *J. R. Soc. Interf.*, **5**(18), 85.
274 Schröder-Turk, G.E., Wickham, S., Averdunk, H. et al. (2011) *J. Struct. Biol.*, **174**(2), 290.
275 Wohlgemuth, M., Yufa, N., Hoffman, J., Thomas, E.L. (2001) *Macromolecules*, **34**(17), 6083.
276 Fratzl, P., Jakob, H.F., Rinnerthaler, S. et al. (1997) *J. Appl. Crystallogr.*, **30**(5, Pt. 2), 765.
277 Glatter, O. (1977) *J. Appl. Crystallogr.*, **10**, 415.
278 Glatter, O. (1980) *J. Appl. Crystallogr.*, **13**, 7.
279 Donaldson, L., Xu, P. (2005) *Trees*, **19**(6), 644.
280 Salmen, L., Burgert, I., Salmén, L. (2009) *Holzforschung*, **63**(2), 121.
281 Fritz-Popovski, G. (2015) *J. Appl. Crystallogr.*, **48**, 44.
282 Feigin, L.A., Svergun, D.I. (1987) *Structure Analysis by Small-Angle X-Ray and Neutron Scattering*, Plenum Press, New York, pp. 219–273.
283 Ota, T., Takahashi, M., Hibi, T. et al. (1995) *J. Am. Ceram. Soc.*, **78**, 3409.
284 Ota, T., Imaeda, M., Takase, H. et al. (2000) *J. Am. Ceram. Soc.*, **83**(6), 1521.
285 Sun, B.H., Fan, T.X., Zhang, D., Okabe, T. (2004) *Carbon*, **42**, 177.
286 Wood, L.L., Messina, P., Frisch, K.C. (1974) US Patent 3,833,386A.
287 Peng, H.X., Fan, Z., Evans, J.R.G., Busfield, J.J.C. (2000) *J. Eur, Ceram. Soc.*, **20**, 807.
288 Byrne, C.E., Nagle, D.C. (1997) *Mat. Res. Innovat*, **1**(3), 137.
289 Chakrabarti, O., Weisensel, L., Sieber, H. (2005) *J. Am. Ceram. Soc.*, **88**, 1792.
290 Greil, P., Lifka, T., Kaindl, A. (1998) *J. Eur. Ceram. Soc*, **18**, 1974.
291 Vogli, E., Zollfrank, C., Sieber, H. et al. (2002) *Ceram. Transact.*, **129**, 33.
292 Zollfrank, C., Kladny, R., Sieber, H., Greil, P. (2001) In *Verbundwerkstoffe und Werkstoffverbunde* (Eds: B. Wielage, G. Leonhardt), Wiley-VCH, Weinheim, Germany, pp. 274–278.
293 Brandt, B., Zollfrank, C., Franke, O. et al. (2010) *Acta BioMater.*, **6**, 4345.
294 Paris, O., Zollfrank, C., Zickler, G.A. (2005) *Carbon*, **43**, 53.
295 Greil, P., Lifka, T., Kaindl, A. (1998) *J. Eur. Ceram. Soc.*, **18**(14), 1975.
296 Xu, Y., Cheng, L., Zhang, L. (1999) *Carbon*, **37**, 1179.
297 Singh, M., Martınez-Fernández, J., de Arellano-López, A.R. (2003) *Curr. Opin. Solid State Mater. Sci.*, **7**, 247.
298 Weiner, S. Addadi, L. (1997) *J. Mater. Chem.*, **7**, 689.
299 Zhang, X., Yu, M., Liu, J., Li, S. (2012) *Langmuir*, **28**, 3690.
300 Fritz-Popovski, G., Morak, R., Schöberl, T. et al. (2014) *Bioinspired, Biomim. Nanobiomater.*, **3**, 160.
301 Wang, J. He, S., Xie, S. et al. (2007) *Mater. Lett.*, **61**, 917.
302 Zhang, X., Yu, M., Liu, J., Li, S. (2012) *Langmuir*, **28**, 3690.
303 Götze, J., Möckel, R., Langhof, N. et al. (2008) *Ceram. Silikáty*, **52**, 268.
304 Wang, T.-C., Fan, T-X., Zhang, D., Zhang, G.-D. (2006) *Mater. Lett.*, **60**, 2695.
305 Wang, T.-C., Fan, T-X., Zhang, D., Zhang, G.-D. (2006) *Carbon N, Y.*, **44**, 900.
306 Haas, D., Fey, T., Greil, P. (2007) *Adv. Eng. Mater.*, **9**(10), 892.
307 Wang, T.-C., Fan, T-X., Zhang, D. et al. (2007) *Mater. Lett.*, **61**, 1849.
308 Arellano-López, A.R., Martínez-Fernandez, J., González, P. et al. (2004) *Int. J. Appl. Ceram. Technol.*, **67**, 56.
309 Beck-Candanedo, S., Roman, M., Gray, D.G. (2005) *Biomacromolecules*, **6**, 1048.
310 Padalkar, S., Capadona, J.R., Rowan, S.J. et al. (2010) *Langmuir*, **226**, 8497.
311 Zhang, T., Wang, W., Zhang, D. et al. (2010) *Adv, Funct. Mater.*, **20**, 1152.
312 Gruber, S., Klupp Taylor, R.N., Scheel, H. et al. (2011) *Mater. Chem Phys.*, **129**, 19.
313 Gruber, S., Zollfrank, C. (2012) *Bioinspired, Biomim. Nanobiomater.*, **1**(2), 95.
314 Scheel, H., Zollfrank, C., Greil, P. (2009) *J. Mater. Res.*, **24**, 1709.
315 Nogi, B.M., Yano, H., Nogi, M. (2008) *Adv. Mater.*, **20**, 1849.
316 Olsson, R.T., Azizi Samir, M.A.S., Salazar-Alvarez, G. et al. (2010) *Nat. Nanotechnol.*, **5**, 584.
317 Korhonen, J.T., Hiekkataipale, P., Malm, J. et al. (2011) *ACS Nano.*, **5**, 1967.

318 Ago, M., Jakes, J.E., Johansson, L.S. et al. (2012) *ACS Appl. Mater. Interfac.*, **4**, 6849.
319 Eichhorn, S.J., Dufresne, A., Aranguren, M. et al. (2009) *J. Mater. Sci.*, **45**, 1.
320 Rosenau, T., Potthast, A., Adorjan, I. et al. (2002) *Cellulose*, **9**, 283.
321 Rauch, M.W., Dressler, M., Scheel, H. et al. (2012) *Eur. J. Inorg. Chem.*, **2012**, 5192.
322 Van Opdenbosch, D., Maisch, P., Fritz-Popovski, G. et al. (2012) *Carbohydr. Polym.*, **87**, 257.
323 Kondo, T., Togawa, E., Brown Jr., R.M. (2001) *Biomacromolecules*, **2**, 1324.
324 Shopsowitz, K.E., Qi, H., Hamad, W.Y., MacLachlan, M.J. (2010) *Nature*, **468**, 422
325 Moon, R.J., Martini, A., Nairn, J. et al. (2011) *Chem. Soc. Rev.*, **40**, 3941.
326 Barreiro, A.M., Recouvreux, D.O.S., Hotza, D. et al. (2010) *J. Mater. Sci.*, **45**, 5252.
327 Rambo, C.R., Sieber, H., Genova, L.A. (2008) *J. Porous Mater.*, **15**, 419.
328 Cao, J., Rambo, C.R., Sieber, H. (2004) *J. Porous Mater.*, **2**, 11(3), 163.
329 Rambo, C., Cao, J., Sieber, H. (2004) *Mater. Chem. Phys.*, **87**, 345.
330 Sieber, H., Rambo, C., Cao, J., Vogli, E., Greil, P. (2002) *Key Eng. Mater.*, **206**, 2009.
331 Pickenäcker, O., Pickenäcker, K., Wawrzinek, K. et al. (1999) *Keram. Z.*, **51**, 108.
332 Joannopoulos, J.D., Johnson, S.G., Winn, J.N., Meade, R.D. (2008) *Photonic Crystals: Molding the Flow of Light*, Princeton University Press, Princeton, NJ.
333 Simonis, P., Vigneron, J.P. (2011) *Phys. Rev. E*, **83**(1), 11908.
334 Colomer, J-F., Simonis, P., Bay, A. et al. (2012) *Phys. Rev. E*, **85**(1), 011907.
335 Burresi, M., Cortese, L., Pattelli, M. et al. (2014) *Scientific Reports*, **4**.
336 Pouya, C., Stavenga, D.G., Vukusic, P. (2011) *Optics Express.*, **19**(12), 11355.
337 Galusha, J.W., Richey, L.R., Jorgensen, M.R. et al. (2010) *J. Mater. Chem.*, **20**, 1277.
338 Parker, A.R., Townley, H.E. (2007) *Nat. Nanotech.*, **2**(6), 347.
339 Watanabe, K., Hoshino, T., Kanda, K. et al. (2005) *Jpn. J. Appl. Phys.*, **44**(1), 48.
340 Van Opdenbosch, D., Johannes, M., Wu, X. et al. (2012) *Photonics Nanostruct. Fundam. Appl.*, **10**(4), 516.
341 Maradudin, A.A, (1991) *Photonic Band Struct.*, **80**(3), 199.
342 Wu, X., Erbe, A., Raabe, D., Fabritius, H. (2013) *Adv. Funct. Mater.*, **23**, 3615.
343 Busch, K., John, S. (1998) *Phys. Rev. E*, **58**, 3896;.
344 Huang, J., Wang, X., Wang, Z.L.Z.L. (2006) *Nano Lett.*, **6**, 2325.
345 Aliev, A.E., Zakhidov, A.A., Baughman, R.H., Yablonovitch, E. (2006) *Int. J. Nanosci.*, **5**(01), 157.
346 Galusha, J.W., Jorgensen, M.R., Bartl, M.H., Galusha, B.J.W. (2010) *Adv. Mater.*, **22**, 107.
347 Huang, J., Wang, X., Wang, Z.L. (2006) *Nano Lett.*, **6**, 2325.
348 Zhu, S., Zhang, D., Chen, Z., Zhou, G. (2010) *J. Nanoparticle Res.*, **12**, 2445.
349 Schröder-Turk, G.E., Wickham, S., Averdunk, H. et al. (2011) *J. Struct. Biol.*, **174**(2), 290.
350 Wilts, B.D., Michielsen, K., De Raedt, H., Stavenga, D.G. (2011) *Interfac. Focus*, **1**.
351 Saranathan, V., Osuji, C.O., Mochrie, S.G.J. et al. (2010) *Proc. Nat. Acad. Sci.*, **107**(26), 11676.
352 Dolan, J.A., Wilts, B.D., Vignolini, S. et al. (2015) *Adv. Opt. Mater.*, **3**, 12.
353 Scherer, M.R.J. (2013) *Double-Gyroid-Structured Functional Materials*, Springer T., Springer International Publishing, Cham, Switzerland.
354 Ullal, C.K., Maldovan, M., Thomas, E.L. et al. (2004) *Appl. Phys. Lett.*, **84**, 5434.
355 Maldovan, M., Thomas, E.L. (2004) *Nat. Mater.*, **3**(9), 593.
356 Mille, C., Tyrode, E.C.,Corkery, R.W. et al. (2011) *Chem. Comm.*, **47**, 9873.
357 Dolan, J.A., Wilts, B.D., Vignolini, S. et al. (2015) *Adv. Opt. Mater.*, **3**, 12.
358 Nam, S., Park, W.H., Ihm, D., Hudson, S.M. (2010) *Carbohyd. Poly.*, **80**, 80291.
359 Dolgopyatova, N.V., Novikov, V.Y., Konovalova, I.N., Putintsev, N.N. (2011) *Russ. J. Appl. Chem.*, **84**(6), 1026.
360 Gao, H., Liu, Z., Zhang, J. et al. (2007) *Appl. Phys. Lett.*, **90**, 123115.
361 Xie, G., Zhang, G., Lin, F. et al. (2008) *Nano. Technol.*, **19**, 9560.
362 Cheng, X., Lu, C., Yang, B. (2011) *Rec. Pat. Mater. Sci.*, **4**, 15.
363 Istrate, E., Sargent, E.H. (2002) *J. Opt. A*, **4**, 242.
364 Li, M., Chaiko, D.J., Schlossman, M.L. (2003) *J. Phys. Chem. B*, **107**, 9079.
365 Jiang, W., Gu, L., Chen, X., Chen, R.T. (2007) *Solid-State Electron.*, **51**, 1278.
366 Kostovski, G., White, D.J., Mitchell, A. et al. (2012) *Biosens. Bioelec.*, **24**, 1531.
367 King, B.H., Wong, T., Sailor, M.J. (2011) *Langmuir*, **27**, 8576.
368 Ge, J., Yin, Y. (2011) *Angew. Chem.*, **123**, 1530.
369 Zhu, S., Zhang, D., Chen, Z. et al. (2009) *Nanotechn.*, **20**, 315303/8.
370 Gutbrod, K., Greil, P., Zollfrank, C. (2011) *Appl. Catal. B, Environ.*, **103**(1–), 240.
371 Li, X., Fan, T., Zhou, H. et al. (2009) *Adv. Funct. Mater.*, **19**, 45.
372 Sommer, A.P. Gente, M. (2006) *Energy & Fuels*, **20**(5), 2189.

373 Zhang, W., Zhang, D., Fan, T. et al. (2009) *Mater. Sci. Eng. C*, **29**(1), 92.
374 Dorval Courchesne, N-M., Steiner III, S.A., Cantú, V.J. et al. (2015) *Chem. Mater.*, **27**, 5361.
375 Biró, L.P., Kertész, K., Vértesy, Z., Bálint, Z. (2008) In *Photonic Nanoarchitectures Occurring in Butterfly Scales as Selective Gas/Vapor Sensors*, Vol. 7057, Creath, K. (Ed.), spie, San Diego, CA, USA
376 Potyrailo, R.A., Ghiradella, H., Vertiatchikh, A. et al. (2007) *Nat. Photonics*, **1**, 123.
377 Zhu, S., Liu, X., Chen, Z. et al. (2010) *J. Mater. Chem.*, **20**(41), 9126.
378 Liu, Z.T., Fan, T.X., Zhang, D. et al. (2009) *Sensors Actuators B Chem.*, **136**, 499.
379 Fudouzi, H., Sawada, T. (2006) *Langmuir*, **22**(3), 1365.
380 Gao, L., McCarthy, T.J. (2006) *Langmuir*, **22**(7), 2966.
381 Barthlott, W., Neinhuis, C. (1997) *Planta*, **202**(1), 1.
382 Feng, L., Li, S., Li, Y et al. (2002) *Adv. Mater.*, **14**, 1857.
383 Fürstner, R., Barthlott, W., Neinhuis, C., Walzel, P. (2005) *Langmuir*, **21**(3), 956.
384 Sun, M., Luo, C., Xu, L. et al. (2005) *Langmuir*, **21**(19), 8978.
385 Saison, T., Peroz, C., Chauveau, V. et al. (2008) *Bioinspiration Biomimetics*, **3**(4), 046004/1.
386 Kandarakis, I., Cavouras, D., Panayiotakis, G. et al. (1996) *Phys. Med. Biol.*, **41**(2), 297.
387 Schweizer, S. (2001) *Phys. Status Solidi*, **187**, 335.
388 Nagarkar, V.V., Tipnis, S.V., Gupta, T.K. (1999) *Nucl. Sci.*, **46**(3), 232.
389 Schmitt, B., Fuchs, M., Hell, E. et al. (2002) *Nucl. Instrum. Meth. B*, **191**, 800.
390 Rowlands, J.A. (2002) *Phys, Med. Biol.*, **47**, R123.
391 Nagarkar, V.V., Tipnis, S.V., Gupta, T.K. (1999) *Nucl. Sci.*, **46**(3), 232.
392 Martin, T., Koch, A. (2006) *J. Synchrotron Radiat.*, **13**, 180.
393 Kostova, M.H., Zollfrank, C., Batentschuk, M. et al. (2009) *Adv. Funct. Mater.*, **19**(4), 599.
394 Van Opdenbosch, D., Kostova, M.H., Gruber, S. et al. (2010) *Wood Sci. Technol.*, **44**, 547.
395 Kostova, M.H., Batentschuk, M., Goetz-Neunhoeffer, F. et al. (2010) *Mater. Chem. Phys.*, **123**(1), 166.
396 Mann, S. (1995) *J. Mater. Chem.*, **5**(7), 935.
397 Faatz, M., Gröhn, F., Wegner, G. (2004) *Adv. Mater.*, **16**, 996.
398 Levi-Kalisman, Y., Raz, S., Weiner, S. et al. (2002) *Adv. Funct. Mater.*, **12**, 43.
399 Loste, E., Meldrum, F.C. (2001) *Chem. Comm.*, **1**, 901.
400 Donners, J.J.J.M., Heywood, B.R., Meijer, E.W. et al. (2002) *Chemistry (Easton)*, **8**, 2561.
401 Beniash, E., Aizenberg, J. Addadi, L., Weiner, S. (1997) *Proc. R. Soc. B Biol. Sci.*, **264**(1380), 461.
402 Deshpande, A.S., Burgert, I., Paris, O. (2006) *Small*, **2**(8–9), 994.
403 He, G., Gajjeraman, S., Schultz, D. et al. (2005) *Biochemistry*, **44**, 16140.
404 Ciobanu, M., Bobu, E., Ciolacu, F. (2010) *Chemistry, Cellul. Chem. Technol.*, **44**, 379.
405 Persson, P.V., Hafrén, J., Fogden, A. et al. (2004) *Biomacromolecules*, **5**(3), 1097.
406 Pauling, L. (1960) *The Nature of the Chemical Bond and the Structure of Molecules and Crystals*, Cornell University Press: Ithaca, NY, USA
407 Liu, Z., Fan, T., Zhang, D, (2006) *J. Am. Ceram, Soc.*, **89**(2), 662.
408 Zhang, Y., Liu, X., Huang, J. (2011) *ACS Appl. Mater. Interfac.*, **3**, 3272.
409 Huang, J., Kunitake, T. (2003) *J. Am. Chem. Soc.*, **125**, 11834.
410 Huang, J., Kunitake, T., Onoue, S-Y. (2004) *Chemical, Chem. Commun.*, **1008**.
411 Huang, J., Matsunaga, N., Shimanoe, K. et al. (2005) *Chem. Mater.*, **17**, 3513.
412 Aoki, Y., Huang, J., Kunitake, T. (2006) *J. Mater. Chem.*, **16**, 292.
413 Liu, Z., Fan, T., Ding, J. et al. (2008) *Ceram. Int.*, **34**, 69.
414 Fan, T., Li, X., Liu, Z. et al. (2006) *J. Am. Ceram. Soc.*, **89**, 3511.
415 Li, X., Fan, T., Liu, Z. et al. (2006) *J. Eur, Ceram. Soc.*, **26**(16), 3657.
416 Liu, Z., Fan, T., Zhang, W., Zhang, D. (2005) *Microporous Mesoporous Mater.*, **85**(1–2), 82.
417 Liu, Z., Fan, T., Zhang, D. (2006) *J. Am. Ceram. Soc*, **89**, 662.
418 Zhu, S., Zhang, D., Chen, Z., Zhou, G. (2010) *J. Nanoparticle Res.*, **12**, 2445.
419 Luo, M., Gao, J.-Q., Yang, J-F. et al. (2007) *J. Am. Ceram. Soc.*, **90**, 4036.
420 Zheng, Z., Huang, B., Ma, H. et al. (2007) *Cryst. Growth Des.*, **7**, 1912.
421 Maca, K., Dobsak, P., Boccaccini, A.R. (2001) *Ceram. Int.*, **27**, 577.
422 Miao, X., Sun, D. (2010) *Materials*, **3**, 26.
423 Triantafillidis, C., Elsaesser, M.S., Husing, N. (2013) *Chem. Soc. Rev.*, **42**, 3833.
424 Fischer, S.F., Thielen, M., Loprang, R.R. et al. (2010) *Adv. Eng. Mater.*, **12**(12), B658.
425 Van Opdenbosch, D., Thielen, M., Seidel, R. et al. (2011) *Bioinsp., Biomim. Nanobiomater.*, **1**, 117.
426 Schnepp, Z., Yang, W., Antonietti, M., Giordano, C. (2010) *Angew. Chem. Int. Ed.*, **49**(37), 6564.
427 Del-Gallo, P., Gary, D., Chartier, T. et al. (2010) European Patent 2141140,
428 Greil, P. (2010) *MRS Bull.*, **35**, 145.
429 Zampieri, A., Mabande, G.T.P., Selvam, T. et al. (2006) *Mater. Sci. Eng. C*, **26**(1), 130.
430 Zampieri, A., Schwieger, W., Zollfrank, C., Greil, P. (2007) *Handbook of Biomineralization: Biomimetic*

and Bioinspired Chemistry, Wiley-VCH, Weinheim, Germany, 255.
431 Ren, Z., Zhai, T.,Wang, Z. et al. (2008) *Adv. Mater.*, **20**, 2337.
432 Frölich, A.,. Fischer, J., Zebrowski, T. et al. (2013) *Adv. Mater.* **25**, 3588.
433 Ball, P. (2008) *Nat. Mat.*, **7**, 102.
434 Zampieri, A., Sieber, H., Selvam, T. et al. (2005) *Adv. Mat.*, **17**(3), 344.
435 Bhushan, B. (2009) *Phil. Trans. A*, **367**(1893), 1445.
436 Fan, T-X., Chow, S-K., Zhang, D. (2009) *Prog. Mat. Sci.*, **54**(5), 542.
437 Pouget, E., Dujardin, E., Cavalier, A. et al. (2007) *Nat. Mater.*, **6**(6), 434.
438 Yoo, P.J., Nam, K.T., Qi, J. et al. (2006) *Nat. Mater.*, **5**(3), 234.
439 Vignolini, S., Rudall, P.J., Rowland, A.V. et al. (2012) *Proc. Nat. Acad. Sci. U.S.A.*, **109**, 15712.
440 Simon, R., Holderied, M.W., Koch, C.U., von Helversen, O. (2011) *Science*, **333**(6042), 631.

2

Tubular Tissue Engineering Based on Microfluidics

Lixue Tang, Wenfu Zheng, and Xingyu Jiang*

Beijing Engineering Research Center for BioNanotechnology and CAS Key Lab for Biological Effects of Nanomaterials and Nanosafety, National Center for NanoScience and Technology, 11, BeiYiTiao, ZhongGuanCun, Beijing, 100190, China

2.1 Introduction

There are many tubular structures in the human body, such as blood vessels, lymphatic vessels, tracheae, guts, and the renal tubules. These tubular structures play important roles in supporting human life. However, diseases, injury, and complications may lead to large-scale lesions of these tissues. For example, atherosclerosis may cause stenosis of the whole aorta and subsequent high mortality. Without medical help, these lesions cannot repair themselves. Autografts, allografts, and xenografts, although they have been used in clinics for some time, face many problems, such as limited donor source, immunological rejection, and risk of infection. Tissue engineering, as an emerging tissue-repair technology, is a promising approach in treating large-scale lesions in tubular structures. Microfluidics technology offers promising tools to precisely control the cellular microenvironment. Various cell manipulation techniques in microfluidics have been applied and been successful in tissue engineering. In this review, after a short introduction to the system of natural tubular structures and microfluidics, we focus on the combination of microfluidics, cells, and biomaterials in tubular tissue engineering (Figure 2.1).

2.2 Natural Tubular Structures

Tubular systems are found everywhere in the human body and interconnect the cells, tissues, and organs to realize delivery and exchange of materials and signaling transmission. Among various tubular structures we will introduce blood vessels, lymph vessels, vessels in the digestive system, and vessels in the respiratory system in this section due to their vital functions in the body and their relevant roles in tissue engineering.

2.2.1 Blood Vessels

Blood vessels refer to a series of different sized and variously structured pipes through which blood flows and nutrients, wastes, and other metabolic materials are transported. Blood vessels can be divided into arteries, veins, and capillaries, according to their different configurations and functions. According to the role of the blood vessels in tissue engineering, we will only introduce arteries in this review.

Classified by diameter, arteries can be divided into four categories: large artery, medium-sized artery, small artery, and arteriole. The wall of an artery comprises three layers: (1) the tunica intima; (2) the tunica media; and (3) the tunica adventitia. The tunica intima is composed of the endothelial and the sub-endothelial layers. The tunica media consists of many layers of elastic membrane and has different thicknesses, depending on its type. Medium-sized arteries and small arteries also contain layers of circular smooth muscle fibers in their tunica media. The tunica adventitia, the outermost layer of arteries, contains fibroblasts and collagen fibers [1].

2.2.2 Lymphatic Vessels

Lymphatic vessels are parts of the lymphatic system which can transport redundant interstitial fluid away from tissues and finally return the fluid into the venous circulation. Lymphatic vessels also serve as part of the immune system and involve the migration and maturation of immune cells. Lymphatic vessels are typically composed of three layers: (1) the tunica intima; (2) the tunica media; and (3) the tunica adventitia. The tunica intima is the inner layer of the lymph vessel and is composed of endothelium [2]. The tunica media is the middle layer of the lymph vessel which is composed of circularly arranged smooth muscle and elastic fibers.

* Email: zhengwf@nanoctr.cn

Bioinspired Materials Science and Engineering, First Edition. Edited by Guang Yang, Lin Xiao, and Lallepak Lamboni.

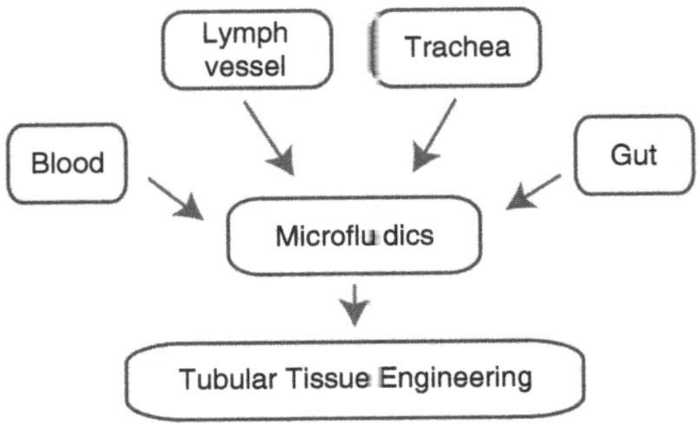

Figure 2.1 Schematic illustration of tubular tissue engineering on microfluidics by mimicking various natural tubular structures.

The tunica media alters the diameter of the lumen and provides the propulsion for the lymph movement by contracting or relaxing [3]. The tunica adventitia is the outer layer of the lymph vessel which is composed of connective tissue, collagen, and elastic fibers.

2.2.3 Vessels in the Digestive System

Vessels in the digestive system, also called the gastrointestinal tract, are tubes for transferring, consuming, and digesting foodstuffs, absorbing nutrients, and expelling waste. These vessels usually are divided into four parts: (1) the esophagus; (2) the small intestine; (3) the large intestine; and (4) ducts, such as the bile duct. The wall of the vessels in the digestive system, with a similar structure, consists of four parts from the innermost lumen to the outermost wall: (1) the mucosa; (2) the submucosa; (3) the muscularis externa; and (4) the adventitia. The mucosa is made up of the epithelium, the lamina propria, and the muscularis mucosae. Epithelium secretes the digestive juice for digestion and to absorb nutrients. The lamina propria is a layer of connective tissue which contains the capillaries and the lymphatic capillaries. The muscularis mucosae is a layer of smooth muscle responsible for the bowel movement. The submucosa consists of connective tissue with large blood vessels, lymphatics, and nerves. The muscularis externa consists of muscular layers for preventing food from traveling backward and shortening the tract. The adventitia consists of several layers of connective tissue [1].

2.2.4 Vessels in the Respiratory System

Vessels of the respiratory system, responsible for passing and exchanging gas, can be divided into three parts: (1) the trachea; (2) the bronchi; and (3) the alveolus. The trachea, a wide hollow tube, branches into the bronchi, leading to the lobes of the lungs which are filled with small, spongy sacs called alveoli. The alveolus is composed of epithelial cell monolayers, and its function is to exchange oxygen and carbon dioxide. The walls of the trachea are made up of four layers from the innermost to the outermost: (1) the mucosa; (2) the submucosa; (3) the hyaline cartilage; and (4) the adventitia. The mucosa consists of ciliated pseudostratified columnar epithelium with many goblet cells. The submucosa layer is composed of connective tissue where collagen, elastin, and reticular protein fibers give soft support and elasticity to the trachea, while blood vessels and nerves support the nutrient supply and sensing of the tracheal wall. Longitudinal smooth muscle fibers are present in the posterior trachea between the ends of the cartilage rings, allowing the trachea to adjust its diameter. The hyaline cartilage forms the supportive rings of the trachea and provides flexible structure that maintains an open airway and is resistant to external stresses. The adventitia is composed of areolar connective tissue that loosely anchors the trachea to the surrounding soft tissues [1].

2.2.5 The Features of the Natural Tubular Structures

The same features of different vessels and their difference have been summarized here to direct how to prepare biomaterials and scaffolds that will satisfy the requirements of tissue engineering for these different tissues. Generally speaking, the natural tubular structures have two distinguishing features: (1) they have 3D tubular shapes; and (2) they all have a multi-layered structure, where different types of cells are distributed at specific locations [4]. So, in order to bioengineer tubular structural tissues, we first of all need to fabricate a tubular scaffold that mimics the natural tubular tissues, and, second, incorporate the appropriate cells in the tubular scaffold.

2.3 Microfluidics

2.3.1 An Introduction to Microfluidics

Microfluidics is a set of techniques that can precisely control minute amounts of fluids at a micrometer scale [5]. The concept of microfluidics dates back to the 1970s [6] and increasingly became popular in the 1990s [7, 8]. As a form of miniaturized platforms, microfluidics creates numerous new opportunities for tissue engineering [9]. The inherent advantages of microfluidics for tissue engineering include:

1) *a suitable size scale*: Ranging from several to hundreds of micrometers, the size of the structures in microfluidics is close to single cell or a cell-scaffold complex which allows convenient manipulations of cultured cells and tissues.

2) *a fluidic microenvironment*: Various fluidic conditions, such as laminar flow, micro-droplets, and the vortex can be recapitulated on the microscale to satisfy research under specific conditions.
3) *multiple function integration*: The ability to integrate multi-structures at a microscale shows the huge potential of microfluidics to integrate multiple functions in parallel or cascaded.
4) *a large surface area*: The large surface-to-volume ratio of microfluidic chips facilitates surface-dependent operations, such as capturing cells by affinity. All these advantages make microfluidics a compelling platform for bio-analysis and tissue engineering [10–12]

2.3.2 Microfluidics to Manipulate Cells

Microfluidics is an unique toolbox for biological research, especially cell manipulation due to its inherent merits in matching the requirements of the cellular microenvironments. Hence, cell research focusing on various themes has been applied to microfluidics. Here, we introduce cell patterning and cell mechanics on microfluidics, which are closely related to tissue engineering.

2.3.2.1 Cell Patterning

Cell patterning refers to positioning cells on a substrate with controllable size, orientation, distribution, and spatial arrangement. By confining cell adhesion to a rational, pre-designed geometrical shape, biologists can study various scientific issues, such as cell viability, cell migration, cell proliferation, cell differentiation, and cell-cell assembly. The techniques typically adopted to pattern cells are known as "soft lithography," which was developed by Whitesides and colleagues. Soft lithographic techniques include self-assembled monolayers (SAMs), microcontact printing (μCP), and microfluidic printing (μFP). SAMs are 1–3 nm-thick regularly self-assembled nanostructures formed by organic molecules onto a surface. The most extensively used SAMs are the SAMs of alkanethiolates on a gold surface which permit control over the interfacial structure and properties. μCP is a technique that uses topographic patterns on the surface of an elastomeric polydimethylsiloxane (PDMS) stamp to form the same patterns on the surfaces of various substrates. The stamp, inked with a solution containing patterning components, can be brought into contact with a substrate to transfer the components and form corresponding patterns on the substrate with high spatial definition (Figure 2.2a). μFP is a technique that uses microfluidic channels to restrict the flow of fluids to desired regions to form patterns on various substrates. Ligands, proteins, or cells could be patterned on various substrates such as gold, glass, a Petri dish, and PDMS (Figure 2.2b). These key techniques of soft lithography provide a convenient, effective, and low-cost method for the formation and manufacturing of micro patterns of cells on various substrates [13–15].

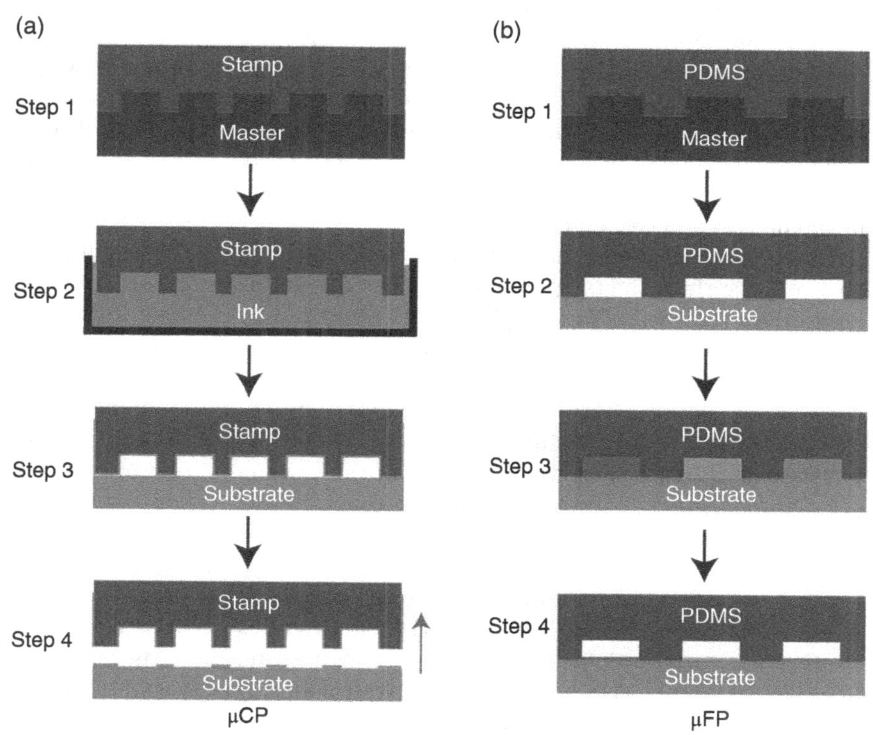

Figure 2.2 The principle of μCP, μFP, and LFP. (a) Schematic demonstration of the process of μCP. Step 1, pour PDMS prepolymer over master and cure it to form a PDMS stamp; Step 2, dip the PDMS stamp into patterning solution; Step 3, press PDMS stamp onto the substrate for seconds to minutes; Step 4, Lift up the stamp and leave patterns on the substrate. (b) Schematic demonstration of the process of μFP. Step 1, pour PDMS prepolymer over master and cure it to form a PDMS mold containing microchannels; Step 2, put the PDMS mold and substrate together to form sealed microchannels; Step 3, introduce different solutions into different microchannels to deposit components onto the substrate; Step 4, wash away the solution and leave components on the substrate to form patterns in confined areas.

To realize the cell patterning, most mammalian cells must first of all attach themselves to a surface though cell-surface interactions in order to survive, proliferate, and function. To enhance the cell adhesion, necessary surface modification techniques such as SAMs formation, Langmuir-Blodgett deposition, and layer by-layer assembly have been employed to fabricate a thin layer of composites whose properties are similar to an extracellular matrix (ECM), which provides structural integrity and a suitable chemical environment for the cells. Among these modification methods, SAMs are widely implemented because the nature and density of functional groups can easily be tuned on SAMs to tune cell adhesion [16].

The cell-surface interactions can be divided into two categories: nonspecific and specific interactions. Nonspecific interactions are mediated by physical attachments, including electrostatic force, van der Waals interaction, and hydrophobic forces. By employing nonspecific interactions, Lee et al. used dopamine self-polymerization to form thin, surface-adherent polydopamine films on a wide range of inorganic and organic materials [16]. Jiang et al. reported that polydopamine is a versatile nonspecific adhesive material for cell adhesion [17].

Specific cell-surface interactions refer to the involvement of specific receptor-ligand bonding formations in cell adhesion. The RGD (Arg-Gly-Asp) sequence is an excellent mediator to promote the specific attachment of cells on surfaces. Its ability to bind a variety of cells through ligand-receptor interactions makes RGD an exceptionally useful sequence for cell patterning [18]. The surface density of RGD peptides has been demonstrated to elicit different cellular responses and a minimum RGD density of 10 fmol/cm^2 is needed for cell adhesion [19]. The clustering of RGD peptides at the nanoscale level can significantly reduce the average ligand density and cell adhesion [20].

The physical properties such as topographical and stiffness of the surface, the temperature, light can also influence the cell adhesion. Topographical cues generated by the ECM, independent of biochemistry, have direct effects on cell behavior such as adhesion, migration, cytoskeletal arrangements, and differentiation [21–24]. Teruo Okano et al. introduced the Poly(N-isopropylacrylamide) (PIPAAm) onto the surface of polystyrene cell culture dishes to control the cell adhesion by adjusting the temperature of the polymer [25]. Cells attached on PIPAAm-grafted surfaces at 37 °C, and detached from the surfaces at 20 °C without the usual damage associated with trypsinization. The cell sheets fabricated using this strategy have been widely used in the tissue engineering field [26, 27]. Light also has been used to dynamically control cell adhesion on substrates. However, existing techniques have mostly relied on irreversible control, that is, once the original structure for controlling cell adhesion on the surface is altered, it cannot be regenerated for further use [28–31]. Jiang et al. reported a photochemical method to reversibly control cell adhesion using azobenzene-embedded self-assembled monolayers. In this method, reversible cell adhesion was realized depending on the conformational transition of the azobenzene group embedded in the SAMs, which were elicited by different wavelengths of lights (Figure 2.3a) [32].

Patterning cells in 3D to simultaneously realize the combination of ordered cells arrangement and functions still remains challenging. Jiang et al. reported an on-chip method to construct a kind of long-term 3D neuronal network with ordered somata patterns and guided neurite connections as well as recordable neuronal activities in functions [33]. In this strategy, gravity-driven, the microchamber-assisted assembly of borosilicate glass microspheres is employed to fabricate scaffolds onto which dissociated neurons develop 3D neuronal networks (Figure 2.3b). Compared with the existing strategies which can just realize either ordered pattern of neuronal somata [34, 35] or controllable guidance of neurites [36, 37], this strategy is more promising for applications in the fields of fundamental neuroscience, tissue engineering, and brain-machine interfaces.

2.3.2.2 Cell Mechanics

Mechanics play key role in tissue development, tissue repair, and microenvironment maintenance. The mechanical cues, typically including stretching, shear stress, and matrix rigidity, drastically affect cell proliferation, differentiation, and functions, and consequently influence the process of tissue regeneration. Compressive loads have been shown to increase cell differentiation and extracellular matrix synthesis [38]. Dynamic fluid perfusion can induce higher cell proliferation and differentiation than static cultures [39]. Shear stress has been reported to influence cell adhesion, cytoskeletal alignment, as well as gene expressions [40]. Tensile forces can stimulate cell proliferation and vascular remodeling in living skin [41]. The rigidity of the substrate can also influence the behavior of the cells. Chen et al. observed cells prefer to migrate to stiffer regions of substrate [42]. LeDuc et al. employed microfluidic channels to pattern alternative rigidities on substrate. Their results show that localized rigidity would restrict the outgrowth and orientation of neurite in differentiating neuroblasts [40]. All these discoveries imply that mechanical stimuli is a very important component in the field of tissue engineering.

Microfluidics provides unique tools for mimicking various mechanical environments to investigate cells behavior *in vitro*. Douville et al. developed a microsystem

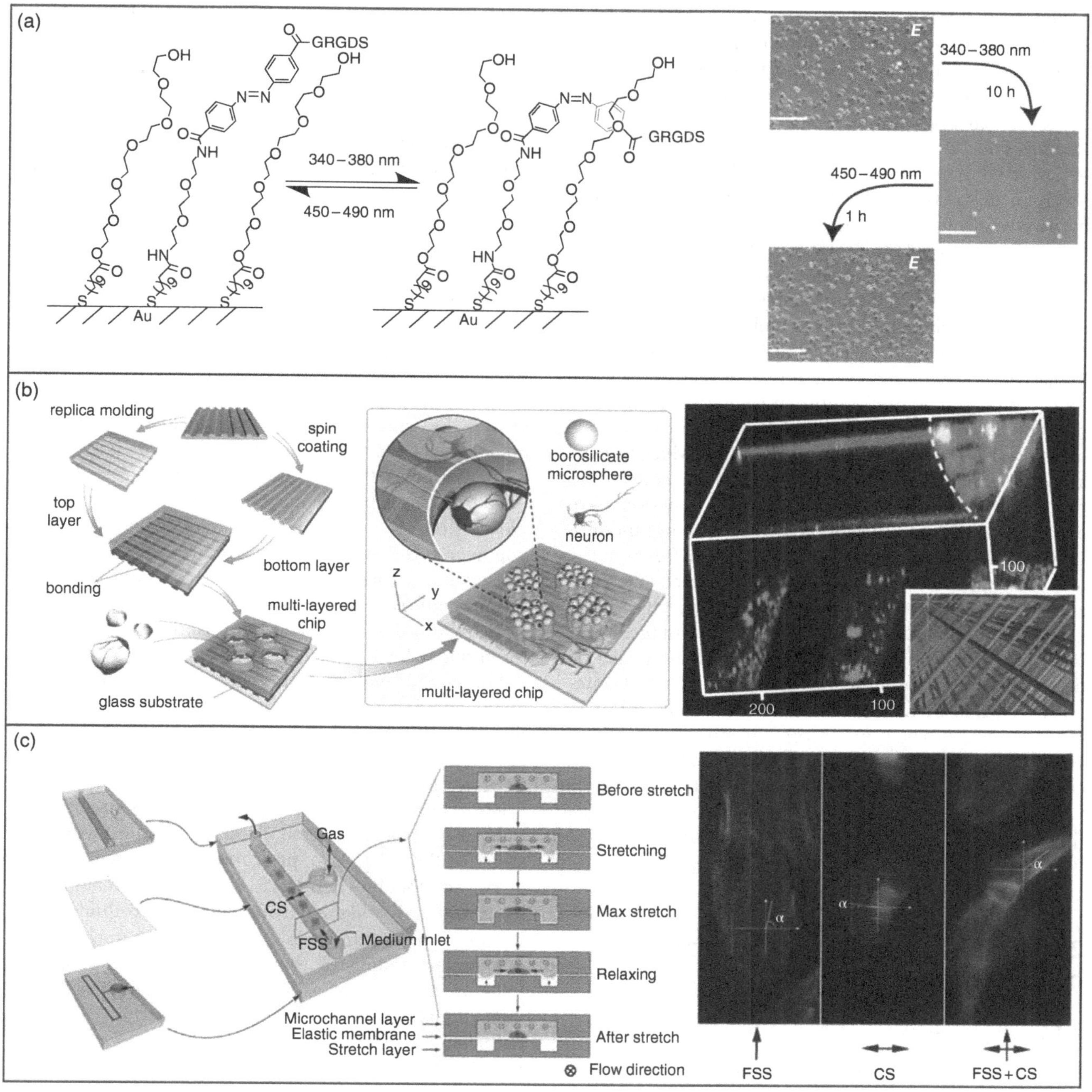

Figure 2.3 (a) The azobenzene moiety can be converted photochemically between the E and Z configurations to either present or mask the RGD ligand and hence modulate cell adhesion. Cells adhered onto SAMs with the azobenzene group in the E configuration. Few cells adhered to the same SAMs with azobenzene in the Z configuration. Cells adhered to the SAMs again when the conformation of azobenzene was changed from Z to E. (b) Schematics of fabrication processes of the multilayered microchip and 3D reconstruction of the multilayered neurite connections. The inset represents the orthogonal neurite pathways in the brain. Copyright 2012, American Association for the Advancement of Science. (c) Schematic of the microfluidic flow-stretch chip. The stress fibers tend to align in the direction parallel to that of the FSS, CS and the resultant force of the FSS and CS. *Sources:* (a) Reproduced with permission from [29]. (b) Reproduced with permission from [30]. (c) Reproduced with permission from [1]. (*See insert for color representation of the figure.*)

to mimic the mechanical environment in the alveolus and demonstrated significant morphological differences between alveolar epithelial cells exposed to a combination of mechanical and surface tension stresses compared to cell populations exposed solely to the cyclic stretch. This "alveoli-on-a-chip" research describes new tools for studying the combined effects of two kinds of mechanical stress on cells [43]. Ingber et al. developed a bioinspired "lung-on-a-chip" that reconstitutes the functional alveolar capillary interface of the human lung. The device reproduces complex, integrated organ-level responses to bacteria, inflammatory cytokines, and nanoparticles introduced into the alveolar space [44]. Based on a similar device, Ingber's group presented an *in vitro* living cell-based model of the intestine that mimics the mechanical, structural, absorptive, transport and pathophysiological properties of the human gut, i.e. a biomimetic "human gut-on-a-chip" [45]. Mechanically active "organ-on-a-chip" microdevices that reconstitute tissue/tissue interfaces critical to organ function have expanded the capabilities of *in vitro* cell culture models. In the cardiovascular system, hemodynamic forces in the form of fluid shear stress (FSS) and cyclic stretch (CS) are important for the maintenance of the physiological condition of the cardiovascular system and the development of cardiovascular diseases. Jiang et al. [1] developed a microfluidic flow-stretch chip that integrates FSS and CS for cultured endothelial cells, mimicking the hemodynamic microenvironment of blood vessels *in vivo*. Under FSS-only, CS-only, and FSS + CS conditions, the alignment of the cellular stress fibers varied with the types of stimulations (Figure 2.3c). The flow-stretch chip is a reliable tool for simulating the hemodynamic microenvironment.

2.4 Fabrication of Tubular Structures by Microfluidics

2.4.1 Angiogenesis

Angiogenesis is the process of the growth of new microvessels from pre-existing blood vessels. This process occurs either naturally during development or wound healing or as a consequence of a pathological injury, and is orchestrated by a range of pro- and anti-angiogenic factors [46]. Various diseases [47], including cancer [48], rheumatoid arthritis [49–51], and atherosclerosis [52, 53], involve angiogenesis.

Tissue engineering has made tremendous progress in recent years, however, most applications have been restricted to avascular or thin tissues, such as cartilage, skin, or bladder [54, 55]. In these tissues, the delivery of nutrients and oxygen to cells and the transport of waste from cells can be realized by diffusion. However, for thick tissues, without supplies from capillaries, diffusion cannot satisfy the survival condition of the cells, which would undergo hypoxia and apoptosis [56]. Therefore, the formation of vascular networks that are able to deliver nutrients and oxygen though the engineered tissues is a main principle in the engineering of complex tissues. Angiogenic growth factors play an important role and regulate each stage of the angiogenesis process. For example, the vascular endothelial growth factor (VEGF) contributes to the initiation of angiogenesis, and induces endothelial cell proliferation and migration. The platelet-derived growth factor (PDGF) is a mitogen and chemotactic agent that recruits pericytes and smooth muscle cells (SMCs). Transforming growth factor-β (TGF-β causes ECM deposition for stabilization of new vessels [57–59]. Therefore, the incorporation of an angiogenic factor into the bioengineered tissue before implantation with slow and sustained-release kinetics enhances new capillary ingrowth from the host's vascular plexus [60, 61]. As an alternative, cells within the engineered tissue can be genetically modified to secrete angiogenic factors [62, 63] to recruit circulating endothelial progenitor cells (EPCs) or mesenchymal stem cells (MSCs) into the scaffold and induce the angiogenesis of the tissue [64, 65].

The microfluidics offer enormous opportunities in angiogenesis because they provide flexibility in controlling the critical biochemical and biomechanical factors, such as the establishment of biochemical gradients in two- or three-dimensional geometries [66]. For example, scientists can recapitulate FSS, CS, and pressure differentials to control the metabolic stress or the mechanical stress of the cells. Also, pro- and anti-angiogenic molecules and other biochemical factors could be manipulated to mimic a vivo-like environment in microfluidic chips for cells to form microvessels. The research may promote our understanding of the mechanism of angiogenesis [67–71]. Chrobak et al. [72] generated endothelialized microtubes within which perfusion can be realized by culturing endothelial cells on the surfaces of open channels that span entire collagen gels. Zheng et al. introduced networks of endothelialized microchannels within matrices of type I collagen to study angiogenesis in healthy and pathological scenarios, and thrombosis under quiescent and inflammatory conditions [73].

2.4.2 Tissue Engineering of Natural Tubes

2.4.2.1 A Brief Review of Vascular Tissue Engineering

The development of artificial blood vessels in general can be divided into three stages, ranging from autograft substitute, to synthetic vascular alternatives, and the tissue engineering of blood vessels. Autografts, the graft from the patient's own body, are the gold standard grafts used to treat vascular occlusions. However, the limited source of the autografts hindered their application in clinics. The synthetic grafts such as expanded polytetrafluoroethylene (Teflon®) and polyethylene terephthalate (Dacron®) grafts, due to their sufficient source supply, play a big part in the production of the artificial blood vessels. However, although the synthetic grafts have been widely used in clinics, their drawbacks are apparent: immunogenicity, occlusion, and non-degradation. Also, the performance of synthetic blood vessels at small diameters (inner diameter <6 mm) are disappointing [74, 75]. Therefore, scientists nowadays are beginning to focus on producing tissue-engineered blood vessels.

The purpose of vascular tissue engineering is to design biocompatible, responsive, and degradable blood vessels similar to the natural tissues [76]. Engineered vascular grafts should have sufficient burst pressure (at least 1,700 mmHg, equal to that of a saphenous vein), enough fatigue resistance to CS, negligible dilatation, fabrication consistency, and a functional endothelial cell layer which is similar to the intimal in a natural vessel. Also, the vascular grafts should not cause chronic inflammatory responses [77]. Mechanical and biochemical environments are usually needed to promote the cell growth, differentiation, and ECM secretion of the engineered blood vessel [78]. Emerging vascular surgery, in general, has evolved from "replacement" to "regeneration" of the vascular tissue.

Generally, vascular tissue engineering can be broadly divided into two groups: (1) scaffold-free vascular tissue engineering; and (2) scaffold-guided vascular tissue engineering [79].

2.4.2.1.1 Scaffold-free Vascular Tissue Engineering

In 1986, pioneering work was performed by Weinberg and Bell, who produced the world's first tissue-engineering blood vessels [80]. A culture medium mixed with collagen and smooth muscle cells was cast into a mold to prepare a tube which corresponds to the tunica media of blood vessel. Adventitial fibroblasts were cast between the tunica media and a Dacron mesh sleeve to form a layer corresponding to the tunica adventitia. The lumen was lined with endothelial cells by rotating the vessel. This structure has many characteristics of a real blood vessel, such as layered structures and electrophysiological responses. However, it needs at least one month to form the whole structure and also needs a Dacron mesh sleeve to strengthen the structure to provide enough mechanical strength.

The successful application of epithelial cell sheets in the treatment of burns patients inspired the application of a cell sheet to vascular tissue engineering. The study found that the smooth muscle cells and fibroblasts of the human umbilical vein, under the stimulation of ascorbic acid, can secrete large amounts of extra ECM to form a cohesive cellular sheet which proved to be a milestone in tissue engineering. Current tissue engineering blood vessel reconstruction can be dated back to 1998, when L'Heureux et al. used the cell sheet-based method to produce a blood vessel with excellent performance [81]. Human vascular smooth muscle cells (hvSMC) cultured with ascorbic acid will produce a cohesive cellular sheet. This sheet can be placed around a tubular support to prepare the media of an artificial blood vessel. A similar sheet of human fibroblasts can be wrapped around the media to form an adventitia. After maturation, the tubular support can be removed and endothelial cells can be seeded in the lumen to form an integrated blood vessel. In 2006, L'Heureux's group further improved their method using the original basis. They simplified the former three-layer structure in which the smooth muscle cell layer was removed, and between the intima and the media, an acellular layer from a decellularized fibroblast cell sheet was added to strengthen the vessel as well as provide a barrier against cell migration toward the lumen [82].

The results of L'Heureux's studies challenge the presumption that smooth muscle cells and scaffolds are actually necessary for vascular tissue fabrication. Cell sheets have been used to create an arteriovenous shunt for hemodialysis access in patients with end-stage renal disease in clinical trials. Although they offer good performance, these methods are limited by the long-term process of fabrication. Therefore, shortening the fabrication time has become a priority for researchers.

2.4.2.1.2 Scaffold-guided Vascular Tissue Engineering

Scaffold-guided vascular tissue engineering is essentially a continuation of the study of synthetic vascular grafts. The difference is that, in the scaffold-guided vascular tissue engineering, the vascular cells will infiltrate into the materials *in vitro* and become maturated before implantation. The scaffold is a three-dimensional template to support the entire structure of the organization, provide guidance during the development of tissues, and influence the cellular functions, such as cell adhesion, migration, proliferation, differentiation, and secretion of the extracellular matrix [83]. Therefore, the scaffold should be biocompatible and reproducible, at the same time, it should have the appropriate mechanical properties. Also, the adaptability of the scaffold material is one of the most important factors that affect vascular remolding. To stimulate cell migration, information exchange, and tissue growth, the scaffold structure must be highly porous and interconnected [84]. This porous network of scaffolds provides channels for cell migration, signaling molecule diffusion, nutrient delivery, and waste removal. Scaffolds can also be divided into two types, one has a good biological stability which provides long-term support in the human body; the other is made of biodegradable materials whose degradation rate equals that of tissue regeneration. Scaffolds of both types can be made from synthetic biomaterials or be naturally occurring.

Synthetic materials have great potential in terms of being vascular scaffolds because they are widely available, and their cost and performance are controllable. However, the biocompatibility of the synthetic materials influences their performance in vascular tissue engineering. Zella et al. lined ePTFE grafts with functional autologous endothelium to overcome the problems associated with acute thrombosis [85]. However, this approach introduces a potential risk to the body due to the permanent presence of a synthetic scaffold. In 1999, Niklason et al. fabricated tissue engineering blood vessels with biodegradable polyglycolic acid (PGA). SMCs were seeded into the tubular PGA scaffolds and cultured in a bioreactor under conditions of pulsatile radial stress for eight weeks, and finally endothelial cells were lined in the luminal surface. Their vessels had high patent and burst pressure with contractile responses to pharmacological agents [78]. However, problems, such as the long period of construction time and the complex fabrication process, limited this method. With advances in technology, some materials with excellent performance are being developed, and the acellular tissue engineering scaffolds are attracting increasing attention. Wang et al. developed a cell-free scaffold from poly (glycerol sebacate) (PGS), a biodegradable material that has an appropriate degradation time [86]. The inner layer of the scaffold is porous PGS for host cell infiltration, and over 99.99% of the pores are interconnected. In the luminal surface, heparin is coated to reduce the platelet adhesion to prevent thrombosis. To enhance the mechanical strength and prevent blood cell loss, the grafts were enclosed in a dense, nonwoven poly (ε-caprolactone) PCL sheath. The entire scaffold was directly connected to rat abdominal aorta and exposed to the hemodynamic environment without seeding any cells *in vitro*. Under the interaction of the scaffold degradation and host remolding, new arteries, which were almost free of foreign materials, were yielded after three months.

There are also some non-mainstream methods for vascular remolding. Some researchers inserted silastic, polypropylene or polyglycolic acid tube into the peritoneal cavity, in which macrophage-derived granulation tissue will form on the surface of the tube, the vessel yielded after the tube was removed. This living vessel, though with good patency and mechanical strength, is still risky in clinical trials owing to the invasive requirement of inserting foreign materials into the human body [87, 88].

2.4.2.2 Rapid Prototyping of 3D Tubular Structures

Tissues with tubular structures, such as blood vessels, lymph vessels, trachea, and intestines are abundant in the bodies of higher-order animals. Tubular tissues have two distinguishing features: (1) they have specific 3D shapes; and (2) they have different types of cells at specific locations. Mimicking both of these features is a prerequisite in fabricating functional tubular tissues. Furthermore, most of the present methods used on 3D tubular structure fabrication need several months of *in vitro* culture before implantation, a rapid 3D tubular structure fabrication method is desirable. Jiang and colleagues provide three efficient and facile strategies to fabricate tubular structures. Briefly, based on the stress-induced rolling membrane (SIRM) technique, combined with the microfluidic and cell patterning technology, tissue-engineered blood vessels with highly controllable distribution and orientation of vascular cells were fabricated.

The first strategy, based on the SIRM technique, provided a self-assembly method to fabricate a series of complex 3D structures, such as tubular structures, tubes-in-a-tube structures, spirals, and tubes with wrinkled walls [89]. Using the SIRM techniques to fabricate the tubular structure, the SIRM is fabricated by binding two elastic membranes together. The top membrane of the two binding elastic membranes is a piece of cured PDMS which is stretched by a mechanical stretcher to produce stress, and the bottom membrane is relaxed and semi-cured PDMS. After the two layers of membranes bond together, the SIRM is obtained. if the top and the bottom membrane were released at the same time, the bi-layer membrane will roll up and turn into a tubular structure because of the internal stress. By using this technique, the tubes-in-a-tube structure, which refers to the structure where several small tubes are embedded in a big tube, could be fabricated with only two steps (Figure 2.4a): first, a rectangular array was cut in the middle part of the SIRM with a scalpel, with one side of each rectangle left uncut. The rectangular array will roll up to form the tubular array. Next, one end of the entire SIRM was cut off along one side of the tubular array, then the entire SIRM is rolled toward another end to form a big tube. The two ends of the big tube were cut to generate the tubes-in-a-tube structure. In addition, spiral structures could be formed using the SIRM technique. It is known that closing the gaps of the spiral ribbons can lead to the formation of rigid and straight tubes [90]. Conversely, if a rectangular strip were cut with a certain angle (?° from

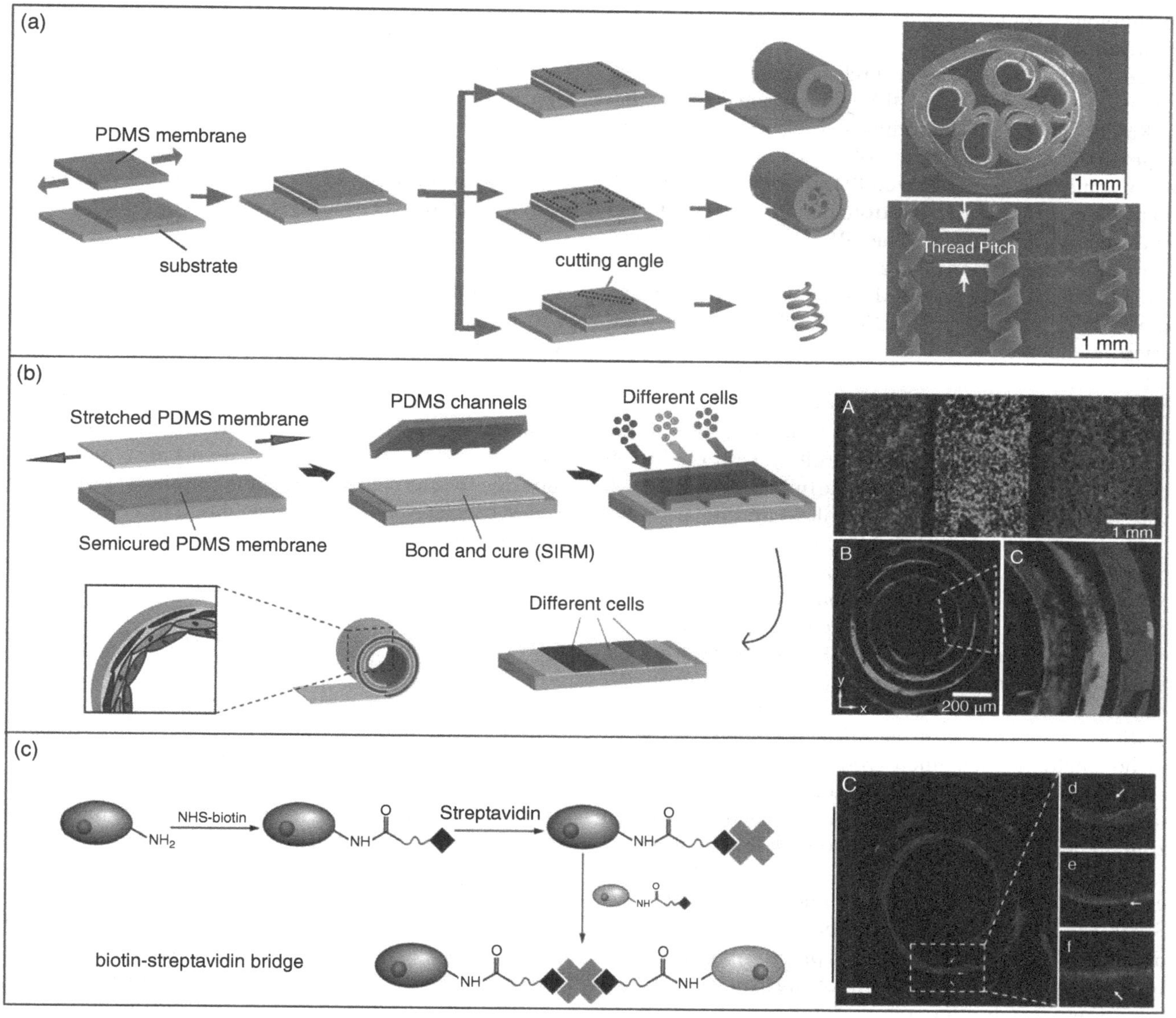

Figure 2.4 (a) Schematic illustration of the fabrication process of tubes, tubes-in-a-tube and spirals. (b) Schematic illustration of a stress-induced rolling membrane (SIRM) and tubes with multiple types of cells as the walls. A thin PDMS membrane is stretched as the top layer of the SIRM and covers a semi-cured PDMS membrane. After curing the two layers to cause adhesion, a SIRM is obtained that rolled up when the ends were released. Microfluidic channels cover the surface of the SIRM, different cells are delivered via microfluidic channels to the surface of the SIRM. One end of the SIRM is released by cutting its edge. The SIRM rolls up into a tube and each type of cell is delivered to a designated position as the tubular wall. The structure of the tube is similar to that of tubular tissues, such as small-diameter vessels. (c) Schematic diagram of cell surface modification and stepwise formation of multicellular structures. Confocal image of a 3D reconstruction of the bilayer on the SIRM after rolling. *Sources:* (a) Reproduced with permission from [85]. (b) Reproduced with permission from [87]. (c) Reproduced with permission from [88]. (*See insert for color representation of the figure.*)

the stretching direction on the SIRM, the narrow strips can self-assemble into the spiral structure which mimics the alignment of SMCs in the blood vessel (Figure 2.4a). Compared with tubes with smooth walls, wrinkles can prevent the closure and clogging of the inner cavity if the tubes are bent. In order to obtain tubes with wrinkled walls, liquid PDMS instead of solid PDMS as the bottom layer was used. If a stretched PDMS membrane were put on the liquid PDMS, after curing the bi-layer, wrinkles will be formed on the bi-layer structure. When this kind of bi-layer PDMS membranes is released, tubes with wrinkled walls will be formed. Therefore, by tuning the SIRM structure and shape, various tubular structure with specific features could be fabricated to satisfy clinic applications.

The second strategy [91], combined with the SIRM technique, the cell patterning technique, and microfluidic technology, contains two steps: (1) delivering and patterning different types of cells on a 2D SIRM using microfluidic channels; and (2) releasing the SIRM to roll up into a 3D tube (Figure 2.4b). As mentioned above, the SIRM is fabricated by binding two elastic membranes together. The top membrane of the two binding elastic membranes is a piece of cured PDMS which is stretched to produce stress, and the bottom membrane is a relaxed and semi-cured PDMS. The SIRM tends to roll up because of the internal stress; to avoid this, the two ends of the SIRM were fixed inside a Petri dish. Then the SIRM is covered by a PDMS mold embedded with microfluidic channels. These microfluidic channels are separated, and different types of cells are delivered into different channels. After the cells have attached to the membrane, the mold is removed. When one end of the SIRM was cut, the cell sheet will roll up with the SIRM due to the internal stress, finally, the tube with the layered wall with different types of cells will form, similar to the natural vessel structure. In addition, the diameter of the tube and the thickness of tubular wall can be controlled. What is more, this method could also control the distribution and orientation of the cells in the 3D scaffold because many vascular cells *in vivo* such as SMCs are polarized: they orientate either longitudinally or circumferentially. In order to obtain patterned cells in the scaffold, the top layer of the SIRM was patterned with orthogonally aligned microgrooves. On the patterned region, SMCs would grow along the grooves due to contact guidance. When the membrane rolls up, the horizontally aligned SMCs become longitudinally oriented and the vertically aligned ones are circumferentially oriented. To encourage the communication of cells, the SIRM can be fabricated with a porous structure. This method provides a new method of precisely fabricating a tubular structure with expectant cells. However, in these tubes there is only a monolayer of the same type of cells between the two layers of PDMS, and the cells between the different PDMS layers cannot directly communicate with each other. To overcome these problems, cell surface modification is needed.

Biotinylation of cell surfaces is one of the commonest methods used for cell surface modification through the direct chemical reaction of primary amine groups presented on the cell membrane with the amine-reactive biotin, such as N-hydroxy-succinimide biotin derivatives. In order to obtain a three-dimensional, multilayered structure, different types of cells were stacked using biotin-SA (streptavidin) interaction [92]. The whole procedure of cell surface modification consists of two steps: first, treating the cell surface amines with sulphonated biotinyl-N-hydroxy-succinimide (NHS–biotin), and then followed by biotin-SA interaction. The adherent cells on the substrates were modified with NHS-biotin and SA successively. The population of another type of cells modified with NHS-biotin was added on the top of the monolayer of the SA-coated adherent cells. So the biotinylated cells can make contact with the SA-coated cells through the biotin-SA interaction to form a bi-layer (Figure 2.4c). The structures can also be continuously coated with SA, followed by adding another population of biotinylated cells to form structures of multiple types of cells.

Based on the SIRM technique, a monolayer of NHS-biotin modified smooth muscle cells (SMCs) on the SIRM was fabricated, then the monolayer was modified with SA, then the suspension of the biotinylated HUVECs was added on the surface of the monolayer. HUVECs connected with the monolayer of SMCs to form the bi-layer (Figure 2.4c). When one end of the SIRM was released, the cell-covered SIRM began to roll up into a tube due to internal stress, and the cell sheet was layered as the tubular wall. In this way, different types of cells make contact with each other, and they will have direct material and signal communication.

2.4.3 Tissue Engineering of Other Tubular Structures

2.4.3.1 Tissue Engineering and Regeneration of Lymphatics

The lymphatics play an instrumental role in the regulation of immunologic, circulatory, and metabolic systems. However, the lymphatic system can be damaged by trauma, tumor metastasis, genetic abnormalities, or infections. So, the repair of lymphatic system by tissue engineering is necessary and meaningful. The engineering of the lymphatic system can be broadly classified into two categories: (1) the growth factor approach; and (2) the cellular strategy. The growth factor approach is used to promote the regeneration of lymphatic vessels, whereas the cellular strategy is designed to deliver lymphatic endothelial cells (LECs) or stem cells to the scaffold. Some complex approaches may combine growth factors, cells, and scaffolds together.

2.4.3.1.1 Growth Factor-based Approaches to Tissue Engineering

The discovery of lymphatic-specific cellular markers was a significant advance in the study of the lymphatic system especially in the field of lymphatic tissue engineering. These critical molecules, such as vascular endothelial growth factor-C (VEGF-C), VEGF-D, fibroblast growth factor (FGF) [93, 94], hepatocyte growth factor (HGF) [95, 96], insulin-like growth factor (IGF) [97], and platelet-derived growth factor (PDGF) [98, 99], regulate lymphatic development and regeneration. For example, intradermal ear injection of recombinant VEGF-C-expressing an adeno-associated virus has been shown to be an effective means of promoting lymphatic sprouting and growth in a congenital model of primary lymphedema that lacks capillary lymphatics [100].

2.4.3.1.2 Cellular Approaches to Tissue Engineering and Regeneration

Isolation and analysis of LECs *in vitro* have increased our understanding of the lymphatic system and have served as a key step in lymphatic tissue engineering. Swartz seeded enterocytes and LECs to a Transwell dish and created a model of an intestinal lacteal to study lymphatic-enterocyte interphase and lipid transport [101]. LECs form 3D tubular structures when placed on matrigel or other scaffolds. This response is greatly increased with exposure of the LECs to lymphangiogenic cytokines or inhibition of antilymphangiogenic growth factors [102]. Dai et al. took advantage of this phenomenon by culturing LECs in a PGA scaffold *in vitro* and then transferring these constructs *in vivo* to tissue engineer a lymphatic vascular network [103]. Lymphatic networks can be also created using scaffolds that promote LEC ingrowth. Based on this concept, Wong et al. evaluated the potential for acellular dermal matrices as a biological scaffold for lymphatic vessel ingrowth, and found that both blood and lymphatic vessel ingrowth occur rapidly in decellularized dermis [104].

2.4.3.2 Tissue Engineering and Regeneration of the Digestive System

2.4.3.2.1 The Small Intestine

Tissue engineering and regeneration of the small intestine are necessary in cases such as the loss of intestinal segments caused by diseases such as the short bowel syndrome, inflammation, and cancers. The effective treatment for these disease is intestinal transplantation, however, the autogenous grafts from the patient's own body are limited and difficult to obtain. Therefore, tissue engineering provides solutions to treat these diseases. Similar to the tissue engineering of blood vessels, the small intestine tissue engineering also should be guided by scaffolds. The PGA scaffold, because of its good biocompatibility and biodegradability, was successfully applied by Vacanti et al. They transplanted organoid units on the PGA scaffold into the omentum of a Lewis rat, and tissue engineered the improved recovery of the small intestine after a massive small bowel resection in a rat model [105]. Small intestinal submucosa (SIS) has been reported as a scaffold for small intestine regeneration in rat [106, 107], rabbit [108], and canine models [109]. The use of an SIS patch produced three regenerated layers, including mucosa, smooth muscle, and a serosal covering [110]. Hori et al. used material of autologous origin and a collagen sponge as a scaffold [110] to regenerate the small intestine, however, the regenerated intestine lacked a smooth muscle layer, which is essential for functional peristalsis and maintenance of the lumen caliber. In an attempt to regenerate the smooth muscle layer, autologous smooth muscle cells were seeded into collagen sponge scaffolds, which have been successfully implanted as a patch graft in a canine model [111]. This patch graft regenerated the mucosal and intestinal epithelial layers along with the villi structures. The major problem encountered with these grafts was shrinkage. Lee et al. used pseudotubular structures formed from collagen sponge scaffolds seeded with intestinal smooth muscle cells, which were neovascularized within a month after prevascularization in the omentum [112].

2.4.3.2.2 The Colon

The colon has important functions, such as fatty acid production, sodium and water absorption, and storage. Loss of colonic segments by surgical resections significantly alters colonic motility, which may cause constipation or diarrhea.

Vacanti et al. generated a tissue-engineered colon on composite poly lactic and glycolic acid (PLGA) polymers from organoid units isolated from the sigmoid colon [113]. The tissue engineered conduits have significant absorptive capacity when implanted. Hecker et al. bioengineered 3D fibrin-based models of a colon that allow self-alignment of circular smooth muscle layers concentrically around a patent lumen. The bioengineered tissues mimic the native smooth muscle alignment and maintains aspects of colonic physiology, such as peristalsis, contraction, and relaxation [114]. Metzger et al. [115] demonstrated that adult human gut-derived enteric progenitor cells can repopulate segments of human aganglionic colon grown in organotypic cultures. Transplantation of enteric neuronal progenitor cells therefore offer successful alternate routes to regeneration and improvement of the gut function.

2.4.3.3 Tissue Engineering in the Tracheal System

Extended lesions on the tracheal may drastically shorten the sufficient length of the natural airway for breathing. Tissue engineering is a promising way to address this issue. Prerequisites for the ideal tracheal graft include lateral rigidity, longitudinal flexibility, biocompatibility, nonimmunogenicity, and the ability to support resurfacing of ciliated respiratory epithelium [116, 117].

Vacanti et al. [118] isolated the chondrocytes from the shoulders of newborn calves and seeded them onto a PGA scaffold. After incubation and implantation over 4 weeks, the chondrocytes produced cartilage of selected shape and size, vascularized connective tissue surrounded the implants and lined the inner surface. The structure can resist collapse and was used to replace segments of trachea in rats. In further experiments, an attempt was made to seed the inner surface with tracheal epithelial cells. Moreover, a decellularized human donor trachea, colonized by the recipient's epithelial cells and chondrogenic mesenchymal stem cells (MSC), was implanted into a patient with end-stage airway disease [119]. More than one year after the tissue-engineered tracheal transplantation, the function parameters of the implanted tracheal are all within the normal range. This provides compelling evidence that tissue engineering with a well-designed strategy is an ideal way to treat large tracheal lesions.

2.5 Conclusion

In this review, we first introduced the main natural tubular structures in the human body: the blood vessel, the lymphatic vessel, the vessels in the digestive system, and the vessels in the respiratory system and their key structural features such as 3D tubular shapes and multi-layered structure. This knowledge may help us to understand the importance of the structural and functional mimicking in the construction of tissue engineering scaffold.

Microfluidics, an emerging set of technologies for manipulating cells, cell-cell interactions, and cell-material interactions, are powerful tools for tubular structure tissue engineering. By bonding and stretching bi-layer membrane with different Young modulus, a two-dimensional membrane could become a three-dimensional tubular structure with multi-layers. By microfluidic printing (μ CP) different cell types with different micro-channels on the two-dimensional membrane, the corresponding three-dimensional tubular structure could be a multi-layered construction where different types of cells are distributed at specific locations. This three-dimensional rapid prototyping technology is a promising method for the construction of advanced, complex, tubular structures.

Although tubular tissue engineering has achieved exciting progress, there are still many problems that need to be addressed. For example, the elasticity, stiffness, and degradation rate of the scaffold materials still cannot fully match the requirement of the operations, the mechanical properties of the original tissues, and the tissue repair processes. The source of the pre-loaded cells in the scaffold is also limited at present. To solve these problems, we may focus on the improvement of biomaterials and the utilization of stem cells.

Acknowledgments

Financial support was provided by the Ministry of Science and Technology of China (2011CB933201, 2012AA030308), the National Science Foundation of China (31170905, 81361140345, 31470911, 51373043, and 21222502), the CAS (XDA09030305, XDA09030307), and the CAS/SAFEA International Partnership Program for Creative Research Teams.

References

1 Papka, R.E. (1995) *Encyclopedia of Neuroscience.*
2 Pepper, M.S., Skobe, M. (2003) *J. Cell Biol,*. **163**, 209–213.
3 Shayan, R., Achen, M.G., Stacker, S.A. (2006) *Carcinogenesis*, **27**, 1729–1738.
4 Nerem, R.M., Seliktar, D. (2001) *Annu. Rev. Biomed. Eng.*, **3**, 225–243.
5 Whitesides, G.M. (2006) *Nat.*, **442**, 368–373.
6 Terry, S.C., Jerman, J.H., Angell, J.B. (1979) *IEEE Trans. Electron. Dev.*, **26**, 1880–1886.
7 Manz, A., Miyahara, Y., Miura, J. et al. (1990) *Sensor Actuator B: Chem.*, **1**, 249–255.
8 Harrison, D.J., Fluri, K., Seiler, K. et al. (1993) *Sci.*, **261**, 895–895.
9 El-Ali, J., Sorger, P.K. Jensen, K.F. (2006) *Nat.*, **442**, 403–411.
10 Meyvantsson, I., Beebe, D.J. (2008) *Annu. Rev. Anal. Chem,*. **1**, 423–449.
11 Khademhosseini, A., Langer, R., Borenstein, J., Vacanti, J.P. (2006) *Proc. Natl. Acad. Sci. U.S.A.*, **103**, 2480–2487.
12 Weibel, D.B., DiLuzio, W.R., Whitesides, G.M. (2007) *Nat. Rev. Microbiol.*, **5**, 209–218.
13 Whitesides, G.M., Ostuni, E., Takayama, S. et al. (2001) *Annu. Rev. Biomed. Eng.*, **3**, 335–373.
14 Kane, R.S., Takayama, S., Ostuni, E. et al. (1999) *Biomat.*, **20**, 2363–2376.
15 Mrksich, M. (2009) *Acta Biomaterialia*, **5**, 832–841.
16 Lee, H., Dellatore, S.M., Miller, W.M., Messersmith, P.B. (2007) *Sci.* **318**, 426–430.
17 Sun, K., Song, L., Xie, Y. et al. (2011) *Langmuir*, **27**, 5709–5712.
18 Humphries, J.D., Byron, A., Humphries, M.J. (2006) *J. Cell Sci.*, **119**, 3901–3903.
19 Massia, S.P., Hubbell, J.A. (1991) *J. Cell Biol.*, **114**, 1089–1100.

20 Maheshwari, G., Brown, G., Lauffenburger, D.A. et al. (2000) *J. Cell Sci*, 2000, **113**, 1677–1686.
21 Yang, M.T., Fu, J., Wang, Y-K., et al. (2011) *Nat. Protoc.*, **6**, 187–213.
22 Agudelo-Garcia, P.A., De Jesus, J.K., Williams, S.P., et al. (2011) *Neoplasia*, **13**, 831–IN822.
23 Doyle, A.D., Wang, F.W., Matsumoto, K., Yamada, K.M. (2009) *J. Cell Biol.*, **184**, 481–490.
24 Diehl, K.A., Foley, J.D., Nealey, P.F., Murphy, C.J. (2005) *J. Biomed. Mater. Res. Part A*, **75**, 603–611.
25 Okano, T., Yamada, N., Okuhara, M., Sakai, H. (2011) *Biomat.: Silver Jubilee Compendium*, **16**, 109.
26 Shimizu, T., Yamato, M., Kikuchi, A., Okano, T. (2003) *Biomat.*, **24**, 2309–2316.
27 J. Yang, J., Yamato, M., Kohno, C. et al. (2005) *Biomat.*, **26**, 6415–6422.
28 Yamaguchi, S., Yamahira, S., Kikuchi, K. et al. (2012) *Angew. Chem.*, **124**, 132–135.
29 Rechenmacher, F., Neubauer, S., Polleux, J. et al. (2013) *Angew. Chem. Int. Ed.*, **52**, 1572–1575.
30 Ng, C.C.A., Magenau, A., Ngalim, S.H. et al. (2012) *Angew. Chem. Int. Ed.*, **51**, 7706–7710.
31 Siltanen, C., Shin, D.S., Sutcliffe, J., Revzin, A. (2013) *Angew. Chem.*, **125**, 9394–9398.
32 Liu, D., Xie, Y., Shao, H., Jiang, X. (2009) *Angew. Chem. Int. Ed.*, **48**, 4406–4408.
33 Huang, Z., Sun, Y., Liu, W., et al. (2014) *Small*, **10**, 2530–2536.
34 Pautot, S. Wyart, C., Isacoff, E.Y. (2008) *Nat. Methods*, **5**, 735–740.
35 Kato-Negishi, M., Tsuda, Y., Onoe, H., Takeuchi, S. (2010) *Biomat.*, **31**, 8939–8945.
36 Luo, Y., Shoichet, M.S. (2004) *Nat. Mater.*, **3**, 249–253.
37 Ladd, C., So, J.H., Muth, J., Dickey, M.D. (2013) *Adv. Mater.*, **25**, 5081–5085.
38 Meyer, U., Büchter, A., Nazer, N., Wiesmann, H.P. (2006) *Brit. J. Oral Max. Surg.*, **44**, 134–140.
39 Fassina, L., Visai, L., Asti, L. et al. (2005) *Tiss. Eng.*, **11**, 685–700.
40 Chou, S-Y., Cheng, C-M., Chen, C-C., LeDuc, P.R. (2011) *Soft Matt.*, **7**, 9871–9877.
41 Pietramaggiori, G., Liu, P., Scherer, S.S. et al. (2007) *Ann. Surg.*, **246**, 896–902.
42 Gray, D.S., Tien, J., Chen, C.S. (2003) *J. Biomed. Mater. Res. Part A*, **66**, 605–614.
43 Douville, N.J., Zamankhan, P., Tung, Y-C. et al. (2011) *Lab. Chip*, **11**, 609–619.
44 Huh, D., Matthews, B.D., Mammoto, A. (2010) *Sci.*, **328**, 1662–1668.
45 Huh, D., Torisawa, Y-S., Hamilton, G.A. et al. (2012) *Lab. Chip*, **12**, 2156–2164.
46 Folkman, J. (2007) *Nat. Rev. Drug. Discov.*, **6**, 273–286.
47 Carmeliet, P., Jain, R.K. (2000) *Nat.*, **407**, 249–257.
48 Folkman, J. (2002) *Canc. Cell*, **1**, 113–115.
49 Colville-Nash, P.R., Scott, D.L. (1992) *Ann. Rheum. Dis.*, **51**, 919–925.
50 Koch, A. E. (1998) *Arthritis Rheum.*, **41**, 951–962.
51 Weber, A.J., De Bandt, M. (1999) *Joint, Bone, Spine: Revue du Rhumatisme*, **67**, 366–383.
52 O'Brien, E.R., Garvin, M.R., Dev, R. et al. (1994) *Am. J. Pathol.*, **145**, 883.
53 Winter, P.M., Morawski, A.M., Caruthers, S.D. et al. (2003) *Circul.* **108**, 2270–2274.
54 Atala, A. (2008) *Cur.r Stem. Cell. Res. The.r*, **3**, 21–31.
55 Naughton, G. K. (2002) *Ann. New York Aca. Sci.*, **961**, 372–385.
56 Patel, Z.S., Mikos, A.G. (2004) *J. Biomater. Sci. Polym. Edit.*, **15**, 701–726.
57 Silvestre, J-S., Lévy, B.I., Tedgui, A. (2008) *Cardiovas. Re.s*, **78**, 201–202.
58 Chen, R.R., Silva, E.A., Yuen, W.W., Mooney, D.J. (2007) *Pharm. Res.*, **24**, 258–264.
59 Darland, D-C., D'Amore, P.A. (1999) *J. Clin. Invest.*, **103**, 157.
60 Laschke, M.W. Harder, Y., Amon, M. et al. (2006) *Tiss. Eng.*, **12**, 2093–2104.
61 Shen, Y.H., Shoichet, M.S., Radisic, M. (2008) *Acta Biomaterialia*, **4**, 477–489.
62 Yang, F., Cho, S-W., Son, S.M. et al. (2010) *Proc. Natl. Acad. Sci. USA*, **107**, 3317–3322.
63 Vallbacka J.J., Sefton, M.V. (2007) *Tiss. Eng.*, **13**, 2259–2269.
64 Hoenig, M.R. Bianchi C., Sellke, F.W. (2008) *Cur. Drug. Targets*, **9**, 422–435.
65 Matsumoto, T., Kuroda, R., Mifune, Y. et al. (2008) *Bone*, **43**, 434–439.
66 Chung, S., Sudo, R., Vickerman, V. et al. (2010) *Ann. Biomed. Eng.*, **38**, 1164–1177.
67 Chung, S., Sudo, R., Mack, P.J. et al. (2009) *Lab. Chip*, **9**, 269–275.
68 Frisk, T., Rydholm, S., Liebmann, T. et al. (2007) *Electrophor.*, **28**, 4705–4712.
69 Mack, P.J., Zhang, Y., Chung, S. et al. (2009) *J. Biol. Chem.*, **284**, 8412–8420.
70 Saadi, W., Rhee, S.W., Lin, F. et al. (2007) *Biomed. Microdevices*, **9**, 627–635.
71 Vickerman, V., Blundo, J., Chung, S., Kamm, R. (2008) *Lab. Chip*, **8**, 1468–1477.
72 Chrobak, K.M., Potter, D.R., Tien, J. (2006) *Microvas. Res.*, **71**, 185–196.
73 Zheng, Y., Chen, J., Craven, M. et al. (2012) *Proc. Natl. Acad. Sci. USA*, **109**, 9342–9347.
74 Abbott, W.M., Callow, A., Moore, W. et al. (1993) *J. Vas. Surg.*, **17**, 746–756.
75 Bennion, R. S., Williams, R.A., Stabile, B.E. et al. (1985) *Surg. Gynecol. Obstet.*, **160**, 239–242.
76 Shin'oka, T., Imai, Y., Ikada, Y. (2001) *New Eng. J. Med.*, **344**, 532–533.

77 L'Heureux, N., Dusserre, N., Marini, A. et al. (2007) *Nat. Clin. Pract. Cardiovas. Med*, **4**, 389–395.
78 Niklason, L.E., Gao, J., Abbott, W.M. et al. (1999) *Sci.*, **284**, 489–493.
79 Seifu, D.G., Purnama A., Mequanint, K., Mantovani, D. (2013) *Nat. Rev. Cardiol.*, **10**, 410–421.
80 Weinberg, C.B., Bell, E. (1986) *Sci.*, **231**, 397–400.
81 L'Heureux, N., Pâquet, S., Labbé, R. et al. (1998) *FASEB J.*, **12**, 47–56.
82 L'Heureux, N., Dusserre, N., Konig, G. et al. (2006) *Nat. Med.*, **12**, 361–365.
83 Zhang, W.J., Liu, W., Cui, L., Cao, Y. (2007) *J. Cell. Mol. Med.*, **11**, 945–957.
84 Berglund, J.D., Galis, Z.S. (2003) *Brit. J. Pharm.*, **140**, 627–636.
85 Meinhart, J.G., Deutsch, M., Fischlein, T. et al. (2001) *Ann. Thor. Surg.*, **71**, S327–S331.
86 Wu, W., Allen, R.A., Wang, Y. (2012) *Nat. Med.*, **18**, 1148–1153.
87 Campbell, J.H., Efendy, J.L., Campbell, G.R. (1999) *Circ. Res.*, **85**, 1173–1178.
88 Chue, W-L., Campbell, G.R., Caplice, N. et al. (2004) *J. Vas. Surg.*, **39**, 859–867.
89 Jin, Y., Wang, N., Yuan, B. (2013) *Small*, **9**, 2410–2414.
90 Zhang, X., Zou, J., Tamhane, K. et al. (2010) *Small*, **6**, 217–220.
91 Yuan, B., Jin, Y., Sun, Y. et al. (2012) *Adv. Mater.*, **24**, 890–896.
92 Gong, P., Zheng, W., Huang, Z. et al. (2013) *Adv. Func. Mater.*, **23**, 42–46.
93 Chang, L.K., Garcia-Cardeña, G., Farnebo, F. et al. (2004) *Proc. Natl. Acad. Sci. U.S.A.*, **101**, 11658–11663.
94 Cao, R., Eriksson, A., Kubo, H. et al. (2004) *Circ. Res.*, **94**, 664–670.
95 Kajiya, K., Hirakawa, S., Ma, B. et al. (2005) *EMBO J.*, **24**, 2885–2895.
96 Saito, Y., Nakagami, H., Morishita, R. et al. (2006) *Circul.*, **114**, 1177–1184.
97 Björndahl, M., Cao, R., Nissen, L.J. et al. (2005) *Proc. Nat. Acad. Sci. U.S.A.*, **102**, 15593–15598.
98 Cao, Y. (2005) *Cell Cycle*, **4**, 231–233.
99 Onimaru, M., Yonemitsu, Y., Fujii, T. et al. (2009) *Am. J. Physiol. Heart Circ. Physiol.*, **297**, H1685–H1696.
100 Karkkainen, M.J., Saaristo, A., Jussila, L. et al. (2001) *Proc. Nat. Acad. Sci. U.S.A.*, **98**, 12677–12682.
101 Dixon, J.B., Raghunathan, S., Swartz, M.A. (2009) *Biotechnol. Bioeng.*, **103**, 1224–1235.
102 Avraham, T., Daluvoy, S., Zampell, J. (2010) *Am. J Pathol.*, **177**, 3202–3214.
103 Dai, T., Jiang, Z.H., Li, S.L. et al. (2010) *J. Biotechnol.*, **150**, 182–189.
104 Wong, A.K., Schonmeyer, B.H., Singh, P. et al. (2008) *Plast. Reconstr. Surg.*, **121**, 1144–1152.
105 Grikscheit, T.C., Siddique, A., Ochoa, E.R. et al. (2004) *Ann. Surg.*, **240**, 748.
106 Parnigotto, P.P., Marzaro, M., Artusi, T. et al. (2000) *J. Pediatr. Surg.*, **35**, 1304–1308.
107 Wang, Z.Q., Watanabe, Y., Toki, A. (2003) *J. Pediatr. Surg.*, **38**, 1596–1601.
108 Demirbilek, S., Kanmaz, T., Özardalı İ. (2003) *Pediatr. Surg. Int.*, **19**, 588–592.
109 Chen, M.K., Badylak, S.F. (2001) *J. Surg. Res.*, **99**, 352–358.
110 Hori, Y., Nakamura, T., Matsumoto, K. et al. (2001) *Int. J. Artif. Organs*, **24**, 50–54.
111 Nakase, Y., Hagiwara, A., Nakamura, T. et al. (2006) *Tiss. Eng.*, **12**, 403–412.
112 Lee, M., Wu, B.M., Stelzner, M. et al. (2008) *Tiss. Eng. Part A*, **14**, 1395–1402.
113 Grikscheit, T.C., Ochoa, E.R., Ramsanahie, A. et al. (2003) *Ann. Surg.*, **238**, 35.
114 Hecker, L., Baar, K., Dennis, R.G., Bitar, K.N. (2005) *Am. J. Physiol. Gastrointest. Liver Physiol*, **289**, G188–G196.
115 Metzger, M., Caldwell, C., Barlow, A.J. et al. (2009) *Gastroent.*, **136**, 2214–2225.
116 Belsey, R. (1950) *Brit. J. Surg.*, **38**, 200–205.
117 Neville, W.E., Bolanowski, J P., Kotia, G.G. (1990) *J. Thor. Cardiovas. Surg.*, **99**, 604–612.
118 Vacanti, C.A., Kim, W.S., Sakata, J. et al. (1994) *J. Pediatr. Surg.*, **29**, 201–205.
119 Macchiarini, P., Jungebluth, P., Go, T. et al. (2008) *Lancet*, **372**, 2023–2030.

3

Construction of Three-Dimensional Tissues with Capillary Networks by Coating of Nanometer- or Micrometer-Sized Film on Cell Surfaces

Michiya Matsusaki[1,2], Akihiro Nishiguchi[1,3], Chun-Yen Liu[1,4], and Mitsuru Akashi[1,5,]*

[1] *Graduate School of Engineering, Osaka University, Osaka, Japan*
[2] *PRESTO, Japan Science and Technology Agency (JST), Saitama, Japan*
[3] *DWI Leibniz-Institute for Interactive Materials, RWTH Aachen University, Aachen, Germany*
[4] *University of California, Berkeley, California, USA*
[5] *Graduate School of Frontier Biosciences, Osaka University, Osaka, Japan*

3.1 Introduction

The surface of a cell is a coating covering a nanometer- or micrometer-sized polymer thin film, such as liquids, gases, or solids on the surfaces of substrates, such as paper or fabric. This is used for different applications in industrial fields because functional materials may change the surface properties of the substrates, such as their wettability [1], adhesion [2–4], physical or chemical resistance [5, 6]. In the case of a chemical coating, there are many methodologies, such as vapor deposition [7–9], spraying [10–12], roll-to-roll coating [13, 14], and chemical/electrochemical techniques to modify and control the surface properties of the substrates for further applications. The concept of coating is also important in the development of biological and tissue engineering fields. For example, the surface control of protein or polymer to control cell adhesion [15], growth direction [16, 17], or even to kill specific bacteria, or make a biosensor chip [18].

One of these different coating methods is called "Layer-by-Layer" (LbL), developed by G. Decher, in 1981 to coat polymer and proteins onto substrate surfaces by dipping or spraying [19, 20]. Recently, this simple and novel approach has attracted a lot of attention for cell surface modification, due to its advantages of easy preparation, versatility, and tunable composition on cell surfaces under normal ambient and physiological conditions.

Tissues in the human body usually consist of different types of cells surrounded by various amounts of extracellular matrix (ECM) fibers. There are various kinds of ECMs in the body, for example, hyaluronic acid, laminin, heparin, fibronectin, gelatin, and collagen, etc., which play an important role *in vivo* to regulate cellular activities and contribute to cell regulation [21, 22], adhesion [23], communication [24], and migration [25, 26]. But in general *in vitro* cell culture, we can only obtain monolayer structures even after the culture dish is confluent. This may be due to the insufficient extracellular matrices expressed from the cell to form three-dimensional (3D) tissues in a short time. Accordingly, in order to construct 3D-tissue models with similar structures and functions similar to natural tissues, many research groups have developed their own special approaches to construct 3D-tissue models for various purposes, for example, cell sheet engineering [27–29], decellularization [30], cell-containing gels [31–33], magnetic liposome [34], and cell beads [35, 36]. These approaches provide the possibility of constructing target tissue models. However, these techniques need complicated procedures and take a lot of time to construct, and, precise control of the cell location, density, and the components inside the constructs still remains a central challenge.

Recently, we have developed a simple and unique bottom-up approach called the "cell accumulation technique" by preparing Fibronectin-Gelatin (FN-G) nanometer-sized films on cell surfaces [37, 38], based on the concept of the LbL assembly method. About 6 nm-thick FN-G films were fabricated on cell surfaces by FN and G alternatively for nine steps. Until now, over 100 μm-thick 3D-tissue models with capillary networks [39, 40] have been successfully constructed for further advanced studies such as drug delivery systems studies, cancer cell invasion mechanism observation, or even manufacturing a biosensor chip [41] for specific detection purposes, etc.

Some tissues, for example, cartilage tissues [42] or skin tissues [43–45], consist of cells with a lot of ECMs, where

* Email: akashi@chem.eng.osaka-u.ac.jp

Bioinspired Materials Science and Engineering, First Edition. Edited by Guang Yang, Lin Xiao, and Lallepak Lamboni.

the cells are surrounded by a lot of ECMs. In such a situation, the cell distance in the tissues is of a micrometer-sized level. Although the LbL FN-G coating methodology provide possibilities to construct functionalized 3D-tissue models in one day, the fabricated films on the cell surfaces are only nanometer-sized, so it is hard to obtain thick films in a short time. Hence, we developed another novel approach called the "collagen coating method" and the "multiple coating method" to achieve the aim of construction of 3D-tissue models with lower cell density [46]. This chapter gives an overview of our attempts to construct artificial 3D-tissue models possessing similar structures and functions as a natural tissue in the body by using nano- and micro-coating technologies (Figure 3.1).

3.2 Fabrication of Nanometer- and Micrometer-Sized ECM Layers on Cell Surfaces

Tissues and organs consist of various cell types and densities, for example, heart tissues and muscle tissues are composed of higher cell densities, and skin tissues and cartilage tissues are composed of lower or gradient cell densities with aging or even injury. To develop a 3D-tissue model similar to natural tissues, direct fabrication on cell surfaces of nanometer- and micrometer-sized ECM fibrous scaffolds onto a cell membrane is crucial, as insufficient ECMs are secreted during the early cell culture to form 3D-tissue models. In this study, we focus on the fabrication of ECMs films on a cell membrane, based on specific interactions between ECMs and the integrin receptor on the cell membranes. There are many different types of ECM in the body, for example, collagen, laminin, hyaluronic acid, fibronectin, elastin, and gelatin, etc. Selection of the appropriate natural ECM components is important when fabricating ECM films onto cell surfaces without cytotoxicity. A typical ECM has specific cell-adhesive moieties like RGD (arginine-glycine-aspartic acid) [47].

The cell membrane is a biological membrane which is selectively favorable to ions or organic molecules in and out of cells, which consist of a phospholipid bi-layer with various kinds of embedded proteins. Proteins on cell membranes play an important role as they are responsible for various biological activities. Among these proteins, integrin receptors are one of the most important proteins on cell membranes. Integrin receptors consist of 18 types of α subunits and 8 types of β subunits. The combination of α and β subunits determines the specificity for extracellular ligands as well as intracellular signaling processes [48, 49]. In these different kinds of integrin receptors, $\alpha5\beta1$ integrin receptor [50] and other integrin receptors can recognize FN since it contains an RGD sequence. Binding of the RGD motif to initiate fibril formation to integrin receptors may be the initial event which triggers the subsequent reactions. In the case of G, this is a mixture of peptides and proteins produced from partial hydrolysis of collagen extracted from the skin, bones, and connective tissues. FN and G can interact with each other because FN has a collagen-binding domain, indicating these two proteins may connect to each other similar to a nanometer-sized film on cell surfaces. Thus, the FN-G nanofilms are expected to provide a suitable cell-adhesive surface similar to the natural ECMs for multilayered structures without any cytotoxicity. As shown in Figure 3.1, a 3D multilayer composed of cells and FN-G nano films was fabricated onto cell surfaces by using an LbL assembly based on the their specific interactions.

3.2.1 Control of Cell Surface by FN Nanofilms

In order to fabricate micrometer-sized films onto cell surfaces, another natural molecule is needed. Collagen is the most abundant structural protein found in the human body and is the major and most abundant component of animal connective tissues. Among different types of collagen, type I is the most widely used material in biomedical and tissue engineering fields, because it is easily denatured by environment temperatures and pH [51, 52]. Different from $\alpha5\beta1$ integrin receptor used for the recognition of FN on cell membranes [47], another integrin receptor called the $\alpha2\beta1$ integrin receptor can recognize collagen molecules with an association constant of $6.7 \times 10^4 M^{-1}$ [53]. The fabrication of micrometer-sized layers on cell membranes was performed according to the process shown in Figure 3.1. Recently, we published a new coating method called "collagen coating method" and "multiple coating method" to fabricate micrometer-sized films onto cell surfaces based on specific recognition abilities between the $\alpha2\beta1$ integrin receptors and collagen fibers.

In order to discover the thickness of the ECM films coated onto cell surfaces, in the case of FN-G assembly films, the film thickness was analyzed quantitatively by a quartz crystal microbalance (QCM) as the assembly substrate, and with a phospholipid bi-layer membrane as a model cell membrane (Figure 3.2). A phospholipid bi-layer was prepared by 1, 2-dipalmitoyl-*sn*-glycero-3-phosphatidylcholine (DPPC) and 1, 2-dipalmitoyl-*sn*-glycero-3-phosphate (DPPA) on a base layer constructed by LbL assembly of poly(diallyldimethyl-ammonium chloride) (PDDA) and poly(sodium styrenesulfate) (PSS) as stated in Krishna's report [54]. About 2.3, 6.2,

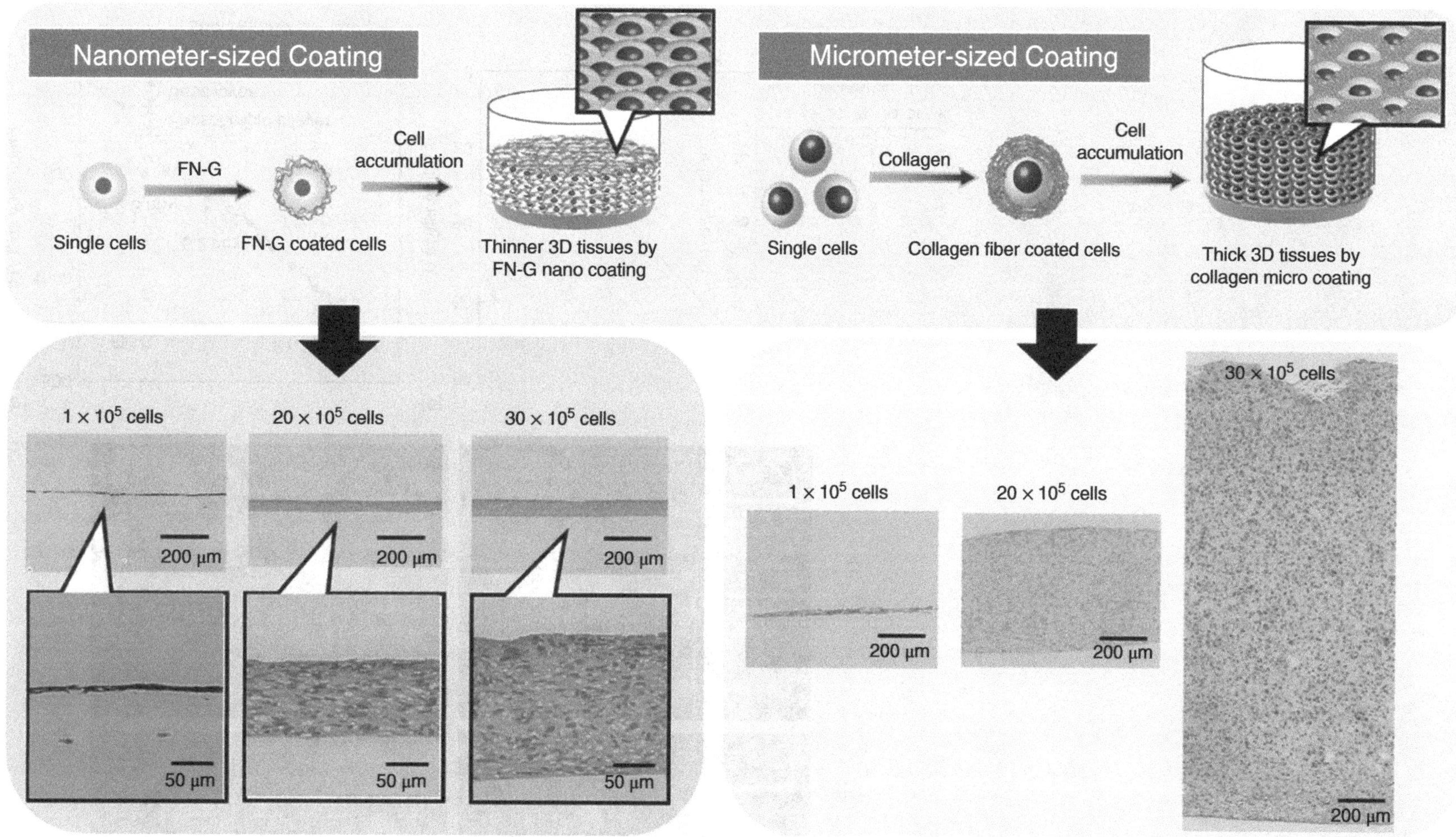

Figure 3.1 Schematic illustration of fabrication of ECM layers on cell surfaces by (left) nanometer-sized FN/G films and (right) micrometer-sized collagen nanofiber matrices to construct 3D- tissue models with higher and lower cell densities.

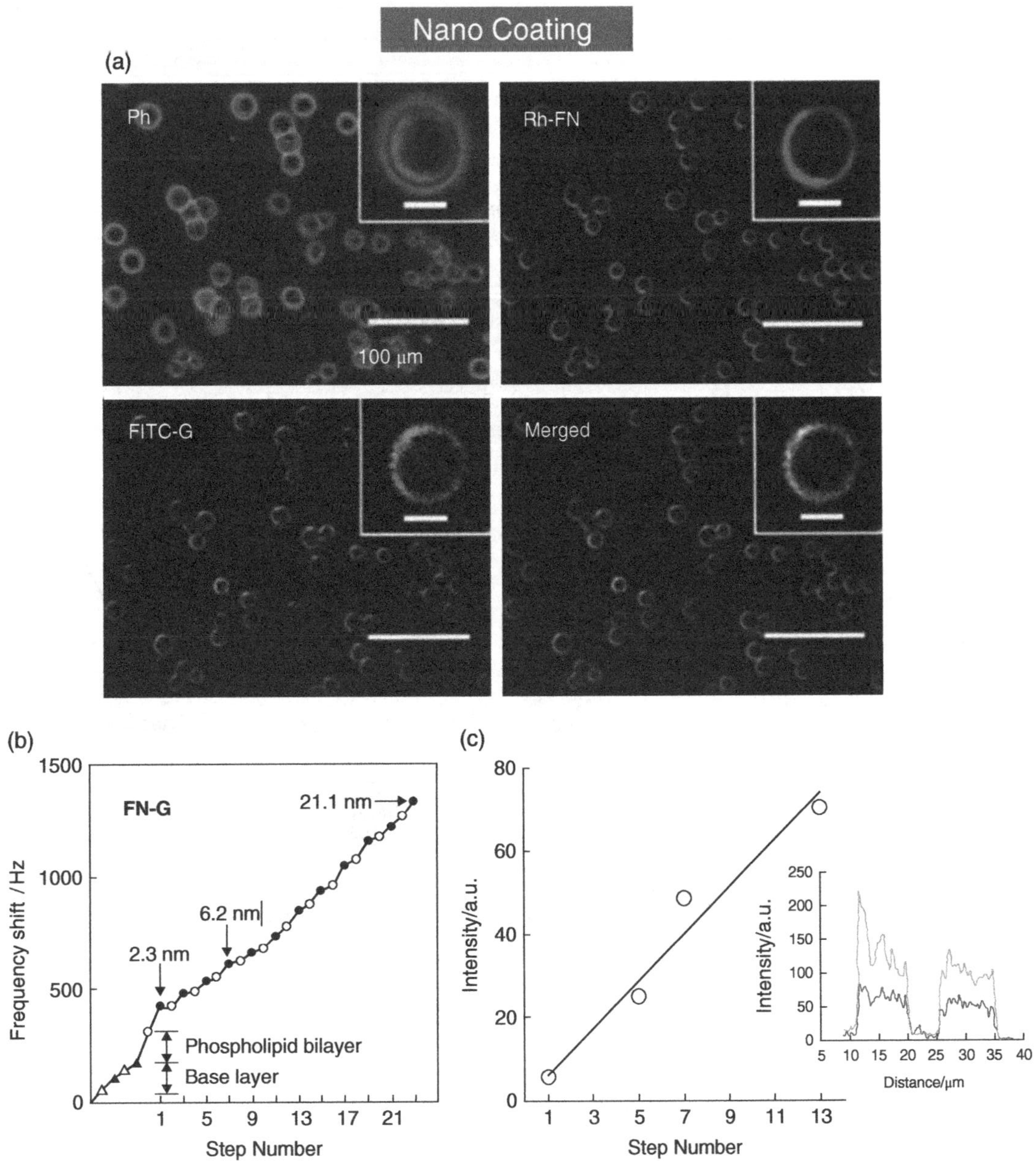

Figure 3.2 (a) Phase and fluorescent microscopic images of L929 mouse fibroblast cells coated with nanometer-sized films by 9-step assembly of Rh-FN and FITC-G. Scale bars in the figure are 10 mm. (b) Frequency shift of the quartz crystal microbalance (QCM) LbL assembly of FN-G nanofilms onto a phospholipid bilayer. Closed (●) and open circles (○) represent the assembly steps of FN and G, respectively. The phospholipid bilayer was fabricated on a base layer prepared by PDDA and PSS. (c) Fluorescence intensity of the cells and 7-step-assembled Rh-FN-G nanofilms on cell surfaces observed by CLSM. (d) Schematic illustration of cells after coating for collagen nanofiber matrix once, twice, and three times with rhodamine or FITC-labeled collagen. (e) From left to right are the coating results of once, twice, and three-times- coated cells observed by CLSM. (f) line scanning results of collagen-coated cells with different sizes of collagen microlayers on cell surfaces. *Source:* Reprinted with permission from [40, 64]. (*See insert for color representation of the figure.*)

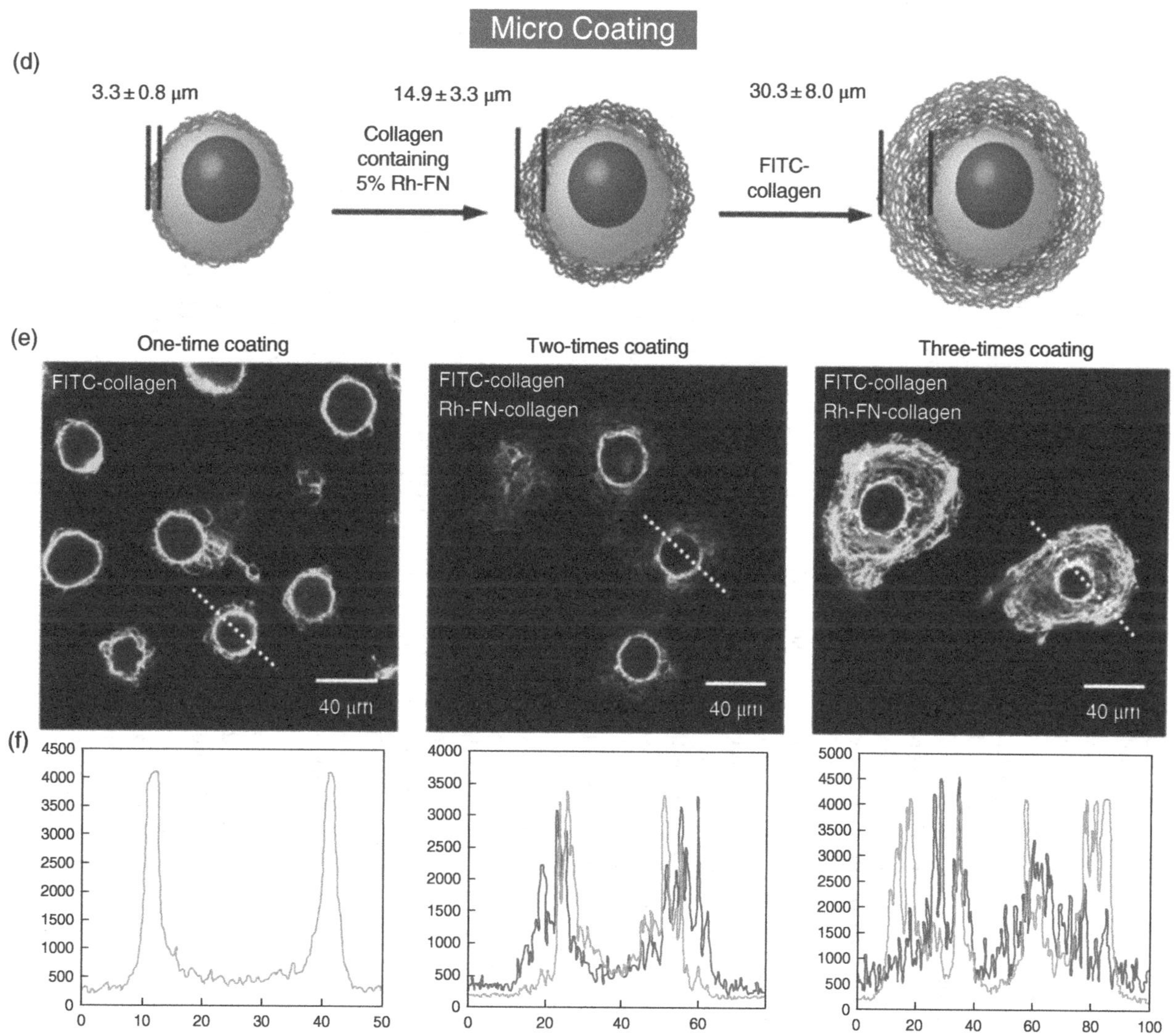

Figure 3.2 (Continued)

and 21.1 nm-thick FN-G nanofilms were fabricated on the substrates after 1, 7, and 23 assembly steps. By using the fluorescently labeled method, deposition of rhodamine-labeled FN (Rh-FN) on cell surfaces was observed by using a confocal laser scanning microscope (CLSM). The thickness of FN increased after 7-step-assembled Rh-FN-G procedures (Figure 3.2b) and also can be calculated by detecting changes in the fluorescence intensity. By using the fluorescent labeling method, the coating results of Rh-FN and FITC-G nanofilms on the floating (Figure 3.2a) and the adhered (Figure 3.2c) cell surfaces also can be visualized by CLSM.

Other compounds such as sodium dextran sulfate (DS) were also applied to coat a nanometer-sized films with the combination of FN on the cell surfaces, although FN and DS have a negative charge under physiological conditions [55]. Furthermore, we also successfully applied PE films, for example, FN-ε-Lys and ε-Lys-PSS onto the cell surfaces based on their common electrostatic interactions between the positive and negative components. To evaluate the effect of LbL nanometer-sized films on cell functions, mouse L929 cells with and without different kinds of LbL films were used for proliferation experiments. Figure 3.3a shows the morphologies of cells after 24 h of incubation with and without different LbL films on the cell surfaces. From the results, the FN-G and FN-DS nanofilms-coated mouse L929 fibroblasts revealed almost the same morphologies as shown in the control group. Proliferation curves of FN-G and FN-DS groups were also evaluated. The results shown in Figure 3.3b indicate that proliferation curves of FN nanofilm groups agreed well with the control group (without coating) during 72 h of incubation, suggesting no cytotoxicity to cell functions. To understand the details of the FN nanofilms coating, mouse L929 fibroblasts with FN nanofilms coating were also observed by fluorescence microscopy and scanning electron microscopy (SEM). As shown in Figures 3.3a, b, the results of nanometer-sized meshwork structures of the FN films spread on culture dishes and the cell surfaces were observed after 24 h of incubation, which are similar to natural ECMs in the human body.

Similarly, proliferation experiments of polyelectrolyte (PE) nanofilms-coated cells (FN-ε-Lys and ε-Lys-PSS) were also evaluated during 72 h of incubation. The results show that the coating affects the cell adhesion morphology (Figure 3.4a) and their proliferation (Figure 3.4b). This might be due to the strong non-specific adsorption of cationic polymers onto the cell surfaces. The results showed that the FN nanofilms did not affect the cell functions, and can form a nano-meshwork on the cell surfaces, which increases cell adhesion on culture dishes.

3.2.2 Control of Cell Surface by Collagen Microfilms

Apart from the FN nanofilm coating methodology, we also developed a microfilm coating approach called the "collagen coating method" as mentioned in Section 3.2.1 [46]. Based on the specific interactions between collagen fibers and α2β1 integrin receptors on the cell membranes, we successfully coated collagen microfilms on the cell surfaces. To modulate cell density precisely, we tried to control the thickness of the collagen microfilms on the cell surfaces by multiple coating approach as shown in Figure 3.2d, NHDFs were coated with collagen molecules once, twice, and three times. To visualize the thickness changing of each layer, fluorescein isothiocyanate (FITC)-labeled collagen was applied to distinguish each time the coating of the collagen microlayer on the cell surfaces, and rhodamine (Rh)-labeled FN was added to the collagen solution to clarify the second layer. Figure 3.2e clearly presents the green fluorescent collagen microfilms on the cell surfaces by CLSM. When the second coating solution contains Rh-FN, red fluorescence corresponding to Rh-FN was also clearly observed. We also tried to fabricate a third layer using a FITC-collagen solution without Rh-FN. CLSM images clearly revealed three distinct layers consisting of collagen fibers and FN. The average thicknesses of these three layers are about 3, 15, and 30 μm, respectively. The line scanning results shown in Figure 3.2f also indicate each coated layer after different coating procedures. The results demonstrated the fabrication of collagen microfilms on the cell surfaces.

To confirm the collagen microfilm coating results under wet conditions, we used a low-voltage SEM instrument (LVSEM, Miniscope Tm3030, Hitachi). A high density of meshwork structures was observed on surfaces with collagen coating in Figure 3.3d, whereas no fiber structure was coated on the cell surfaces (Figure 3.3c) [56]. The meshwork structures coated on the cell surfaces are similar to natural ECMs. Other detail measurements such as FTIR and XRD were also applied to evaluate the coating effect on the structures of collagen and cells.

To evaluate the effects of cells on the functional groups of collagen molecules, FTIR spectra analyses of cells and collagen micro-coated cells were measured. From previous reported studies [57, 58], the amide I band of collagen molecules, which exhibits a high sensitivity to conformational changes, is a frequently used spectral region for IR spectroscopic analysis of protein secondary structure. This band of $\nu_{C=O}$ is centered in the range 1650–1660 cm^{-1}. The amide II band is located at ~ 1540–1550 cm^{-1}, which represents the complex nature comprising both the N-H bending vibration (δ^{NH}) [59], whereas the stretching

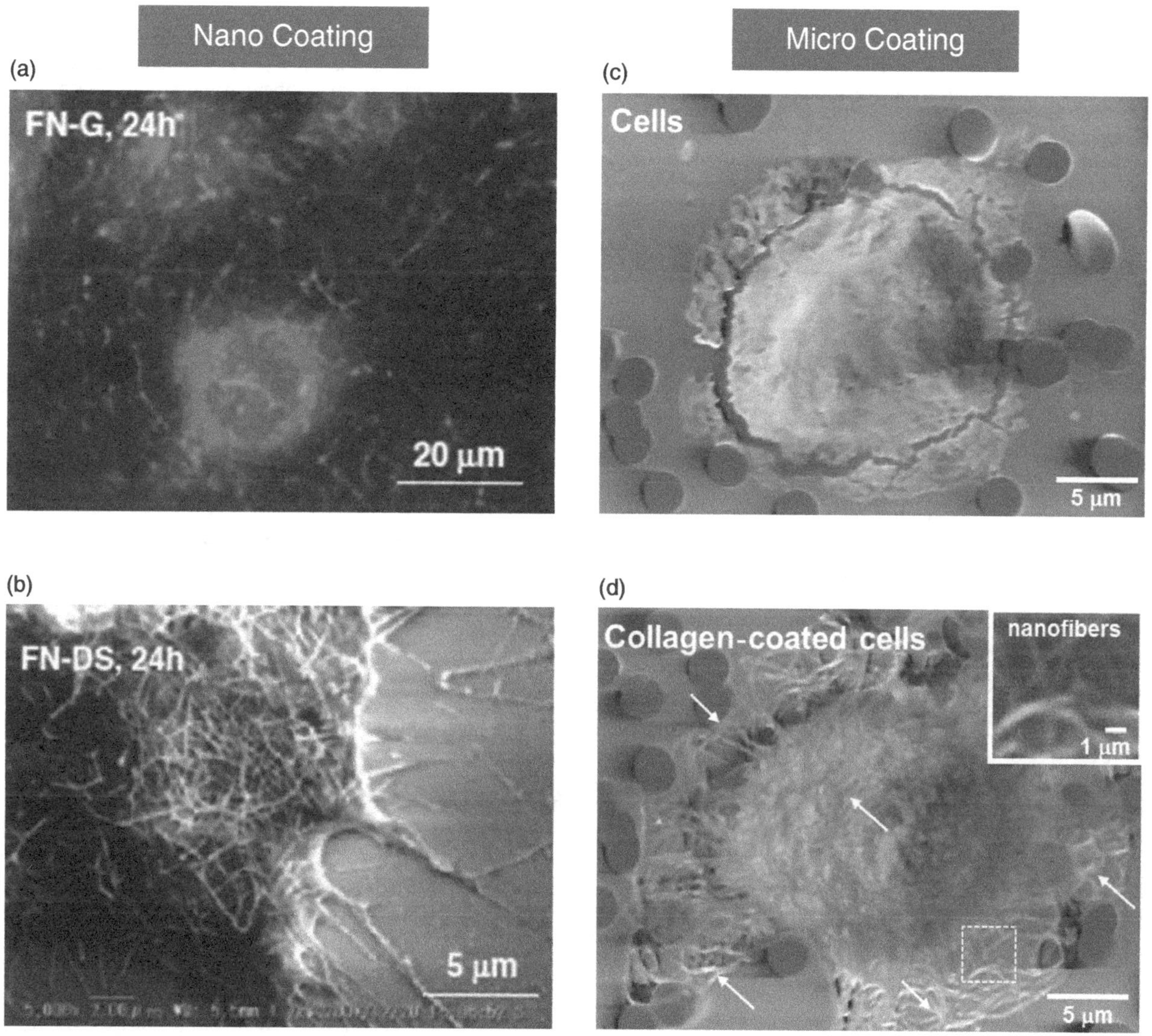

Figure 3.3 (a) Fluorescence microscopic image Rh-FN-FITC-G films prepared on L929 cells after 24 h of incubation. (b) SEM image of the L929 cells with FN-DS films after 24 h of incubation. NHDF cells (c) with and (d) without micrometer-sized collagen fiber coating procedures were observed by LV-SEM. *Source:* Reprinted with permission from [55]. (*See insert for color representation of the figure.*)

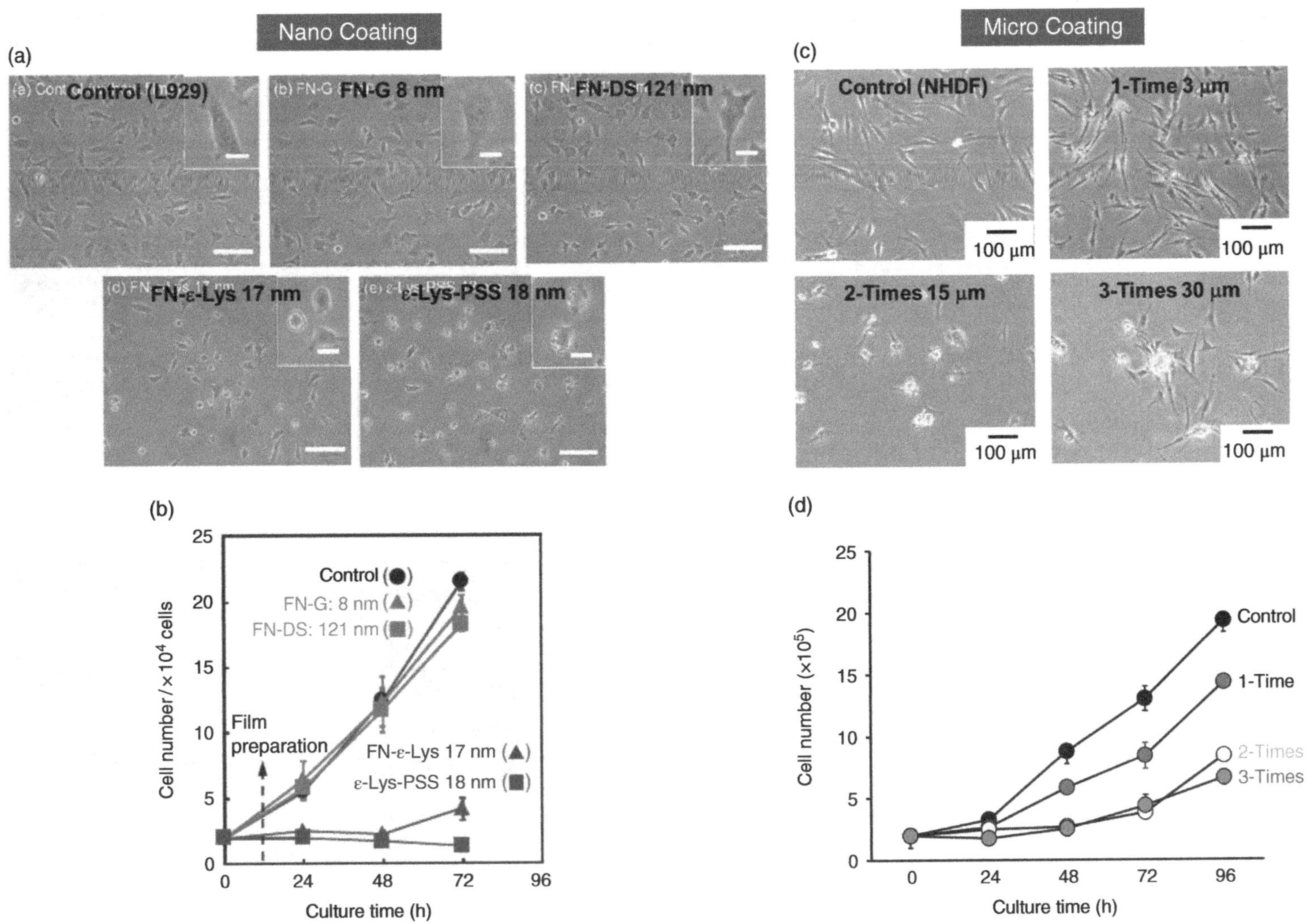

Figure 3.4 Phase contrast images of (a) the L929 fibroblasts without or with various nanofilms on the cell surfaces and (c) cells after no coating, and cells coated with 0.03 wt% once, twice, and three times after 24 h of incubation. Cell proliferation in (b) various nanofilms prepared on the L929 cell surfaces and (d) various micrometer-sized collagen fiber matrix on the cell surfaces during 72 h of incubation were also evaluated. *Source:* Reprinted with permission from [46, 68].

vibration of C-N groups of amide III (ν_{C-N}) was found at ~ 1400–1410 cm^{-1} [60]. Additionally, peaks at ~ 1550–1800 cm^{-1} are assigned to proteins of collagen: the peaks at 1652 cm^{-1} have been assigned to α-helix, 1626 cm^{-1} to β-sheets, 1634 cm^{-1} to triple helix and 1646 cm^{-1} to random coils [61]. In the cases of living cells and cells covered by collagen fibers, a band centered at 1236 cm^{-1} originates from asymmetric phosphate stretching vibrations, which can be attributed to phosphodiester groups of nucleic acid in DNA [62]. This specific band can be observed in the results from living cells and cells covered by collagen microfilms, indicating the existence of living cells after the coating procedures. Specific peaks belonging to both dried living cells and collagen fiber were observed, indicating that collagen coating procedures did not affect the functional groups and amide bands of collagen molecules.

To understand the effect of cells on collagen molecular orientation, XRD analysis was performed. As in the previous examination groups, both dried living cells and collagen-coated cells were measured. Both the FTIR results show a broad pattern from $2\theta = 20°–30°$, which represent typical peaks of cells. Due to the huge imbalance between cells and collagen molecules, the XRD diffraction patterns of typical collagen fibers centered at $2\theta = 23°$ were not obvious. The strong peak at $2\theta = 27°$ was assigned to inorganic salts, $CaCl_2$ and NaCl, in the culture medium [63, 64]. From these results, we confirmed that living cells show no effect on collagen structures and their functional groups.

In order to compare the morphology changing of the micro-coated cells, coatings were applied once, twice, and three times. All these different cells were seeded in the culture mediums filled with DMEM and the growing situation was continuously observed for 4 days. The culture mediums of all samples changed every day. As shown in Figures 3.4c, d, in the cells coated with collagen layers, the cell number was stable in the first 48 h, and then started to grow at a normal speed after 72 h (especially the cells that were coated two and three times). This may show that the cells needed time to break out of the coated collagen layers on the cell surfaces, and then to start proliferating. These results indicate the coating steps are harmless to the cells.

3.3 3D-Tissue with Various Thicknesses and Cell Densities

To evaluate the coating effect on the construction of 3D-tissue models, cells without coating, cells coated with FN-G nanofilms or collagen microfilms were employed. The cells of each group were added in FN-coated culture inserts with sufficient DMEM supplement for 24 h. After incubation, all the constructs were washed with PBS and fixed by 10% formalin solutions for about 24 h. Hematoxylin and eosin (HE) staining was applied to stain the fixed specimens to understand the structures inside each group. As shown in Figure 3.5a, heterogeneous and porous thin structures were obtained in the tissues constructed by cells without coating. On the other hand, cells coated with collagen microfilms produced nearly 80 μm-thick tissues after 1 day of incubation (Figure 3.5c). This was about double the high cell density 3D-tissues with 33 μm constructed by the nano-coating, although the cell number was 1.67-fold higher (Figure 3.5b). The results suggest about 1/3 the cell number was sufficient to fabricate 3D-tissues of an equal thickness by a collagen microfilm coating method. Apart from the cell number, we also calculated the cell density in both 3D-tissues. Cell densities obtained via the "collagen microfiber coating methods" and the "cell accumulation techniques" were about 2.3 and 8.0 × 10^8 cells cm^{-3}, respectively, suggesting a 3.5-fold lower cell density in 3D-tissues using a collagen microfiber coating method. Cell–cell distances in these 3D-tissues were calculated as 15.6 ± 4.0 μm and 8.8 ± 3.1 μm, respectively.

To understand the relationship between the thickness of the 3D-tissues and seeding cell number, 1, 5, 10, 20, 25, and 30 × 10^5 cells coated with nano- or micro-films were seeded in the 24-microwell culture inserts, and then a histological evaluation was performed [65]. From the results shown in Figure 3.6, the thickness of all samples increased with the increase in the seeding cell number. In the case of 3D-tissues composed of nano-coated cells, 3D-tissues with 8–154 μm thickness were obtained, whereas collagen micro-coated cells revealed 25–1871 μm-thick tissues (Figure 3.7). Some significant pores or cracks were found inside all tissues, and this indicates that we can easily control the thickness of 3D-tissues by employing two coating methods.

With different sizes of collagen microfilms, we can control the cell distance in the two-dimensional (2D) view easily. Cells without coating resulted in a compact distribution in the certain area, the distance between the cells is really small, whereas in cells coated with collagen microfilm once, the distance between cells is about 3 times greater than for the cells without coating. If we look from a 2D view to a 3D view, we can see that the cell density inside the -tissues could be controlled by different sizes of collagen micro-coated cells. The cell-cell distance in 3D tissues with one, two, and three times micro-coatings were 15 ± 3, 23 ± 5, and 30 ± 7 μm, respectively. Magnified histological HE staining photographs showed that cell – cell distances in 3D tissues constructed with collagen micro-coated cells remained stable and depended on the thickness of the layers on the cell surfaces. These results

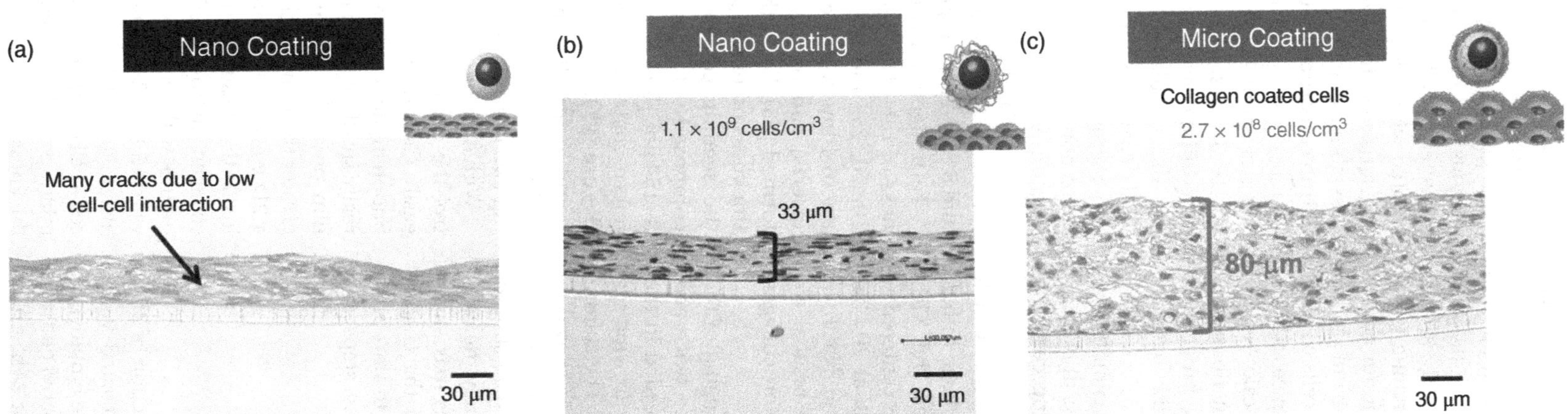

Figure 3.5 Histological staining images and schematic illustration of cell densities inside 3D tissues constructed by (a) cells without coating, (b) FN-G coated cells, and (c) collagen-coated cells (coated once), using the cell accumulation technique. *Source:* Reprinted with permission from [46].

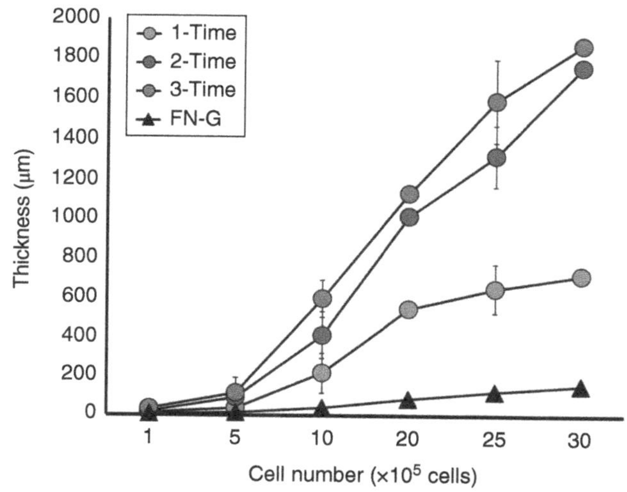

Figure 3.6 Thickness of constructed 3D tissues made of cells coated once, twice, or three times with collagen nanofibers with various numbers of seeded cells after 24 h of incubation. *Source:* Reprinted with permission from [64].

showed that cells were surrounded by ECM homogenously, without any cracks inside the tissues.

3.4 Fabrication of Vascularized 3D-Tissues and Their Applications

Finally, we tried to construct functional 3D-tissue models using nano- and micro-coating methods. In the case of FN-G nanofilm coating, we constructed multilayered tissues with endothelial tube networks by embedding human umbilical vein endothelial cells (HUVECs) in 3D-tissues composed of NHDFs. The formation of the capillary networks of HUVECs on the scaffolds or tissues by 3D-culture has been previously reported [66, 67]. The development of 3D-vascularized thick tissues possessing high density blood capillary networks is still an important issue not only for tissue engineering, but also for angiogenesis models for pharmaceutical assays. In this experiment, we performed a sandwich culture with one layer of HUVECs between 4 or 10 layers of FN-G nanofilm-coated NHDFs for one week. We observed the highly developed capillary networks and tubular morphology of the HUVECs by confocal laser scanning microscopy (CLSM) analyses as shown in Figure 3.8a (HUVECs were fluorescently labeled with an anti-CD31 antibody). Amazingly, a dense and homogenous vascularized morphology in the multilayered tissues of 1 cm width and 50 μm height was confirmed. The occupied area percentage and distance of this capillary network were calculated to be approximately 63 ± 12% and 50–150 μm in 3D-multilayered tissues by image analysis. Immunohistological results of 3D-tissues constructed by one layer on HUVECs between two layers of FN-G-coated NHDFs layers were also evaluated. From the results shown in Figure 3.8c, both two samples with different thicknesses showed brown color-labelled structures inside the 3D-tissues, indicating vascularized HUVEC network structures are well grown during the incubation, suggesting that there was sufficient oxygen and nutrients supplement to the cells inside the tissues, which is essential to the structures with tissue thickness over 50–100 μm [68]. We also applied micrometer-sized collagen fiber-coated cells to construct 3D-tissues with vascularized network structures by using the sandwich culture method and mixing method (Figures 3.8b, d). Vascularized network structures in these samples with different construction methods can be observed by the immunostaining method, suggesting that the collagen coating method also provides the possibility to construct vascularized thick 3D-tissues, and to control the specific location of vascularized structures inside the tissues [69].

Other functional tissue models constructed by FN-G nanofilm-coated cells, such as induced pluripotent stem cells-derived cardiomyocytes (iPS-CMs) or lymph epithelial cells (LECs), were also successfully constructed. Recently, we tried to coat FN-G nanofilm by using the filtration method instead of the centrifugation method, which may increase the cell viability after the coating procedures. Also, different ratios of FN-G-coated iPS-CMs and Normal Human Cardiac Fibroblast (NHCF) were evaluated by tissue thicknesses, proteins, and beating speeds, etc. In the case of FN-G-coated iPS-CM, about 1×10^6 cells were seeded in the 24-well culture insert. Strong beating phenomenon can be observed during incubation. The HE and Azan staining results indicate 3D-tissues with FN-G-coated iPS-CM cells were successfully constructed [70].

In the same way as the vascularized 3D-tissues, we also succeeded in constructing functional 3D-tissue models by using the collagen microfilm coating method. Similar to the FN-G nanofilm coating method, iPS-CMs were also used to coat with collagen fibers following the previously reported procedures. About 2×10^6 cells with two coats of collagen fiber microfilms were seeded in 24-microwell culture insert for 4 days of incubation. Constructed iPS-CM 3D-tissues can be seen with the naked eye directly on the insert membranes. Furthermore, the constructed 3D-tissues show stable and strong beating (80 times/min) and high cell viability (>90%) after 4 days of incubation. Histological HE staining results showed that 800 μm-thick 3D-tissues composed of collagen-coated iPS-CM and NHCF (2×10^6 cells in total) were obtained. Four immunostaining assays such as actinin antibody, troponin T antibody, connexin antibody, and Azan stains were applied in this study to evaluate the

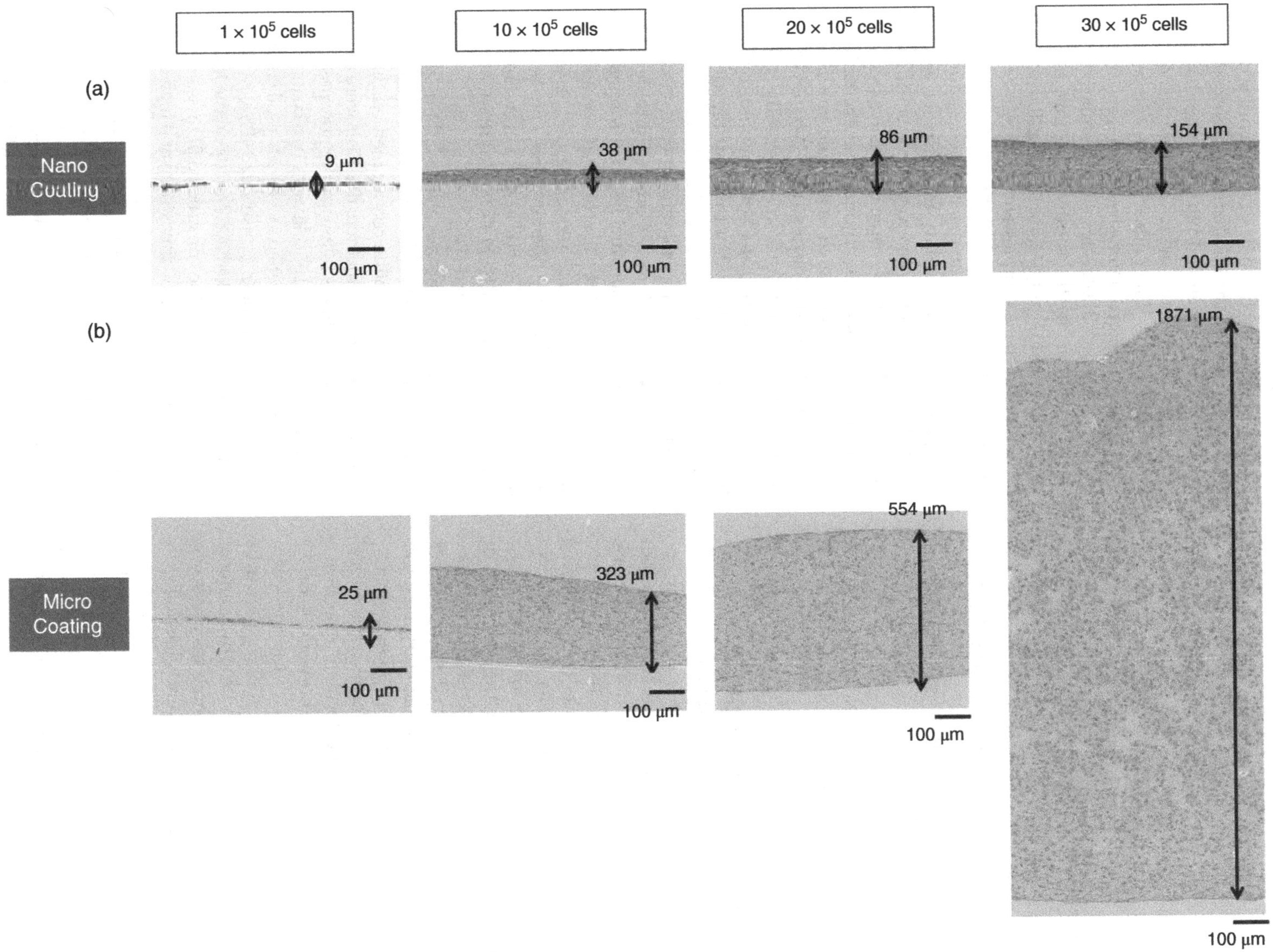

Figure 3.7 HE staining results of 3D tissues constructed with (a) FN-G nanofilm-coated cells and (b) micrometer-sized collagen fiber-coated cells with various seeding cell numbers after a 24 h of incubation.

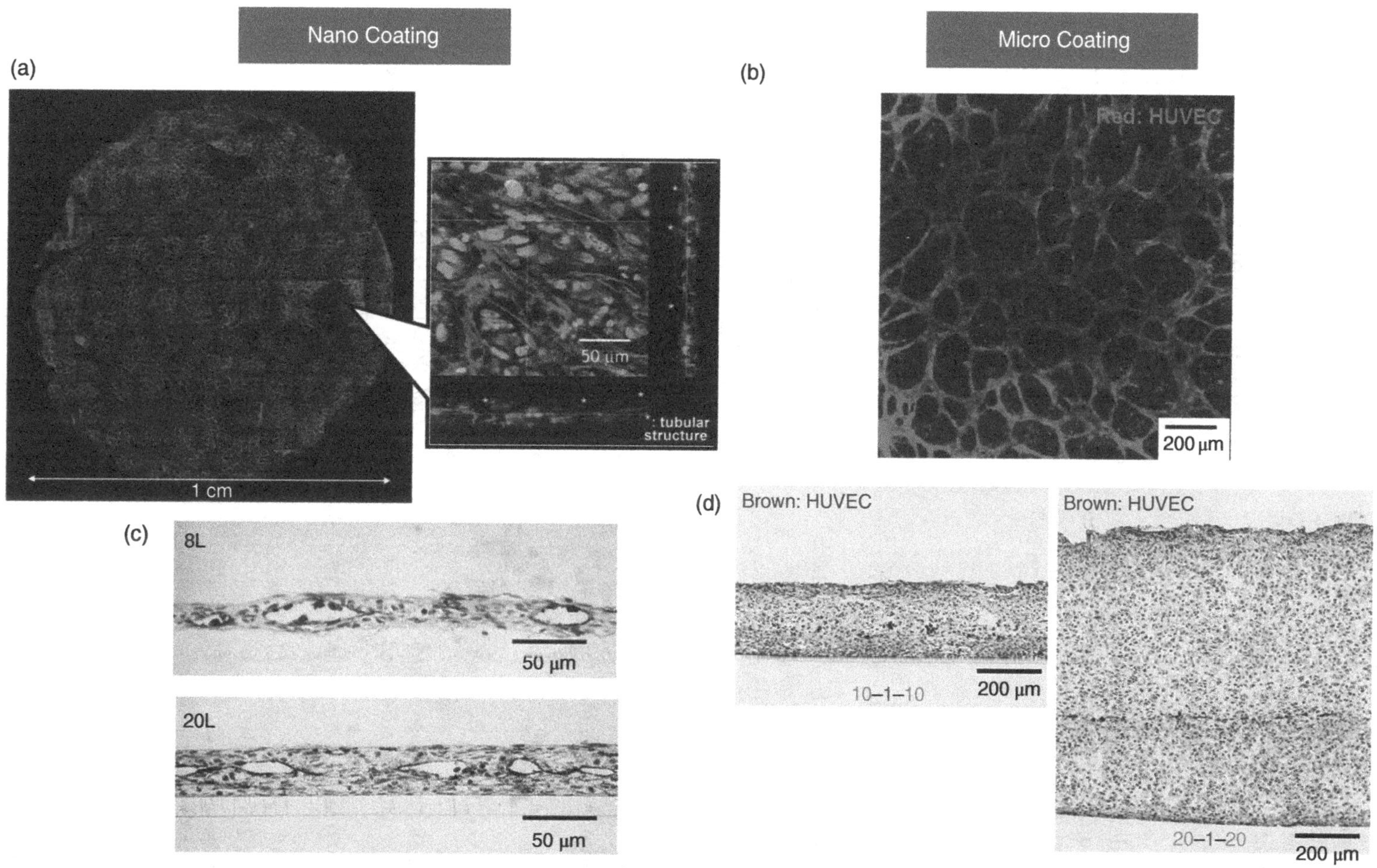

Figure 3.8 Images of confocal laser scanning microscopy (CLSM) of vascularized 3D -tissue models constructed by (a) nano- and (b) micro-coatings by sandwich culture. Immunohistological staining images using anti-CD31 antibody of the 3D tissue models by (c) nano- and (d) micro coating methods. *Source:* Reprinted with permission from [39, 40].

expression of cardiac myoblasts in the 3D iPS-CM tissues. Protein staining such as actin staining identified the cytoskeleton, which contributes to muscle contraction, troponin T staining proved the modulation of strained muscle, connexin staining marked the typical gap junction at virtually every interface between the cultured cardiomyocytes in the tissues, and Azan staining was applied to distinguish the collagen matrix inside the tissues [65]. These results suggested that collagen microfilm coating method can be applied to various types of cells for different purposes.

3.5 Conclusion

We have successfully developed nanofilm and microfilm coating method to coat different sizes of cells to construct 3D-tissue models for different purposes. Both coating methods showed no obvious effect on cell functions. A bottom-up approach is valuable to develop 3D-tissue architectures using various types of cells with controlled location, functions, and cell densities. These two approaches could be the solutions to construct functional 3D-tissue models similar to natural tissues. We would like to apply these methods to contribute to the development of tissue engineering and pharmaceutical fields.

Acknowledgments

This chapter was supported by the NEXT Program (LR026), a Grant-in-Aid for Scientific Research (S), a Grant-in-Aid for Scientific Research on Innovative Areas (26106717), and a Grant-in-Aid for Scientific Research (B) (26282138) from MEXT of Japan, and JST-PRESTO (10825).

References

1 Takao, H., Hasegawa, H. (2004) *Mater. Trans.*, **45**, 747–753.
2 Guzman, L., Miotello, A., Voltolini, E., Adami, M. (2000) *Thin Solid Films*, **377–378**, 760–765.
3 Shulz, U. Munzert, P.. Kaiser N. (2001) *Surf. Coat. Technol.*, **45**, 507–511.
4 Nakamori, T., Adachi, Y., Arai, M., Shibuya, A. (1995) *ISIJ Int.*, **35**, 1494–1501.
5 Nikravesh, B., Ramezanzadeh, B., Sarabi, A.A., Kasiriha, S.M. (2011) *Corros. Sci.*, **53**, 1592–1603.
6 Kalenda, P., Kalendová, A. (1995) *Dye. Pigment.*, **27**, 305–312.
7 Da Silva Sobrinho, A (1998) *J. Vac. Sci. Technol.*, **16**, 3190.
8 Quinto, D.T., Wolfe, G.J., Jindal, P.C. (1987) *Thin Solid Films*, **153**, 19–36.
9 Sproul, W.D. (1996) *Surf. Coatings Technol.*, **81**, 1–7.
10 Chen, H., Ding, C.X. (2002) *Surf. Coatings Technol.*, **150**, 31–36.
11 Li, C. J., Li, W.Y. (2003) *Surf. Coatings Technol.*, **167**, 278–283.
12 Bakshi, S.R., Singh, V., Balani, K. et al. (2008) *Surf. Coatings Technol.*, **202**, 5162–5169.
13 Søndergaard, R., Hösel, M., Angmo, D. et al. (2012) *Mater. Today*, **15**, 36–49.
14 Krebs, F.C. (2009) *Sol. Energy Mater. Sol. Cells*, **93**, 465–475.
15 Pierschbacher M.D. (1987) *J. Biol. Chem.*, **262**, 17294–17298.
16 Recknor, J. B., Sakaguchi, D.S., Mallapragada, S.K. (2006) *Biomaterials*, **27**, 4098–4108.
17 Chang, P.-C., Liu, B.-Y., Liu, C.-M. et al. (2007) *J. Biomed. Mater. Res. A*, **81**, 771–780.
18 Liss, M., Petersen, B., Wolf, H., Prohaska, E. (2002) *Anal. Chem.*, **74**, 4488–4495.
19 Decher, G. (1997) *Science*, **277**, 1232–1237.
20 Decher, G., Hong, J. (1991) *J. Makromol. Chem., Macromol. Symp.*, **46**, 321–327.
21 Roberts, A.B., McCune, B.K., Sporn, M.B. (1992) *Kidney Int.*, **41**, 557–559.
22 Vukicevic, S., Kleinman, H.K., Luyten, F.P. et al. (1992) *Exp. Cell Res.*, **202**, 1–8.
23 Gumbiner, B.M. (1996) *Cell*, **84**, 345–357.
24 Werb, Z. (1997) *Cell*, **91**, 439–442.
25 Cursio, R., Mari, B., Louis, K. et al. (2002) *FASEB J.*, **16**, 93–95.
26 Wolf, K., te Lindert, M., Krause, M. et al. (2013) *P. J. Cell Biol.*, **201**, 1069–1084.
27 Sasagawa, T., Shimizu, T., Sekiya, S. et al. (2010) *Biomaterials*, **31**, 1646–1654.
28 Sekine, H., Shimizu, T., Sakaguchi, K. et al. (2013) *Nat. Comm.*, **4**, 1399.
29 Yang, J., Yamato, M., Kohno, C. et al. (2005) *Biomaterials*, **26**, 6415–6422.
30 Uygun, B.E., Soto-Gutierrez, A., Yagi, H. et al. (2010) *Nat. Med.*, **16**, 814–820.
31 Themistocleous, G. S., Katopodis, H., Sourla, A. et al. (2004) *In Vivo*, **18**, 687–696.
32 Cheema, U., Brown, R.A. (2013) *Adv. Wound Care*, **2**, 176–184.
33 Choi, N.W., Cabodi, M., Held, B. et al. (2007) *Nat. Mater.*, **6**, 908–915.

34 Akiyama, H., Ito, A., Kawabe, Y., Kamihira, M. (2010) *Biomaterials*, **31**, 1251–1259.
35 Morimoto, Y., Tanaka, R., Takeuchi, S. (2013) *Adv. Healthcare Mater.*, **2**, 261–265.
36 Matsunaga, Y.T., Morimoto, Y., Takeuchi, S. (2011) *Adv. Mater.*, **23**, H90–H94.
37 Matsusaki, M., Amemori, S., Kadowaki, K., Akashi, M. (2011) *Angew. Chem. Int. Ed.*, **50**, 7557–7561.
38 Matsusaki, M., Kadowaki, K., Nakaha, Y., Akashi, M. (2007) *Angew. Chem. Int. Ed.*, **46**, 4689–4692.
39 Nishiguchi, A., Matsusaki, M., Asano, Y. et al. (2014) *Biomaterials*, **35**, 4739–4748.
40 Nishiguchi, A., Yoshida, H., Matsusaki, M., Akashi, M. (2011) *Adv. Mater.*, **23**, 3506–3510.
41 Matsusaki, M., Akashi, M. (2014) *Polym. J.*, **46**, 524–536.
42 Stockwell, R.A. (1967) *J. Anat.*, **101**, 753–763.
43 Kanitakis, P. (2002) *J. Eur. J. Dermatol.*, **12**, 390–401.
44 Lang, I.M., Shaker, R. (1997) *J. Dermatol. Nurses Assoc.*, **3**, 50–55.
45 Bouwstra, J.A., Honeywell-Nguyen, P.L. (2002) *Adv. Drug Deliv. Rev.*, **54**, 41–55.
46 Liu, C.-Y., Matsusaki, M., Akashi, M. (2014) *RSC Adv.*, **4**, 46141–46144.
47 Ruoslahti, E., Pierschbacher, M.D. (1987) *Science*, **238**, 491–497.
48 Campbell, I.D., Humphries, M.J. (2011) *Cold Spring Harbor Perspect. Biol.*, **20**, 3, 1–14.
49 Arnaout, M. a; Mahalingam, B., Xiong, J.-P. (2005) *Ann. Rev. Cell Dev. Biol.*, **21**, 381–410.
50 Nagae, M., Re, S., Mihara, E. et al. (2012) *J. Cell Biol.*, **197**, 131–140.
51 Harris, J.R., Reiber, A. (2007) *Micron*, **38**, 513–521.
52 Harris, A.K., Wild, P., Stopak, D. (1980) *Science*, **208**, 177–179.
53 Calderwood, D.A, Tuckwell, D.S., Eble, J. (1997) *J. Biol. Chem.*, **272**, 12311–12317.
54 Krishna, G., Shutava, T., Lvov, Y. (2005) *Chem. Commun.*, **345**, 2796–2798.
55 Kadowaki, K., Matsusaki, M., Akashi, M. (2010) *Langmuir*, **26**, 5670–5678.
56 Liu, C.-Y., Matsusaki, M., Akashi, M. (2015) *Polym. J.*, **47**, 391–399.
57 Dong, A., Huang, P., Caughey, W.S. (1990) *Biochemistry*, **29**, 3303–3308.
58 De Campos Vidal, B., Mello, M.L.S. (2011) *Micron*, **42**, 283–289.
59 Vileno, B., Jeney, S., Sienkiewicz, A. et al. (2010) *Biophys. Chem.*, **152**, 164–169.
60 Baia, L., Baia, M., Danciu, V. et al. (2008) *J. Opto. Adv. Mater*, **10**, 933–936.
61 Lai, J.-Y., Lue, S.J., Cheng, H.-Y., Ma, D.H.-K. (2013) *J. Biomed. Nanotechnol.*, **9**, 2048–2062.
62 Bongda, M., Vileno, B., Runne, H. (2011) *Anal. Chem.*, **83**, 7712–7720.
63 Sugino, A., Miyazaki, T., Ohtsuki, C. (2008) *J. Mater. Sci. Mater. Med*, **19**, 2269–2274.
64 Lu, J.Q., Luo, M.F., Lei, H. et al. (2002) *J. Catal.*, **211**, 552–555.
65 Liu, C.-Y., Matsusaki, M., Akashi, M. (2015) *ACS Biomater. Sci. Eng.*, **1**, 639–645.
66 Unger, R. E., Sartoris, A., Peters, K. (2007) *Biomaterials*, **28**, 3965–3976.
67 Armulik, A., Abramsson, A., Betsholtz, C. (2005) *Circ. Res.*, **97**, 512–523.
68 Pittman, R.N. (2011) *Acta Physiol.*, **202**, 311–322.
69 Liu, C.-Y., Matsusaki, M., Akashi, M. (2016) *Biochem. Biomed. Res. Commun.*, **472**, 131–136.
70 Amano, Y., Nishiguchi, A., Matsusaki, M. et al. (2016) *Acta Biomater*, **33**, 110–121.

4

Three-dimensional Biofabrication on Nematic Ordered Cellulose Templates

*Tetsuo Kondo**

Graduate School of Bioresource and Bioenvironmental Sciences, Kyushu University, Fukuoka, Japan

4.1 Introduction

The unique interaction of biological systems with entirely synthetic molecular assemblies has prompted the consideration of a new generation of approaches to controlled nano-assembly. Quite some time ago, we found a unique phenomenon that the molecular tracks of nematic ordered cellulose (NOC) [1–3] substrate regulated the direction of the fiber secretion of *Gluconacetobacter xylinus*, a Gram-negative bacterium that synthesizes and extrudes a cellulose nanofiber. This resulted also in regulating the direction of the bacterial movement caused by the inverse force of the secretion. An epitaxial deposition of the nanofibers was induced due to the strong interaction between the nascent nanofibrils, and influenced the order of molecules of the template [4]. This indicated that the conjunction of directed biosynthesis and the ordered fabrication from the nano to the micro scales could lead to new methodologies for the design of functional materials with the desired nanostructures.

In the unique phenomenon, the most influential factor for the regulated bacterial movement was the surface properties due to the nano/micro structure of the template. The NOC substrate where cellulose molecules are highly ordered, but not assembled to form the crystalline state, was prepared by uniaxial stretching of water-swollen and fixed cellulose gel from the dimethyl acetamide/LiCl solution. Under the bacterial culture conditions, the cellulose molecular chains tend to be still oriented towards the stretching axis. Simultaneously, the hydroxyl groups at the C-6 position that are equatorially bonded to the anhydroglucose unit in the cellulose molecules are automatically oriented at a certain angle against the surface and also aligned along the same axis as the tracks [2, 3]. On the contrary, the lateral order of the OH groups is not well coordinated because the molecular chains slip against each other in the NOC due to the uniaxial stretching. Therefore, the OH groups tend to be oriented as molecular tracks only in one direction across the entire NOC surface [1–3]. These ordering tracks of the OH groups on the NOC surface appear to induce the epitaxial deposition of the biosynthesized cellulose nanofibers as they are secreted from *Gluconacetobacter xylinus* along the same axis as the tracks in the NOC. In addition, the regulated movement of the bacterium was not observed in the synthetic polymers with OH groups such as poly (vinyl alcohol) [4]. Therefore, when the nematic ordered state is prepared using other natural polymers, it is of interest and importance to know whether the similar regulated bacterial movement may occur in order to extend this phenomenon into a methodology for the fabrication of new types of functional bio-based materials.

In this chapter, after the properties of the NOC have been described, the behavior of the cellulose production of two types of *Gluconacetobacter xylinus* bacteria (NQ-5: ATCC 53582, and AY-201: ATCC 23769) is first shown in Schramm-Hestrin media on NOC templates [2–5].

Second, we employed chitin and cellulose/chitin blends as the components for the above templates instead of NOC [6]. Chitin is a repeating carbohydrate polymer of (1,4) linked 2-deoxy-2-acetamide-β-D-glucose. Namely, as the primary structure, chitin has only a replacement of hydroxyl groups into acetamide groups at the C-2 position of the anhydroglucose unit when compared with cellulose. However, when treated in the same manner as the NOC, the chitin exhibited a variety of different micro/nano hierarchical nematic ordered states [7]. The NOC-like ordered templates of cellulose/cellulose

* Email: tekondo@agr.kyushu-u.ac.jp

Bioinspired Materials Science and Engineering, First Edition. Edited by Guang Yang, Lin Xiao, and Lallepak Lamboni.

derivatives blends were also employed for the bacterial cultures. Furthermore, the autonomous bottom-up fabrication of three-dimensional honeycomb cellulose structures was established [8] using *G. xylinus* as a bacterial nanoengine, on cellulose honeycomb templates prepared [9] by casting water-in-oil emulsions on glass substrates.

As an extended concept for the NOC, attempts are discussed to propose a sort of biomimic mineralization of calcium phosphate using modified NOC templates. This process was initially mediated by the modified NOC template in the roles of the ion supply sources and scaffolds for 3D-ordering architecture of the calcium phosphate as a biomineral [10]. In addition, the NOC template successfully established a three-dimensional, hierarchical structure of epidermal cells by cell attachment and subsequent culture [11].

In this way, this chapter will provide results on the unique surface properties of NOC templates in patterning the bacterial movement and deposition of their secreted cellulose nanofibers, on the patterned deposition of biominerals and the induction of the formation of three-dimensional, hierarchical structures of epidermal cells. Then, the regulated fabrication could lead to the design of 3D architecture from the nano to the micro scales of functional bio-based materials with the desired patterned nanostructures, which also opens the door to the preparation of potentially versatile organic-inorganic as well as organic-organic 3D order-patterned composites with less energy consumption.

4.2 What Is Nematic Ordered Cellulose (NOC)?

4.2.1 Nematic Ordered Cellulose

Before we discuss NOC it is necessary to know the characteristic features of a cellulose molecule (Figure 4.1). Cellulose has an extended structure with a 2_1 screw axis composed by the β-1,4 glucosidic linkages between anhydroglucose units. Thus, it would be natural to accept the dimer called "cellobiose" as a repeating unit. The present three kinds of hydroxyl groups within an anhydroglucose unit exhibit different polarities, which contribute to the formation of various kinds of inter- and intramolecular hydrogen bonds among secondary OH at the C-2, secondary OH at the C-3 and primary OH at the C-6 positions. In addition, all the hydroxyl groups are bonded to a glucopyranose ring equatorially. This causes the appearance of a hydrophilic site parallel to the ring plane. On the contrary, the CH groups are bonded to a glucopyranose ring axially, causing a hydrophobic site perpendicular to the ring as shown in the upper image of Figure 4.1. These effects lead to the formation of hydrogen bonds in parallel direction to a glucopyranose ring, and to Van der Waals interaction perpendicular to the ring.

Another important feature for the hydroxyl groups is the type of hydroxymethyl conformation at the C-6 position, because the conformation of C(5)–C(6) and the resulting interactions, including inter- and intramolecular hydrogen bonds in the present cellulose structure, may differ from that in crystallites and also it is assumed to make up the extent of crystallization, as well as the final morphology of cellulose. In the noncrystalline regions, the rotational position of hydroxymethyl groups at the C-6 position may be considered indeterminate or totally non-oriented, which are not identical to those in the crystallites. Therefore, it was important to confirm the type of O(6) rotational position with respect to the O(5) and C(4) in a β-glucan chain, by employing CP/MAS^{13} and C NMR [12]. The types of hydroxymethyl conformations are gauche-trans (*gt*), trans-gauche (*tg*) or gauche-gauche (*gg*) at the C-6 positions in carbohydrates. As for the noncrystalline states, they are considered the *gg* conformation (see Figure 4.2).

Now, NOC is prepared by uniaxial stretching of water-swollen cellulose from the N, N-dimethylacetamide (DMAc)/LiCl solution at the draw ratio of 2.0 to provide highly oriented β-1,4-glucan molecular chains of cellulose toward the stretching axis [13].When the dissolved cellulose molecules in DMAc/LiCl system are self-aggregated in water vapor to form a gel, presumably by a minimum amount of restricted engagement between the hydrophobic sites of the ring above described, it is stretched to reach NOC. It is noted that the drawability requires entanglements such as cross-linking, but they are supposed to be minimized. In the present DMAc/LiCl system for NOC preparation, however, the slower coagulation using water vapor yielded the stable noncrystalline cellulose. The key to maintaining the amorphous or noncrystalline state should be the formation of small amounts of the three-dimensional cross-linking network. When the cellulose solution spread on the Petri dish is exposed by a saturated water vapor, it turns to a gel state with shrinking. The shrinkage is probably due to the interchain hydrogen bonds forming during its coagulation process of cellulose molecules into a sort of network structure. After the solvent was exchanged into water, a transparent water-swollen fixed gel of cellulose (WSC) was obtained. The WSC is composed of approximately 93 wt% water and 7 wt% cellulose. During the drawing of the WSC, the water retained inside of the WSC appeared to bleed out from its surface. The expulsion of the excess water from the drawn WSC gel may be attributed to the cellulose chains linking between the

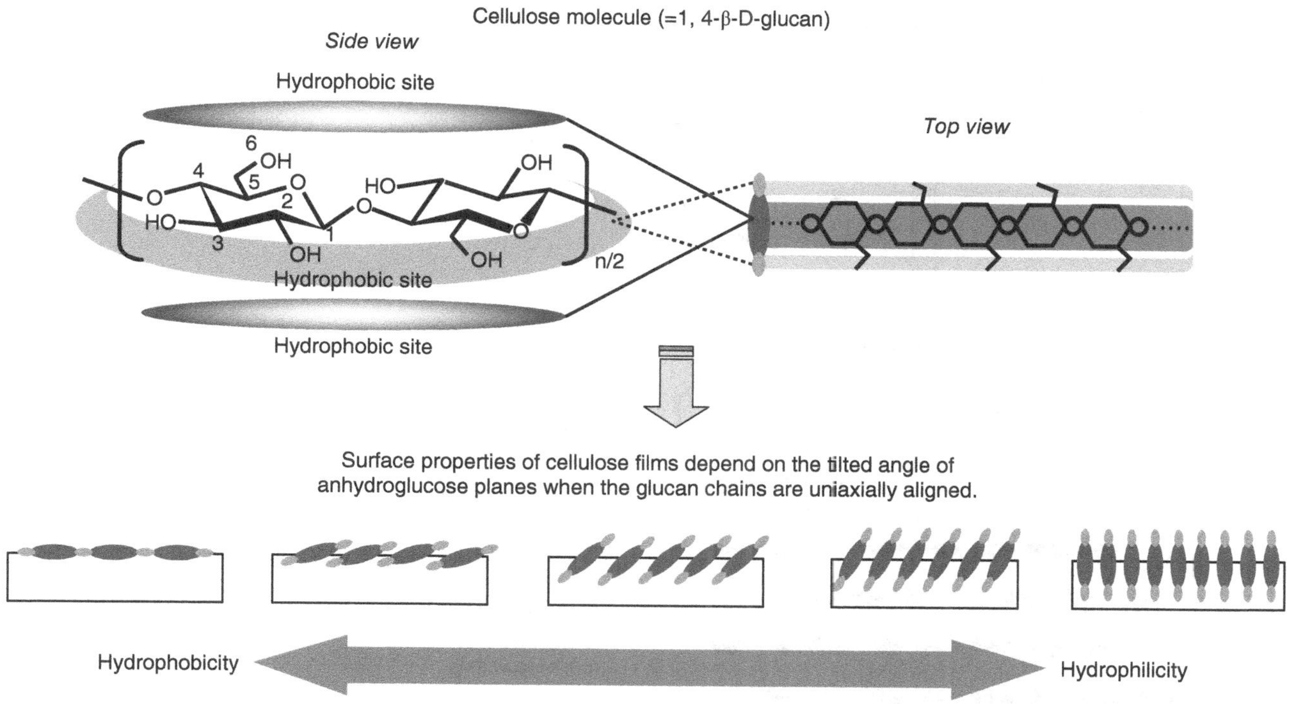

Figure 4.1 Site-specific, amphiphilic nature in cellulose chemical structure, leading to unique surface properties of cellulose films, depending on the tilted angle of anhydroglucose planes when the glucan chains are uniaxially aligned.

Figure 4.2 Schematic diagram of the hydroxymethyl conformations at the C-6 position, namely, the orientation of the C6-O6 bond, gauche-trans (*gt*), trans-gauche (*tg*) or gauche-gauche (*gg*) with a cellobiose unit.

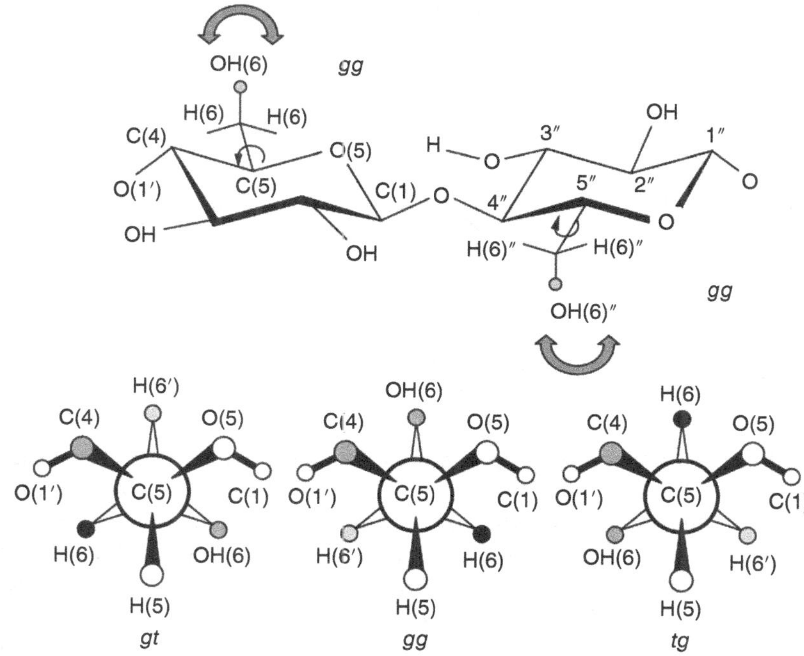

junction zones accompanying the orientation process, as shown in Figures 4.3a and b. The junction zone in Figure 4.3a represents the strongly hydrogen-bonded area where even D_2O vapor could not penetrate [14, 15]. The zone probably plays a role as the cross-linking point in the WSC fixed gel of an initial state for NOC. After drying of the drawn WSC under a stressed state, it had decreased in thickness more than 80% and shrunk in

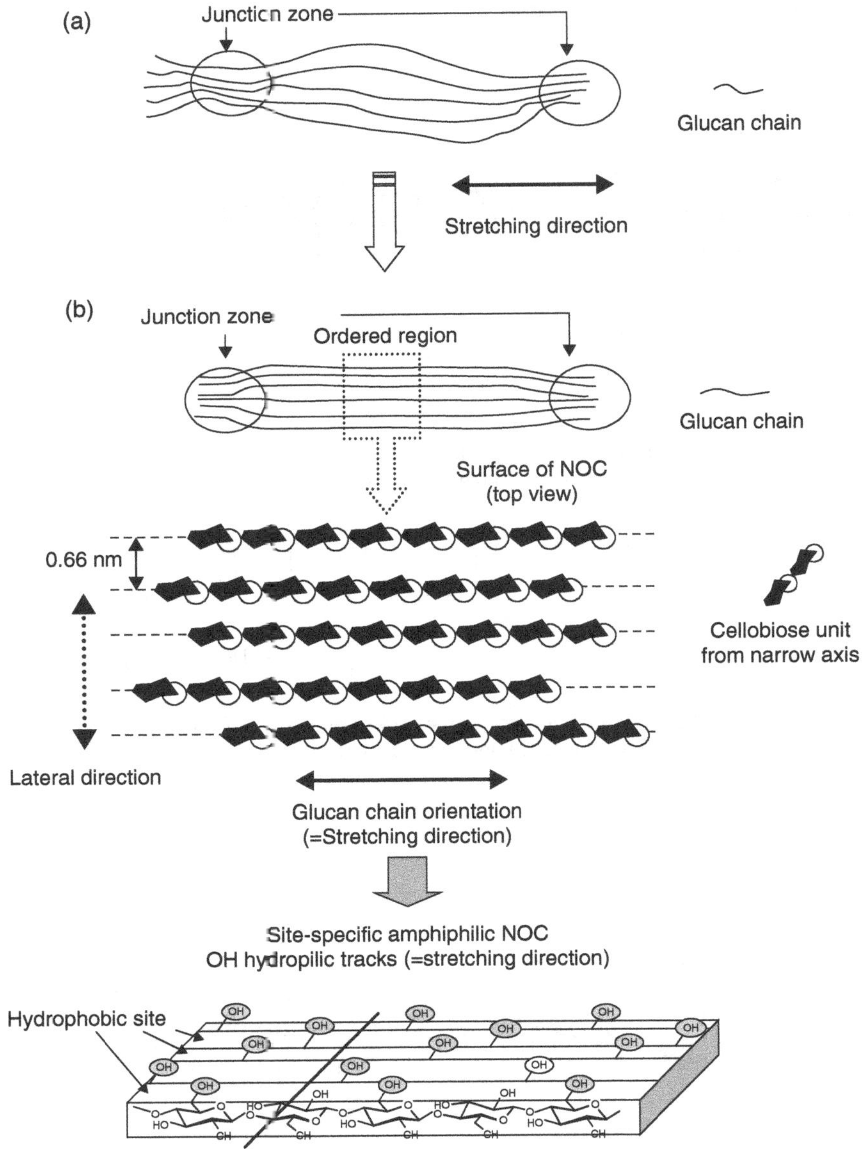

Figure 4.3 Schematic images of NOC: (a) the initial state before stretching; (b) after stretching; the inserted view is the arrangement of cellulose molecules on the NOC surface.

width over 60%, when compared with the starting sizes of WSC. The NOC films thus obtained exhibited almost perfect transparency like the coagulated cellulose solution under water vapor atmosphere. This transparency suggests that the NOC consists of noncrystalline regions.

Figure 4.3 also illustrates structural characteristics of NOC in relation to the situation of OH groups explained in the following: When NOC is prepared by uniaxial stretching of WSC, the cellulose molecular chains tend to be oriented toward the stretching axis, as already described. Further, the hydroxymethyl groups at the C-6 position of the anhydroglucose unit are vertically standing up against the surface, which indicates that the neighboring anhydroglucose ring planes are facing each other. Simultaneously the standing-up OH groups in the individual molecular chains are also aligned like tracks along the stretching axis. On the contrary, the lateral order of the OH groups among the neighboring chains is not well coordinated because the molecular chains are slipping against each other. Therefore, the hydrophilic and polarized OH groups are completely oriented as molecular tracks only in the stretching direction across the entire NOC surface. Between the hydrophilic molecular tracks, the hydrophobic phase due to the anhydroglucose plane also appeared alternately, resulting in both hydrophilic and hydrophobic tracks next to each other across the NOC surface. The amphiphilic molecular tracks enhance the unique surface properties of NOC as described later.

The NOC structure was basically indicated by the high-resolution TEM image (see Figure 4.4). The cross-section image of the NOC template with molecular ordering is also shown in Figure 4.4 [1–3]. The high-resolution TEM image shows a preferentially oriented direction on the surface and suggests tilting of the glucose planes of a cellulose molecule with an angle of 29.3° to the vertical axis against the surface of NOC. It should be noted that the static sessile contact angle of a drop of water on the NOC surface is ~72°, indicating it is fairly hydrophobic [11]. The surface condition is totally altered by stretching to provide tilting of the glucose planes to expose the specific hydrophobic site. Namely, the hydrophobic phases due to the anhydroglucose plane appear alternately next to the hydrophilic OH tracks, resulting in unique amphiphilic molecular tracks.

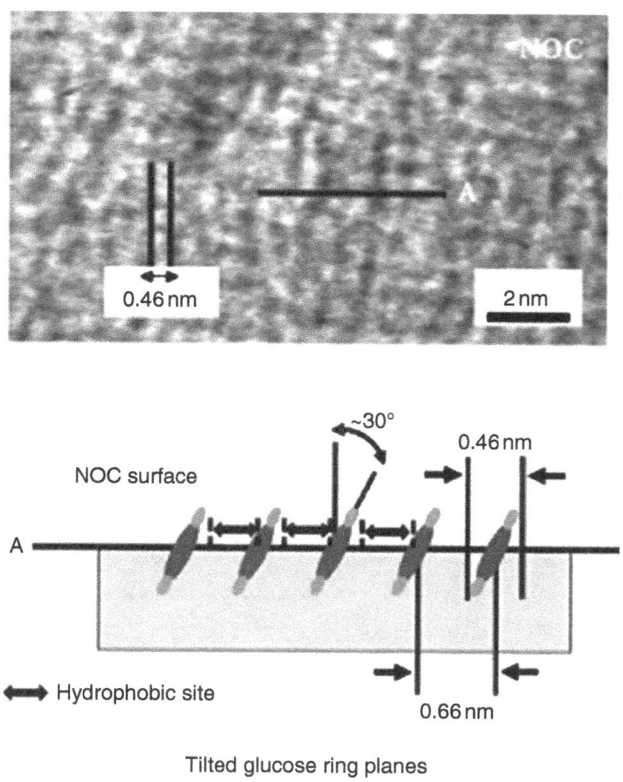

Figure 4.4 High-resolution TEM image of NOC, the molecular ordering template, together with the schematic cross-section image along the perpendicular A (= lateral direction) to the stretching direction indicating the tilted arrangement of the glucan chains. The film was negatively stained by uranyl acetate and spans a region of the copper support grid. Note the individual glucan chains that are separated by an average distance of 0.66 nm (arrows). The arrangement of glucan chains is clearly resolved with 0.46 nm in width.

The orientation parameter calculated from wide angle X-ray diffraction (WAXD) photographs was 0.88, which indicated a high degree of orientation. However, the crystallinity did not significantly follow the increase of the orientation by the stretching [1]. The crystallinity was ca. 14.8% and 16.8% before and after stretching, respectively [1, 13]. IR spectra of deuterated samples also supported the low crystallinity of the film, based on the ratio of the remaining hydroxyl groups in the drawn films that corresponds to the crystallinity index [14, 15]. Thus, we should consider the states of the structure for NOC to be ordered states that are neither crystalline nor amorphous.

The structural stability of NOC in water was examined, and it was found that the noncrystalline character based on the specific chain orientation of NOC was retained after the treatment in water even at 50 °C for long periods [16].

4.2.2 Various Nematic Ordered Templates and Modified Nematic Ordered Cellulose

4.2.2.1 Nematic Ordered Chitin and Chitin/Cellulose Blends

The chitin template prepared in the same way as the NOC template also has apparent orientation of molecular chains [17]. The distance between individual chains (1.6 nm for TEM, 0.54 nm for AFM) is greater than the crystalline lattice constant of the a axis of α-chitin (a: 0.474 nm, b: 1.886 nm, c: 1.032 nm, $\alpha = \beta = \gamma = 90$ degrees) [18]. Similar to NOC, in which the glucan chain dimensions are 0.4–0.5 nm viewed from its narrow axis, AFM images with the chain width of 0.45 nm clearly exhibit single molecular chains in the narrow axis on the surface of the chitin template. This means that nematic ordered chitin was achieved, similar to our previous work with cellulose [1]. High-resolution TEM images of the molecular assembly in the stretched samples of chitin/cellulose blend with a composition of 75/25, which was prepared in the same way as NOC, also exhibit the occurrence of the orientation of molecular aggregation, but are not resolved at the individual molecular chain scale [17]. By image analyses of the TEM photograph, the average width and distance between two parallel lines were found to be 1.38 ± 0.18 nm and 1.65 ± 0.27 nm, respectively. The average line width is narrower than that (1.51 ± 0.27 nm) from the TEM image of the nematic ordered pure chitin, indicating that the intermolecular interaction between cellulose and chitin may be engaged. Possibly each molecular chain is facing each other against the surface through a hydrophobic interaction such as a van der Waals force. On the other hand, the average distance between two parallel lines was not significantly different between the two stretched films in the pure chitin (1.62 nm ± 0.21) and the cellulose/chitin blend (1.65 ± 0.27 nm). Therefore, it is considered that the cellulose/chitin molecular aggregates in the stretched film are aligned similarly to the nematic ordered pure chitin. In addition, these results indicate the presence of another nematic ordered state with a different scale, for example, nano/micro hierarchical structures, in chitin and the blends with cellulose. (See the more detailed study using WAXD on the nematic ordered pure chitin and the blends with cellulose in [6] and [17].)

4.2.2.2 Modified Nematic Ordered Cellulose

Other types of templates have been explored and the first successful fabrication of cellulose and cellulose acetate films with honeycomb-patterned pores was reported [9]. These were produced by casting water-in-oil emulsions on glass substrates in a saturated water vapor atmosphere. Micropatterned honeycomb films are formed more easily using this technique than by using lithographic techniques such as photolithography. In the film preparation, a honeycomb-patterned cellulose acetate was fabricated and then deacetylated to yield a cellulose film, without deformation of the honeycomb pores; the micropore sizes ranged from 1–100 μm depending on the preparation conditions. As the volatile solvent in the polymer solution evaporated, the polymer precipitated to form a film, which simultaneously caused stretching by natural drying. The honeycomb-patterned frames are therefore expected to have specific arrangements of β-glucan chains, based on the NOC, which functions as a template.

A novel type of biomineralization process was proposed by developing a bi-functional cellulose template [5]. Namely, the NOC template was modified into a bi-functional template having a pair of the functions required for biomineralization, resulting in inducing oriented deposition of the biominerals as well as supplying anions in the synthesis. In this case, the new NOC template that is prepared to contain phosphate anions (termed p-NOC) is expected to mediate a unique order patterned deposition of calcium phosphates on the surface by reaction with calcium cations during immersion in a buffer solution.

The modification of NOC is performed as follows: A transparent water-swollen fixed gel of cellulose (WSC) is followed by the procedure using DMAc/LiCl solvent system as described in Section 4.2.1 [1, 13]. The WSC is immersed in phosphate buffered saline (PBS, Sigma-Aldrich co., Ltd.) for one week to prepare a water-swollen cellulose gel containing phosphate anions. Starting with the gel-like films containing phosphate anions thus prepared, p-NOC is obtained by uniaxial elongation at a draw ratio more than 2.0 at room temperature. In the drawing process, the specimen is kept in a wet state by pouring the PBS on the surface of the film. The specimen is then air-dried for 24 hours before being vacuum-dried at 25 °C for more than 48 hours. The AFM images of the p-NOC surface containing phosphate anions together with the original NOC substrate are shown in Figure 4.5. As a reference, a non-ordered cellulose template containing phosphate anions is also shown in Figure 4.5 [10].

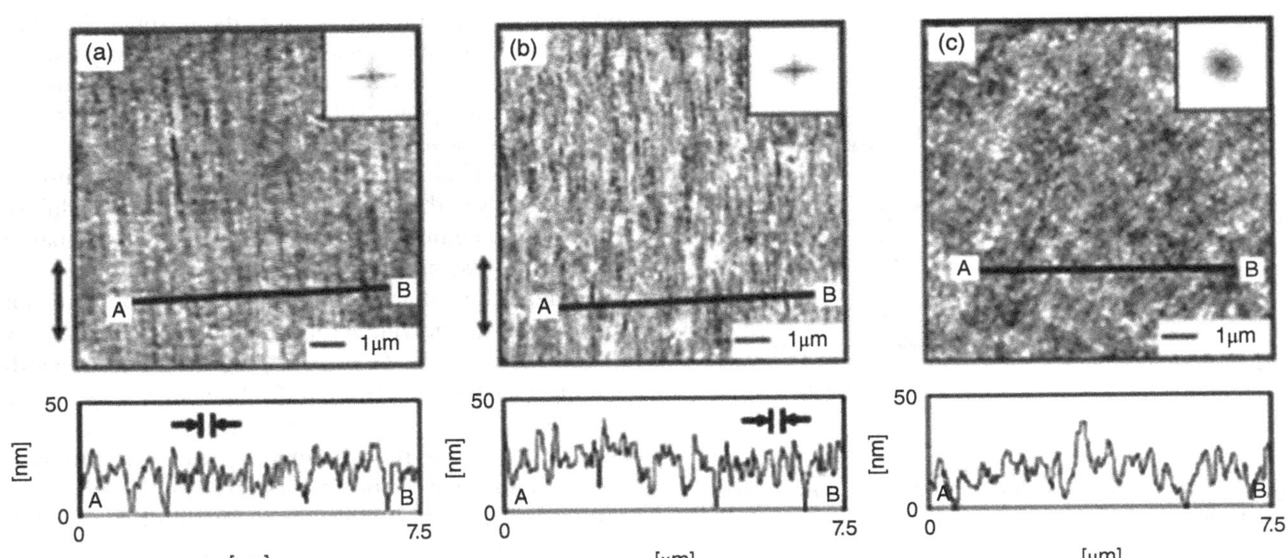

Figure 4.5 AFM images of (a) the NOC template, (b) the p-NOC template and (c) the non-ordered cellulose template as a reference for the p-NOC before the reaction with calcium cations. Insets in the upper right indicate the fast Fourier transformed images. The cross-section profiles along the perpendicular A–B are shown in the bottom of each figure. In figures (a) and (b), the average distances between two parallel lines were (a) ca. 0.17 μm and (b) ca. 0.14 μm. The double arrows indicate the stretching direction.

Comparing Figure 4.5a and Figure 4.5b, the AFM images of height mode demonstrate that the p-NOC containing the phosphate anions possessed a well-ordered surface morphology along the stretching direction similar to the original NOC substrate. Furthermore, the polarized distribution of spots in the two-dimensional fast Fourier transformed (2D-FFT) image inserted in Figure 4.5b confirms a high degree of uniaxial orientation on the surface of the p-NOC. The average line width was calculated from the cross-section analyses of the height along the A–B line perpendicular to the stretching direction (see the bottom figures in Figure 4.5). The average distance between the two parallel cellulose chains was $0.14 \pm 0.03\,\mu m$ in the p-NOC template surface (Figure 4.5b). The line width was similar to that of the original NOC template $0.17 \pm 0.06\,\mu m$ as shown in Figure 4.5a, indicating that the physical association of cellulose chains is not upset even if there are phosphate anions in the p-NOC template. In addition, the negatively polarized surface due to the tracks of OH groups along the molecular orientation is supposed to appear to some extent on the surface of the original NOC [1–4]. Thus, if phosphate anions are contained in NOC templates, then they are likely to be immobilized on the glucose planes only between the OH tracks in the surface that is negatively charged. This means that anions with negative charges could be introduced as the second minus charge into the initially negative-charged film surface. Accordingly, the phosphate anions are also supposed to be aligned along the molecular tracks in the newly prepared p-NOC template.

Without stretching the gel containing the phosphate anions, a non-ordered structure was observed on the surface, which is confirmed by the 2D-FFT image as shown in Figure 4.5c. By uniaxial stretching of the cellulose gel, it is shown that the p-NOC has the unique surface morphology similar to NOC that is totally different from the non-ordered cellulose template containing the phosphate anions. The root-mean-square (RMS) volumes were calculated from the obtained AFM height images to compare the surface roughness of each template. The RMS values for p-NOC and the non-stretched cellulose corresponded to 8.3 nm and 6.2 nm, respectively. Namely, p-NOC exhibited a rougher surface than cellulose without stretching.

4.3 Exclusive Surface Properties of NOC and Its Unique Applications

4.3.1 Bio-Directed Epitaxial Nano-Deposition on Molecular Tracks of the NOC Template

As already described, NOC exhibits several unique properties as a template having the exclusive surface of a certain nano/micro structure. Namely, the OH groups tend to be oriented as molecular tracks only in one direction across the entire NOC surface. In addition, as shown in Figure 4.4, the hydrophobic sites appear alternately next to the OH tracks, resulting in amphiphilic tracks in one direction.

The ordering amphiphilic tracks of hydrophilic OH groups and the hydrophobic polarity in NOC can induce an epitaxial deposition of biosynthesized cellulose nanofibers secreted from the gram-negative bacterium, *Gluconacetobacter xylinus* (= formerly *Acetobacter xylinum*), along the same axis of the tracks in NOC [2–4]. When active *Gluconacetobacter xylinus* cells are transferred to the oriented surface, they synthesize cellulose ribbons parallel to the molecular orientation of the substrate. This is evidenced by direct video imaging of the motion of the bacteria as they synthesize the cellulose ribbon. The movement of *Gluconacetobacter xylinus* in relation to cellulose biosynthesis was reported in 1976 [19]. The cell movement (at a constant rate of 4.5 μm/min at 24 °C) is the result of an inverse force imposed by the directed polymerization and crystallization of the cellulose. It is also well known that the bacterium rotates on its own axis. The bio-directed epitaxial nano-deposition of secreted fibers was studied over time and under varied conditions, including different substrates. Basically, the cells on the NOC template move straight along the molecular tracks. Midway through this series, the cell "jumps off" of the oriented substrate, and continues to secrete its cellulose ribbon, but now in the form of a spiral or tunnel structure that is the normal pattern of formation when not in contact with any organized substrate [20]. When the interaction between the bacterium and the surface of NOC is strong enough, the bacterium follows the track of the molecular template. As the polymer orientation of the NOC is not necessarily perfect, the bacterium may jump off the track at a structural defect. After the bacterium leaves the track, two forces, the inverse force of the secretion and the close interaction of adjacent microfibrils, affect the bacterium, possibly resulting in the synthesis of a spiral ribbon of microfibrils. In addition, the bacterium begins to rotate on its own axis. This rotation is the visible result of ribbon twisting which occurs to relieve strain induced by the absence of interaction with the substrate. When the ribbon is assembled in direct contact with the oriented molecular NOC substrate, ribbon twisting is prevented, thus suggesting a control of this oriented solid surface over the final physical interaction of polymer chains immediately after synthesis and during the early stages of crystallization.

Ribbon interaction is more clearly shown by field emission scanning electron microscopy (FE-SEM) in Figures 4.6a and b. In this experiment, the motion of the same sample is observed using light microscopy and

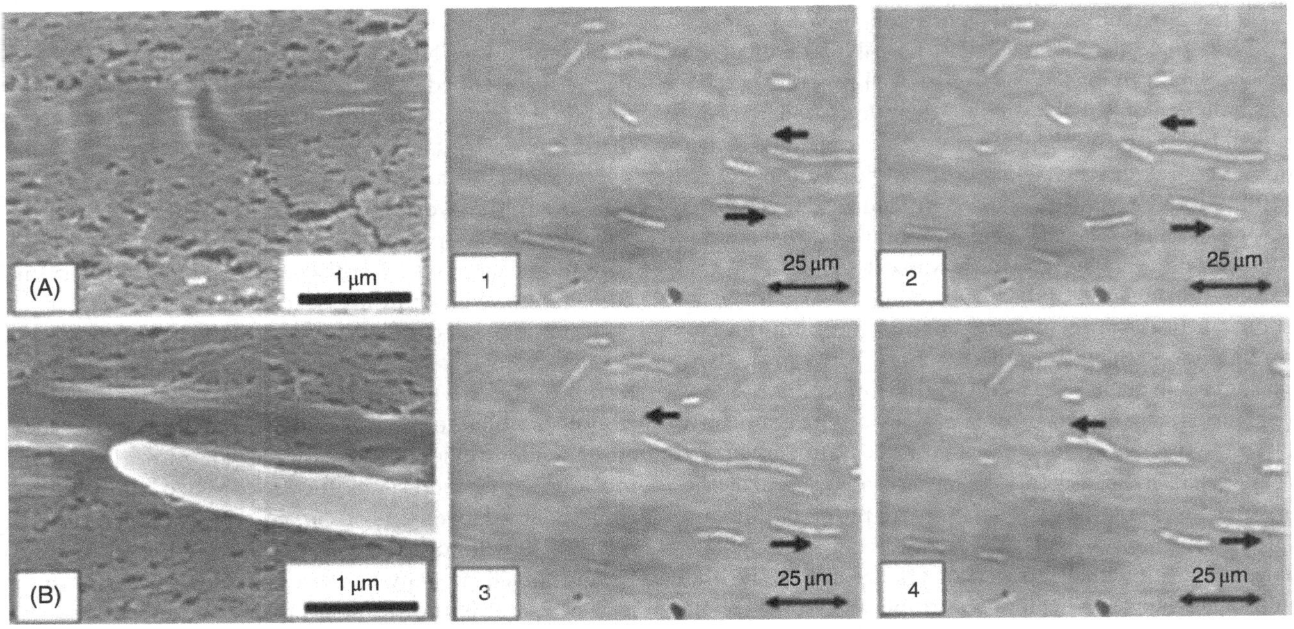

Figure 4.6 FE-SEM images of the cellulose ribbon deposition process. (A and B) are examples of bacteria synthesizing cellulose ribbons on the oriented molecular track of NOC. Figure 4.6A demonstrates the tight association between the molecular track and the cellulose ribbon. In 4.6B, a bacterium shows a flat ribbon immediately behind its site of synthesis. (1–4) are successive images showing the motion of a bacterium on nematic ordered blended film of cellulose and cellulose acetate which has been uniaxially stretched in the same manner for NOC.

then the sample is prepared for FE-SEM observations. The ribbon structure shown by the FE-SEM images in Figures 4.6a and b reveal no twisting and a more perfect alignment with the substrate. Likewise, when the bacterium "jumps off" the track, the cellulose ribbon is twisted (data not shown). Figures 1–4 in Figure 4.6 show the active *Gluconacetobacter* cells cultured on the oriented surface of a blend of cellulose and *O*-methylcellulose with DS 1.6 (50/50 w/w) that is uniaxially stretched in the same manner as for NOC. The bacteria also tend to follow the molecular tracks on the surface of the oriented cellulosic blend template, similar to the NOC case. In addition, the bacteria exhibit the similar linear movements along the molecular tracks on nematic ordered cellulose/cellulose acetate (DS 1.7) templates like the NOC. Namely, the nematic ordered cellulosic blends as well as NOC are supposed to show the same unique properties as templates that direct the bacterial movement. When *Gluconacetobacter* cells are placed on an agar surface as a control, their interaction seems to be reduced, and twisted ribbons are always deposited, also in a random fashion. It should be noted that this regulated movement of the bacterium is not observed in the synthetic polymers having OH groups such as poly (vinyl alcohol), but only in nematic ordered states of carbohydrate polymers described above [4].

4.3.2 Critical Factors in Bio-Directed Epitaxial Nano-Deposition on Molecular Tracks

4.3.2.1 Hierarchical Order

Figure 4.7 compares AFM images before and after incubation of *Gluconacetobacter* cells, showing the fabrication of a 3D architecture along the molecular tracks of NOC as a scaffold. After the 10 hours of culture, the height increased by 10 times to ca. 350 nm selectively in the stripes on the NOC (see Figure 4.7b). This is evidence of the directed close contact between the secreted nanofibers and the molecular tracks on NOC templates. Moreover, the elastic modulus increased slightly from 6.2 ± 0.1GPa to 6.9 ± 0.0GPa by the patterned fabrication with the secreted cellulose nanofibers on the NOC surface. Presumably, the change is mostly influenced by the sufficient strength of NOC as a scaffold when compared with the deposited nanofibers on the above. Here one can observe that additional layers secreted over time continue to mimic the original orientation initiated in the first deposited layer (e.g. the epitaxial deposition).

4.3.2.2 Epitaxial Deposition

The adhesion occurs between biosynthesized cellulose microfibrils and the C(6) primary hydroxyl groups and/or hydrophobic tracks described in Section 4.2.1. Namely, the important factor in controlling bacterial movements may be close contact and strong interaction between the molecular tracks on NOC and the newly synthesized cellulose microfibrils. Figure 4.8 explains this concept. Normally, *Gluconacetobacter xylinus* secretes a cellulose nanofiber that comprises a microfibril of ca. 3.5 nm in width, as shown in Figure 4.8 left. The microfibrils from the subunit of terminal complexes are assembled by interfacial interactions including hydrogen bonding and van der Waals force between the component molecules, resulting in forming a ribbon-like cellulose nanofiber of ca. 50 nm in width and 10 nm in thickness, respectively [21]. On the other hand, a cellulose microfibril is synthesized along the molecular track on the NOC by the strong adhesive effect (Figure 4.8 right). This epitaxial deposition is clearly confirmed by the FE-SEM image of the width of the NOC-altered cellulose ribbons having 500 nm with splayed microfibrils 7–8 nm apart (see also Figures 4.6A and 4.6B [2–4]).

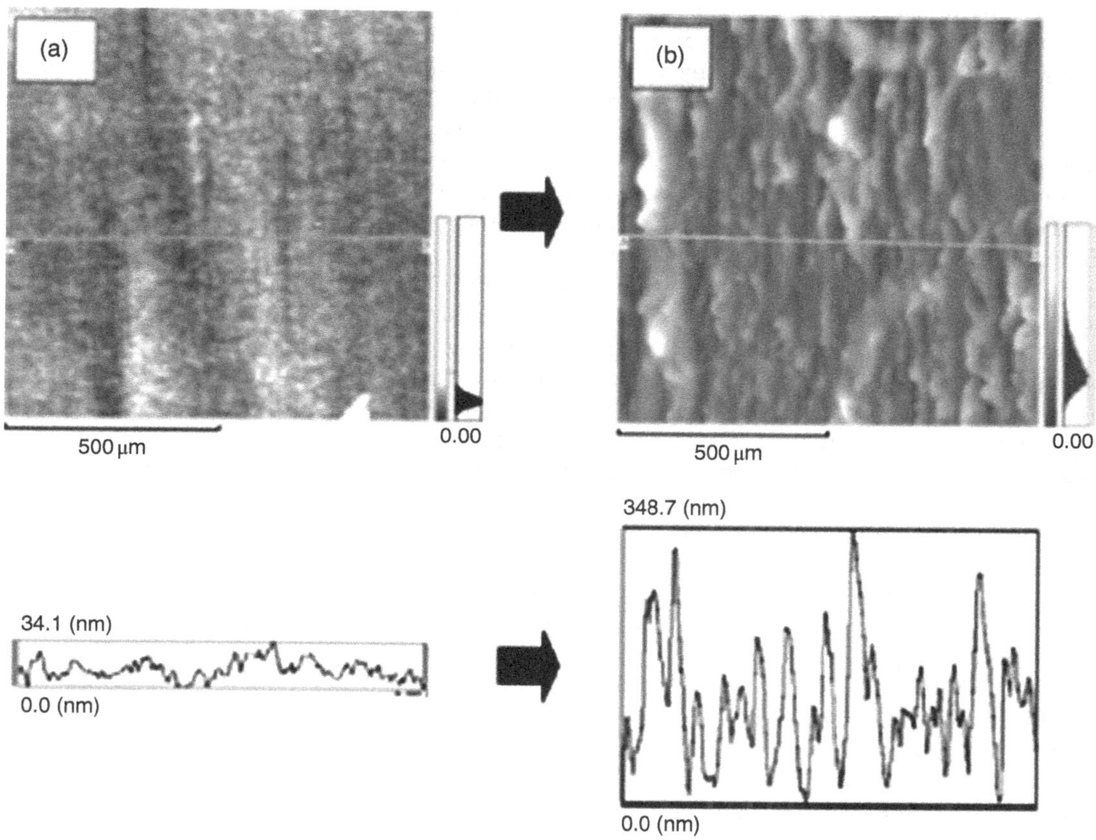

Figure 4.7 AFM images before (a) and after (b) incubation of *Gluconacetobacter xylinus* on NOC for 10 hours in SH media.

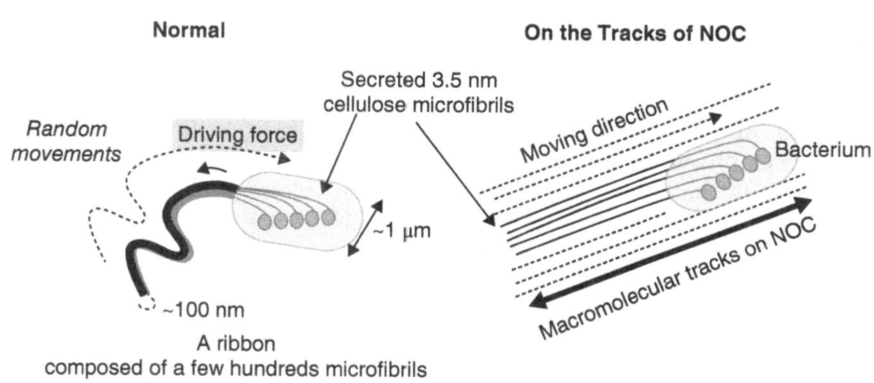

Figure 4.8 Schematic figures of self-assembly of cellulose microfibrils secreted from *G. xylinus* in normal fashion (left) and on the tracks of NOC templates (right). The circles indicate cellulose-synthesizing enzyme subunits linearly arranged across the cell body.

The amphiphilic molecular tracks on NOC allow close contact with the glucan chains of microfibrils, as they are being secreted and crystallized. Thus, in theory, this interaction is predicted to alter the self-assembly of glucan chains in the crystalline microfibril [21–25]. In fact, alteration of the aggregation of cellulose microfibrils on NOC was demonstrated as described above. Such interactions with the substrate alter the production of the cellulose ribbon and also the motion of the bacterium, and provide support for the epitaxial nature of the cellulose deposition on NOC. Thus, nano-scale interactions induce micro-scale structural changes. Such interactions could be used on a larger scale to selectively modify biopolymers for many applications.

4.3.2.3 Critical Factors

The adhesion for epitaxial growth of microfibril deposition requires several different critical factors relating to the molecular "order" of the substrate. This order is represented by three factors: (1) the molecular chain orientation; (2) the specific polymer chain conformation; and (3) O(6) rotational position with respect to the O(5) and C(4) in a β-glucan chain. The importance of molecular chain orientation is supported by the observation that unstretched NOC does not cause epitaxial deposition. As already described, we consider that the C(6) hydroxymethyl conformation of NOC is also of importance. Its role is like an anchor for the molecular tracks to direct the epitaxial deposition of the cellulose nanofibers. The conformation of C(6) hydroxyl groups is believed to contribute to the crystalline forms depending on either *gt* (cellulose II) or *tg* (native cellulose). When the NOC with *gg* of hydroxymethyl conformation is crystallized, it is converted to ordered films containing cellulose II crystalline forms with *gt* hydroxymethyl conformation [16]. When *Gluconacetobacter* is incubated on the crystallized NOC, there is no adhesion of the secreted microbial cellulose nanofibers. The crystallized films do not exhibit the molecular track effect to direct bacterial movements. The same lack of adhesion is observed with substrates prepared from native crystalline cellulose I from *Valonia* (data not shown). In the crystalline samples, the specific polymer chain conformation is changed, and the rotational conformation at C(6) is *gt* in crystallized NOC and *tg* in *Valonia* cellulose. Considering the above, we conclude that the role of primary hydroxyl groups at the C(6) position is greater than those of hydroxyl groups at the C(2) and C(3) positions. In addition, it is also noted that not only the OH tracks but the hydrophobic tracks are supposed to appear on the NOC surface, and thus the amphiphilic molecular tracks would be critical in regulating the epitaxial deposition of the cellulose nanofibers.

Ordered, non-crystalline cellulose (NOC) and related blends have been proved to work very well as the templates. Surfaces of nematic ordered chitin also show the adhesive effect, but in an unusual fashion.

The rate and direction of the movement correspond to those of the fiber production. The fibers were reported to produce at a rate of 2 μm/min at 25 °C [19]. However, *Gluconacetobacter xylinus* produced the fiber faster on NOC templates at a rate of 4.5 μ/min at 24 °C [4]. This gap is considered due to the difference in the strength of the interaction between the biosynthesized fiber and the NOC surface. This also indicated that the NOC surface promoted the secretion rate of the fibers.

The optimal condition on the templates containing the additives for the culture medium was studied in order to understand the influential factor in the rate of the movement of *G. xylinus* accompanied by the secretion of a nanofiber [26]. The template with a molecular orientation at a higher draw ratio was more effective at enhancing the moving rate. Further, in the midway to the maximum draw ratio, the moving rate may depend on the constituents of the template. The additive, CMC, with a lower DP, was effective in preventing secreted cellulose microfibrils from self-assembling, which resulted in an increase in the rate of the movement on the agar substrate. In fact, the SH medium containing 1% of CMC with a lower DP enhanced the moving rate of the bacterium cultured on the template without orientation. Furthermore, on a NOC template, the SH medium containing 1% of CMC with a high DP was more effective in increasing the moving rate. Presumably, the alignment of the CMC molecules onto the oriented cellulose molecules in the NOC by a strong interaction caused the increase in the moving rate. From the results, the condition necessary to increase the moving rate is proposed to employ a flat drawn template with an orientation that can engage with other materials due to a strong interaction. Further, the additive to the medium, which prevents self-assembly of the secreted cellulose microfibrils, is also effective. Thus, understanding how to prevent the self-assembly of the secreted microfibrils on the oriented templates could be at present the easiest way for us to find an optimal condition to enhance the fiber production. More importantly, the regulated movement of the bacteria due to the ordered surface of NOC may trigger the development of a 3D structure of hierarchical architecture from the nano to the micro levels. Therefore, we have been attempting to regulate the 3-D architecture of materials using nanofiber secreted by *Gluconacetobacter xylinus* as a building block.

4.3.3 Regulated Patterns of Bacterial Movements Based on Their Secreted Cellulose Nanofibers Interacting Interfacially with Ordered Chitin and Honeycomb Cellulose Templates

4.3.3.1 On Ordered Chitin Template

When active *Gluconacetobacter xylinus* cells are transferred to the nematic ordered chitin surface, they synthesize cellulose nanofibers (= cellulose ribbons) that are not parallel to the molecular orientation of the substrate, unlike the secreted fibers deposited on NOC templates. Instead, the bacteria begin to follow the molecular tracks, but soon they "jump off" the track, are diverted and then synthesize their cellulose parallel to neighboring tracks, once again soon jumping off. This "wavy" pattern was repeated across the template, as evidenced by direct video imaging of the motion of the bacteria as they secrete the cellulose nanofiber, as shown in Figure 4.9. Figure 4.9 shows the time course (120 s between each frame) of the wavy pattern of movement for a *G. xylinus* cell having 1 μm in width and 10 μm in length, respectively, together with the deposited fiber in the same waving pattern. The time lapse observation also proved that the cell movement was at a constant rate of 2.05 μm (standard deviation (SD) = ± 0.14 μm)/min at 24 °C, which is considered the result of an inverse force imposed by the directed polymerization and crystallization of the cellulose in the bacterium. The amplitude and periodicity of the waving movement as indicated in Figure 4.9 were 6.11 ± 0.53 μm and 17.85 ± 0.04 μm, respectively. As seen in the minimum standard deviation in the periodicity of the wavy pattern, the movement of the bacterium is fairly periodical and regularized. However, the rate of motion on the nematic ordered chitin template was the same as that for cell movements without templates. Furthermore, there was no difference in the rate between movement on the track and off the track, indicating that the interaction of the nascent cellulose with the nematic ordered chitin template was not as strong as that with the NOC template [2–4]. In nematic ordered chitin

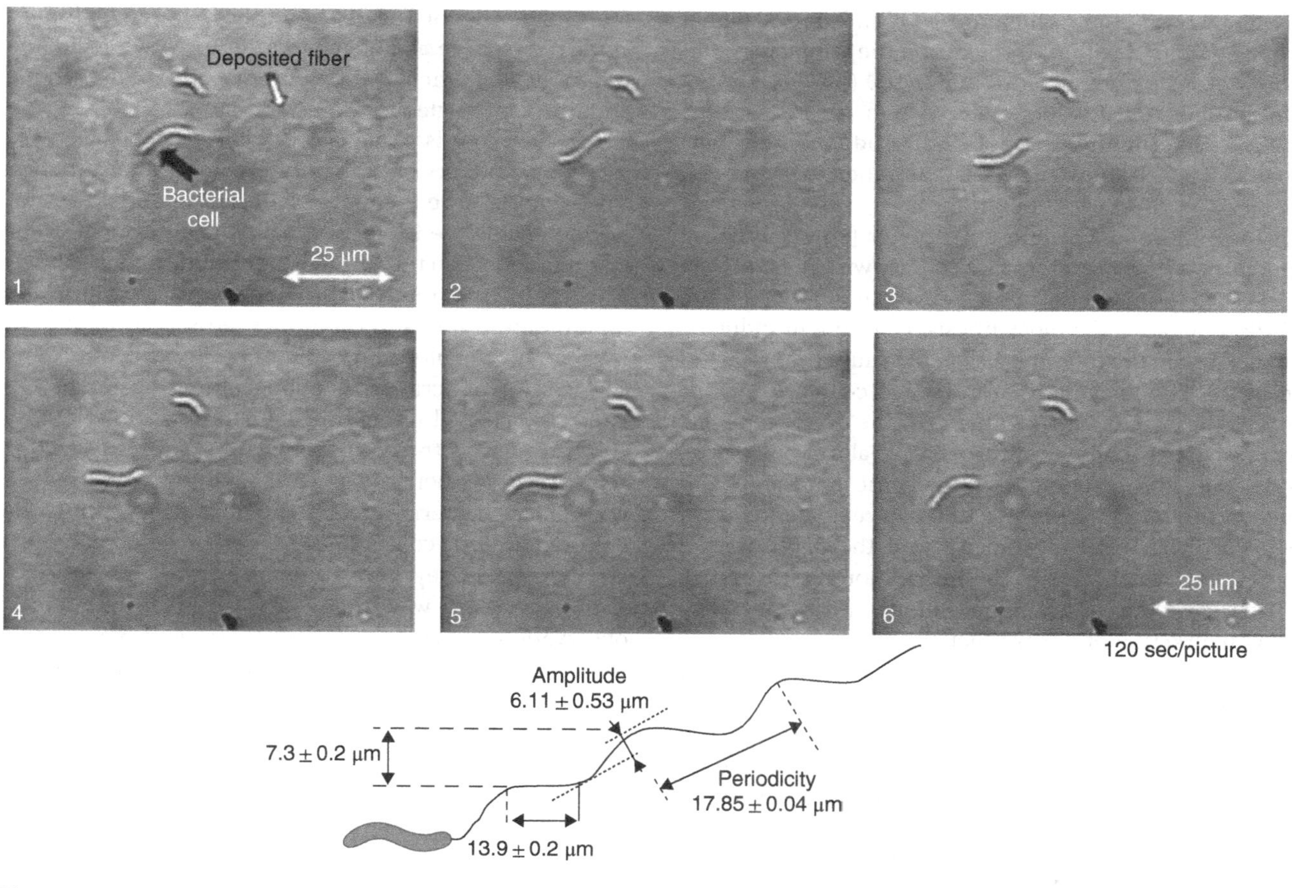

Figure 4.9 Successive images showing the motion using real-time video analysis of a bacterium as it secretes a cellulose nanofiber on the nematic ordered chitin template. The bacterium is attached, and thereafter synthesizing the fiber on the molecular tracks in the template to show a "wavy" moving pattern.

templates, hydrophilic hydroxyl groups at the C-6 position of the glucopyranose ring and the relatively hydrophobic acetamide groups at the C-2 position appeared alternately on the surface because of the 2_1 screw axis of the molecular chain, whereas the NOC template provides only hydroxyl groups at the C-6 position on the surface. The balance of the presence of the two substituents strongly affects the strength of the interaction between the surface and the secreted cellulose microfibrils. When the contact with the chitin template is not as strong as that for NOC, the biosynthesized individual sub-elementary fibrils (of a few nm in width) usually tend to self-assemble to form a ribbon-type nanofiber of 40–60 nm in width. The situation when the bacteria secrete cellulose nanofibers cannot be changed without such a strong interaction between the interfaces. Accordingly, the rate of the deposition of the fibrils on the nematic ordered chitin template that corresponds to the rate of the bacterial movement did not change from the initial rate without the template (both the moving rates were 2.05 μm ± 5% at 24 °C). On the contrary, when the interaction was strong enough as in the case for NOC, the rate of movement on NOC was much higher when compared with movements on the templates composed of 100/0 (pure chitin) and 0/100 (pure cellulose). One explanation for this phenomenon is to determine the rate-determining step for both production of fibers and the bacterial movements. It is assumed to be the self-assembly process of biosynthesized sub-elementary fibrils in forming a cellulose nanofiber [26]. Namely, in this case, the strong interfacial contact with the template prevented the assembly of the individual sub-elementary fibrils soon after the biosynthesis, resulting in inducing an increase in the rate, and also could cause the release of stress for the enzymatic synthesis of cellulose. As the result, the production of fibers on the NOC template increases so to match the rate of bacterial movement (see the details in [6]). These results suggest that the magnitude of the interfacial interaction can regulate the wavy pattern of bacterial movements and the deposition of nanofibers, which leads to the autonomous fabrication of 3D structures with a unique pattern and a controlled growth direction of the product.

With the addition of cellulose as a component for nematic ordered chitin template, the molecular ordering was altered, as described previously in Section 4.2.2.1. It was found that application of the preparative method for NOC to chitin and cellulose/chitin blends did not permit the formation of nematic ordered states like NOC on the molecular scale, but instead, induced a variety of hierarchical nematic ordered states on various scales [17]. It was expected to alter the magnitude of the interfacial interactions that occurred with the microfibrils by the addition of cellulose as a component. In the two templates of nematic ordered chitin/cellulose blends of 50/50 and 25/75, the moving patterns were still "wavy". However, a remarkable difference appeared depending on the cellulose component ratio. The more cellulose was added, the less the amplitude became, namely, getting close to a linear pattern [6]. This is interpreted that more cellulose that is contained, the more the moving amplitudes into a straight manner are reduced. The moving rate for the individual blend ratios seemed to be within a non-significant range.

4.3.3.2 On Honeycomb Cellulose Template

A honeycomb-patterned cellulose film was used as a scaffold for the culture of *G. xylinus* [10]. When the molecular orientation on the film surface exhibits amphiphilic molecular tracks similar to those on NOC, the bacterial cells autonomously fabricate 3D honeycomb structures by the deposition of secreted cellulose nanofibers on templates [2–4, 6, 8]. According to the detailed examination of NOC (see Section 4.2.1), the hydrophobic phases due to the anhydroglucose plane appear alternately next to the hydrophilic OH tracks, resulting in unique amphiphilic molecular tracks. This molecular order on the surface of the honeycomb frames appeared to have a similar ordering to NOC, including the molecular orientation along the honeycomb frame and the tilted angle of the glucose plane [1–4, 6].

The first step is to induce the bacteria to selectively secrete nanofibers onto the honeycomb frame surfaces. The bacteria were therefore incubated under slow rotation to enable the bacteria to secrete and deposit cellulose nanofibers on the surfaces of the honeycomb frames. This means that once bacteria apparently attach to the honeycomb frame surface, they keep secreting or producing nanocellulose onto the same honeycomb frame surface. In this critical rotation culture, bacteria were naturally selected to deposit the nanofiber onto honeycomb frames in the entire template. As a result of this selective attachment of bacterial nanofibers, which does not indicate bacterial attachment, during incubation, the cultured bacteria continuously deposit cellulose nanofibers along the honeycomb frame surfaces. Then the selective deposition was found by the real-time video observation to occur in the entire areas of honeycomb frame surfaces. Namely, this rotational culturing of the bacteria successfully achieved a preferential location on the entire honeycomb frame surface, presumably by anchoring as a result of selective interactions between the secreted nanofibers and the surfaces, as shown in Figure 4.10. It should be added that before anchoring, the bacterial cells were just floating and not observable in the same frame as in Figure 4.10.

After the attached movement of *G. xylinus* along the honeycomb frame surfaces, the bacterial cells began to synthesize cellulose nanofibers, enabling further selective

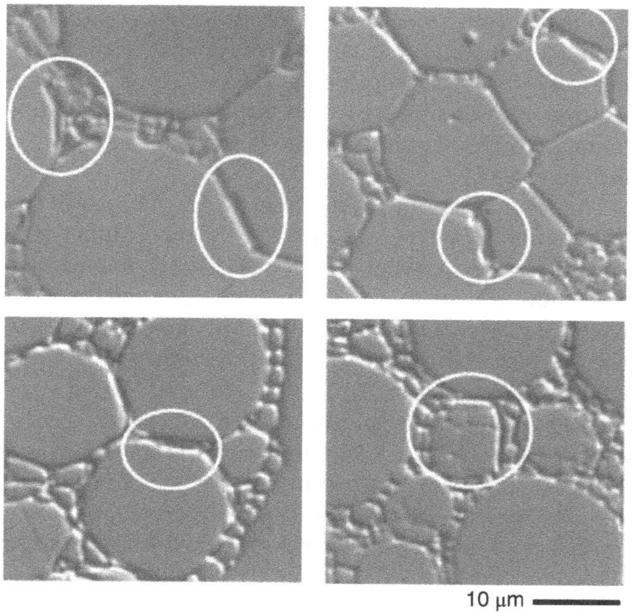

Figure 4.10 Differential interference light microscopic images of selective attachment of bacteria (inside circles) on surfaces of honeycomb frames under slow rotation during culture.

movements along the honeycomb frames of the substrate, as shown in Figure 4.11. In contrast, there were no cellulose fibers inside the pore spaces. The bacterial movement can therefore only be regulated on the surfaces of the honeycomb frames across the entire template, as shown by direct video imaging of the motion of a bacterium as it secreted the cellulose nanofibers. Presumably, the rest of the bacteria, which are not attached to the honeycomb frame, are floating in the media, and not capable of fixing themselves onto any scaffold.

Another method of fabricating 3D honeycomb cellulose structures was reported by Uraki et al. [27, 28]. The honeycomb film was prepared by a transcription method [26], for use as a honeycomb groove for *G. xylinus* movement to give a 3D architecture. This method was based on physical mechanical control of the bacterial movements; this is totally different from the present study, which was based on physicochemical interfacial interactions between cellulose nanofibers and templates.

As described above [2–4], the bacteria followed the orientation direction of the NOC and moved at 4.5 μm/min at 24 °C. In the present study, the cells moved at a

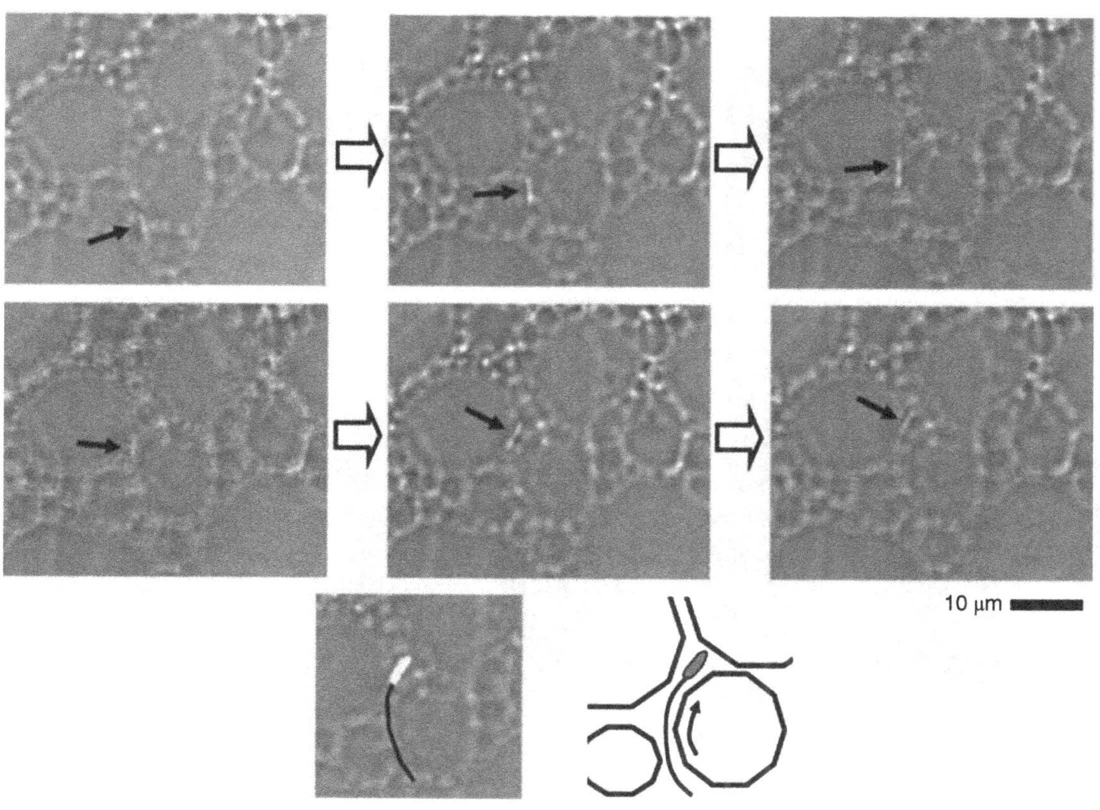

Figure 4.11 Successive images showing motion of a bacterium as it secretes a cellulose nanofiber, using real-time video analysis. The bacterium was selectively attached to the honeycomb frame and synthesized its cellulose fiber on the molecular tracks on the honeycomb frame. The interval of each image was set to be 1 min.

constant rate of 2.79 ± 0.46 µm/min at 24 °C when forming the first layer on the top of the honeycomb frame. This was considered to be the result of interactive forces between the surface of the honeycomb film and the secreted microbial cellulose fibers. For the second and subsequent layers, interactions between the cellulose nanofibers contributed to the movement rate; it became 2.92 ± 0.26 µm/min at 24 °C, which was faster than that for the first layer. The experimental results could be achieved by keeping observation of bacterial deposition all through the beginning to the end under an optical microscope using real-time video system (3 frames/min). Furthermore, both rates are faster than that of 2.0 µm/min at 25 °C [19] for cell movement without templates. The increase in the movement rate for NOC and the honeycomb templates may be caused by the inhibition of the self-assembly of the nascent cellulose microfibrils by the strong interactions between the biosynthesized fibers and the template [26]. The increase in the number of contact points on the microfibrils then causes strong interactions with the scaffold surface. This phenomenon is the rate-determining step for both production of fibers and bacterial movement [2, 3, 6, 25]. In this case, the strong interfacial contact with the template prevented the assembly of individual cellulose microfibrils of width 3.5 nm soon after biosynthesis, resulting in an increase in the rate, and could also cause stress release in the enzymatic synthesis of cellulose [3, 6]. As a result, the production rate of fibers on the surface of an NOC-like template increases with the increasing rate of bacterial movement.

The two-dimensional (2D) microscopic images of the honeycomb frames before and after the incubation of the bacteria showed no change in the 2D honeycomb pattern after culture. However, AFM observations revealed the phenomenon shown in Figure 4.12. In this study, an AFM with a voice coil motor was used to obtain topographic images with very rough surfaces. Normal AFM instruments equipped with a piezoelectric scanner could not be used to observe such rough honeycomb frames, because the piezo element cannot scan the large height changes. The changes in the heights in the surface topographic images of a honeycomb-patterned cellulose film are shown in Figure 4.12. Before and after incubation, microscopic observations with a light microscope and AFM were performed for the same sample, the same position, and the same measurement spot. It should be noted that great skill was needed to perform the experiments. This method was used to prove the 2D deformation of the honeycomb pattern. However, the cross-sectional AFM

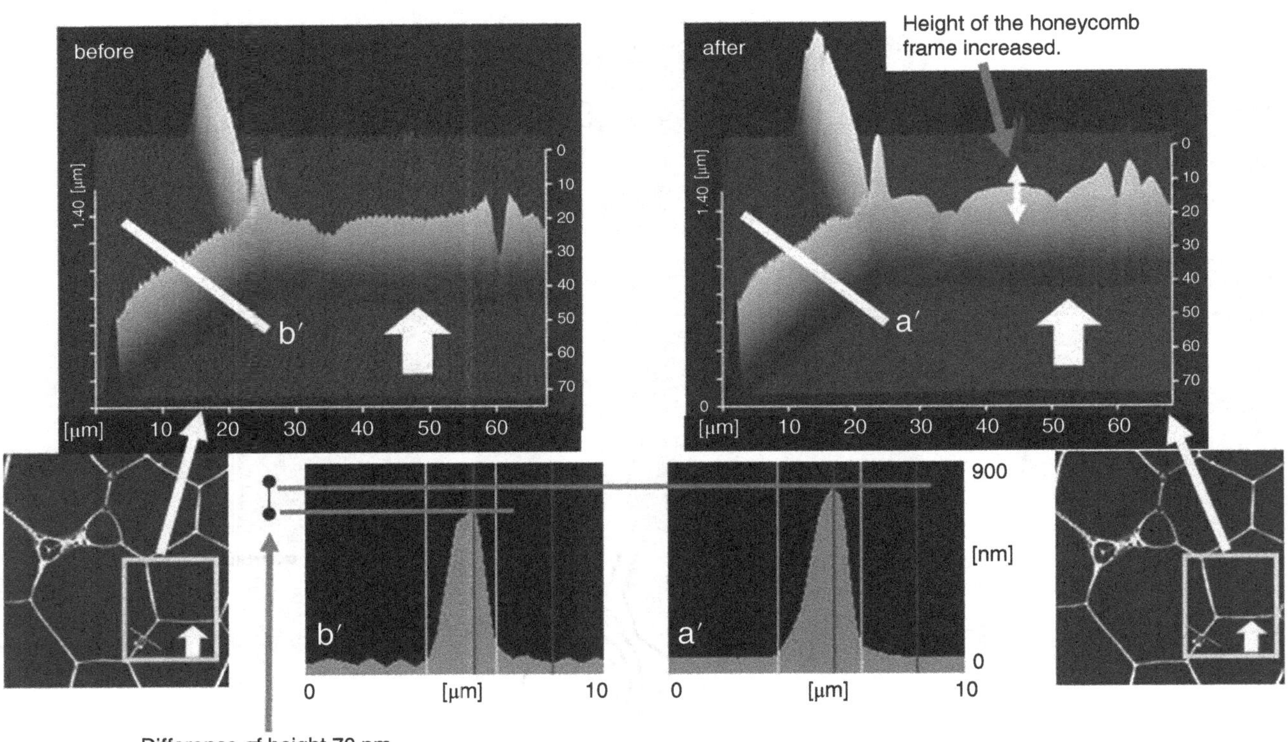

Figure 4.12 Morphological changes in the honeycomb frame on film, observed using atomic force microscopy, after culture. The square areas were focused and cross-sectioned, and show the 3D images before and after culture. Twenty layers of bacterial cellulose nanofibers were accumulated in 10 h (one layer = 3.5 nm).

image (right) after incubation of the bacterial cells clearly showed an increase in only the height direction of the honeycomb frames. In culture for 10 h, the average height increased by 70 nm, as shown in the height image at the bottom right of Figure 4.12. If the thickness of the individual microbial cellulose microfibrils is assumed to be 3.5 nm [6, 8], 20 layers of fibers were accumulated. It should be noted that we kept observing bacterial deposition from the beginning to the end under an optical microscope using time lapse video system (3 frames/min). Soon after the optical observation, the height was measured in the same location of the honeycomb frame. Therefore, the substance accumulated on the honeycomb frame should be cellulose nanofibers secreted by the bacteria. These results indicate that the honeycomb pattern could be used as a template for autonomous building up of 3D structures from 2D structures, with cellulose nanofibers as the building blocks, using a bacterial nanoengine.

The aim of this study is to develop a method to create novel 3D nanostructures using biological systems combined with templates with specific molecular orientation patterns. We expect that the relationship between directed biosynthesis by bacteria and ordered fabrication from the nano- to the micro-scale will lead to sustainable bottom-up methods for the design of functional materials with desired nanostructures, with low energy consumption.

4.3.4 NOC Templates Mediating Order-Patterned Deposition Accompanied by Synthesis of Calcium Phosphates as Biomimic Mineralization

This section briefly introduces an advanced possibility of NOC templates having bi-function (p-NOC; see Section 4.2.2.2) in terms of a novel bio-mineralization [10]. As described above, the NOC exhibits a unique property as a scaffold for the fabrication of the organic hierarchical structure as seen in living things. Recently, the authors have extended the above concept of NOC further to the fabrication of an organic-inorganic hierarchical structure using less energy consumption. The NOC was at first modified into a bi-functional template containing phosphate anions for bio-mineralization, resulting in the induction of the oriented deposition of calcium phosphates as well as supplying the counter-phosphate anions.

The surfaces of both p-NOC and non-ordered p-cellulose templates as a reference for the p-NOC, after reaction with the calcium cations in the buffer solution, were analyzed by using SEM observation. Figures 4.13a–c exhibit the surface morphological change of the p-NOC as the elapse of the reaction time with calcium cations. Large amounts of the calcium phosphates were formed all over the surface of the p-NOC template after the reaction with calcium cations. In addition, the assembled pattern of the deposited minerals was well arranged along the molecular orientation of the p-NOC surface. The inset FFT images corresponding to each SEM image in Figures 4.13a–c indicate that the longer the reaction time proceeded, the poorer the orientation of deposition of calcium phosphate aggregates became. The average distance between the oriented mineral lines ($n = 50$) was approximately $0.23 \pm 0.07\,\mu m$ (a), $0.19 \pm 0.04\,\mu m$ (b) and $0.30 \pm 0.08\,\mu m$ (c), respectively. These values were larger than that between the apparent tracks due to cellulose molecular chains on the p-NOC surface (see Figure 4.5b: ca. $0.14 \pm 0.03\,\mu m$ ($n = 50$)). The FFT image with a polar distribution confirms better orientation of the lines as shown in the inset images in Figures 4.13a–c, when compared with the inset of the non-ordered cellulose template in the same figure as described below.

The calcium phosphates did not take any ordered shape on the surface of the non-ordered p-cellulose template as a reference for p-NOC, although large amounts of calcium phosphates were also formed all over the surface (see the SEM image in the bottom right image in Figure 4.13). In addition, the inserted FFT image of the non-ordered cellulose template exhibited randomly distributed spots, indicating that the deposition pattern of the calcium phosphates on the non-ordered template was not induced in a preferable manner. Accordingly, this result strongly suggests that the p-NOC surface encourages the uniaxially oriented deposition accompanied by synthesis of the calcium phosphates. It was obvious that the surface orientation of the p-NOC template mediated the orientation of the calcium phosphates, because of the correspondence with each other. However, the orientation of the inorganic deposits on the p-NOC surface tended to be reduced with the longer reaction time through a–c in Figure 4.13.

Furthermore, the particle size of the calcium phosphates on the p-NOC template was smaller than those on the non-ordered p-cellulose template. It was presumably because a three-dimensional mineral deposition and its growth tended to be inhibited by the close contact of the uniaxially orientated cellulose molecular chains on p-NOC. The deposition of calcium phosphates was mostly limited to the orientation direction of the cellulose molecular chain, and thereby the shape of calcium phosphates was supposed to be also arranged along the stretching direction of p-NOC. This was a similar mechanism to a bio-mineralization with the inhibition of the three-dimensional crystalline growth and the arrangement of the mineral shape mediated by orientated scaffolds such as collagen or chitin [29, 30]. In this way, the

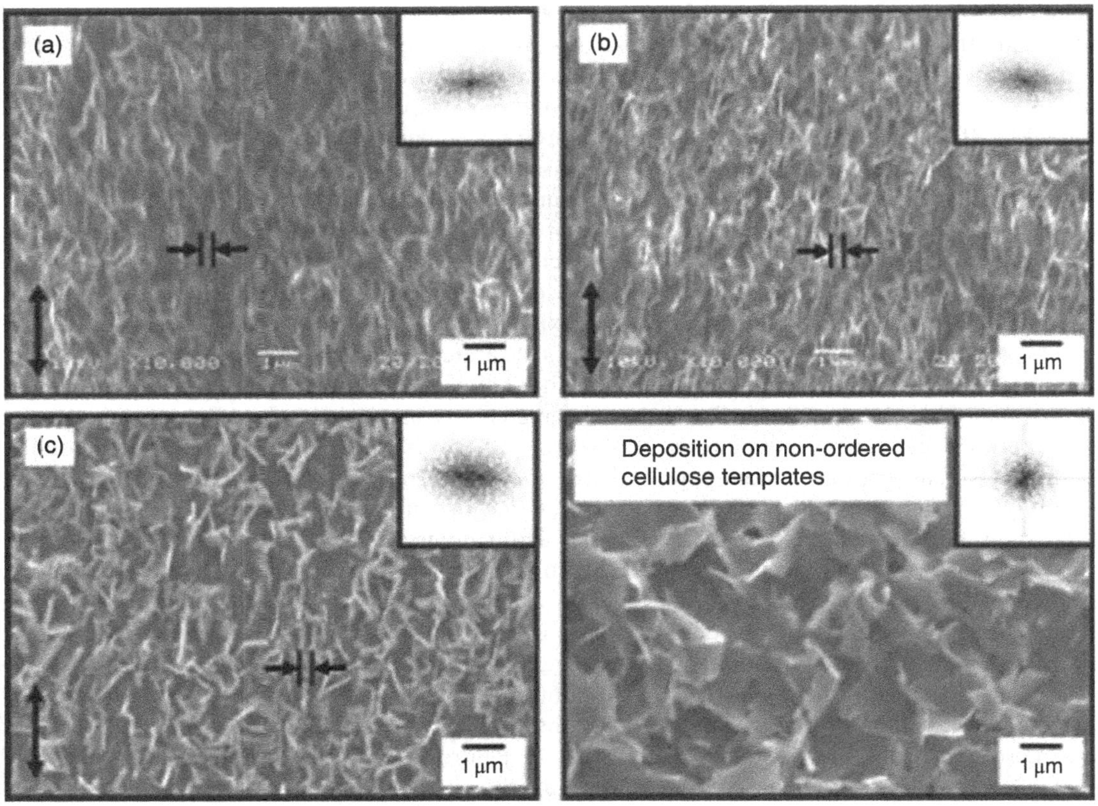

Figure 4.13 SEM images of the surface of p-NOC templates after reactions with calcium cations for (a) 0.5, (b) 5.0 and (c) 10 min with the fast Fourier transformed images as insets on the right. The morphology of calcium phosphate deposition on the non-ordered cellulose template after the same reaction for 10 min is also displayed as a reference for the p-NOC template. In (a)–(c), the line widths of calcium phosphates were 0.23 ± 0.07 μm, 0.19 ± 0.04 μm and 0.30 ± 0.08 μm, respectively. The double arrows indicate the stretching direction.

p-NOC template is expected as a bio-mimic template in terms of inducing the orientation of calcium phosphates, as well as the phosphate anion supplier [10].

4.3.5 Three-Dimensional Culture of Epidermal Cells on NOC Scaffolds

As so far described, amphiphilic molecular tracks of both hydrophilic and hydrophobic tracks next to each other alternating across the NOC surface are capable of enhancing the unique surface properties, which include serving as a template or scaffold for ordered deposition of various substances, such as nanofibers or inorganics [2–4, 6, 8, 10, 11].

The interfacial affinity between cells and culture plates is in general affected by differences in the plate surface properties, because a relatively hydrophobic culture plate is believed to induce the proliferation of cells following cell adhesion and subsequent settlement [31]. Recently, some scaffolds for cell culture have been proposed to establish a technique for regeneration medicine. Among available scaffolds, collagen [32–36], hyaluronic acid [37–39], chitosan [40], silk fibroin, and spider silk [41, 42] are major biopolymers playing a bioactive role in such medical purposes. Although cotton cellulose has not been thought to be biocompatible, it was recently clarified that bacterial cellulose pellicle produced from *Gluconacetobacter xylinus* has biocompatibility [43] and could be used for medical materials and tissue engineering [43, 44]. The latest research reported that cell divisions were induced toward the direction of cellulose nanofiber as the scaffold [45, 46]. Here, the authors succeeded in culturing epidermal cells on the noncrystalline cellulose template (NOC) with ordered cellulose molecules, instead of crystalline cellulose nanofibrils.

Initially, surface characterization of NOC templates was carried out by contact angle measurements and atomic force microscopy (AFM) analyses. Then, cell behaviors when cultured on NOC templates were observed using inversed phase contrast microscopy, followed by attempts to produce a three-dimensional culture of epidermal keratinocytes on a newly modified NOC system proposed.

The possibility of cell fixation onto a culture template depends on the appropriate surface hydrophobicity. A contact angle range of 60–80 ° has been proposed as the most suitable for cell fixation and proliferation [31]. The contact angle of the NOC template is indeed 71.9 ± 1.3 °. This value is in the middle of the 60–80 ° range and indicates suitability for cell culture.

Epidermal cells were then cultured on both the T-25 flask and the NOC template and, after attachment, observed by inversed phase contrast microscopy. The cell shapes were observed to change from spherical to spreading several hours after introduction to both the T-25 flask and NOC templates. Spreading shapes indicated that cells were successfully attaching onto both culture plates and such a behavior of cells also indicated the preference of the cells for particular plates.

Following adherence to the template or culture flask surfaces, cell division began. It should be noted that T-25 flask surfaces are usually modified by negative electric charges to induce stable cell attachment. On NOC templates, it was observed that the cells successfully attached and proliferated on both these templates and T-25 flasks; the calculated growth ratio showed similar tendencies to proliferate (Figure 4.14). At the beginning of the culture periods, cells were scattered across the NOC template. After 8 days of cultivation, cells completed covered the surface and were considered confluent and stably attached. Taken together, these behaviors indicated that the NOC template surface was capable of supporting these cells. Growth observation results also verified the relationship between cell attachment and the contact angle value of 71.9 ± 1.3 °.

On the other hand, the cells appeared to be randomly distributed after proliferation on the NOC surface. This may have indicated that the cellulose molecular orientation on the NOC surface was not effective in controlling the direction of the cell growth. In this case, it appeared more important for cell attachment and proliferation to have surface hydrophobicity because of the cellulose molecular orientation.

This study also attempted to establish a three-dimensional cell culture using a NOC template by confirming it was a suitable culture template, as in Section 4.3.4. A three-dimensional formation of epidermal cells is important as a bio-mimicking sample for medical and cosmetic analyses without involving living organisms. Stability of cell adherence and the continuous introduction of nutrients are required for three-dimensional cell culture. As NOC templates were swollen in the medium, forming a semipermeable membrane because of its dominant non-crystalline domains, they served as a means to constantly provide fresh nutrient for the growing cells (Figure 4.15). Using the new system after 3 days of cultures, the cells proliferated and formed one layer over the entire NOC

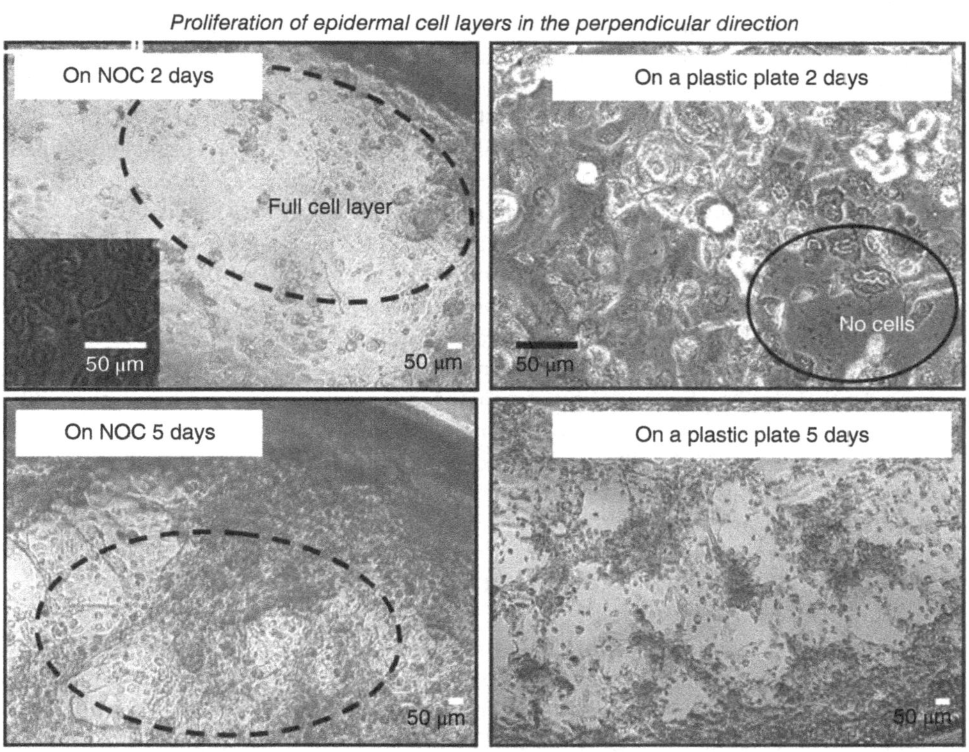

Figure 4.14 Culturing epidermal cells on NOC scaffolds and a plastic plate.

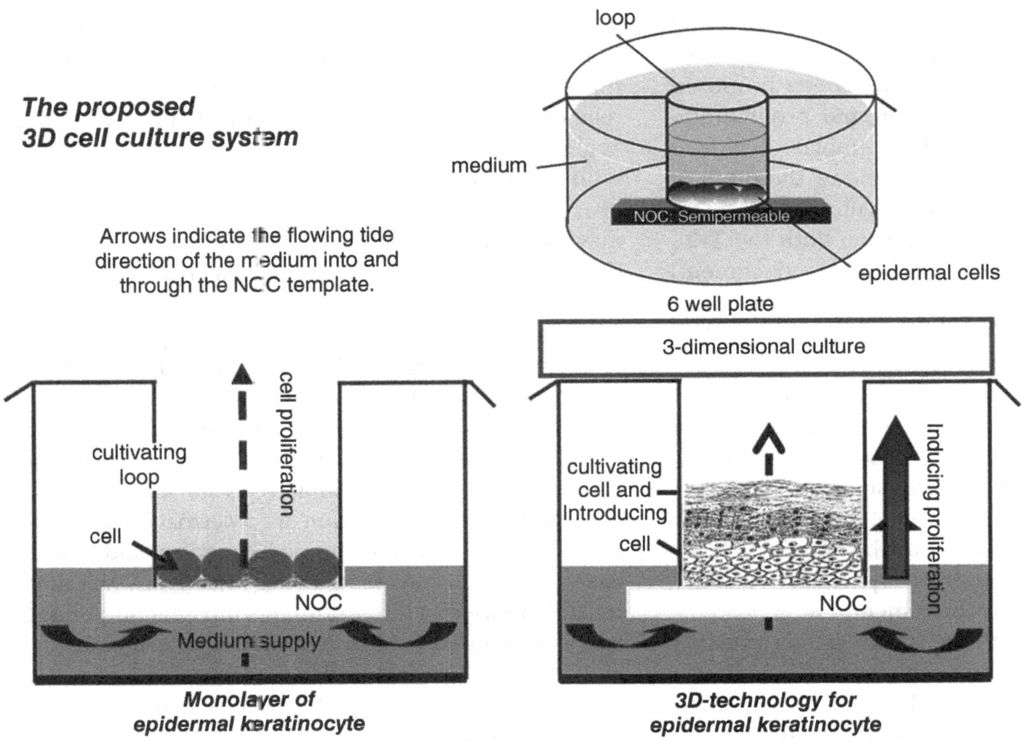

Figure 4.15 Schematic image of three-dimensional cultivation on the semipermeable NOC template. Broken arrow indicates possible proliferation direction of epidermal cells and arrows indicate the flowing tide direction of the medium into and through NOC template.

template surface (Figure 4.16). Furthermore, after 5 days of cultures, some colonies with 36 μm differences in height were observed on top of the previously formed cell layer. This indicated that cell division was induced upward, perpendicular to the NOC template bottom. Specifically, this effect indicated that the film played a role as a semipermeable membrane in providing and maintaining fresh culture conditions (Figure 4.15). To support the 3-D features mentioned above, the height of a single cell was measured by AFM in a cultivation that was allowed to build only a single cell layer on the NOC surface in the loop bottom. The cross-sectional image was assumed to indicate the height of a single fixed cell, at 1.42 ± 0.13 μm. Although measurement of the height of some layered cells, as a colony, was attempted, it was too high to detect by AFM. Alternatively, the height of these layers was measured at ~36 μm via a through-focus series of images using light microscopy. The colony was concluded to consist of ~30 layered cells. These results also suggested that the very bottom layer of cells did not detach from the NOC template during culture; therefore, the formerly established layered cells were considered to fulfill the role of a scaffold for the next newly-formed cell layer.

This study served to establish that NOC, as a cellulose film having a unique surface structure and properties, could be employed as a 3-D culture template for epidermal cells. Initial light microscopic observations of cell attachment and growth (or proliferation) on NOC templates indicated that the template was suitable as a cultivating scaffold for animal cells. It also indicated that the template was suitable for inducing construction processes leading to the formation of a three-dimensional cell accumulation because the NOC structure played a role as both a scaffold and a semipermeable membrane. The functions of the scaffold in supporting cell adherence and proliferation were enhanced by the amphiphilic molecular tracks on the unique NOC surface. However, the semipermeable membrane function was attributed to the dominant noncrystalline domains of the NOC film, providing fresh nutrient for the cells from the bottom to the top, similar to the process of human skin formation.

4.4 Conclusion

In biological systems, skeletal materials such as cell walls, bones, and shells are made primarily of a nano-scale building block of polysaccharides, proteins, and inorganic salts. The assembly of these building blocks facilitates the production of a hierarchical framework

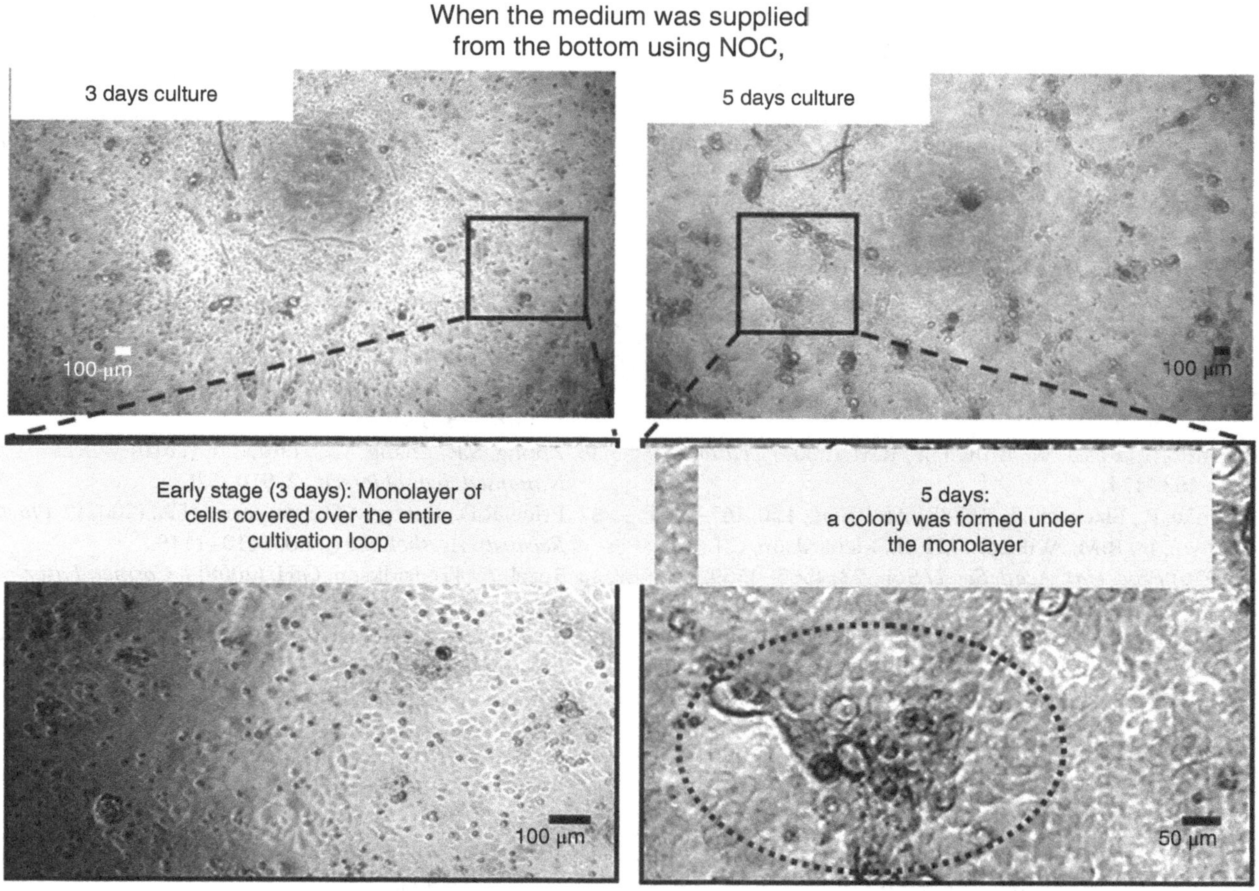

Figure 4.16 Cell proliferation on the provided medium of the NOC templates from the loop bottom to the top through the semipermeable NOC template. Culture times, (a, a′) 3 days and (b, b′) 5 days cultivation, and circle, colony. On NOC, the cells kept covering the surface after 5 days culture, indicating that the former layers could be a scaffold for the next generation coming from the bottom.

structure. The formation dynamics observed in this NOC could be applicable to the design of nano-scale controlled, hierarchically structured materials with specific properties. We have employed a biological system combined with a polymer platform with an NOC-like surface in order to directly fabricate hierarchically ordered materials from the nano level up to the micron level. Thus, if a nano/micro-patterned film with similar surface characteristics as NOC can be fabricated, the 3D-patterned materials will be built up by mediation of the cellulose or related scaffolds, and thereby widely appreciated and used.

References

1 Kondo T., Togawa E., Brown, Jr., R.M. (2001) *Biomacromolecules*, **2**, 1324–1330.
2 Kondo, T. (2007) In *Cellulose: Molecular and Structural Biology*, R.M. Brown, Jr., I.M. Saxena (eds.), Springer, Dordrecht, pp. 285–306.
3 Kondo, T. (2012) In *Bacterial Nano Cellulose: A Sophisticated Multifunctional Material*, M. Gama, P. Gatenholm, D. Klemm (eds.), CRC Press, Boca Raton, FL, pp. 113–142.
4 Kondo, T., Nojiri, M., Hishikawa, Y. et al. (2002) *Proc. Natl. Acad. Sci. U.S.A.*, **99**, 14008–14013.
5 Hesse, S., Kondo, T. (2005) *Carbohy. Polym.*, **60**, 457–465.
6 T. Kondo, W. Kasai, M. Nojiri, et al. (2012) *J. Biosci. Bioeng.*, **114**, 113–120.
7 Kondo, T., Kasai, W., Brown, Jr., R.M. (2004) *Cellulose*, **11**, 463–474.
8 Kondo, T., Kasai, W. (2014) *J. Biosci. Bioeng.*, **118**, 482–487.

9 Kasai, W., Kondo, T. (2004) *Macromol. Biosci.*, **4**, 17–21.
10 Higashi, K., Kondo, T. (2012) *Cellulose*, **19**, 81–91.
11 Seyama, T., Suh, E-Y. Kondo, T., (2013) *Biofabrication*, **5**, 025010.
12 Horii, F., Hirai, A., Kitamaru, R. (1983) *Polym. Bull.*, **10**, 357–361.
13 Togawa, E., Kondo, T. (1999) *J. Polym. Sci., B:Polym. Phys*, **37**, 451–459.
14 Hishikawa, Y., Togawa, E., Kataoka, Y., Kondo, T. (1999) *Polymer*, **40**, 7117–7124.
15 Hishikawa, Y., Togawa, E., Kondo, T. (2010) *Cellulose*, **17**, 539–545.
16 Togawa, E., Kondo, T. (2007) *J. Polym. Sci., B:Polym. Phys.*, **45**, 2850–2859.
17 Kondo, T., Kasai, W., Brown, Jr., R.M. (2004) *Cellulose*, **11**, 463–474.
18 Minke, R., Blackwell, J. (1978) *J. Mol. Biol.*, **120**, 167–181.
19 Brown, Jr., R.M., Willison, J.H.M., Richardson, C.L. (1976) *Proc. Nat. Acad. Sci. U.S.A.*, **73**, 4565–4569.
20 Thompson N.S., Kaustinen, H.M., Carlson, J.A. Uhlin, K.I. (1988) *Int. J. Biol. Macromol.*, **10**, 126–127.
21 Tokoh, C.K.M., Takabe, M., Fujita Saiki, H. (1998) *Cellulose*, **5**, 249–261.
22 Benziman, M., Haigler, C.H., Brown, Jr., R.M. et al. (1980) *Proc. Nat. Acad. Sci. U.S.A.*, **77**, 6678–6682.
23 Haigler, C.H., Brown, Jr., R.M., Benziman, M. (1980) *Science*, **210**, 903–906.
24 Haigler C.H., White A.R., Brown, Jr., R.M., Cooper, K.M. (1982) *J. Cell. Biol.*, **94**, 64–69.
25 Yamamoto H., Horii F., Hirai A. (1996) *Cellulose*, **3**, 229–242.
26 Tomita, Y., Kondo, T. (2009) *Carbohydr. Polym.*, **77**, 754–759.
27 Nemoto, J., Uraki, Y., Kishikoto, T. et al. (2005) *Biotech.*, **96**, 1955–1958.
28 Uraki Y., Nemoto J., Otsuka H. et al. (2007) *Carbohydr. Polym.*, **69**, 1–6.
29 Rho, J.Y. Kuhn-Spearing L., Zioupos, P. (1998) *Med. Engine. Phys.*, **20**, 92–102.
30 Kato, T., Sugawara, A., Hosoda, N. (2002) *Adv. Mater.*, **14**, 869–877.
31 Ikada, Y., Tamada Y. (1986) In *Polymers in Medicine II*, E. Chiellini, P. Giusti, C. Migliaresi, L. Nicolais (eds.), Plenum, New York, pp. 101–115.
32 Doillon, C.J. (1992) *Clin. Mater.*, **9**, 189–193.
33 Hutmacher, D.W., Vanscheidt, W. (2002) *Drugs of Today*, **38**, 113–133.
34 Brännvall, K., Bergman, K., Wallenquist, U. et al. (2007) *J. Neurosci., Res.*, **85**, 2138–2146.
35 Uchino, T., Takezawa, T., Ikarashi, Y. (2009) *Toxicology in Vitro*, **23**, 333–337.
36 Zhong, S.P., Zhang, Y.Z., Lim, C.T. (2010) *WIREs Nanomed. Nanobiotech.*, **2**, 510–525.
37 Price, R.D., Berry, M.G., Navsaria, H.A. (2007) *J. Plast. Reconstr. Aesthet. Surg.*, **60**, 1110–1119.
38 Tezel, A., Fredridkson, G.H. (2008) *J. Cosmet. Laser Ther.*, **10**, 35–42.
39 Scuderi, N., Anniboletti, T., Carlesimo, B., Onesti, M.G. (2009) *Our Experience In Vivo*, **23**, 991–1004.
40 Tchemtchoua, V.T. et al. (2011) *Biomacromolecules*, **12**, 3194–3204.
41 Okabayashi, R., Nakamura, M., Okabayashi, T. et al. (2009) *J. Biomed. Mater. Res. B. Appl. Biomat.*, **90B**, 641–646.
42 Wendt, H., Hillmer, A., Reimers, K. et al. (2011) *PLoS ONE*, **6**, e21833.
43 Klemm, D, Heublein, B., Fink, H.P., Bohn, A. (2005) *Angew. Chem. Int. Ed.*, **44**, 3358–3393.
44 Helenius, G., Bäckdahl, H., Bodin, A. et al., (2006) *J. Biomed. Mater. Res.*, **76A**, 431–438.
45 Dugan J.M., Collins R.F., Gough J.E., Eichhorn, S.J. (2013) *Acta Biomater.*, **9**, 4707–4715.
46 Dugan J.M., Gough J.E., Eichhorn, S.J. (2010) *Biomacromolecules*, **11**, 2498–2504.

5

Preparation and Application of Biomimetic Materials Inspired by Mussel Adhesive Proteins

*Heng Shen, Zhenchao Qian, Ning Zhao, and Jian Xu**

Beijing National Laboratory for Molecular Sciences, Chinese Academy of Sciences, Beijing, China

5.1 Introduction

Mussels are a kind of common marine organisms. They exhibit strong adhesive ability and can attach themselves to the surface of boats, rocks, and other organisms, bearing the impact of storm and waves. Surprisingly, mussels can adhere to almost all kinds of substrates, even including the surface of Teflon which is considered a non-sticky surface due to its low surface energy. Their versatile, strong, and long-term adhesion in a wet environment attracts a lot of research attention.

It is found that the mussels adhere to the solid surface through byssus in a radial pattern, and every byssus forms an adhesive disk of several square millimeters with substrate. Allen and Wilker [1, 2] measured the adhesion strength of a single adhesive disk of different mussels to the rock surface and found it was between 0.3–6 MPa. There are 25–30 kinds of proteins in byssus, and 7 or 8 proteins exist in the area of the adhesive disk. These specific proteins are called mussel foot proteins (mfp) by Waite and co-workers [3]. Among these mfps, mfp-1 [4, 5] is mainly distributed in the cuticle and it is related to the high strength and high toughness of the byssus but is not involved in the interface adhesion. Mfp-2 [6, 7] and mfp-4 [8] are mainly distributed in the interior of the adhesive disk and they are related to the cross-linking and solidification, which increase the cohesive energy. Mfp-3 [9] and mfp-5 10] are the major proteins distributed on the interface of the byssus and the substrate. Although mfp-6 [11] is also distributed on the interface, it is responsible for connecting proteins on the interface and the interior. After analyzing the amino acid sequence of mfp, the fact that all of these mfps contain a kind of amino acid called 3, 4-dihydroxyphenyl-L-alanine (Dopa) is revealed. Especially in mfp-3 and mfp-5, the content of Dopa is as high as 20% and 28%, respectively.

Figure 5.1 shows the possible interactions between Dopa and various substrates, including metal ions, metal oxides and organic groups [12]. The extensive reactivity of mussels is responsible for their versatile adhesion. In the 1980s, Koide et al. found that the byssus of mussel could enrich metal ions from waste water [13] and the content of the inorganic component (Fe, Mn, Cu, Zn, V, Al and Ti) was up to 2 wt% [14]. Taylor et al. [15, 16] studied the coordination interaction between peptide containing Dopa and Fe^{3+} using Raman spectra. This revealed that the bis- or tris-coordination complex was formed by changing the ratio of Dopa and Fe^{3+}. When the mole ratio of Dopa to Fe^{3+} was 3:1, the stability constant of the formed tris-coordination catechol/Fe complex in a neutral solution with 0.1 M NaCl exceeded ^{1040}. Further research found that the association constants of catechol group with other metal ions were also very high [17]. The stability constants of Dopa with solid metals or metal oxides are high as well. Dalsin et al. [18] detected that charge transfer complexes were formed when a model compound containing Dopa made contact with the surface of titanic oxide and iron oxide by different detection methods. The coordination bond between Dopa and the metal oxides which can replace the hydration on the interface is the key factor in adhesion for the mussel adhesive proteins. For organic substrates, the high combining capacity of Dopa with substrates is due to the high reactivity of oxidized dopaquinone. The dopaquinone can react with Dopa to form free radicals and couple to form di-Dopa structure [19, 20]. It also can react with amino, thiol and imidazole via Michael-addition or Schiff base reaction to form covalent bonds [21, 22]. These reactions not only enhance the interactions between mfps and organic substrates but also lead to the cross-linking of mfps, resulting in the enhancement of the cohesive strength. In addition, the catechol of Dopa

* Email: jxu@iccas.ac.cn

Bioinspired Materials Science and Engineering, First Edition. Edited by Guang Yang, Lin Xiao, and Lallepak Lamboni.

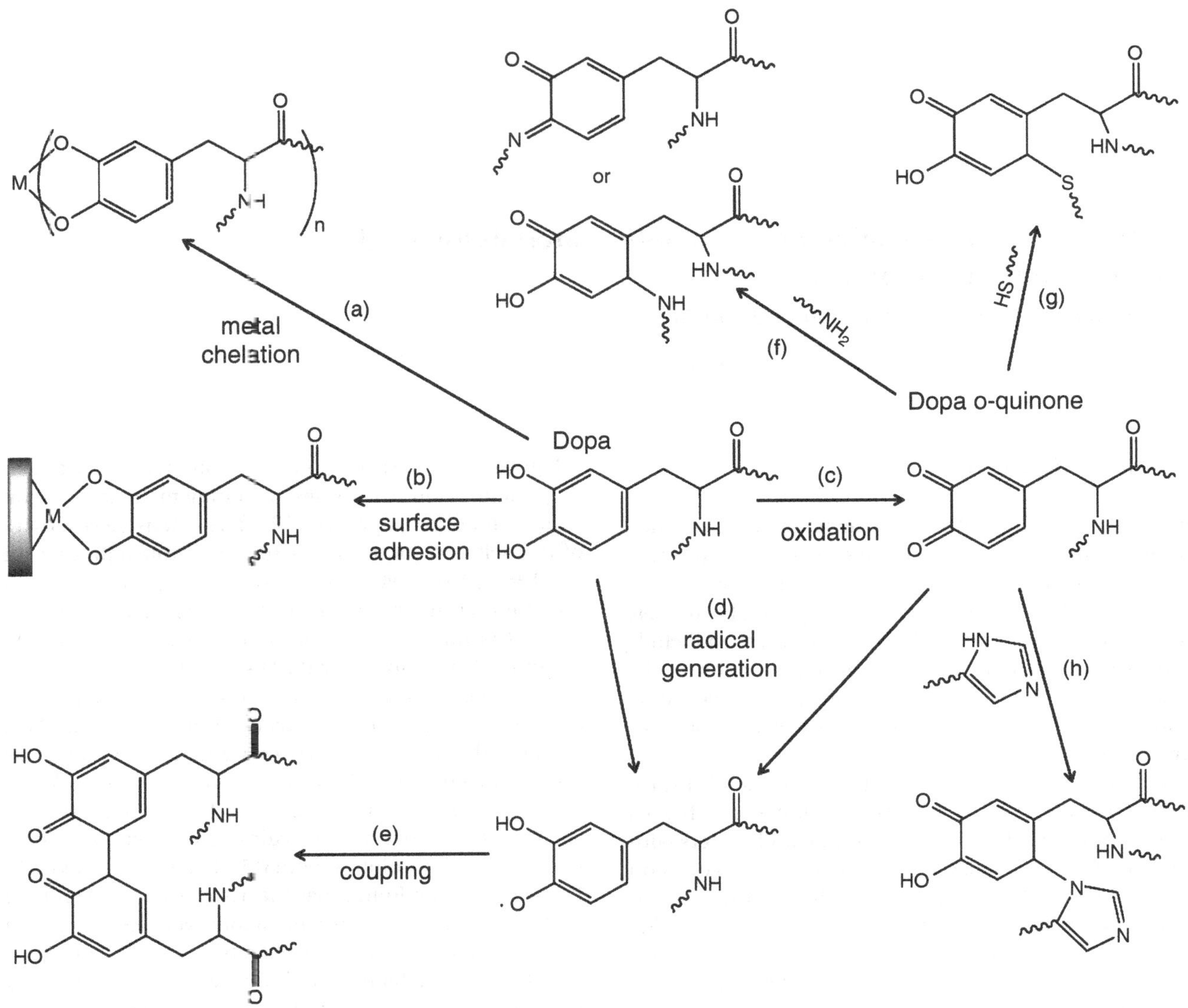

Figure 5.1 Hypothetical adhesion, oxidation and cross-linking reaction pathways between Dopa and different substrates [12].

can easily form a hydrogen bond with polar groups, and a benzene ring may form π-π stacking with other conjugated structures. Recently, researchers found that the mfps usually formed a complex with polyanionic proteins via electrostatic force [23, 24]. These protein fluids, with high concentration (1–2 g/mL) formed by phase separation, have low surface energy and are easy to spread out on the surface. This is an advantage for underwater adhesion.

5.2 Various Research Studies

Inspired by the strong and long-term adhesion of mussels in a wet environment, great efforts have been made to synthesize polymers with the catechol group as novel adhesives in order to improve adhesion in a wet environment since most conventional adhesives show poor adhesion when they become wet. Deming et al. [25] synthesized the copolymer composed of Dopa and lysine and studied the adhesive ability between aluminum and steel. It was found that the copolymer showed strong adhesion under water and the adhesive strength was related with the content of Dopa, the component of oxidant and the oxidation condition. Payne et al. [26] mixed chitosan and dopamine, oxidized by enzyme, and the shear adhesive strength under water of the prepared adhesive was as high as 400 kPa. Messersmith et al. [27] prepared a PMMA-PMAA-PMMA triblock copolymer with DOPA groups and investigated the adhesive properties to the surface of TiO_2 and tissue. The adhesive strength decreased after several recycles, because of the residue of adhesive on the surface due to the strong interaction between Dopa and the contacted surface. In addition, the addition of oxidants reduced the adhesion for TiO_2 while it enhanced that for tissue. It was speculated that after Dopa was oxidized from catechol to quinone, the non-covalent bond with TiO_2 reduced while the covalent bond formed with tissue.

Wilker and et al. [28–30] synthesized a series of poly[(3,4-dihydroxystyrene)-co-styrene] and studied the adhesive properties under different modes of cross-linking. By adjusting the component of the monomer, the molecular weight of the polymer and the cross-linking condition, the highest shear adhesive strength on aluminum reached 7 MPa, which was comparable with cyanoacrylate. The authors added the cationic styrene monomer to synthesize the terpolymer in order to investigate the effect of the electrostatic interaction on the adhesion, and it revealed that the wet adhesion was preferable to the commercial adhesive. The optimal content of the cationic styrene monomer was 7%.

Lee et al. [31] combined the properties of physical dry adhesion of geckos resulting from the micro- and nano-structure and the chemical wet adhesion of mussels resulting from the Dopa to fabricate a hybrid adhesive. They employed arrays of gecko-mimetic nanoscale pillars coated with a thin mussel-mimetic polymer film, which was composed of poly(dopamine methacrylamide-co-methoxyethyl acrylate) (p(DMA-co-MEA)) (Figure 5.2a). The wet adhesion of the hybrid adhesive was enhanced 15-fold compared with that without the mussel-mimetic polymer coating, and the property had only slightly decreased after over 1000 adhesion cycles (Figure 5.2b, c).

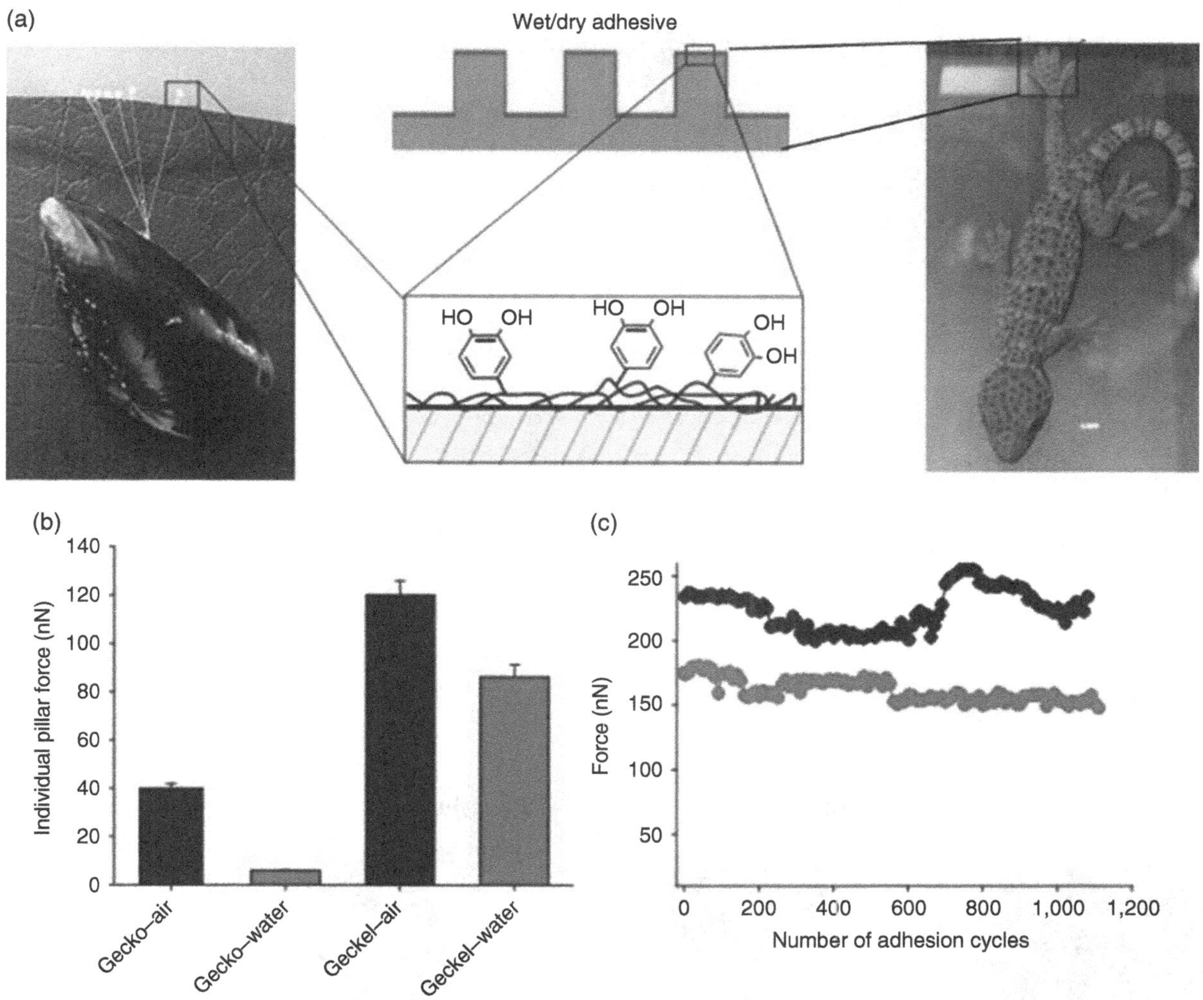

Figure 5.2 (a) Scheme of the gecko- and mussel-inspired wet/dry adhesive. (b) Adhesion forces of the adhesive with and without mussel-mimetic polymer in wet and dry condition. (c) Performance of geckel adhesive during multiple contact cycles in water (red) and air (black) [31].

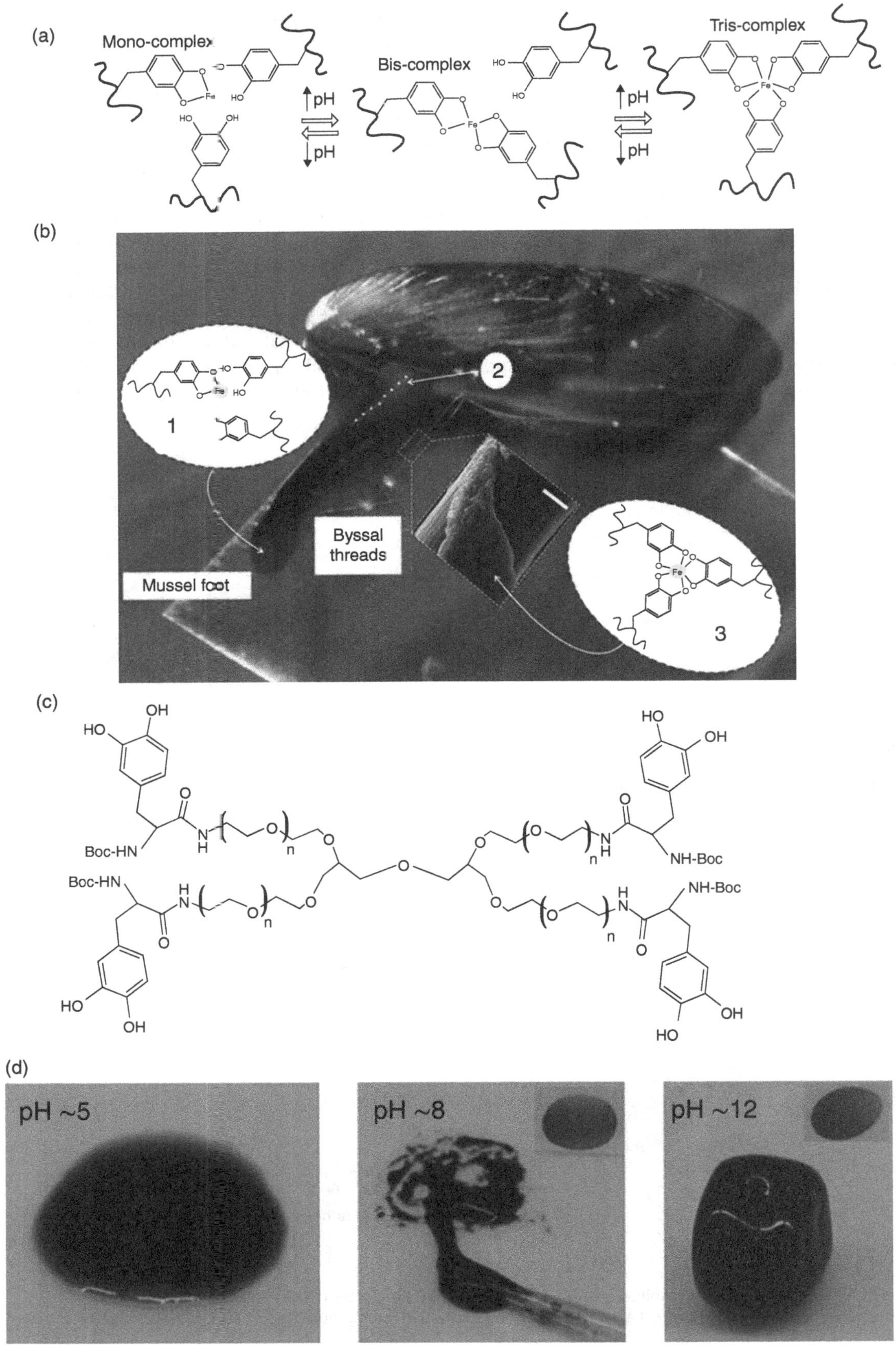

Figure 5.3 (a) The pH-dependent stoichiometry of Fe^{3+}-catechol complexes. (b) Schematic of proposed cross-linking mechanism of byssal thread cuticle. (c) PEG-Dopa$_4$. (d) Physical state and color of PEG-Dopa$_4$ gels in different pH [40].

Glass et al. [32, 33] further increased the wet adhesion by optimizing the size of the pillars, the thickness of the mussel-mimetic polymer coating, and the degree of cross-linking.

Special prepared mussel-mimetic polymers are qualified to be medical adhesives. In this field, they usually must meet several special demands: (1) be soluble in biological fluids; (2) offer sufficient wet adhesion; (3) have controllable rapid gelation in seconds; (4) have adjustable viscoelasticity; and (5) have excellent biocompatibility.

Messersmith et al. [34, 35] designed various adhesive hydrogels with line, star and branching PEGs which were modified with Dopa. The effects of the cross-linking condition, the molecular weight of PEGs and the topological structure of the polymer chain on the gelation time and gel structure were investigated. They also synthesized the adhesive hydrogel via light-induced polymerization. This strategy, without extra oxidants, was efficient enough to avoid the decrease of the adhesion caused by oxidization, and the elastic modulus of the gel was controllable by adjusting the dosage of the cross-linking agent to match that of biological tissue [36]. They applied these kinds of bio-mimetic adhesives to the application of living organism. For instance, they synthesized a kind of adhesive precursor polymer consisting of a four-armed poly(ethylene glycol) (PEG) core, whose end groups were derived from catechol, and the *in vivo* performance was assessed in a murine model of extrahepatic syngeneic islet transplantation. It was found that the catechol-derived PEG (cPEG) solutions gelled in 30 seconds under oxidation conditions and no obvious inflammatory reaction was elicited in one year [37]. This adhesive polymer was also successfully used for the closure of iatrogenic membrane defects [38]. They also synthesized a series of adhesive polymers composed of Dopa, PEG and polycaprolactone (PCL). These polymers were cast onto the bovine pericardium and used in hernia repair. It was revealed that the adhesive strength was significantly higher than fibrin glue and the mechanical strength, swelling degree, or degradation rate could be tailored by the composition of the PCL [39].

Metal ions play another critical role in mussel adhesion. Taking advantage of the strong and reversible coordination interaction of Dopa with Fe^{3+}, various self-healing hydrogels have been developed. Waite et al. [40] developed a simple strategy to control catechol-Fe^{3+} interpolymer cross-linking via pH (Figure 5.3). They prepared the self-healing hydrogel by mixing synthetic four-armed PEG-$Dopa_4$ with Fe^{3+}. With the increase of the pH from 5 to 12, the complex turned to tris-complex from mono-complex, gradually leading to the increase of the cross-linking degree. The mechanical strength was comparable to that of the hydrogel formed by covalent cross-linking. Similar self-healing hydrogels with different macromolecules as a skeleton, such as chitosan [41], gelatin [42], polyallylamine [43], and polyacrylamide [44], were also reported. Campo et al. [45] synthesized PEG-ND_4 using nitrodopamine (ND) instead of Dopa, and fabricated self-healing hydrogel by cross-linking with Fe^{3+} as well. At the same time, the nitrodopamine was cleaved under ultraviolet irradiation, resulting in the degradation of the hydrogel. This evidenced underwater bonding and unbonding on demand and had potential applications in multiple cell and medical fields. Kohane et al. [46] synthesized PEG4-dopmanine with low molecular weight of four-armed PEG (2000 Da). This liquid polymer changed to an elastic solid with good biocompatibility and biodegradation after being mixed with a small amount of Fe^{3+} solution and the elastic property could be tailored by adjusting the Fe^{3+} concentration. Both hydrophilic and hydrophobic drugs could be incorporated into the elastomer and released slowly, indicating its potential application in drug delivery and other biomedical fields.

Besides the reversible interaction between catechol and Fe^{3+}, boric acid can be used as the reversible cross-linker with the catechol group as well. Messersmith et al. [47] fabricated pH-responsive and self-healing hydrogels based on a boronate-catechol complexation (Figure 5.4a). When 1,3-benzenediboronic acid was added, the liquid synthetic cPEG catechol transformed into hydrogel under basic aqueous conditions due to the cross-linking by forming a tetrahedral borate ester with the catechol end groups, and this cross-linking network could be dissociated in and acidic condition. Stadler et al. [48] found that covalent gels could be formed by adding $NaBH_4$ to a copolymer containing catechol groups in DMF (Figure 5.4b). The gels could adsorb humidity and the $NaBH_4$ would be hydrolyzed to H_3BO_3, which made the covalent bonds from irreversible to reversible. They proved that water played a key role in reversible cross-linking between boric acid and catechol.

Recently, Waite et al. [49] developed a new strategy to fabricate self-healing polymers through a surface-initiated hydrogen bond between interfacial catechol groups without the assistance of Fe^{3+} or boric acid (Figure 5.5). The polymer was synthesized from triethylsilane-blocked catecholacrylate. The polymer rod was cut into two parts, and after being immersed in the low pH buffer, the silyl-blocked catechol was hydrolyzed and the catechol groups at the interface were exposed. Complete healing was realized after rejoining the cut rods in water due to the interfacial hydrogen bonding from the exposed catechol groups. These self-healing polymeric materials without additional additives, such as metal ions or boric acid under wet conditions, show significant promise in biomedical and environmental applications.

Metal oxides nanoparticles have been widely used in various fields, including materials, energy, catalysts, and

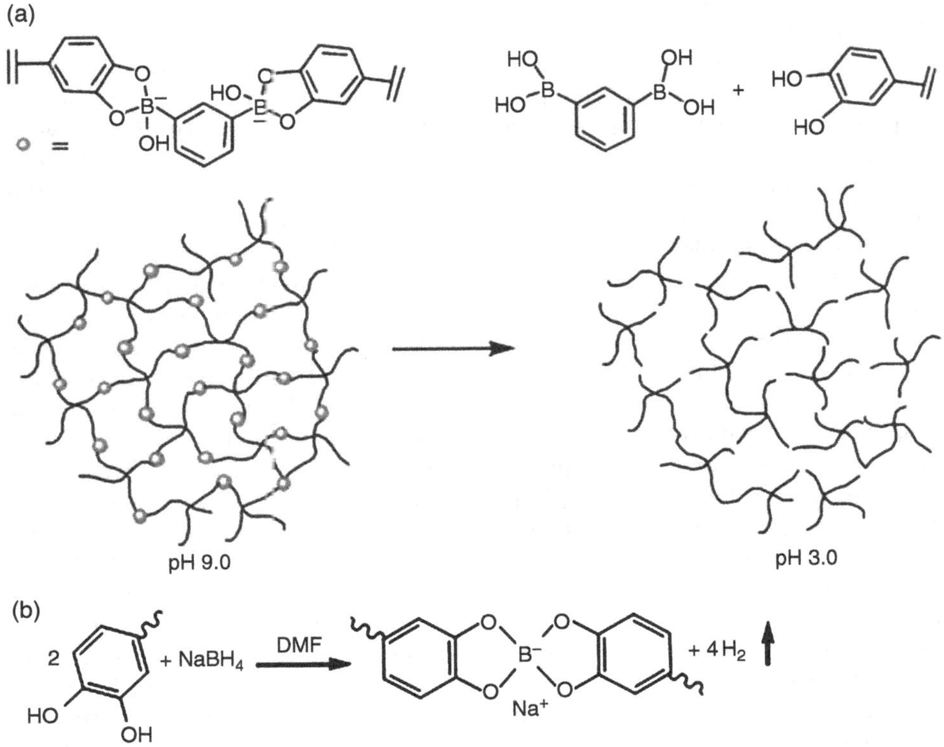

Figure 5.4 (a) pH-responsive hydrogel based on cPEG and 1,3-benzenediboronic acid [47]. (b) gel structure in DMF [48].

biomedicine. However, nanoparticles tend to aggregate as a result of their small size and high surface energy. This disadvantage usually limits their applications, thus surface modification of nanoparticles is essential not only to avoid aggregation but also to introduce the unique properties of the modifying agent. Catechol derivatives are efficient for bidentate coordination with metal ions, especially with ferric ions. Various stabilizers with bioactive molecules and catecholic anchors that modified NPs have been explored. Their long-lasting stability in a physiological environment and their biocompatibility make the NPs potentially useful in biomedical applications such as MRI imaging [50–52] and cancer therapy [53, 54].

Of these successful research studies, Xu et al. are one of the first research groups to apply superparamagnetic nanoparticles for potential therapeutic work with catecholic anchors. The authors developed an easy method to fabricate a stable conjugate of porphyrin and magnetite nanoparticles with dopamine as a surface anchor. The modified nanoparticles could induce the cell apoptosis when exposed to yellow light after penetrating the cytoplasm of the cancer cell line HeLa cells [53]. Similarly, Park et al. immobilized hyaluronic acid (HA) with dopamine as the surface anchor onto the magnetite nanocrystals [54]. Bawendi et al. coated superparamagnetic Fe2O3 nanoparticles with a synthetic zwitterionic dopamine sulfonate (ZDS) ligand [51]. Mattoussiet et al. coated iron oxide nanocrystals with multidentate catechol-PEG anchors [52]. Schmalz et al. employed PDMAEMA chains as the coating of Fe2O3 via ATRP using a dopamine-functionalized initiator as an anchor [55]. Frey et al. coated MnO nanoparticles with synthetic PEG-catechol ligands through oxyanionic polymerization [56]. These hybrid nanoparticles were suitable for biomedical applications because of their long-term stability and biocompatibility.

Hyeon et al. [50] synthesized a new kind of multiple-interaction ligand (MIL), consisting of methoxypoly(ethylene glycol) (mPEG), hyperbranchedpolyethylenimine (bPEI) and poly(L-3,4-dihydroxyphenylalanine) (polyDOPA). Different nanoparticles, for instance, Fe_3O_4, MnO and Au, coated with one kind of MIL called MIL2 (the molar ratio of each segment PEG: PEI: DOPA = 8: 1: 15) were highly stable in harsh aqueous environments, such as highly acidic and basic media, a highly concentrated NaCl solution, and even boiling water. The authors suggested this ultra-stability resulted from the cooperative work of different binding

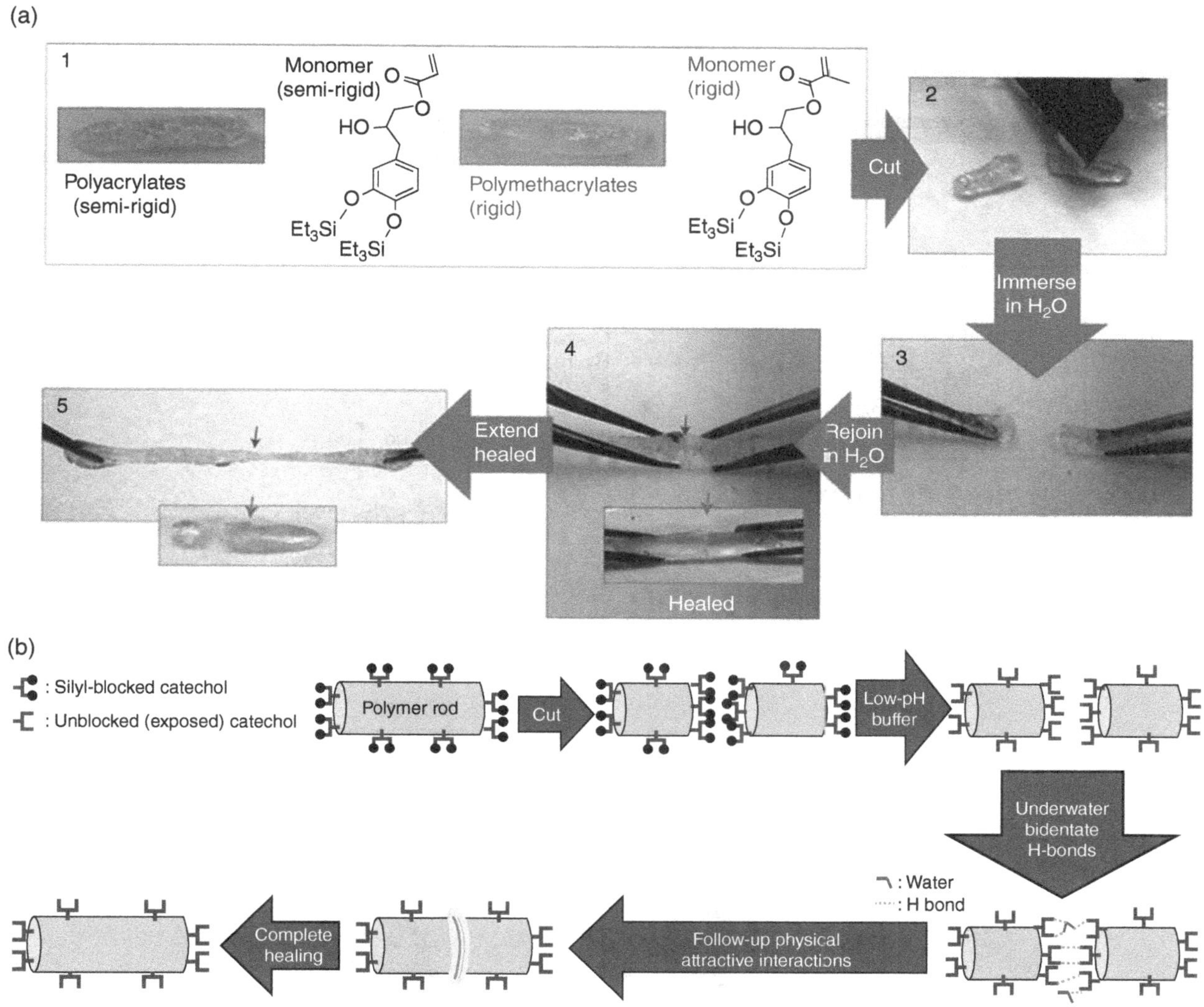

Figure 5.5 (a) Photos and (b) schematic diagram of the self-healing polymer [49].

modes, including simultaneous catechol binding and amine binding, micelles formation because of the amphiphilic hyperbranched block copolymer structure which could encapsulate nanoparticles, and the electrostatic interaction between positively charged bPEI moiety and negatively charged nanoparticles (Figure 5.6a). The MIL2-coated Fe_3O_4 nanoparticles were successfully employed *in vivo* MRI application, which confirmed these types were suitable for biomedical applications.

Most of the catechol derivatives offer an active end group, like carboxyl or the amino group, which result in being easily conjugated to other functional moieties. When functional molecules are introduced, the coated NPs may exhibit more unique characteristics.

Klajn et al. [57] coated superparamagnetic Fe_3O_4 nanoparticles with a synthetic ligand azobenzene-catechol (AC), giving the hybrid nanoparticles a dual-responsive property due to the superparamagnetic cores and photoresponsive ligands as a result of the introduction of azobenzene. The functional nanoparticles could self-assemble into larger spherical clusters in solvents of low dielectric constant when exposed to UV light, due to the attractive electric dipole interactions between the nanoparticles resulting from a rapid *trans* → *cis* isomerization of the azobenzene. This aggregation was reversible. Taking the small diameter of the cores (11.3 nm) into consideration, the magnetic dipole interaction was too weak to overcome the thermal energy, so the NPs did not aggregate in an external magnetic field. The magnetic dipole interaction proportional to r^6 increased significantly with the increasing diameter when spherical clusters formed upon exposure to UV light. In that situation, the spherical clusters could assemble into threads in an external magnetic field. Once the threads formed, the

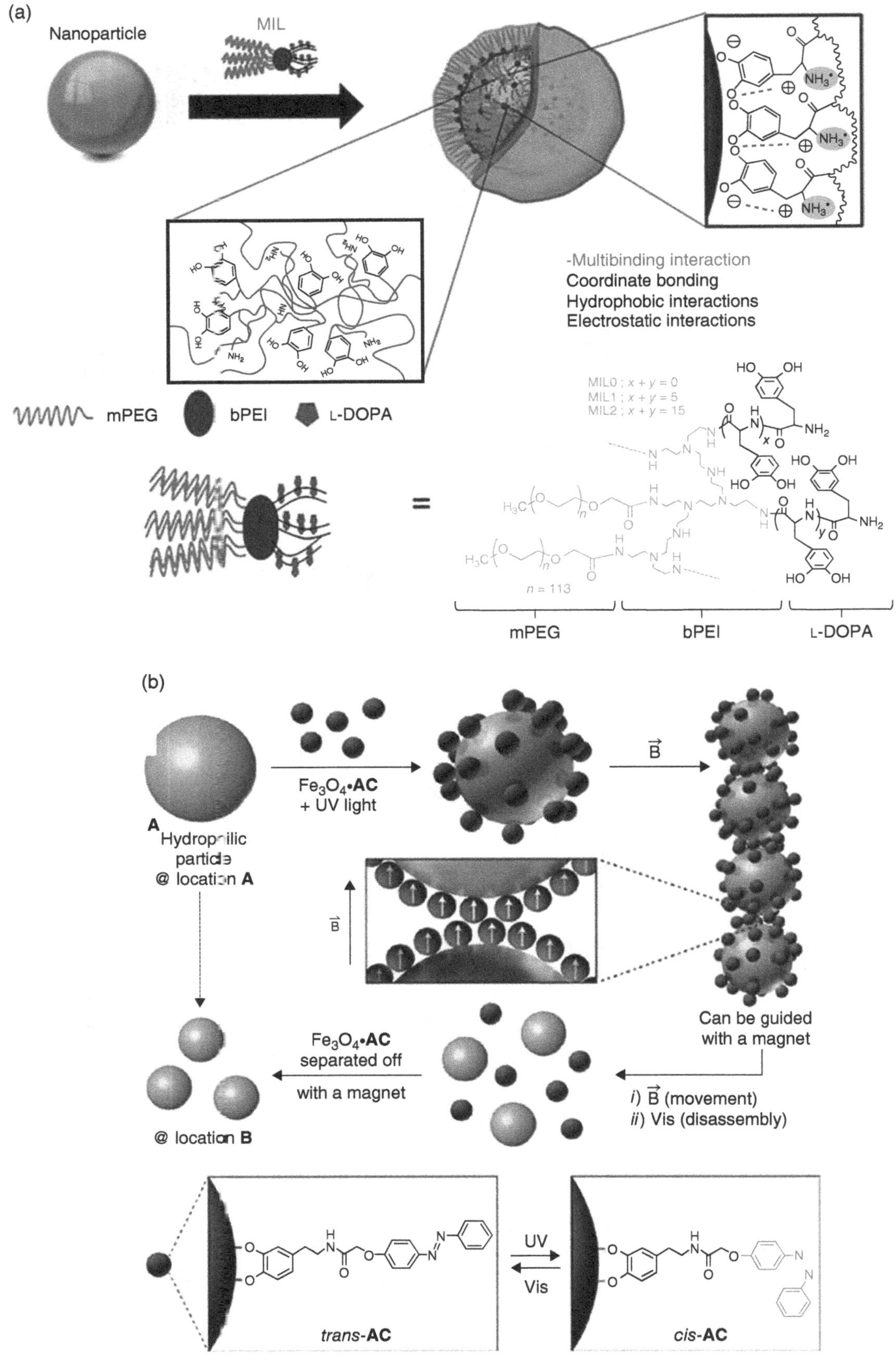

Figure 5.6 Schematic drawing of (a) water-dispersible nanoparticles coated with multiple-interaction ligand (MIL) [50]. (b) delivery of diamagnetic particles with magnets by the reversible, light-controlled formation of a dynamic paramagnetic coating on their surfaces, resulting from the isomerization of azobenzene [58].

removal of the magnetic field had no influence on them, while the threads disassembled when UV irradiation stopped. In a follow-up study, these dual-responsive hybrid paramagnetic nanoparticles were applied to transport diamagnetic particles. In a mixture of solvents with intermediate polarity, the hybrid nanoparticles formed a thin coating on the surface of diamagnetic particles under UV light irradiation, the aggregates could be guided with a magnet to the desired location. After that, the aggregates disintegrated to release the diamagnetic particles when exposed under visible light instead of UV light (Figure 5.6b) [58].

Morey et al. reported on the iodide squaramide ammonium-functional iron NPs with dopamine moiety as an anchor. The hybrid NPs could be employed to detect mono and dicarboxylates in water but not tricarboxylates due to the selective coordination [59].

These synthetic polymers containing catechol groups inspired by mussel adhesive proteins show an excellent performance and wide applications. However, the synthetic process is usually complicated and expensive. Messersmith et al. [60] proposed that a dopamine molecule could be considered one kind of mussel-inspired molecule because it contained catechol and amino groups simultaneously as shown in Figure 5.7. They found that dopamine could self-polymerize to polydopamine (PDA) in an alkaline solution and was deposited on almost all kinds of substrates, including PTFE, which was just like the mussel adhesive proteins, and functionalization could be conducted easily followed by the PDA deposition. This strategy has attracted extensive attention due to its simple process and versatile applications.

Nevertheless, PDA has a complex structure, and it is not soluble or can be melted. The clear structure of PDA has not been demonstrated so far due to the difficulties in characterization. Figure 5.8 shows the possible polymerization process and structure model of PDA. Messersmith et al. [60] found a trimeric conjugate of 5,6-dihydroxyindole (DHI) and fragments of hydroxyl and phenyl groups by time-of-flight secondary ion mass spectrometry

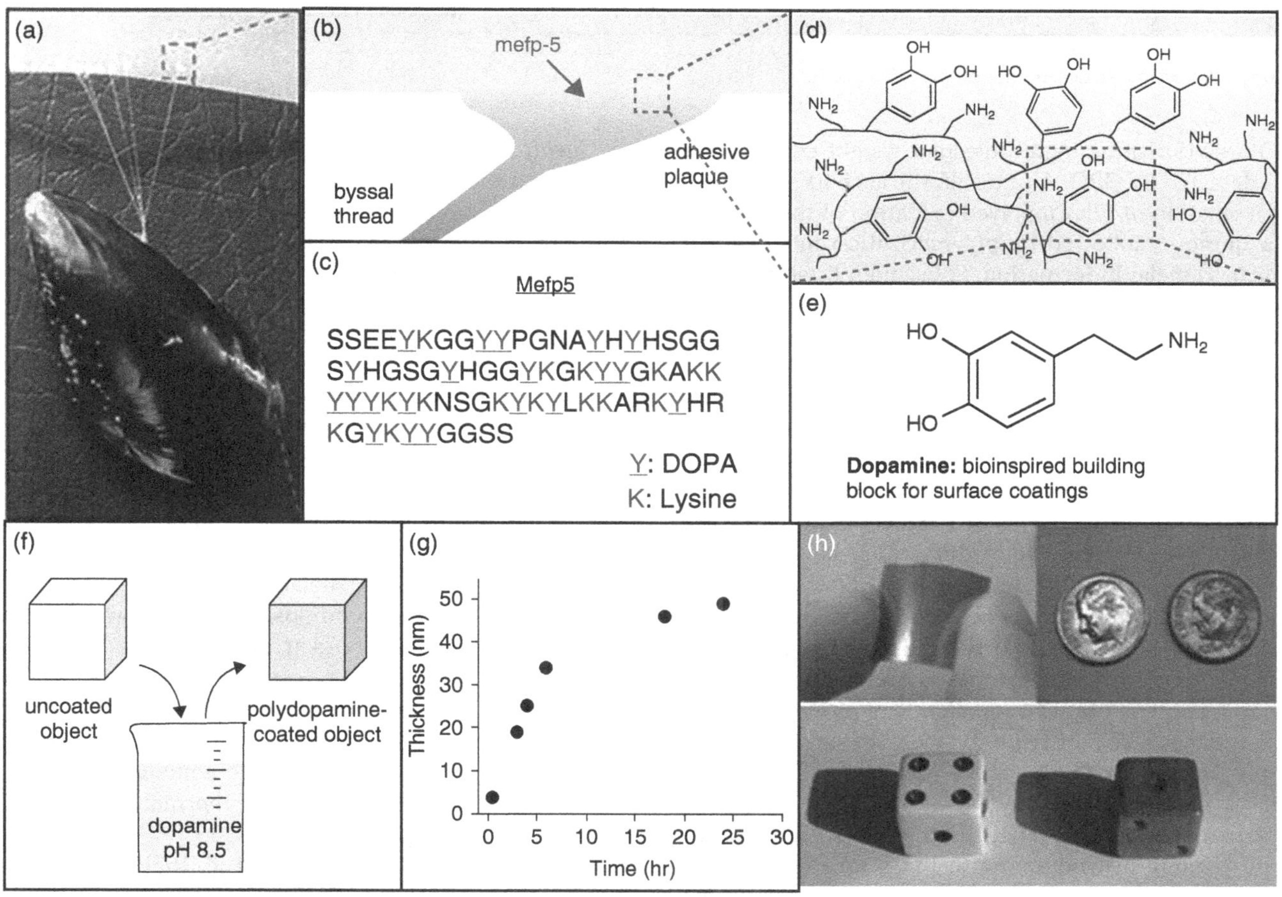

Figure 5.7 (a–d) Photos of mussel and adhesive protein mfp-5. (e) The molecule structure of dopamine. (f) Modification process of PDA coating. (g) The thickness of PDA coating with reaction time. (h) Electroless metallization of PDA-coated substrates [60].

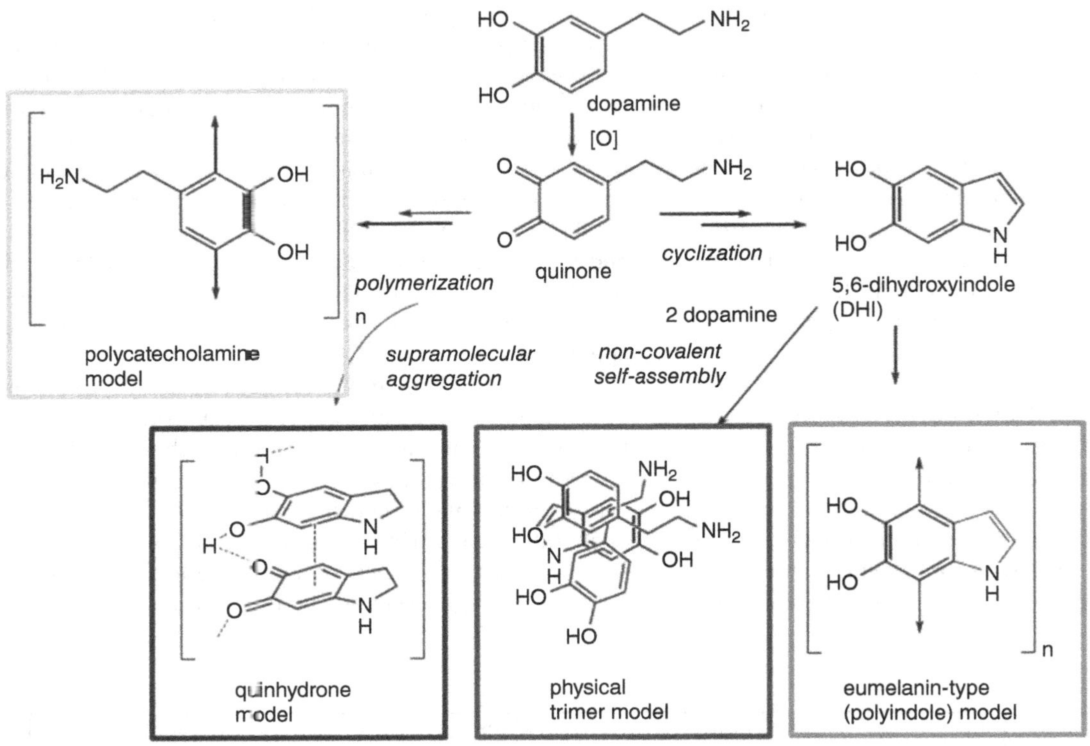

Figure 5.8 Possible structure models of PDA [61, 62].

(TOF-SIMS), and a high molecular weight composition was detected by GPC. As a result, the authors speculated the formation of PDA involved oxidation of the catechol to a quinone, followed by polymerization in a manner similar to melanin formation. Subsequently, after in situ monitoring the polymerization process by ^{1}H NMR and analyzing with the results of high performance liquid chromatography (HPLC), and mass spectrum (MS), they amended the covalent model and proposed that the non-covalent self-assembled trimer of (dopamine)$_2$/DHI was another important composition of PDA besides the conjugates of DHI [61]. Dreyer et al. [62] considered that the structure unit of DHI was self-stacking through non-covalent interactions, including multiple hydrogen bonds, π-π stacking and charge transfer by the results of ^{15}N, ^{13}C NMR, XRD and FTIR.

Although the clear structure of PDA has not yet been demonstrated, it has been widely used in various fields due to its unique physical and chemical properties. PDA is considered a kind of synthetic melanin and it has similar properties to melanin:

1) *Reactivity*: there are a large number of structure units of quinone/hydroquinone in PDA, as a result, PDA processes redox ability. It can easily react with amino and thiol groups and reduce novel metal ions, such as Au, Ag and Pt.
2) *Chelating ability*: PDA has a strong chelating ability with various metal ions such as Fe^{3+}, Ti^{4+}, Ca^{2+}, Mn^{3+}, Zn^{2+}, Cu^{2+}, etc., due to its amino and catechol groups.
3) *pH stimuli-responsibility*: because of the different pKa of amino and catechol groups, PDA has the stimuli-responsibility. Yu et al. [63] revealed that the PDA film possessed permeation selectivity of different ions under different pH conditions, the anion could permeate under acidic condition, while the cation could permeate under basic conditions. The isoelectric point is approximately 4 [64].
4) *Free radical property*: although the free radical property has not been studied a lot, some research has demonstrated that PDA has the ability to eliminate free radicals similar to melanin. Phua et al. [65] fabricated a polypropene composite with clay, which was modified with PDA, and it improved the ability of UV resistance of the composite.
5) *Stability*: PDA has superior structural and thermal stability. It is not soluble in common solvents and only can be dissolved in strong basic aqueous (pH > 12). The carbon residue of PDA after thermal decomposition under inert gas is up to 60% [66].
6) *Optical property*: Bernsmann et al. [67] studied the UV-visible spectra of PDA. They revealed that the absorption spectra covered the UV to visible area, and the absorbance decreased with the increase of

wavelength, which was similar to melanin. However, different from melanin, the absorbance of PDA was not linear in wavelength in logarithmic coordinates, indicating that the heterogenization degree of PDA was less than melanin and it could not ensure the complete absorption of light. Lu et al. [68] found that the light absorption of PDA particles could be extended to the near-infrared area, and the high energy infrared laser could be transformed into heat energy. Recently, some research has found the photoluminescence of PDA nanoparticles after appropriate treatment. Wei et al. [69] prepared the wormlike nanoparticles sized 10–100 nm by treating with hydrogen peroxide for 5 h after the polymerization of dopamine in a tris buffer for 15 min. The obtained nanoparticles exhibited fluorescence after excitation by UV-visible light.

7) *Biocompatibility*: PDA has been demonstrated to be biocompatible. This property makes it preferable in bio-medical fields.

Because of these unique properties of PDA, it is widely used in surface modification and functionalization. Surface wettability is one of the most important research issues in surface science. PDA is hydrophilic and its reactivity makes it easy to adjust the surface wettability. Lee et al. [70] modified the low surface energy substrates including PTFE, graphene and Au, and realized the surface patterning of a block copolymer via photolithography. The authors also modified the superhydrophobic surface by this method, and fabricated the hydrophilic/superhydrophobic patterning surface inspired by beetles [71] and a micro-channel device driven by gravity with high flux [72]. PDA can also be used to modify porous membranes to improve hydrophilicity and flux [73–75]. Park and Choi et al. [76, 77] modified the PP separator for Li battery with PDA, and it not only improved the wettability and adsorbing capacity of the polar electrolyte but it also decreased the resistance of the membrane and enhanced the discharge current density of the battery.

The PDA-modified surface can also be used in the preparation of a conductive layer. Xu et al. [78] developed a novel strategy to fabricate a conductive silver film on different substrates (Figure 5.9a). By depositing a PDA layer on substrates, they obtained a layer of Ag nano-seeds taking advantage of the high reducing ability of the PDA. After the deposition of more Ag nanoparticles and immersion in electrolyte solutions, the Ag nanoparticles would sinter at room temperature to form a conductive thin film. Zhang et al. applied a similar method to deposit a conductive layer with Ag nanoparticles on the surface of rubber [79] and Kevlar fibers [80]. Akter et al. [81] sprayed the Ag nano wires on the surface of PDMS modified with PDA to fabricate a flexible and transparent conductive coating, making use of the strong interaction between PDA and Ag nano wires. PDA can induce the growth of an inorganic layer on the surface as well. Ou et al. [82] successfully deposited the TiO_2 film on the surfaces of PTFE, PE and PET modified with PDA via the sol-gel method to improve the biocompatibility of the substrates. Park et al. [83, 84] induced the mineralization of hydroxylapatite on various planes and fiber substrates by taking advantage of the universal adhesion and chelating ability of PDA. Rai et al. [85] developed a universal strategy to prepare the mineralization surface of silicon. They modified the surface of Au, PS and a silicon wafer by PDA and decapeptide, and silicatein was applied to fabricate a uniform silica film (Figure 5.9b). Kang et al. [86] used 2-dimethylaminoethanethiol instead of silicatein to fabricate a biomimetic silica coating to the PE separators. The modified porous membrane was used as the separator of a Li ion battery and it was found that the wettability of the electrolyte and the thermal stability were greatly enhanced.

The bioactive component is usually attached to the surface modified with PDA to fabricate the bioactive surface, improving the affinity between materials and cell, protein, or bacteria. Wang et al. [87] attached bovine serum albumin to the surface of a diamond modified with PDA, and it showed that this modification decreased the cytotoxicity and enhanced the blood compatibility efficiently. Park et al. [88] immobilized heparin on a metal surface coated with PDA, and a therapeutic layer-by-layer multilayer composed of paclitaxel (PTX) encapsulated poly(lactic-coglycolic acid) grafted hyaluronic acid (HA-g-PLGA) micelles, heparin, and poly-L-lysine (PLL) was built up, which was efficient enough to inhibit the formation of thrombus and enhance blood compatibility. Microcontact printing [89] or microfluidic patterning [90] is usually applied to deposit PDA regionally to realize the cell patterning, which is an optional method for the fabrication of a biological chip.

Elimelech et al. [91] modified the AFM tip with PDA and attached a single live lactobacillus to study the interaction between the cell and the material surface. Liu et al. [92] found that the PDA-modified sensor of Au/polypyrrole could enrich bacteria and enhance the adhesion.

The PDA modification is also widely used in preparing antibiosis and an antiseptic surface. Rahimipour et al. [93] deposited PDA on various substrates and grafted the lipopetide with low toxicity and antibacteria, and demonstrated that the modified surface showed antibacterial property for *Escherichia coli*. Xu et al. [94] fabricated an antibacterial composite fiber with washing durability by *in situ* reduction of Ag nanoparticles on the surface of cotton fiber which was modified with PDA.

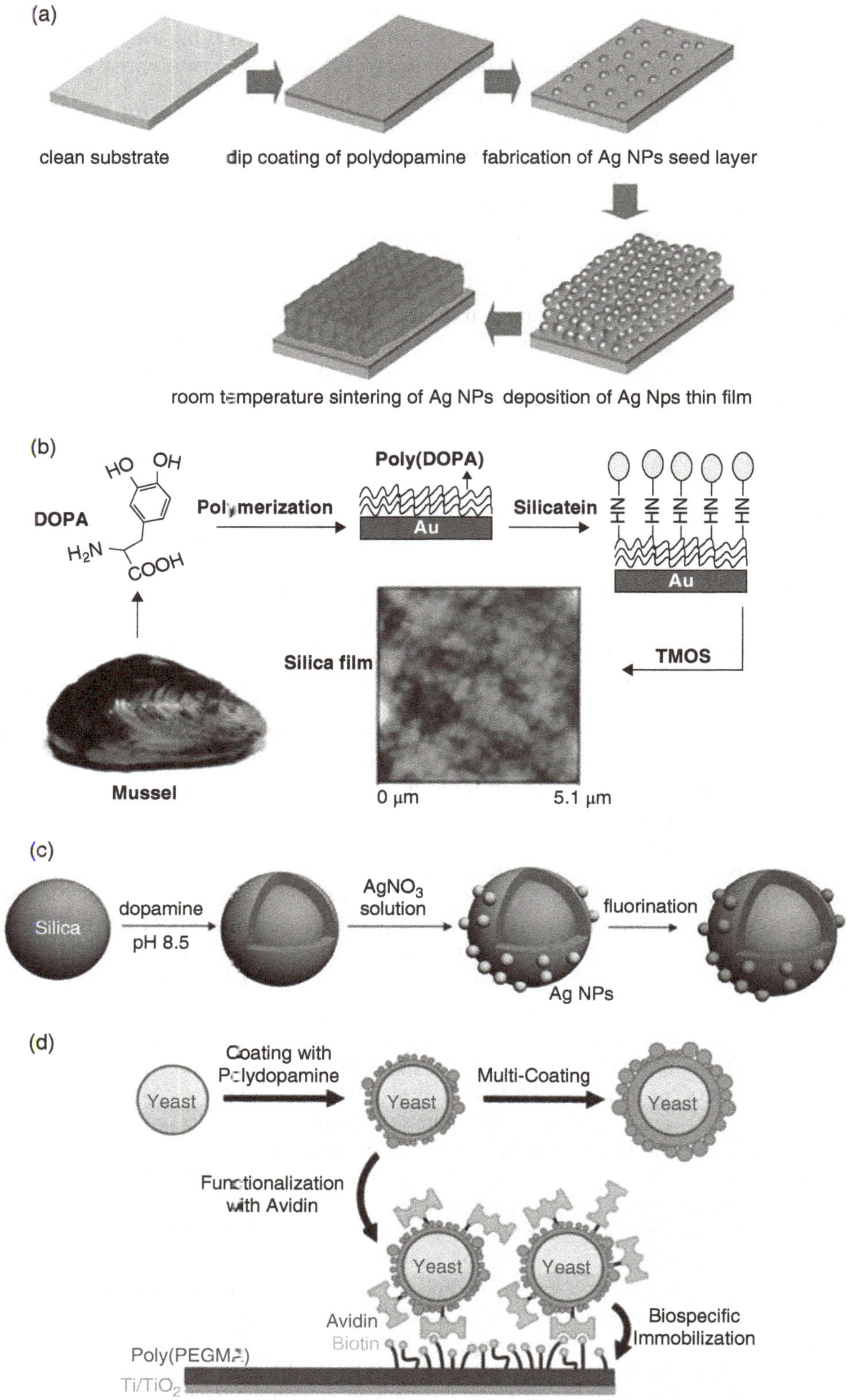

Figure 5.9 The schematic illustration of the fabrication of (a) Ag conductive film [78], (b) silica film [85], (c) superhydrophobic particles [58], and (d) functionalized yeast cell [101].

Similarly, the bacterial cellulose [95] and a polyimide film [96] modified with PDA/Ag exhibited excellent antibacterial property.

Recently, Yang et al. [97] deposited a PDA layer on an aluminum surface modified with a silane coupling agent with sulfydryl, and the corrosion resistance of the coating enhanced twofold. If a layer of tetradecylamine was further attached to the surface, the corrosion resistance would enhance threefold [98]. However, this protective coating was unstable, its performance decreased significantly after two days. The authors fabricated a PDA/ZrO_2 composite coating on a silicon substrate by LbL, and it enhanced both the surface hardness and the corrosion resistance [99].

Because the modification of PDA is independent of the shape and size of the substrates, it is widely used in the modification of various microparticles and nanoparticles. Xu et al. [58] reported a versatile method to synthesize superhydrophobic microparticles. PDA was deposited on the surface of targeted particles and then the Ag nanoparticles were reduced, due to the reducibility of PDA, resulting in the formation of a hierarchical structure similar to that of the micromorphology of a lotus leaf. Followed by modification with thiolfluoroalkanes, superhydrophobic particles were fabricated (Figure 5.9c). When magnetic particles were chosen as the core materials, the hybrid particles had potential applications in oil/water separation and transportation. Jung et al. [100] modified TiO_2 nanoparticles with PDA and applied them to a dye-sensitized solar cell, and the conversion efficiency of light energy was up to 1.2%. Choi et al. [101] deposited a PDA coating on a yeast cell and grafted the avidin molecule. After this modification, the yeast cell could be attached to the surface of TiO_2 which was modified with biotin, and the PDA coating protected the yeast cell from harm by the enzyme and inhibited the differentiation of the yeast cell (Figure 5.9d).

PDA has significant applications in the modification of one- and two-dimensional nanomaterials as well. Carbon nanotubes (CNTs) are easily modified by PDA. The initiator can be grafted onto CNTs to form polymer brushes [102] or hydrophilic/hydrophobic molecules can be grafted to improve the dispersion ability in solvents [103]. It also can be used to load novel metal nanoparticles such as Ag, Au, Pt or Pd. These nanohybrids are usually sensors and catalysts with a high performance [104, 105]. Kaolin nanotubes were modified with PDA and loaded $Ru(bpy)_3^{2+}$, which was used as the electrode for the electro-chemiluminescence sensor [106]. Thakur modified boron nitride nanotubes with PDA to improve the dispersion ability in solvents and fabricate high performance composites [107]. In addition, fibers such as PLLA, PCL, PVA, PS, etc. can also be modified with PDA to improve the biocompatibility, antibacterial ability, and wettability [108–110]. For clay, the modification of PDA is mainly to improve the dispersion ability in organic solvents and fabricate epoxy resin [111] or polyurethane [112] composites with high performance. For graphene, the graphene oxide can be reduced to graphene during the polymerization of dopmanine [113, 114], and the modification can give the graphene more functionality [115–117].

In addition to being used as an efficient surface anchor, dopamine can self-polymerize to form nanoparticles easily under air oxidation in alkaline media. The structure of PDA nanoparticles is similar to melanins which are widely distributed in living organisms. Lee et al. [118] reported on a method to synthesize melanin-like nanoparticles by polymerization of dopamine hydrochloride under air oxidation using NaOH to adjust the pH. The NPs below 100 nm with long-term stability could be generated by controlling the concentration of NaOH, dopamine, and the temperature. Thiol-terminated methoxy-poly(ethylene glycol) (mPEG-SH) was conjugated effectively to the NPs by Michael-addition. The modified NPs were demonstrated to be biocompatible and have free radical scavenging activity.

Lu et al. reported on a versatile strategy to fabricate carbon sub-micrometer spheres (carbon SMSs) by using PDA nanoparticles as a carbon resource. The SMSs were monodispersed and size-controlled by tuning the ratio of ammonia to the dopamine used during polymerization in the mixed solution of water and ethonal. This kind of carbon SMSs was N-rich and more sp^2 C, resulting in the higher electroconductivity and catalytic activity of oxygen reduction reaction [119]. The PDA nanoparticles were also severed as a new photothermal therapeutic agent due to their biocompatibility, strong NIR absorption, and photothermal conversion efficiency [68].

Lee et al. [120] synthesized polydopamine spheres in a mixed solvent of alcohol and Tris-buffer solution. The PDA spheres were monodispersed with tunable diameters. MnO_2 hollow spheres or PDA/Fe_3O_4 and PDA/Ag core/shell nanostructures were successfully synthesized by using the PDA spheres as active templates because of the existence of abundant surface functional groups, e.g. -OH and -NH_2, which could coordinate with the metal ions efficiently.

As templates are introduced into the dopamine aqueous solution, the in-situ polymerization of dopamine occurs on the surface of the templates. Various materials with specific structures and functionalities can be fabricated in this method. Dai et al. [66] used SiO_2 nanospheres as templates to get SiO_2@PDA spheres in a dopamine aqueous solution. After carbonization and templates removal, hollow carbon spheres were synthesized. When Au@SiO_2 spheres were used instead of

SiO_2, Au@C yolk-shell nanocomposites with high catalytic ability and stability were fabricated using the same strategy.

5.3 Conclusion

Researchers usually find inspiration in nature. The bioinspired materials often exhibit novel and superior functions compared to conventional materials. The polymer inspired by mussel adhesive proteins, including polymer containing catechol groups and PDA, is one of the most successful examples. The mussel adhesive proteins mimetic polymer has received increasing attention in recent years because of its universal adhesive property and easy post-functionalization, which make the process simple and the materials multi-functional. Catechol and the derivatives, especially dopamine, may be employed to functionalize almost all kinds of materials from zero to three dimensions, due to the multiple physical and chemical interactions including chelation, hydrogen bonding, π-π stacking, and electrostatic interaction. Although plenty of successful examples have been reported, all benefits from the catechol chemistry, some issues have not yet been addressed. The interactions between catecholic anchors and different substrates are varied. For instance, the affinity of catecholic anchors to silica is weak while that to metal oxides is highly stable. This difference is worthy of attention when substrates are modified. As the polymerized mechanism of dopamine is ambiguous at present, the side reactions should not be ignored. During the process of modifying the substrates, especially nanoparticles, how to initiate an optimized condition to avoid the formation of PDA particles and their sticking to each other needs to be considered. These issues are undergoing research and we look forward to these being resolved so as to broaden the fields of the applications of the nano objects themselves and also as building blocks for the construction of sophisticated structured materials.

References

1 Allen, J.A., Cook, M., Jackson, D.J. et al. (1976) *J. Moll. Stud.*, **42**, 279.
2 Burkett, J.R., Wojtas, J.L., Cloud, J.L., Wilker, J.J. (2009) *J. Adhes.*, **85**, 601.
3 Waite, J.H., Lichtenegger, H.C., Stucky, G.D., Hansma, P. (2004) *Biochemistry*, **43**, 7653.
4 Hwang, D.S., Zeng, H., Masic, A. et al. (2009) *J. Biol. Chem.*, **285**, 25850.
5 Harrington, M.J., Masic, A., Holten-Andersen, N. et al. (2020) *Science*, **328**, 216.
6 Rzepecki, L.M., Hansen, K.M., Waite, J.H. (1992) *Biol. Bull.*, **183**, 123.
7 Inoue, K., Takeuchi, Y., Miki, D., Odo, S. (1995) *J. Biol. Chem.*, **270**, 6698.
8 Zhao, H., Waite, J.H. (2006) *Biochemistry*, **45**, 14223.
9 Zhao, H., Robertson, N.B., Jewhurst, S.A., Waite, J.H. (2006) *J. Biol. Chem.*, **281**, 11090.
10 Waite, J.H., Qin, X. (2001) *Biochemistry*, **40**, 2887.
11 Zhao, H., Waite, J.H. (2006) *J. Biol. Chem.*, **281**, 26150.
12 Deming, T.J. (1999) *Curr. Opin. Chem. Biol.*, **3**, 100.
13 Koide, M., Lee, D.S., Goldberg, E.D. (1982) *Estuarine, Coast. Shelf. S.*, **15**, 679.
14 Coombs, T.L., Keller, P.J. (1981) *Aquat. Toxicol*, **1**, 291.
15 Taylor, S.W., Luther III, G.W., Waite, J.H. (1994) *Inorg. Chem.*, **33**, 5819.
16 Taylor, S.W., Chase, D B., Emptage, M.H. (1996) *Inorg. Chem.*, **35**, 7572.
17 Waite, J.H., Andersen, N.H., Jewhurst, S., Sun, C. (2005) *Adhes.*, **81**, 297.
18 Dalsin, J.L., Lin, L., Tosatti, S. et al. (2005) *Langmuir*, **21**, 640.
19 McDowell, L.M., Burzio, L.A., Waite, J.H., Schaefer, J. (1999) *J. Biol. Chem.*, **274**, 20293.
20 Haemers, S., Koper, G.J., Frens, G. (2003) *Biomacromol.*, **4**, 1098.
21 Land, E.J., Ramsden, C.A., Riley, P.A. (2004) *Method Enzymol.*, **378**, 88.
22 Burzio, L.A., Waite, J.H. (2000) *Biochem.*, **39**, 11147.
23 Zhao, H., Sun, C., Stewart, R.J., Waite, J.H. (2005) *J. Biol. Chem.*, **280**, 42938.
24 Shao, H. (2010) *Adv. Mater.*, **22**, 729.
25 Yu, M., Deming, T.J. (1998) *Macromol.*, **31**, 4739.
26 Yamada, K., Chen, T., Kumar, G. et al. (2000) *Biomacromol.*, **1**, 252.
27 Guvendiren, M. (2009) *J. Adhes.*, **85**, 631.
28 Westwood, G., Horton, T.N., Wilker, J.J. (2007) *Macromol.*, **40**, 3960.
29 White, J.D., Wilker, J.J. (2011) *Macromol.*, **44**, 5085.
30 Matos-Perez, C.R. (2012) *J. Am. Chem. Soc.*, **134**, 9498.
31 Lee, H., Lee, B.P., Messersmith, P. B. (2007) *Nat.*, **448**, 338.
32 Glass, P., Chung, H., Washburn, N.R., Sitti, M. (2009) *Langmuir*, **25**, 6607.
33 Glass, P., Chung, H., Washburn, N.R., Sitti, M. (2010) *Langmuir*, **26**, 17357.
34 Lee, B.P., Huang, K., Nunalee, F.N. et al. (2004) *J. Biomater. Sci., Polym. Ed.* **15**, 449.
35 Lee, B.P., Dalsin, J.L., Messersmith, P.B. (2002) *Biomacromol.*, **3**, 1038.

36 Lee, B.P., Chao, C.Y., Nunalee, F.N. et al. (2006) *Macromol.*, **39**, 1740.
37 Brubaker, C.E., Kissler, H., Wang, L.J. et al. (2010) *Biomaterials*, **31**, 420.
38 Bilic, G., Brubaker, C., Messersmith, P.B.J. (2010) *Am. J. Obstet. Gynecol.*, **202**, 2009.
39 Murphy, J.L., Vollenweider, L., Xu, F., Lee, B.P. (2010) *Biomacromol.*, **11**, 2976.
40 Holten-Andersen, N., Harrington, M.J., Birkedal, H. et al. (2011) *PNAS* **108**, 2651.
41 Yavvari, P. S., Srivastava, A. (2015) *J. Mater. Chem.* **B3**, 899.
42 Choi, Y.C., Choi, J.S., Jung, Y.J., Cho, Y.W. (2014) *J. Mater. Chem. B*, **2**, 201.
43 Krogsgaard, M., Behrens, M.A., Pedersen, J.S., Birkedal, H. *Biomacromol.*, **14**, 297.
44 White, E.M., Seppala, J.E., Rushworth, P.M. et al. (2013) *Macromol.*, **46**, 8882.
45 Shafiq, Z., Cui, J., Pastor-Pérez, L. et al. (2012) *Angew. Chem. Int. Ed.*, **51**, 4332.
46 Mizrahi, B., Shankarappa, S.A., Hickey, J.M. et al. (2013) *Adv. Funct. Mater.*, **23**, 1527.
47 He, L., Fullenkamp, D.E., Rivera, J.G., Messersmith, P.B. (2011) *Chem. Commun.*, **47**, 7497.
48 Vatankhah-Varnoosfaderani, M., Ghavami Nejad, A., Hashmi, S., Stadler, F.J. (2013) *Chem. Comm.*, **49**, 4685.
49 Ahn, B.K., Lee, D.W., Israelachvili, J.N., Waite, J.H. (2014) *Nat. Mater.*, **13**, 867.
50 Ling, D., Park, W., Park, Y.I. et al. (2011) *Angew. Chem. Int. Ed.*, **50**, 11360.
51 Wei, H., Insin, N., Lee, J. et al. (2012) *Nano Lett.*, **12**, 22.
52 Na, H. B., Palui, G., Rosenberg, J. T. et al. (2012) *ACS Nano*, **6**, 389.
53 Gu, H., Xu, K., Xu, C., Xu, B. (2005) *Chem. Comm.*, **89**, 4270.
54 Lee, Y., Lee, H., Kim, Y. B. et al. (2008) *Adv. Mater.*, **20**, 4154.
55 Majewski, A.P., Schallon, A., Jérôme, V. et al. (2012) *Biomacromol.*, **13**, 857.
56 Wilms, V.S., Bauer, H., Tonhauser, C. et al. (2013) *Biomacromol.*, **14**, 193.
57 Das, S., Ranjan, P., Maiti, P.S. et al. (2013) *Adv. Mater.*, **25**, 422.
58 Zhang, L., Wu, J., Wang, Y. et al. (2012) *J. Am. Chem. Soc.*, **134**, 9879.
59 López, K.A., Piña, M.N., Morey, J. (2013) *Sens. Actuators, B*, **181**, 267.
60 Lee, H., Dellatore, S.M., Miller, W.M., Messersmith, P.B. (2007) *Science*, **318**, 426.
61 Hong, S., Na, Y.S., Choi, S. et al. (2012) *Adv. Funct. Mater.*, **22**, 4711.
62 Dreyer, D.R., Miller, D.J., Freeman, B.D. et al. (2012) *Langmuir*, **28**, 6428.
63 Yu, B., Wang, D.A., Ye, Q. et al. (2009) *Chem. Comm.*, **85**, 6789.
64 Ball, V. (2010) *Colloids and Surfaces A: Physicochem. Eng. Aspects*, **363**, 92.
65 Phua, S.L., Yang, L., Toh, C.L. et al. (2013) *ACS Appl. Mater. Interfaces*, **5**, 1302.
66 Liu, R., Mahurin, S.M., Li, C. et al. (2011) *Angew. Chem. Int. Ed.*, **50**, 6799.
67 Bernsmann, F., Ponche, A., Ringwald, C. et al. (2009) *J. Phys. Chem. C.*, **113**, 8234.
68 Liu, Y., Ai, K., Liu, J. et al. (2013) *Adv. Mater.*, **25**, 1353.
69 Zhang, X., Wang, S., Xu, L. et al. (2012) *Nanoscale*, **4**, 5581.
70 Kim, B.H., Lee, D.H., Kim, J.Y. et al. (2011) *Adv. Mater.*, **23**, 5618.
71 Kang, K., Lee, S., Kim, R. et al. (2012) *Angew. Chem. Int. Ed.*, **51**, 13101.
72 You, I., Kang, S.M., Lee, S. et al. (2012) *Angew. Chem. Int. Ed.*, **51**, 6126.
73 McCloskey, B.D., Park, H.B., Ju, H. et al. (2010) *Polymer*, **51**, 3472.
74 Xi, Z.Y., Xu, Y.Y., Zhu, L.P. et al. (2009) *J. Membr. Sci.*, **327**, 244.
75 Cheng, C., Li, S., Zhao, W. et al. (2012) *J. Membr. Sci.*, **417**, 228.
76 Ryou, M.H., Lee, Y.M., Park, J.K., Choi, J.W. (2011) *Adv. Mater.*, **23**, 3066.
77 Ryou, M.H., Lee, D.J., Lee, J.N. et al. (2012) *Adv. Energy Mater.*, **2**, 645.
78 Long, Y., Wu, J., Wang, H. et al. (2011) *J. Mater. Chem.*, **21**, 4875.
79 Du, W., Zou, H., Tian, M. et al. (2012) *Polym. Adv. Technol.*, **23**, 1029.
80 Wang, W., Li, R., Tian, M. et al. (2013) *ACS Appl. Mater. Interfaces*, **5**, 2062.
81 Akter, T., Kim, W.S. (2012) *ACS Appl. Mater. Interfaces*, **4**, 1855.
82 Ou, J., Wang, J., Zhang, D. et al. (2010) *Colloid Surface B*, **76**, 123.
83 Ryu, J., Ku, S.H., Lee, H., Park, C.B. (2010) *Adv. Funct. Mater.*, **20**, 2132.
84 Kim, S., Park, C.B. (2010) *Biomaterials*, **31**, 6628.
85 Rai, A., Perry, C.C. (2012) *J. Mater. Chem.*, **22**, 4790.
86 Kang, S.M., Ryou, M.H., Choi, J.W., Lee, H. (2012) *Chem. Mater.*, **24**, 3481.
87 Tao, C., Zhang, J., Yang, S. (2011) *J. Nanosci. Nanotechno*, **11**, 5068.
88 Kim, T.G., Lee, H., Jang, Y., Park, T.G. (2009) *Biomacromol.*, **10**, 1532.
89 Sun, K., Xie, Y., Ye, D. et al. (2012) *Langmuir*, **28**, 2131.
90 Sun, K., Song, L., Xie, Y. et al. (2011) *Langmuir*, **27**, 5709.
91 Kang, S., Elimelech, M. (2009) *Langmuir*, **25**, 9656.

92 Liu, A., Zhao, L., Bai, H. et al, (2009) *ACS Appl. Mater. Interfaces*, **1**, 951.
93 Shalev, T., Gopin, A., Bauer, M. et al. (2012) *J. Mater. Chem.*, **22**, 2026.
94 Xu, H., Shi, X., Ma, H. et al. (2011) *Appl. Surf. Sci.*, **257**, 6799.
95 Yang, C., Liang, G.L., Xu, K.M. et al. (2009) *J. Mater. Sci.*, **44**, 1894.
96 Sureshkumar, M., Siswanto, D.Y., Lee, C.K. (2010) *J. Mater. Chem.*, **20**, 6948.
97 Ou, J., Wang, J., Zhou, J. et al. (2010) *Prog. Org. Coat.*, **68**, 244.
98 Ou, J., Wang, J., Liu, S. et al. (2009) *J. Phys. Chem. C*, **113**, 20429.
99 Ou, J., Wang, J., Qiu, Y. et al. (2011) *Surf. Interface Anal*, **43**, 803.
100 Nam, H.J., Kim, B., Ko, M.J. et al. (2012) *Chem-Eur. J.*, **18**, 14000.
101 Yang, S.H., Kang, S.M., Lee, K.B. et al. (2011) *J. Am. Chem. Soc.*, **133**, 2795.
102 Hu, H., Yu, B., Ye, Q. et al. (2010) *Carbon*, **48**, 2347.
103 Zhang, X., Liu, M., Zhang, Y. et al. (2012) *RSC Adv.*, **2**, 12153.
104 Çelen, B., Ekiz, D., Pişkin, E. et al. (2011) *J. Mol. Catal. A-Chem.*, **350**, 97.
105 Ye, W., Hu, H., Zhang, H. et al. (2010) *Appl. Surf. Sci.*, **256**, 6723.
106 Xing, B., Yin, X.B. (2009) *PloS One*, **4**, e6451.
107 Thakur, V.K., Yan, J., Lin, M.F. et al. (2012) *Polym. Chem.*, **3**, 962.
108 Yang, H., Lan, Y., Zhu, W. et al. (2012) *J. Mater. Chem.*, **22**, 16994.
109 Rim, N.G., Kim, S.J., Shin, Y.M. et al. (2012) *Colloid Surface B*, **91**, 189.
110 Xie, J., Michael, P. L., Zhong, S. et al. (2012) *J. Biomed. Mater. Res. A*, **100A**, 929.
111 Yang, L.P., Phua, S.L., Teo, J.K.H. et al. (2011) *ACS Appl. Mater. Interfaces*, **3**, 3026.
112 Phua, S.L., Yang, L., Toh, C.L. et al. (2012) *ACS Appl. Mater. Interfaces*, **4**, 4571.
113 Xu, L.Q., Yang, W.J., Neoh, K.G. et al. (2010) *Macromol.*, **43**, 8336.
114 Kang, S.M., Park, S., Kim, D. et al. (2011) *Adv. Funct. Mater.*, **21**, 108.
115 Cheng, C., Nie, S., Li, S. et al. (2013) *J. Mater. Chem. B*, **1**, 265
116 Zhang, Z., Zhang, J., Zhang, B., Tang, J., (2013) *Nanoscale*, **5**, 118.
117 Liu, H,Y. (2012) *J. Phys. Chem. C*, **116**, 3334.
118 Ju, K.Y., Lee, Y., Lee, S. et al. (2011) *Biomacromol.*, **12**, 625.
119 Ai, K., Liu, Y., Ruan, C. et al. (2013) *Adv. Mater.*, **25**, 998.
120 Yan, J., Yang, L., Lin, M.F. et al. (2013) *Small*, **9**, 596.

6

Self-assembly of Polylactic Acid-based Amphiphilic Block Copolymers and Their Application in the Biomedical Field

*Lin Xiao, Lixia Huang, Li Liu, and Guang Yang**

College of Life Science and Technology, Huazhong University of Science and Technology, Wuhan, China

6.1 Introduction

Polylactide (PLA) and its copolymers, such as poly(lactide-co-glycolide) (PLGA), are hydrophobic aliphatic polyesters possessing superior biocompatibility and biodegradability. They show no cytotoxicity *in vitro* and very low immunogenicity *in vivo* and thus have been approved by the US Food and Drug Administration (FDA) for clinical use. They can be degraded by hydrolysis or enzymolysis under physiological conditions with a variable degradation rate depending on their molecular weights and crystallinity. Moreover, they have rather good mechanical properties and processability. These unique properties have facilitated their applications in biomedical field as drug delivery systems, surgery sutures, bone fixation implants, and tissue engineering scaffolds [1–6]. However, the hydrophobicity of PLA and its copolymers has impeded their applications in the biomedical field, especially as functional drug/cell delivery systems in the body, because of the nonspecific interaction between the PLA particles and serum components, leading to their severe aggregation and rapid clearance from the circulation by the reticuloendothelial system following systemic administration rapid clearance [7–9]. The introduction of hydrophilic polymers such as polyethylene glycol (PEG), polypeptide and polysaccharides into the hydrophobic PLA and its copolymers can provide the resultant polymers with amphiphilic features and thus a number of special properties [10–12].

The amphiphilic block polymers containing a hydrophobic block of PLA or its copolymer and one or more hydrophilic blocks of other polymers are attractive in a variety of fields because of their ability to self-assemble in aqueous solution. As is well known, micellar structures, including spherical micelles, rod-like micelles, and vesicles can be prepared from PLA-based amphiphilic block copolymers in dilute solution (e.g. 0.5 wt% PLA-*b*-PEG in an aqueous solution), while hydrogels can be formed when the polymer concentration increases to a rather high level (e.g. 20 wt% PLA-*b*-PEG in an aqueous solution). The self-assembled morphologies of micellar structures can be affected by many factors, such as the copolymer composition, the copolymer concentration, and solvents among others. Properties such as the tensile modulus and stimulus responsiveness of the hydrogels can be also tailored by tuning the copolymer compositions and concentration. Because of their ordered structures and superior properties, the self-assembled micellar structures and hydrogels have been widely explored in biomedical applications, including drug/gene/cell delivery systems, imaging, tissue engineering scaffolds, tissue augmentation, and more [13–15].

This chapter will begin with the discussion of the mechanisms of the self-assembly behaviors of PLA-based amphiphilic block copolymers in an aqueous solution which produce micellar structures and hydrogels respectively, analyzing the possible factors that affect the process. Further, the properties of the assemblies containing PLA, especially the stimulus responsiveness, will be the focus, with an emphasis on the design considerations. The final section will deal with the applications of these assemblies that contain PLA in biomedical fields and will provide an outlook for the future.

6.2 Micellar Structures from PLA-based Amphiphilic Block Copolymers

PLA-based amphiphilic block copolymers self-assemble into nanoparticles with micellar structures in a dilute aqueous solution. The hydrophobic PLA or its copolymers form the core of the spherical or rod-like micelles or the wall of vesicles, while the hydrophilic chain segments stretch in water to act as the corona of the spherical or rod-like micelles or the cavity of vesicles. These morphologies

* Email: yang_sunny@yahoo.com

Bioinspired Materials Science and Engineering, First Edition. Edited by Guang Yang, Lin Xiac, and Lallepak Lamboni.

can be transformed between each other in specific conditions that obey some uniform mechanisms.

6.2.1 Preparation and Mechanism of Micellar Structures

6.2.1.1 Methods for Preparation of Micellar Structures

Several methods have been developed to prepare micellar aggregates from PLA-based amphiphilic block copolymers in a dilute aqueous solution. For most amphiphilic block copolymers containing a glassy hydrophobic block at room temperature, a co-solvent method is frequently employed to produce micellar aggregates [16, 17]. Typically, the amphiphilic block copolymers are dissolved in a common solvent, such as DMF, dioxane, or THF, which are good solvents for all blocks. Then a selective solvent, such as water (which is a nonsolvent for the hydrophobic block), is slowly added to the solution until the water content (generally 25–50 wt%) is much higher than that at which aggregation starts. Then the aggregates are usually quenched in an excess of water to freeze the kinetic processes and morphologies. Finally, the common solvent is removed by dialysis of the resulting solution against water. It is noted that with regard to PLA-based amphiphilic block copolymers containing a highly hydrophilic block or a relatively very long hydrophilic block, it is possible to form micellar aggregates through a simple dissolution method because of the good water solubility of the copolymers [18, 19]. It is well known that both PLA-*b*-PEG and PEG-*b*-PLA-*b*-PEG can form micelles by dissolving them in water with very low concentration. Other methods such as microfludics techniques, layer-by-layer, solvent evaporation, among others, have also been developed to prepare micellar aggregates from PLA-based amphiphilic block copolymers [20–23].

The formation of micellar aggregates can be confirmed by cryogenic transmission electron microscopy (cryo-TEM) and dynamic light scattering (DLC) and also can be reflected by using dye solubility experiments [24, 25].

6.2.1.2 Mechanism of Micellar Structures

Generally, the formation of micellar aggregates from amphiphilic block copolymers is driven by hydrophobic interactions, that is, by packing of the hydrophobic chain segments [26]. Mai and Eisenberg [27] described the possible factors affecting the self-assembled morphologies of amphiphilic block copolymers. It was believed that the formation of thermodynamically stable block copolymer aggregates of various morphologies is governed by the free energy of the system, which is made up of three aspects: (1) the degree of stretching of the core-forming blocks; (2) the interfacial tension between the micelle core and the solvent outside the core; and (3) the repulsive interactions among corona-forming chains [28, 29]. Therefore, the self-assembled aggregates could be controlled by the factors that may affect the three contributions. As for PLA-based amphiphilic block copolymers, the affecting factors include the copolymer composition (i.e. the length of each block and ratio of different blocks) and concentration, the water content in the solution, the nature of the common solvent, or the presence of additives, such as ions or homopolymers, etc. [30].

The copolymer composition plays a key role in determining the self-assembled morphologies of the amphiphilic block copolymers, the changes of which can induce the transition of aggregate morphologies from one to another. This is because the chain length of both hydrophobic block and hydrophilic block can affect the three aspects described above contributing to the free energy of the system. The cone-column concept, first raised by Discher and his colleagues [31], is used to explain the mechanism of morphological transitions caused by copolymer composition (Figure 6.1) [27]. It is believed that block copolymers with relatively long hydrophilic chains usually tend to form

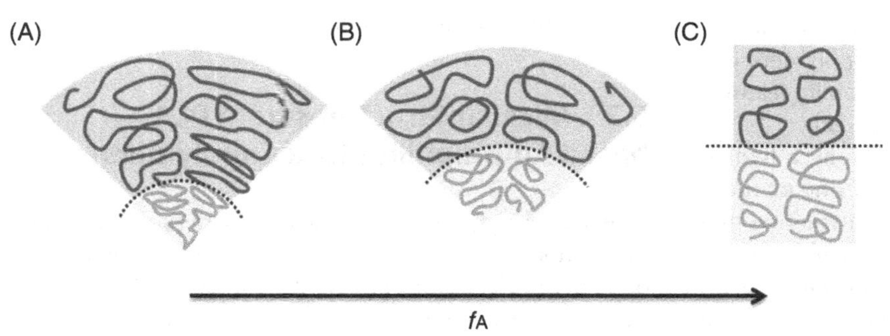

Figure 6.1 Schematic illustration of the possible polymer chain arrangements in different morphologies of diblock copolymers changing from sphere (A) to cylinder (B) and to lamella (C), as the volume fraction (f_A) of the hydrophobic block (orange) increases. The dashed curve in each morphology represents a part of the interface between hydrophobic and hydrophilic domains. *Source*: The concept of this figure originates from [31].

spherical micelles. With an increase in the length of the hydrophobic block, it is generally considered that the aggregate morphologies may be transformed from spherical micelles, via rod- or wormlike micelles into bilayer structures or vesicles. Typical examples of this tendency are polystyrene-block-polyacrylic acid (PS-b-PAA), polystyrene-b-poly(4-vinylpyridine) (PS-b-P4VP) and polystyrene-block-polyethylene oxide (PS-b-PEO) [32–34].

PLA-based amphiphilic block copolymers have been used to prepare spherical micelles for drug carriers by different research groups. However, there are a few examples of other aggregate morphologies such as rod-like micelles and bilayer structures (i.e. lamella and vesicles). The first report of a biodegradable vesicle of PLA-based amphiphilic block copolymers was by Feijen et al. [35]. In their study, a series of PLA-*b*-PEG copolymers with variable molecular weight formed vesicles in the chloroform/water system. The diameter of the vesicles covered a broad range from 70 nm to 50 μm. It should be noted that these PLA-*b*-PEG copolymers possessed relatively long PLA blocks (3–8 times as long as PEG blocks), which seems in accordance with the mechanism discussed above. Park et al. [36] also prepared self-assembled vesicles from PLA-*b*-PEG copolymers with relatively long PLA blocks using the emulsion solvent evaporation method. The vesicles were of various sizes from several hundred nanometers to a few micrometers, which could be readily controlled by altering the relative hydrodynamic volume fraction ratio between hydrophilic and hydrophobic blocks in the copolymer structure. Moreover, it was found that the vesicles became more stable when increasing the molecular weight of the PEG block within a certain scope. Nevertheless, results showing different tendency were also reported on the self-assembly of PLA-based amphiphilic block copolymers considering the effect of hydrophobic blocks on the micellar morphologies. Du et al. [37] studied the self-assembly behaviors of PLA-*b*-PEG copolymers with a fixed PEG length and a variable PLA length. It was found in their study that by increasing the hydrophilic PEG fractions (f_{EO}) of the diblock copolymers, the morphology of PLA-*b*-PEG micelles shifted from spherical micelles ($f_{EO} < 20\%$) to vesicles ($f_{EO} \sim 20–42\%$) or wormlike micelles ($f_{EO} > 42\%$). However, it was also found that when increasing the hydrophobic PLA block from $x = 56$ to $x = 132$ and to $x = 212$ for PLA_x-*b*-PEG_{44} block copolymers, the self-assembled micellar morphologies transformed from worm-like micelles via open lamellae into closed lamellae (vesicles) as shown in Figure 6.2. These results are in accordance with the general view found in many other diblock copolymers, which is the

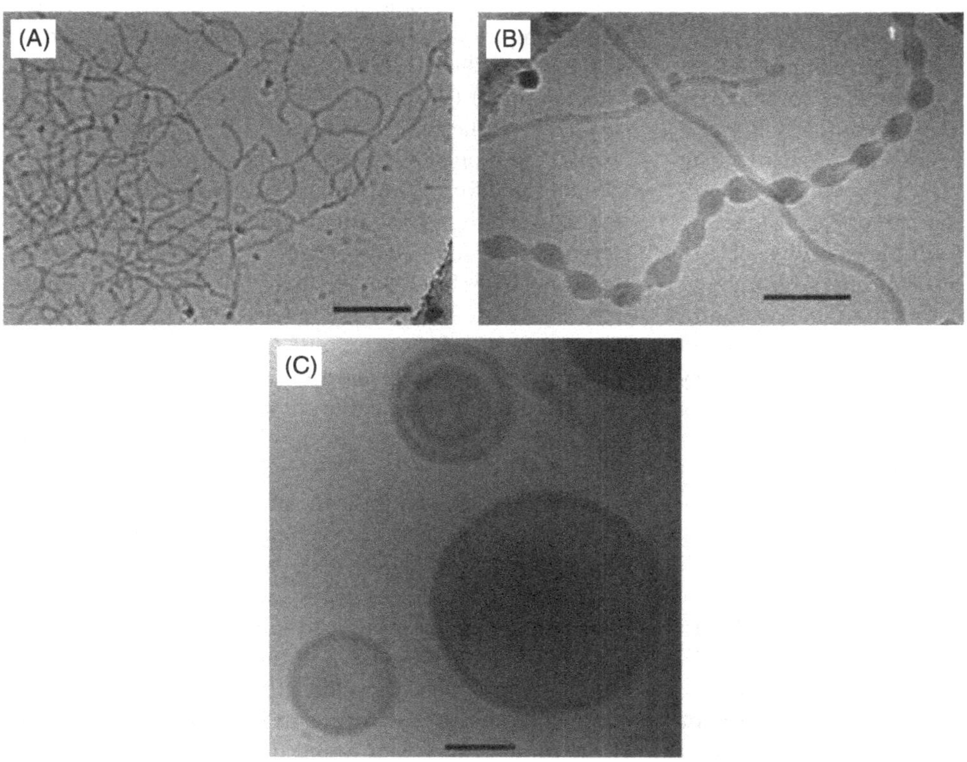

Figure 6.2 Cryo-TEM images of PLA_x-*b*-PEG_{44} micelles in aqueous solution. The concentrations of initial PLA_x-*b*-PEG_{44} THF solutions were 0.45 wt%. (A) PLA_{56}-*b*-PEG_{44}, (B) PLA_{134}-*b*-PEG_{44}, and (C) PLA_{212}-*b*-PEG_{44}. The scale bars in the images present 200 nm. *Source:* Reproduced with permission from [37].

result of the energy balance among the stretching of the core-forming block, the interfacial tension between the core and the solvent outside the core region, and the repulsion among the corona chains.

The self-assembled morphology of aggregates can be also affected by the common solvent used to dissolve the PLA-based amphiphilic block copolymers, since the common solvent directly affects the dimensions of both the hydrophobic domains and the hydrophilic corona chains of the aggregates. Feijen et al. [35] prepared vesicles from PLA-*b*-PEG using different solvent systems. It was found that the block copolymer formed vesicles coexisting with other bilayer structures, e.g. tubules when using the chloroform/water system, while it produced only vesicles when using the acetone/water or tetrahydrofuran (THF)/water system. Moreover, the vesicles were of a smaller size and had lower polydispersity compared with those prepared by the chloroform/water system. Besides the copolymer composition and solvent factors, other factors such as copolymer concentration, water content, and additives are also crucial in the formation of self-assembled aggregates from PLA-based amphiphilic block copolymers.

6.2.2 Stability and Stimuli-Responsive Properties: Molecular Design and Biomedical Applications

Self-assembled aggregates from PLA-based amphiphilic block copolymers have great potential in biomedical applications because of their high biocompatibility and superior comprehensive properties. Considering drug delivery systems, the non-fouling hydrophilic blocks in PLA-based amphiphilic block copolymers usually acting as the corona of micellar aggregates play a crucial role in stabilizing the nanoparticles *in vivo* through minimizing the nonspecific interactions between nanoparticles and serum components, and thus leading to a prolonged circulation time [38–40]. Among various drug delivery systems made from PLA-based amphiphilic block copolymers, PLA-*b*-PEG is considered one of the most promising species with bright prospects in clinical applications. Recently, a paclitaxel-(PTX-)-loaded nanoparticle based on PLA-*b*-PEG diblock copolymer was approved for cancer chemotherapy in South Korea and entered Phase II clinical trials in the United States [41, 42]. However, the stability of PLA-*b*-PEG nanoparticles in blood circulation remains a major challenge [43]. Another issue is the lack of responsive ability to specific biological microenvironments, which limits the treatment efficacy of the nanoparticles. Especially, the non-fouling hydrophilic blocks such as PEG, could reduce the cellular internalization of nanoparticles in disease location, leading to a greatly discounted therapeutic efficacy [44–46]. Ideally, the drug delivery system from PLA-based amphiphilic block copolymers should be colloidally stable in circulation with the protection of hydrophilic blocks, while simultaneously becoming active at the disease site to promote cellular uptake through responsive loosening of hydrophilic blocks [47].

It has been proved that the formation of poly(lactic acid) stereo complexes by blending enantiomeric poly(L-lactic acid) (PLLA) and poly(D-lactic acid) (PDLA) can effectively improve the stability of micellar aggregates of PLA-based amphiphilic block copolymers in physiological conditions [48]. This improved stability stems from the H-bonding force from specific $CH_3 \cdots O=C$ and $C\alpha H \cdots O=C$ interactions between both PLA stereoisomers from a combined study with FT-IR spectroscopy and molecular modeling by Sarasua et al. [49]. Using this interesting behavior of stereo complexation, researchers have developed drug-loaded micelles from PLA-based amphiphilic block copolymers with higher stability [50–52]. As an example, Leroux et al. [50] prepared a stereo complex micellar system from PLLA-*b*-PEG and PDLA-*b*-PEG copolymers, which showed much higher kinetic stability compared to micelles from isotactic or racemic PLA alone. Moreover, the stereo complexation can be employed as a type of noncovalent driven force to prepare stereo complex micellar structures from PLA-based amphiphilic block copolymers with different hydrophilic blocks. Hedrick et al. [53] reported a stable complex micellar system prepared from the mixture of PDLA-*b*-PEG and poly(L-lactide)-block-poly(N-isopropylacrylamide) (PLLA-*b*- PNIPAAM). The corona of the complex micelle was composed of PEG and PNIPAAM chains, the repulsion between which was regarded to be compensated for by the peculiarly strong stereo complexation between the PDLA and the PLLA in the core. Besides stereo complexation, researchers are exploring other effective ways to enhance the stability of micellar structures from PLA-based amphiphilic block copolymers. Recently Ding et al. [43] developed a PLA-*b*-PEG micellar system for sustained intracellular drug delivery which could be controlled by dual regulation of stereo complexation and host-guest interactions (Figure 6.3). In their study, 4-arm PLLA-PEG and PDLA-PEG with the PLA end modified by cholesterol were synthesized and employed to prepare complex micelles. The stereo complexation between PLLA and PDLA could significantly improve the stability of the micelles. Moreover, poly(β-cyclodextrin) (PCD) was introduced into this stereo complex system, which could further enhance the stability of the micelles by host-guest interactions between β-cyclodextrin and cholesterol, while endowing the micellar core with increased hydrophilicity at the same time. It was found that this complex PLA-*b*-PEG micellar system had higher drug-loading capacity and a controllable drug release rate, which is promising in cancer therapy.

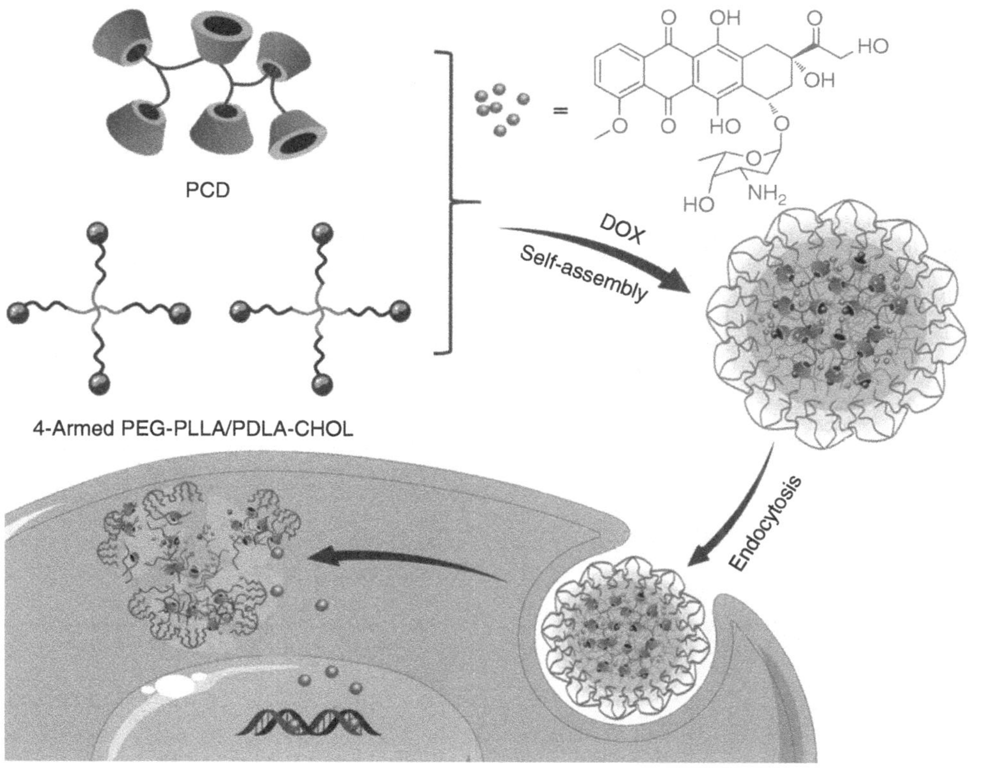

Figure 6.3 PLA-*b*-PEG micellar system for sustained intracellular drug delivery regulated by stereo complexation and host-guest interactions. *Source:* Reproduced with permission from [43]. (*See insert for color representation of the figure.*)

Micellar nanoparticles from PLA-based amphiphilic block copolymers have been widely used as drug carriers for tumor-targeting treatment in the past two decades. As is well known, in contrast to normal tissues, tumoral tissues possess a microenvironment characterized by enhanced rate of glycolysis, hypoxia, intracellular reducibility, high interstitial fluid pressure (IFP), low pH, and the presence of tumor-specific expressed proteins (such as matrix metalloproteinase-2, MMP-2), which provide potential targets for drug or gene delivery in cancer therapy [54–56]. It could be an efficient way to develop tumor-targeting delivery systems using the features of the tumoral microenvironment.

With respect to PLA-based amphiphilic block copolymers, the introduction of a tumoral microenvironment-responsive ability, such as acidity-sensitivity and redox-responsiveness, could significantly improve the anti-tumor efficacy of the drug-loaded micellar nanoparticles, which represents the cutting-edge in this field. To fulfill the pH-sensitive shedding of hydrophilic chains, pH-sensitive functional groups are applied as links between the core-forming hydrophobic chains and the coating hydrophilic chains. pH-sensitive functional groups including the diorthoester, the orthoester, the phosphoramidate, the vinyl ether, the hydrazone, and the b-thiopropionate usually undergo protonation in the low pH environment, leading to hydrolysis of the sensitive bond and therefore to detach the coating hydrophilic chains [57–60]. Recently we synthesized a pH-sensitive PEG-*b*-PLA block copolymer through introducing an acid-labile acetal group as link between the PLA and the PEG chain segments [61]. A biodegradable micellar drug delivery system with a pH-responsive sheddable PEG shell was developed and applied to the tumoral release of paclitaxel (PTX). The micelles, with a diameter of ca. 150 nm, were stable in PBS at pH 7.4 but aggregated at pH 5.5 after shedding the PEG chains. Drug release profiles in simulated human blood and tumoral environments demonstrated the sustained and acidity-triggered release of PTX from the micelles. Compared with its non-cleavable counterpart, this pH-sensitive micellar drug delivery system exhibited higher anti-tumor efficacy *in vitro* and *in vivo* [61]. Besides the pH-sensitive groups by covalent bondings, non-covalent interactions can be also applied to design pH sensitive micellar nanoparticles. Zhang et al. [62] reported pH-responsive supramolecular micelles based on benzimidazole-terminated poly(ethylene glycol) (PEG-BM) and β-cyclodextrin-modified poly(L-lactide) (CD-PLLA) (Figure 6.4a). The pH sensitivity of this micellar system was due to the pH-dependent association/

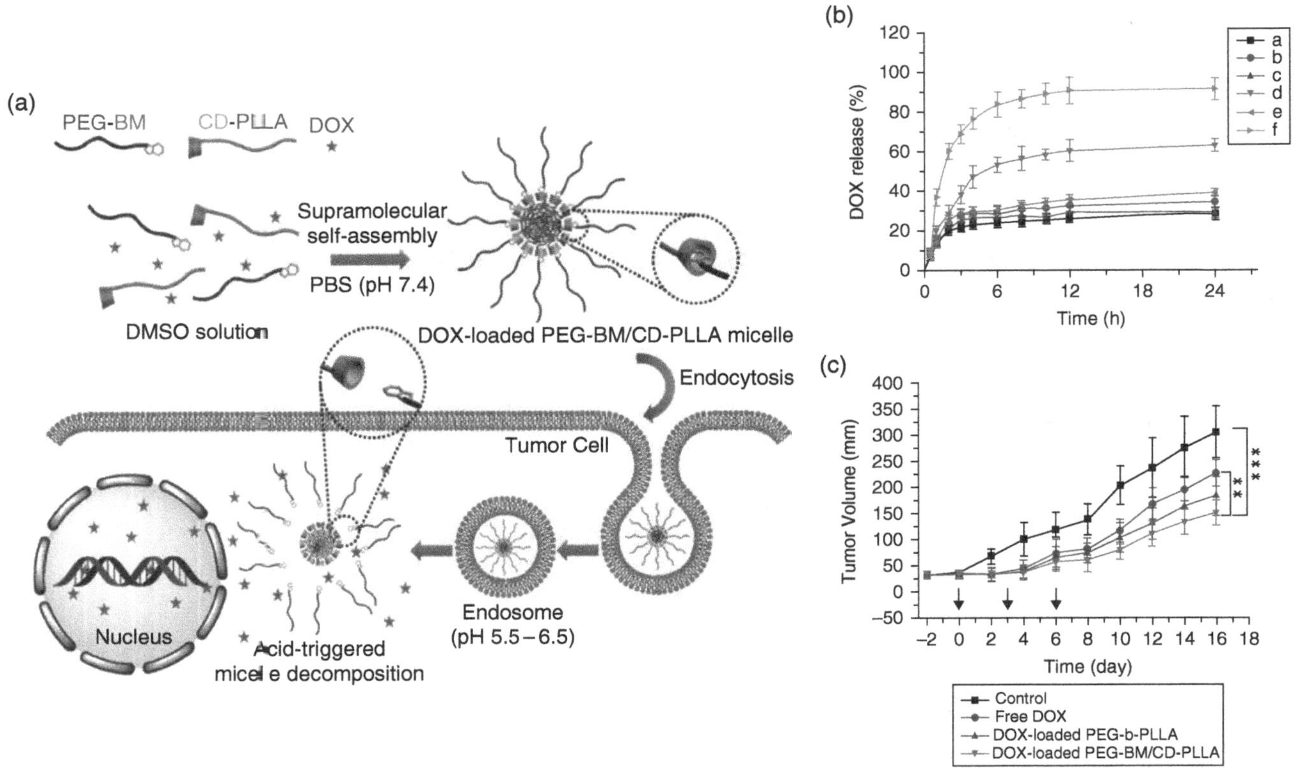

Figure 6.4 (a) Schematic illustration of formation and triggered drug release from DOX-loaded PEG-BM/CDPLLA supramolecular micelles in response to the intracellular microenvironment; (b) *In vitro* DOX release profiles of DOX-loaded micelles in PBS at 37 °C and different pH values: (a,c,e) DOX-loaded PEG-*b*- PLLA and (b,d,f) DOX-loaded PEG-BM/CD-PLLA at pH (a,b) 7.4, (c,d) 6.5, and (e,f) 5.5; (c) *In vivo* antitumor efficacies after tail-vein injection of PBS (control), free DOX, DOX-loaded PEG-b-PLLA, and DOX-loaded PEG-BM/CD-PLLA into male BALB/c nude mice bearing HepG2 xenografts. *Source:* Reproduced with permission from [62]. (*See insert for color representation of the figure.*)

dissociation of the complexes formed by PEG-BM and CD-PLLA via host-guest interactions between the benzimidazole and β-cyclodextrin groups. The supramolecular micelles dissociated in response to the acidic intracellular microenvironment because the link by host-guest interactions between benzimidazole and β-cyclodextrin groups was broken in a weak acidic environment. It was found that the release of DOX from the supramolecular micelles was significantly accelerated as the pH was reduced from 7.4 to 5.5 (Figure 6.4b). *In vivo* studies revealed that the drug-loaded PEG-BM/CD-PLLA micelles could inhibit tumor growth more efficiently than its non-responsive counterpart (Figure 6.4c).

Redox-responsive design is also a powerful way to gain the desired anti-tumor efficacy. The reductive microenvironment is attributed to the high glutathione (GSH) concentration (2 – 10 mM) in cell cytosol, which is responsive to disulfide linkages through redox reactions. Inspired by this fact, one can design redox-responsive micelles for tumor-targeted drug delivery by using disulfide linkages as the link between the hydrophobic block and the hydrophilic block. Koul et al. [63] prepared a redox-responsive biodegradable vesicles comprising of poly(ethyleneglycol)-polylacticacid-poly(ethyleneglycol) [PEG-s-s-PLA-s-s-PLA-s-s-PEG] triblock copolymer with multiple disulfide linkages, which could improve the intracellular delivery and enhance the chemotherapeutic efficacy of doxorubicin in breast cancer with minimal cardiotoxicity. Enhanced doxorubicin release was achieved in a simulated tumor environment at pH 5.0 and in the presence of 10 mM glutathione (GSH). With the active tumor-targeting ability of folic acid, this redox-responsive vesicle system showed improved cellular uptake and enhanced apoptosis in breast cancer cell lines *in vitro*, and enhanced antitumor efficacy and minimal cardiotoxicity *in vivo* conducted with Ehrlich ascites tumor-bearing Swiss albino mice. In addition, Chen et al. [64] reported a redox-responsive nano delivery system by a double emulsion method from an amphiphilic copolymer composed of a hydrophobic poly (lactic-co-glycolic acid) (PLGA) head and a hydrophilic hyaluronic acid (HA) segment linked by a disulfide bond. Dual drug-loaded particles were fabricated by incorporating doxorubicin (DOX) and cyclopamine (CYC) in the system. The resultant drug-loaded nanoparticles demonstrated a redox-responsive drug release profile *in vitro*, and achieved a remarkable synergistic anti-tumor effect and prolonged survival *in vivo*.

6.3 Hydrogels from PLA-based Amphiphilic Block Copolymers

Unlike the micellar structures formed in an aqueous solution with rather low polymer concentration, hydrogels are usually obtained at rather high polymer concentration. Although PLA is considered a typical hydrophobic polymer, it can form hydrogels through the introduction of hydrophilic chain segments such as PEG, polypeptides, and polysaccharides in specific conditions. The first hydrogel-containing PLA was reported by Jeong et al. [65] in 1997, when a triblock copolymer PEG-PLLA-PEG was employed. Combining the superior biocompatibility and biodegradability of the polymer skeleton and the high water content of hydrogel, hydrogels from PLA-based amphiphilic block copolymers are attractive materials in drug or cell delivery and tissue engineering.

6.3.1 Mechanism of Hydrogel Formation from PLA-based Amphiphilic Block Copolymers

The driven forces for hydrogel formation from PLA-based amphiphilic block copolymers mainly include hydrophobic entanglement, stereo complexation, ionic interaction, and photo-induced cross-linking polymerization [66–72].

The self-assembled micelles from PLA-based amphiphilic block copolymers are formed by hydrophobic interactions, that is by the packing of hydrophobic PLA chains at low polymer concentrations. When increasing the polymer concentration to a rather high level (e.g. 20 wt%), the polymer aqueous solution can form hydrogel probably due to the association of micelles by a chain exchange mechanism shown in Figure 6.5a [73]. The hydrophobic interactions are related tightly to heat, thus the hydrophobic entanglement in PLA-based amphiphilic block copolymers could be regulated by temperature. An aqueous solution of PLA-based amphiphilic block copolymers is in free-flowing sol phase at room temperature or lower, while it could become a gel at higher temperature. This is because at high temperatures the large aggregates of micelles may form and the hydrophobic interaction of the PLA segments is enhanced, leading to the loss of solubility and physical gelation [74, 75]. Stereo complexation has also been employed to obtain hydrogels from PLA-based amphiphilic block copolymers because of the stereo complex formation of PLLA and PDLA (Figure 6.5b) [73]. It was reported the mixture of PLLA–PEG–PLLA and PDLA–PEG–PDLA forms hydrogel in an aqueous solution at body temperature (37 °C) [76]. Hydrogels were also prepared from star block copolymers PEG–$(PLLA)_8$ and triblock copolymers of PDLA–PEG–PDLA through stereo complexation [77, 78]. In addition, other strategies such as ionic interaction and photo-induced cross-linking can be also used as driven forces for hydrogel formation. To obtain multifunctional hydrogels, multiple driven forces were applied in a single system. Feijen et al. [79] developed a rapidly *in situ* forming biodegradable robust hydrogel by combining stereo complexation and photo-polymerization from methacrylate-functionalized PEG – PLLA and PEG – PDLA star block copolymers. The authors found that stereo-complexed hydrogels could be rapidly formed (within 1 – 2 min.) in a polymer concentration range of 12.5 – 17.5% (w/v), while the methacrylate group hardly affected the stereo complexation. After that, photo-polymerization of methacrylate was performed, and the resultant hydrogels showed higher storage moduli than those of the corresponding hydrogels that were cross-linked by stereo complexation or photo-polymerization only.

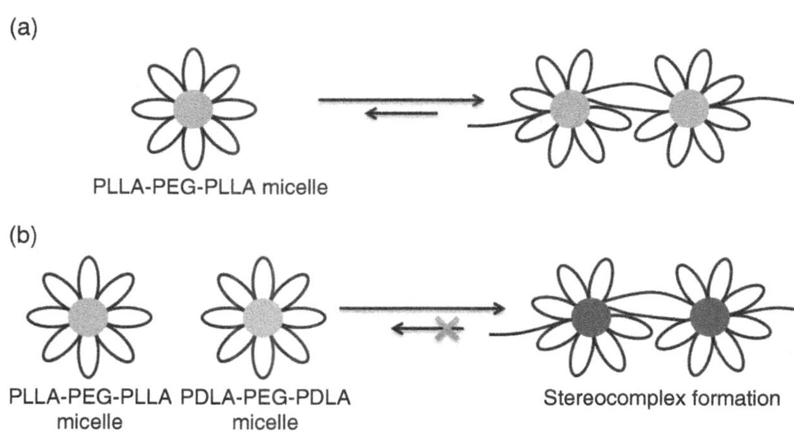

Figure 6.5 Hydrogel formation from PLA-PEG copolymers (a) driven by hydrophobic interactions or (b) stereocomplex driven. Micelles containing either PLLA or PDLA cores are designated as orange or green, respectively. However, micelles containing both PLLA and PDLA within a single core are designated by the color brown to emphasize the co-crystallization and formation of a new stereo-complexed crystal. *Source:* The concept of this figure originates from [73].

Besides the driven forces, there are several other factors determining the hydrogel formation from PLA-based amphiphilic block copolymers such as the concentration of copolymer, molecular weight and chain length of both hydrophobic and hydrophilic blocks. Jeong et al. [65] for the first time prepared hydrogels from PLLA-PEG diblock copolymer and PEG-PLLA-PEG triblock copolymer in an aqueous solution. The authors studied the gelation behavior of the polymer solutions with variable polymer concentrations and chain length of the PLLA block. It was found that the sol-gel transition temperature of the polymer solutions increased as the polymer concentration and chain length of PLLA block increased both for PLLA-PEG and PEG-PLLA-PEG (Figure 6.6). In other words, increasing the PLLA block length increases the aggregation tendency of a block copolymer in water, resulting in a steepening of the gel-sol transition curve slopes and the onset of gelation at a lower concentration. Conversely, a higher concentration of polymer is needed as the PEG block length increases. Ding et al. [74, 75, 80] studied the effect of molecular weight distribution of PEG chains on the thermo-gelling behavior of a PLA-PEG aqueous solution and provided a possible mechanism for the gelation of PEG–PLA based hydrogel systems. They proposed that water molecules surrounded the PEG chains in order of hydrogen bondings, which could lead to a significant entropy loss of water. The increased temperature could enhance this entropy effect and thus increase the hydrophobicity of PEG chains, which facilitated the aggregate formation from the micelles. On the other hand, the hydrophobic interaction of PLA segments was also enhanced at higher temperature. These phenomena could result in the physical gelation and loss of solubility.

6.3.2 Properties and Biomedical Applications of Hydrogel from PLA-based Amphiphilic Block Copolymers

Thermo-responsiveness is one of the most attractive properties of hydrogels from PLA-based amphiphilic block copolymers. As an example, the aqueous solution of PLA-PEG-polyurethane copolymer remained soluble at 45 °C, while it formed a gel at 37 °C [65]. This hydrogel has the potential to be used in the body as a drug/cell carrier and tissue engineering scaffold by injection, where it can be injected in a liquid state, but quickly form a gel inside the body. Another example is the multi-block PLA-PEG copolymer consisting of alternating PLA and PEG blocks, which was synthesized by polycondensation of PEG with α,ω-dicarboxylic acid-terminated PLA. At low temperatures, self-assembled micelles were formed from this multiblock copolymer with a diameter of 20 nm in an aqueous solution; but hydrogel was formed as the temperature increased, which was ascribed to the aggregation of the micelles [81]. Additionally, star-like PLA-PEG copolymers also exhibited thermos-responsive properties. It was reported that an 8-Arm PEG-b-PLLA copolymer conjugated with cholesterol could form temperature-induced hydrogels by self-assembly, which could be used as an injectable scaffold for tissue engineering [82]. Hydrogels of PLA with polysaccharides (such as chitosan) have been reported to have thermos-responsiveness and be used in drug delivery. Seo et al. [83] prepared thermo-sensitive hydrogels from PLGA and chitosan by graft copolymerization, which were then applied for the entrapment and release of drugs, using vancomycin hydrochloride and betamethasone sodium phosphate as model drugs.

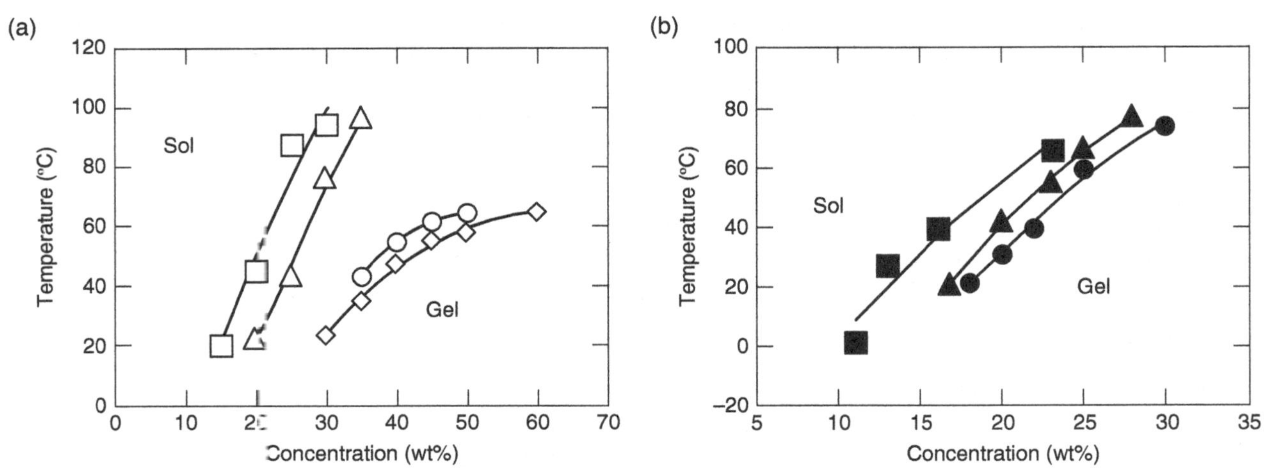

Figure 6.6 Gel-sol transition curves: (a) PEO-PLLA diblock copolymers with Mr values as follows: diamonds, 5000–720; circles, 5000–1000; triangles, 5000–1730; squares, 5000–1960; (b) PEO-PLLA-PEO triblock copolymers with Mr values as follows: filled circles, 5000–2040–5000; filled triangles, 5000–3000–5000; filled squares, 5000–5000–5000. The gel–sol transition temperature was determined as follows. *Source:* Reproduced with permission from [65].

Mechanical properties of hydrogels are very important, especially when used as a scaffold for tissue engineering. The mechanical properties, including strength and elastic modulus, are determined by many factors for PLA-based amphiphilic copolymers. The employment of driven forces is the dominant factor deciding the mechanical properties of hydrogels. As for the hydrogel driven by stereo complexation, the hydrogel strength and elastic modulus usually increase as the mixing time of the stereo complex increases in scope and as the stereo complex concentration increases [84]. The introduction of multiple driven forces is an effective strategy to obtain hydrogels with superior mechanical properties. Fan et al. [85] reported a PLA-PEG-based hydrogel which was prepared from acrylate-PEG-PLLA and acrylate-PEG-PDLA by dual cross-linking of radical polymerization of acrylate and stereo complexation of PLLA and PDLA. Compared to its non-stereo complex counterpart, this dual co-network was stronger. Moreover, it was observed that the storage and loss moduli of this co-network hydrogel increased with the increasing amount of PLA stereo complex.

The pH responsiveness is one of the most attractive properties for hydrogels used for biomedical applications. Hydrogel from copolymerization of L-lactic acid and citric acid followed by grafting with chitosan was reported to have pH responsiveness [86]. This hydrogel could support fibroblast cells better than chitosan film alone. Hydrogels with pH responsiveness can be prepared through host-guest interactions. Using the acid-labile of host-guest interaction between benzimidazole and β-cyclodextrin, Zhang et al. [87] fabricated a pH-responsive hydrogel system based on benzimidazole-terminated poly(ethylene glycol) and β-cyclodextrin-modified poly(L-lactide). *In vitro* drug release profile showed that the release of doxorubicin was triggered in a simulated tumoral acidic environment where the pH is about 5.5.

6.4 Conclusion

PLA-based amphiphilic copolymers can form micellar aggregates, including spherical micelles, rod-like micelles, or vesicles by self-assembly in aqueous solution when the polymer concentration is low. The morphologies of formed aggregates are mainly affected by the copolymer composition and concentration, the water content in the solution, or the nature of the common solvent, among others. The self-assembled aggregates of PLA-based amphiphilic copolymers show great prospects in applications in the biomedical field, such as drug delivery, especially when the nanoparticles possess environment stimuli-responsive properties. On the other hand, hydrogels are formed from the aqueous solution of PLA-based amphiphilic copolymers when the polymer concentration is rather high under specific conditions. The driven force, copolymer concentration, molecular weight and distribution of copolymer blocks, and the temperature among others, are the factors determining the hydrogel formation. Since they have good biocompatibility, high water content, and superior physical and chemical properties such as tunable mechanical properties and stimuli-responsiveness, hydrogels based on PLA copolymers have promising potential in developing biomedical materials especially for tissue engineering scaffolds.

Acknowledgments

The authors acknowledge the China Postdoctoral Science Foundation (General Program, No. 2015M580640) and the National Natural Science Foundation of China (General Program, No. 21574050) for financial support.

References

1 Basu, A., Kunduru, K.R., Doppalapudi, S. et al. (2016) *Adv. Drug Deliv. Rev.*, **107**, 192–205.
2 Tsuji, H. (2016) *Adv. Drug Deliv. Rev.*, **107**, 97–135.
3 Uhrich, K.E., Cannizzaro, S.M., Langer, R.S., Shakesheff, K.M. (1999) *Chem. Rev.*, **99**(11), 3181–3198.
4 Drumright, R.E., Gruber, P.R., Henton, D.E. (2000) *Adv. Mat.*, **12**(23), 1841–1846.
5 Dechy-Cabaret, O., Martin-Vaca, B., Bourissou, D. (2004) *Chem. Rev.*, **104**(12), 6147–6176.
6 Jacobson, G.B., Shinde, R., Contag, C.H., Zare, R.N. (2008) *Angew. Chem. Int. Ed.*, **47**(41), 7880–7882.
7 Nomoto, T., Matsumoto, Y., Miyata, K. et al. (2011) *J. Cont. Rel.*, **151**(2), 104–109.
8 Ko, Y.T., Bhattacharya, R., Bickel, U. (2009) *J. Cont. Rel.*, **133**(3), 230–237.
9 Jeong, J.H., Kim, S.W., Park, T.G. (2007) *Prog. Polym. Sci.*, **32**(11), 1239–1274.
10 Abebe, D.G., Liu, K.Y., Mishra, S.R. et al. (2015) *RSC Adv.*, **5**(116), 96019–96027.
11 Sahoo, S., Chung, C., Khetan, S., Burdick, J.A. (2008) *Biomacromol.*, **9**(4), 1088.
12 Motulsky, A., Lafleur, M., Couffin-Hoarau, A.C. et al. (2005) *Biomat.*, **26**(31), 6242–6253.
13 Wu, X.L., Kim, J.H., Koo, H. (2010) *Bioconj. Chem.*, **21**(2), 208–213.
14 Yang, J., Lee, C.H., Ko, H.J. (2007) *Angew. Chem., Int. Ed.*, **46**, 8836.

15 Wang, Q., Wang, L., Detamore, M.S., Berkland, C. (2008) *Adv. Mat.*, **20**(2), 236–239.
16 Zhang, L., Eisenberg, A. (1996) *J. Am. Chem. Soc.*, **118**(13), 3168–3181.
17 Zhang, L., Eisenberg, A. (1998) *Polym. Adv. Technol.* **9**(10–11), 677–699.
18 Yang, L., Wu, X., Liu, F. et al. (2009) *Pharm. Res.*, **26**(10), 2332–2342.
19 Alexandridis, P., Holzwarth, J.F., Hatton, T.A. (1994) *Macromol.* **27**(9), 2414–2425.
20 Decher, G. (1997) *Science*, **277**(5330), 1232–1237.
21 Caruso, F. (2001) *Adv. Mat.*, **13**(1), 11–22.
22 Shum, H.C., Kim, J.W., Weitz, D.A. (2008) *J. Am. Chem. Soc.*, **130**(29), 9543–9549.
23 Tyrrell, Z.L., Shen, Y., Radosz, M. (2010) *Prog. Polym. Sci.*, **35**(9), 1128–1143.
24 Jain, S., Bates, F.S. (2003) *Science*, **300**(5618), 460–464.
25 Mai, Y., Eisenberg, A. (2010) *J. Am. Chem. Soc.*, **132**(29), 10078–10084.
26 Jeong, B., Bae, Y.H., Lee, D.S., Kim, S.W. (1997) *Nature*, **388**(6645), 860.
27 Mai, Y., Eisenberg, A. (2012) *Chem. Soc. Rev.*, **41**(18), 5969–5985.
28 Zhang, L., Eisenberg, A. (1996) *J. Am. Chem. Soc.*, **118**(13), 3168–3181.
29 Zhang, L., Eisenberg, A. (1998) *Polym. Adv. Technol.*, **9**(10-11), 677–699.
30 Oh, J.K. (2011) *Soft Matt.*, **7**(11), 5096–5108.
31 Discher, D.E., Ahmed, F. (2006) *Annu. Rev. Biomed. Eng.*, **8**, 323–341.
32 Walther, A., Goldmann, A.S., Yelamanchili, R.S. et al. (2008) *Macromol.*, **41**(9), 3254–3260.
33 Yu, Y., Zhang, L., Eisenberg, A. (1998) *Macromol.*, **31**(4), 1144–1154.
34 Tung, P.H., Kuo, S.W, Chen, S.C. et al. (2007) *Polym.*, **48**(11), 3192–3200.
35 Meng, F., Hiemstra, C., Engbers, G.H., Feijen, J. (2003) *Macromol.*, **36**(9), 3004–3006.
36 Lee, Y., Chang, J.B., Kim, H.K., Park, T.G. (2006) *Macromol. Res.*, **14**(3), 359–364.
37 Du, B., Mei, A., Yin, K. (2009) *Macromol.*, **42**(21), 8477–8484.
38 Poon, Z., Chang, D., Zhao, X., Hammond, P.T. (2011) *ACS Nano*, **5**(6), 4284.
39 Davis, M.E., Shin, D.M. (2008) *Nat. Rev. Drug Disc.*, **7**(9), 771–782.
40 Hatakeyama, H., Akita, H., Harashima, H. (2011) *Adv. Drug Del. Rev.* **63**(3), 152–160.
41 Oh, J.K. (2011) *Soft Matt.*, **7**(11), 5096–5108.
42 Inkinen, S., Hakkarainen, M., Albertsson, A.C., Södergård, A. (2011) *Biomacromol.*, **12**(3), 523–532.
43 Feng, X., Ding J., Gref, R., Chen, X. (2017) *Chin. J. Polym. Sci.*, **35**(6): 693–699.
44 Gratton, S.E., Ropp, P.A., Pohlhaus, P.D. et al. (2008) *Proc. Nat. Acad. Sci.*, **105**(33), 11613–11618.
45 Edinger, D., Wagner, E. (2011) *Wiley Interdisc. Rev.: Nanomed Nanobiotech*, **3**(1), 33–46.
46 Romberg, B., Hennink, W.E., Storm, G. (2008) *Pharm. Res.*, **25**(1), 55–71.
47 Yang, X.Z., Du, J.Z., Dou, S. (2011) *ACS Nano*, **6**(1), 771–781.
48 Ikada, Y., Jamshidi, K., Tsuji, H., Hyon, S.H. (1987) *Macromol.*, **20**(4), 904–906.
49 Sarasua, J.R., Rodríguez, N.L., Arraiza, A.L., Meaurio, E. (2005) *Macromol.*, **38**(20), 8362–8371.
50 Kang, N., Perron, M.È., Prud'Homme, R.E., et al. (2005) *Nano Lett.*, **5**(2), 315–319.
51 Chen, L., Xie, Z., Hu, J. (2007) *J. Nanopart. Res.*, **9**(5), 777.
52 Riley, T., Stolnik, S., Heald, C.R. (2001) *Langmuir*, **17**(11), 3168–3174.
53 Kim, S.H., Tan, J.P., Nederberg, F. (2008) *Macromol.*, **42**(1), 25–29.
54 Gupta G.P, Massague J. (2006) *Cell*, **127**(4), 679–695.
55 Fidler I.J. (2003) *Nat. Rev Canc.*, **3**(6), 453–458.
56 Sung S, Johnstone P.A.S. (2007) *Curr, Prob. Canc,.* **31**(2), 36–100.
57 Choi, J.S., MacKay, J.A., Szoka, F.C. (2003) *Bioconj. Chem.*, **14**(2), 420–429.
58 Masson, C., Garinot, M., Mignet, N. et al. (2004) *J. Cont. Rel.*, **99**(3), 423–434.
59 Sawant, R.M., Hurley, J.P., Salmaso, S. et al. (2006) *Bioconj. Chem.*, **17**(4), 943.
60 Oishi, M., Nagasaki, Y., Itaka, K., et al. (2005) *J. Am. Chem. Soc.* **127**(6), 1624–1625.
61 Xiao, L., Huang, L., Moingeon, F. et al. (2017) *Biomacromol.*, **18**(9), 2711–2722.
62 Zhang, Z., Lv, Q., Gao, X., et al. (2015) *ACS App. Mat. Interf.*, **7**(16), 8404–8411.
63 Lale, S.V., Kumar, A., Prasad, S. et al. (2015) *Biomacromol.*, **16**(6), 1736–1752.
64 Hu, K., Zhou, H., Liu, Y. et al. (2015) *Nanosci.*, **7**(18), 8607–8618.
65 Jeong, B., Bae, Y.H., Lee, D.S., Kim, S.W. (1997) *Nature*, **388**(6645), 860–862.
66 Lee, J., Bae, Y.H., Sohn, Y.S., Jeong, B. (2006) *Biomacromol.*, **7**(6), 1729–1734.
67 Mason, M.N., Metters, A.T., Bowman, C.N., Anseth, K.S. (2001) *Macromol.*, **34**(13), 4630–4635.
68 Slager, J., Domb, A.J. (2003) *Adv. Drug Deliv. Rev.*, **55**(4), 549–583.
69 Abebe, D.G., Fujiwara, T. (2012) *Biomacromol.*, **13**(6), 1828–1836.
70 Wang, Q., Wang, L., Detamore, M.S., Berkland, C. (2008) *Adv. Mater.*, **20**(2), 236–239.
71 Hiemstra, C., Zhou, W., Zhong, Z. et al. (2007) *J. Am. Chem. Soc.*, **129**(32), 9918–9926.

72 Censi, R., Vermonden, T., Deschout, H. et al. (2010) *Biomacromol.*, **11**(8), 2143–2151.

73 Abebe, D.G., Liu, K.Y., Mishra, S.R. et al. (2015) *RSC Adv.*, **5**(116), 96019–96027.

74 Chen, L., Ci, T., Li, T. et al. (2014) *Macromol.*, **47**(17), 5895–5903.

75 Chen, L., Ci, T., Yu, L., Ding, J. (2015) *Macromol.*, **48**(11), 3662–3671.

76 Mukose, T., Fujiwara, T., Nakano, J. et al. (2004) *Macromol. Biosci.*, **4**(3), 361–367.

77 Hiemstra, C., Zhong, Z., Dijkstra, P.J., Feijen, J. (2005) In *Macromolecular Symposia* (Vol. 224, No. 1, pp. 119–132). WILEY-VCH Verlag, Mannheim.

78 Hiemstra, C., Zhong, Z.Y., Jiang, X. et al. (2006) *J. Cont. Rel.*, **116**(2), e17–e19.

79 Hiemstra, C., Zhou, W., Zhong, Z., Wouters, M., Feijen, J. (2007) *J. Am. Chem. Soc.*, **129**(32), 9918–9926.

80 Yu, L., Ding, J. (2008) *Chem. Soc. Rev.*, **37**(8), 1473–1481.

81 Lee, J., Bae, Y.H., Sohn, Y.S., Jeong, B. (2006) *Biomacromol.* **7**(6), 1729–1734.

82 Nagahama, K., Ouchi, T., Ohya, Y. (2008) *Adv. Funct. Mat.*, **18**(8), 1220–1231.

83 Ganji, F., Abdekhodaie, M.J. (2010) *Carbohyd. Polym.*, **80**(3), 740–746.

84 Li, S., El Ghzaoui, A., Dewinck, E. (2005). In *Macromolecular Symposia* (Vol. 222, No. 1, pp. 23–36). WILEY-VCH Verlag, Mannheim.

85 Fan, X., Wang, M., Yuan, D., He, C. (2013) *Langmuir*, **29**(46), 14307–14313.

86 Wu, Y., Zheng, Y., Yang, W., et al. (2005) *Carbohyd. Polym.*, **59**(2), 165–171.

87 Zhang, Z., Lv, Q., Gao, X., et al. (2015) *ACS Appl. Mat. Interf.*, **7**(16), 8404–8411.

Part II

Biomacromolecules

7

Electroconductive Bioscaffolds for 2D and 3D Cell Culture

Zhijun Shi[1,2], Lin Mao[1,2], Muhammad Wajid Ullah[1,2], Sixiang Li[1,2], Li Wang[1,2], Sanming Hu[1,2], and Guang Yang[1,2,]*

[1] *College of Life Science and Technology, Huazhong University of Science and Technology, Wuhan, China*
[2] *National Engineering Research Center for Nano-Medicine, Huazhong University of Science and Technology, Wuhan, China*

7.1 Introduction

Bioscaffolds are artificial structural materials which can potentially be used to promote tissue regeneration and injury recovery after their implantation into the human body by various methods. The ideal scaffolds should demonstrate certain characteristic features such as:

1) They should be highly biocompatible to support the cell adhesion, proliferation, differentiation, and migration.
2) They should possess a fine three-dimensional (3D) structure. The fine 3D structure of scaffolds has a good influence on the cell activity and growth of tissues. The fine 3D structure of the scaffold is achieved through special process technology by controlling the specific parameters, such as surface properties, geometry, and porosity, etc.
3) Different tissues have different mechanical properties, thus the scaffold used for the regeneration of new tissues must possess the appropriate mechanical features and be degraded at a suitable rate compared to the host tissues.
4) The scaffold should direct and control the behavior of the cells and tissues during the interaction between the host cells.
5) The scaffolds must have appropriate biochemistry and nano-/micro-scale surface topographies and form desirable binding sites [1].

In addition to these characteristic features, nowadays bioscaffolds have been used for regenerative medicine and tissue engineering to effect cell activity and function owing to their diverse biological, structural, chemical, and physiological functions. Such scaffolds can direct the microenvironments (including stiffness, pore size, cell-matrix interactions, or sensitivity to the electrical signal) of cells, and promote cell adhesion, proliferation, spreading, migration, and differentiation in response to electrical signals [2–4]. Many of those reaction processes involve different extracellular transmissions. For example, endogenous DC electric fields (EFs) are one such important signaling, which, if present in electrically conductive pathways, can induce polarized ion transport and current flow to influence the correlations between the adjacent cells for their changed voltage and ion flux so as to mimic this kind of electrically conductive pathway and to build this system [5, 6]. The electroconductive bioscaffold is a biocompatible electrically conductive pathway to facilitate electrical communication among the cells (Figure 7.1). Once combined with electrical stimulation, bioscaffolds can influence cell performances, such as cell attachment, viability, migration, and differentiation, and further help tissue regeneration, especially in neuron, muscular, bone, and skin tissue regeneration [7].

This chapter focuses on methods and approaches used for the development of electroconductive bioscaffolds. Further, it details the phenomenon of how to culture various cells, such as neurons, myoblasts, osteoblasts, and fibroblasts. It also overviews the role of stimulation with electrical signals which directs neuron, muscular, bone, and skin tissue regeneration.

7.2 Electrical Stimulation

An electric field (EF) at a given point is defined as the amount of force exerted on a static particle of unit charge by electromagnetic forces. It gives an estimation of the total potential difference between the two given points located at a known distance and is measured in volts per

* Email: yang_sunny@yahoo.com

Bioinspired Materials Science and Engineering, First Edition. Edited by Guang Yang, Lin Xiao, and Lallepak Lamboni.

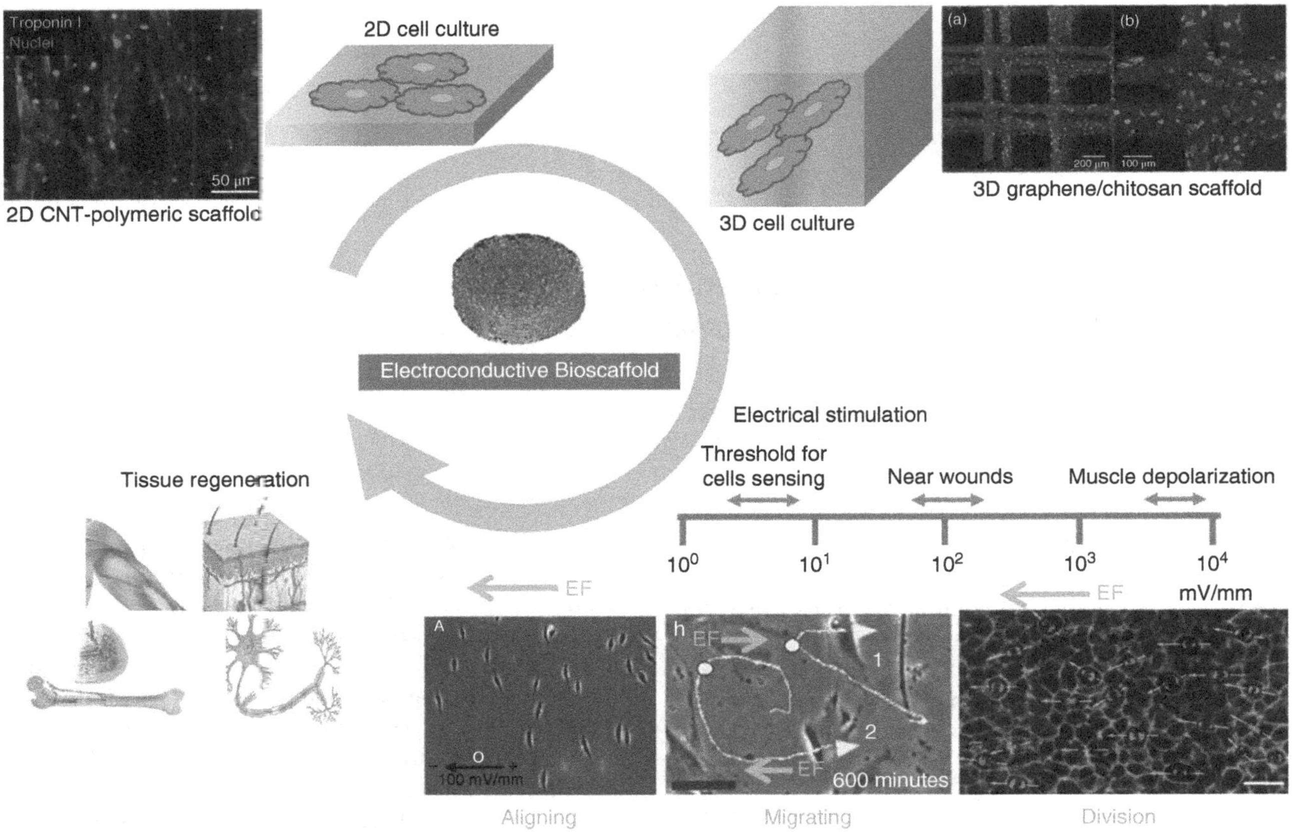

Figure 7.1 Electroconductive bioscaffold for 2D and 3D cell cultures. Following stimulation with electrical signals, vertebrate cells usually respond to EF by promoting cell alignment, division, migration, and growth parallel to the EF lines. An electroconductive bioscaffold could be used as a conductive, biomimetic scaffold for the regeneration of skin, nerves, and skeletal muscles. *Source:* Reproduced with permission from [7, 8].

distance (V/m). The intensity of the EF largely depends on a voltage source (usually a battery) and a medium (known as a conductive pathway) [9]. In biological systems, an endogenous EF is generated by two components: a polarized ion transport and a conductive extracellular pathway. In nature, biological systems possess two inherent voltage sources: the trans-plasma membrane electrical potential and trans-epithelial potential (TEP). A conductive extracellular pathway is created when a plasma membrane or tight-junctional epithelium is injured or becomes dysfunctional, which cause short-circuiting of voltage source and results in a current flow at the injury site. A study carried out to investigate wound healing in human and rodent cornea revealed the continuous generation of electric current. Specifically, a continuous current of $3\mu Acm^{-2}$ was generated at the wound edges of rat rodent cornea skin which was sustained for several hours [10]. Similarly, the extracellular EF in the vicinity of damaged neurons spread to a distance of several millimeters from the site of injury and sustained for several days. For example, injury to the spinal cord of th guinea pig generated an enormous ($\leq mA/cm^2$) bioelectric current at the site of the injury. The intensity of the endogenous current decreased with time and distance from the site of injury, however, it maintained a lower but significant value of $<50\ \mu A/cm^2$ [11]. EFs are also generated during embryonic development where the electro-conductive pathways are naturally developed and not initiated by some injury [12, 13].

Endogenous direct current electric fields (DC-EFs) play a vital role in injury development and wound recovery. Removal of endogenous EFs, replacing them with some pharmacological agents, or changing the polarity of EFs significantly affects injury development and wound recovery. However, increasing the intensity of endogenous EFs enhances the overall rate of wound healing and extent of their regeneration capacity [9]. EFs have been applied to humans in several clinical trials to provide a devastating signal for the enhanced healing potential of chronic wounds. These clinical practices and several other studies led to a positive coverage decision by the U.S. Department of Health and Human Services in 2002 to approve electrical stimulation (ES) for the treatment of various chronic wounds, for example, pressure ulcers, arterial ulcers, diabetic ulcers, and venous stasis ulcers [14]. ES has also been successfully used to promote the healing of the spinal cord and during peripheral nerve regeneration [15]. Successful clinical trials using functional ES systems have enabled several individuals with spinal cord injury to effectively use their paralyzed arms, hands, arms, legs, trunk, and diaphragm for normal functioning and attained a degree of control over bladder and bowel evacuation [16]. In addition, bone, because it possesses natural electrical conductivity, such as piezoelectricity [17]. can create an endogenous electrical field in response to strains that alter cell proliferation. Thus, ES can significantly influence the bone healing process [18]. Furthermore, ES can also be used for cancer therapy. For example, a study has shown the effect of neuromuscular electrical stimulation in patients suffering from dysphagia following the treatment for head and neck cancer. The results demonstrated that neuromuscular electrical stimulation combined with traditional swallowing training was much more effective than traditional swallowing training alone in these patients [19].

Studies have shown that various types of ES can normalize the physiological functions of cells, including cell division, migration, differentiation, and apoptosis. Generally, the cultured vertebrate cells show a weak response to DC-EFs through alignment, migration, and growth along a specific direction with respect to EF lines. Generally, the movement of cells is observed to be perpendicular to the EF lines and directed either toward the cathode or the anode (see Table 7.1).

ES can regulate cell physiological activities which play a vital role in the development and regeneration of organisms. Thus, if bioscaffolds incorporated with electroconductive properties are developed, these can be effectively used for improved tissue regeneration under an EF that mimics the natural environment of a living system, and thus can further be used in the fields of bone, muscle, neuron regeneration, and wound healing, etc.

7.3 Electroconductive Bioscaffolds

7.3.1 Conductive Polymers-based Electroconductive Bioscaffolds

Conductive polymers (CPs) are a series of functional polymers possessing good electrical conductivity and demonstrate reversible chemical, electrochemical, and physical properties through a doping/de-doping process on a delocalized π-system backbone along the polymeric chains [28]. These have received immense consideration in the last two decades owing to the various physical and chemical properties of organic polymers while maintaining the electrical characteristics of metals [29] and are receiving growing interest for the construction of electronics, photonics, phononics, and biomedical devices.

By making use of conducting polymers, researchers have developed diversified methods to fabricate electroconductive bioscaffolds. Various methods used for the preparation of conducting polymers are briefly described in the following:

1) CPs are blended with a scaffold solution and precursors or monomers through a variety of sol-gel transition processes, including cross-linking, casting, drying or electrospinning to form an electroconductive hydrogel, aerogel, nanofiber, or membrane, etc. For instance, polyaniline (PAni) and polypyrol (PPy) are

Table 7.1 Some examples of vertebrate cells' responses to weak DC-EFs.

Cells' response to EFs	Intensity of electrical stimulation	Cell type	Ref.
Aligned perpendicular to the EF lines	20–200mV/mm	Fibroblast cell	[20]
	106 mA cm^{-2} at 245 mV mm^{-1}	Schwann cells	[21]
	50 ~ 600 mV/mm	Retinal pigment epithelial cells	[22]
Cell division is aligned perpendicular to EF lines	0 ~ 500% transcorneal potential difference(40mV)	Corneal epithelia cells	[23]
Drift toward the cathode	150 ~ 400 mV/mm	Human microvascular endothelial cells	[24]
	~7 V/cm	T lymphocytes	[25]
	3 V/cm	Lung cancer cells	[26]
Migrate toward the anode	1.5 ~ 5 V/cm	Breast cancer cells	[27]
	150 ~ 400 mV/mm	Vascular fibroblasts, human umbilical vein endothelial cells, smooth muscle cells	[24]

widely used CPs which can be blended with PCL, poly(lactic acid) (PLA) or gelatin to form electroconductive nanofibers through electrospinning. Even without ES, cells' adhesion, proliferation, and spreading can be achieved and enhanced by using these electroconductive nanofibers, which has been shown to be effective for different types of cells, including fibroblasts, myoblasts, neural cells, etc. [30–32]. Moreover, the electrospinning technique is used to make aligned or oriented nanofibers to guide cells' differentiation. For example, highly aligned and electrically conductive PCL/PAni nanofibers are prepared in a magnetic field by electrospinning, which can serve as functional scaffolds and guide myoblast orientation and accelerate myotube formation and maturation [33].

2) Electroconductive bioscaffolds (Figure 7.2) can be constructed via the polymerization of CP precursors or monomers into an existing nanostructure of nonconductive scaffolds. To this end, the CPs are introduced into the matrix through *in situ* chemical or electrochemical polymerization. Natural hydrogels with a fine 3D structure such as bacterial cellulose (BC), offer an alternative matrix for CPs' polymerization [34, 35].

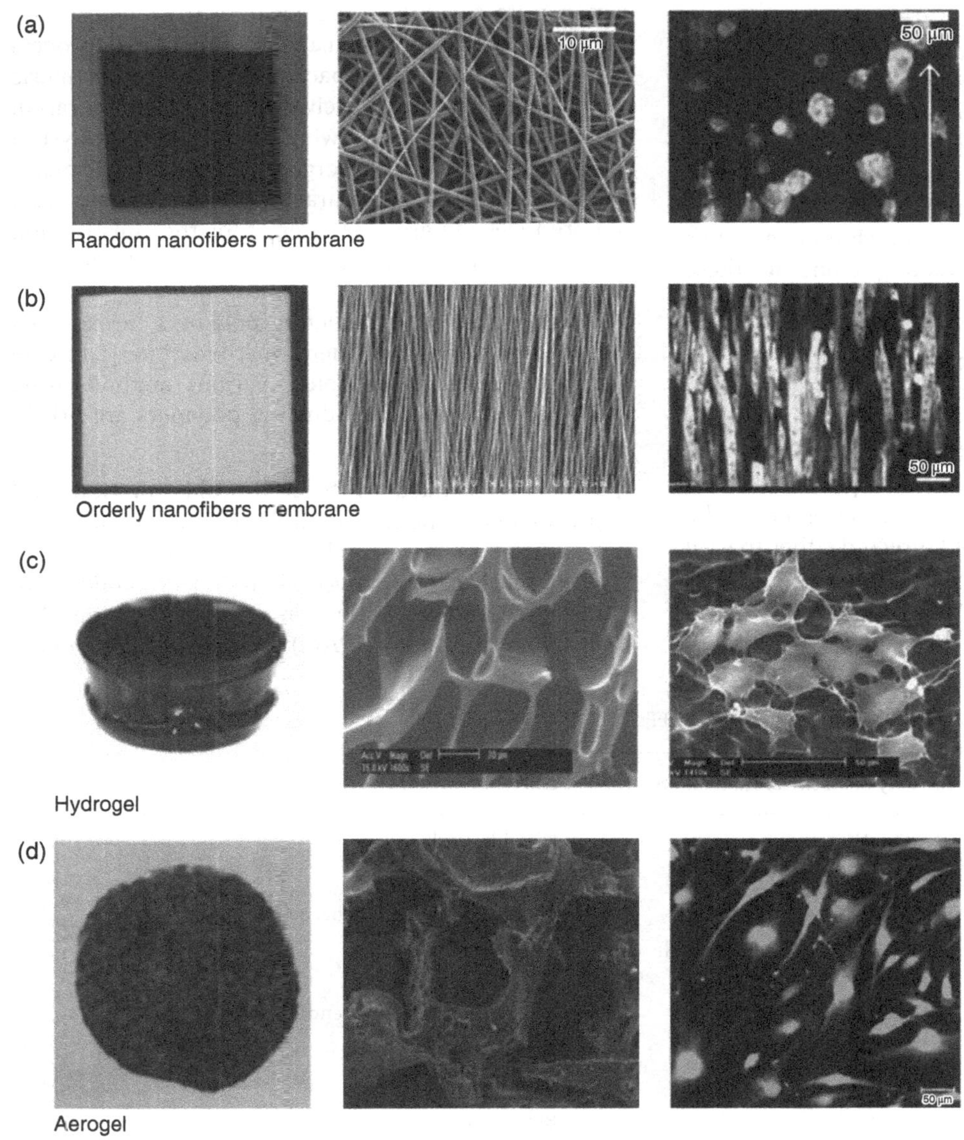

Figure 7.2 Diversified electroconductive bioscaffolds based on conductive polymers. *Source:* Reproduced with permission from [33, 37–39].

3) Conductive oligomers, such as aniline pentamer, combined with other active molecules could be used as a functional branch for electroconductive bioscaffolds through cross-linking and graft reactions. For example, aniline pentamer (AP) was grafted into carboxymethyl chitosan in the presence of glutaraldehyde, which formed a cytocompatible and electrically conductive hydrogel that supported the adhesion and proliferation of C2C12 myoblast cells and reduced the burst release of bioactive molecules [36].

Under electrical stimulation, CPs-based electroconductive bioscaffolds showed promising properties for 2D and 3D cell cultures. Some recent studies are summarized in Table 7.2 to support the potential use of electrically stimulated CPs-based electroconductive bioscaffolds in 2D and 3D cell culturing.

1) With proper electrical stimulation, CPs-based electroconductive bioscaffolds are capable of promoting axonal regeneration in nerve injury repair. Applying ES through conductive PPy/chitosan combined with olfactory ensheathing cells (OECs) significantly enhanced the proliferation of cells besides enhancing the expression and secretion of mRNA and protein levels in OECs. These results indicated the increased proliferation and neurotrophin secretion by ES in conductive scaffolds likely contributed toward the enhancement of nerve regeneration [40]. In another study, a composite of PPy and poly(D, L-lactic acid) (PDLLA) was fabricated to serve as a nerve conduit and used to culture PC12 cells. The cells exhibited enhanced and prolonged neurite when stimulated with 100mV for 2 h. Utilization of a nerve conduit to fill a 10 mm-long injury in the sciatic nerve demonstrated excellent recovery. Its efficacy was comparable to gold standard autologous nerve graft and much more efficient than the conventional PDLLA-based conduits. This illustrates the great potential of the conduit for neural tissue regeneration [41]. In another study, a nanometer-thick coating of PPy was spread on submicrometer-scale poly(lacticacid-co-glycolic acid) (PLGA) fibers presenting abundant carboxylic groups. After immobilizing the nerve growth factor (NGF) onto the surface of the fibers, the scaffold effectively enhanced the PC12 neurite formation and neurite outgrowth beside improving neurite development and length under electrical stimulation [42].
2) Endogenous electric currents that are generated instantly at skin wounds guide the migration of epithelial cells [10], hence the electrical signal plays an important role in the wound healing process, for which CPs-based electroconductive bioscaffolds have incomparable prevalence in promoting wound healing rather than using normal ones. After applying 100 mV DC stimulation on an electroconductive PPy/PDLLA membrane, the growth of fibroblasts cultured on it was significantly up-regulated [43]. CPSA-PAni/PLCL nanofibers were prepared using an electrospinning technique, which controlled the growth of NIH-3T3 fibroblasts through the application of various direct ES.
3) Bone possesses natural electrical properties such as piezoelectricity [17]. An endogenous electrical field can be generated in response to strains to alter cell proliferation [18]. Thus, electroconductive bioscaffolds can be helpful in regulating the cellular activities for bone tissue regeneration [44]. A study showed that a PPy–alginate blend incorporated with chitosan, when employed as a conducting scaffold, effectively supported the adhesion of human osteoblast-like cells (MG63) [45]. In another study, conductive scaffolds containing PLA, heparin, and PPy were applied to culture Saos-2 cells for the study of osteoblast's proliferation under ES. The results showed that important genes of osteoblast markers (including alkaline phosphatase and osteocalcin) and relevant protein production were modulated with specific ES parameters [46].
4) Muscle tissue regeneration is another interesting area where CPs-based electroconductive bioscaffolds function by regulating the cellular activities. Under ES, myo-nuclear density and myoblast differentiation were significantly enhanced when used with an aligned MWNTs with *para*-toluene sulphonic acid doped PPy (PPy-*p*TS) platforms [47]. To recover from myocardial infarct, the CPs-based electroconductive bioscaffolds are used to improve the cardiac function after bioscaffold implantation. By grafting PPy to chitosan, Mihic et al. [48] fabricated a PPy-chitosan hydrogel which demonstrated high biocompatibility and showed enhanced Ca^{2+} signal conduction in neonatal rat cardiomyocytes, thus it effectively improved the electric coupling between skeletal muscles. In the *in vivo* test, the injection of PPy-chitosan turned out to be an improved cardiac function after myocardial infarction compared to a nonconductive scaffold. These studies provide a base for future studies to improve the impulse propagation and heart function.

7.3.2 Carbon Nanotubes-based Electroconductive Bioscaffolds

Carbon nanotubes (CNTs) are well-ordered, hollow nanostructures comprised of carbon atoms bonded to each other via sp2 bonds that are stronger than sp and sp3 bonds. There are two main types of carbon nanotubes: single-walled carbon nanotubes (SWCNTs) and multi-walled carbon nanotubes (MWCNTS) which have

Table 7.2 Conductive polymers-based bioscaffolds used in cell culture and tissue engineering, based on the literature.

Bioscaffolds	Fabrication method	Conductivity	*In vitro* or *in vivo* test				Ref.
			Spatial	Electrical stimulation	Cell	Biocompatibility	
PPy/chitosan membranes	Blending	—	2D	100 mV/mm for 4 h	Schwann cells	Supporting cell adhesion, spreading, and proliferation, and enhancing the expression and secretion of NGF and BDNF	[49]
Polypyrrole/ chitosanlongitudinally oriented micro-channels	Cross-linking and freezing drying in a mold	—	3D	3 V, 20 Hz for 1 h	*In vivo* (Sprague-Dawley rats)	Accelerating nerve regeneration	[50]
PPy/chitosan	Blending	$(1.3 \pm 0.1) \times 10^{-3}$ S/cm	2D	Constant potential gradients (50 ~ 1000mV/mm)	Olfactory ensheathing cells	Supporting cell adhesion, spreading, and proliferation with or without ES	[40]
PPy/PDLLA composite	*In situ* oxidative polymerization	$10^3 \Omega$/sq	2D	100mV	Fibroblasts	Inducing cell growth with current stimulation	[43]
PPy/PDLLA conduit	Blending	5.65 ~ 15.56 mS/cm	2D/3D	100 mV	PC12 cells/*in vivo*	*In vitro*: lengthening neurite, *in vivo*: recovering function.	[41]
PCLF–PPy conductive nerve conduits	*In situ* oxidative polymerization	—	2D	10 µA direct current or 20 Hz frequency for 1 h every 24 h	PC12 cells	ES can direct neurite extension and increase the number of neurites	[51]
NGF-immobilized PPy-coated PLGA fibers	Electrospinning	<500 KΩ/ square	2D	10 mV/cm	PC12 cells	Supporting PC12 neurite formation and neurite outgrowth.	[42]
PPy/Cellulose aerogels	*In situ* synthesis of PPy in cellulose gels	0.08 S/cm	2D	100 mV (frequency of 100 Hz for 30 min every 24 h, repeated 3 times)	PC12 cells	Enhancing the adhesion and proliferation	[52]
Chitosan-gelatin/Ppy cryogel scaffold	Blending	10^4S/m	3D	100 mV for 2 h every 24 h	Neuro2a cells, C2C12 cells	Electrical stimulation increasing the cell proliferation of neural cells	[53]
PPy/Heparin/PLLA	Blending	—	2D	100, 200, 300 and 400 mV/mm for 2, 4, 6 and 8 h	Osteoblast	ES promoting osteoblast expression and production of osteocalcin	[46]

Aligned PPy-coated PLA fibrous scaffold	Electrospinning	$10^{-2} \sim 10$S/cm	2D	100 mV for 2 h continuing with biphasic 100 mV for 4 h per day	Rat hippocampal neural cells	Supporting attachment and migration. And extended neurite outgrowth under electrical stimulation	[54]
Ppy/PLGA nanofibers	Electrospinning and *in situ* oxidative polymerization	$10^{3} \sim 10^{4}\Omega$/sq	2D	10 mV/cm	PC12 cells	Longer neurites and more neurite formation with applied potential	[37]
Aligned MWNT/ PPy-pTS platforms	Electro-deposition	—	2D	0.125 mA cm^{-2} bipolar square wave 2 ms pulses separated by a 2 μs pause in the applied current, at a frequency of 10 Hz, 8 h/ 24 h for 3 days	Myoblasts	Marked improvement in total myo-nuclear density and myoblast differentiation	[47]
PAni free standing films	Oxidative polymerization and casting	$10^{-9} \sim 10$ S/cm	2D	100 mV/cm for 10 min every 24 h	hMSCs	Enhancing cytoskeleton elongation	[55]
CPSA-PAni/PLCL nanofibers	Blending and electrospinning	10^{-2} S/cm	2D	2 ~ 200 Mv for 2 days	Human dermal fibroblasts, NIH-3T3 fibroblasts, and C2C12 myoblasts	Higher adhesion, and enhancing the growth of NIH-3T3 fibroblasts under the stimulation of various direct current flows	[56]
Chondroitin sulfate doped PPy film	Doping	10^{-3} S/cm	2D	Biphasic electric current of ±0.2 Vamplitude, 2.5 ms pulse width and 100 Hz pulse repetition frequency	Human adipose stem cells (hASCs)	Supporting early osteogenic differentiation, and strengthening mineralization o fhASCs under ES	[57]

highly ordered and regular structures. A single graphite sheet seamlessly rolled into a cylindrical tube can form a SWCNT, while an array of such graphene sheets that are concentrically wrapped like rings of a tree trunk create MWCNTs.

CNTs have attracted increasing attention owing to their distinctive mechanical, electrical, thermal, optical, and structural properties. The high mechanical strength and excellent electrical and thermal conductivities of CNTs can be exploited in biological applications to design an ideal electrically conductive scaffold. An earlier study corroborated that CNT substrates can boost neuronal electrical signaling transduction, support the dendrite elongation, and cell adhesion, thus effectively improving the growth of neuronal circuits on a CNT grid, which was accompanied by a significant increase in network activity [58]. In another study, Malarkey et al. [59] used conductive nanotubes with steadily varying electrical conductivity in order to investigate how electrical conductivity influences the growth of neurons. The results demonstrated that nanotubes promoted the growth of neurites while decreasing the number of growth cones, in a narrow range of electrical conductivity, however, and increased the area of cell body. However, these effects disappeared at higher conductance value.

Despite the extensive advantages offered by CNTs, there are several disadvantages associated with them, which limits their broad spectrum biomedical applications. In particular, toxicity of CNTs toward live cells remains a major issue that restricts their biological applications. Therefore, in order to improve the biocompatibility of CNTs, polymers have been extensively employed as a more permissive and versatile platform to integrate CNTs in biological systems. Likewise, due to their naturally chemical inertness and poor dispersion in common solvents, additional features are imparted to CNTs through functionalization by adding additional functional groups such as carboxyl or alcohol groups to the walls and ends of the nanotubes. For example, it was reported that carboxyl functional group functionalized MWCNT (f-MWCNT) could be uniformly dispersed in a chitosan matrix which usually involves a chemical interaction between chitosan and CNTs. The composite scaffolds of chitosan and f-MWCNT significantly enhanced the adhesion and proliferation of human osteo-sarcoma cells, besides increasing the protein content, alkaline phosphatase activity, and mineralization compared to chitosan scaffold alone [60]. In another study, carboxyl acid group functionalized CNTs at varying concentrations were incorporated into aligned poly(glycerol sebacate):gelatin (PG) electrospun nanofibers by using an electrospinning technique to form hybrid scaffolds. The results showed that the addition of functionalized CNTs markedly improved the fiber alignment and enhanced the electrical conductivity, besides enhancing the toughness and flexibility of scaffolds. Furthermore, the hybrid scaffolds not only maintained the viability, retention, alignment, and contractile features of cardiomyocytes (CMs) cultured on scaffolds but also caused spontaneous and synchronous beating behavior of the cells [8].

CNTs can be embedded within the matrix of several polymers to fabricate nanofibres, membranes, aerogels, and hydrogels, etc. for their applications as electroconductive bioscaffolds through various techniques, such as electrospinning, freeze-casting, and sol-gel transition, etc. (see Table 7.3). Under ES, CNTs-based electroconductive bioscaffolds could accelerate skin, bone, muscle, and neuron regeneration. For instances, electrospun scaffolds of poly(e-caprolactone) with CNTs have been proven to promote the *in vitro* cardiac differentiation of hMSCs by extrinsic ES [61]. In addition, Koppes et al. [62] used SWCNT and collagen to develop an electrically conductive 3D composite hydrogel. In the SWCNT/collagen scaffold, the SWCNTs were affirmed in order to electrically activate the cells even in the absence of endogenous ES. The results showed that SWCNT/collagen hydrogel markedly improved the total neurite outgrowth and prolonged the neurite as compared to non-ES nanofiller-free controls. Further, the dorsal root ganglia outgrowth was found to be greater when exogenous ES (50 mV/mm) was imposed in combination with the conductive hydrogels.

It is interesting to know that ES applied to an ordered microstructure can further facilitate the directional differentiation of cells. A study has reported the vertical alignment of CNTs within the methacrylated gelatin hydrogels using the dielectrophoresis technique (Figure 7.3). The results indicated that the vertically aligned CNT hydrogels significantly contributed to increasing the number of functional myofibers of skeletal muscle cells grown on them, compared to randomly distributed and horizontally aligned CNTs. Further, it resulted in more profound elevation in the expression of myogenic gene and protein expression when ES stimulation (voltage, 8 V; frequency, 1 Hz; duration, 10 ms) was applied along the direction of the aligned CNTs. This could be due to the anisotropic conductivity of CNTs [63].

7.3.3 Graphene-based Electroconductive Bioscaffolds

Graphene is a 2D carbon material which possesses remarkable features, such as atomic thickness, high surface area, remarkable thermal conductivity, excellent electrical conductivity, and high mechanical properties. These unique features and excellent performance has made graphene an attractive material in recent years for

Table 7.3 CNTs-based bioscaffolds used in cell culture and tissue engineering, based on the literature.

				In vitro or *in vivo* test				
Bioscaffolds	**Fabrication method**	**Functionalization of CNTs**	**Conductivity**	**Spatial**	**Electrical stimulation**	**Cell**	**Biocompatibility**	**Ref.**
Non-woven SWCNTs films	Floating chemical vapor deposition method	—	—	2D	No	Mouse fibroblasts	Enhanced cells adhesion and proliferation	[64]
Chitosan/ functionalized-MWCNTs scaffold	Freeze drying	Carboxyl functional group functionalized	—	2D	No	Human osteosarcoma cell line	Higher cell proliferation, alkaline phosphatase activity, protein concentration and mineralization	[60]
Gelatin-MWCNT scaffold	Electrospinning technique	—	—	3D	Yes	C2C12 myoblast cells	Enhanced the maturation of the myotubes and the amplitude of the myotube contractions under ES	[65]
CNT-poly(glycerol sebacate):gelatin scaffolds	Electrospinning technique	Carboxyl acid group functionalized	—	3D	No	Cardiomy-ocytes	Stronger spontaneous and synchronous beating behavior	[8]
SWCNT-collagen hydrogel	Sol-gel	Functionalization with carboxyl groups	1300 ~ 2400 ms/m	3D	Yes	Dorsal root ganglia	Electrical stimulation enhanced neurite outgrowth	[62]
Peptides-PLLA-MWCNT scaffold	Drop-cast	Phome functionalized	130 kΩ	2D	No	Human SH-SY5Y cells	Positively modulated neuronal differentiation	[66]
CNTs-interfaced glass fiber scaffold	Aligned PGF bundle interfaced with CNTs	Carboxylated	10^{-5}–10^{-6} S cm^{-1}	3D	No	PC12 cell and dorsal root ganglion cells, transected rat sciatic nerve model	Good cell viability and neuronal interactions	[67]
Carbon-based hierarchical scaffolds	CNT grow on microporous carbon foams and aligned carbon fiber mats	—	—	3D	No	C2C12 cells	Facilitated myoblast adhesion, growth and differentiation into myocytes	[68]

(*Continued*)

Table 7.3 (Continued)

Bioscaffolds	Fabrication method	Functionalization of CNTs	Conductivity	*In vitro* or *in vivo* test: Spatial	Electrical stimulation	Cell	Biocompatibility	Ref.
CNTs/PLLA aligned nanofiber scaffolds	Electrospinning and plasma-treated	—	5 ~ 6 ms/cm	2D	No	Mouse embryonic stem cells	Enhanced the expression of mature neuronal markers in differentiated MESC, MESC aligned with the underlying nanofiber substrate	[69]
Plasma-treated chitin/CNT composite scaffolds	Sol-gel casting	Coated with carboxymethyl cellulose	0.73 (±0.28) ~ 2.89 (±0.54) $S\ m^{-1}$	2D	No	Cortices from embryonic day 18 (E18) Wistar rats	Supported synaptic function	[70]
SWCNT/PLLA nanofibrous scaffolds	Electrospinning and plasma-treated	—	—	2D	No	Olfactory ensheathing glial cell	Promoted the adhesion, growth, survival and proliferation of OEC	[71]
MWCNTs-coated collagen sponge	Doping	Carboxylated	$8 \times 10^3 \Omega m$	3D	No	Mouse osteoblast cells	Strong cell adhesion	[72]
3D macroporous all-carbon scaffolds	Radical initiated thermal crosslinking	—	—	3D	No	MC3T3 preosteobl-ast cells	Good cell viability, attachment, proliferation, and cell infiltration	[73]
SWCNTs-chitosan hydrogel	Sol-gel casting	—	—	3D	No	Cardiomy-ocytes	Enhanced electrical coupling, synchronous beating, and cardiomyocytes function	[74]
Poly(e-caprolactone)/ CNTs scaffolds	Electrospinning	—	5 ~ 40 ms/cm	2D	Yes	Human mesenchy-mal tem cells (hMSCS)	Enhanced cardiac differentiation of human mesenchymal stem cells	[61]
Methacrylated gelatin-aligned CNT hydrogels	Dielectrophore-sis to vertically align CNTs within the hydrogels	Acid treatment	Anisotropic electrical conductivity	3D	Yes	C2C12 myoblasts	High number of functional myofibers, increased gene and protein expression upon ES	[75]

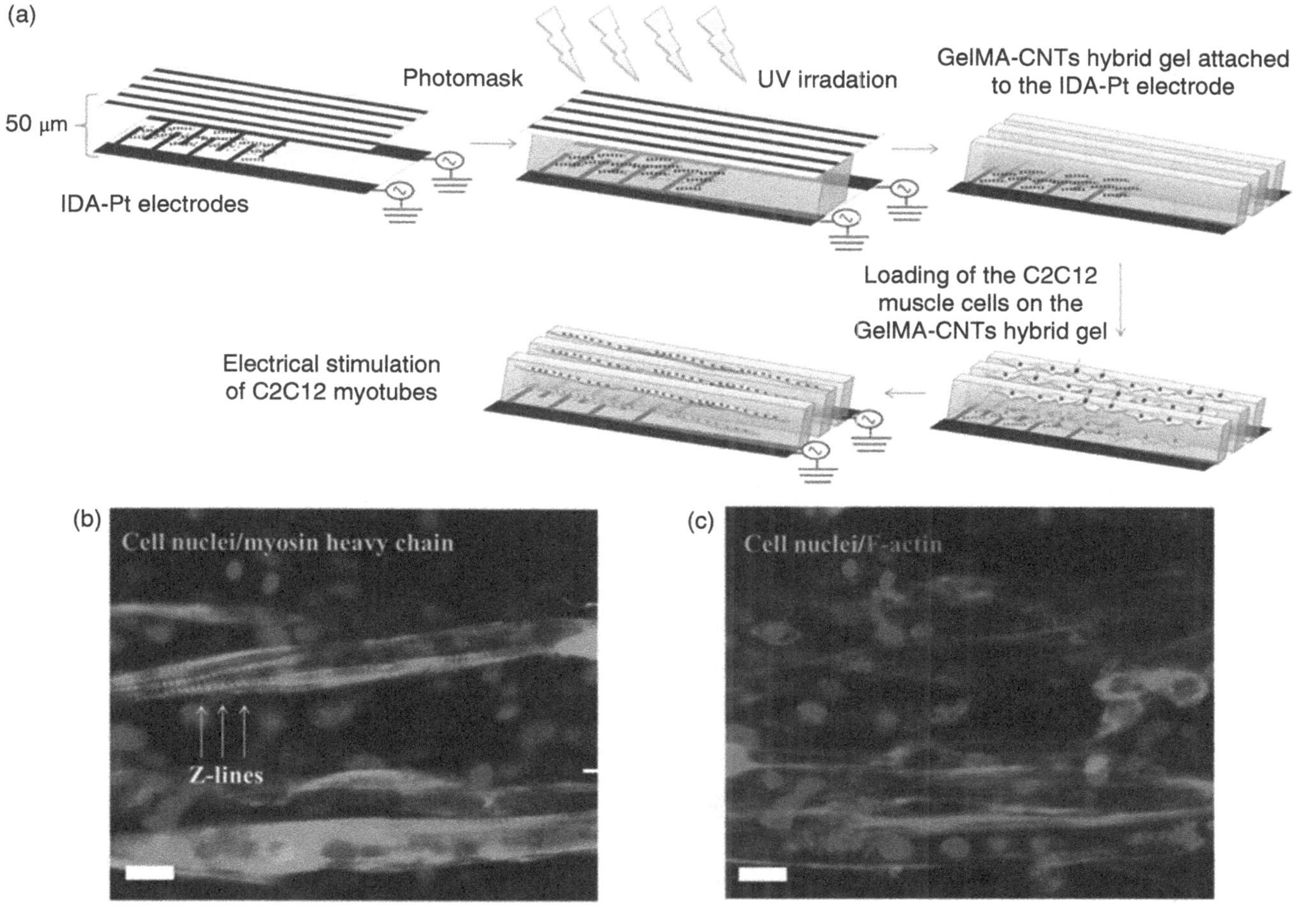

Figure 7.3 Schematic illustration of the procedure for the fabrication of a groove ridge topography within the GelMA-CNT hybrid gel (0.3 mg/mL CNTs) with (+ES) or without (−ES) the electrical stimulation at day 10 of the culture; (b)–(c) Immunostaining of cell nuclei/myosin heavy chain (b), and cell nuclei/F-actin (c). Scale bar: (A) 30 μm, (B) 50 μm. *Source:* Reproduced with permission from [63]. (*See insert for color representation of the figure.*)

broad spectrum applications in the fields of materials engineering, electronics, sensors, energy storage/conversion, and tissue engineering [76].

In the field of tissue engineering, graphene has been shown to be a useful material for the development of bioscaffolds owing to its potential to impart topographical, chemical, and electrical features to the scaffold and provide an optimal environment for tissue regeneration, hence it can be more effective than conventional inert biomaterials. It has been reported that 2D graphene possesses unique surface properties which can induce stem cells to preferentially differentiate into a specific lineage. For example, GO nano-sheets were coated onto aligned and aminolyzed poly-L-(lactic acid) (PLLA) nanofibrous scaffolds. The results showed that the GO coating significantly enhanced the various biomedical properties of the aligned PLLA nanofibrous scaffolds, such as their surface porosity and hydrophilicity and promoted the interaction of cells with material which ultimately resulted in the promotion of Schwann cells' (SCs) growth as well as regulating their orientation. Further, these induced PC12 cells differentiation and neurite growth [77]. In another study, Wang et al. [78] used fluorinated graphene on mesenchymal stem cells (MSCs) where it promoted proliferation and induced stronger polarization of MSCs. Furthermore, the MSCs can be controllably patterned on fluorinated graphene which points toward a scaling up and development of a robust strategy to align stem cells and their ability to provoke neuronal lineage, even in the absence of a chemical inducer.

Graphene-based bioscaffolds have been fabricated through various methods such as the template-directed chemical vapor deposition method [79, 80], the nucleate boiling method [81], the hydrothermal method [82], the chemical reduction method [83], the freeze drying method [84, 85], and so on. The 3D graphene-based scaffolds find different applications in various fields such as electronics, photonics, sensor, energy generation, and storage [86]. Recently, extensive research has been carried out to explore

Table 7.4 Graphene and graphene oxide-based bioscaffolds used in cell culture and tissue engineering.

			In vitro or *in vivo* test				
Bioscaffolds	**Fabrication method**	**Conductivity**	**Spatial**	**Electrical stimulation**	**Cells type**	**Biocompatibility**	**Ref.**
Fluorinated graphene sheets	Exposing graphene to a fluorinating agent	—	2D	—	MSCs	Induced higher proliferation and stronger polarization and induced neuronal lineage	[78]
Aligned PLLA/GO scaffold	GO coated onto aligned PLLA scaffolds	—	2D	—	PC12	Promoted SCs growth and regulated cell orientation, and induced PC12 cell differentiation and neurite growth	[77]
GO–PCL meshes	Electrospinning	—	2D	—	CB-hMSCs	Excellent myoblast differentiation and promoted self-aligned myotubes formation similar to natural orientation	[90]
Graphene films	Large-scale pattern growth	280 Ω/sq	2D	1 ~ 10 of 500 mV monophasic/ cathodic voltage pulses with 1 ~ 100 ms duration in a second	hNSCs	Good electrical coupling with the differentiated neurons for electrical stimulation	[91]
Graphene/PET	Chemical vapor deposition	—	2D	A train of balanced biphasic stimulus charges of 9 mV voltage	Human neuroblas-toma cells	Promoted new cell-to-cell coupling and strengthened existing cell-to-cell coupling	[92]
Graphene-heparin/ poly-L-lysine	Layer-by-layer (LbL) deposition	362 to 17.1 KΩ/sq	2D/3D	—	Primary cortical neurons (PCNs)	Supported neuron cell adhesion and neurite outgrowth	[93]
Graphene oxide	Freeze casting	—	3D	—	ENPC	Formation of neural networks containing neurons, glial cells, dendrites, axons and synaptic nerve bridges were formed, containing neuronal cells	[84]
Graphene hydrogel	Hydrothermal approach	—	3D	—	MG63 cell line	Guided filopodia protrusions of MG63 on the hydrogel.proving the compatible bio-application of the grapheme hydrogel structure.	[94]
GO/PVA hydrogel	Freeze/thaw method	—	3D	—	Osteoblast cells	The incorporation of certain amount GO into PVA hydrogels does not affect the toxicity of PVA to osteoblast cells	[85]
GO/PVA nanofibrous scaffolds	Electrospinning	—	3D	—	Osteoblast cells	Supported cells' adhesion and migration	[95]
Grapheme/ polylactide-co-glycolide	3D printing	800 S/m	3D	—	hMSCs	Adhesion, viability, proliferation, and neurogenic differentiation with significant up-regulation of glial and neuronal genes	[89]
Rolled graphene oxide foams	GOF layers were rolled to obtain 3D-scaffolds	~170Ω/ sq	3D	100 ms voltage pulses in intervals of 10s, currents ~20 mA	hNSCs	Proliferation of cells and differentiation into neurons	[96]

the potential of 3D graphene as cell culture platforms or their ability to guide the fate of stem cell [87]. The 3D graphene could provide 3D microenvironments to cells, mimicking their *in vivo* counterparts. Additionally, the 3D graphene foams have been reported to support the attachment and viability of hMSCs and induced spontaneous osteogenic differentiation [88]. Another study showed the 3D GOx-based scaffolds developed by the biocompatible freeze casting procedure to investigate the adhesion, morphology, viability, and neuronal/glial differentiation of embryonic neural progenitor cells (ENPC). The results showed that highly viable and interconnected neural networks were developed on these 3D scaffolds containing both neurons and glial cells. The neuronal cells were rich in dendrites, axons, and synaptic connections [84]. In another study, Adam et al. [89] used a composite consisting of majority graphene and minority poly(lactide-co-glycolide) as 3D printable ink to create a multi-layered 3D bioscaffold. The 3D printable graphene demonstrated high mechanical properties and good conductivity. Also, it supported the adhesion, viability, proliferation, and neurogenic differentiation with significant up-regulation of glial and neuronal genes.

Graphene-based bioscaffolds could provide an environment for tissue regeneration by integrating the electrical cues. As a result of the outstanding performance of pathways for charge transport, such as the enormous interface, 3D multiplexing, and high conductivity could be obtained for the excellent electrical conductivity of graphene bioscaffolds. The behavior of electro-active cells can be modulated by ES. For examples, graphene sheets showing good electrical coupling with electrically stimulated and differentiated neurons were applied to 2D electro-conductive templates to differentiate human neural stem cells (hNSCs) [91]. In another study, a graphene/polyethylene terephthalate (PET) film was prepared to serve as a stimulator to manage the neural cell-cell interactions through non-contact EF stimulation [92]. This stimulator produced a weak non-contact EF by applying a series of balanced biphasic stimulus charges (9 mV). The results showed that the electric field promoted the coupling of new cells and strengthened the existing cell-to-cell coupling. When ES was applied, the electrically conductive 3D-scaffolds based on graphene promoted the adhesion and proliferation of the hNSCs. Graphene oxide foams (GOFs) as electrically conductive 3D-scaffolds have been effectively used to direct the growth of neural fibers and differentiation of hNSCs into neurons under an ES following the method devised by Omid Akhavan et al. [85]. In this study, chemically exfoliated GO sheets were precipitated to fabricate GOFs layers in an aqueous suspension under UV irradiation. Thereafter, the 3D-scaffolds with potential directed proliferation and the differentiation of neurons along the main axis of the rolls were obtained by rolling the GOFs layers. The electrical sheet resistance of the GOFs was about 170Ω/sq, when adding the mall voltage (<5V) to generate the stimulation current (~20mA). Apparently, the proliferation of cells was induced and stimulated the differentiation into neurons (rather than glia) upon application of the ES.

7.4 Conclusion

Electroconductive bioscaffolds have been extensively studied for their application in regenerative medicines owing to their excellent structural, mechanical, electrical, and biocompatible properties. Electroconductive bioscaffolds are prepared by various methods for their applications in 2D and 3D cell cultures. Combining natural or synthetic polymers with conductive polymers such as carbon nanotubes and grapheme can result in the preparation of different types of electroconductive bioscaffolds which can be implanted into the body to promote tissue regeneration and injury recovery. Further, these electroconductive bioscaffolds induce biocompatible and electrically conductive pathways to stimulate electrical communication among the cells. It has been demonstrated that incorporating bioscaffolds with an electroconductive property can be used to enhance tissue regeneration under an EF that mimics the natural environment of human body. These can be further used in the fields of skin, bone, muscle, and neural regeneration, etc.

Acknowledgments

This work was supported by the National Natural Science Foundation of China (21574050, 51603079), and the China Postdoctoral Science Foundation (2015M572132).

References

1 Albano, J.J., Alexander, R.W. (2011) *Clin. J. Sport Med.*, **21**, 35.
2 Thiele, J., Ma, Y., Bruekers, S.M.C. et al. (2014) *Adv Mater*, **26**, 125.
3 Annabi, N. et al. (2014). *Adv Mater*, **26**, 85.
4 Li, Y. et al. (2014). *Biotechnol Adv.*, **32**, 347.
5 Frohlich, F., McCormick, D.A. (2010). *Neuron.*, **67**, 129.
6 Song, B. et al. (2007) *Nat Protoc.*, **2**, 1479.

7 Shi, Z. et al. (2016) *Biomater.*, **111**, 40.
8 Kharaziha, M. et al. (2014) *Biomater.*, **35**, 7346.
9 Messerli, M. A., Graham, D. M. (2011) *The Biol. Bull.*, **221**, 79.
10 Guo, A. et al. (2010) *J. Invest. Dermatol.*, **130**, 2320.
11 Zuberi, M., Liusnyder, P., Haque, A., et al. (2008) *J. Biol. Engin.*, **2**, 17.
12 Hotary, K. B., Robinson, K.R. (1994) *Dev. Biol.*, **166**, 789.
13 Hotary, K.B., Robinson, K.R. (1990) *Dev Biol.*, **140**, 149.
14 Kloth, LC. (2005) *Int. J. Lower Extrem Wounds.*, **4**, 23.
15 Willand, M.P., Nguyen, M., Borschel, G.H., Gordon, T. (2016) *Neurorehab Neural Res.*, **30**, 490.
16 Ragnarsson, K.T. (2007) *Spin. Cord*, **46**, 255.
17 Fukada, E., Yasuda, I. (1957) *J. Phys. Soc. Jpn.*, **12**, 1158.
18 Isaacson, B.M., Bloebaum, R.D. (2010) *J. Biomed. Mater. Res., A.* **95**, 1270.
19 Ryu, J. S. et al. (2009) *Oral Oncol.*, **45**, 665.
20 Gibson, I.R., McCaig, C.D. (2004) *Mrs Proc.*, **845**.
21 Nguyen, H.T. et al. (2013) *J. Neural Eng.*, **10** 46011.
22 Gamboa, O.L. et al. (2010) *Exp. Eye Res.*,**91** 195.
23 Song, B., Zhao, M., Forrester, J.V., McCaig, C.D. (2002) *Proc. Natl. Acad. Sci. USA*, **99**, 13577.
24 Bai, H., McCaig, C.D., Forrester, J.V., et al. (2004) *Arteriosc., Throm., Vasc. Biol.*, **24**, 1234.
25 Li, J. et al. (2011) *Lab Chip*, **11** 1298.
26 Yan, X. et al. (2009) *Bioelectromag.*, **30**, 29.
27 Wu, D., Ma, X., Lin, F. (2013) *Cell Biochem. Biophys.*, **67**, 1115.
28 Zhao, Y., Liu, B., Pan, L., Yu, G. (2013) *Energy Environm. Sci.*, **6** 2856.
29 Macdiarmid, A.G. (2001) *Angew. Chem.*, **40**, 2581.
30 Sharma, Y. et al. (2012) *Int. J. Biol. Macromol.*, **51**, 627.
31 Li, M., Guo, Y., Wei, Y., et al. (2006) *Biomater.*, **27**, 2705.
32 Kai, D., Prabhakaran, M.P., Jin, G., Ramakrishna, S. (2011) *J. Biomed. Mater. Res. A.*, **99**, 376.
33 Chen, M., Sun, Y., Chen, Y. (2013) *Acta Biomater.*, **9**, 5562.
34 Shi, Z. et al. (2012) *RSC Adv.*, **2**, 1040.
35 Shi, Z., Li, Y., Chen, X., et al. (2014) *Nanoscale*, **6**, 970.
36 Zhang, L., Li, Y., Li, L. et al. (2014) *Reactive Funct. Polym.*, **82**, 81.
37 Lee, J.Y., Bashur, C.A., Goldstein, A.S., Schmidt, C.E. (2009) *Biomater.*, **30**, 4325.
38 Mawad, D. et al. (2012) *Adv.Funct. Mater.*, **22**, 2692.
39 Guarino, V., Alvarezperez, M.A., Borriello, A. et al. (2013) *Adv. Healthcare Mater.*, **2**, 218.
40 Qi, F. et al. (2013) *Biomaterials*, **34**, 1799.
41 Xu, H. et al. (2014) **35**, 225.
42 Lee, J. Y. et al. (2012) *IEEE T Nanobiosci.*, **11**, 15.
43 Shi, G., Rouabhia, M., Wang, Z. (2004) *Biomater.*, **25**, 2477.
44 Shahini, A. et al. (2014) *Int. J. Nanomed.*, **9**, 167.
45 Sajesh, K.M., Jayakumar, R., Nair, S.V., Chennazhi, K.P. (2013) *Int. J. Biol. Macromol.*, **62**, 465.
46 Meng, S., Rouabhia, M., Zhang, Z. (2013) *Bioelectromag.*, **34**, 189.
47 Quigley, A.F. et al. (2012) *Adv. Healthcare Mater.*, **1**, 801.
48 Mihic, A. et al. (2015) *Circul.*, **132**, 772.
49 Huang, J. et al. (2010) *J. Biomed Mater Res., A*, **93**, 164.
50 Huang, J. et al. (2012) *Plos One*, **7**.
51 Moroder, P. et al. (2011) *Acta Biomater.*, **7**, 944.
52 Shi, Z. et al. (2014) *Angew. Chem.*, **53**,5380.
53 Vishnoi, T., Kumar, A., (2012) *J. Mater. Sci.: Mater. Med.*, **24**, 447.
54 Sudwilai, T. et al. (2014) *J. Biomat, Sci-Polym. E*, **25**, 1240.
55 Thrivikraman, G., Madras, G., Basu, B. (2014) *Biomaterials.*, **35**, 6219.
56 Jeong, S. I. et al. (2008) *Macromol. Biosci.*, **8**, 627.
57 Bjorninen, M. et al. (2014) *Ann. Biomed. Eng.*, **42**, 1889.
58 Lovat, V. et al. (2005) *Nano Lett.*, **5**, 1107.
59 Malarkey, E.B. et al. (2009) *Nano Lett.*, **9**, 264 (2009).
60 Venkatesan, J., Ryu, B., Sudha, P.N., Kim, S. (2012) *Int. J. Biol. Macromol.*, **50**, 393.
61 Crowder, S.W. et al. (2013) *Nanomed.-UK.*, **8**, 1763.
62 Koppes, A.N. et al. (2016) *Acta Biomater.*, **39**, 34.
63 Ramonazcon, J. et al. (2013) *Adv Mater.*, **25**, 4028.
64 Meng, J. et al. (2006) *J. Biomed Mater Res A.*, **79A**, 298.
65 Ostrovidov, S. et al. (2014) *Biomater.*, **35**, 6268.
66 Scapin, G. et al (2015) *Nanomed.: Nanotech., Biol. Med.*, **11**, 621.
67 Ahn, H. et al. (2015) *Acta Biomat.*, **13**, 324.
68 Patel, A. et al. (2016) *Acta Biomat.*, **32**,77.
69 Kabiri, M. et al. (2012) *Biotechnol. Lett.*, T34, 1357.
70 Singh, N. et al. (2016) *Nanoscale*, **8**, 8288.
71 Kabiri, M. et al. (2015) *Excl. J.*, **14**, 851.
72 Hirata, E. et al. (2009) *J. Biomed. Mater. Res. Part B: App. Biomat.*, **90B**, 629.
73 Lalwani, G. et al. (2015) *J. Biomed. Mater. Res. A*, **103**, 3212.
74 Pok, S. et al. (2014) *Acs. Nano.*, **8**, 9822.
75 Ahadian, S. et al. (2014) *Sci, Rep-UK*, **4**, 4271.
76 Li, N. et al. (2014) *Chin. Sci. Bull.*, **59**, 1341.
77 Zhang, K., Zheng, H., Liang, S., Gao, C. (2016) *Acta Biomater.*, **37**, 131.
78 Wang, Y. et al. (2012) *Adv. Mater.*, **24**, 4285.
79 Chen, Z. et al. (2011) *Nat. Mater*, **10**, 424.
80 Cao, X. et al. (2011) *Small*, **7**, 3163.
81 Ahn, H. S. et *al.* (2013) *Sci. Rep-UK*, **3**.
82 Li, Y. et al. (2014) *Adv. Mater*, **26**, 4789.
83 Tang, G. et al. (2014) *Carbon*, **77**, 592.
84 Serrano, M. C. et al. (2014) *J. Mater. Chem. B*, **2**, 5698.
85 Zhang, L. et al. (2011) *J. Mater. Chem.*, **21**, 10399.

86 Novoselov, K.S. et al. (2012) *Nature*, **490**, 192.
87 Krishna, K.V., Ménard-Moyon, C., Verma, S., Bianco, A. (2013) *Nanomed.-UK*, **8**, 1669.
88 Crowder, S.W. et al (2013) *Nanoscale*, **5**, 4171.
89 Jakus, A. E. et al. (2015) *ACS Nano.*, **9**, 4636.
90 Chaudhuri, B., Bhadra, D., Moroni, L., Pramanik, K. (2015) *Biofabrication*, **7**, 15009.
91 Park, S.Y. et al. (2011) *Adv. Mater.*, **23**, 123.
92 Heo, C. et al. (2011) *Biomater.*, **32**, 19.
93 Zhou, K. et al. (2012) *Appl. Mater. Inter*, **4**, 4524.
94 Ming, H.N. (2011) *Int. J. Nano. Med.*, **6**, 1954.
95 Qi, Y. Y. et al. (2013) *J. Appl. Polym. Sci*, **127**, 1885.
96 Akhavan, O., Ghaderi, E., Shirazian, S.A., Rahighi, R. (2016) *Carbon*, **97**, 71.

8

Starch and Plant Storage Polysaccharides

Francisco Vilaplana[1,2], Wei Zou[2,3], and Robert G. Gilbert[2,3,]*

[1] *Joint International Research Laboratory of Agriculture and Agri-Product Safety, College of Agriculture, Yangzhou University, Yangzhou Jiangsu Province 225009, China*
[2] *KTH Royal Institute of Technology, Stockholm, Sweden*
[3] *University of Queensland, Centre for Nutrition and Food Sciences, Brisbane, Australia*

8.1 Starch and Other Seed Polysaccharides: Availability, Molecular Structure, and Heterogeneity

8.1.1 Molecular Structure and Composition of Seeds and Cereal Grains

Starch and other seed polysaccharides have been essential to mankind ever since the development of agriculture, about 10 000 years ago. Their main role is to supply the largest single component of our food energy, and they also provide functional food and nutrients, many health benefits as dietary fibers, and an ever-expanding range of materials for packaging, food processing, pharmacological applications, and many more.

Starch is the glucose storage molecule in plants. Transient starch is present in all leaves. It is synthesized during the day as a result of the photosynthetic process and degraded at night: this is transient, or transitory, starch. Storage starch is found in the storage organs of grains, legumes, and tubers, where it remains undegraded over extended periods for subsequent use in the plant, such as providing glucose for the germinating seed. Storage starch provides about half of the food energy for humanity, the actual amount depending on geography, economic status, and custom.

The most important sources of starch for human and animal nutrition, and for processing into food ingredients and other industrial products, are grains, legumes, and tubers. These have complex structures, as illustrated for a rice grain in Figure 8.1. Starch can be obtained from these botanical structures in various ways. In rice, for example, after harvesting, the whole grain is de-hulled and polished, leaving a white rice grain largely comprising the starchy endosperm. This process has developed through thousands of years of rice cultivation by humans, resulting in food products that are preferred by a vast number of consumers. It also results in various by-products, such as rice bran, which has been, up to this point, simply used as fuel, animal feed or soil nutrient, or in some higher-value product such as processing to rice-bran oil. Although such refined products have arisen from, and have sometimes caused, cultural preferences, they can result in the loss of some advantageous components, such as nutritionally useful proteins and non-starch polysaccharides, in the case of white rice. Additionally, such agricultural by-products as the bran, husk, and straw nowadays constitute a fundamental renewable resource for the production of bio-based materials and platform chemicals through biorefinery processes.

Other non-starch polysaccharides can be found as storage biomolecules in the cell wall of seeds of many important plant families [1].

8.1.2 Starch Hierarchical Structure from Bonds to the Granule

Starch is very simple from the chemical point of view: a homopolymer of glucose, with linear $(1\longrightarrow4)$-α links and $(1\longrightarrow6)$-α links at the branch points, as shown in Figure 8.2.

However, it has a hierarchical ultrastructure, as shown in Figure 8.3, and now elaborated.

8.1.3 Crystalline Structure

The crystalline lamellae in starch are formed by amylopectin branches: both A and B chains are confined into one cluster. Because of steric hindrance, most branch points are located in the amorphous lamellae, resulting in the semicrystalline structure of starch. Amylose, intertwined with amylopectin, is mostly

* Email: b.gilbert@uq.edu.au

Bioinspired Materials Science and Engineering, First Edition. Edited by Guang Yang, Lin Xiao, and Lallepak Lamboni.

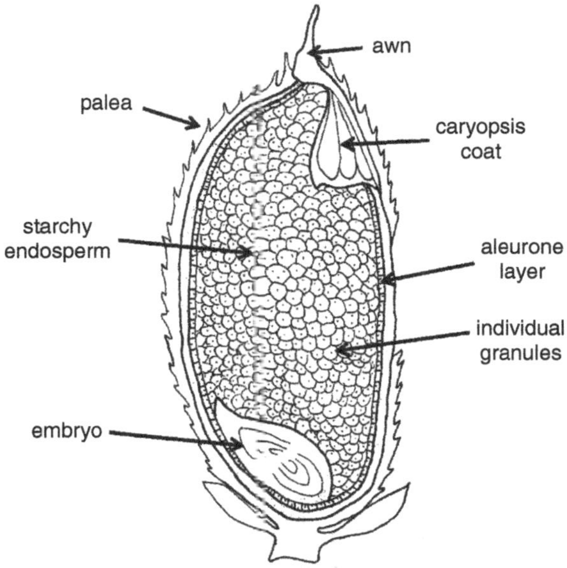

Figure 8.1 Sketch of the structure of a rice grain.

confined to the amorphous lamellae. The crystalline and amorphous lamellae are arranged alternatively with a similar repeat distance of ~ 9 nm, being conserved among different botanical origins.

8.1.4 Granular Structure

Starch granules from different botanical origins exhibit differences in shapes and size varying from 1–2 to ~100 μm [2]. Starch granule morphologies include spherical, oval, disk, polygonal, elongated, kidney, and lobe. Moreover, some cereals contain granules of different sizes within the same plant. For example, wheat, barley, rye, and triticale produce larger A-granules and smaller B-granules [3]; rice and oat have compound granules containing granules of different sizes, which are arranged to form irregular shapes in the endosperm.

8.1.5 Mannans, Galactomannans, and Glucomannans

Mannans are the main energy-reserve polysaccharide in the seeds of different plant families (e.g. *Leguminosae*, *Palmae*) and they can also be found as structural components in the secondary plant cell wall of softwoods [4–6]. Normally mannans in secondary cell walls have much lower molar mass than those extracted from seed tissues. The general structure of mannans (Figure 8.4), involves a backbone of (1→4)-β-linked D-mannopyranose (Man*p*) solely or in combination with D-glucopyranose (Glc*p*). This backbone may be occasionally substituted by single D-galactopyranosyl (Gal*p*) groups in the (1⟶6)-α positions, by (1⟶6)-β-D-glucopyranosyl (Glc*p*) units (in glucomannans), and also may be acetylated in the *O*-2 or *O*-3 positions [7, 8]. Depending on the backbone and substitution pattern, mannans can be further classified into linear mannans, galactomannans, glucomannans, and galactoglucomannans. Linear mannans refer to polysaccharides with an unsubstituted backbone of (1→4)-β-Man*p* units, as, for example, in the endosperm of ivory nut *Phytelephas aequatorialis*. These hemicelluloses, due to the stiffness of the unsubstituted (1→4)-β-linked backbone, similar to that of cellulose, confer structural stability to the plant tissues. Galactomannans involve a backbone of (1→4)-β-Man*p* units with (1⟶6)-α–Gal*p* substitutions, widely found in the seeds from the *Leguminosae* family. The ratio of Gal*p* substitution (usually described as the Man to Gal ratio or M:G) varies from different species, from fully substituted in fenugreek gum (M:G ratio of approximately 1:1), to progressively less substituted galactomannans as in guar gum (M:G of approximately 2:1), tara gum (M:G of approximately 3:1), and locust bean gum (M:G of approximately 4:1). A detailed list of the occurrence of galactomannan polysaccharides in different Leguminosae and non-Leguminosae species, together with the M:G ratio, is provided in [9, 10]. The degree (M:G ratio) and pattern of substitution have a fundamental effect on the

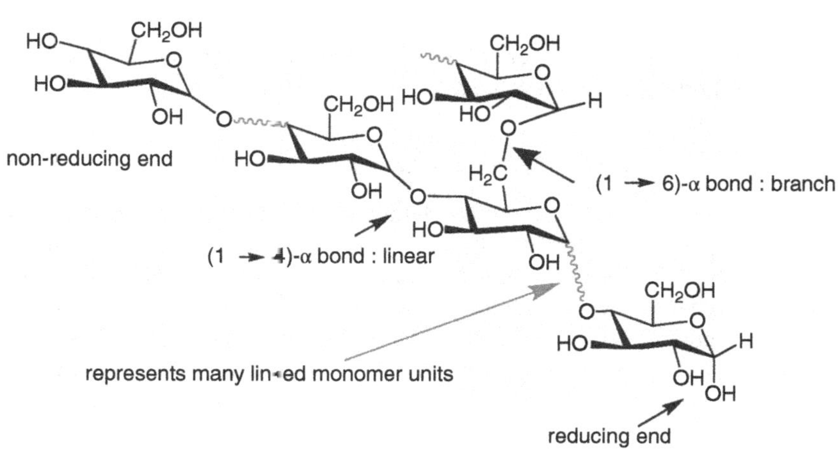

Figure 8.2 The chemical structure of starch.

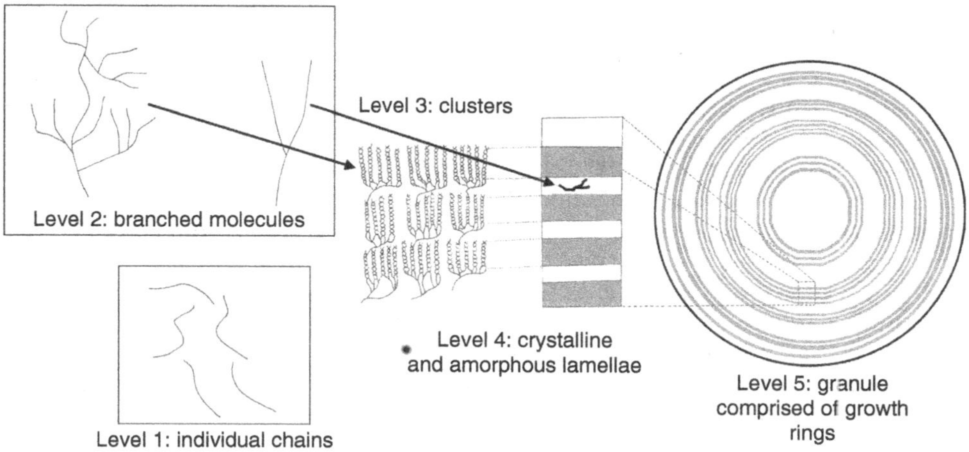

Figure 8.3 Starch structural levels. Level 1 is the individual bonds in a single chain: (1⟶4)-α linkages (see Figure 8.2) between anhydroglucose units, with (1⟶6)-α branch points. Level 2 is branched molecules. Level 3 comprises clusters of crystalline helices made of shorter amylopectin chains and of portions of longer amylopectin chains which span more than one lamella. Level 4 comprises alternating amorphous and crystalline lamellae, with clusters of shorter amylopectin chains largely confined to a single crystalline lamella, while the amorphous region largely contains amylose molecules, longer amylopectin chains (which span two or more lamellae), branch points and "dangling" ends of amylopectin chains. Level 5 is the granules, made up of growth rings of alternating lamellae.

Figure 8.4 Molecular structure of mannans: Linear mannan; carob galactomannan (redrawn after [9, 10]); konjac glucomannan (redrawn after [12]); and softwood galactoglucomannan. (redrawn after [7, 13]).

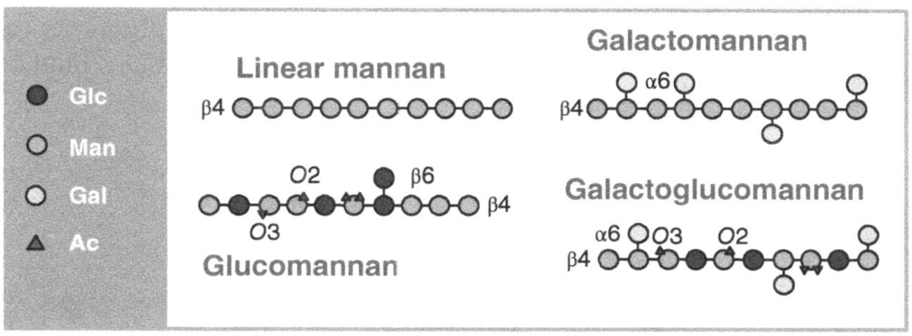

hydrodynamic properties and the solubility of the polysaccharides [11], and therefore on the subsequent material properties. Glucomannans have a backbone of β-(1→4)-linked-Man*p* and Glc*p* units, with the presence of minor (1→6)- β-Glc*p* substitutions in the glucopyranosyl units of the backbone (8% reported branching points in Konjac glucomannan), and some acetylation modification in the *O*-2 and/or *O*-3 positions [12]. The copolymer distribution sequence of (1→4)- β-Man*p* and Glc*p* units in the glucomannan backbone is still not understood, but it can be assumed that is not random due to the biosynthetic mechanisms for hemicelluloses. Finally, galactoglucomannans involve a backbone of (1→4)- β-Man*p* and Glc*p* units, with single D-galactopyranosyl (Gal*p*) substitutions in the (1⟶6)-α positions and eventually acetylation in the *O*-2 or *O*-3 positions [7, 13].

8.1.6 Xyloglucans

Xyloglucans (XyGs) are a complex and structurally diverse family of polysaccharides that can be found as a fundamental hemicellulose component in the primary cell wall of all land plants, especially in flowering (dicot) plants in what has been defined as type I primary cell wall [14]. Additionally, XyGs can be found as storage polysaccharides in the seeds of some plants, such as tamarind (*Tamarindus indica*) and nasturtium (*Tropaeolum majus*) [1, 15]. Xyloglucans share a general structure with a backbone composed of β-(1→4)-D-glucopyranosyl (Glc*p*) units, same as cellulose, but regularly substituted with α-(1→6)-linked D-xylopyranosyl residues (Xyl*p*). These Xyl*p* units can themselves carry further glycosyl substitutions, including D-galactopyranose, D-xylopyranose, L-arabinopyranose, L-arabinofuranose, D-galacturonic acid, L-fucose and L-galactopyranose. In addition to this, XyG sidechains can be further modified by acetylation and methylation. These substitution patterns create repeating structural motifs in the xyloglucan backbone that are specific to different species, and are believed to contribute to their structural and signaling functions in cell walls. A systematic nomenclature based on nature of the substitutions in the XyG backbone was initially proposed by Fry et al. [16], and has been more recently expanded by Tuomivaara et al. [17]. As an example, an unsubstituted backbone Glc*p* residue is normally represented by the letter **G**, a single Xyl*p*-α – (1→6)-Glc*p* unit is portrayed as **X**, a Gal*p*-β-(1→2)-Xyl*p*-α-(1→6)-Glc*p* unit is represented as **L**, and so on (Figure 8.5). For a comprehensive library of the structural diversity and the nomenclature of the XyG sidechain substitutions, recent reviews are recommended [17, 18].

Seed storage XyGs are preferably used for different industrial applications in the food and pharmaceutical sectors rather than their primary cell wall counterparts, due to their higher relative abundance in such tissues, and their easier extractability and purification procedures. The basic structure of storage XyGs is similar to the primary cell wall XyGs, but they do not possess such diversity in sidechain substitution [1]. Storage XyGs are normally heavily galactosylated and they do not possess further modifications (e.g. fucosylation, acetylation). The substitution pattern of XyGs is normally analyzed using a XyG-specific *endo*-glucanase (XGase), which hydrolyzes the backbone glycosidic linkage between an unsubstituted Glc*p* unit and a Xyl*p*-substituted Glc*p* residue [19]. This enzymatic depolymerization yields specific xyloglucan oligosaccharides (XGOs) that can be further analyzed by mass spectrometry, thus providing an oligomeric fingerprinting of the sidechains in XyG considering the specific mass profiles of the different substituents [20, 21]. The substitution pattern of tamarind XyG obtained by XGase treatment and analysis by matrix-assisted laser desorption/ionization mass spectrometry (MALDI-MS) is provided in Figure 8.2. As can be observed, tamarind XyG shows galactosylated oligosaccharides XLXG/XXLG and XLLG, together with XXXG, as the main building blocks that conform the polymeric chains. Longer building blocks, such as XXXXG, XLXXG, XXLXG, and XXXLG, have been reported as well in seeds of Hymenaea courbaril, which may prevent aggregation and improve rheological properties when compared with the XyG extracted from tamarind seeds [22, 23].

The differences in sidechain substituents and the pattern of substitution along the backbone influence the conformation, solubility, and aggregation of storage XyG, which in turn has a large impact on the macroscopic properties, such as rheological properties and viscoelasticity, fundamental for their material applications.

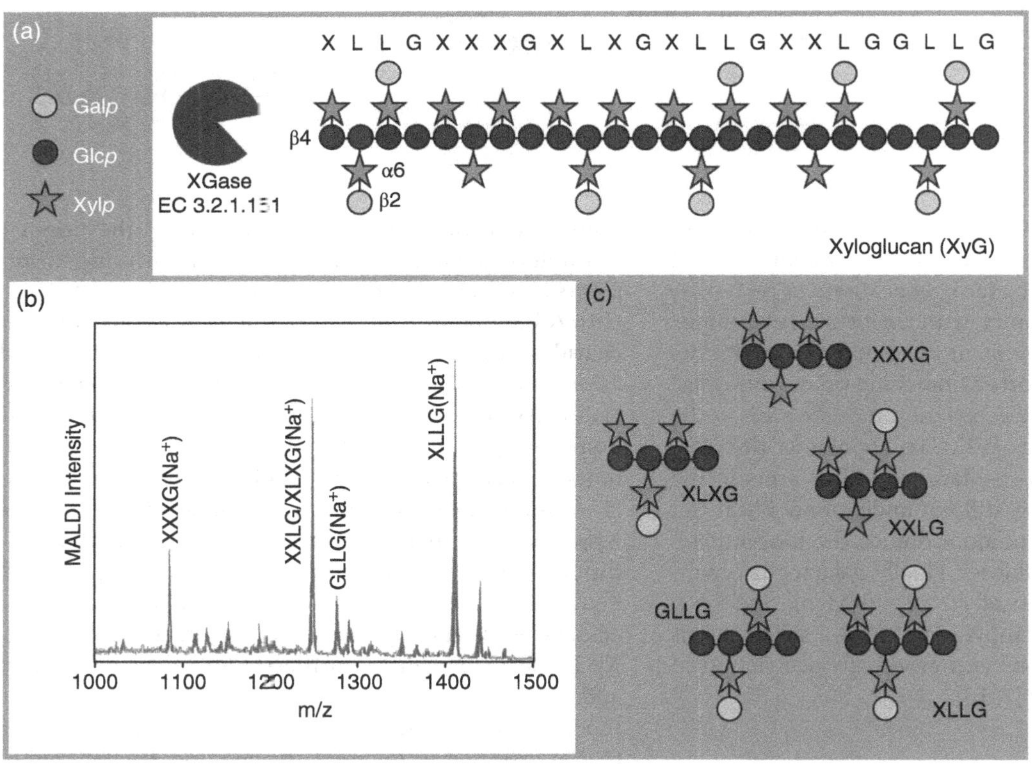

Figure 8.5 Mass profiling of the substitution pattern of tamarind XyG by enzymatic depolymerization using a XyG-specific endoglucanase (XGase) and MALDI-MS analysis. The substitutions in the XyG backbone are named following the nomenclature initially proposed by Fry et al. [16].

8.1.7 Xylans. Arabinoxylans, Glucuronoxylans, and Glucuronoarabinoxylans

Xylans are the most abundant polysaccharides in plant biomass after cellulose, since they are the main hemicellulose in the cell wall of annual plants (including cereal grains) and hardwoods, and they are also present in softwoods [5, 24]. Xylans (Figure 8.6) are heteropolysaccharides that share a (1→4)-linked β-D-xylopyranosyl (Xyl*p*) backbone, but they differ in the nature, degree, and pattern of substitutions along the backbone, not only from one plant species to another, but also spatially and temporally within the different tissues of a plant species. The main β-(1→4)-D-Xyl*p* chain can be substituted at positions C-2, C-3 or both by α-L-arabinofuranosyl (Ara*f*), D-glucuronic acid (or its 4-*O*-methyl ether), or in minor abundance by other glycosyl substituents (e.g. D-Xyl*p* or D-Gal*p* residues). Some naturally occurring xylans also carry *O*-acetyl and feruloyl substituents [25, 26]. Depending on the nature of the substituents, xylans can be classified into glucuronoxylan (GX), arabinoxylan (AX), glucuronoarabinoxylan (GAX), and arabinoglucuronoxylan (AGX), as depicted in Figure 8.5. Glucuronoxylans are the main hemicellulose in hardwoods; they are typically substituted by 4-*O*-methyl α-(1→2)-D-glucuronic acid (MeGlcA), and they can be acetylated in the *O*-2 and *O*-3 positions [27]. Softwood glucuronoarabinoxylans (GAX) are slightly more complex than hardwood GX, since the β-(1→4)-Xyl*p* backbone is not only substituted by α-(1→2)-linked MeGlcA but also by arabinofuranosyl (Ara*f*) residues via α-(1→3) linkages [28, 29]. Xylans in grasses and specially in cereal grains are somewhat different in structure than those in higher plants, since the cell walls in the grain endosperm have negligible load-bearing functions during development and germination. [30] Arabinoxylan (AX) constitutes the most abundant non-starch cell wall polysaccharide in cereal grains. Arabinoxylans are mainly mono- or disubstituted by L-arabinofuranosyl units (Ara*f*) by (1⟶2)-α and/or (1⟶3)-α linkages to different extents and intramolecular distributions, depending on the biological source and cereal tissue (e.g. endosperm, aleurone layer, bran) [26, 30, 31]. The arabinose decorations may be further substituted by ester links in the C5 position by phenolic compounds, mainly ferulic acid; these phenolic compounds can dimerize with other ferulic acid moieties in other arabinoxylan molecules, creating a complex cross-linked covalent structure in some tissues like the bran [32]. In some tissues and species, as in maize, arabinoxylans exhibit an extremely complex macromolecular architecture, where additional glycosyl substitutions and sidechains can be found, such as (1⟶2)-α-MeGlcA, (1⟶3)-β-Xyl*p*, (1⟶2)-α- and (1⟶3)-α-Xyl*p*, and (1⟶2)-α-L-Gal*p* [33, 34]. The function of such complex substitution pattern in bran xylans may be to prevent enzymatic degradation or to modulate tighter intermolecular interactions with other cell wall components [33].

From a technological point of view, the arabinoxylans from different cereals and tissues have traditionally been classified into water-extractable arabinoxylans (WE-AX) and water-unextractable arabinoxylans (WU-AX), depending on their potential aqueous extractability. Again, the content of each arabinoxylan fraction and its macromolecular structure depend on the cereal source and the tissue. As a rule of thumb, WE-AXs are mainly present in the cereal endosperm and are moderately substituted with Ara*f*. On the other hand, WU-AXs show great heterogeneity in their structure, with fractions containing very divergent Ara*f* substitution patterns and further decorated with MeGlcA and ferulic acid, which contribute to cross-linking with other arabinoxylan molecules and other cell wall components. There is broad variability in the literature about the availability and composition of each arabinoxylan fraction in each cereal and tissue (Table 8.1). It is therefore necessary to study individually the content and structure of each

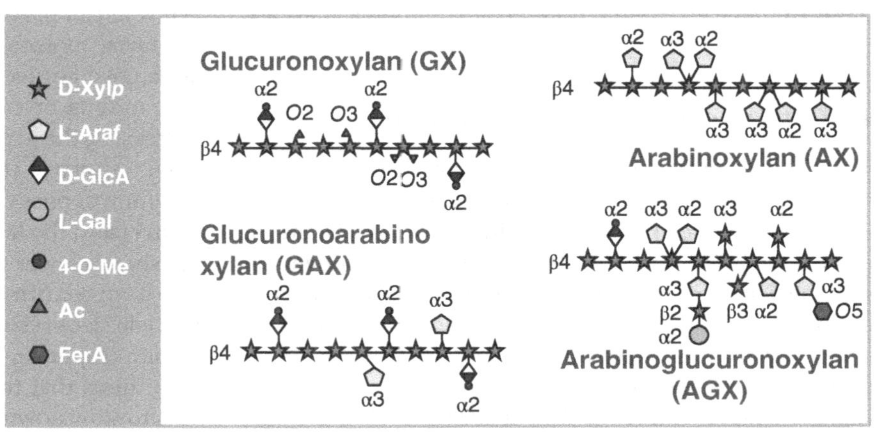

Figure 8.6 Molecular structure of xylans: Hardwood glucuronoxylan (redrawn after [27]); softwood arabinoglucuronoxylan (redrawn after [29]); cereal endosperm arabinoxylans and bran arabinoglucuronoxylan (redrawn after [33, 34]).

Table 8.1 Composition and relative amounts (in % dry weight) of AX and mixed-linkage β-glucans in different cereal tissues.

		Arabinoxylans (AX)		Mixed-linkage β-glucans
Cereal tissue		WE-AX (%)	Total (%)	Total (%)
Wheat	Grain [31, 35]	0.5	6	0.4–1.5 [36]
	Endosperm (flour) [31, 35]	0.5	2.2	
	Bran [31, 35]	0.8	20–33	
Barley	Grain [35]	0.5–1	4–7	2.5–11.3 [36]
	Brewers' spent grain [35, 36]	0.5	13.5	
Rye	Grain [37, 38]	2.4–8	7–12	2–3 [36]
	Flour [37, 38]	2.5	3.5–9	
	Bran [37]	2.1	12.6	
Oats	Grain [35]	0.2	2.7	2.2–7.8 [36]
	Bran [35]	0.33	3.5	

individual cereal industrial stream for an optimized exploitation of these valuable polysaccharide fractions for the development of materials.

8.2 Effect of the Molecular Structure of Starch and Seed Polysaccharides on the Macroscopic Properties of Derived Carbohydrate-based Materials

The complex branching structure of starch at the macromolecular level is defined by tight biosynthetic constraints, and determines its hierarchical organization in the grain. Similarly, the structural heterogeneity of seed polysaccharides in terms of substitution pattern and molecular weight is fundamental for their specific functionality and environmental adaptability. However, this structural complexity imposes major difficulties for improving the digestibility of starch-based foods in the biotechnological exploitation of such polysaccharides into bio-based materials.

8.2.1 Factors Affecting Starch Digestibility

Since the invention of agriculture, starch-based foods have been the major energy source for humans, and arguably have been key factors in the development of civilization [39]. However, this essential component of our diets is now responsible for major public health problems, as more and more of humanity adopts a less physically arduous lifestyle. This is because most of the popular starch-based foods are such that the starch therein is rapidly digested. This tends to result in a spike in blood sugar, less "resistant" starch reaching the lower bowel, and reduced satiety shortly after food consumption. All these factors are significant contributors to the frightening levels of obesity, diabetes, and colo-rectal cancers: for example, the level of diabetes in China is now higher, and growing more quickly, than in most developed countries [40], and there are increasing problems of severe obesity [41]. A partial answer is the development and acceptance of starch-containing foods wherein the starch is digested more slowly, but which are both palatable and have cultural acceptance.

Plant cell walls are a common native plant structure. The plant cells that encapsulate starches in the endosperm of whole grains are surrounded by the cell wall matrix, which is composed of polysaccharides that cannot be digested by non-ruminant animals' digestive enzymes, even though they can readily be degraded by ruminant microorganisms in the digestive tract. Therefore, the cell walls potentially protect the entrapped starches from endogenous amylase activity in non-ruminant animals' small intestine. Moreover the intact cell wall can prevent the complete swelling and leaching of starch molecules from granules, thus inhibiting the access of digestive enzymes. Some processing procedures (e.g. milling or cooking) can disrupt cell walls, thereby increasing the accessibility of starch to hydrolysis. It has been reported that higher blood glucose levels and insulin responses result when whole rice is ground [42]. Conversely, the larger particle sizes of wheat, maize, and oat show a lower *in vitro* starch digestion rate, which has also been confirmed by an *in vivo* experiment, with larger particle sizes resulting in lower blood-glucose levels and insulin responses [43]. All these phenomena support the hypothesis that the plant cell wall is an effective barrier able to slow down the rate of starch digestion.

Fiber is a common component in flour and is also able to influence starch digestion. Fiber can be classified into the water-soluble (non-starch polysaccharide) and water-insoluble fractions (cell wall, bran, etc.); there are other technical definitions of fiber, but here we use the soluble/insoluble categorization. Inclusion of high amounts of insoluble fibers can disrupt the protein matrix, resulting in a porous structure; this may lead to an increase in the accessibility of starch granules to degradative enzymes [44]. Conversely, the inclusion of soluble fibers will enhance the resistance of starch to enzymatic degradation, as it can induce the formation of a viscous protein-fiber-starch network that can entrap the starch granules, reducing glucose release [44]. Many studies using human subjects and animal models have provided evidence that added soluble fibers can significantly reduce the glycaemic response and plasma insulin levels [45, 46]. Apart from this *in vivo* experimental evidence, the rate and extent of starch hydrolysis by amylases can also be shown to be reduced when soluble fiber is incorporated into a food matrix and subjected to a digestion medium.

Soluble fiber can alter the viscosity of the food matrix, affecting the rate of starch digestion. For example, polysaccharide-based gums, a type of water-soluble non-starch polysaccharide, are thought to be beneficial as they reduce postprandial glycaemia [47]. This has been confirmed with the experimental result that guar gum in pig meals caused an increase of the zero-shear viscosity of jejuna digesta, along with a significant decrease in the rate of glucose absorption [48]. This beneficial postprandial effect results from an increase in the digesta viscosity inside the gastrointestinal tract, caused by the swelling of fully hydrated galactomannan chains; this reduces the digestion rate and absorption of carbohydrates, leading to a lowering of postprandial rises in blood glucose. The presence of hydrated galactomannan chains restricts the swelling of starch granules during gelatinization. As a result, some granules may not be gelatinized completely, reducing the water available to the starch granules [49]. This incomplete gelatinization of starch may also contribute to the starch's resistance to digestion. It is noted that some viscous fiber derived from guar, tragacanth, can increase the viscosity even at a low polymer concentration, which increases the overall viscosity of digesta in the gastrointestinal tract. Moreover an increase in the viscosity of digesta can also affect gastric behavior and reduce the propulsive and mixing effects from peristalsis, decreasing the interaction frequency between substrates and digestive enzymes. In addition to the lower rate of starch digestion, the transport of some intermediate products (e.g. maltose, α-limit dextrins) to intestine mucosal surfaces would also be slowed down [50]. Although the addition of other cereal-based viscous fibers can also decrease starch digestion and glucose absorption to some extent, the mechanism behind this reduction in digestion may be slightly different, as these fibers are physically and functionally different from viscous gums.

Not only can soluble fibers decrease the digestion rate of starch by increasing the viscosity, but they can also play a role as a physical barrier, reducing the interactions between enzyme and substrate [50, 51]. Moreover some soluble fibers like galactomannans can even act as a non-competitive inhibitor on the activity of α-amylase [52]. The resulting complex formed by galactomannans and enzyme has diminished activity, decreasing the activity of pancreatic amylase.

Native α-amylase inhibitors, a type of anti-nutrient, commonly occur in natural plants. These inhibitors occur in varying quantities among a wide variety of crops, e.g. beans, rye, wheat, and oats. The most common types of α-amylase inhibitors are either protein or glycoprotein in nature, are non-competitive, non-dialyzable and are effected by heat treatment [53]. However, there have also been non-protein amylase inhibitors reported (e.g. polyphenolic compounds or phenolic acids, acarbose, isocarbose and acarviosine-glucose) [54]. For example, amylostatin (extracted from the fungi *Streptomyces*) and acarbose (extracted from the family Actinoplanaceae) are two types of oligosaccharides that can inhibit amylase activity. In addition, some natural molecules extracted from plants can also inhibit the activity of α-amylase. For example, natural inhibitors from white beans have been found to reduce the peak of postprandial glucose in healthy and type 2 diabetic subjects [55]; some low molecular weight molecules derived from plants such as luteolin, strawberry extracts, and polyphenols in green tea have also been observed to prohibit the activity of α-amylase and lower postprandial hyperglycemia [56, 57]. It has been reported that there are compounds from wheat that do not affect the wheat's amylase activity but can inhibit the activity of mammalian salivary and pancreatic α-amylase [58].

In general, amylase inhibitors tend to inhibit the activity of certain specific amylases. These inhibitors can interact with amylase to form a complex, leaving the enzyme inactive. According to their specific functions, inhibitors from plant sources can be divided into two groups: those that inhibit the activity of exogenous α-amylase from predators, thereby protecting the seeds or other organs, and those that inhibit endogenous α-amylases to regulate their activity during seed development or germination [59]. All amylase inhibitors are pH- and temperature-sensitive. For example, α-amylase inhibitors from the kidney bean, while having the highest activity, are pH dependent; α-amylases inhibitors from legume crops are generally inactivated at temperatures above 100 °C. It has been shown that some amylase inhibitors can permanently lose their activity during processing. For instance, natural amylase that has been isolated from wheat germ fractions can be destroyed when being passed through a roller mill [60]. Although many products have become available to block the activity of amylases since the early 1940s, many of them were found to be ineffective *in vivo*, despite being reported to be promising during *in vitro* experiments. It is interesting that amylase inhibitors are unstable in the stomach and can only be active when being incubated with amylase prior to the reaction with starch [61].

Another common amylase inhibitor from plant sources is phytic acid, one of the most important reserves for phosphate in many plants. It is able to form complexes with proteins or metal ions and thereby reduce their biological availability [62]. Phytic acid was shown to decrease the rate of sugar release when incubated with saliva prior to reacting with starch, whereas there is no evident inhibition if the phytic acid and amylase are added into the starch solution simultaneously. Phytic acid is able to affect starch digestibility by interacting with amylase proteins as well as by binding with salivary minerals like calcium that are known to increase amylase activity. Moreover, processing can also affect phytic acid, resulting in a change in the activity of amylase [63, 64]. For example, phytic acid levels can be significantly reduced when the seeds are allowed to germinate for 72 h; extrusion cooking has also been found to reduce phytic acid, as well as lead to condensed tannins and polyphenolics; thermal processing can lead to the degradation of anti-nutrient molecules including phytic acid, changing their activity in the formation of insoluble complexes [64].

The size of starch granules has been shown to affect the amylolysis, with several studies showing that larger granules have a lower starch digestion rate than their smaller counterparts [65–70], even though differences in granule composition may also partially affect the amylolysis rate. The primary factor controlling the starch hydrolysis rate is thought to be the number of starch molecules available for the absorption and binding, with the lower number leading to a slower rate [71]. This has been supported by the experimental observation that small potato granules digested almost three times faster than their larger counterparts [72]. There is also a size difference between the channels in A-polymorphic starches and the pores (opening of channels). The former is thought to be lined with proteins and lipids [73–75], with diameters ranging from 0.007–0.1 μm, while the latter has larger diameters, ranging from 0.1–0.3 μm, and this is large enough to allow the amylase to diffuse inside the maize starch granules [76, 77]. The enzymatic digestion of starch granules with few pores and channels, however, can only occur from the outside towards the inner part of the granule, as the high local concentration and tight packing of amylose (AM) and amylopectin (AP) make the granule surface impermeable to amylase [78, 79]. As well as the architectural features of the native starch granules, the proteins [80] and lipids [81] on the granule surface can also affect the starch digestion rate by blocking the accessibility of the starches and disturbing the enzyme binding [78]. Conversely the less-ordered starch chains [71, 82] and damaged surfaces favor the binding of enzyme to the granule surface [78].

Seed protein has a significant effect on the rate of starch digestion. The protein usually surrounds or extends into the endosperm structure containing the starch granules, forming a continuous protein matrix. The protein factors that affect starch digestion can generally be divided into two groups: exogenous effects including the interaction of protein with non-protein components, like polyphenols, starch, non-starch polysaccharides, and lipids; and endogenous influences arising from the proteins' natural properties and organization in the grains [83, 84]. The amount of protein in a grain is approximately 10% protein, compared to starch which accounts for approximately 70–80% of the dry weight of most cereals. Protein affects the functional properties of starch, e.g. gelatinization and starch digestion rate [84–89] as the protein composition may form a contiguous layer surrounding the starch granules. This phenomenon has been confirmed for many cereals and is especially evident for sorghum [90].

In addition to native protein components, lipids are another kind of non-starch component that associate with starch granules and affect the starch digestion rate (raw or gelatinized starch granules). Similar to proteins, lipids also reduce the accessibility and binding of enzymes to the starch molecules. In general, cereal starch granules contain internal lipids that are mainly free fatty acids and lysophospholipids existing independently [91], as well as amylose-lipid complexes [92]. The lipids existing independently can form complexes with amylose after thermal processing, restricting the swelling and solubility of starch granules [93]. Both the native amylose-lipids complexes and those produced by processing are much more resistant to digestion than normal single helical amylose chains [94], as the active site of the amylases are hindered.

The activities of enzymes and amylose-lipid complexes are determined by the structural features of these complexes. Generally speaking, fatty acids less than 10 carbon units long are water-soluble, with the chains becoming insoluble when exceeding 12 carbon units in length. Therefore the shorter, water-soluble fatty acids are less effective at forming complexes than the longer water-insoluble fatty acids. Furthermore, amylose-lipid complexes can be classified into two forms based on their melting temperatures: form I consists of amorphous amylose-lipid complexes that have a lower melting temperature and are more susceptible to enzyme hydrolysis; form II consists of crystalline complexes with higher melting temperatures [95–98].

The catalytic reaction between amylase and starch molecules can be affected by many barriers, with the structural parameters of the native starch being important in determining the starch's digestibility. The main structural parameters that control starch digestibility are usually considered to be: the amylose (AM)/amylopectin (AP) ratio, amylopectin chain length, the amount of crystallites caused by the helical structure of the starch, the packing of the crystallites, and the interaction of the starch molecules with non-starch components. Recently, amylose fine structure has also been revealed as a significant factor [99, 100]. These factors can individually, or in combination, affect the binding and catalysis of starch granules with amylase.

The crystalline and amorphous phases provide barriers that inhibit the accessibility of enzymes to the interior of starch granules, which can greatly affect the rate and extent of the enzymatic hydrolysis of starch [101]. Therefore, processing of the starch granules, for example, by hydrothermal treatment, is introduced to swell the amorphous and ordered regions, making them more readily available for amylase-catalyzed hydrolysis, allowing the kinetics to be analyzed [82].

The type of crystallinity also affects the rate of starch digestion. With the exception of extremely high amylose variants (e.g. Gelose 80) that are bred specifically for their amylose content (but have poor palatability in food [102]), A-type starches from normal cereals are more readily hydrolyzed than B-type starches from high amylose cereals or potatoes. The effect of the crystallinity type is also dependent on the chain length of amylopectin [103–106]: specifically, B-type starches contain amylopectin with more long branch chains, which can extend through several crystals to make the internal structure stable; conversely A-type starches have more short branch chains, leading to an unstable lamellar structure [107, 108]. The structural difference makes B-type starches more resistant to enzymatic digestion compared with A-type starches. There are some reports [104] regarding this aspect: e.g. A- and B-type polymorphs from the same source (amylose spherocrystals, DP 15 – 20) were subjected to enzymatic digestion *in vitro*, with the B-type polymorphs being more resistant to α-, β-amylase and amyloglucosidase digestion. However another study found that A-type crystals from debranched waxy starches were more resistant to enzymatic digestion than the B-type polymorphs from the same starch source [109, 110]. Although the double helical structures are almost identical for the two polymorphic forms (A- and B-types) [111–113], the double helices in the A-type monoclinic unit cells are packed more compactly (with a lower water content) than B-type crystals which are packed in a hexagonal unit cell [106, 114]. This might explain why in this study the A-type starches were more resistant to enzymatic hydrolysis. Considering the conflicting results mentioned above, even though the distribution of crystallites and their influence on local granule organization probably affects the overall activity of amylase enzymes on starch hydrolysis [115], it is likely amylose or debranched amylopectin crystals are not the sole determinants of amylase activity on starch granules. It seems likely that a combination of factors regarding granule structure at all structural levels come together to determine the overall hydrolysis rate.

Supramolecular organization is also thought to affect the digestion of less organized amorphous regions more than the ordered semi-crystalline regions, as observed using scanning electron microscopy (SEM) [106, 116] and reported in molecular structural analysis studies [116]. A-type starches have been reported to be more accessible to enzymes than B-type starches [117, 118]; however, in this study both crystalline and amorphous regions were evenly degraded. This phenomenon may be explained if the glucan chains of the amorphous regions are not as free and mobile as previously thought, in which they are densely packed and arranged tightly adjacent to the crystalline regions. This special structural feature would be effective in inhibiting the enzymatic hydrolysis of the starch [104, 117]. Further, the non-crystalline fractions contain amorphous regions entrapped within imperfect crystals and double helical structures that are not a part of crystallites. It has been hypothesized that these fractions in the non-crystalline regions lead to an increase in resistance to enzymatic hydrolysis [119]. This explanation has been supported in other studies: for instance, amylopectin lamellae are thought to form spherical blocklets on the surface regions of B-type starches, making these starches granules comparatively larger (200–500 nm) than in A-type starches (20–120 nm) [106, 120, 121]. The organization of amylopectin lamellae can inhibit the accessibility of enzymes into the inside of starch granules. Also the AM in starches with high AM contents (e.g. high amylose maize starches, HAMS), can reinforce and stabilize amorphous rings, prohibiting amylase attack, an extra barrier that starches with higher AM contents (e.g. normal maize starches, MS) do not have. However, other studies have shown that potato starch with similar AM contents to MS had a slower digestion rate more similar to the digestion pattern of HAMS. This different digestion rate might be attributed to the size of the granules' architecture, in which the higher levels of structure in the range of ~100 nm (e.g. pores and channels, damage to granule surface) are the primary aspect determining the digestion rate, where the AM/AP ratio, the chain lengths of AP and the amount and types of crystallinity only play minor or secondary roles [118, 122].

All cereal-based food products undergo a series of processing procedures, any of which can result in changes in the size of the starch granules or affect the protein network which can potentially lead to changes in the digestion of the starch. Wide ranges of technologies have been employed by industry to process food materials, altering the food structure and thus the nutritional qualities, e.g. starch digestibility. The most common procedures used include milling, mixing, sheeting, extrusion, drying, cooking, and chemically modifying.

Both milling and grinding of grains can break the cell wall structures and protein networks, leading to an increased activity of amylase digestion. It has been observed that the grinding of barley and sorghum grains led to an approximate doubling of the amylase digestibility with the particle size decreasing from ~500 to ~250 μm [123]. Because the size of individual endosperm cells range typically from 50–50 μm in diameter, starch granules in particles of ~500 μm in size are primarily contained within intact cell walls; however, particles of ~250 μm are likely to have more broken cells, exposing the intra-cellular starches to adsorption and catalytic action by amylases. Moreover, the smaller particle size means a larger reaction area available for higher catalytic efficiency and rate of hydrolysis [72, 82].

The size of cereal-based food products can also influence starch digestion. For example, the digestibility of two pastas with different sizes have been compared, with starch in the larger-sized spaghetti being less susceptible to α-amylase than starch in the smaller-sized vermicelli, presumably due to the higher surface to weight ratio of vermicelli compared to spaghetti [124]. Pasta shape was also found to affect the rate of starch digestion, with macaroni (smaller particle size) producing a higher *in vivo* glucose response than spaghetti (larger particle size) [125]. It was also shown that spaghetti after being ground had a higher susceptibility to enzymatic digestion, possibly because the pasta structure after destruction facilitated the diffusion of α-amylase [126]. One experiment showed that the disintegration of spaghetti into coarse porridge greatly increased the glucose response over that of intact spaghetti [93]. All of these results highlight the importance of food structure, including particle size.

8.2.2 Structural Aspects of Seed Polysaccharides Affecting Configuration and Macroscopic Properties

The presence of different substitutions along the polymeric backbone (e.g. neutral sugars, uronic acids, and chemical modifications such as acetylation) of seed polysaccharides, their pattern of substitution, and molecular weight determine the macroscopic properties of seed polysaccharides. These structural effects are complex and sometimes even conflicting. An early review [127] pointed out the role of the substituents in complex polysaccharides in altering the inter- and intramolecular hydrogen bonding networks and the water interaction. This in turn influences the macromolecular conformation in solution, including the ability to form ordered structures (e.g. helices) by cooperative interactions, the stiffness of the molecular chains and finally the (in)solubility/aggregative behavior in the specific solvent. Finally, the macroscopic properties in both solution and the solid state, including viscosity, gel forming, rheology and viscoelasticity, crystallinity, film forming, barrier properties, and mechanical properties are largely influenced by these structural features. It is therefore fundamental to assess thoroughly the molecular structure of the polysaccharides to understand and control the final macroscopic properties of the derived materials.

A good example to understand the effect of the glycosyl substitutions on the macromolecular conformation and the solution properties in water of seed polysaccharides is the case of seed galactomannans with different ratios of galactosyl substitutions (M:G ratios). Molecular dynamic studies of galactomannans, from a purely linear mannan to galactomannans with increasing Gal substitutions (M:G from 5:1 to 1:1) pointed out the influence of substitutions in chain flexibility [128]. Computational studies were used to predict spatial characteristics of the (galacto)mannan chains, including the persistence length, which is a mechanical property related to the stiffness of a polymeric chain. Linear mannan exhibits a rigid backbone chain with an estimated persistence length of 14 nm, which is progressively reduced by the addition of Gal side chains. This influence of side substitutions of chain flexibility reaches a plateau for high levels of substitution (between 3:1 to 1:1), with estimated values of persistence length of 9 nm. In these cases, it is the substitution pattern rather than the content of substitutions that plays a major role. These observations are in agreement with the experimental values obtained for tara, locust bean and guar gum obtained by multiple-detection size-exclusion chromatography, differential viscometry, light scattering, and sedimentation studies [129–132].

Multiple-detection size-exclusion chromatography (SEC) is a very powerful technique to investigate the influence of fine structural details on the conformational properties of seed polysaccharides. The combined use of multi-angle laser light scattering (MALLS) and a concentration detector (typically differential refractometry for polysaccharides) hyphenated to SEC provides both the size dependence of the absolute weight-average molecular weight ($\bar{M}_W$) and the root-mean-square z-average radius of gyration $R_{g,z}$, given the known values of the differential refractive index increment (d*n*/d*c*) for the polysaccharide in the specific solvent and temperature of separation [133]. The direct representation of molecular weight versus macromolecular size, also known as a conformational plot, offers a robust estimation of the conformation of the polysaccharides in solution. This dependence is well described by a simple power equation usually known as the de Gennes scaling law, [134] whose exponent directly indicates the fractal dimensions of the polymer chain in solution (Eq. 8.1) [135]:

$$R_g = K_g \bar{M}_W{}^{\nu_g} \quad (8.1)$$

Here K_g is a constant and the exponent ν_g indicates the fractal dimensions, which depend on the macromolecular architecture, the temperature and polymer-solvent interactions. A value of ν_g between 0.5–0.6 corresponds to a random coil conformation in solution, a value of $\nu_g = 0.33$ indicates a compact hard sphere, and a value of $\nu_g = 0.1$ corresponds to a rigid rod of homogeneous density.

Molecular weight is one of the structural parameters that directly influence the conformation in solution of polysaccharide chains and, in turn, overall macroscopic properties. The unique influence of the molecular weight can be clearly illustrated for the case of mixed-linkage β-glucan. Indeed, mixed-linkage β-glucan can be envisaged as a linear copolymer of cellotriosyl and cellotetraosyl units linked by (1⟶3)-β-linkages without any side substitution or branching. In this case, a direct correlation between molecular weight, conformation in solution, viscosity, and rheological properties was established for different barley mixed-linkage β-glucans [136]. Increasing molecular weight distributions results in a slightly stiffer conformation in solution, as evidenced by the increasing fractal dimensions (ν_g) in the conformational plots of radius of gyration versus weight-average molecular weight (Figure 8.7). This stiffer conformation promotes aggregative behavior among the β-glucan chains, which in turn enhance the viscosity and rheological properties of β-glucan solutions.

Another example that shows the complex relation between molecular structure and properties, which can in turn be revealed by multiple-detection SEC, is the influence of the pattern of substitution in cereal arabinoxylans. Rye and wheat endosperm arabinoxylans have distinct substitution pattern, with rye AX Xyl*p* backbone

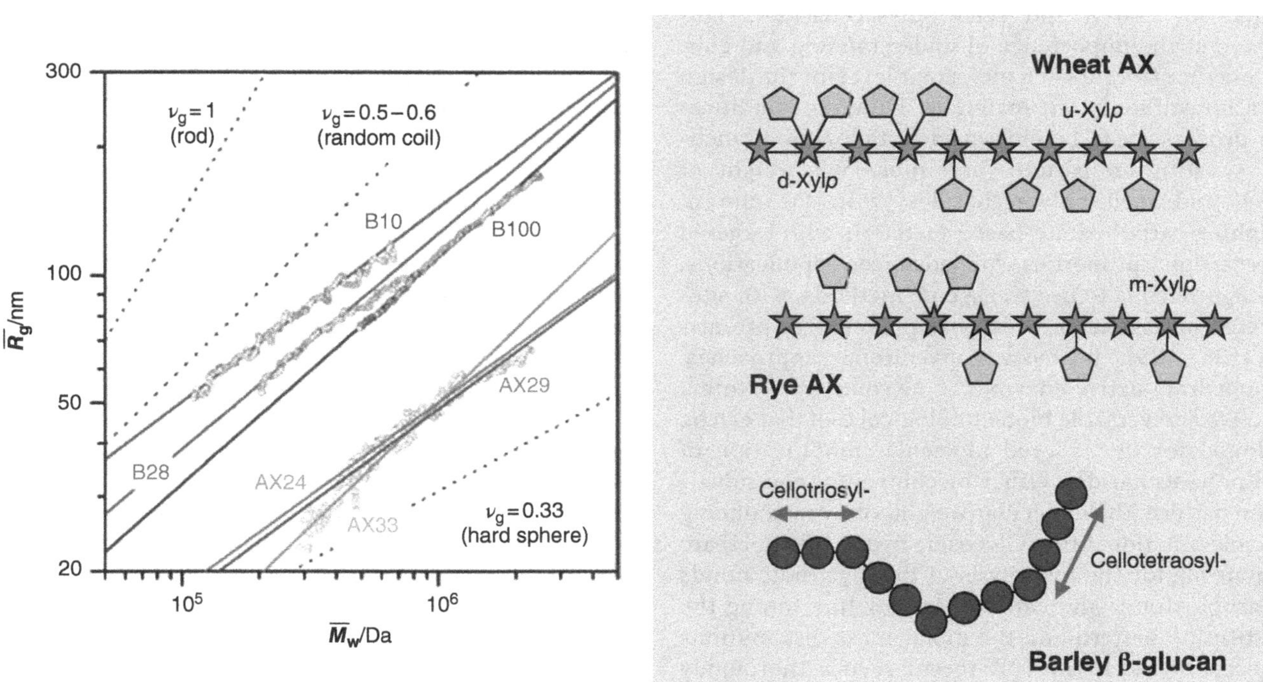

Figure 8.7 Effect of substitution pattern and molecular weight on the hydrodynamic properties and conformation in solution of cereal arabinoxylans and β-glucans.

units being preferably mono-substituted by Ara*f* residues (m-Xyl*p*) and wheat AX having a larger ratio of double-substituted Xyl*p* units (d-Xyl*p*) by Ara*f* residues [137]. The complex and sometimes conflicting effect of the substitution pattern and the molecular weight on the macroscopic properties of AX has been elucidated (Figure 8.7). Rye AX (AX33) showed unexpectedly similar intrinsic viscosity and promoted higher diffusion of a macroscopic probe in solution compared to wheat AXs (AX29 and AX24) with lower nominal molecular weights. In this case, the conformational plots revealed again the differences in fractal properties between rye and wheat AXs, with rye AX exhibiting a more rigid and stiff polymeric conformation and wheat AX samples a more flexible and compact conformations. This was directly correlated with the substitution pattern in AX, with rye AX having more mono-substituted Xyl*p* backbone, inducing chain stiffness and promoting interchain aggregation in solution, therefore influencing the rheological and diffusional properties.

8.3 Chemo-enzymatic Modification Routes for Starch and Seed Polysaccharides

In Section 8.2, the complex and sometimes conflicting correlation among molecular structure, intermolecular association, and macroscopic properties was highlighted for starch and seed polysaccharides. This points out the importance of understanding and controlling the structure at a molecular level for the design of carbohydrate-based materials. Different modification processes can be envisaged to tailor the branching, substitution pattern, and molecular weight of starch and seed polysaccharides, with the aim of designing carbohydrate-based materials with targeted macroscopic properties for advanced applications. One approach is to use transgenic methods, with site-directed mutagenesis, which has proved very promising [138, 139]. In more conventional approaches, carbohydrate-active enzymes (CAZymes) constitute a selective and versatile biotechnological tool that can be exploited for the tailored molecular modification of seed polysaccharides, either by controlling the substitution pattern, the molecular weight, or by introducing valuable functionalities. Glycoside hydrolases (GH) are responsible for the hydrolysis of the glycosidic bonds in carbohydrates and can be used for fine-tuning the substitution pattern and the molar mass distributions of seed polysaccharides. A recent review thoroughly described the use and mode of action of different GHs for the extraction and modification of different plant polysaccharide families from lignocellulosic biomass, establishing important correlations between changes in molecular structure and macroscopic properties [140].

Different studies have focused on tailoring the substitution pattern and molar mass distributions in xylans by enzymatic means, in order to modify the macroscopic properties for enhanced preparation and application of xylan film barriers [141–145]. *Endo* β-xylanases cleave the β-Xyl*p*-(1→4)-linked units of the backbone, whereas *exo*-α-arabinofuranosidases and *exo*-α-glucuronosidases remove the Ara*f* and GlcA substitutions, respectively. The variation in arabinose substitution has a significant impact on the solubility and material properties of cereal xylans. Different α-L-arabinofuranosidases acting on polymeric xylans (arabinoxylan arabinofuranohydrolases, AXH) can be exploited depending on their substrate specificities [25, 146]. Some AXHs act on (1→2)- and (1→3)-linked α-L-arabinofuranosyl units on monosubstituted xylopyranosyl residues (AXH-m), whereas others release solely (1→3)-linked α-L-arabinofuranosyl units from disubstituted xylopyranosyl residues (AXH-d) [147]. Progressive enzymatic hydrolysis of the Ara*f* substitutions xylan by AXH-m causes agglomeration of the polysaccharide in water. [143]. This reduction in Ara*f* side groups also affects the macroscopic properties of AX materials, by increasing the crystallinity and the oxygen barrier properties of the prepared films; however, the mechanical properties were not enhanced by such treatment, causing a decrease of the elongation at break without a concomitant improvement of the tensile strength [24, 143–145, 148]. The degree of Ara*f* substitution is not only responsible for the changes in macroscopic properties but also the pattern of such substitutions, as can be observed in Figure 8.7. The combined use of AXH-m and AXH-d3 allowed the production of tailored AX with similar Ara*f*/Xylp ratios but different patterns of unsubstituted, mono- and disubstituted Xyl*p* units, which controls the crystallinity, elongation at break, and barrier properties of xylans [149].

Seed mannans can be selectively modified by the enzymatic hydrolysis of the (1→4)-β-Man*p* backbone by *endo* β-mannanases, by the selective removal of the (1⟶6)-α-Gal*p* substitutions by *exo*-α-galactosidases, and finally by the selective oxidation of the hydroxyl group in the C-6 position of Gal*p* by galactose oxidases (Figure 8.8). On one hand, enzymatic backbone-cleaving of mannans by *endo* β-mannanases drastically reduces the viscosity of the polysaccharides, which may enable processivity of these solutions [150]. On the other hand, progressive reduction of the degree of Gal*p* substitution

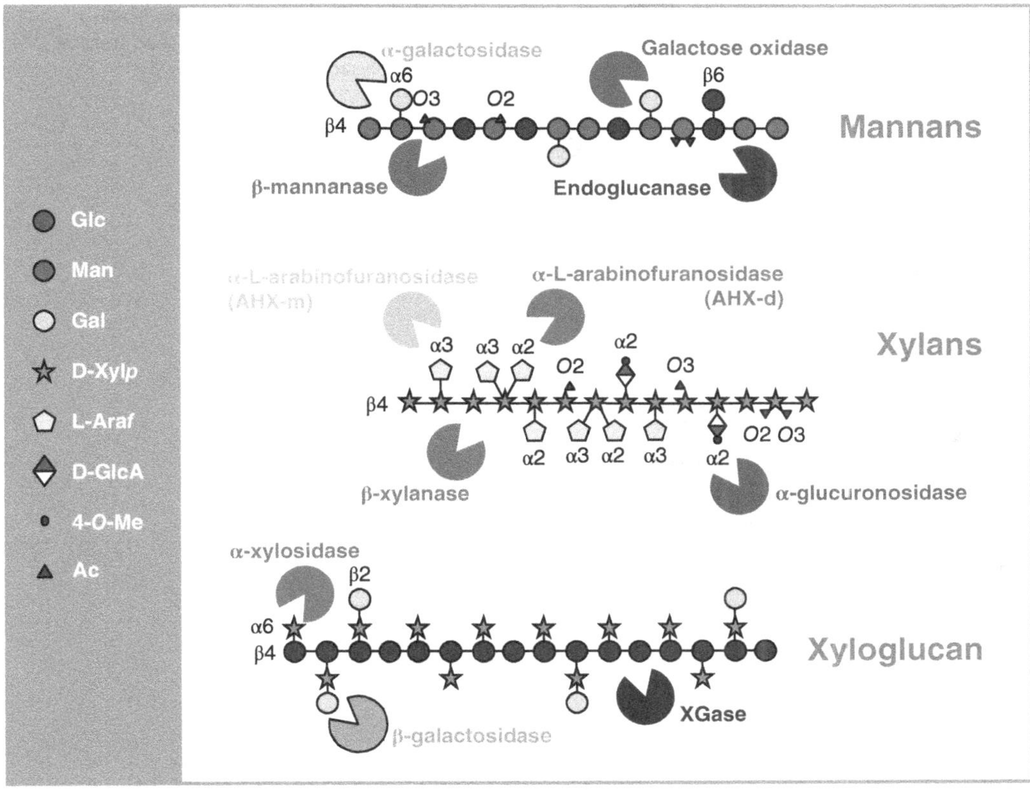

Figure 8.8 Use of selected carbohydrate-active enzymes (CAZymes) for structural modification of starch and seed polysaccharides (redrawn after [140]).

by α-galactosidases is responsible for an increase in the mechanical properties (elongation at break and tensile strength) of films produced from locust bean and guar gums [151]. Finally, oxidation of the Gal*p* residues by galactose oxidation can be exploited for the preparation of hydrogels and aerogels from high molecular galactomannans [152–154]. Such partially-oxidized mannans exhibit improved film-forming properties due to the increased viscosity of the solutions, which results in enhanced thermal stability and mechanical performance.

A similar influence of enzymatic "trimming" by *exo*-β-galactosidase on the properties of seed xyloglucans has been reported. Progressive removal of Galp decorations induces aggregation of the polysaccharide chains in solution and influences the gelling behavior [155]. Moreover, the mechanical and the oxygen barrier properties, together with the moisture sensitivity of XyG films, can be significantly improved by the enzymatic removal of Gal*p* residues in tamarind XyG films, which is fundamental for their application in food packaging under large humidity conditions [156].

Chemical modification of starch and seed polysaccharides constitutes another valuable route for structural modification and fine-tuning the macroscopic properties of the derived materials. The rich palette of carbohydrate chemistry offers almost unlimited possibilities for the modification and substitution of the hydroxyl groups in polysaccharides. This can be implemented by oxidation, esterification (acetylation, succinoylation, benzoylation), etherification (carboxymethyl, methylation, hydroxylakylation, benzylation), grafting from, grafting to, and cross-linking (Figure 8.9) [157, 158].

8.4 Conclusion

Starch and seed storage polysaccharides have many advantages for a wide range of uses. The greatest advantages are that they are from renewable resources, and their structural diversity and complexity. However, the latter is also a major disadvantage. A future challenge for scientists in this field is to overcome the disadvantages to capitalize on the advantages.

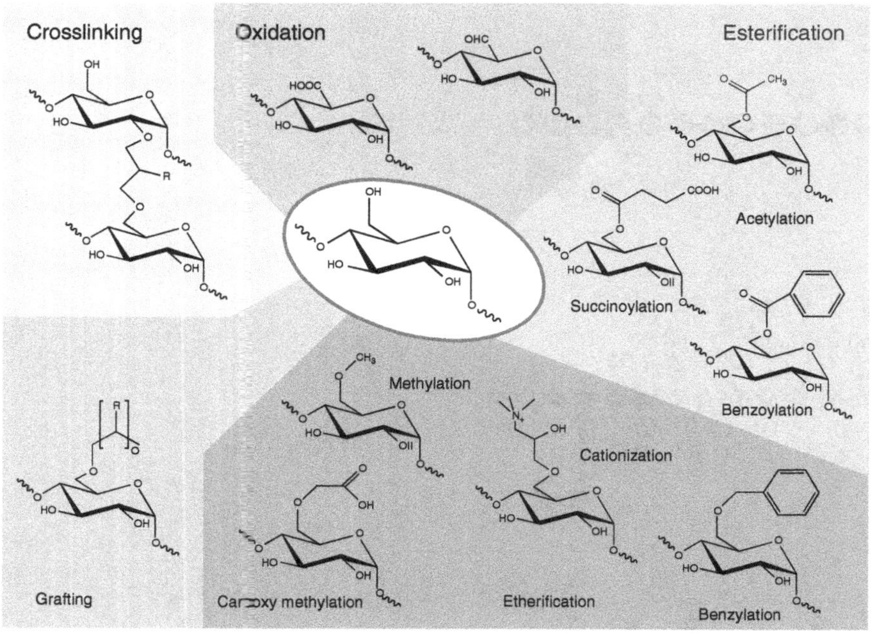

Figure 8.9 Chemical modification methods for starch and seed storage polysaccharides.

References

1 Buckeridge, M.S. (2010) *Plant Physiology*, **154**, 1017–1023.
2 Jane, J.L., Kasemsuwan, T., Leas, S. et al. (1994) *Starch-Stärke*, **46**, 121–129.
3 Ao, Z., Jane, J.L. (2007) *Carbohyd. Polym.*, **67**, 46–55.
4 Gidley, M.J., Grant Reid, J.S.,(2006) In *Food Polysaccharides and their Applications*, Stephen, A.M., Phillips, G.O., Williams, P.A. (eds.), Taylor and Francis: Boca Raton, FL, pp. 181–205.
5 Ebringerova, A., Heinze, T. (2000) *Macromol. Rapid Comm.*, **21**, 542–556.
6 Moreira, L.R.S., Filho, E.X.F. (2008) *Appl. Microbio. Biotech.*, **79**, 165–178.
7 Lundqvist, J., Teleman, A., Junel, L. et al. (2002) *Carbohyd. Polym.*, **48**, 29–39.
8 Willför, S., Sundberg, K., Tenkanen, M., Holmbom, B. (2008) *Carbohyd. Polym.* **72**, 197–210.
9 Prajapati, V.D., Jani, G K., Moradiya, N.G. et al. (2013) *Int. J. Biol. Macromol.*, **60**, 83–92.
10 Buckeridge, M.S., Dietrich, S.M.C., de Lima, D.U. (2000).In *Developments in Crop Science*, Anil Kumar, G., Narinder, K. (eds.), Elsevier, Oxford. Vol. 26, pp. 283–316.
11 McCleary, B.V., Amado, R., Waibel, R., Neukom, H. (1981) *Carbohyd. Res.*, **92**, 269–285.
12 Katsuraya, K., Okuyama, K., Hatanaka, K. et al. (2003) *Carbohyd. Polym*, **53**, 183–189.
13 Hannuksela, T., Hervé du Penhoat, C. (2004) *Carbohyd. Res.*, **339**, 301–312.
14 Carpita, N.C., Gibeaut, D.M. (1993) *The Plant J.*, **3**, 1–30.
15 Edwards, M., Dea, I.M., Bulpin, P., Reid, J.S.G. (1985) *Planta*, **163**, 133–140.
16 Fry, S.C., York, W.S., Albersheim, P., et al. (1993) *Physiologia Plantarum*, **89**, 1–3.
17 Tuomivaara, S.T., Yaoi, K.,O' Neill, M.A., York, W.S. (2015) *Carbohyd. Res.* **402**, 56–66.
18 Schultink, A., Liu, L., Zhu, L., Pauly, M. (2014) *Plants*, **3**, 526.
19 Hilz, H., de Jong, L.E., Kabel, M.A. et al. (2006) *J. Chromatog. A*, **1133**, 275–286.
20 Gunl, M., Gille, S., Pauly, M. O.(2010) *J. Vis. Exp.*, e2046.
21 Obel, N., Erben, V., Schwarz, T.,et al. (2009) *Mol. Plant*, **2**, 922–932.
22 Buckeridge, M.S., Crombie, H.J., Mendes, C.J.M. et al. (1997) *Carbohyd. Res.*, **303**, 233–237.
23 Freitas, R.A., Martin, S., Santos, G.L. et al. (2005) *Carbohyd. Polym.*, **60**, 507–514.
24 Mikkonen, K., Pitkänen, L., Liljeström, V. et al. (2012) *Cellulose*, **19**, 467–480.
25 Höije, A., Gröndahl, M., Tämmeraas, K., Gatenholm, P. (2005) *Carbohyd. Polym.*, **61**, 266–275.
26 Izydorczyk, M.S., Biliaderis, C.G. (1995) *Carbohyd. Polym.*, **28**, 33–48.
27 Sjostrom, E. (1993) In *Wood Chemistry* (2nd edn), Sjöström, E. (ed.), Academic Press: San Diego, pp. 51–70.

28 Escalante, A., Goncalves, A., Bodin, A. (2012) Flexible oxygen barrier films from spruce xylan. *Carbohydr. Polym.*, **87**.

29 McKee, L.S., Sunner, H., Anasontzis, G.E., et al. (2016) *Biotech.Biofuels*, **9**, 1–13.

30 Burton, R.A., Fincher, G.B. (2014) *Frontiers Plant Sci.*, **5**, 456.

31 Saulnier, L., Guillon, F., Sado, P.-E., Rouau, X. (2007) In *Comprehensive Glycoscience*, Kamerling, J., Boons, G.-J., Lee, Y., et al. (eds.), Elsevier: Amsterdam, Vol. 2, pp. 653–68.

32 Nino-Medina, G., Carvajal-Millan, E., Rascon-Chu, A. et al. (2010) *Phytochem. Rev.*, **9**, 111–120.

33 Saulnier, L., Marot, C., Chanliaud, E., Thibault, J.F. (1995) *Carbohyd. Polym.*, **26**, 279–287.

34 Rogowski, A., Briggs, J.A., Mortimer, J.C. et al. (2015) *Nat. Commun.*, **6** 7481.

35 Hashimoto, S., Shogren, M.D., Pomeranz, Y. (1987) *Cer. Chem.*, **64**, 30–34.

36 Izydorczyk, M.S., Dexter, J.E. (2008) *Food Res. Inter.*, **41**, 850–868.

37 Vinkx, C.J.A., Delcour, J.A. (1996) *J. Cer. Sci.*, **24**, 1–14.

38 Andersson, R., Fransson, G., Tietjen, M., Aman, P. (2009) *J. Agric, Food Chem.*, **57**, 2004–2008.

39 Hardy, K., Brand-Miller, J., Brown, K.D. et al. (2015) *Quart. Rev. Biol.*, **90**, 251–268.

40 Xu, Y., Wang, L., He, J., et al. (2013) *J. Amer. Med. Assoc.*, **310**, 948–959.

41 Zhang, Y.-X., Wang, Z.-X., Zhao, J.-S, Chu, Z.-H. (2016) *Eur. J. Prev. Card.*, 2047487316643830.

42 O'Dea, K., Nestel, P., Antonoff, L. (1980) *Amer. J. Clin. Nutr.*, **33**, 760–765.

43 Heaton, K.W., Marcus, S., Emmett, P., Bolton, C. (1988) *Amer. J. Clin. Nutr.*, **47**, 675–682.

44 Tudorica, C., Kuri, V., Brennan, C. (2002) *J. Agric.Food Chem.*, **50**, 347–356.

45 Ellis, P., Apling, E., Leeds, A., Bolster, N. (1981) *Brit. J. Nutr.*, **46**, 267–276.

46 Jenkins, D.J.A., Wolever, T.M.S., Leeds, A.R. et al. (1978) *BMJ.*, **1**, 1392–1394.

47 Kaur, L., Singh, J., Govil, J., Singh, V. (2009) *Standardization Herbal/Ayurvedic Formul.*, **67**, 329–343.

48 Ellis, P., Roberts, F., Low, A., Morgan, L.(1995) *Brit. J. Nutr.*, **74**, 539–556.

49 Kaur, L., Singh, J., Singh, H., McCarthy, O.J. (2008) *Food Chem.*, **111**, 1–10.

50 Brennan, C., Blake, D., Ellis, P., Schofield, J. (1996) *J. Cer. Sci.*, **24**, 151–160.

51 Ellis, P.R., Dawoud, F.M., Morris, E.R. (1991) *Brit. J. Nutr.*, **66**, 363–79.

52 Slaughter, S.L., Ellis, P.R., Jackson, E.C., Butterworth, P.J. (2002) *Biochimica et Biophysica Acta (BBA)-General Subjects*, **1571**, 55–63.

53 Dreher, M.L., Dreher, C. J., Berry, J.W., Fleming, S.E. (1984) *Crit. Rev. Food Sci & Nutr.*, **20**,47–71.

54 Farias, L.R., Costa, F.T., Souza, L. (2007) *Pesticide Biochem. Physiol.*, **87**, 255–260.

55 Boivin, M., Flourie, B., Rizza, R.A. et al. (1988) *Gastroenterology*, **94**, 387–394.

56 He, Q., Lv, Y., Yao, K. (2007) *Food Chem.*, **101**, 1178–1182.

57 McDougall, G.J., Shapiro, F., Dobson, P. et al. (2005) *J. Agric. Food Chem.*, **53**, 2760–2766.

58 Lankisch, M., Layer, P., Rizza, R.A., DiMagno, E.P. (1998) *Pancreas*, **17**, 176–181.

59 Singh, J., Dartois, A., Kaur, L. (2010) *Trends in Food Sci. Techn.*, **21**, 168–180.

60 Snow, P., O'Dea, K. (1981) *Amer. J. Clin. Nutr.*, **34**, 2721–7.

61 Lajolo, F.M., Genovese, M.I. (2002) *J. Agric.Food Chem.*, **50**, 6592–6598.

62 Yoon, J.H., Thompson, L.U., Jenkins, D. (1983) *Amer. J. Clin.Nutr.*, **38**, 835–842.

63 Alonso, R., Aguirre, A., Marzo, F. (2000) *Food Chem.*, **68**, 159–165.

64 Rehman, Z.-U., Shah, W. (2005) *Food Chem.* **91**, 327–331.

65 Franco, C.M., Ciacco, C.F., Tavares, D.Q.(1998) *Starch-Stärke*, **50**, 193–198.

66 Franco, C.M., do Rio Preto, S.J., Ciacco, C.F. (1992) *Starch-Stärke*, **44**, 422–426.

67 Kasemwong, K., Piyachomkwan, K., Wansuksri, R., Sriroth, K. (2008) *Starch-Stärke*, **60**, 624–633.

68 Noda, T., Takigawa, S., Matsuura-Endo, C. et al. (2005) *Carbohyd. Polym.*, **60**, 245–251.

69 Vasanthan, T., Bhatti, R.S. (1996) *Cereal Chem.*, **73**, 199–207.

70 Naguleswaran, S., Li, J., Vasanthan, T. et al. (2012) *Carbohyd. Polym.*, **88**, 864–874.

71 Warren, F.J., Butterworth, P.J., Ellis, P.R. (2012) *Carbohyd. Res.*, **358**, 67–71.

72 Dhital, S., Shrestha, A.K., Gidley, M.J. (2010) *Carbohyd. Polym*, **82**, 480–488.

73 Benmoussa, M., Hamaker, B.R., Huang, C.P. et al. (2010) *J. Cer. Sci.*, **52**, 22–29.

74 Han, X.Z., Benmoussa, M., Gray, J.A. et al. (2005) *Cer. Chem.*, **82**, 351–355.

75 Naguleswaran, S., Li, J., Vasanthan, T., Bressler, D. (2011) *Cer. Chem.*, **88**, 87–94.

76 Planchot, V., Colonna, P. (1995) *Carbohyd. Res.*, **272**, 97–109.

77 Payan, F., Haser, R., Pierrot, M. et al. (1980) *Acta Crystallogr., Sect. B*, **B36**, 416–21.

78 Oates, C.G. (1997) *Trends in Food Sci. Techn.*, **8**, 375–382.

79 Jane, J.-L. (2006) *J. Appl. Glycosci*, **53**, 205–213.

80 Baldwin, P.M. (2001) *Starch-Stärke*, **53**, 475–503.

81 Baldwin, P., Davies, M., Melia, C. (1997) *Int. J. Biol. Macromol.*, **21**, 103–107.
82 Tahir, R., Ellis, P.R., Butterworth, P.J. (2010) *Carbohyd. Polym.*, **81**, 57–62.
83 Belton, P., Delgadillo, I., Halford, N., Shewry, P. (2006) *J. Cer. Sci.*, **44**, 272–286.
84 Duodu, K., Nunes, A., Delgadillo, I., et al. (2001) *J. Cerl Sci.*, 2002, **35**, 161–174.
85 Chandrashekar, A., Kirleis, A. (1988) *Cer. Chem.*, **65**, 457–462.
86 Ezeogu, L.I., Duodu, K.G., Taylor, J.R.N. (2005) *J. Cer. Sci.*, **42**, 33–44.
87 Ezeogu, L.I., Duodu, K.G., Emmambux, M.N., Taylor, J.R. (2008) *Cer. Chem.*, **85**, 397–402.
88 Zou, W., Sissons, M., Gidley, M.J. et al. (2015) *Food Chem.*, **188**, 559–68.
89 Zou, W., Sissons, M., Warren, F.J. et al. (2016) *Carbohyd. Polym.*, DOI 10.1016/j.carbpol.2016.06.016.
90 Wong, J.H., Lau, T. Cai, N. et al. (2009) *J. Cer. Sci.*, **49**, 73–82.
91 Morrison, W. (1981) *Starch-Stärke*, **33**, 408–410.
92 Morrison, W.R. (1988) *J. Cer. Sci.*, **8**, 1–15.
93 Björck, I., Granfeldt, Y., Liljeberg, H. et al. (1994) *Amer. J. Clin. Nutr.*, **59**, 699S–705S.
94 Holm, J., Björck, I., Ostrowska, S. et al. (1983) *Starch-Stärke*, **35**, 294–297.
95 Tufvesson, F., Wahlgren, M., Eliasson, A.C. (2003) *Starch-Stärke*, **55**, 61–71.
96 Tufvesson, F., Wahlgren, M., Eliasson, A.C. (2003) *Starch-Stärke*, **55**, 138–149.
97 Yotsawimonwat, S., Sriroth, K., Kaewvichit, S. et al. (2008) *Int. J. Biol. Macromol.*, **43**, 94–99.
98 Gelders, G.G., Goesaert, H., Delcour, J.A. (2005) *Biomacromol.*, **6**, 2622–2629.
99 Syahariza, Z.A., Sar S., Tizzotti, M. et al. (2013) *Food Chem.*, **136**, 742–749.
100 Syahariza, Z.A., Sar S., Warren, F.J. et al. (2014) *Food Chem.*, **145**, 617–618.
101 Colonna, P., Leloup, V., Buleon, A. (1992) *Eur. J. Clin. Nutr*, **46**, S17–S32.
102 Nakamura, S., Satoh, H., Ohtsubo, K. (2011) *J. Agric. Food Chem.*, **59**, 10665–10676.
103 Sujka, M., Jamroz, J. (2010) *Starch/Starke*, **62**, 229–35.
104 Planchot, V., Colonna, P., Buleon, A. (1997) *Carbohyd. Res.*, **298**, 319–326.
105 Jane, J.-l., Ao, Z., Duvick, S. A. et al. (1992) *J. Appl. Glycoscience*, **50**, 167–172.
106 Gallant, D., Bouchet, B., Buleon, A., Perez, S. (1992) *Eur. J. Clin. Nutr.*, **46**, S3–S16.
107 McPherson, A.E., Jane, J. (1999) *Carbohyd. Polym.*, **40**, 57–70.
108 Jane, J., Chen, Y.Y., Lee, L.F. et al. (1999) *Cer. Chem.*, **76**, 629–637.
109 Cai, L., Shi, Y.-C. (2010) *Carbohyd. Polym.*, **79**, 1117–1123.
110 Cai, L., Shi, Y.-C., Rong, L., Hsiao, B.S. (2010) *Carbohyd. Polym*, **81**, 385–393.
111 Gidley, M.J. (1987) *Carbohydrate Res.*, **161**, 301–304.
112 Imberty, A., Chanzy, H., Perez, S. et al. (1988) *J. Mol. Biol.*, **201**, 365–78.
113 Imberty, A., Perez, S. (1988) *Biopolym.*, **27**, 1205–1221.
114 Tester, R.F., Karkalas, J., Qi, X. (2004) *J. Cer. Sci.*, **39**, 151–165.
115 Gerard, C., Colonna, P., Buleon, A., Planchot, V. (2001) *J. Sci. Food Agric.*, **81**, 1281–1287.
116 Blazek, J., Gilbert, E. (2010) *Biomacromol.*, **11**, 3275–3289.
117 Zhang, G., Ao, Z., Hamaker, B.R. (2006). *Biomacromol.*, **7**, 3252–3258.
118 Shrestha, A.K., Blazek, J., Flanagan, B.M. et al. (2012) *Carbohydr. Polym.*, **90**, 23–33.
119 Shrestha, A.K., Ng, C.S., Lopez-Rubio, A. et al. (2010) *Carbohyd. Polym.*, **80**, 699–710.
120 Gallant, D.J., Bouchet, B., Baldwin, P.M. (1997) *Carbohyd. Polym.*, **32**, 177–191.
121 Tang, H., Mitsunaga, T., Kawamura, Y. (2006) *Carbohyd, Polym.*, **63**, 555–560.
122 Dhital, S., Shrestha, A.K., Gidley, M.J. (2010) *Food Hydrocoll.*, **24**, 152–163.
123 Al-Rabadi, G.J., Gilbert, R.G., Gidley, M. (2009) *J. Cer. Sci.*, **50**, 198–204.
124 Granfeldt, Y., Bjorck, I., Hagander, B. (1991) *Eur. J. Clin. Nutr.*, **45**, 489–499.
125 Wolever, T., Jenkins, D., Kalmusky, J. et al. (1986) *Diabet. Car*, **9**, 401–404.
126 Colonna, P., Barry, J.-L., Cloarec, D. et al. (1990) *J. Cer. Sci.*, **11**, 59–70.
127 Rinaudo, M. (2004) *Biomacromol.*, **5**, 1155–1165.
128 Petkowicz, C.L.O., Reicher, F., Mazeau, K. (1998) *Carbohyd. Polym.*, **37**, 25–39.
129 Picout, D.R., Ross-Murphy, S.B., Errington, N., Harding, S.E. (2001) *Biomacromol.*, **2**, 1301–1309.
130 Picout, D.R., Ross-Murphy, S.B., Jumel, K., Harding, S.E. (2002) *Biomacromol.*, **3**, 761–767.
131 Morris, G.A., Patel, T.R., Picout, D.R. (2008) *Carbohyd. Polym.*, **72**, 356–360.
132 Patel, T.R., Picout, D.R., Ross-Murphy, S.B., Harding, S.E. (2006) *Biomacromol.*, **7**, 3513–3520.
133 Vilaplana, F., Gilbert, R.G. (2010) *J. Separation Sci.*, **33**, 3537–54.
134 de Gennes, P.-G. (1979) *Scaling Concepts in Polymer Physics*. Cornell University: Ithaca, NY.
135 Burchard, W. (1999) *Adv. Polym. Sci.*, **143**, 113–194.
136 Shelat, K.J., Vilaplana, F., Nicholson, T.M. et al. (2011) *Carbohyd. Polym.*, **86**, 1732–8.

137 Pitkanen, L., Virkki, L., Tenkanen, M., Tuomainen, P. (2009) *Biomacromol.*, **10**, 1962–1969.

138 Li, C., Gilbert, R.G. (2016) *Planta*, **243**, 13–22.

139 Li, C., Wu, A.C., Go, R. et al. (2015) *PLoS One*, **10** e0125507.

140 Martínez-Sanz, M., Gidley, M.J., Gilbert, E.P. (2015) *Carbohyd. Polym.*, **2015**, 125, 120–134.

141 Mikkonen, K.S., Heikkilä, M.I., Willför, S.M., Tenkanen, M. (2012) *Int. J. Polym. Sci.*, **18**, 356–370.

142 Mikkonen, K.S., Pitkanen, L., Liljestrom, V. (2012) *Cellulose*, **19**, 467–480.

143 Höije, A., Stememalm, E., Heikkinen, S. et al. (2008) *Biomacromol.*, **9**, 2042–2047.

144 Stevanic, J. S., Joly, C., Mikkonen, K. S. et al. (2011) *J. Appl. Polym. Sci.*, **122**, 1030–1039.

145 Pitkänen, L., Tuomainen, P., Mikkonen, K.S., Tenkanen, M. (2011) *Carbohyd. Polym.*, **86**, 1230–1235.

146 Tenkanen, M. (2003) *Enzymatic Tailoring of Hemicelluloses*. American Chemical Society: Washington, DC.

147 Van Laere, K.M.J., Voragen, C.H.L., Kroef, T. et al. (1999) *App. Microbiol. Biotech.*, **51**, 606–613.

148 Zhang, Y., Pitkänen, L., Douglade, J. et al. (2011) *Carbohyd. Polym.*, **86**, 852–859.

149 Heikkinen, S.L., Mikkonen, K.S., Pirkkalainen, K. et al. (2013) *Carbohyd. Polym*, **92**, 733–740.

150 Tayal, A., Pai, V., Kelly, R.M., Khan, S.A (2002) *Enzymatic Modification of Guar Solutions*. Springer US, New York, pp. 41–49.

151 Mikkonen, K.S., Rita, H., Helén, H. et al. (2007) *Biomacromol.* **8**, 3198–3205.

152 Mikkonen, K.S., Parikka, K., Suuronen, J.P. et al. (2014) *RSC Adv.*, **4**, 11884–11892.

153 Parikka, K., Ansari, F., Hietala, S., Tenkanen, M. (2012) *Food Hydrocolloids*, **26**, 212–220.

154 Parikka, K., Leppänen, A.-S., Pitkänen, L. et al. (2010) *J. Agric. Food Chem.*, **58**, 262–271.

155 Shirakawa, M., Yamatoya, K., Nishinari, K. (1998) *Food Hydrocolloids*, **12**, 25–28.

156 Kochumalayil, J.J., Berglund, L.A. (2014) *Green Chem.*, **16**, 1904–1910.

157 Chen, Q., Yu, H., Wang, L. et al. (2015) *RSC Adv.*, **5**, 67459–67474.

158 Soderqvist Lindblad, M., Albertsson, A.-C. (2004) In *Polysaccharides*, Dumitriu, S. (ed.), CRC Press: Baton Rouge, LA.

9

Conformational Properties of Polysaccharide Derivatives

*Ken Terao and Takahiro Sato**

Osaka University, Toyonaka, Osaka, Japan

9.1 Introduction

Polysaccharide is one of the most abundantly available polymer materials. Humans have used polysaccharides in daily life as food, clothing, and shelter for thousands of years. Although many synthetic polymer materials have been produced since the last century, the importance of polysaccharides as polymer materials is still increasing, because polysaccharides are carbon-neutral materials [1], i.e. the plants produce polysaccharides from carbon dioxide in the air.

Polysaccharides possess unique physical properties, due to the chemical structures of their monomer units with the rigid ring structure and hydroxyl groups. To improve their physical properties and also processability, polysaccharides are often chemically modified, by substituting various functional groups for the hydroxyl groups. Those polysaccharide derivatives have also various uses in polymer industry.

After Staudinger established the concept of the macromolecule in chemistry, dilute solution properties of polysaccharides and their derivatives (Figure 9.1) were extensively studied to determine molar mass as well as dimensional and hydrodynamic properties by means of viscometry, ultracentrifuge, light, X-ray, and neutron scattering, and so on [2]. The aim of the solution studies is to establish the relationship between the chain conformation and physical properties for polysaccharides and their derivatives. However, because many native polysaccharides except for pullulan (3. in Figure 9.1) have solubility problems, dilute solution properties have been more extensively examined for polysaccharide derivatives, e.g. cellulose tris(phenylcarbamate) (CTPC, 4. in Figure 9.1) and amylose tris(phenylcarbamate) (ATPC, 5. in Figure 9.1), since the 1960s [3–26]. Some of the data were analyzed in terms of the wormlike chain model [27] to determine the chain stiffness. Both CTPC and ATPC were found to behave as stiff chains in solution. More recently, the main chain of native cellulose was also demonstrated to be stiff in *N,N'*-dimethylacetamide [28] and in metal complexing water containing solvents [29–32]. Amylose has a more flexible main chain in dimethyl sulfoxide than that of ATPC [33, 34] but they behave like rather stiff chains in iron-sodium tartrate [31]. The higher chain stiffness of ATPC than amylose in the former solvent was considered to be due to the intramolecular hydrogen bonds between NH and C=O groups of the neighboring carbamate groups and the steric hindrance of the internal rotation [3, 12, 17]. Thus, polysaccharides carbamate derivatives may have rigid helical structure in solution, which provides important solution and solid state properties, e.g. lyotropic liquid crystallinity, high viscosity, and toughness.

Another characteristic feature of cellulose and amylose is the perfect stereo-regularity of the main chain. In the 1980s, these phenylcarbamate derivatives were found to be extremely useful as chiral stationary phase (CSP) [35]. Okamoto et al. [36] examined the efficiency of the chiral resolution of various racemic compounds by polysaccharide-based CSPs. The efficiency of the CSPs strongly depends on the chemical structure of the side group. Actually, amylose tris(3,5-dimethylphenylcarbamate) (ADMPC) and cellulose tris(3,5-dimethylphenylcarbamate) (CDMPC) [37] (Figure 9.2) are the most effective CSPs while less performance was reported for amylose and cellulose alkylcarbamate derivatives [36, 38–41]. This difference may be related to the chain conformation of the derivatives in the mobile phase as well as the spatial arrangement of the side groups around the polar NH and C=O groups because the interaction of the objective substance with the polar groups of CSPs should be important. While the helical structure can be estimated from the crystal structure of CTPC, no reports are available

* Email: tsato@chem.sci.osaka-u.ac.jp

Bioinspired Materials Science and Engineering, First Edition. Edited by Guang Yang, Lin Xiao, and Lallepak Lamboni.

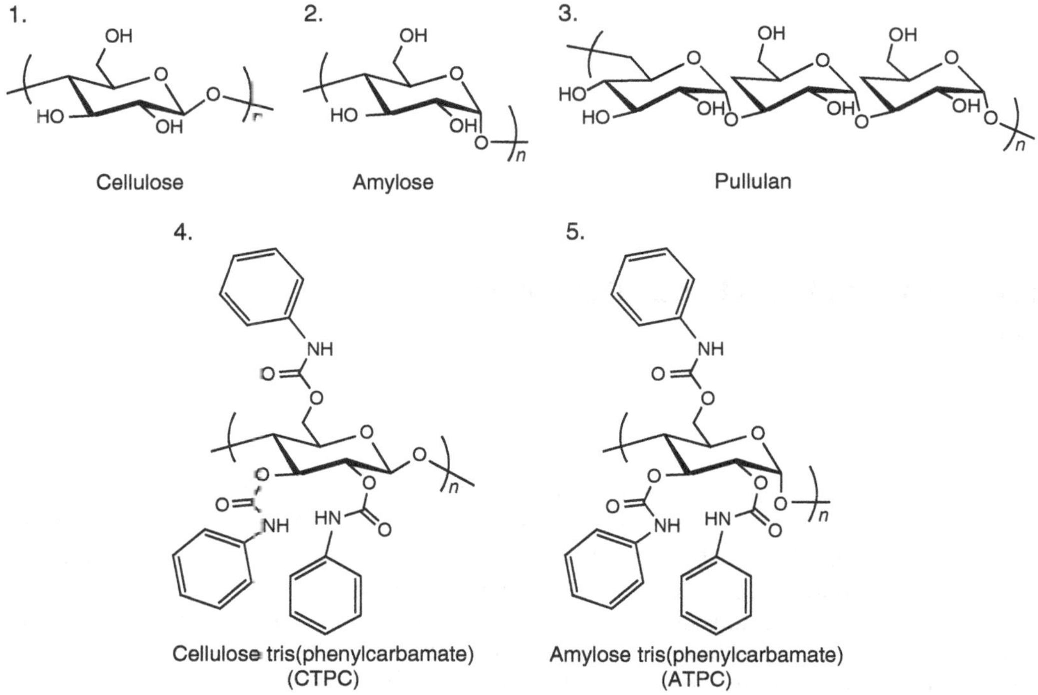

Figure 9.1 Chemical structures of 1. cellulose, 2. amylose, 3. pullulan, 4. cellulose tris(phenylcarbamayte) (CTPC), and 5. amylose tris(phenylcarbamayte) (ATPC).

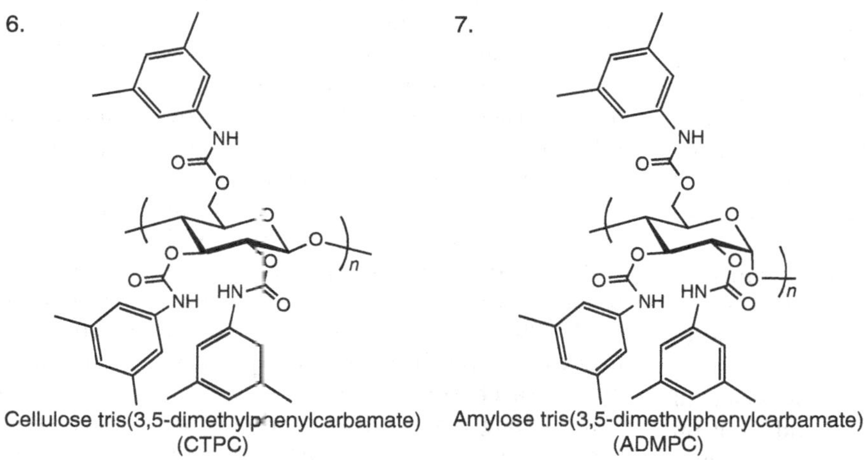

Figure 9.2 Chemical structures of 6. cellulose tris(3,5-dimethylphenylcarbamayte) (CDMPC) and 7. amylose tris(3,5-dimethylphenylcarbamayte) (ADMPC).

for amylose carbamates due to the difficulty of obtaining suitable crystalline samples to determine the helical structure. Yamamoto et al. therefore estimated the helical structure of ADMPC from multi-dimensional NMR and molecular modeling [42]. On the other hand, Kasat et al. [43] investigated the interaction between ADMPC and eluent molecules by means of infrared absorption, X-ray diffraction, and NMR for solid state CSPs. Dilute solution properties also give us important information about the helical structure of polysaccharides and their derivatives.

Recently, we investigated systematically dilute solution properties of carbamate derivatives of amylose, cellulose, and curdlan to determine the chain stiffness and local helical structure in various solvents by light and small-angle X-ray scattering, viscometry and infrared absorption measurements. We found that the chain stiffness and the local helical structure are governed by the chemical structures of the main and side chains, the intramolecular hydrogen bonds, and the interaction with solvent molecules. In this chapter, we summarize our recent results for conformations of polysaccharide carbamate derivatives in solution to discuss the performance as CSPs. We also present the conformation and properties of cyclic amylose carbamate derivatives, which is quantitatively obtained from cyclic amylose. High purity cyclic amylose sample may be obtained by enzymatic synthesis [44]. These polymers are one of the class of rigid cyclic polymers which is rarely obtained other than in cyclic comb polymer but they have potential use as a component of nanomachines [45].

9.2 Theoretical Backbone to Determine the Chain Conformation of Linear and Cyclic Polymers from Dilute Solution Properties

As described in Section 9.1, the wormlike chain model [27], a special limit of the helical wormlike chain [46, 47], is one of the most successful models to explain the dilute solution properties of polymers obtained from static and dynamic light scattering, small-angle X-ray and neutron scattering, and viscometry for many stiff and flexible polymers. Although the helical wormlike chain model [48] is more general, the wormlike chain model is enough to describe dilute solution properties of polysaccharide carbamate derivatives discussed in this chapter. According to the theoretical expression of the wormlike chain model, the chain conformation of polymer molecules is characterized by the contour length L, the Kuhn segment length λ^{-1}, and the chain diameter d. The second parameter λ^{-1} is the indicator of the chain stiffness and it is equal to twice the persistence length which is proportional to the characteristic ratio C_∞ for the infinitely long chain. The stiffness parameter λ^{-1} can also be related to the bending force constant ε of the polymer chain viewed as the elastic wire by

$$\lambda^{-1} = 2\varepsilon / k_B T \tag{9.1}$$

where k_B is the Boltzmann constant and T is the absolute temperature.

In what follows, we will present the helical conformation of polysaccharides and their derivatives. It is noted that the wormlike chain contour corresponds to the helix axis of the helical polymers, not to their helical chain backbone. Thus, the contour length L of the wormlike chain must be calculated from the helix pitch or the helix rise h per monosaccharide unit and the degree of polymerization N_0 by $L = hN_0$. The pitch h is not usually equal to the length of the monosaccharide unit but its projection length onto the helix axis.

When all the internal rotation angles of the polymer chain are identical, the chain takes the helical conformation. If the polymer chain takes a perfect helical conformation, the stiffness parameter λ^{-1} becomes infinitely large. The finite value of λ^{-1} indicates the polymer chain will not be the perfect helix. The imperfection comes from the fluctuation of the internal rotation angle, and the fluctuation is determined by the steepness of the internal potential minimum. Although we will use the term the *helical conformation* or the *helical structure* in the following, the reader should not imagine the perfect helix taking the rigid rod-like conformation from this term. In this chapter, the helical conformation means a locally helical structure but globally the one viewed as the wormlike chain model.

The radius of gyration $\langle S^2 \rangle$ for linear thin wormlike chains is calculated by the Benoit-Doty [49] equation as

$$\langle S^2 \rangle = \lambda^{-2} \left[\frac{1}{6}\lambda L - \frac{1}{4} + \frac{1}{4\lambda L} - \frac{1}{8(\lambda L)^2}\left(1 - e^{-2\lambda L}\right) \right] \tag{9.2}$$

While L is proportional to the molar mass M of the polymer, the molar mass per unit contour length M_L ($\equiv M/L$) or the helix pitch per residue h ($\equiv M_0/M_L$; M_0: the molar mass of the residue) are preferably used as the wormlike chain parameter independent of M. For non-helical chains, the latter parameter h may be called as the contour length per residue. Figure 9.3 illustrates plots of $\lambda\langle S^2 \rangle_z^{1/2}$ vs the Kuhn segment number n_K ($\equiv \lambda L$, the reduced chain length) for ATPC in 1,4-dioxane [50], 2-ethoxyethanol [50], methyl acetate [51], ethyl acetate [51], 4-methyl-2-pentanone [51], and 2-butanone [52]. Most of the data are well fitted by the theoretical values calculated in terms of Eq. 9.2. From the curve-fitting procedure, the two parameters h and λ^{-1} can

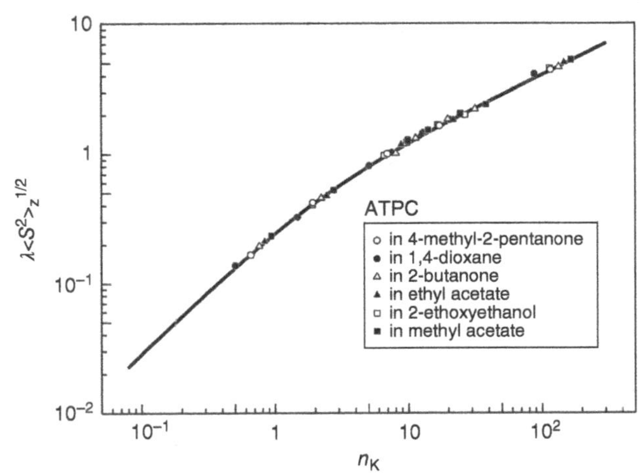

Figure 9.3 Universal plots of $\lambda\langle S^2 \rangle_z^{1/2}$ vs the Kuhn segment number n_K ($\equiv \lambda L$) for amylose tris(phenylcarbamate) (ATPC) in 4-methyl-2-pentanone at 25 °C [51] (unfilled circles), in 1,4-dioxane at 25 °C [50] (filled circles), in 2-butanone at 25 °C [52] (unfilled triangles), in ethyl acetate at 33 °C [51] (filled triangles), in 2-ethoxyethanol at 25 °C [50] (unfilled squares), and in methyl acetate at 25 °C [51] (filled squares).

be unequivocally determined. Some data points at a high n_K range are slightly larger than the theoretical solid curve. This is most likely due to the intramolecular excluded-volume effects and the discrepancy can be explained by the established theory, that is, the quasi-two-parameter (QTP) theory [48, 53, 54] for the wormlike chain with the Domb-Barrett equation for $\langle S^2 \rangle$ [55]. It should be noted that the chain thickness effects on $\langle S^2 \rangle$ is only operable if the chain length is quite short, that is, less than ten times of the chain length, thus negligible for the current data. A similar analysis may also be done for molar mass dependence of the intrinsic viscosity $[\eta]$. Theoretical values can be calculated with h, λ^{-1}, and d in terms of the Yamakawa-Fujii-Yoshizaki theory [48, 56, 57] for the cylindrical wormlike chain or the Yoshizaki-Nitta-Yamakawa [48, 58] theory for the touched-bead wormlike chain, since their chain thickness affects the hydrodynamic properties including $[\eta]$, sedimentation coefficients, and diffusion coefficients even for high molar mass polymers in solution unlike the case of $\langle S^2 \rangle$. For good solvent systems, the intramolecular excluded-volume effect was taken into consideration by the QTP theory with the Barrett equation for $[\eta]$ [59].

Similar information is also obtained from the particle-scattering function $P(q)$. The theoretical values for the linear wormlike chain can be calculated from the Nakamura-Norisuye expression [60, 61]. Figure 9.4 shows two examples of Holtzer plots for ATPC having different chain lengths in 1,4-dioxane. The three parameters λ^{-1}, h, and d mainly affect the shape of the scattering function at low, middle, and higher q region, respectively, thus, the three parameters can be unequivocally determined from the curve fitting procedure. Specifically, the height of the plateau region around $q = 0.5\,\text{nm}^{-1}$ for ATPC50K mostly determines the contour length L. Appreciable peak height, that is, the difference between the dashed line for the theoretical values of the rigid cylinder and the data points at the low q region, is mainly due to the chain flexibility, thus, the chain stiffness parameter can be estimated. It should be noted that sufficient long main chain is mandatory to determine the chain stiffness because the actual shape of the polymer chain becomes similar to the rigid cylinder with decreasing chain length. The difference between the solid and dashed curves for ATPC20K is indeed much smaller than that for ATPC50K. Furthermore, a slight decrease with increasing q at high q region reflects the chain thickness. Thus, the chain diameter d may be determinable. Since the obtained d reflects the electron density profile [48, 62], a different d value might be obtained from hydrodynamic properties, such as intrinsic viscosities. For example, the d value for ATPC in 1,4-dioxane was determined to be 1.5 nm from $P(q)$ while 2.9 nm was estimated from the viscosity data. We determined the two wormlike chain parameters h and λ^{-1} to explain both dimensional and hydrodynamic properties consistently and the resultant parameters for CTPC in tetrahydrofuran (THF) and ATPC in 1,4-dioxane are summarized in Table 9.1 [50, 63]. We use these values in the following sections whereas slightly different parameters were reported for ATPC by other researchers [24] because side group and solvent dependence should be discussed using the data by the same group with similar methods. It should be noted that the research group difference might be due to the analysis methods of static light scattering since the Zimm plot [64] was used to analyze light scattering data instead of the Berry plot [65]. The former plot may overestimate both the weight-average molar mass M_w and $\langle S^2 \rangle_z$ for linear semi-flexible chains.

On the one hand, for the thin cyclic wormlike chain, Shimada and Yamakawa's approximate expression [67] is available to calculate $\langle S^2 \rangle$ as

$$\langle S^2 \rangle = \frac{L^2}{4\pi^2}\begin{bmatrix} 1 - 0.1140\lambda L - 0.0055258(\lambda L)^2 \\ +0.0022471(\lambda L)^3 - 0.00013155(\lambda L)^4 \end{bmatrix} \quad for \quad \lambda L \le 6$$
$$= \frac{L}{12\lambda}\left\{1 - \frac{7}{6\lambda L} - 0.025\exp\left[-0.01(\lambda L)^2\right]\right\} \quad for \quad \lambda L \ge 6 \tag{9.3}$$

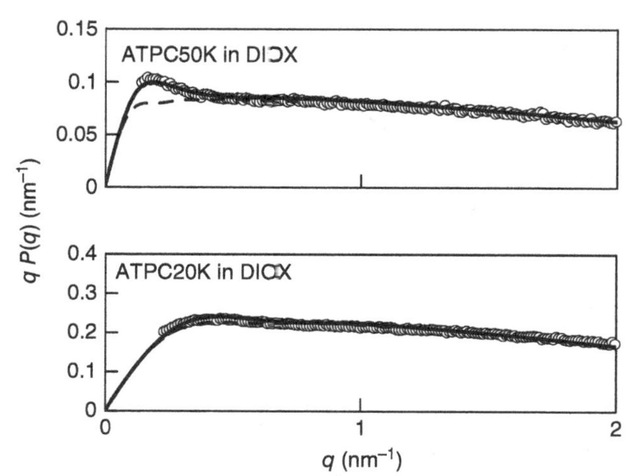

Figure 9.4 Holtzer plots [$qP(q)$ vs q] for ATPC50K ($M_w = 5.5 \times 10^4$) and ATPC20K ($M_w = 1.9 \times 10^4$) in 1,4-dioxane at 25 °C [50]. Solid curves, theoretical values for the unperturbed wormlike cylinder. Dashed curves, theoretical values for the rigid cylinder limit ($\lambda^{-1} = \infty$).

Table 9.1 The Kuhn segment length λ^{-1} and the helix pitch per residue h for ATPC in 1,4-dioxane, CTPC in THF, and curdlan tris(phenylcarbamate) (CdTPC) in THF at 25 °C.

Polymer	Solvent	T (°C)	h (nm)	λ^{-1} (nm)	Ref
ATPC	1,4-dioxane	25	0.34	22	[50]
CTPC	THF	25	0.50	21	[63]
CdTPC	THF	25	0.39	57	[66]

while those for the particle-scattering function and hydrodynamic properties are reported for limited range for the chain stiffness [68, 69] and therefore we mainly analyzed $\langle S^2 \rangle$ data for cyclic chains. It should be noted that the recent simulation method developed by Ida et al. [69] may be available to analyze the data for cyclic polymers with rather high chain stiffness in the near future.

9.3 Chain Conformation of Linear Polysaccharides Carbamate Derivatives in Dilute Solution

9.3.1 Effects of the Main Chain Linkage of the Polysaccharides Phenylcarbamate Derivatives

As mentioned in Section 9.1, intramolecular hydrogen bonds play an important role for the stiffness of cellulose and amylose phenylcarbamates, that is, CTPC and ATPC. In this section, we demonstrate the relationship between intramolecular hydrogen bonds and the main chain structures for CTPC, ATPC, and curdlan tris(phenylcarbamate) (CdTPC). Recently, Kasat et al. [43] reported that hydrogen bonding state including intramolecularly and intermolecularly with solvent molecules for amylose tris(3,5-dimethylphenylcarbamate) can be evaluated from the amide I (1700–1750 cm^{-1}) band of the infrared absorption (IR) spectra corresponding to the C=O stretch mode. We determined the IR spectra for ATPC in 1,4-dioxane, 2-ethoxyethanol, and their mixtures. Double peaks were observed as shown in Figure 9.5 [50]. The peaks at 1754 cm^{-1} and 1706 cm^{-1} were assigned as free and hydrogen bonding C=O groups, respectively. It is noted that another absorption band from the C=O groups having intermolecular hydrogen bonds with solvent molecules should be between the two peaks. The composition of the intramolecular hydrogen bonding increases with increasing the 1,4-dioxane content and indeed, the determined Kuhn segment length λ^{-1} (22 nm) for ATPC in 1,4-dioxane is appreciably larger than that in 2-ethoxyethanol (16 nm) summarized in Table 9.2. This difference becomes more significant for the "bis" phenylcarbamate derivative, that is, amylose-2-acetyl-3,6-bis(phenylcarbamate) (AAPC) [70] of which chemical structure is shown in Figure 9.6. In actuality, λ^{-1} for AAPC in 1,4-dioxane is substantially the same as that for ATPC in the same solvent whereas that in 2-ethoxyethanol is 12 nm, which is 25% smaller than that for ATPC in the same solvent. The difference of intramolecular hydrogen bonds between the two solvents was more clearly seen for AAPC in the IR spectra.

While the chain stiffness parameter for ATPC in 1,4-dioxane is substantially close to that for CTPC in THF, that for CdTPC was determined to be 57 nm, which is much larger than those for CTPC and ATPC. Chemical structures of curdlan and CdTPC are shown in Figure 9.6. The FT-IR spectra for the three phenylcarbamate derivatives are illustrated in Figure 9.7. A broad peak found at 1735 cm^{-1} for CdTPC may be assigned to be intramolecular hydrogen bonding C=O group, while major peaks for ATPC and CTPC at 1753 cm^{-1} are "free" C=O groups. Another smaller peak for CTPC and ATPC around 1710 cm^{-1} should be due to the intramolecular hydrogen bonding C=O groups. Therefore, we may conclude that CdTPC has a weak but a greater content of hydrogen bonding C=O groups than CTPC and ATPC, while those for ATPC are slightly larger than that for CTPC. On the other hand, CdTPC has much smaller chain stiffness (λ^{-1} = 17 nm) with much longer helix pitch (or helix rise) per residue (h = 0.51 nm) in 4-methyl-2-pentanone while the h value in THF was determined to be 0.39 nm, suggesting that the solvent molecules may form hydrogen bonds with NH groups of

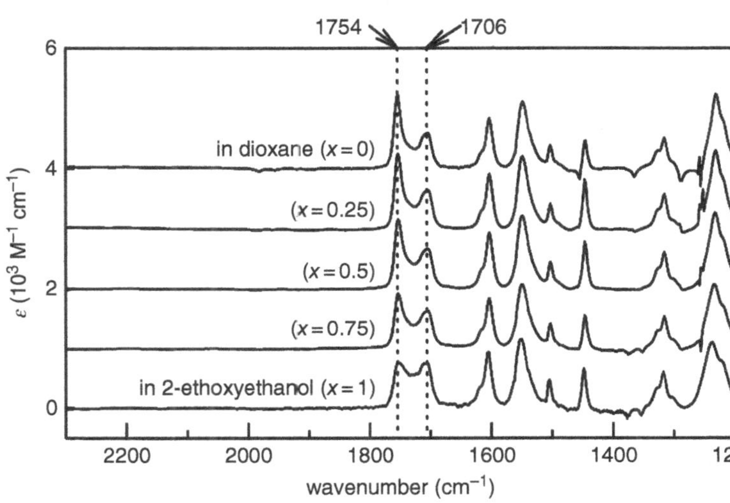

Figure 9.5 Infrared absorption spectra (molar extinction coefficient ε vs wavenumber) for an ATPC sample ($M_w = 2.8 \times 10^5$) in 1,4-dioxane, 2-ethoxyethanol, and their mixtures of which volume fraction of 2-ethoxyethanol is shown as x [50].

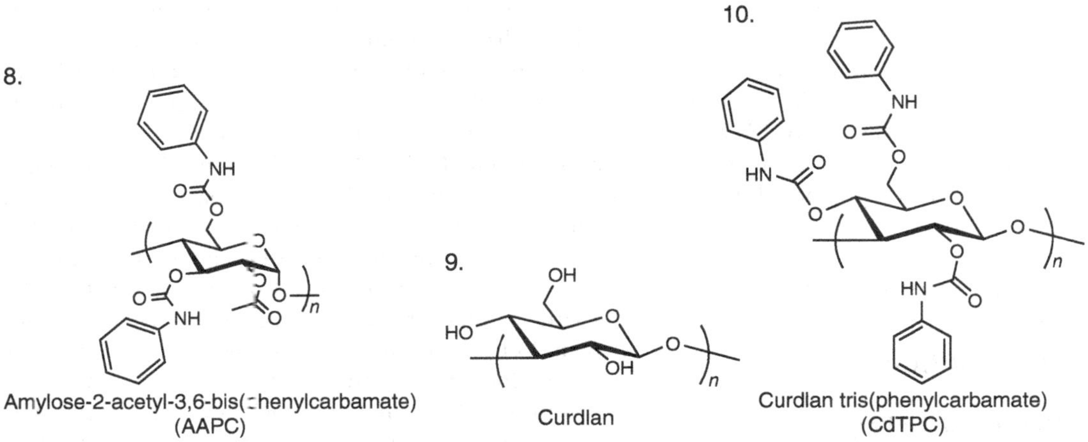

Figure 9.6 Chemical structures of 8. amylose-2-acetyl-3,6-bis(phenylcarbamate) (AAPC), 9. curdlan, and 10. curdlan tris(phenylcarbamate) (CdTPC).

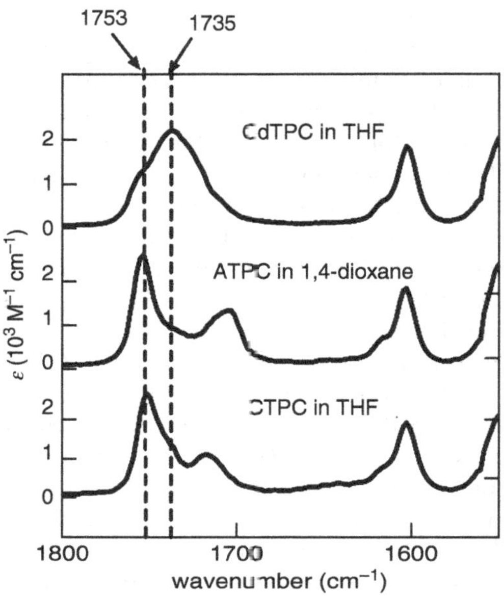

Figure 9.7 Infrared absorption spectra for CdTPC ($M_w = 1.5 \times 10^6$) in THF, ATPC ($M_w = 3.3 \times 10^6$) in 1,4-dioxane, and CTPC ($M_w = 2.6 \times 10^6$) in THF [66].

CdTPC and the content of intramolecular hydrogen bonds of CdTPC are substantially smaller in 4-methyl-2-pentanone than in THF. This supports the importance of intramolecular hydrogen bonds to stabilize the rigid helical main chain structure of CdTPC in THF.

9.3.2 Effects of Hydrogen Bonds to Stabilize the Helical Structure

As mentioned in Section 9.3.1, the main chain stiffness as well as the helical structure of polysaccharide carbamate derivatives may be strongly influenced by the intramolecular hydrogen bonds. To clarify this point, both the dimensional properties and the intramolecular hydrogen bonds should be determined in solvents covering a wide polarity range. While we could not find suitable solvents for polysaccharide tris(phenylcarbamate) systems, we found amylose tris(*n*-butylcarbamate) (ATBC) and amylose tris(ethylcarbamate) (ATEC) have sufficient solubility in large numbers of alcohols and ethers in which the amide I band of side carbamate groups enables detection of the hydrogen bonding state. Figure 9.8 illustrates the chemical structures of ATEC, ATBC, and amylose tris(*n*-hexylcarbamate) (ATHC), which has good solubility in THF but limited solubility in alcohols. Fortunately, the chain stiffness substantially depends on the solvent from 9 nm for ATEC in methanol (MeOH) [71] to 75 nm for ATBC in THF [72]. Figure 9.9 displays the IR spectra for ATBC in six solvents, that is, MeOH, 2-propanol (2PrOH), 1-propanol (1PrOH), 2-ethoxyethanol (2EE), 2-butanol (2BuOH), and THF [73]. While the amide I band in MeOH shows a broad single peak which is fitted by the normal distribution with a certain width, that in THF is bimodal and is explained by the summation of two normal distributions with a rather narrow width. We did waveform separation for the other solvent systems to determine the content f_{hyd} of intramolecular hydrogen bonding C = O groups at 1698 cm^{-1}. The obtained f_{hyd} for ATBC in THF, MeOH, and their mixed solvents fairly fitted by the specific optical rotation $[\alpha]_{280}$ at the wavelength of 280 nm, as shown in Figure 9.10. Furthermore, we can obtain a possible 3D structure with a large number of intramolecular hydrogen bonds by a certain method, as illustrated in Figure 9.11. We thus concluded that the values of f_{hyd} should be proportional to that for helix content stabilized by intramolecular hydrogen bonds. Furthermore, no molar mass dependence was observed both for f_{hyd} and $[\alpha]_{280}$, indicating that cooperation in the solvent-induced conformational change is not important

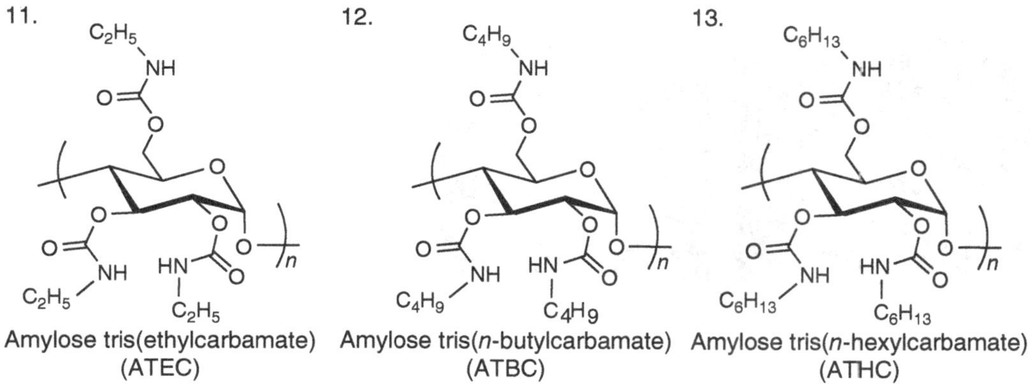

Figure 9.8 Chemical structures of 11. amylose tris(ethylcarbamate) (ATEC), 12. amylose tris(n-butylcarbamate) (ATBC), and 13. amylose tris(n-hexylcarbamate) (ATHC).

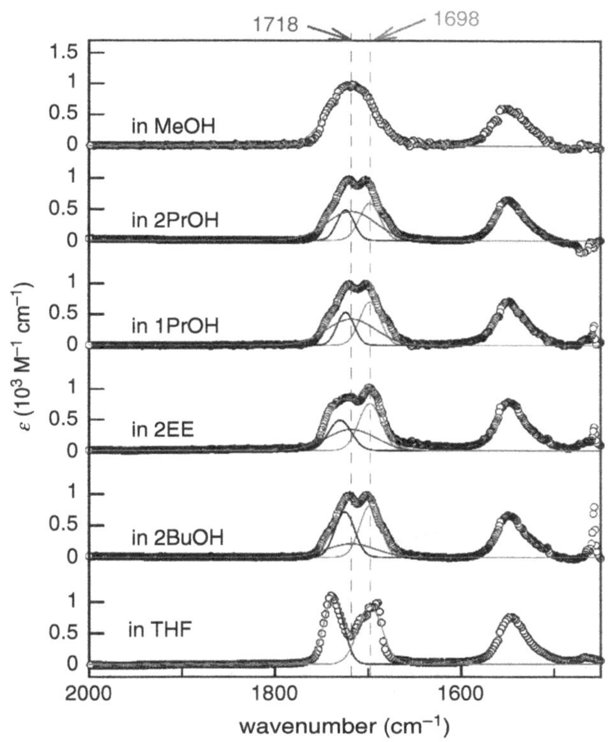

Figure 9.9 Infrared absorption spectra for ATBC ($M_w = 4.6 \times 10^5$) in methanol (MeOH), 2-propanol (2PrOH), 1-propanol (1PrOH), 2-ethoxyethanol (2EE), 2-butanol (2BuOH), and THF [73].

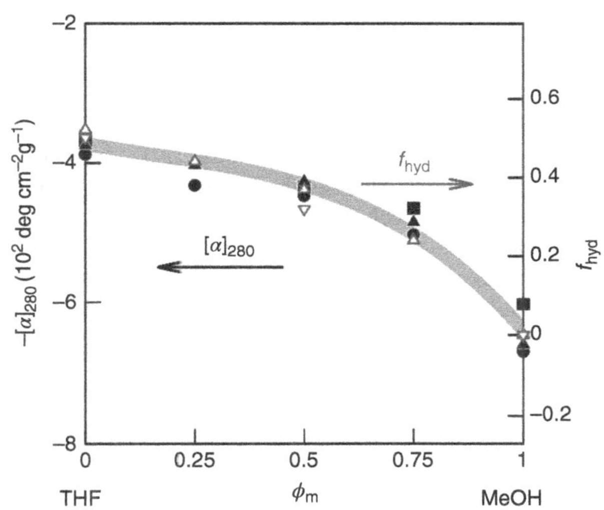

Figure 9.10 Dependence of $-[\alpha]_{280}$ (filled symbols) and f_{hyd} (unfilled symbols) on the volume fraction of methanol (ϕ_m) for ATBC900K ($M_w = 8.9 \times 10^5$, circles), ATBC460K ($M_w = 4.6 \times 10^5$, triangles), ATBC55K ($M_w = 5.5 \times 10^4$, squares), and ATBC17K ($M_w = 1.7 \times 10^4$, inverted triangles) [72].

for both ATBC and ATEC systems while it is definitely important for α-helix – random coil transition for polypeptides in solution [74]. This intramolecular hydrogen bonds may be broken by including a small amount of water. Indeed, the intrinsic viscosity of an ATBC sample with the weight-average molar mass M_w of $8.6 \times 10^5\,\mathrm{g\,mol^{-1}}$, in THF including 10% water (v/v) is 25% smaller than that in pure THF. This reduction of $[\eta]$ corresponds to that in 25% MeOH – 75% THF solution, suggesting that the water molecules break the intramolecular hydrogen bonds of ATBC, but including the water molecules as an impurity in pure THF (<0.1%) is not significant.

If the chain conformations of ATBC and ATEC are determined only by the intramolecular hydrogen bonds, solvent dependence of the wormlike chain parameters should be a function of the helix content, that is, f_{hyd}. If so, we may model the polymer chains consisting of two different monomer units (F and R),

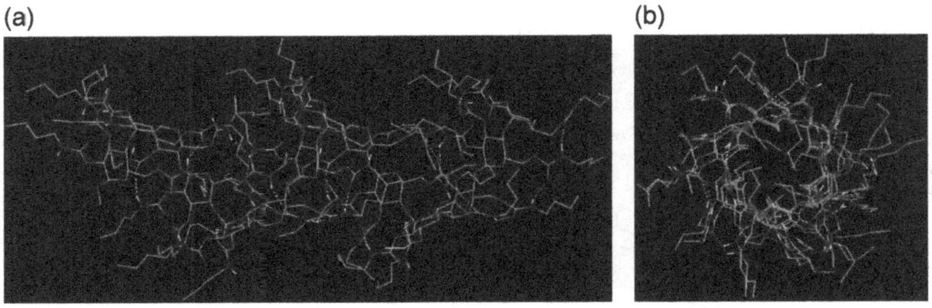

Figure 9.11 A possible 3D structure of rigid helical ATBC (6-fold helix and $h = 0.25$ nm): (a) side view; (b) top view [73].

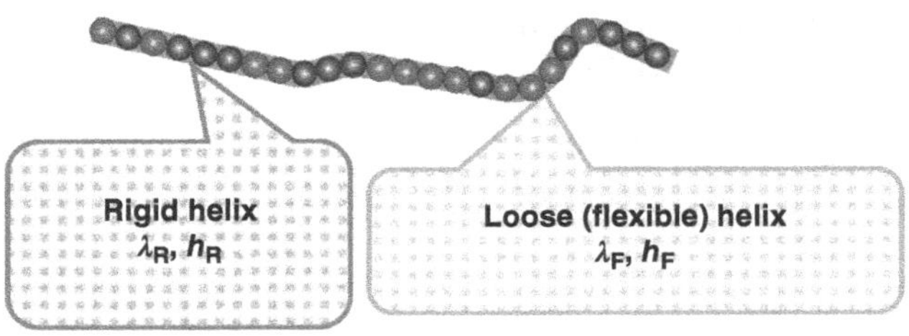

Figure 9.12 Schematic representation of the two-state wormlike chain (TSWC) model.

having different helical structure and chain stiffness (the two-state wormlike chain mode, TSWC), which is originally developed for a random copolymer [75], and the schematic illustration is shown in Figure 9.12. We thus assumed the Kuhn segment lengths λ_F^{-1} and λ_R^{-1} and the helix pitches per residue h_F and h_R for F and R monomers, respectively. The "average" wormlike chain parameters h and λ^{-1} estimated from the analysis of the theoretical expression of the conventional wormlike chain model can be expressed with the content of the rigid portion F_R as:

$$h = F_R h_R + (1 - F_R) h_F \tag{9.4}$$

$$\lambda h = F_R \lambda_R h_R + (1 - F_R) \lambda_F h_F \tag{9.5}$$

where F_R is proportional to f_{hyd}. From these two equations, we may eliminate F_R as

$$h^{-1} - h_F^{-1} = -\frac{h_R^{-1} - h_F^{-1}}{\lambda_F - \lambda_R}(\lambda - \lambda_F) \tag{9.6}$$

This relationship allows us to compare the data when we cannot observe the amide I band, for example, due to the absorption of solvent molecules. The corresponding plots for the above equations are shown in Figure 9.13 for ATEC, ATBC, and ATHC. Linear relationships were observed for ATEC and ATBC, indicating that the TSWC model works well for the two polymer–solvent systems. In other words, chain conformations of the two carbamate derivatives of amylose are determined only by the content of intramolecular hydrogen bonds.

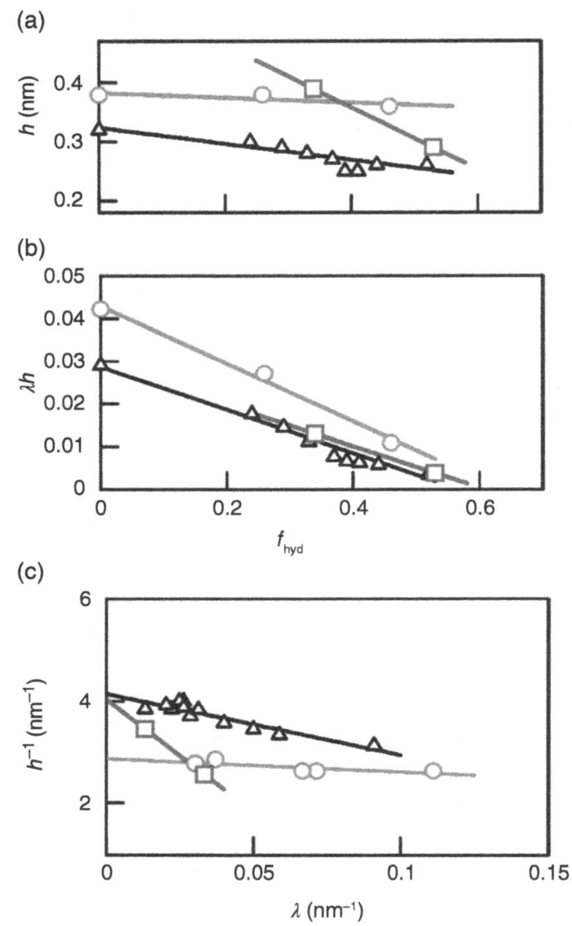

Figure 9.13 Plots of (a) h vs f_{hyd}, (b) λh vs f_{hyd}, and (c) h^{-1} vs λ for ATEC (circles), ATBC (triangles), and ATHC (squares) [71].

Interestingly, the side group difference of h for ATEC and ATBC, however, becomes more significant with decreasing f_{hyd}, indicating that h_R for ATBC is appreciably smaller than that for ATEC. We summarize the h and λ^{-1} values for ATEC, ATBC and ATHC in THF or for ATPC in 1,4-dioxane in Figure 9.14. If we assume the phenyl group of ATPC may be regarded as longer (bulky) alkyl groups, we can see that the helical structure significantly depends on the length of the side alkyl groups, and if the alkyl side chain length is intermediate, the h value may have the

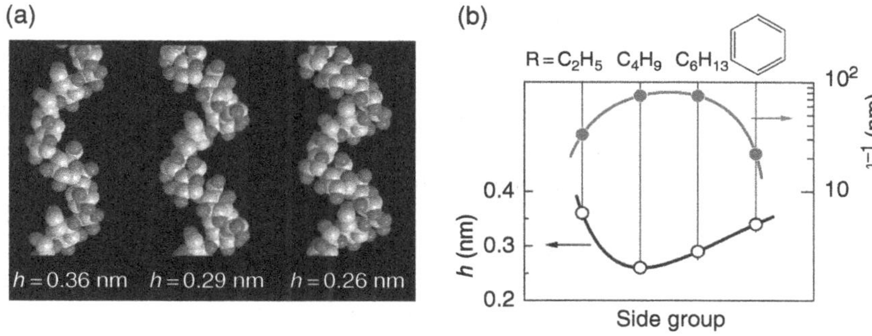

Figure 9.14 (a) Schematic representation of the amylosic main chain for $h = 0.36$, 0.29, and 0.26 nm; (b) h and λ^{-1} values for ATEC, ATBC, ATHC, and ATPC in THF or 1,4-dioxane [71].

lowest value, which would be around 0.25 nm, and furthermore the chain stiffness would be the greatest. This is most likely because some side alkyl groups may be located inside the helical structure and stabilize it with the hydrogen bonds. This suggests that for amylosic chains, *n*-butyl or *n*-hexyl carbamate groups allow the most tightly wound helical structure with a high chain stiffness.

9.3.3 Enantiomeric Composition Dependent Chain Dimensions: ATBC and ATEC in D-, DL-, L-ethyl lactates [72, 76]

As mentioned in Section 9.1, some polysaccharide carbamate derivatives are very useful as a chiral stationary phase to separate or to analyze enantiomers since left- and right-handed molecules interact differently with such polysaccharide derivatives. The solvent-dependent conformation of polysaccharide carbamate derivatives described above makes us predict that different chain dimensions can be observed between right- and left-handed solvent molecules. While optically active solvents are generally extremely expensive, lactate esters are rather inexpensive since both L- and D-lactates can be synthesized by the fermentation methods. From our preliminary viscosity measurements, we found that intrinsic viscosities of ATBC and ATEC in D-ethyl lactate are appreciably larger than those in L-ethyl lactate. These discrepancies were more significant than those for CTPC and CdTPC; we note that ATPC is not soluble in the lactic esters we investigated. We thus determined the dimensional and hydrodynamic properties of ATBC and ATEC in ethyl lactates by means of SAXS, light scattering, and viscometry to determine the wormlike chain parameters, λ^{-1} and h. The resultant chain stiffness parameters λ^{-1} for ATBC and ATEC in D-ethyl lactate are 1.53 and 1.80 times larger than those in L-ethyl lactate, respectively. In spite of being unable to observe amide I band in these solvents, we can compare the determined "average" wormlike chain parameters with the TSWC model with Eq. 9.6 as illustrated in Figure 9.13c. Figure 9.13 includes the data points for ATBC and ATEC in D-, L-, and DL-ethyl lactates. All data points in ethyl lactate systems obey the corresponding straight lines for ATBC and ATEC, indicating the applicability of the TSWC model for the solvent systems. The content of the intramolecularly hydrogen bonding C = O groups can therefore be estimated from Eq. 9.5. The obtained number fraction of hydrogen bonding C = O group f_{hyd} for ATBC in D-ethyl lactate was 15% larger than that in L-ethyl lactate, and furthermore the

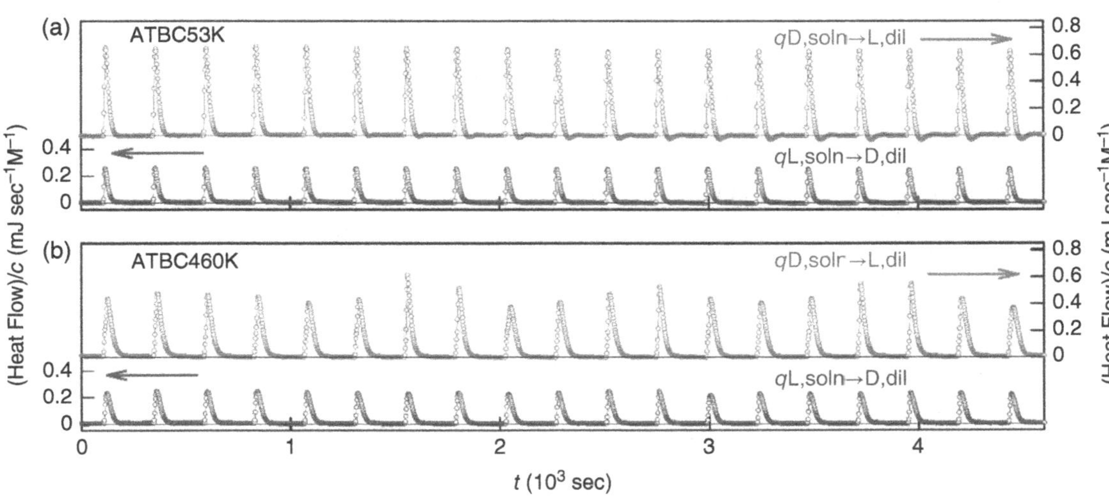

Figure 9.15 Isothermal titration calorimetric data for ATBC53K (a, $M_w = 5.3 \times 10^4$) and ATBC460K (b, $M_w = 4.6 \times 10^5$) in ethyl lactates at 25 °C. 2 μL solution was dropped at every 2 min. [76].

difference is more significant for ATEC systems. This discrepancy in the intramolecular hydrogen bonds was observed by using isothermal calorimetry. Figure 9.15 indicates the time course of $q_{\mathrm{D,soln}\rightarrow\mathrm{L,dil}}$ and $q_{\mathrm{L,soln}\rightarrow\mathrm{D,dil}}$ for two ATBC samples in ethyl lactates where $q_{\mathrm{D,soln}\rightarrow\mathrm{L,dil}}$ denotes the amount of heat at each dripping (2 μm) of ATBC – D-ethyl lactate solution to pure L-ethyl lactate (2 mL). It should be noted that the heat of dilution ($q_{\mathrm{D,soln}\rightarrow\mathrm{D,dil}}$ and $q_{\mathrm{L,soln}\rightarrow\mathrm{L,dil}}$) are negligibly smaller than those for $q_{\mathrm{D,soln}\rightarrow\mathrm{L,dil}}$ and $q_{\mathrm{L,soln}\rightarrow\mathrm{D,dil}}$. Heats of mixing for the two chiral solvents $q_{\mathrm{D,dil}\rightarrow\mathrm{L,dil}}$ and $q_{\mathrm{L,dil}\rightarrow\mathrm{D,dil}}$ were close to the average of $q_{\mathrm{D,soln}\rightarrow\mathrm{L,dil}}$ and $q_{\mathrm{L,soln}\rightarrow\mathrm{D,dil}}$. The appreciable difference between red and blue peaks in Figure 9.15 indicates the difference in the heat of dissolution of the two solvents. This was consistent with the energy of hydrogen bonds, if we assume the difference is only due to the number of hydrogen bonds of C=O groups of ATBC estimated from f_{hyd}. While ATBC is reported as the chiral stationary phase for gas chromatography [77], Okamoto et al. [38–41] insisted that amylose tris(methylcarbamate) and amylose tris(isopropylcarbamate), are not useful for the purpose for liquid chromatography. Considering CTPC has a sufficient ability as a CSP, this may suggest that the main chain structure should not be affected by the enantiomer excess of the chiral solvents to attain higher dissolution of enantiomers as the chiral stationary phase.

9.3.4 Solvent-Dependent Helical Structure and the Chain Stiffness of Amylose Phenylcarbamates in Polar Solvents

In Section 9.3.3, we showed that the intramolecular hydrogen bonds between neighboring NH and C=O groups on the carbamate groups of polysaccharide carbamates decisively influence the polymer conformation in solution, and furthermore solvent-dependent wormlike chain parameters for ATEC and ATBC can be explained only by the content of intramolecular hydrogen bonds. On the other hand, we found that amylose carbamate, having rather bulky side groups in bulky polar solvents, have a different solvent dependence since hydrogen bonding solvent molecule may play an important role in the conformation.

Figure 9.16 shows the plots for h^{-1} vs λ for ATPC in various solvents along with those for ATBC. Interestingly, the data for ATPC do not obey a straight line, indicating that the TSWC model (Eq. 9.6) is not applicable for ATPC in the solvents investigated. The h^{-1} data for ATPC in ketone and esters, shown as filled triangles in Figure 9.16, are appreciably smaller than those in the other solvents (unfilled triangles), 1,4-dioxane and 2-ethoxyethanol. In the former solvents, both h and λ^{-1} increase with an increase in the molar volume of the solvents v_{M}, as shown in Figure 9.17. This is the opposite tendency found for ATBC and ATEC, in that h decreases and λ^{-1} increases with increasing intramolecular hydrogen bonds. This is most likely because hydrogen bonding ketone or ester molecules both extend and stiffen the ATPC main chain, suggesting that this characteristic feature is something to do with the molecular recognition ability of this polymer. While the solvent dependence of λ^{-1} was found to be more significant for amylose tris(3,5-dimethylphenylcarbamate) (ADMPC), one of the most excellent materials as the chiral stationary phase (CSP), the other parameter h is mostly irrespective of solvent, showing that the local helical structure of ADMPC is more stable than ATPC and thus the hydrogen bonding solvent molecules just stiffen the main chain of ADMPC more significantly than that for ATPC.

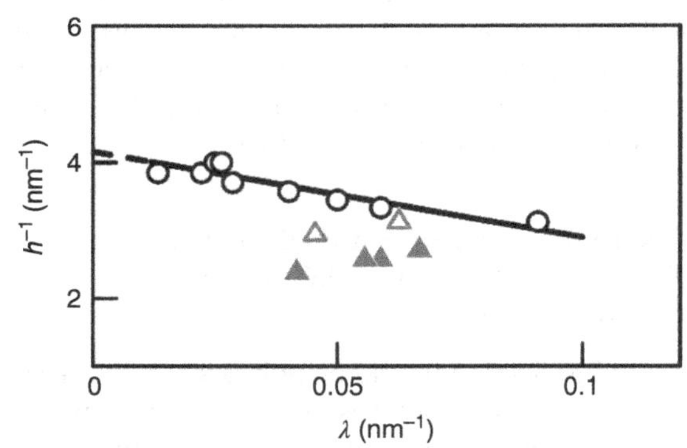

Figure 9.16 Plots of h^{-1} vs λ for ATBC (circles) and ATPC in ketones and esters (filled triangles) and in 1,4-dioxane and 2-ethoxyethanol (unfilled triangles) [50, 51, 72, 73, 76].

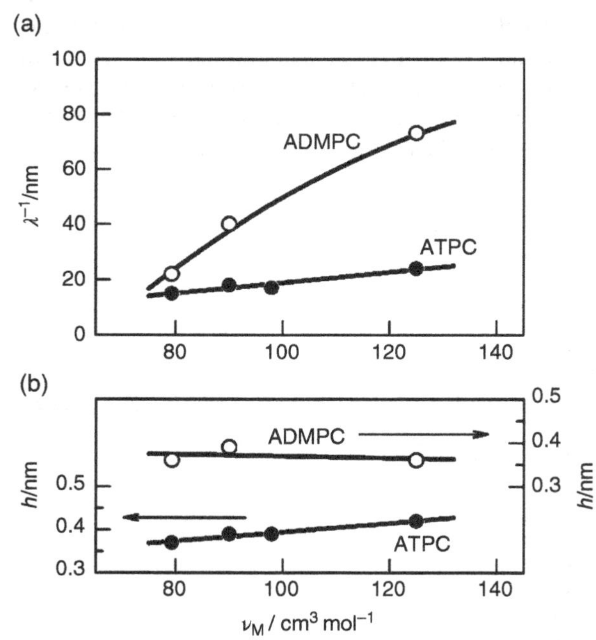

Figure 9.17 Dependences of (a) λ^{-1} and (b) h on the molar volume of the solvent (v_{M}) for ADMPC (unfilled circles) and ATPC (filled circles) in ketones and esters [51, 52].

This stable main chain conformation may be the important characteristics for chiral separation. On the other hand, cellulose derivatives, that is, CTPC [78] and cellulose tris(3,5-dimethylphenylcarbamate) (CDMPC) [79] have a similar chain conformation in 1-methyl-2-purrolidone. While the number of solvents investigated is still a few for cellulosic chains, it seems that, for various solvents, it should be mandatory to measure the restriction of internal rotation by the hydrogen bonding solvent molecules in order to confirm whether the above-mentioned conclusion is applicable to the cellulosic chains. However, this might be difficult as suitable solvents are rarely found for CDMPC because it has less solubility in most organic solvents.

9.4 Lyotropic Liquid Crystallinity of Polysaccharide Carbamate Derivatives

Since most cellulose derivatives have higher chain stiffness than those for typical vinyl polymers, their concentrated solution may form lyotropic liquid crystals [80]. Sato and Teramoto [81] showed that the phase diagram, that is, the phase boundary concentration c_I between the isotropic and the biphasic region and the concentration c_A between the biphasic region and the anisotropic phase, can be explained almost quantitatively for different molar mass samples of hydroxypropyl cellulose, acetoxypropyl cellulose, and cellulose triacetate, schizophyllan, poly(γ-benzyl-*L*-glutamate), poly(*n*-hexylisocyanate), and poly(yne)-platinum polymer in terms of the scaled particle theory, where chain stiffness (or semiflexibility) is taken into consideration. This theory also successfully explained lyotropic liquid crystallinity of CTPC in THF [82] and chitosan phenylcarbamates [83]. More recently, we also found that ATBC, ATEC, ATHC, and CdTPC in THF show lyotropic liquid crystallinity [66, 84]. The phase boundary concentrations of the three amylose derivatives are also successfully explained in terms of the theory with the appropriate molecular parameters, as shown in Figure 9.18. Another feature of the liquid crystals of polysaccharide derivatives is the cholesteric color, as is the case with polypeptides since the polysaccharide main chain has a perfect chiral structure. If the pitch of the cholesteric structure is close to the wavelength of visible light, a so-called structural color can be observed because the corresponding circularly polarized light will reflect, that is,

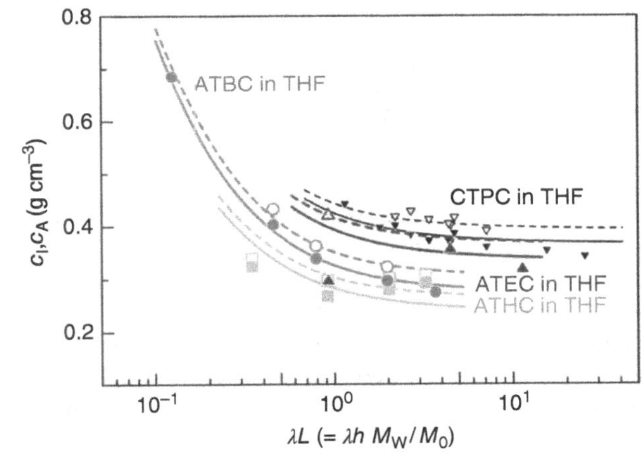

Figure 9.18 Comparison between experimental and theoretical phase boundary concentrations for ATBC (circles), ATEC (triangles), ATHC (squares), and CTPC (inverse triangles) all in THF at 25 °C. Filled and unfilled symbols denote experimental c_I (phase boundary concentration between the isotropic and biphasic region) and c_A (phase boundary concentration between the biphasic and anisotropic region), and solid and dashed curves are theoretical c_I and c_A, respectively [84]. (*See insert for color representation of the figure.*)

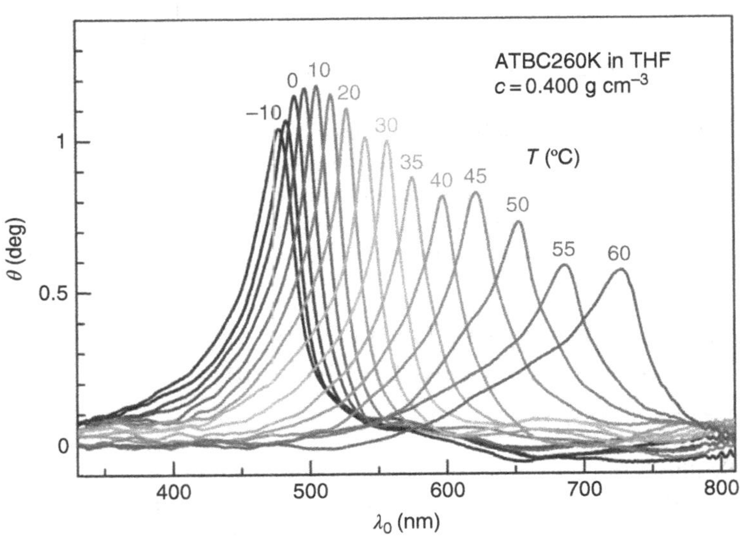

Figure 9.19 Circular dichroism spectra for an ATBC sample ($M_w = 2.6 \times 10^5$) [84].

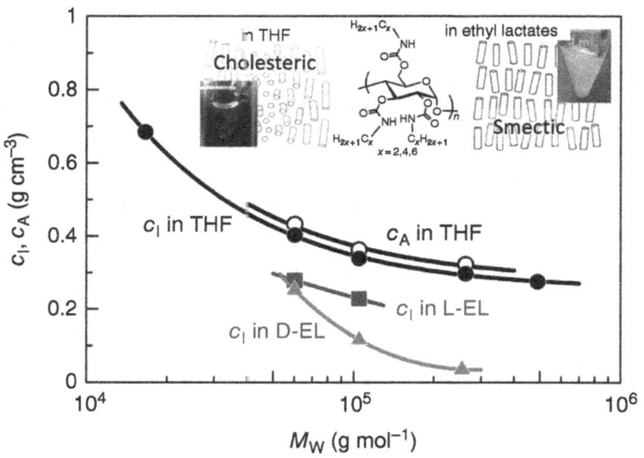

Figure 9.20 Phase diagram of ATBC in L-ethyl lactate (triangles), L-ethyl lactate (squares), and in THF (circles) at 25 °C. Unfilled and filled symbols denote c_A and c_I, respectively [84].

selective reflection. It can be observed as circular dichroism spectra as shown in Figure 9.19 in which the cholesteric pitch appreciably depends on the temperature, thus, it can be used as a simple thermometer.

While most rigid or semiflexible polymer-solvent systems show a phase boundary between the nematic liquid crystal-isotrorpic phase, the smectic phase is rarely found for monodisperse polypeptide systems [85], polydialkylsilane solutions [86], and a polyisocyanide solution [87]. This is probably because molar mass dispersity generally prevents the smectic phase. We recently found this phase for ATBC, ATEC, ATHC in ethyl lactate solution while it cannot be found in THF [84]. The phase boundary concentration c_I is much lower than that in THF and the other boundary concentration c_A was estimated to be much higher, whereas the chain stiffness in ethyl lactates was somewhat lower than that in THF, as shown in Figure 9.20. The above-mentioned scaled particle theory, therefore, does not explain this phenomenon, because this theory does not take into account the smectic order. A significant attractive (or repulsive) interaction between polymer chain ends, as well as substantial anisotropic attractive interactions between two polymer chains, may be also important in addition to the molar mass dispersity, although we have not obtained experimental validation for the origin of the smectic order yet.

9.5 Cyclic Amylose Carbamate Derivatives: An Application to Rigid Cyclic Polymers

Synthesis of cyclic polymers is generally much more difficult than that for linear polymers. For example, both extremely low concentrations and special techniques are required to synthesize cyclic polystyrene [88]. This is because the ring closure probability must be sufficiently less than the probability of attaching two chain ends of different polymer molecules to avoid intermolecular coupling reactions. Furthermore, purification to remove residual linear polymers is also laborious, for example, some special techniques should be applied to purify cyclic polystyrene [89]. Thus, rigid ring polymers are rarely obtained from the coupling reaction of the corresponding linear polymers because of the extreme low ring closure probability. Rather, recently, grafted cyclic chains are reported to visualize the cyclic chain by atomic force microscopy [45, 90, 91]. The main chain of the obtained cyclic polymers should be stiffer than the original cyclic polymer because of repulsive interactions between the grafted side chains. In reality, it is well known that comb polymers with very high grafting densities, e.g. polymacromonomers have appreciably stiffer main chains than those of the corresponding linear polymer [92, 93]. In the biological systems, there are cyclic chains, that is, cyclic DNA and peptides, which have important functions.

Takaha et al. [44] reported that cyclic amylose can be obtained quantitatively by enzymatic synthesis and the obtained cyclic chain has substantially the same main chain stiffness, that is, 4 nm in λ^{-1} estimated from the diffusion coefficients [94]. Taking the high chain stiffness of amylose carbamate derivatives into consideration, we attempted to synthesize cyclic amylose tris(phenylcarbamate) (cATPC) [95] and cyclic amylose tris(n-butylcarbamate) (cATBC) [96] from the enzymatically synthesized cyclic amylose. While cyclodextrin tris carbamate derivatives are rarely obtained probably due to the high steric hindrance, cATPC and cATBC were successfully synthesized from cyclic amylose, where the degree of polymerization ranges from 24 to 290. The obtained cyclic chains should have stiff main chains; indeed, highly concentrated cATBC in THF shows lyotropic liquid crystallinity as shown in Figure 9.21. Figure 9.22 illustrates the plots of $\lambda\langle S^2\rangle_z^{1/2}$ vs n_K for cATPC in 1,4-dioxane and 2-ethoxyethanol and cATBC in methanol, 2-propanol, and THF along with those for the corresponding linear polymer systems.

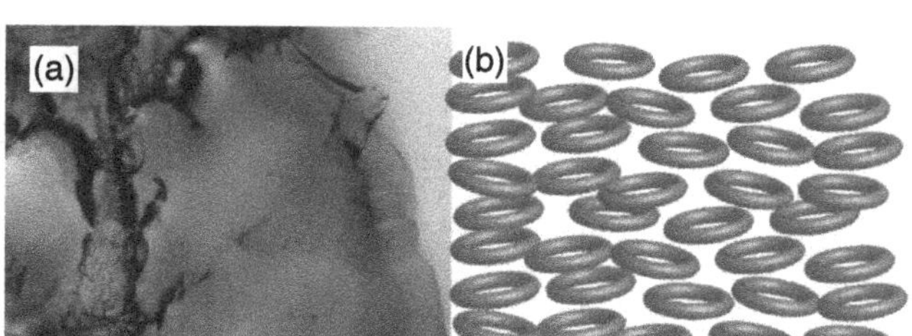

Figure 9.21 (a) Polarized light micrograph of cATBC in THF; (b) schematic representation of the liquid crystal consisting of rigid ring polymers [96]. (*See insert for color representation of the figure.*)

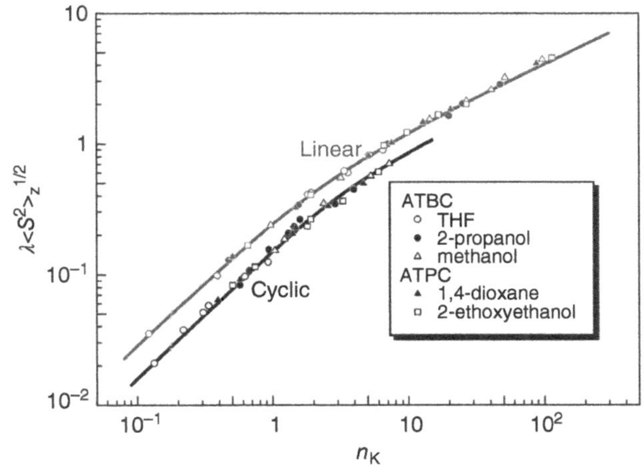

Figure 9.22 Plots of $\lambda\langle S^2\rangle_z^{1/2}$ vs n_K for linear (gray symbols) and cyclic (black symbols) ATBC in THF at 25 °C (unfilled circles), in 2-propanol at 35 °C (filled circles), and in methanol at 25 °C (unfilled triangles) and for linear and cyclic ATPC in 1,4-dioxane at 25 °C (filled triangles) and in 2-ethoxyethanol at 25 °C (unfilled squares). Solid gray and black curves indicate the theoretical values calculated by Eq. 9.2 for linear wormlike chain and by Eq. 9.3 for cyclic wormlike chain, respectively [50, 72, 73, 95, 96].

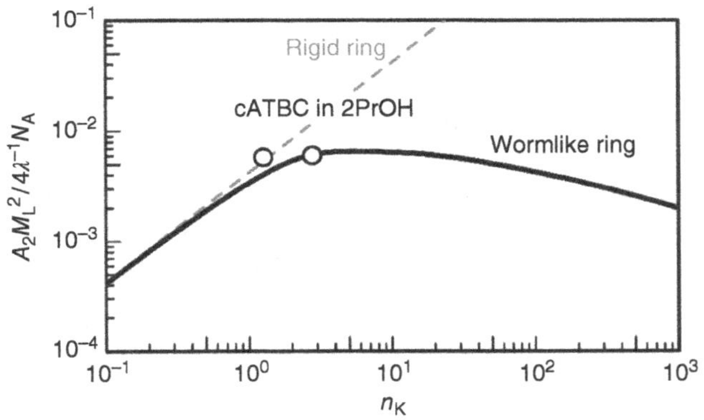

Figure 9.23 The Kuhn segment number n_K dependence of the reduced second virial coefficient ($A_2M_L^2/4\lambda^{-1}N_A$) for cATBC in 2PrOH (unfilled circles) at the theta temperature (35 °C), where M_L is the molar mass per unit contour length and calculated from h and the molar mass of the repeat unit [96]. Solid curve, results from Monte Carlo simulation by Ida et al. [97]. Dashed line, theoretical values for rigid ring [98].

The data points for cyclic chains obey the black curve for the cyclic wormlike chain calculated using Eq. 9.3 with substantially the same wormlike chain parameters as those for the linear polymers. They form a different universal curve which is appreciably smaller than those for the linear one. This indicates that the obtained $\langle S^2\rangle_z^{1/2}$ data for rigid cyclic chains are successfully explained by the molecular parameters obtained from the corresponding linear polymers-solvents systems for which the chain stiffness ranges up to 75 nm. It should be noted that the chain stiffness parameter for cATBC in THF cannot be determined from the current data. Cyclic samples with higher molar mass is desirable to determine the chain stiffness of cyclic chains with high λ^{-1} accurately since the degree of polymerization of the highest cyclic samples were at most 290, while those for linear polymers are about 6000. Another characteristic feature of cyclic chains is the topological interaction between two discrete cyclic chains. This can be seen as an apparent repulsive interaction between two cyclic chains because two discrete cyclic chains cannot be a linked ring, a so-called catenane. Figure 9.23 illustrates the Kuhn segment number n_K dependence of the reduced second virial coefficient $A_2M_L^2/4\lambda^{-1}$ N_A for cATBC in 2-propanol at the theta temperature of the linear ATBC in the same solvent. Thus, the second virial coefficient A_2 of linear ATBC vanishes in the solvent. The appreciable positive A_2 is well explained by the theory by Ida et al. [97] in that, only topological interactions between two rings are considered and they are still substantially close to the theoretical dashed line for the rigid ring by des Cloizeaux [98].

On the other hand, specific phenomena were found for cATPC in ketones and esters. Figure 9.24 displays the molar mass dependence of $\langle S^2\rangle_z^{1/2}$ for cATPC in methyl acetate (MEA), ethyl acetate (EA), and 4-methyl-2-pentanone (MIBK). Unlike the former cases for cATBC and cATPC, the data points for cATPC in ketones and esters are appreciably lower than the theoretical dashed curves calculated for cyclic wormlike chains with the parameters for the linear chains, clearly indicating that the wormlike chain parameters for cATPC in the solvents are different from the corresponding linear one. The obtained parameters, h and λ^{-1}, are plotted against the molar volume of the solvent v_M in Figure 9.25. Both (h and λ^{-1}) the discrepancy between linear and cyclic chains become more significant with increasing v_M, suggesting that an extended main chain structure for ATPC in ketones and esters is easily changeable by the small curvature of the polymers. This tendency is also seen in A_2. Reduced second virial coefficients for cATPC in the theta solvent for linear ATPC are shown in Figure 9.26 along with those for cATBC previously shown in Figure 9.23. Interestingly, the second virial coefficients drastically decrease with an increase in the size of the solvent molecules and become negative in MIBK. We believe

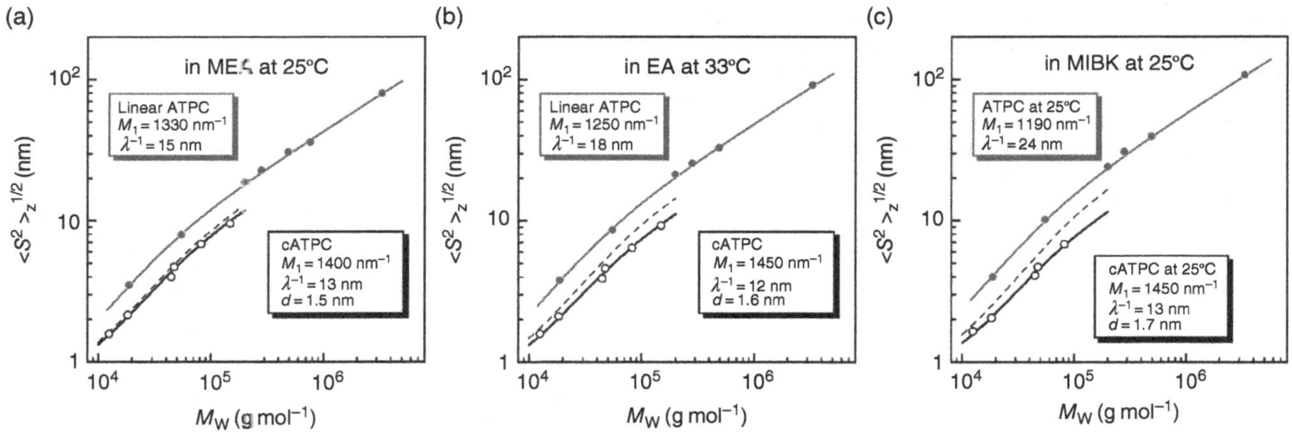

Figure 9.24 Molar mass dependence of $\langle S^2\rangle_z^{1/2}$ for cATPC (unfilled circles) and linear ATPC (filled circles) (a) in MEA at 25 °C, (b) in EA at 33 °C, and (c) in MIBK at 25 °C. Solid black and gray curves, theoretical curves for cylindrical wormlike ring and linear chains, respectively, calculated with the parameters listed in each figure. Dashed lines, theoretical values for wormlike ring with the parameters for linear ATPC [99].

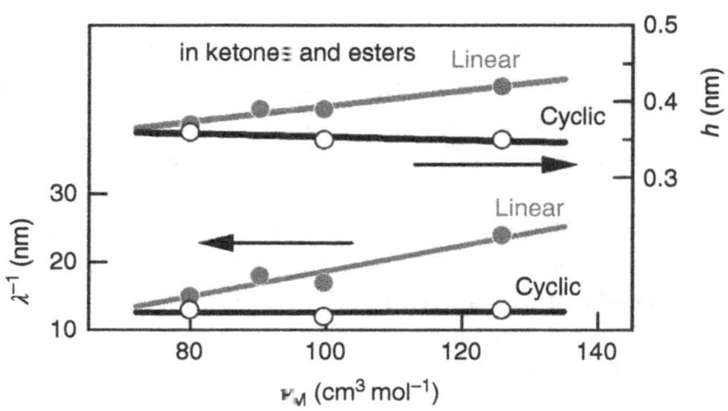

Figure 9.25 Dependences of h (a) and λ^{-1} (b) on v_M for cATPC (unfilled circles) and ATPC (filled circles) in ketones and esters [99].

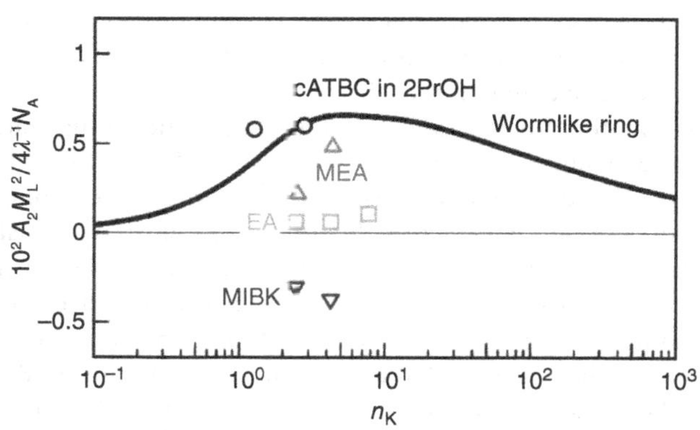

Figure 9.26 The Kuhn segment number n_K dependence of the reduced second virial coefficient ($A_2M_L^2/4\lambda^{-1}N_A$) for cATPC in methyl acetate (MEA, unfilled triangles), ethyl acetate (EA, unfilled squares), and 4-methyl-2-pentanone (MIBK, inverted triangles) at the corresponding Θ temperatures along with those for cATBC in 2-propanol (unfilled circles) at the Θ temperature (35 °C) [99]. Solid curve, results from Monte Carlo simulation by Ida et al. [97].

this is the specific characteristics for rigid cyclic polymers and indeed it is the first finding that the rigid cyclic chains may have different conformation and intermolecular interaction from the corresponding linear chains due to the permanent curvature. From the viewpoint of CSPs, this result suggests that the local chain curvature of amylosic chains on the silica gel may influence the performance as CSP. Hereafter, performance between linear and cyclic amylosic chains as CSP will give us important information even for the preparation of a chiral column with the linear amylose derivatives.

9.6 Conclusion

In this chapter, we have given an overview of the characteristic solution behavior of polysaccharide carbamates derivatives, some of which are useful in the chiral separation of various organic materials. Amylose tris (3,5-dimethylphenylcarbamate) (ADMPC), that is, one of the most useful materials, has a stable helical structure even in polar solvents, in which the helix pitch per residue for other amylose carbamate derivatives appreciably depends on the solvents, suggesting that the stable helical structure might be an indispensable feature in the chiral stationary phase. We also found that very high chain stiffness was attained for some polysaccharide derivatives in rather low polarity solvents, that is, amylose tris(alkylcarbamate)s (ATBC, ATEC, ATHC) and curdlan tris(phenylcarbamate) (CdTPC). Their concentrated solutions have cholesteric liquid crystallinity, which is well known for some α-helical polypeptides. Using this high chain stiffness, rigid cyclic polymer can be quantitatively synthesized from cyclic amylose, which has a potential use as a building block as nanomachines.

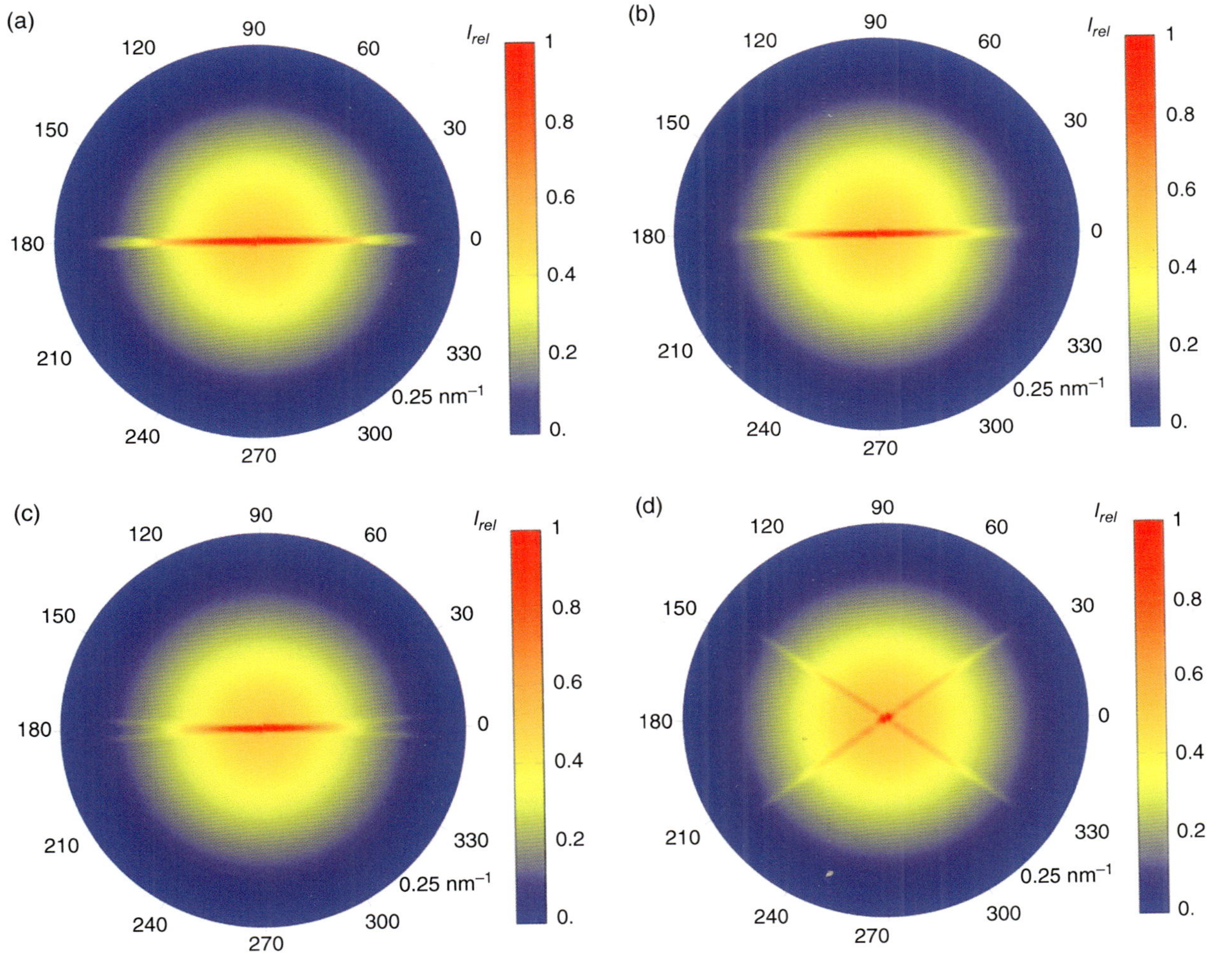

Figure 1.6 Simulated small-angle X-ray scattering patterns from closest-packed arrangements of each 500 fibrillar structures (with angular uniformity of σ_{ang}:0.2 $^{\circ-1}$), 2 nm in diameter, embedded in cylindrical cell walls, which were probed centrally, accounting for isotropic scattering from randomly oriented structures (10 nm) and for a Gaussian beam profile (σ_{beam}:8 nm^{-1}) [282]. The simulated microfibril angles (cylinder front to back) are (a) 0°, (b) 2°, (c) 3° and (d) 35°. The separation of the directional scattering patterns originating from the fibrils can, under such practical conditions, be observed starting at 3°.

Bioinspired Materials Science and Engineering, First Edition. Edited by Guang Yang, Lin Xiao, and Lallepak Lamboni.

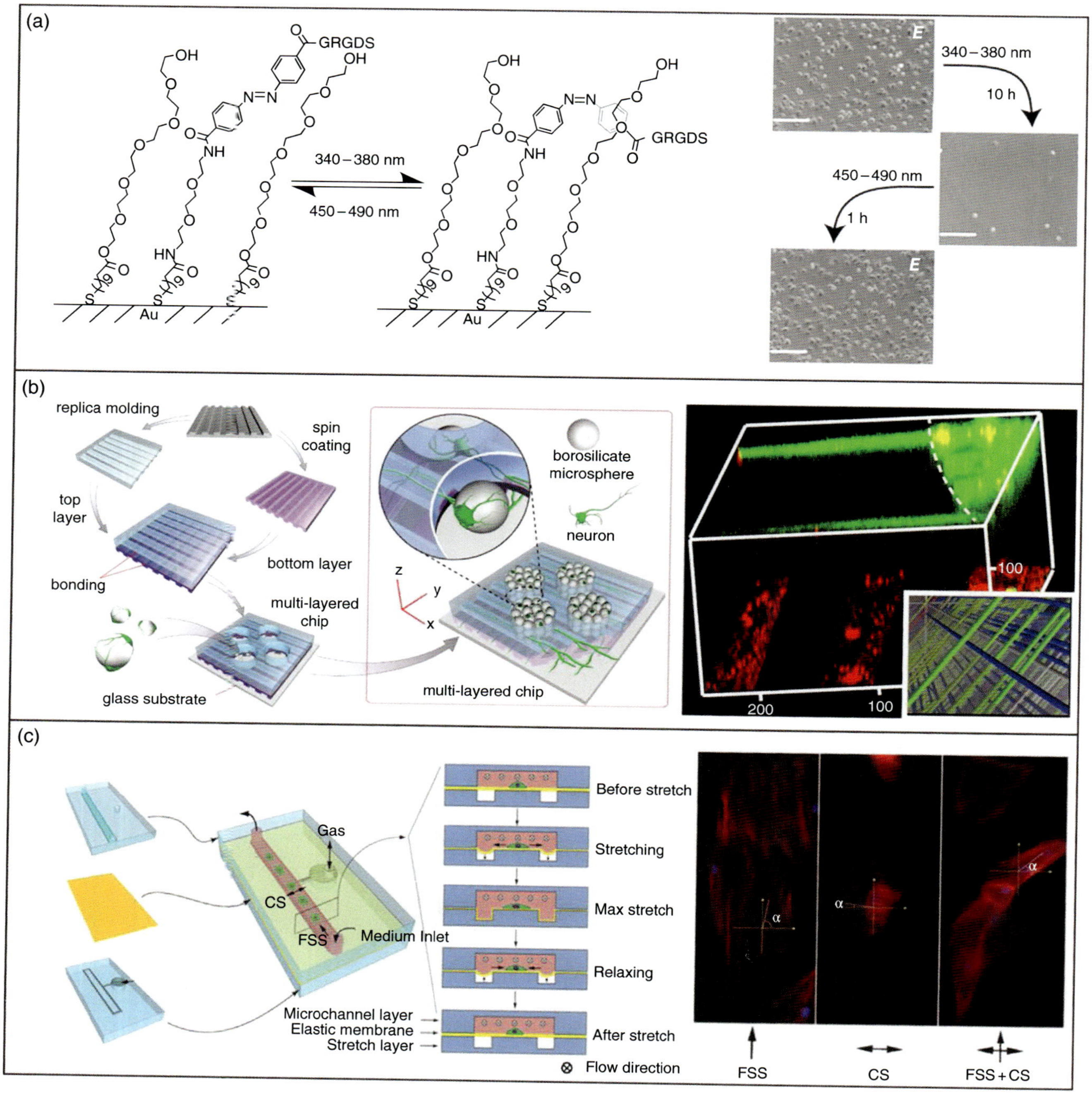

Figure 2.3 (a) The azobenzene moiety can be converted photochemically between the E and Z configurations to either present or mask the RGD ligand and hence modulate cell adhesion. Cells adhered onto SAMs with the azobenzene group in the E configuration. Few cells adhered to the same SAMs with azobenzene in the Z configuration. Cells adhered to the SAMs again when the conformation of azobenzene was changed from Z to E. (b) Schematics of fabrication processes of the multilayered microchip and 3D reconstruction of the multilayered neurite connections. The inset represents the orthogonal neurite pathways in the brain. Copyright 2012, American Association for the Advancement of Science. (c) Schematic of the microfluidic flow-stretch chip. The stress fibers tend to align in the direction parallel to that of the FSS, CS and the resultant force of the FSS and CS. *Sources:* (a) Reproduced with permission from [29]. (b) Reproduced with permission from [30]. (c) Reproduced with permission from [1].

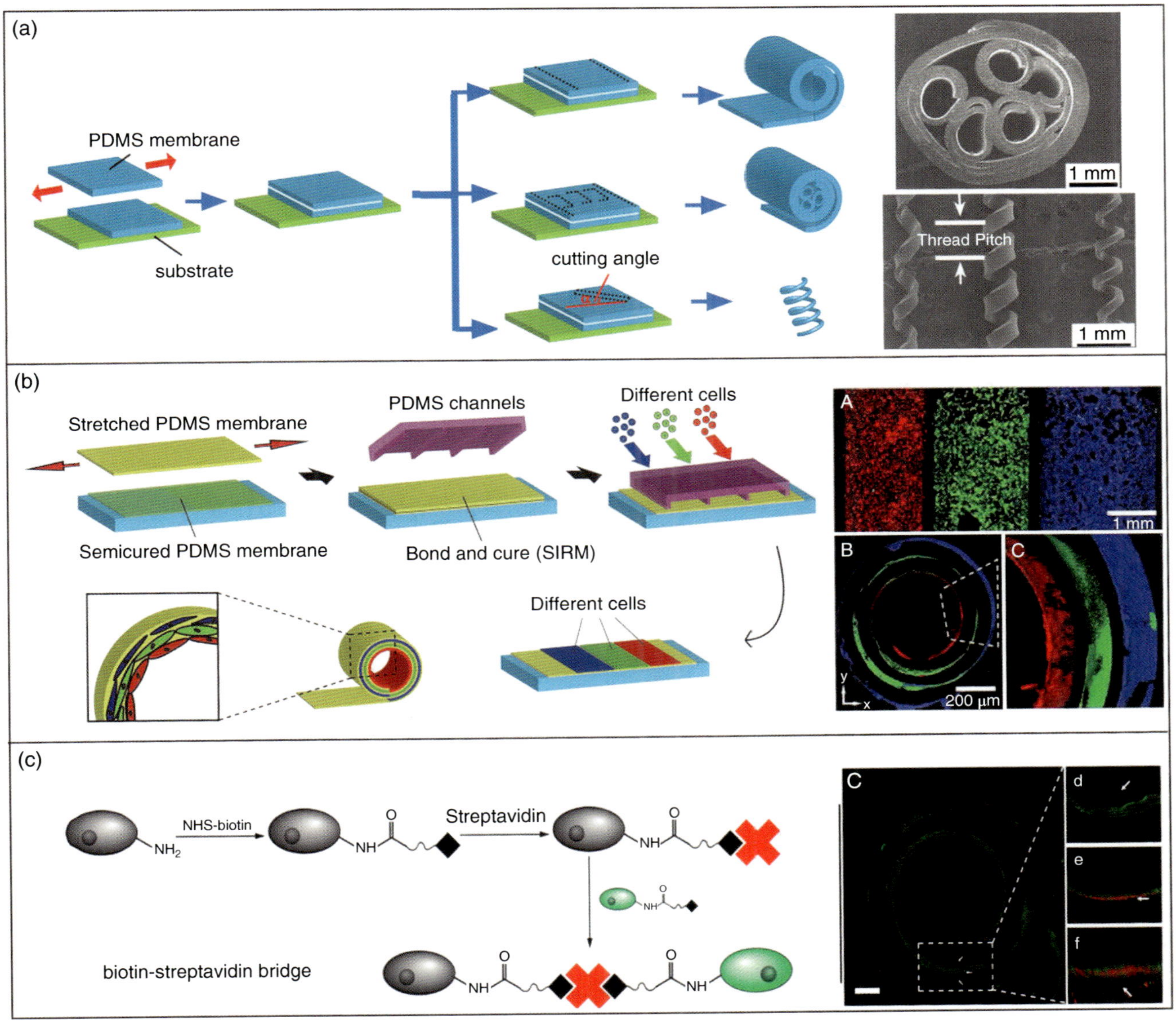

Figure 2.4 (a) Schematic illustration of the fabrication process of tubes, tubes-in-a-tube and spirals. (b) Schematic illustration of a stress-induced rolling membrane (SIRM) and tubes with multiple types of cells as the walls. A thin PDMS membrane is stretched as the top layer of the SIRM and covers a semi-cured PDMS membrane. After curing the two layers to cause adhesion, a SIRM is obtained that rolled up when the ends were released. Microfluidic channels cover the surface of the SIRM, different cells are delivered via microfluidic channels to the surface of the SIRM. One end of the SIRM is released by cutting its edge. The SIRM rolls up into a tube and each type of cell is delivered to a designated position as the tubular wall. The structure of the tube is similar to that of tubular tissues, such as small-diameter vessels. (c) Schematic diagram of cell surface modification and stepwise formation of multicellular structures. Confocal image of a 3D reconstruction of the bilayer on the SIRM after rolling. *Sources:* (a) Reproduced with permission from [85]. (b) Reproduced with permission from [87]. (c) Reproduced with permission from [88].

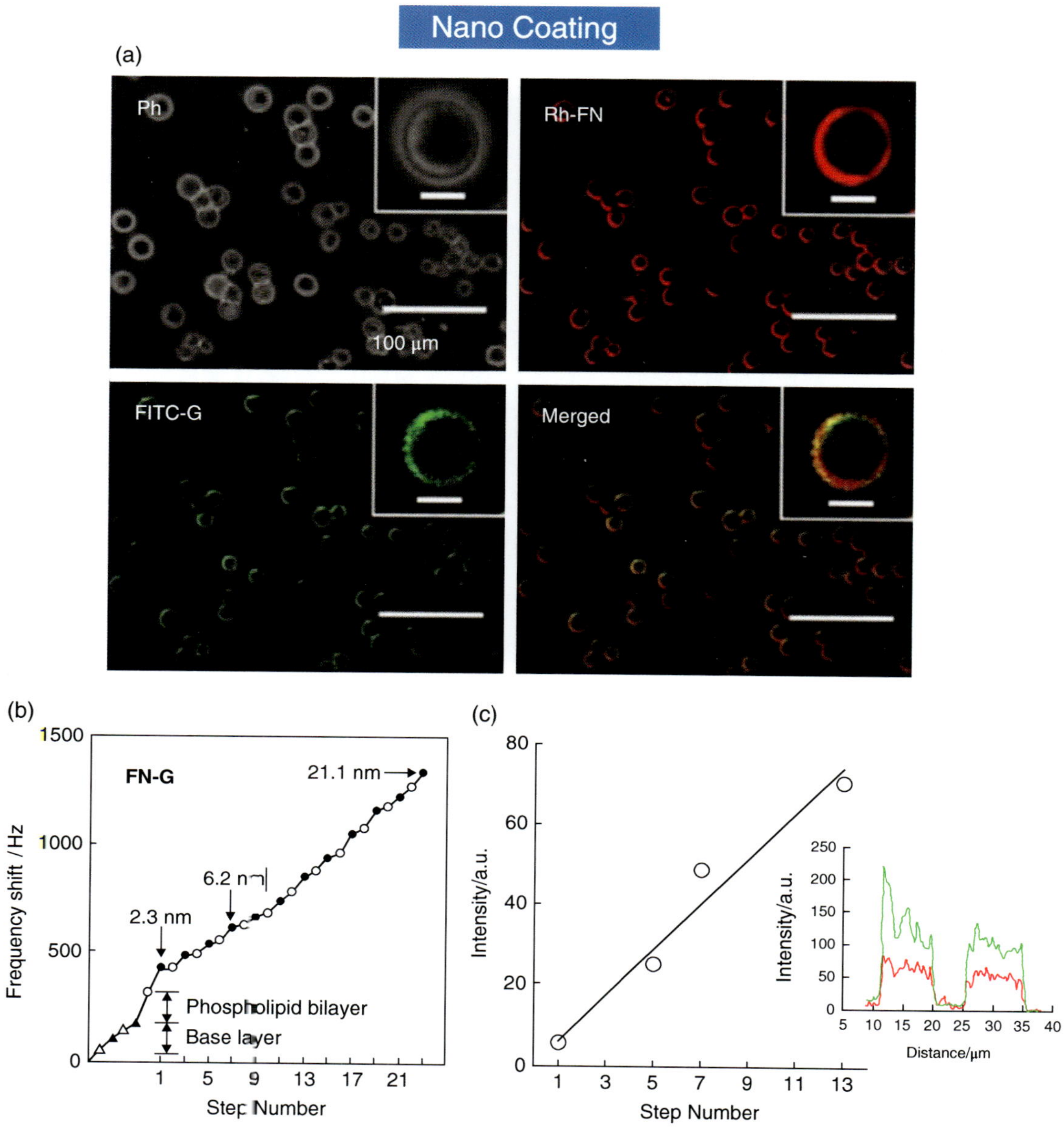

Figure 3.2 (a) Phase and fluorescent microscopic images of L929 mouse fibroblast cells coated with nanometer-sized films by 9-step assembly of Rh-FN and FITC-G. Scale bars in the figure are 10 mm. (b) Frequency shift of the quartz crystal microbalance (QCM) LbL assembly of FN-G nanofilms onto a phospholipid bilayer. Closed (●) and open circles (○) represent the assembly steps of FN and G, respectively. The phospholipid bilayer was fabricated on a base layer prepared by PDDA and PSS. (c) Fluorescence intensity of the cells and 7-step-assembled Rh-FN-G nanofilms on cell surfaces observed by CLSM. (d) Schematic illustration of cells after coating for collagen nanofiber matrix once, twice, and three times with rhodamine or FITC-labeled collagen. (e) From left to right are the coating results of once, twice, and three-times-coated cells observed by CLSM. (f) line scanning results of collagen-coated cells with different sizes of collagen microlayers on cell surfaces. *Source:* Reprinted with permission from [40, 64].

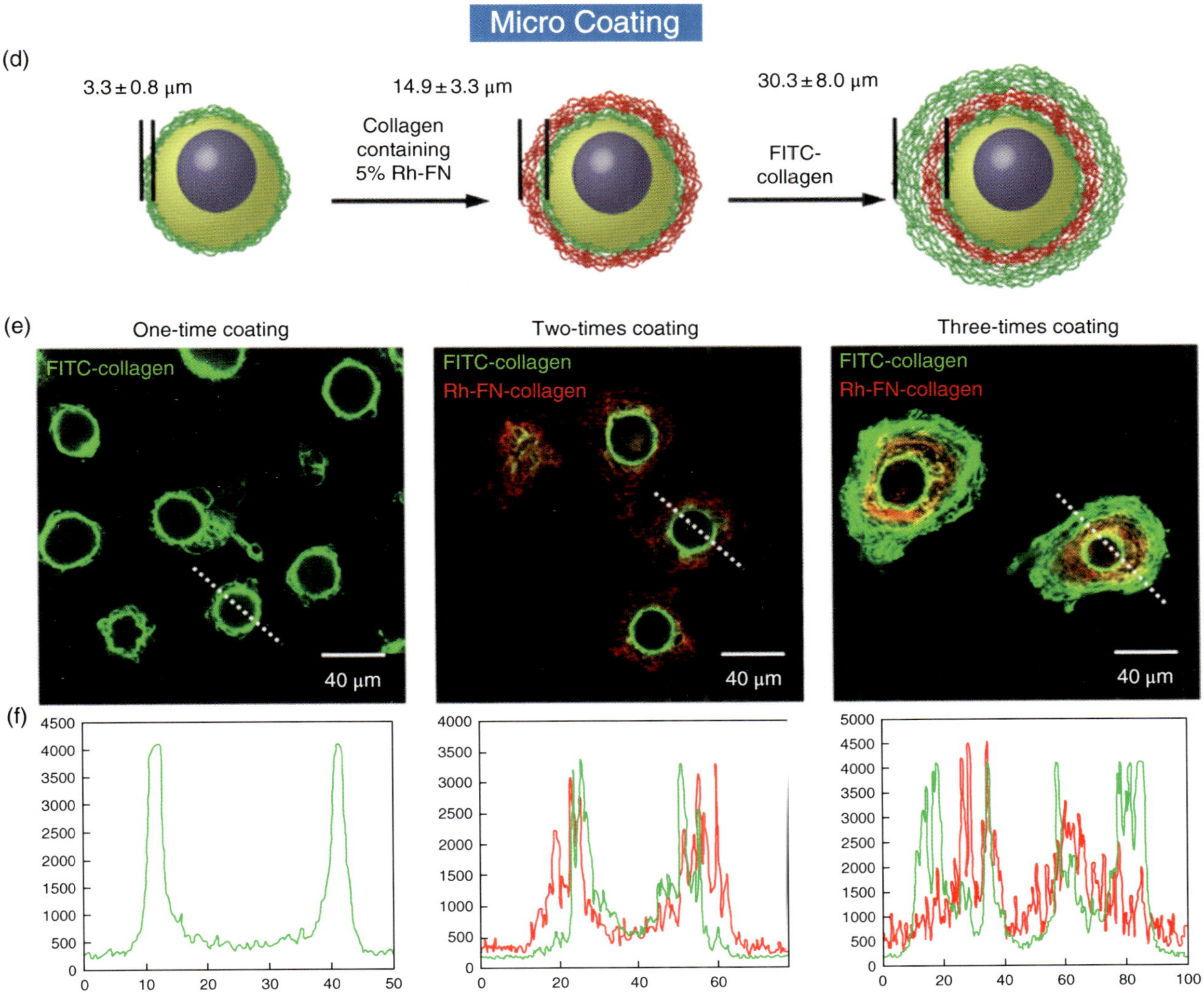

Figure 3.2 (Continued)

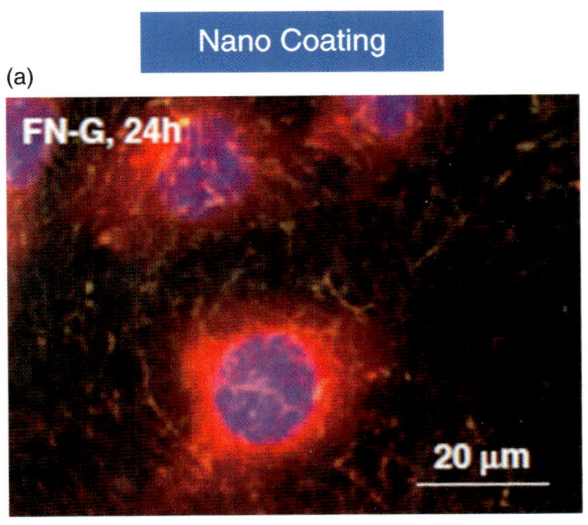

Figure 3.3 (a) Fluorescence microscopic image Rh-FN-FITC-G films prepared on L929 cells after 24 h of incubation. *Source:* Reprinted with permission from [55].

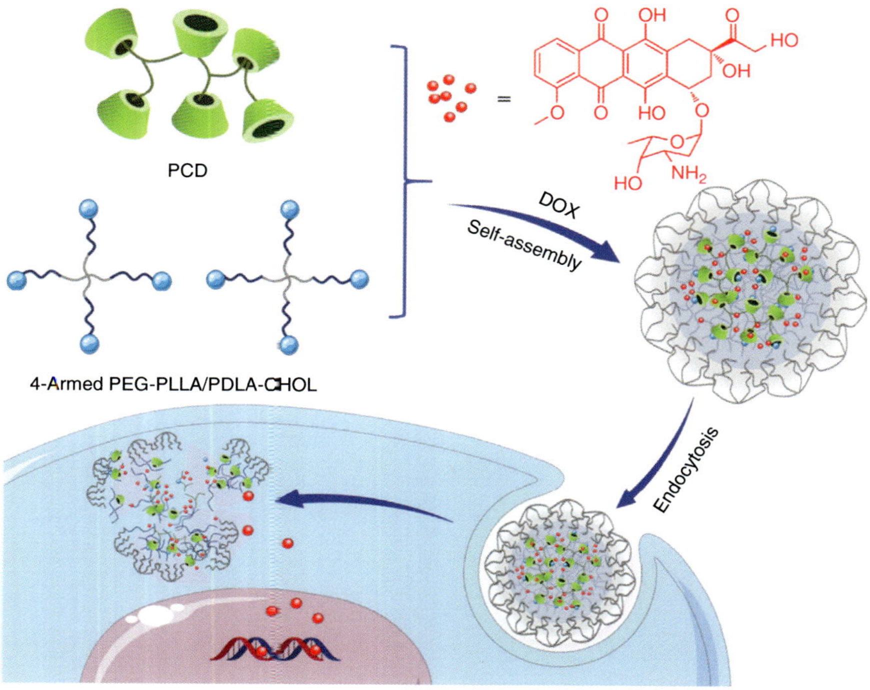

Figure 6.3 PLA-*b*-PEG micellar system for sustained intracellular drug delivery regulated by stereo complexation and host-guest interactions. *Source:* Reproduced with permission from [43].

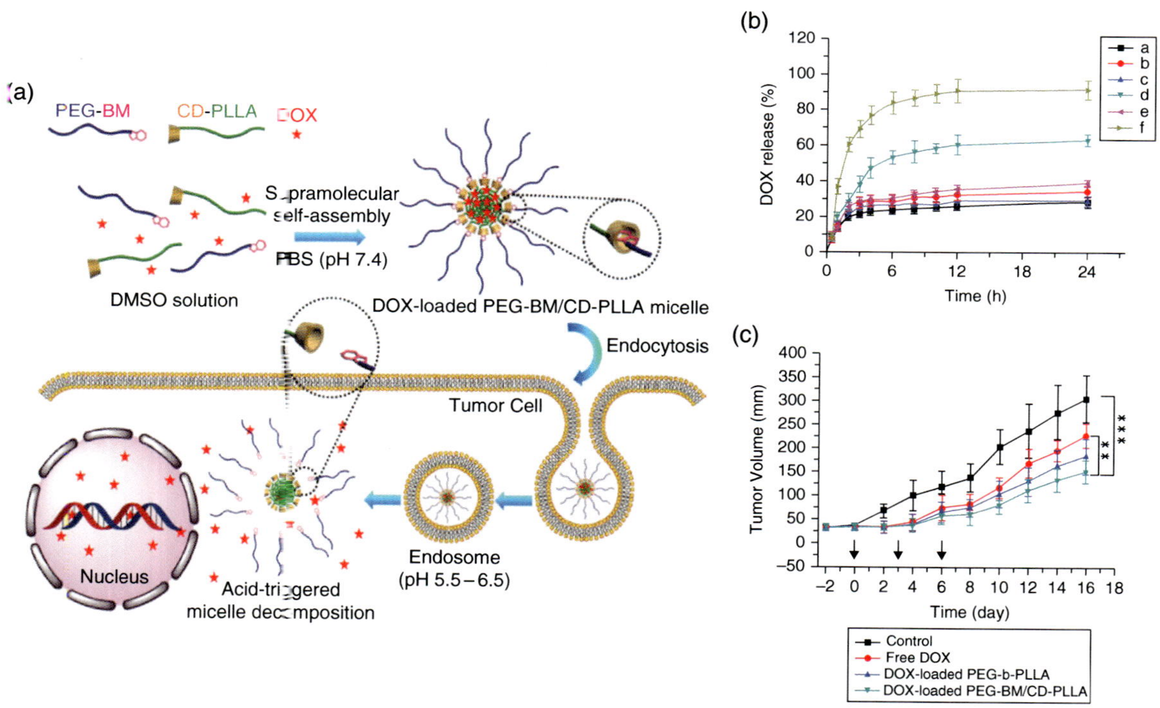

Figure 6.4 (a) Schematic illustration of formation and triggered drug release from DOX-loaded PEG-BM/CDPLLA supramolecular micelles in response to the intracellular microenvironment; (b) *In vitro* DOX release profiles of DOX-loaded micelles in PBS at 37 °C and different pH values: (a,c,e) DOX-loaded PEG-*b*- PLLA and (b,d,f) DOX-loaded PEG-BM/CD-PLLA at pH (a,b) 7.4, (c,d) 6.5, and (e,f) 5.5; (c) *In vivo* antitumor efficacies after tail-vein injection of PBS (control), free DOX, DOX-loaded PEG-b-PLLA, and DOX-loaded PEG-BM/CD-PLLA into male BALB/c nude mice bearing HepG2 xenografts. *Source:* Reproduced with permission from [62].

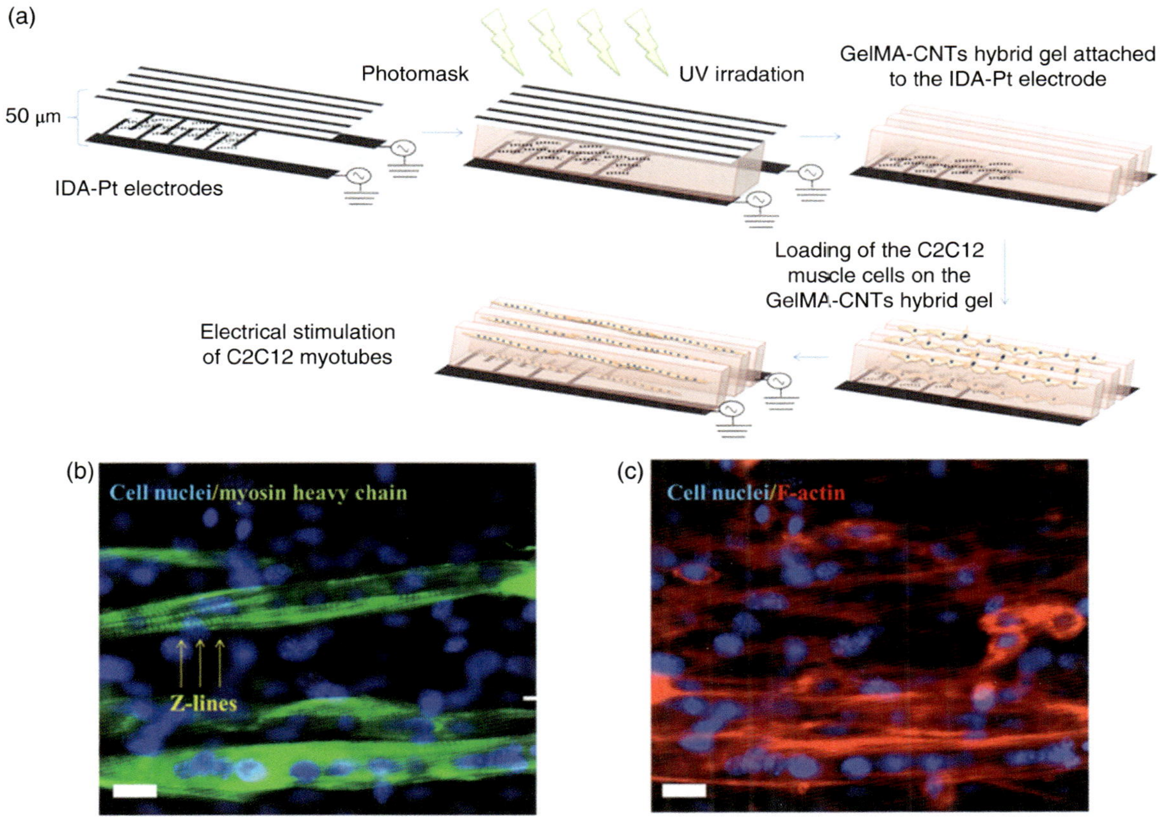

Figure 7.3 Schematic illustration of the procedure for the fabrication of a groove ridge topography within the GelMA-CNT hybrid gel (0.3 mg/mL CNTs) with (+ES) or without (−ES) the electrical stimulation at day 10 of the culture; (b)–(c) Immunostaining of cell nuclei/myosin heavy chain (b), and cell nuclei/F-actin (c). Scale bar: (A) 30 μm, (B) 50 μm. *Source:* Reproduced with permission from [63].

Figure 9.18 Comparison between experimental and theoretical phase boundary concentrations for ATBC (circles), ATEC (triangles), ATHC (squares), and CTPC (inverse triangles) all in THF at 25 °C. Filled and unfilled symbols denote experimental c_I (phase boundary concentration between the isotropic and biphasic region) and c_A (phase boundary concentration between the biphasic and anisotropic region), and solid and dashed curves are theoretical c_I and c_A, respectively [84].

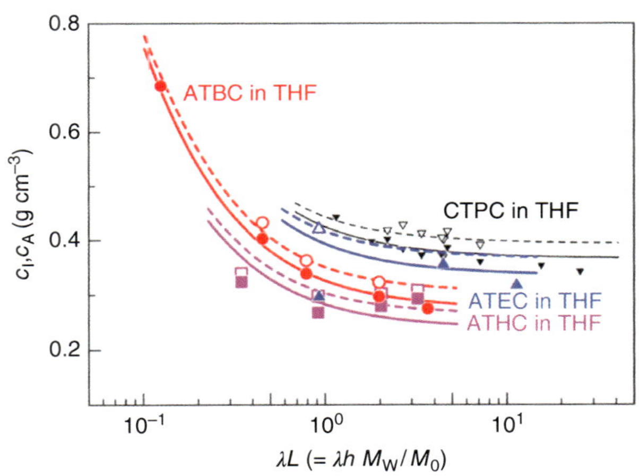

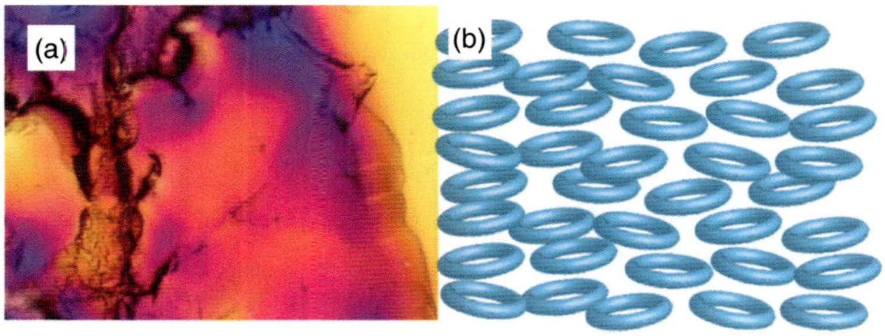

Figure 9.21 (a) Polarized light micrograph of cATBC in THF; (b) schematic representation of the liquid crystal consisting of rigid ring polymers [96].

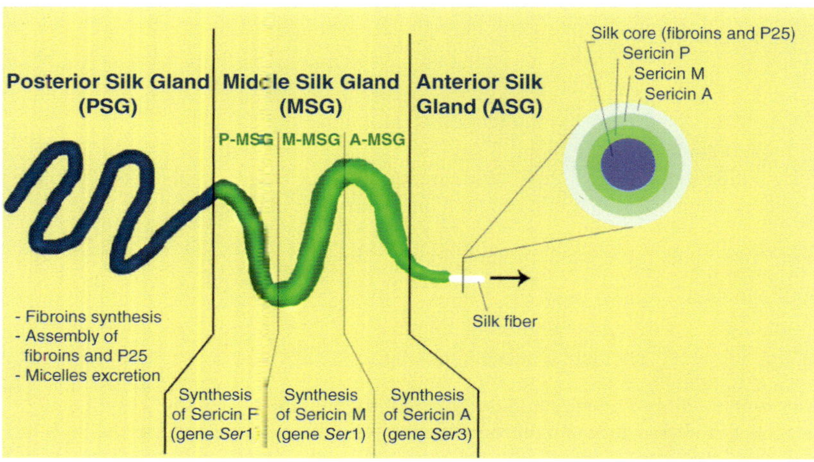

Figure 10.1 Schematic illustration of *Bombyx mori*'s silk gland and silk structure.

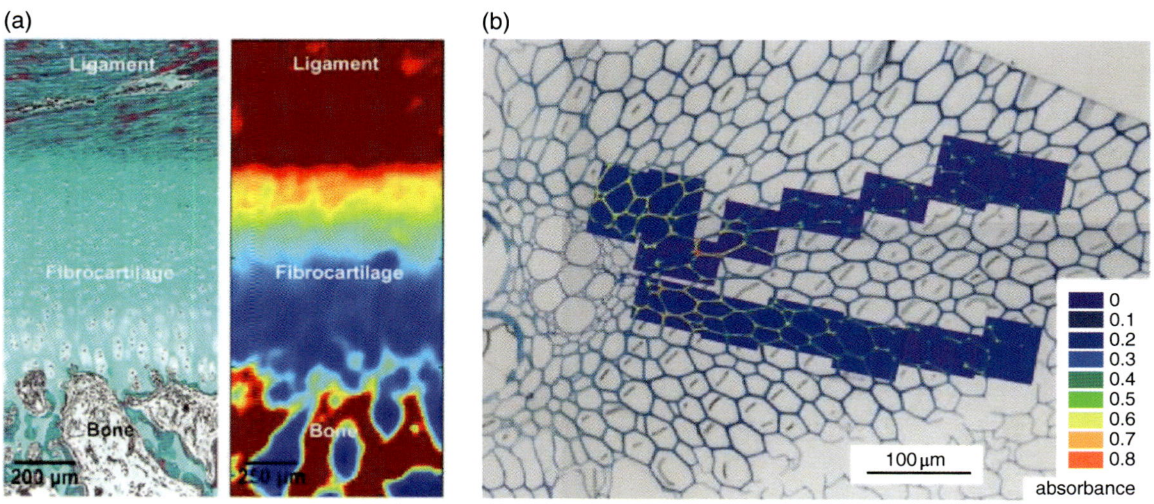

Figure 12.1 Gradient structures in biological systems. (a) The native ACL-bone interface (left) and Fourier Transform Infrared Spectroscopic (FTIR) image (right). (b) Composite image of a fiber cap cross-section. The 2D UV absorbance scans illustrate the degree of cell wall lignification. (red-to-blue color represents high to low levels of lignin). *Sources:* (a) Reproduced from [1] with permission, copyright 2008, Elsevier. (b) Reproduced from [3] with permission, copyright 2008, The Royal Society.

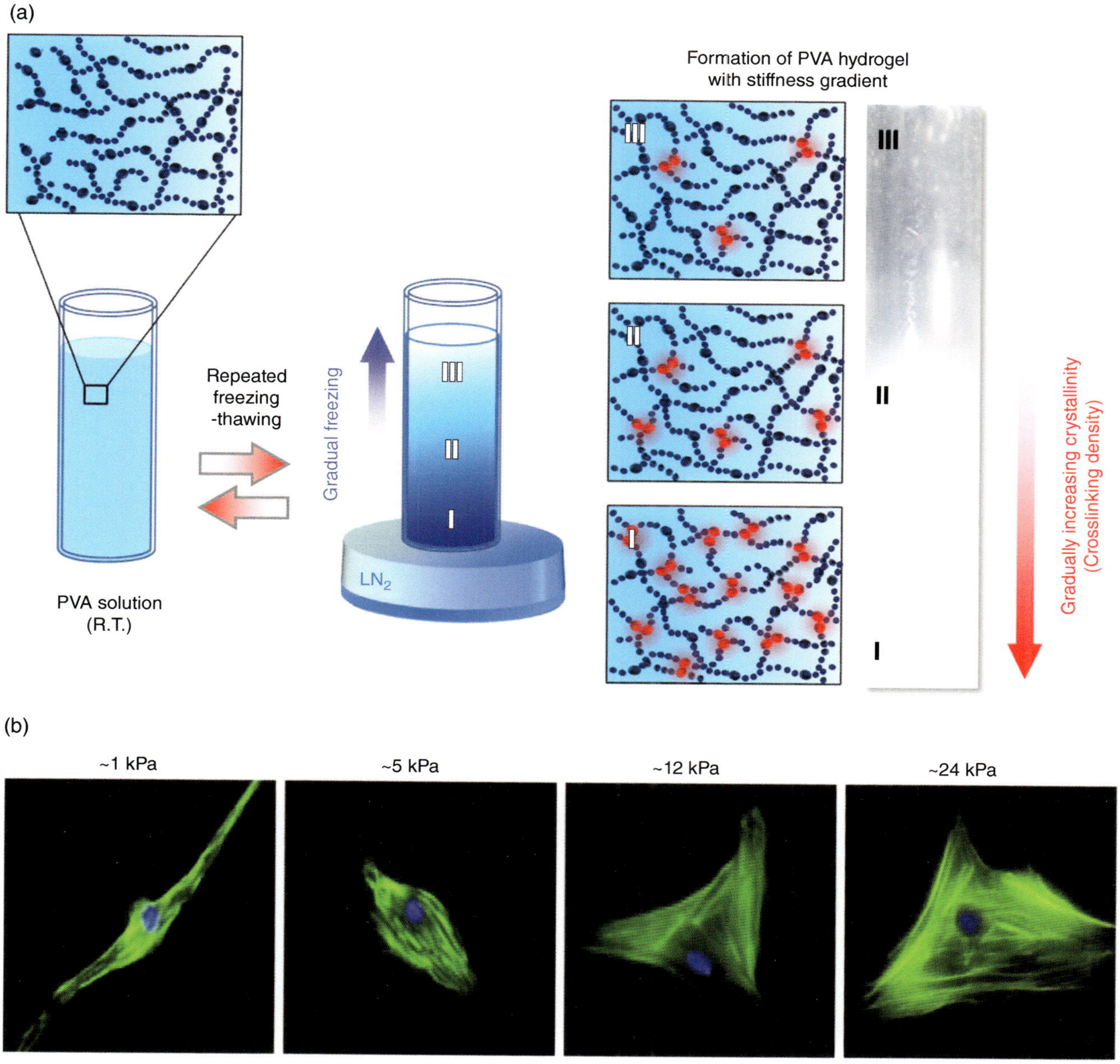

Figure 12.4 Gradient stiffness in hydrogels. (a) Schematic diagrams showing the stiffness gradient PVA hydrogel generated by gradual freezing-thawing method. The red dots represent cross-linking points in PVA hydrogel networks. (b) Immunofluorescent staining (Phalloidin) images of hBMSCs cultured on the stiffness gradient PVA hydrogel after 28 days. (c) Schematic diagrams of the process to prepare an HA-based hydrogel with orthogonal gradients. (d) The orthogonal gradients of both stiffness and ligand density. (e) U373-MG human GBM cells remained rounded and rarely exhibited lamellipodia in soft regions with low fibronectin. In contrast, cells spread extensively and developed large lamillipodia in high stiffness regions with high fibronectin. *Sources:* (a) and (b) reproduced from [25] with permission, copyright 2015, Elsevier. (c)–(e) reproduced from Rape (2015) [27] with permission, copyright 2015, Nature Publishing Group.

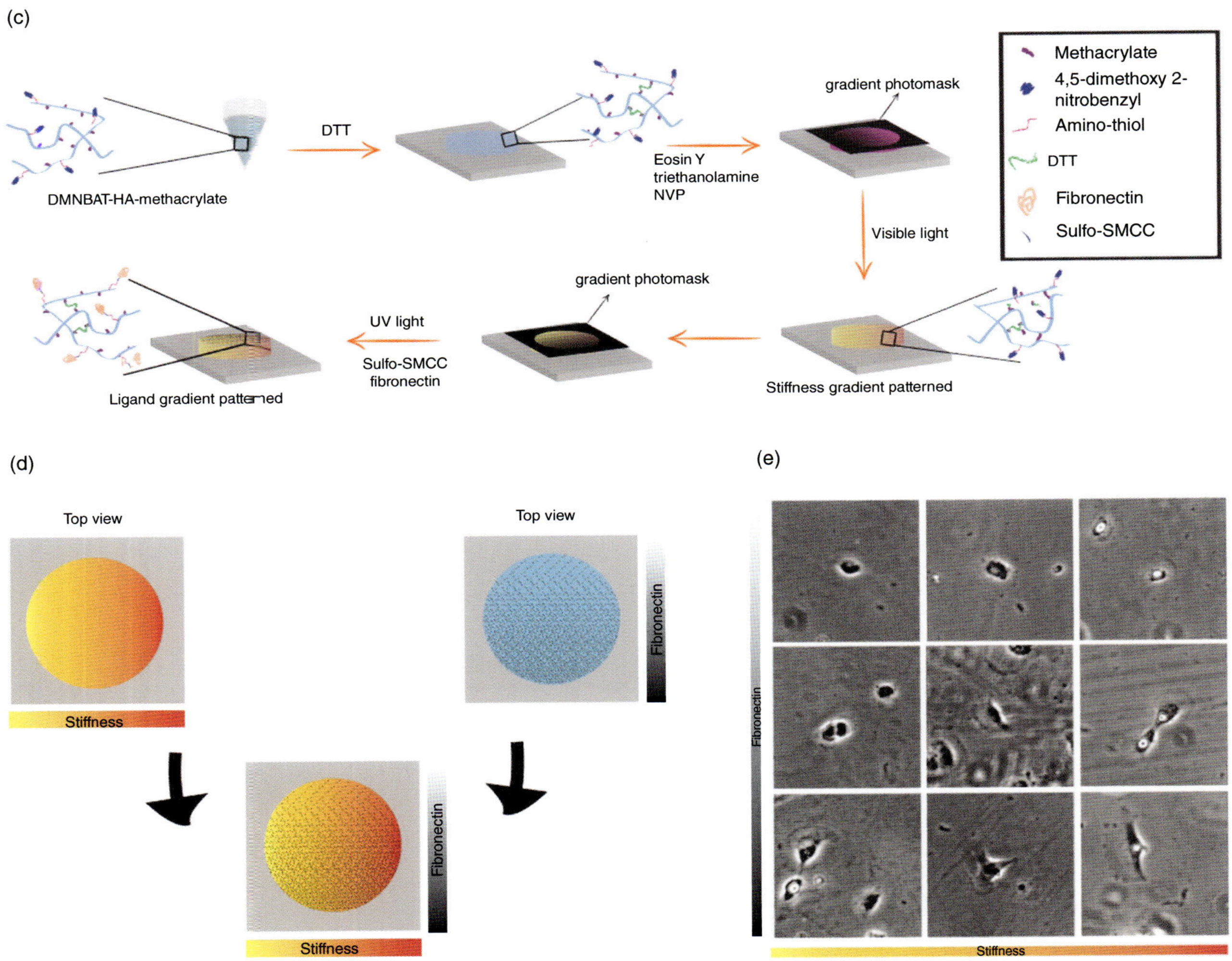

Figure 12.4 (Continued)

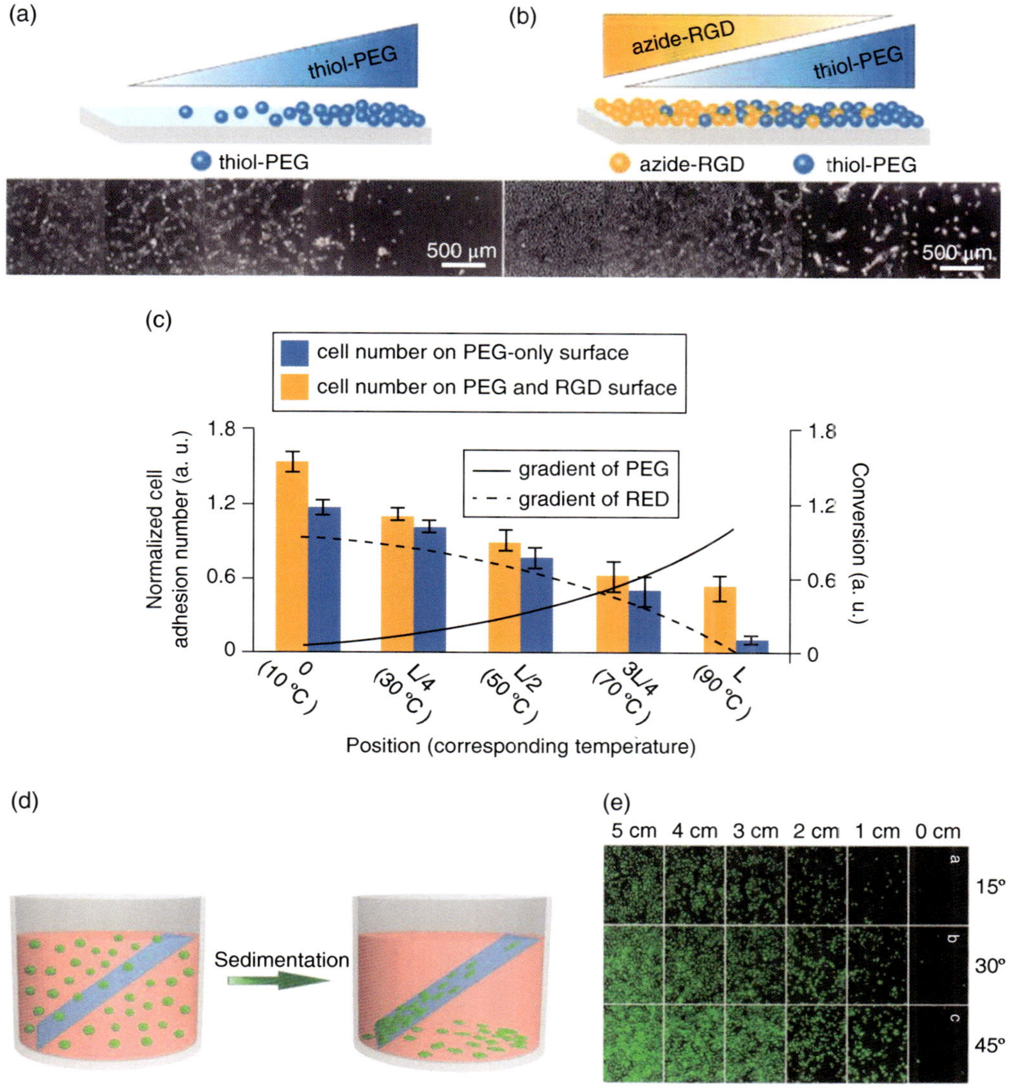

Figure 12.10 Gradient cell adhesion. (a) Schematic of the PEG concentration gradient prepared by click chemistry. 3T3 cells were cultured on this surface to show cell density gradient. (b) Schematic of the countercurrent gradients of PEG and RGD concentration. (c) Quantification of cell attachment on the substrate in (a) and (b). (d) Schematic of the sedimentation method for fabricating the cell density gradient. (e) Fluorescence micrographs showing cell density gradients generated on glass substrate at different tilt angles. (f) Schematic of the rapid generation of a cell gradient. Top view and side view SEM images showing the increased surface roughness and thickness of the nanodendritic gradient, respectively. The fluorescence micrographs showing the gradual increasing cell density along the direction with gradient nanodendrites. (g) The influence of chemical gradients on nanodendrites for generation of cell gradients. *Sources:* (a)–(c) reproduced from [80] with permission, copyright 2016, American Chemical Society. (d) and (e) reproduced from [9] with permission, copyright 2013, John Wiley & Sons. (f) and (g) reproduced from [10] with permission, copyright 2014, John Wiley & Sons.

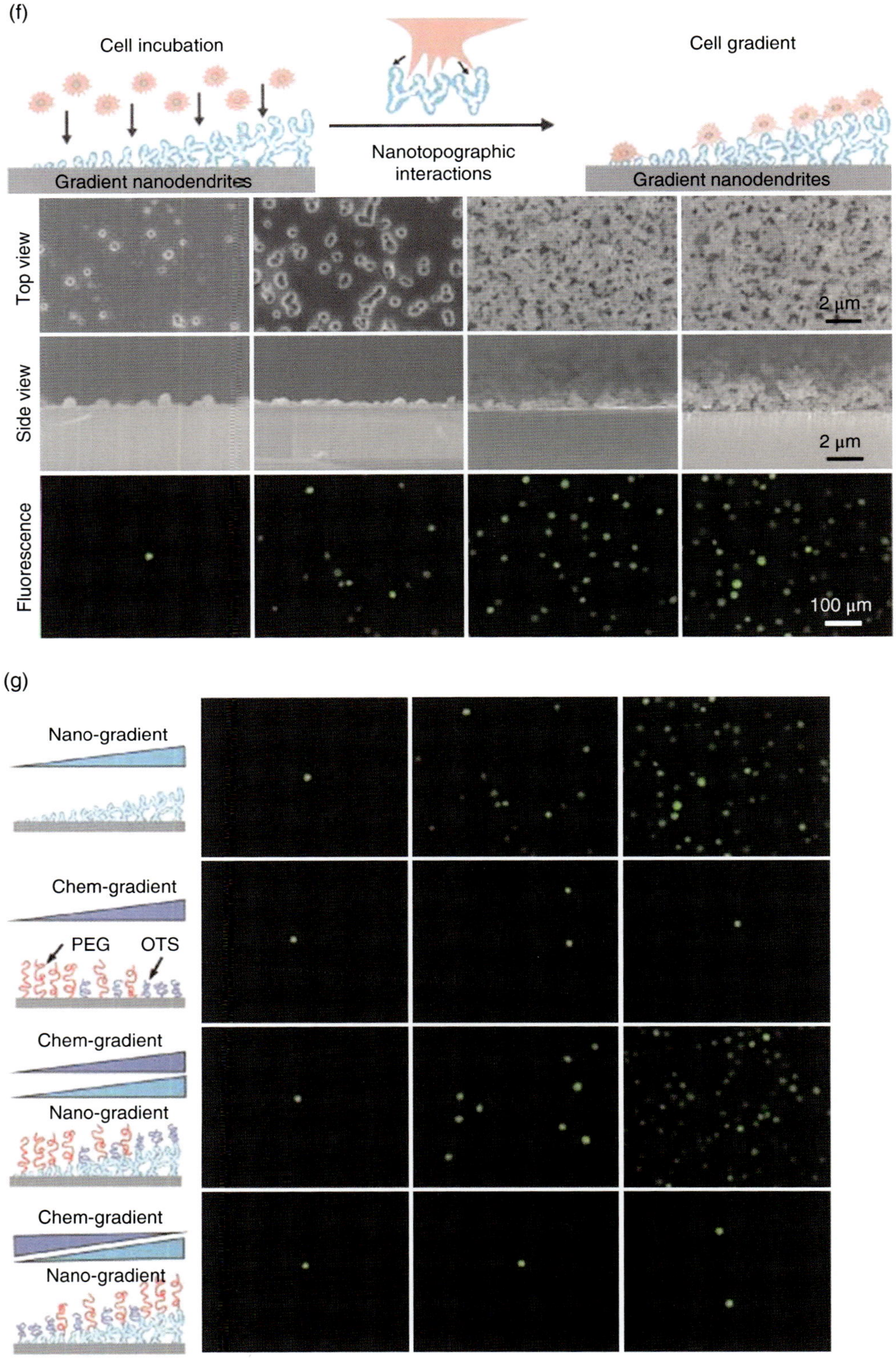

Figure 12.10 (Continued)

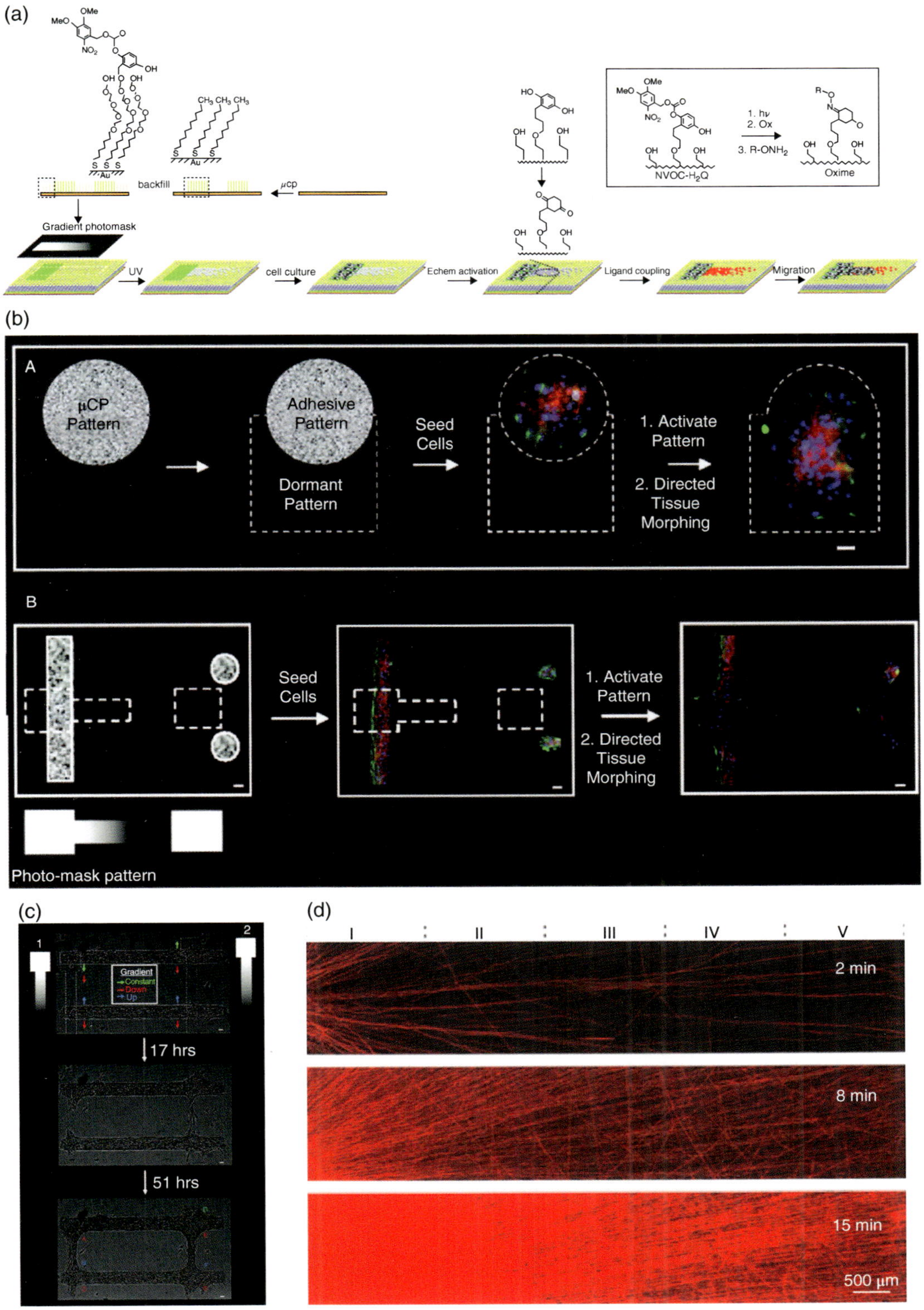

Figure 12.11 Gradient polymeric structures-regulated cell migration. (a) Schematic of the process flow for generation of dynamic ligand surface gradients. (b) Fluorescent micrographs show cells at μCP regions migrated toward unveiled cell-adhesive photo-patterned regions. (c) Time-lapse micrographs showing the different migration rates of patterned cells up or down the RGD peptide gradients. (d) Different gradients of radially aligned fibers with varied collection time of electrospinning. Shallow (2 min), steep (8 min) and moderate (15 min) continuous gradient were generated with the increasing collection time. (e) Fluorescence micrographs showing the migration of NSCs on the CBD-SDF1α immobilized radially aligned fibers. (f) NSCs distribution on radially aligned fibers immobilized with CBD-SDF1α. NAT-SDF1α, PBS, and randomly oriented fibers immobilized with CBD-SDF1α were as the control groups. The dashed lines in all images indicate the border of NSCs seeding. (g) Schematic of the countercurrent density gradients of PHEMA brushes and YIGSR for selective guidance of ECs migration. (h) Migration traces of ECs and SMCs on countercurrent density gradient of PHEMA/YIGSR, single YIGSR gradient and PHEMA gradient. (i) Optical images of ECs and SMCs sheets cultured on PHEMA/YIGSR, YIGSR and PHEMA density gradient surfaces, respectively. *Sources:* (a)–(c) reproduced from [8] with permission, copyright 2011, American Chemical Society. (d)–(f) reproduced from [101] with permission, copyright 2016, John Wiley & Sons. (h) and (i) reproduced from [102] with permission, copyright 2014, American Chemical Society.

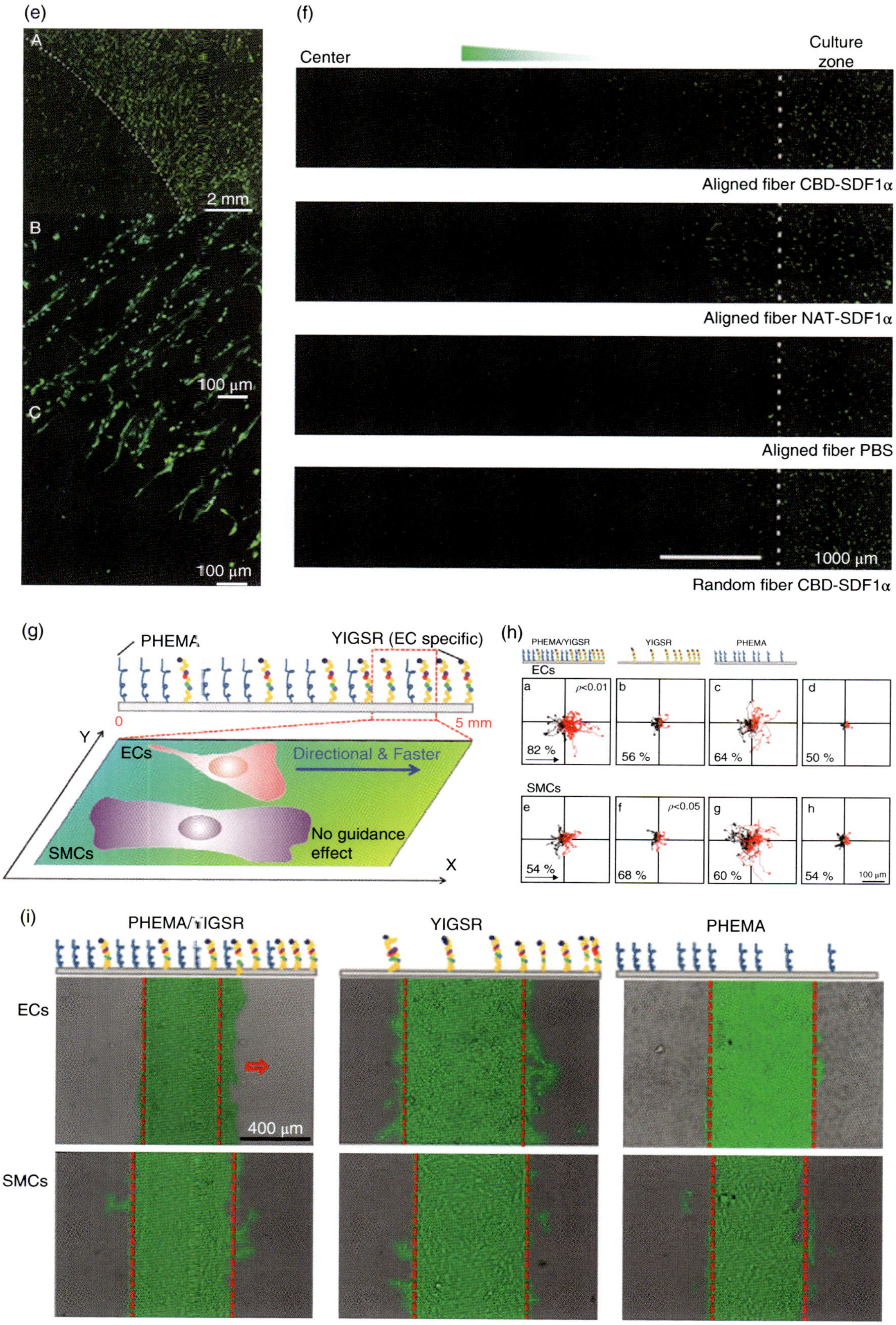

Figure 12.11 (Continued)

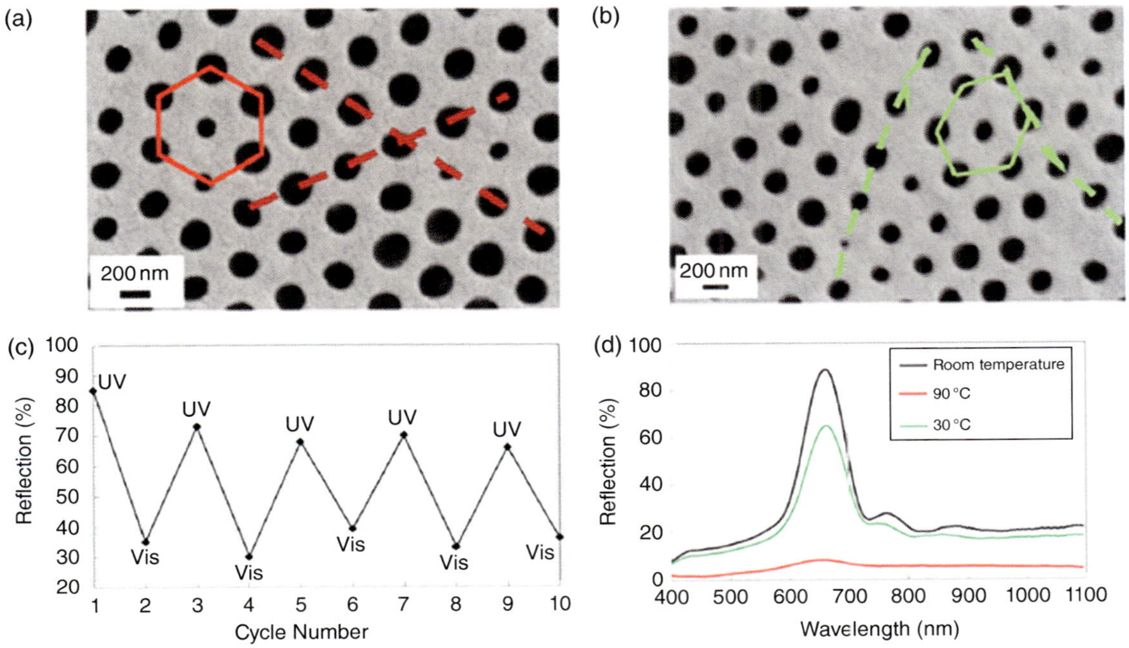

Figure 14.30 SEM images of the inverse opal film (a) before and (b) after UV light irradiation. (c) The maximum intensity of the reflection peak of the inverse opal film under alternating irradiation with UV and visible light at different times. (d) Reflection spectra of the inverse opal film as a function of temperature. *Source:* Reprinted with permission from [21].

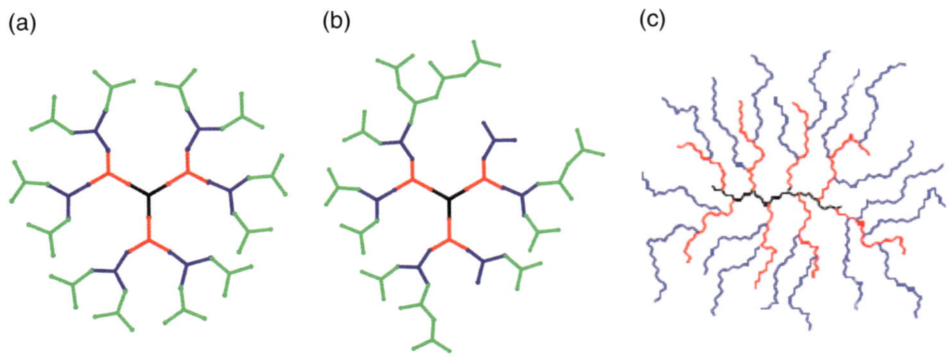

Figure 16.1 Schematic representation of dendritic polymers: (a) dendrimer; (b) hyperbranched polymer; and (c) dendrigraft polymer.

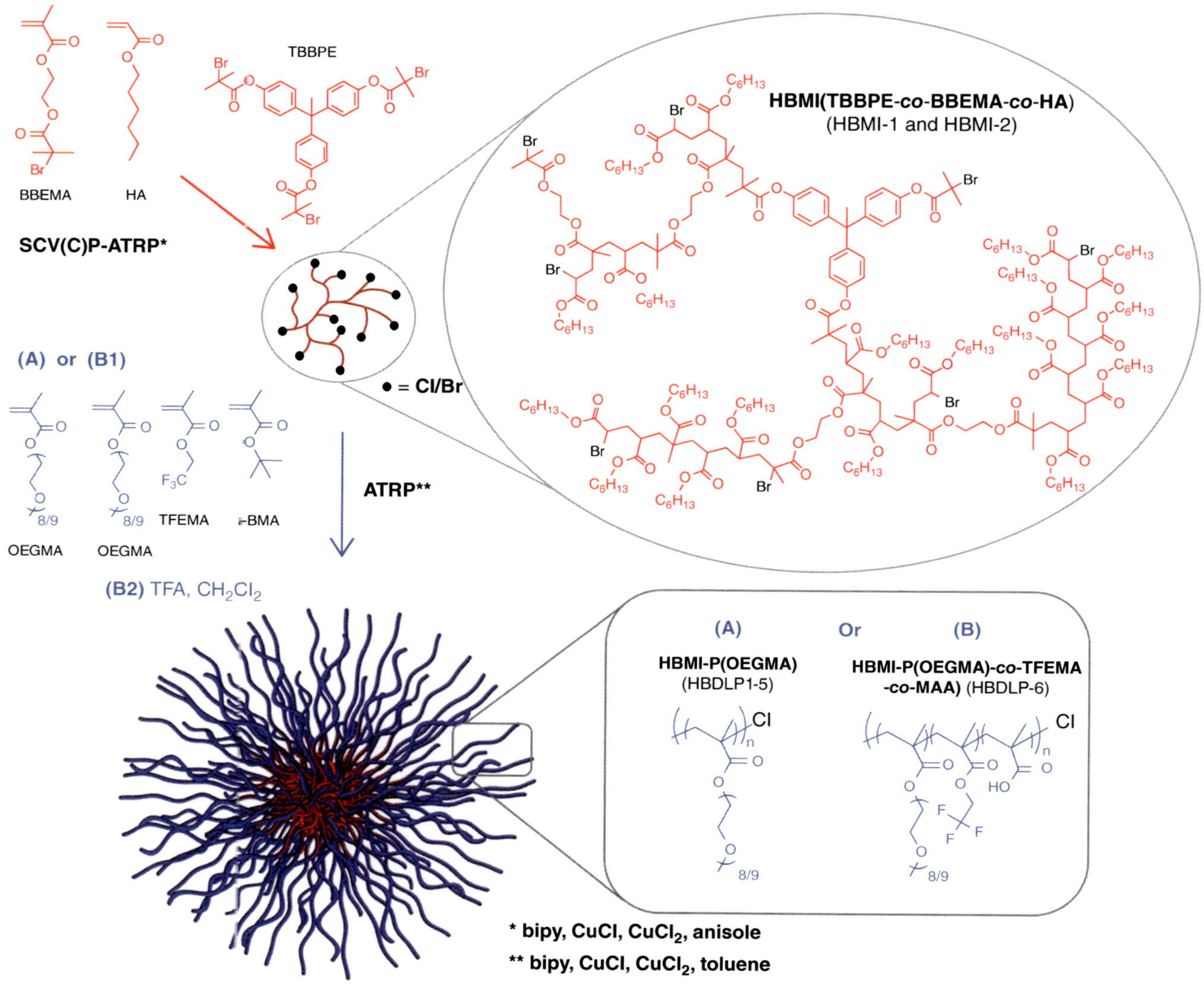

Figure 16.11 Synthesis of hyperbranched dendritic-linear polymers by sequential SCV(C)P-ATRP and ATRP techniques. *Source:* Reprinted with permission from [60]. Copyright 2014 American Chemical Society.

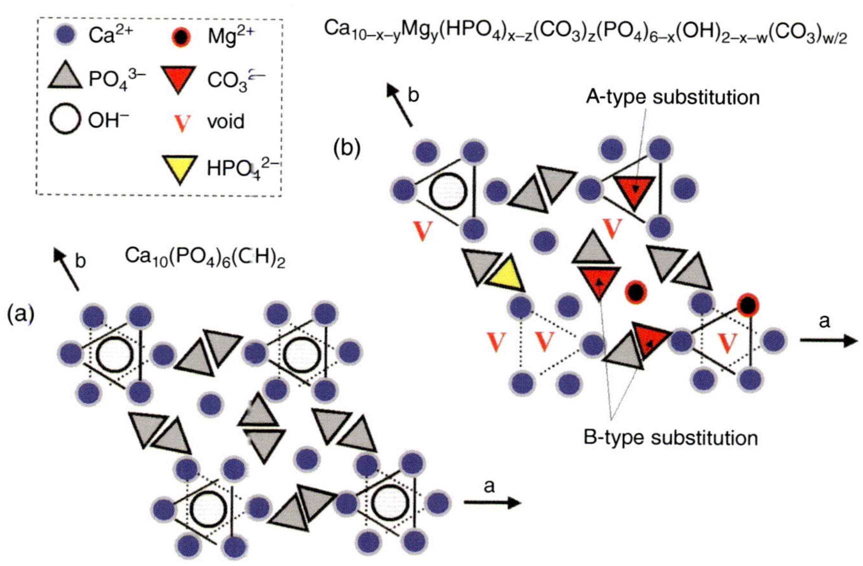

Figure 17.3 Projection of the (001) plane of (a) stoichiometric hydroxy apatite and (b) ionic substitutions in bone apatite or biomimetic apatite.

Appendix: Wormlike Chain Parameters for Polysaccharide Carbamate Derivatives

Table 9.2 Helix pitch per residue *h* and the Kuhn segment length λ^{-1} for polysaccharide carbamate derivatives.

Polymer	Solvent	T/°C	h / nm	λ^{-1} / nm	Ref
Amylose derivatives					
ATEC	THF	25	0.36	33	[71]
	D-ethyl lactate	25	0.35	27	[71]
	L-ethyl lactate	25	0.38	15	[71]
	2-methoxyethanol	25	0.38	14	[71]
	methanol	25	0.38	9	[71]
ATBC	THF	25	0.26	75	[72]
	D-ethyl lactate	25	0.26	49	[76]
	THF / methanol ($\phi_m = 0.25$)	25	0.26	45	[72]
	2-butanol	45	0.25	40	[73]
	DL-ethyl lactate	25	0.26	38	[76]
	2-ethoxyethanol	25	0.25	38	[73]
	THF / methanol ($\phi_m = 0.50$)	25	0.27	35	[72]
	L-ethyl lactate	25	0.26	32	[76]
	1-propanol	40	0.28	25	[73]
	2-propanol	35	0.29	20	[73]
	THF / methanol ($\phi_m = 0.50$)	25	0.30	17	[72]
	methanol	25	0.32	11	[72]
ATHC	THF	25	0.29	75	[71]
	1-propanol	25	0.39	30	[71]
ATPC	4-methyl-2-pentanone	25	0.42	24	[51]
	4-methyl-2-pentanone	58	0.39	17	[99]
	1,4-dioxane	25	0.34	22	[50]
	1,4-dioxane	58	0.34	22	[99]
	2-butanone	25	0.39	18	[52]
	ethyl acetate	33	0.39	17	[51]
	2-ethoxyethanol	25	0.32	16	[50]
	methyl acetate	25	0.37	15	[51]
AAPC	1,4-dioxane	25	0.34	21	[70]
	2-butanone	25	0.36	12	[70]
	2-ethoxyethanol	25	0.35	11	[70]
ADMPC	4-methyl-2-pentanone	25	0.36	73	[52]
	2-butanone	25	0.38	41	[52]
	methyl acetate	25	0.36	22	[52]
Cellulose derivatives					
CTPC	tetrahydrofuran	25	0.50	21	[63]
	1-methyl-2-pyrrolidone	25	0.49	16	[78]
CDMPC	1-methyl-2-pyrrolidone	25	0.52	16	[79]
Curdlan derivative					
CdTPC	THF	25	0.39	57	[66]
	4-methyl-2-pentanone	25	0.51	17	[66]

References

1 Song, E.H. Shang, J., Ratner, D.M. (2012) in *Polymer Science: A Comprehensive Reference* (Eds. Krzysztof, Matyjaszewski, Martin, Möller) Elsevier, Amsterdam, pp. 137.
2 Burchard, W. (2008) in *Soft Matter Characterization*, Redouane, B., Pecora, R. (eds.), Ch. 9, Springer, Dordrecht, pp. 463.
3 Burchard, W., Husemann, E. (1961) *Makromol. Chem.*, **44–6**, 358.
4 Burchard, W. (1964) *Z Phys Chem Neue Fol*, **42**, 293.
5 Cowie,J. M. G. (1965) *Biopolymers*, **3**, 69.
6 Burchard,W. (1965) *Makromol. Chem.*, **88**, 11.
7 Shanbhag, V.P. (1968) *Arkiv for Kemi*, **29**, 1.
8 Shanbhag, V.P. (1968) *Arkiv for Kemi*, **29**, 139.
9 Shanbhag, V.P. (1968) *Arkiv for Kemi*, **29**, 33.
10 Shanbhag, V.P., Ohman, J. (1968) *Arkiv for Kemi*, **29**, 163.
11 Janeschi, H., Burchard, W. (1968) *J Polym. Sci A2*, **6**, 1953.
12 Bittiger, H., Keilich, G. (1969) *Biopolymers*, **7**, 539.
13 Burchard, W. (1971) *Br. Polym. J.*, **3**, 214.
14 Guthrie, J.T., Huglin M.B., Richards, R.W. et al. (1975), *Eur. Polym. J.*, **11**, 527.
15 Noordermeer, J.W.M., Daryanani, R., Janeschitz-Kriegl, H. (1975) *Polymer*, **16**, 359.
16 Gupta, A.K., Marchal, E., Burchard, W. (1975) *Macromolecules*, **8**, 843.
17 Sutter, W., Burchard, W. (1978) *Makromol. Chem. Macromol. Chem. Phys.*, **179**, 1961.
18 Banks, W., Greenwood, C.T., Sloss, J. (1970) *Makromol. Chem.*, **140**, 109.
19 Banks, W., Greenwood, C.T., Sloss, J. (1970) *Makromol. Chem.*, **140**, 119.
20 Banks, W., Greenwood, C.T. (1971) *Makromol. Chem.*, **144**, 135
21 Banks, W., Greenwood, C.T., Sloss, J. (1971) *Eur. Polym. J.*, **7**, 879.
22 Gupta, A.K., Marchal, E., Burchard, W., Pfannemuller, B. (1979) *Macromol.*, **12**, 281.
23 Fried, F., Searby, G.M., Seurin-Vellutini, M.J. et al. (1982) *Polymer*, **23**, 1755.
24 Pfannemuller, B., Schmidt, M., Ziegast, G., Matsuo, K. (1984) *Macromol.*, **17**, 710.
25 Daňhelka, J., Netopilíak, M., Bohdanecký, M. (1987) *J. Polym. Sci., Part B: Polym. Phys.*, **25**, 1801.
26 Muroga, Y., Hayashi, K., Fukunaga, M. et al. (2006) *Biophys. Chem.*, **121**, 96.
27 Kratky, O., Porod, G. (1949) *Recl. Trav. Chim. Pays-Bas*, **68**, 1106.
28 McCormick, C.L., Callais, P.A., Hutchinson, B.H. (1985) *Macromolecules*, **18**, 2394.
29 Valtasaari, L. (1971) *Makromol. Chem.*, **150**, 117.
30 Kamide, K., Saito, M., Suzuki, H. (1983) *Makromol. Chem. Rapid Commun.*, **4**, 33.
31 Seger, B., Aberle, T., Burchard, W. (1996) *Carbohydr. Polym.*, **31**, 105.
32 Saalwachter, K., Burchard, W., Klufers, P. et al. (2000) *Macromolecules*, **33**, 4094.
33 Nakanishi, Y., Norisuye, T., Teramoto, A., Kitamura, S. (1993) *Macromolecules*, **26**, 4220.
34 Norisuye, T. (1994), *Polym. J.*, **26**, 1303.
35 Okamoto, Y., Kawashima, M., Hatada, K. (1984) *J. Am. Chem. Soc.*, **106**, 5357.
36 Shen, J., Ikai, T., Okamoto, Y. (2014) *J. Chromatogr. A*, **1363**, 51.
37 Okamoto, Y., Aburatani, R., Fukumoto, T., Hatada, K. (1987) *Chem. Lett.*, **16**, 1857.
38 Shen, J., Okamoto, Y. (2012) In *Comprehensive Chirality*, Carreira, E.M., Yamamoto, H. (eds.), Elsevier, Amsterdam.
39 Okamoto, Y. (2009) *J. Polym. Sci. Part a-Polymer Chemistry*, **47**, 1731.
40 Ikai, T., Okamoto, Y. (2009) *Chem. Rev.*, **109**, 6077.
41 Yamamoto, C., Okamoto, Y. (2004) *Bull. Chem. Soc. Jpn.*, **77**, 227.
42 Yamamoto, C., Yashima, E., Okamoto, Y. (2002) *J. Am. Chem. Soc.*, **124**, 12583.
43 Kasat, R.B., Zvinevich, Y., Hillhouse, H.W. et al. (2006) *J. Phys. Chem. B*, **110**, 14114.
44 Takaha, T., Yanase,M., Takata, H. et al. (1996) *J. Biol. Chem.*, **271**, 2902.
45 Schappacher, M., Deffieux, A. (2008) *Science*, **319**, 1512.
46 Yamakawa, H., Fujii, M. (1976) *J. Chem. Phys.*, **64**, 5222.
47 Yamakawa, H. (1999) *Polym. J.*, **31**, 109.
48 Yamakawa, H., Yoshizaki, T. (2016) *Helical Wormlike Chains in Polymer Solutions* 2nd ed., Springer, Berlin.
49 Benoit, H., Doty, P. (1953) *J. Phys. Chem.*, **57**, 958.
50 Terao, K., Fujii, T., Tsuda, M. et al. (2009) *Polym. J.*, **41**, 201.
51 Fujii, T., Terao, K., Tsuda, M. et al. (2009) *Biopolymers*, **91**, 729.
52 Tsuda, M., Terao, K., Nakamura, Y. et al. (2010) *Macromol.*, **43**, 5779.
53 Yamakawa, H., Stockmayer, W.H. (1972) *J. Chem. Phys.*, **57**, 2843
54 Shimada, J., Yamakawa, H. (1986) *J. Chem. Phys.*, **85**, 591.
55 Domb, C., Barrett, A.J. (1976) *Polym.*, **17**, 179.
56 Yamakawa, H., Fujii, M. (1974) *Macromol.*, **7**, 128.
57 Yamakawa, H., Yoshizaki, T. (1980) *Macromol.*, **13**, 633.
58 Yoshizaki, T., Nitta, I., Yamakawa, H. (1988) *Macromol.*, **21**, 165.
59 Barrett, A.J. (1984) *Macromol.*, **17**, 1566.

60 Nakamura, Y., Norisuye, T. (2008) In *Soft Matter Characterization*, Borsali, R., Pecora, R. (eds.,) Ch. 5, Springer, Dordrecht.
61 Nakamura, Y., Norisuye, T. (2004) *J. Polym. Sci., Part. B: Polym. Phys.*, **42**, 1398.
62 Hickl, P., Ballauff, M., Scherf, U. et al. (1997) *Macromol.*, **30**, 273.
63 Kasabo, F., Kanematsu, T., Nakagawa, T. et al. (2000) *Macromol.*, **33**, 2748.
64 Zimm, B.H. (1948) *J. Chem. Phys.*, **16**, 1099.
65 Berry, G.C. (1966) *J. Chem. Phys.*, **44**, 4550.
66 Ochiai, T., Terao, K., Nakamura, Y. et al. (2012) *Polymer*, **53**, 3946.
67 Shimada, J., Yamakawa, H. (1988) *Biopolymers*, **27**, 657.
68 Fujii, M., Yamakawa, H. (1975)*Macromo.*, **8**, 792.
69 Tsubouchi, R., Ida, D., Yoshizaki, T., Yamakawa, H. (2014) *Macromol.*, **47**, 1449.
70 Tsuda, M., Terao, K., Kitamura, S., Sato, T. (2012) *Biopolymers*, **97**, 1010.
71 Terao, K,. Maeda, F., Oyamada, K. et al. (2012) *J. Phys. Chem. B*, **116**, 12714.
72 Terao, K., Murashima, M., Sano,Y. et al. (2010) *Macromol.*, **43**, 1061.
73 Sano, Y,. Terao, K., Arakawa, S. et al,. (2010) *Polymer*, **51**, 4243.
74 Teramoto, A.(2001) *Prog. Polym. Sci.*, **26**, 667.
75 Chisaka, S., Norisuye, T. (2001) *J. Polym. Sci., Part. B: Polym. Phys.*, **39**, 2071.
76 Arakawa, S., Terao, K., Kitamura, S., Sato, T. (2012) *Polym. Chem.*, **3**, 472.
77 Schurig, V., Zhu, J., Muschalek, V. (1993) *Chromatographia*, **35**, 237.
78 Norisuye, T., Tsuboi, A., Sato, T., Teramoto, A. (1997) *Macromol Symp.*, **120**, 65.
79 Tsuboi,A., Norisuye, T., Teramoto, A. (1996) *Macromolecules*, **29**, 3597.
80 Gray, D,G., Harkness, B.R. (1994) In *Liquid Crystalline and Mesomorphic Polymers Partially Ordered Systems*, Shibaev, V.P., Lam, L. (eds.) Ch. 9, Springer: New York.
81 Sato, T. Teramoto, A. (1996) *Adv Polym Sci*, **126**, 85.
82 Sato, T., Shimizu, T.,Kasabo, F., Teramoto, A. (2003) *Macromol.*, **36**, 2939.
83 Kuse,Y., Asahina, D., Nishio, Y. (2009) *Biomacromol.*, **10**, 166.
84 Oyamada, K.,Terao, K., Suwa, M., et al. (2013) *Macromol.*, **46**, 4589.
85 Yu,S.M., Conticello, V.P., Zhang, G. et al. (1997) *Nature*, **389**, 167.
86 Okoshi, K., Kamee, H., Suzaki, G. et al. (2002) *Macromol*, **35**, 4556.
87 Onouchi, H., Okoshi, K., Kajitani, T. et al. (2008) *J. Am. Chem. Soc.*, **130**, 229.
88 Roovers, J., Toporowski, P.M. (1983) *Macromol.*, **16**, 843.
89 Lee, H.C., Lee, H., Lee, W. et al. (2000) *Macromol.*, **33**, 8119.
90 Lahasky, S.H., Serem, W.K., Guo, L. et al. (2011) *Macromol.*, **44**, 9063.
91 Zhang, K., Tew, G. N. (2012) *ACS Macro Lett.*, **1**, 574.
92 Wintermantel, M., Schmidt, M., Tsukahara, Y. et al. (1994) *Macromol. Rapid Comm.*, **15**, 279.
93 Terao, K., Nakamura, Y., Norisuye, T. (1999) *Macromol.*, **32**, 711.
94 Nakata, Y., Amitani, K., Norisuye, T., Kitamura, S. (2003) *Biopolym.*, **69**, 508.
95 Terao, K. Asano, N., Kitamura, S., Sato, T. (2012) *ACS Macro Lett.*, **1**, 1291.
96 Terao, K., Shigeuchi, K., Oyamada, K. et al. (2013) *Macromol.*, **46**, 5355.
97 Ida, D., Nakatomi, D,. Yoshizaki, T. (2010) *Polym. J.*, **42**, 735.
98 des Cloizeaux, J. (1981) *J. Phys. Lett.*, **42**, L433.
99 Asano, N., Kitamura, S., Terao, K. (2013) *J. Phys. Chem. B*, **117**, 9576.

10

Silk Proteins: A Natural Resource for Biomaterials

Lallepak Lamboni[1], Tiatou Souho[1,2], Amarachi Rosemary Osi[3], and Guang Yang[1,]*

[1] *Department of Biomedical Engineering, College of Life Science and Technology, Huazhong University of Science and Technology, Wuhan, China*
[2] *University of Kara, Kara, Togo*
[3] *Ningbo Institute of Material Technology and Engineering, Chinese Academy of Sciences, Ningbo, China*

10.1 Introduction

With the advent of tissue engineering, advances in biomedical research are aiming at the development of accurate therapeutic measures, requiring the use of functional materials that can influence biological systems [1]. Increasing progress has been made in the use of biomaterials that potentially not only can repair or replace damaged tissues, but also can serve as implantable supports that induce or accelerate tissue regeneration [2]. In fact, the aim is to generate or stimulate the formation of a defined tissue in a certain location through selection and manipulation of cells, matrices, and biological stimuli [3, 4]. In this context, the choice of biomaterials in biodevices is critical, because they must imitate the architecture of the natural tissue and consider the molecular, structural, and biological compatibility in order to facilitate tissue development [2, 5]. So far, various materials fabricated from natural or synthetic polymers have been tested for tissue engineering purposes. Despite their valuable tailorability in terms of physico-chemical properties, application of synthetic materials tends to be limited by determinant factors such as bioresorbabilty, biocompatibility, and biodegradability [1]. Consequently, natural materials with inherent biocompatibility and potential bio-responsiveness have emerged as desirable candidates for biomedical engineering applications [6]. Among these naturally occurring compounds, silkworm silk distinguishes itself as a versatile and virtuously rich composite that has gained application in many aspects of tissue engineering. In fact, it is made up of two main proteins which are individually applicable as biomaterial, namely silk fibroin (SF) and silk sericin (SS) representing a structural polymer and a functional macromolecule, respectively.

Initially acknowledged for its application in the textile industry but also in medical sciences as a suture material [7, 8], SF has enjoyed many years of research [9] and has recently attracted a lot of attention in biomedical engineering in the quest for improved materials that overcome the limitations of the numerous synthetic materials thus far fabricated [1, 6, 8, 10]. In fact, aside from its irrefutable biocompatibility, SF allows for a versatile processability and a broad application [1], while exhibiting outstanding and tailorable mechanical properties [11] that qualify it as one of the most interesting biopolymers for current tissue engineering strategies. Conversely, SS, the second chief protein composing silk, has for many years been underused, considered a byproduct of the silk industry [12]. However, since the discovery of its various bioactivities in the last few decades, SS has become an important and valuable resource for current tissue engineering research [12, 13] that has increasingly shifted toward the use of biofunctional materials capable of interacting with the biological milieu via biomolecular cues that guide cell behavior and support tissue repair [1]. Therefore, it is increasingly being applied as biomaterial not only for tissue engineering, but also for drug delivery [6]. In the current chapter, silkworm silk proteins are presented. A return to basics is proposed by considering the biosynthesis and production of these macromolecules, before discussing their biomedical properties. Focus is then given to their use in biomaterials, with regard to the respective processing methods.

* Email: yang_sunny@yahoo.com

Bioinspired Materials Science and Engineering, First Edition. Edited by Guang Yang, Lin Xiao, and Lallepak Lamboni.

10.2 Bio-synthesis of Silk Proteins

Silk is a natural fiber of protein polymer produced by a variety of animals and serving in nature different purposes such as pupae protection, web weaving, nest strengthening, etc. [14]. Within the group of Lepidoptera, silk is produced by several species and is spun into cocoons that provide a protective environment for pupae metamorphosis into adult insects [15]. Due to the unique characteristics of silk fibers, cocoons are used in various fields including the textile industry and biomaterials engineering [11, 16]. Particularly, cocoons produced by *Bombyx mori* larvae have been used for thousands of years in the textile industry and their silk is the most abundantly available and most exploited in the world [16, 17].

Bombyx mori silk is composed of several proteins assembled into two main components: fibroin fiber and sericin gum. These protein macromolecules are arranged into fibers, each consisting of a pair of fibroin filaments wrapped with successive layers of sericin which glue the filaments together. The synthesis of silk proteins occurs in caterpillars' mature silk glands during the fifth instars stage At this stage, silkworm larvae produce silk and spin it into cocoons that surround their whole body [18–20].

10.2.1 Silkworm Silk Glands

In silkworms, the silk gland is a long tubular structure lying to the ventral side from the caudal segment to the head. Each silkworm possesses two glands that join at the spinneret from where the silk is exuded [18]. Each gland produces one strand composed of fibroin filament wrapped in sericin gum and both strands assemble into a silk fiber composed of two threads fixed one to the other by sericin. The glands present three distinct regions that differ in diameter and length: the posterior silk gland region (PSG), the middle silk gland region (MSG), and the anterior silk gland region (ASG) [21, 22].

The PSG is the tiniest portion of the gland and is found in the caudal part of the caterpillar. It is highly coiled and measures about 15 cm. The glandular cells in this section synthesize the silk core proteins that they secrete in the lumen [23]. The MSG is the most prominent section of the silk gland. It follows the PSG and presents two bends that divide the region into three distinct sub-regions (the anterior-, the middle-, and the posterior-MSG). The whole MSG measures 7 cm, and its glandular cells are responsible for sericin production. The ASG is the shortest section of the entire silk gland (2 cm). It has no known secretory activity but serves as a canal to transport silk filaments to the spinneret [18, 20]. Right before the spinneret, the ASG is linked to the Filippi's gland, which contributes by lubricating the passage of silk or providing a favorable physiological environment for the silk fiber formation [24].

10.2.2 Regulation of Silk Proteins Synthesis

After hatching out from eggs, silkworms store nutrients by eating leaves throughout their larval evolution until the fifth instars stage. At this stage, silk glands make up about 25% of their body weight and the silk protein synthesis pathways are activated. Glandular cells then begin the synthesis of the proteins using stored nutrients. The protein synthesis is followed by protein secretion in the lumen of the gland allowing the start of silk spinning [25]. Most amino acids derive from the stored nutrients, except for glycine, alanine, serine, and tyrosine (abundant in silk proteins) that are synthesized in silk glands by transamination [18, 26].

The various components of silk (fibroin chains, sericin, and P25 glycoprotein) are produced by different glandular cells, lining along the silk gland wall's inner face. The cells expressing fibroins and P25 are present exclusively in the PSG whereas those producing different types of sericin are present in the MSG. This leads to a specific spatio-temporal protein expression that suggests a complex regulatory process governing silk production [27]. Proteomic studies have revealed the critical role of anterior-MSG in the silk spinning processes. In fact, this section is involved in the production of the initial silk fibers and expresses many protease inhibitors that protect the fiber from enzymatic degradation [28, 29]. Cells in this subsection also produce high levels of hydroxysteroid dehydrogenase, which is involved in insect hormone metabolism, and thus in the growth of the caterpillar and the development of the silk glands [29].

The elements involved in the translation processes are also involved in the regulation of the silk production. Protein Ras1 induces an increase in cell size and stimulates the formation of cellular organelles, such as the ribosome, involved in protein synthesis. Overall, Ras1 acts by increasing the amount of produced silk proteins. This regulatory effect is particularly observed in the middle-MSG [30]. In addition, other regulatory elements also control the secretion of produced proteins throughout their travel from the endoplasmic reticulum to the Golgi and from the Golgi to the gland lumen [29]. Many other regulatory mechanisms have been described. These include *cis* regulatory elements in the upstream region of silk protein genes, or in the intronic regions of some genes that are recognized by specific regulators for gene transcription enhancement [31–34].

Other regulatory mechanisms involve epigenetic elements such as microRNAs and metabolic elements such as amino acids and hormones [35–38].

10.2.3 Synthesis of Fibroin

The core water-insoluble filaments of silk fibers are composed of heavy-chain fibroin (H-fib), light-chain fibroin (L-fib), and glycoprotein P25. The three proteins are synthesized in the same cells in the PSG. After translation, the three proteins undergo intracellular modifications before their combination and secretion in the PSG lumen as micelles. The released protein micelles are combinations of these three proteins in the ratio 6/6/1 (H-fib/L-fib/P25) [39].

H-fib is encoded by *Hfib*, a gene with only one intron of approximately 1 kb. This intron possesses AT-rich sequences that bind the regulatory factors responsible for the *Hfib* transcriptional regulation [40]. Moreover, the promoter region of this gene (sequence −32 to −238) also presents regulatory sequences involved in transcriptional regulation of the gene [32, 34, 38]. Under favorable conditions (availability of nutrients, proper larval development, etc.), *Hfib* is transcribed and translated into a protein that will be glycosylated and transported from the rough endoplasmic reticulum to the Golgi apparatus where it will undergo several modifications, including phosphorylation, cleavage, and formation of intra-molecular disulfide bonds [41–43].

The light-chain fibroin is coded by a long gene (13.5 kb) composed of seven exons and long introns (the first intron represents about 60% of the whole gene) [44]. The promoter region of this gene shows similarities with the promoter region of *Hfib*, showing that these two proteins are regulated by same factors so that they are synthesized in a synchronized way at the same periods under the same conditions [45, 46]. The concomitant synthesis of these two proteins allows their association by an intermolecular disulfide bond (involves Cys-c20 of H-fib and Cys-172 of L-fib) [42, 47]. This dimerization is essential for the proper secretion of fibroin. As a matter of fact, mutant *Bombyx mori* producing a defective L-fib fails to secrete enough amounts of H-fib leading to naked pupa [48].

On the other hand, P25 is a glycoprotein with many N-linked oligosaccharide chains synthesized in PSG cells in the same time frame as H-fib and L-fib [49]. Its peptide sequence indicates the presence of eight Cys residues which are implicated in the glycoprotein compaction through intra-molecular disulfide bonds. The sequence also shows the presence of three sites of N-glycosylation implicated in the protein maturation. It possesses regions with hydrophobic residues that interact with Gly-Ala sequence repeats in the complex formed by H-fib and L-fib [50]. The H-fib-L-fib-P25 complex is excreted in the lumen of PSG as micelles that flow from the PSG to the MSG.

10.2.4 Synthesis of Sericin

The silk gum cementing fibroin filaments in silk cocoon are made of sericin proteins accounting for about 25–30% of silk [51]. These proteins are expressed from three different genes (*Ser1*, *Ser2* and *Ser3*) that generate many mRNAs by alternative splicing, leading to different proteins expressed from a single gene. Sericins encoded by *Ser1* and *Ser3* genes are the most abundant in silk cocoons, forming consecutive layers around fibroin filaments. These sericins are called sericins P, M, and A, and are produced in the posterior, middle, and anterior sections of the MSG [52, 53], respectively. *Ser1* (the gene) is made up of nine exons which are alternatively spliced in at least four different mRNAs, that are translated into four different proteins, among which are sericin P and M [54]. *Ser3* seems to generate a unique protein which is sericin A. It is expressed in the anterior section of MSG and forms the outermost layer of the cocoon [55].

The gene *Ser2* consists of 13 exons and generates by alternative splicing two different mRNAs that are translated into two proteins of 120 and 230 kDa. These proteins are the first sericin to be expressed in the larvae [56]. In fact, their secretion occurs in feeding larvae and continues in post-feeding wandering larvae. While the highest level of *Ser2* products is observed in the anterior-MSG section of feeding larvae, none of the *Ser2* mRNAs is observed in post-feeding stages [56]. It is accepted that *Ser2* is expressed earlier in larvae along the whole MSG and then the expression is restricted to the anterior-MSG in fifth instars before stopping in the post feeding stages [57]. Sericin proteins encoded by this gene are highly adhesive and serve only to fix the first cocoon scaffold to the supporting substrate e.g. tree branches before the cocoon spinning [58]. They are therefore absent from the harvested cocoons [55, 58].

10.2.5 Silk Filament Assembly

All sericins are expressed in the glandular cells and secreted in the lumen of the MSG section where they divert the fibroin gel flowing from the PSG toward the ASG. The flowing gel of fibroin micelles is first covered by sericin P present in the lumen of the posterior-MSG. When the gel covered by sericin P reaches the middle-MSG, a second layer of sericin (sericine M) is added onto the forming filament. In the anterior-MSG, the filament is finally coated with sericin A (Figure 10.1). At the end of the anterior-MSG, the main elements of silk

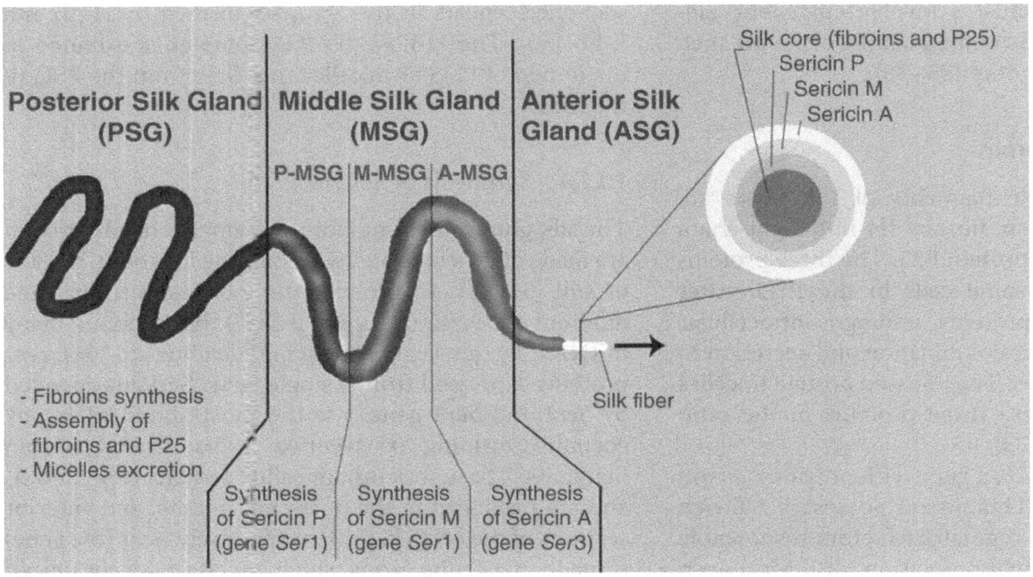

Figure 10.1 Schematic illustration of *Bombyx mori's* silk gland and silk structure. (*See insert for color representation of the figure.*)

filaments are assembled but the filament exists as a jelly substance flowing through the ASG. The filament then solidifies after excretion, when it comes into contact with the air [18, 23].

10.3 Extraction of Silk Proteins

10.3.1 Silk Degumming

Silk degumming is the process consisting in the removal of sericin from silk fibers to obtain the uncovered filaments of fibroin [53, 59]. The process has been known for thousands of years and is used in the textile industry to obtain fibroin fibers. The degumming process is based on the fact that sericin is soluble in water whereas fibroin is not. Traditionally, silk degumming is realized by boiling cocoons in water or aqueous solutions of degumming enhancers,such as salts, soap or enzymes [60, 61]. The process dissolves the sericin gum, yielding clean fibroin fibers that are subsequently spun. In the textile industry, the discarded wastewater obtained after silk degumming is loaded with sericin in concentrations as high as 30g/l [18]. All the extraction protocols of silk proteins commence by degumming, and this step is the determinant of the quality of the purified proteins. In fact, variations in degumming parameters (temperature, pH, pressure, time of incubation, nature of degumming chemicals) lead to the varied quality of both fibroin and sericin [62]. Various chemicals are used to enhance silk degumming. Among these chemicals, sodium carbonate shows the highest ratio of sericin dissolution, followed by sodium hydroxide, urea, and sodium chloride [62, 63].

Enzymatic degumming procedures have also been described. These are based on sericin digestion by proteases that do not digest fibroin, but present the inconvenience of not being adaptable when the degumming procedure is meant to extract sericin with a high molecular weight [64–67]. However, the use of these proteases presents the advantage in textile industry of producing less amounts of polluting wastewater [60].

10.3.2 Fibroin Regeneration

In biomaterials, the regeneration of fibroin is a process in which the degummed fibroin filaments are dissolved in aqueous or organic solvents to generate fibroin solutions that are subsequently processed to recover the protein by dialysis, in order to obtain and design it in desired morphologies [60, 62]. The most widely used methods of fibroin recovery are based on organic solvents systems ($CaCl_2$/ethanol/water, LiBr, LiSCN, NaOH and $Ca(NO_3)_2/MeOH/H_2O$) [47, 68].

The choice of a specific system depends on the expected quality of the to be recovered fibroin. Fibroin fiber contains proteins with molecular weights ranging from 25 to 350 kDa. The dissolution-recovery systems do not always preserve the integrity of proteins. Actually, fibroin recovery systems containing $CaCl_2$ lead to more degraded proteins with a high rate of polypeptides of molecular weight below 60 kDa [69]. However, among the systems of organic solvents, the one based on LiBr shows the highest levels of recovered fibroin with high molecular weight (above 220 kDa) [62, 69]. This system consists in the incubation of degummed fibers in 9.3 M LiBr solution at 60°C for 4 h for dissolution. The recovery of fibroin from the solution can then be realized by dialysis against deionized water for 2–3 days [47, 62, 64, 68, 70]. The recovered fibroin solution can then be lyophilized to obtain powder for later usage or be further processed for biomaterials design [71].

10.3.3 Sericin Recovery

Sericin is composed of proteins with molecular weight ranging from 10 to over 300 kDa. This protein fraction of silk is abundant in the degumming solution and its quality is related to the degumming method. Degumming methods based on protease activity and ultrasound, for example, lead to high sericin degradation and thus to very low amount of high molecular weight sericin [60, 72].

Several methods of sericin recovery have been described. These are based on combinations of techniques of protein precipitation, chemical coagulation, ultrafiltration or nanofiltration [19, 73]. The methods using ultrafiltration show the advantage of recovering high molecular weight sericin from the wastewater [74]. Other approaches have also been used to recover the dissolved sericin. These include water evaporation under moderate hot temperatures, and spray-drying [75, 76].

10.4 Structure and Physical Properties of Silk Proteins

10.4.1 Silk Fibroin

As the main component of the cocoon (70–75%) [12], silk fibroin (SF) is a structural polymer insoluble in water and containing large amounts of Gly, Ala, and Ser in a protein complex composed of 18 amino acids [1, 77]. With an improved ratio of Ala/Gly in non-mulberry silks [7], SF is primarily made up of hydrophobic blocks interspersed with a few hydrophilic domains [78]. Specifically, L-fibroin is hydrophilic, while H-fibroin is mostly composed of hydrophobic domains defined by hydrophobic interactions (mostly hydrogen bonding) that allow for the formation of anti-parallel β-sheets [7, 78]. These further bundle to form highly crystalline regions that endow SF with attractive properties, such as high strength and toughness, together with water stability [79]. The macromolecule also presents remarkable elasticity and extensibility, due to its ability to self-heal and self-assemble by reconstructing after rupture the numerous hydrogen bonds between the crystalline chains [80, 81]. Devoid of L-fibroin, non-mulberry silks present higher mechanical properties in comparison with *B. mori* silk [70] due to improved numbers of poly-Ala blocks generating higher β-sheets contents [7]. Hence, SF mostly constitutes of ordered silk II structures (β-sheets) with few random coils and α-helix (silk I) [82, 83], thereby representing one of the strongest natural fibers [11, 84].

To alter its physical properties, conformational changes may be prompted in the molecular structure of SF using chaotropic agents, which give rise to novel materials with tuned and tailored mechanical properties and related parameters [82, 83]. SF is thus regenerated into a metastable solution made of random coils [85], which is further processed to produce a material of targeted properties. Following regulation of the protein's crystallinity, the material's biodegradability may be controlled [9]. Moreover, the mechanical properties of SF-based materials depend on the molecular weight of the polymer [86], its concentration [87, 88], along with the porosity and pore size within the constructs [85].

Post-processing techniques often used to regulate the conformation and crystallinity of materials prepared with regenerated silk solutions involve the modulation of hydrogen bonds between protein chains, via control of the hydration state within the macromolecule. The classical method for the adjustment and stabilization of SF-based materials consists in the use of chemical cross-linking [83]. Also, alcohol, more frequently methanol but also ethanol, is one of the agents employed for this purpose [43, 89, 180]. This induces dehydration, forcing chains' interactions via hydrogen bonding, which lead to the transition of silk I (amorphous and random coiled) into silk II conformation with increased β-sheet content, crystallinity, and mechanical properties [90, 91]. Water annealing may also be used to modify the physical properties of regenerated silk-based materials [90, 92]. Using water vapor as the active agent induces structural changes by fostering molecular movements that facilitate chains' interactions by hydrogen bonding [9]. Process efficiency may further be improved following heating and applying pressure by autoclaving [90].

10.4.2 Silk Sericin

Sericin, the other protein that forms raw silk, is a mixture of polypeptides also made from 18 amino acids [93], mostly hydrophilic. Serine accounts for 33% of the latter in *B. mori* SS, while representing 19% of the amino acids in *A. mylitta* SS [6]. With glycine, histidine, glutamic acid, aspartic acid, threonine, and tyrosine constituting the other major amino acids in SS [94, 95], the latter exhibits highly polar side chains, rich in functional groups [96]. Additionally to this feature that makes SS hydrophilic, its hydro-solubility strongly lies in its amorphous structure mainly made up of random coils [97, 98], with few β-sheets and turns [99, 100] without α-helices [13]. Thus, due to its possessing low crystallinity, SS often forms fragile materials presenting poor mechanical properties. However, the β-sheets

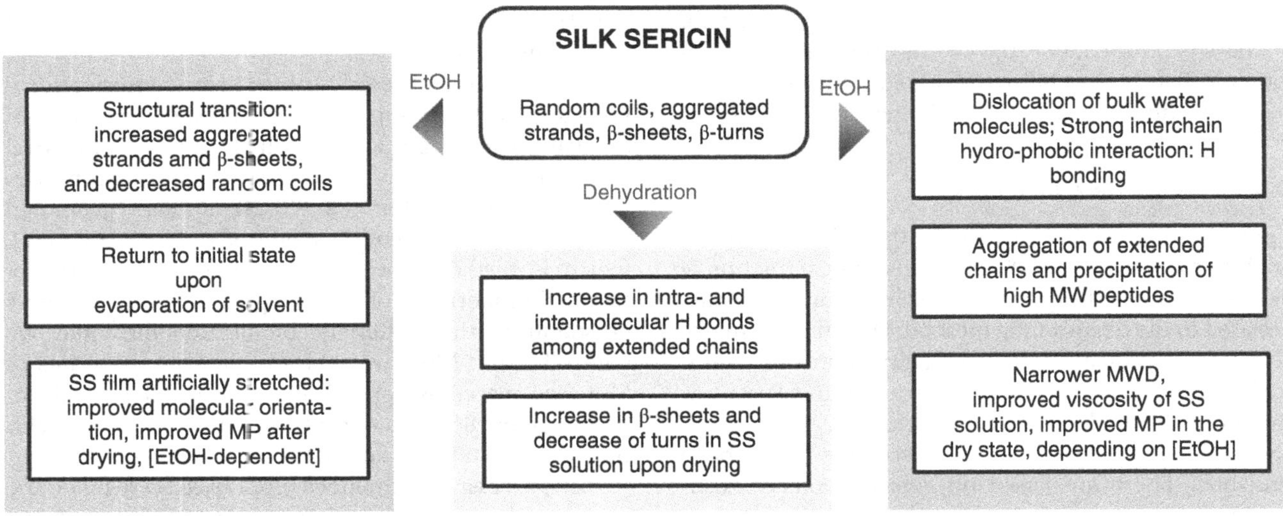

Figure 10.2 Physical properties of sericin (SS) upon ethanol treatment [6].

contents vary with the molecular weight distribution of SS protein, which in turn depends on the extraction conditions employed to obtain the protein [101]. In fact, harsh chemical or physical (temperature) treatments during extraction lead to protein denaturation and degradation that impair the physical stability of the SS-based materials.

Similarly to SF, various methods may be used for the improvement of SS-based materials properties. SS crystallinity may thus be increased by ethanol treatment that enhances the β-sheets content by dehydration and fractionation (Figure 10.2) [6]. An alternative is the use of cross-linking agents. For example, glutaraldehyde is a chemical cross-linker widely used to change the molecular structure in polymers [98]. Polymer blending is another method, that is rather popular, for the improvement of SS-based biomaterials [102]. These modifications are enabled by the numerous functional groups that constitute the side chains of SS and make it chemically highly reactive.

10.5 Properties of Silk Proteins in Biomedical Applications

10.5.1 Silk Fibroin

Certified by the US Food and Drug Administration (FDA) as a biomaterial [78. 103], silk fibroin is a biopolymer with biomedical virtues, as demonstrated by its accepted use in surgery as a suture material [86]. In tissue engineering it is favored for the fabrication of scaffolds [104], while being considered an engineering class material [9]. This attraction primarily is due to the outstanding mechanical properties of SF that are also tailorable, allowing for a versatile and wide range of applications. In fact, for the preparation of biomaterials, SF presents both as a structural and functional polymer [1]. On the other hand, SF displays excellent biocompatibility, with good hemocompatibility [82, 92], and no induction of inflammatory response or cytotoxicity [1, 105–107]. Furthermore, SF has shown the ability to support cell attachment and proliferation when used as a substrate for cell culture [83]. It also has the potential to enhance cell growth by increasing cell adhesion and proliferation as concluded in several studies [82, 87, 88]. Greater cell growth may be observed with non-mulberry SF as a result of the Arg-Gly-Asp sequences (present in the protein) which interact with cells via the integrin system, promoting cell attachment [7, 70, 108].

10.5.2 Biomedical Uses of Silk Sericin

SS protein is a unique biodegradable, abundant, and affordable material which is now recognized as an influential biopolymer desired for diverse applications [6, 109]. Its biophysical and biochemical properties are the driving force behind its applications, both in fundamental and applied biomedicine (Figure 10.3). These are significantly altered by the isolation procedure of the protein, which often involves harsh chemical and/or temperature treatments, leading to a broad molecular weight range and variation in the sequence of amino acids within the protein [109]. Further, SS properties vary according to the source silkworm strain [94]. Meanwhile, different sizes of the protein reflect different secondary structures that may affect the biological

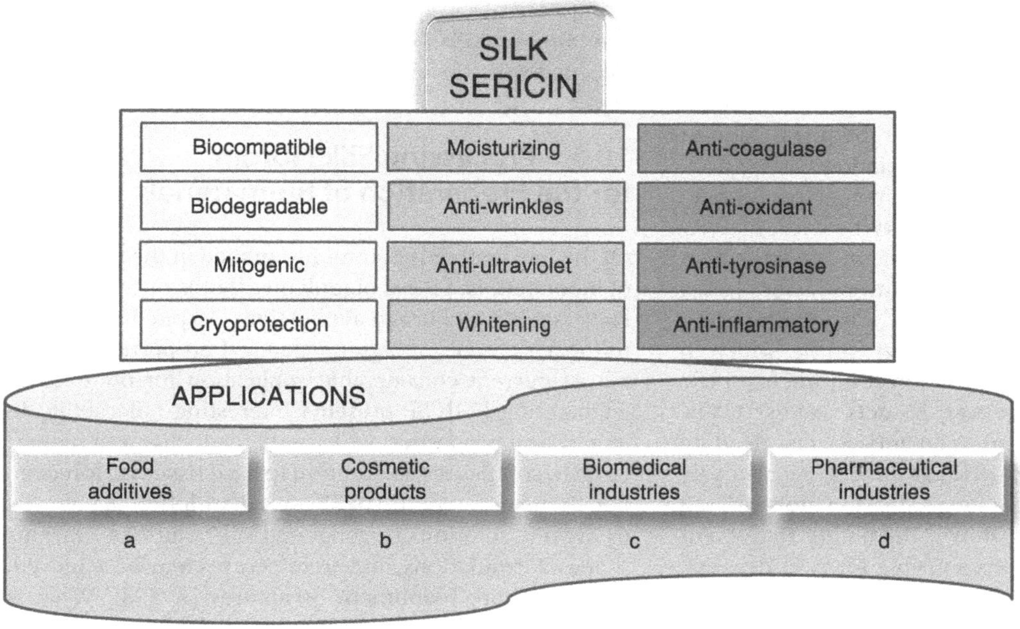

Figure 10.3 Properties of silk sericin (SS) and its biomedical applications. (a) SS reduces oxidative stress caused by free radicals and prevents food browning; (b) SS prevents UV-induced apoptosis and sunburn-related effects on skin; (c) SS is used to develop biomimetic biomaterials and stem cell-based therapeutics; (d) SS possess attractive pharmaceutical effects and is used for targeted drug and cell delivery.

functions of SS [110, 111]. Owing to its highly hydrophilic nature and rich polar side chains, this once underused protein displays functional and biological effects desired for use in the biomedical, food, and pharmaceutical industries [112, 113]. In fact, the biocompatibility [114], anti-inflammatory [95], and mitogenic [115] properties have qualified SS-based materials to be implantable devices in tissue engineering and regenerative medicine [116, 117]. Moreover, the influence of SS on cell integrity and viability when used as a serum alternative in a cell culture medium and as a cryoprotectant for biological tissues demonstrates its advantage for use in biomedical applications [118, 119]. Indeed, investigations to explore the biological response of SS-based materials showed that SS effectively increased cell growth while maintaining cell proliferation and attachment [120–122].

By virtue of its strong antioxidant effect which originates from the presence of hydroxyl amino acid groups (mostly from serine and threonine) [123], SS exhibits the ability to chelate trace elements such as copper and zinc [131]. Hence, SS exerts a defensive effect against hydrogen peroxide-induced oxidative stress in fibroblasts as depicted by the restoration of cell viability and membrane integrity, by elevating antioxidant enzymes [124]. Similarly, prior treatment with SS obstructed hydrogen peroxide-induced lipid peroxidation and increased antioxidant enzymatic activity (elevated levels of catalase, superoxide dismutase, and glutathione peroxidase) in the midgut epithelial cells and hemocytes, thanks to its contents of serine and aspartic acid [125]. Moreover, in addition to scavenging free radicals via its antioxidant activity [126], ingested SS successfully suppresses 1,2-dimethylhydrazine-induced colon cancer, significantly reducing the incidence of colorectal cancer in mice and rats [127]. Colonic tumorigenesis was also inhibited by ingested and yet digested SS which induced apoptosis in cancer cells and hindered tumor cell clones growth by the antioxidant effect [128]. Also, a suppressant effect of SS against skin tumor growth in 7, 12-dimethylbenz α anthracene and 12-*O* tetradecanoylphorbol 13-acetate-activated mice derived from its inhibition of oxidative stress and inflammatory response [129]. Assessment of the photo-protective and antioxidant effect of SS against ultraviolet B (UVB)-irradiated human keratinocytes showed that prior treatment with SS did not elicit apoptosis under oxidative stress; pro-apoptotic factors, such as Bax and caspase-3, were down-regulated, while Bcl-2, an anti-apoptotic factor, was up-regulated [130]. Meanwhile, further studies extended the evaluation of SS to anticoagulant activities displayed by the sulfation of SS using chlorosulfonic acid. Sulfated SS with higher molecular fractions exhibited greater anticoagulant activity [110]. The anticoagulant mechanism of SS in the coagulation cascade involves the interference of sulfated SS with the formation of fibrin fiber by distracting its fibril aggregation [131].

The trending application of SS in the cosmetic industry as a result of its versatile properties is expanding [112]. When used as an anti-tyrosinase agent for the first time, SS inhibited the activity of polyphenol oxidase (tyrosinase), the enzyme that triggers the production of pigmentation, in addition to its association with cancer, neurodegenerative diseases, and browning reaction in several foods [132]. Greater anti-tyrosinase activity was observed in urea-extracted SS due to the large proportions of valine and arginine amino acids [133]. Consequently, SS can be used as a skin whitening agent since it allows for the inhibition of tyrosinase, which in turn reduces skin hyperpigmentation resulting from UV light radiation [134]. Moreover, SS acts against UVB-induced ROS generation and as an anti-elastase, leading to its anti-aging and anti-wrinkle effects, respectively [134]. Also, it is considered a successful skin moisturizing agent [111]. In fact, it has shown the ability to prevent transepidermal water loss responsible for skin dryness, by increasing hydroxyproline and hydration of the epidermal cells when used as a moisturizing agent [135]. Apparently, the high serine content of SS contributes to its high water-absorbing capacity [136]. Furthermore, SS has an amino acid composition similar to that of filaggrin, which with free amino acids constitute the natural skin moisturizing agents [6]. Hence, unsurprisingly, there was promotion of the epidermal levels of hydration as well as free amino acids (from degraded filaggrin) in Nc/Nga mice fed with 1% SS [137].

10.6 Processing Silk Fibroin for the Preparation of Biomaterials

Of the raw materials commonly applied in the fabrication of biomaterials, SF is a biopolymer that is outstanding by its versatility and broad application. Despite its potential to form materials of high mechanical properties predicting its current considerable application for bone tissue engineering [78], SF presents interesting tailorability in its physical properties, which extends its use as a biomaterial to soft tissues, in addition to hard tissues. Moreover, due to the frequent use of SF after dissolution in aqueous or organic solutions (regenerated silk solutions), various shapes, formulations, and architectures may be achieved, enabling more biomimetic structures [1, 138]. What is more, polymer blending may be carried out, with the aim of enhancing functionality and biological performance, or regulating biodegradability. Hence, scaffolds, films, hydrogels, and particle shaped biomaterials are produced for diverse applications (Figure 10.4), following adequate processing methods.

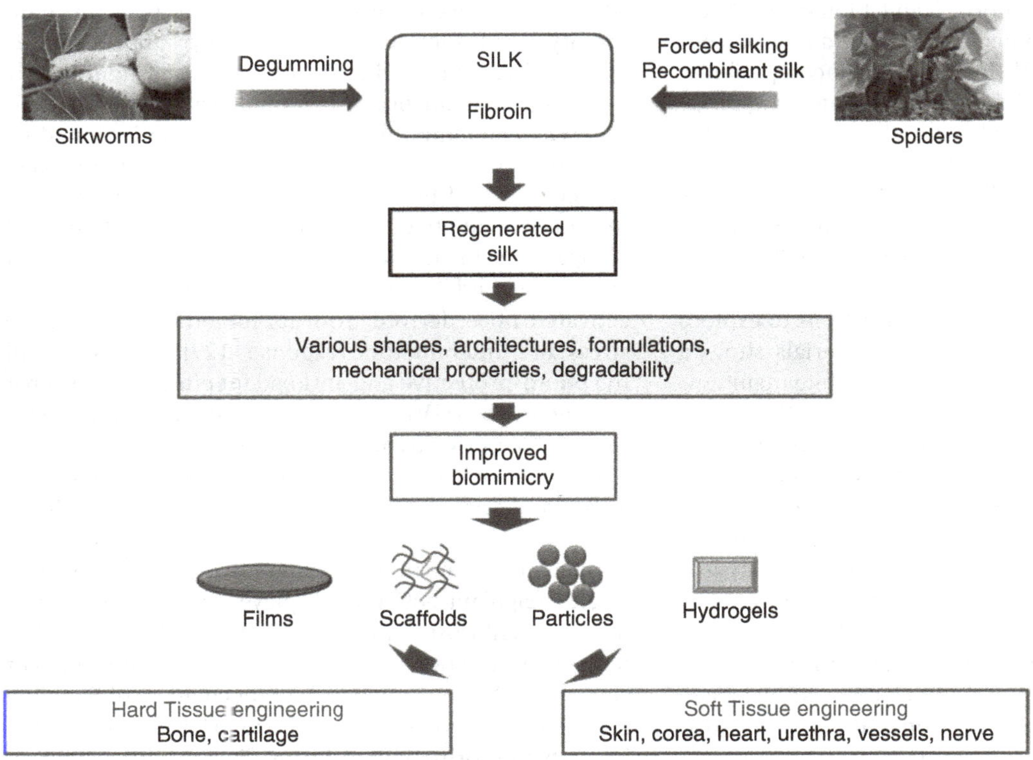

Figure 10.4 Applications of silk fibroin (SF) in tissue engineering and regenerative medicine.

10.6.1 Fabrication of 3D Matrices

In tissue engineering, 3D matrices are of great interest, thanks to their generally high porosity offering a favorable microenvironment for cell adhesion, metabolism, growth, and migration [104]. Furthermore, their spatial organization (3D) provides more *in vivo*-like conditions, an advantage for tissue engineering. These types of materials include sponge-like scaffolds and hydrogels.

For the preparation of SF-based scaffolds, freeze-drying is a simple and common method widely used. It consists of pre-freezing followed by lyophilization of the cast polymer solutions, allowing for the ready fabrication of porous structures by means of simple equipment. By this technique, the porosity and pore size of the resulting materials may be controlled depending on the water and SF content in the initial solutions [90, 139, 140]. In fact, pores are created as a consequence of the sublimation of water crystals formed during the pre-freezing step. Further, the pore morphology may be regulated by both the pre-freezing temperature and the initial solution concentration, parameters that modulate the ice formation [90]. Such methods have been employed to fabricate scaffolds for bone tissue engineering in cleft palate [82], cardiac tissue regeneration [70], and cartilage tissue engineering [80, 141], respectively. Interestingly, controlling the temperature gradient and direction at the stage of ice formation allows for the production of fibrous and lamellar structures which are attracting a lot of attention in the tissue engineering field in the quest for structural mimicry of the natural extracellular matrix [142, 143]. The process is thus termed ice templating or thermally-induced phase separation, and has been applied in order to engineer scaffolds for cartilage reconstruction [139, 144], and the culture of stem cells [90, 142]. Using the same method, blending of SF and SS enables the fabrication of nerve conduits, which, after insertion into a hollow poly(lactic-co-glycolic acid) sheath, successfully promoted nerve regeneration [109]. A derivative of this method consists in freeze-gelling, which involves gelation (in solvent) instead of lyophilization after pre-freezing of the sample, yielding more time and cost efficiency when compared to traditional freeze-drying [104]. Furthermore, similarly to freeze-drying, the porosity obtained by freeze-gelling varies with the pre-freezing temperature. Cytocompatible samples were obtained when ethanol was used as the gelling solvent, supporting the viability and growth of keratinocytes.

Other methods for the fabrication of SF-based porous scaffold structures include the use of porogens (salt leaching) or gas foaming [145, 146] following solution casting, and electrospinning [108, 147, 148]. In particular, the latter is more of a focus in tissue engineering as it enables the manufacture of fibrous structures with controlled fiber diameter from polymer solutions, emulating *in vivo* matrix architectures, thus improving biocompatibility [144, 149]. Fibers are formed on a collection device subsequent to the polymer solution being stimulated by a high voltage [9, 148, 149]. In fact, fibrous surfaces are known to enhance cell attachment and downstream cell proliferation, in comparison with smooth substrates [91]. What is more, SF was interestingly reported to facilitate the electrospinning process [91], which makes it a material of choice for the preparation of fibrous substrates for regenerative medicine applications. Various composite structures incorporating SF have thus been prepared for diverse applications, including bone [83, 91, 149, 186] and cartilage [86] tissue engineering among others.

Hydrogels are another class of materials, highly privileged in regenerative medicine, with a texture similar to that of the natural biological microenvironment (extracellular matrix) [77]. They are made of cross-linked (chemically or physically) very porous networks of polymers, which present high water-absorbing and water-retention abilities without disruptive effects upon their structural integrity. They are mainly useful for cell, drug, or bioactive molecule encapsulation. SF-based hydrogels are known to form from SF or composite solutions following structural transition from random coils to β-sheets [77, 85, 150]. Therefore, controlling the gel formation amounts to regulating the β-sheet formation, which may be induced by various factors [77, 85, 151], including temperature (faster gelation at high temperatures as compared with 37 °C) [152], pH (faster gelation at pH closer to the isoelectric point of SF between 3.6 and 5.2) [153], SF concentration (faster gelation at higher concentrations), shear force (controls the ionic strength by sonication or vortexing) [154, 155], ionic concentration (addition of $CaCl_2$ or KCl) [156], and alcohol treatment [157].

10.6.2 Fabrication of SF-based Films

Fibroin-based films are prepared by conventional materials science methods. Ready to produce, films are obtained following casting of polymer solutions into the desired molds of targeted shape, of a relatively small thickness. These are allowed to dry through solvent evaporation by normal air drying or under laminar flow [158–161]. Different types of molds may be employed, including Teflon molds, PDMS templates, and polystyrene plates among others. Further, for directed cell behavior, patterns may be induced onto the films using negatively patterned molds [158, 159, 162]. Due to the metastable structure of the SF solutions, SF films often require physical stabilization by post-processing as mentioned above (alcohol treatment, water annealing, or cross-linking).

Various applications are attributed to film or membrane substrates in tissue engineering and regenerative medicine. While mostly used as substrates for cell growth, films may also be used as vehicles for the delivery of drugs and active biomolecules, such as growth factors, following grafting or incorporation into the matrix. SF was thus blended with hyaluronic acid [152] or conjugated to heparin [92] to form films used for the controlled release of vascular endothelial growth factor in the field of vascular tissue engineering. The transparency of thin SF films is also an advantage, as it allows it to be used for corneal tissue engineering [162].

10.6.3 Preparation of SF-based Particulate Materials

Particulate materials are composed of compact biomaterials of spherical architectures that have the ability to efficiently entrap cells and/or bioactive signaling molecules which they deliver at specific sites *in vivo*, ensuring longer bioactivity with a targeted response. This characteristic is due to their sub-cellular size (nano- and microparticles) that allows easy circulation and even cellular uptake within the biological milieu. Furthermore, they possess high stability, high carrier capacity as vehicles, and a high surface-to-volume ratio that facilitates interaction with cells [1]. Moreover, using SF presents an additional advantage due to its ability to stabilize bioactive agents, such as enzymes and therapeutics [163]. While microspheres are often prepared by spray-drying or micelle-templating [1, 9, 164, 187], other methods may be used to obtain SF-based nano- and micro-particles, based on the structural properties of SF. In fact, SF gelation may be controlled so as to produce particulate morphologies [165]. For example, nanoparticles were obtained by solidifying SF solution drops in methanol, under vigorous stirring [166]. These materials were further successfully used as vehicles for the delivery of quercetin (*in vitro* studies). The same method was reported elsewhere, with the use of acetone as the gelling solvent [167]. Curcumin was then released in a controlled manner from the obtained microspheres. Inducing gelation by regulating the pH and the ionic strength has also been reported, where K_2HPO_4-KH_2PO_4 solutions were used to promote sphere formation from SF solution, under stirring [168]. Similarly, electrogelation has been used to prepare SF particles [169].

10.7 Processing Silk Sericin for Biomaterials Applications

Initially limited to being a byproduct of the silk industry, the rapid upsurge of interest in SS as a versatile biomaterial has motivated scientists to develop a broad range of SS-based scaffolds. Although reports suggest SS is a suitable biomaterial for fabricating 2D and 3D matrices for biomedical purposes, it is difficult to process pure SS into membranes due to its brittle nature, as a consequence of its amorphous structure and wide range of molecular weight distribution. As a result, useful strategies have been adopted to tailor SS-based materials with improved stability and mechanical properties, as required by the targeted tissues. Based on the presence of strong polar side chains of SS with hydroxyl, carboxyl and amine groups that enable chemical reactivity, SS biomaterials are made from either chemically cross-linked SS and/or SS blended with other organic and inorganic materials. Thus, different SS-based formulations can be initiated from aqueous SS for biomedical applications. For instance, SS-containing nanofibers may be prepared by electrospinning [170, 171]. Hydrogels may be formed by sol-gel transition depending on the β-sheet content and molecular weight distribution within the initial SS sample, usually by use of chemical cross-linking [117, 172, 173], and SS films may be prepared by casting SS solutions with various concentrations [174]. Meanwhile, composites may be formed by mixing polymer powders or solutions in the desired proportions. In most cases, further ethanol treatment is required to increase the crystallinity and physical stability of the SS-based materials.

Mostly acknowledged for its protective and mitogenic effects on mammalian cells and specifically skin cells, as a biomaterial, SS is more frequently used in the fabrication of wound dressings for the acceleration of wound healing and tissue regeneration [102, 172, 175], where few clinical trials have already been reported [176, 177, 180]. Nevertheless, SS has been successfully used for bone [178], cartilage [179], nerve tissue [109], and heart [117] tissue engineering, and as vehicle for the delivery of cells and bioactive molecules [181–185] in blends taking advantage of its chemical reactivity and pH responsiveness.

10.8 Conclusion

As a natural fiber, silk represents one of the leading compounds exploited by human society. Aside from its unprecedented success as a raw material of the textile industry, silk was initially introduced to the biomedical world as a useful material for surgical sutures. With the progress in biomedicine, silk has shown tremendous potential in contributing to human health by providing solutions to tissue engineering and regenerative medicine, a current major and interdisciplinary field making use of biomaterials with the aim of inducing repair and regeneration in damaged tissues. In fact, its major protein components (SF and SS) constitute rich biomaterials which, following a suitable design, may correspond to current tissue engineering strategies. In particular, they allow for biomimicry, one of the main challenges in present tissue engineering research. Considered a bioengineering grade polymer, SF offers extensive tailorability in the preparation of biomaterials, enabling the production of various architectures, formulations, and physical properties that match those of the targeted tissues in order to achieve specific functions. It is thus used as a biomaterial for a wide range of tissues, including bone, cartilage, skin, cornea, urethra, and blood vessels among others. As for SS, it presents mostly as a functional material endowed with regenerative abilities, which may be harnessed to induce and accelerate tissue regeneration. With its many functional groups, it is easily modified with more structural polymers to yield attractive and applicable biomaterials. Its applications also extend to both soft and hard tissues. Thus attracting attention from all sides, silk proteins are anticipated to continue to thrive in tissue engineering applications, used for the development of improved materials with novel and more suitable properties.

Acknowledgments

This work was supported by the National Natural Science Foundation of China (21574050, 31270150).

Abbreviations

2D	2-dimensional
3D	3-dimensional
A. mylitta	*Antheraea mylitta*
ASG	anterior silk gland region
B. mori	*Bombyx mori*
H-fib	heavy chain fibroin
L-fib	light chain fibroin
MSG	middle silk gland region
PDMS	polydimethylsiloxane
PSG	posterior silk gland region
SF	silk fibroin
SS	silk sericin

References

1 Kundu, B., Kurland, N. E., Bano, S. et al. (2014) *Progress in Polymer Sci.*, **39**, 251–267.
2 Mao, A. S., Mooney, D.J. (2015) *Proc. Nat. Acad. Sci.*, **112**, 14452–14459.
3 Sherif, B., Williams, J. K. (2012) *Bioengineering, Biomedical Sci.*, **S2**: 008.
4 Dimmeler, S., Ding, S., Rando, T. A., Trounson, A. (2014) *Nat Med*, **20**, 814–821.
5 Atala, A. (2012) *J. Pediatric Surgery*, **47**, 17–28.
6 Lamboni, L., Gauthier, M., Yang, G., Wang, Q. (2015) *Biotech. Adv.*, **33**, 1855–1867.
7 Jasmine, S., Mandal, B. B. (2014) In *Silk Biomaterials for Tissue Engineering and Regenerative Medicine*, Kundu, S.C.(ed). Woodhead Publishing: Cambridge.
8 Sealy, C. (2015) *Reinforced Plastics*, **59**, 294–301.
9 Lawrence, B.D. (2014) In *Silk Biomaterials for Tissue Engineering and Regenerative Medicine*, Kundu, S.C.(ed.), Woodhead Publishing: Cambridge.
10 Cai, Y., Guo, J., Chen, C. et al. (2017) *Materials Sci. Engineering: C*, **70**, Part 1, 148–154.
11 Murugesh Babu, K. (2017) In *Structure and Properties of High-Performance Fibers*, Gajanan, B.(ed.), Woodhead Publishing: Cambridge.
12 Aramwit, P., Siritientong, T., Srichana, T. (2011) *Waste Management, Res.*, **30**, 217–224.
13 Kundu, S.C., Dash, B.C., Dash, R., Kaplan, D.L. (2008) *Progress in Polymer Sci.*, **33**, 998–1012.
14 Babu, K.M. (2015) In *Textiles and Fashion*. Woodhead Publishing: Cambridge.
15 Epstein, D., Lockshin, R.A. (1981) *J. Insect Physiology*, **27**, 793–798.
16 Kaulenas, M.S., Potswald, H.E., Burns, A.L., Yenofsky, R.L. (1979) *Int. J. Insect Morphology Embryology*, **8**, 33–49.
17 Murugesh Babu, K. (2013) In *Silk*. Woodhead Publishing: Cambridge.
18 Capar, G., Aygun, S.S., Gecit, M. R. (2009) *J. Membrane Sci.*, **342**, 179–189.
19 Chen, F., Porter, D., Vollrath, F. (2012) *Acta Biomaterialia*, **8**, 2620–2627.
20 Ho, M.-P., Wang, H., Lau, K.-T. et al. (2012) *Composites Part B: Engineering*, **43**, 2801–2812
21 Dhawan, S., Gopinathan, K.P. (2003) *Gene Expression Patterns*, **3**, 559–570.
22 Parthasarathy, R., Gopinathan, K.P. (2005) *Gene Expression Patterns*, **5**, 323–339.
23 Borgohain, A. (2015) *J. Academia Industrial Research*, **3**, 626–628.
24 Wang, X., Li, Y., Peng, L. et al. (2016) *Insect Biochem., Molecular Biology*, **68**, 89–99.
25 Wei-Cheng, W.U., Qi-Kang, G.A.O. et al. (2007) *Agricultural Sci.s in China*, **6**, 249–254.
26 Prudhomme, J.-C., Couble, P., Garel, J.-P., Daillie, J. (1985) In *Biochemistry*, Gilbert, L.I. (ed.). Pergamon: Amsterdam.
27 Bovilla, V.R., Padwal, M.K., Siripurapu, P. et al. (2016) *Biochimica et Biophysica Acta (BBA) - Proteins and Proteomics*, **1864**, 860–868.
28 Hinton, H.E. (1981) *Biology of Insect Eggs*. Pergamon: Oxford.

29 Li, J.-Y., Ye, L.-P., Che, J.-Q. et al. (2015) *J. Proteomics*, **126**, 109–120.

30 De Loof, A., Vandersmissen, T., Marchal, E., Schoofs, L. (2015) *Peptides*, **68**, 120–129.

31 Takiya, S., Inoue, H. Kimoto, M. (2011) *Insect Biochem.,Molecular Biology*, **41**, 592–601.

32 Kim, H.J., Um, I.C. (2014) *Inter. J. Biological Macromol.*, **67**, 387–393.

33 Kimoto, M., Tsubota, T., Uchino, K. et al. (2014) *Developmental Biol.* **386**, 64–71.

34 Kimoto, M., Tsubota, T., Uchino, K. et al. (2015) *Insect Biochem., Molecular Biology*, **56**, 29–35.

35 Cao, J., Tong, C., Wu, X.,et al. (2008) *Insect Biochem., Molecular Biology*, **38**, 1066–1071.

36 Chen, Q., Liu, X., Zhao, X. et al. (2015) *Insect Biochem., Molecular Biology*, **57**, 41–50.

37 Qian, P., Wang, X., Jiang, T. et al. (2016) *J. Asia-Pacific Entomology*, **19**, 43?–445.

38 Zhou, C., Zha, X., Shi, P. et al. (2016) *Biochimica et Biophysica Acta (BBA) - General Subjects*, **1860**, 2529–2536.

39 Fabiani, C., Pizzichini, M., Spadoni, M., Zeddita, G. (1996) *Desalination*, **105**, 1–9.

40 Martinez, L., Almagro, J.C., Coll, J.L., Herrera, R J. (2004). *Insect Biochem. Molecular Biology*, **34**, 343–352.

41 Tanaka, K., Kajiyama, N., Ishikura, K. et al. (1999) *Biochimica et Biophysica Acta (BBA): Protein Structure Molecular Enzymology*, **1432**, 92–103

42 Wang, S.P., Guo, T.Q., Guo, X.Y. et al. (2006) *Biochemical Biophysical Res. Communications*, **341**, 1203–1210.

43 Song, J., Che, J., You, Z. et al. (2016) *J. Proteomics*, **148**, 194–201.

44 Kikuchi, Y., Mori, K., Suzuki, S. et al. (1992) *Gene*, **110**, 151–158.

45 Yamaguchi, K., Kikuchi, Y., Takagi, T. et al. (1989) *J. Molecular Biology*, **210**, 127–139.

46 Hui, C.-C., Suzuki, Y., Kikuchi, Y., Mizuno, S. (1990) *J. Molecular Biology*, **213**, 395–398.

47 Kundu, B., Kurland, N.E., Yadavalli, V. K., Kundu, S.C. (2014) *Int. J. Biological Macromol.*, **70**, 70–77.

48 Mori, K., Tanaka, K., Kikuchi, Y. et al. (1995) *Molecular Biology*, **251**, 217–228.

49 Nony, P., Prudhomme, J.-C., Couble, P. (1995) *Biology of the Cell*, **84**, 43–52

50 Tanaka, K., Inoue, S., Mizuno, S. (1999) *Insect Biochem., Molecular Biology*, **29**, 269–276.

51 Murphy, A.R., Romero, I.S. (2014) In *Silk Biomaterials for Tissue Engineering and Regenerative Medicine*, Kundu, S.C.(ed.), Woodhead Publishing: Cambridge.

52 Liu, Y., Yu, L., Guo X. et al. (2006) *Biochemical Biophysical Res. Comm.*, **342**, 273–279.

53 Perea, G. B., Solanas, C., Marí-Buyé, N. et al. (2016) *Eur. Polym. J.*, **78**, 129–140.

54 Garel, A., Deleage, G., Prudhomme, J.-C. (1997) *Insect Biochem., Molecular Biology*, **27**, 469–477.

55 Takasu, Y., Yamada, H., Tamura, T. et al. (2007) *Insect Biochem., Molecular Biology*, **37**, 1234–1240.

56 Kludkiewicz, B., Takasu, Y., Fedic, R. et al. (2009) *Insect Biochem., Molecular Biology*, **39**, 938–946.

57 Couble, P., Michaille, J.-J., Garel, A., et al. (1987) *Developmental Biol.*, **124**, 431–440.

58 Takasu, Y., Hata, T., Uchino, K., Zhang, Q. (2010) *Insect Biochem., Molecular Biology*, **40**, 339–344.

59 Ito, H., Muraoka, Y., Yamazaki, T. et al. (1995) *Textile Research Journal*, **65**, 755–759.

60 Mahmoodi, N.M., Arami, M., Mazaheri, F., Rahimi, S. (2010) *J. Cleaner Production*, **18**, 146–151.

61 Wang, F., Zhang, Y.-Q. (2017). *Materials Sci. Engineering: C*, **74**, 152–158.

62 Ho, M.-P., Wang, H., Lau, K.-T. (2012) *Appl. Surf. Sci.*, **258**, 3948–3955.

63 Ko, J.S., Yoon, K., Ki, C.S. et al. (2013) *Inter. J. Biological Macromol.*, **55**, 161–168.

64 Freddi, G., Mossotti, R., Innocenti, R. (2003) *J. Biotechnology*, **106**, 101–112.

65 Monti, P., Freddi, G., Sampaio, S. et al. (2005) *J. Molecular Structure*, **744–747**, 685–690.

66 Padaki, N.V., Das, B., Thirumalesh, R.M. (2015) In *Advances in Silk Science and Technology*, Basu, A.(ed.), Woodhead Publishing, Cambridge.

67 Pakdel, E., Wang, J., Allardyce, B.J. et al. (2016) *Separation and Purification Techn.*, **170**, 92–101.

68 Aznar-Cervantes, S.D., Vicente-Cervantes, D., Meseguer-Olmo, L. et al, (2013) *Materials Sci. and Engineering: C*, **33**, 1945–1950.

69 Chim-Anage, P., Romsomsa, N. (2009) *J. Biosci. Bioengineering*, **108**, Supplement 1, S119.

70 Patra, C., Talukdar, S., Novoyatleva, T. et al. (2012) *Biomaterials*, **33**, 2673–2680.

71 Wadbua, P., Promdonkoy, B., Maensiri, S., Siri, S. (2010) *Int. J. Biological Macromol.*, **46**, 493–501.

72 Suwannaphan, S., Fufeungsombut, E., Promboon, A., Chim-Anage, P. (2017) *Int. Biodeterioration, Biodegradation*, **117**, 141–149.

73 Li, H., Shi, W., Wang, W., Zhu, H. (2015) *Separation Purification Technology*, **146**, 342–350.

74 Silva, V.R., Ribani, M., Gimenes, M.L., Scheer, A.P. (2012) *Procedia Engineering*, **42**, 833–841.

75 Vaithanomsat, P., Malapant, T., Apiwattanapiwat, W. (2008) *J. Biotechnology*, **136**, Supplement, S335–S336.

76 Oh, H., Lee, J.Y., Kim, M.K.et al. (2011) *Int. J. Biological Macromol.*, **48**, 32–37.

77 Kapoor, S., Kundu, S.C. (2016) *Acta Biomater.*, **31**, 17–32.

78 Melke, J., Midha, S., Ghosh, S., Ito, K., Hofmann, S. (2016) *Acta Biomater.*, **31**, 1–16.
79 Bhattacharjee, M., Ghosh, S. (2014) In *Silk Biomaterials for Tissue Engineering and Regenerative Medicine*, Kundu, S.C(ed.). Woodhead Publishing, Cambridge.
80 Jaipaew, J., Wangkulangkul, P., Meesane, J. et al. (2016) *Materials Sci. Engineering: C*, **64**, 173–182.
81 Lv, X., Li, Z., Chen, S., Xie, M. et al. (2016) *Biomater.*, **84**, 99–110.
82 Sangkert, S., Meesane, J., Kamonmattayakul, S., Chai, W.L. (2016) *Materials Sci. Engineering: C*, **58**, 1138–1149.
83 Singh, B.N., Panda, N.N., Mund, R., Pramanik, K. (2016) *Carbohyd. Polym.*, **151**, 335–347.
84 Römer, L., Scheibel, T. (2008) *Prion*, **2**, 154–161.
85 Numata, K. (2014) In *Silk Biomaterials for Tissue Engineering and Regenerative Medicine*, Kundu, S.C.(ed.), Woodhead Publishing, Cambridge.
86 Mirahmadi, F., Tafazzoli-Shadpour, M., Shokrgozar, M.A., Bonakdar, S. (2013) *Materials Sci. Engineering: C*, **33**, 4786–4794.
87 Bhardwaj, N., Kundu, S.C. (2011) *Carbohydrate Polymers*, **85**, 325–333.
88 Teimouri, A., Azadi, M., Emadi, R. et al. (2015) *Polymer Degradation Stability*, **121**, 18–29.
89 Lan, Y., Li, W., Jiao, Y. et al. (2014) *Acta Biomater.*, **10**, 3167–3176.
90 Oliveira, A.L., Sun, L., Kim, H.J. et al. (2012) *Acta Biomater.*, **8**, 1530–1542.
91 Paşcu, E.I., Stokes, J., Mcguinness, G.B. (2013) *Materials Sci. Engineering: C*, **33**, 4905–4916.
92 Seib, F.P., Herklotz, M., Burke, K.A. et al. (2014) *Biomat.*, **35**, 10.1016/j.biomaterials.2013.09.053
93 Padamwar, M.N., Pawar, A P. (2004) *J. Scientific, Industrial Res.* **63**, 323–329.
94 Dash, R., Ghosh, S.K., Kaplan, D.L., Kundu, S.C. (2007) *Comp. Biochem., Physiology Part B: Biochem., Molecular Biology*, **147**, 129–134.
95 Aramwit, P., Siritientong, T., Kanokpanont, S., Srichana, T. (2010) *Inter. J. Biological Macromol.*, **47**, 668–675.
96 Dash, R., Mukherjee, S., Kundu, S.C. (2006) *Int. J. Biological Macromol.*, **38**, 255–258.
97 Lee, K., Kweon, H., Yeo, J.H. et al. (2003) *Int. J. Biological Macromol.*, **33**, 75–80.
98 Nayak, S., Talukdar, S., Kundu, S. (2012) *Cell and Tissue Res.* **347**, 783–794.
99 Dash, B. C., Mandal, B. B., Kundu, S. C. (2009) *J. Biotechnology*, **144**, 321–329.
100 Siritientong, T., Srichana, T., Aramwit, P. (2011) *AAPS PharmSciTech*, **12**, 771–781.
101 Teramoto, H., Nakajima, K.-I., Takabayashi, C. (2005) *Bioscience, Biotechnology, Biochem.*, **69**, 845–847.
102 Lamboni, L., Li, Y., Liu, J., Yang, G. (2016) *Biomacromol.*, **17**, 3076–3084.
103 Chandrasekaran, A., Novajra, G., Carmagnola, I. et al. (2016) *Ceram. Inter.*, **42**, 13761–13772.
104 Bhardwaj, N., Chakraborty, S., Kundu, S.C. (2011) *Inter. J. Biological Macromol.*, **49**, 260–267.
105 Acharya, C., Ghosh, S.K., Kundu, S.C. (2008) *J. Materials Sci.: Materials in Medicine*, **19**, 2827–2836.
106 Wang, F., Xu, H., Wang, Y. et al. (2014) *Acta Biomater.*, **10**, 4947–4955
107 Lee, K.S., Kim, B.Y., Kim, D.H., Jin, B.R. (2016) *J. Asia-Pacific Entomology*, **19**, 1015–1018.
108 Chouhan, D., Chakraborty, B., Nandi, S.K., Mandal, B.B. (2017) *Acta Biomaterialia*, **48**, 157–174.
109 Rao, J., Cheng, Y., Liu, Y. et al. (2017) *Materials Sci. Engineering: C*, **73**, 319–332.
110 Tamada, Y., Sano, M., Niwa, K. et al. (2004) *J. Biomaterials Sci., Polymer Edition*, **15**, 971–980.
111 Kunz, R.I., Brancalhão, R.M.C., Ribeiro, L.D.F.C., Natali, M.R M. (2016) *BioMed Research Inter.*, **19**.
112 Cao, T.-T., Zhang, Y.-Q. (2016) *Materials Sci., Engineering: C*, **61**, 940–952.
113 Li, L., Wu, J.-H., Zhang, L. et al. (2017) *LWT - Food Sci. and Technology*, **77**, 532–541.
114 Chirila, T.V., Suzuki, S., Bray, L.J. et al. (2013) *Prog. Biomater.*, **2**, 14.
115 Miyamoto, Y., Oishi, K., Yukawa, H. et al. (2012) *Cell Transplan.*, **21**, 617–622.
116 Duan, L., Yuan, J., Yang, X., Cheng, X., Li, J. (2016) *Int. J. Biological Macromol.*, **93**, Part A, 468–475.
117 Song, Y., Zhang, C., Zhang, J. et al. (2016) *Acta Biomaterialia*, **41**, 210–223.
118 Sahu, N., Pal, S., Sapru, S. et al. (2016) *BioMed Research Inter.*, 2016, **13**.
119 Hajarian, H., Aghaz, F., Karami Shabankareh, H. (2017) *Theriogenology*, **92**, 144–148.
120 Dinescu, S., Galateanu, B., Albu, M. et al. (2013a) *Int. J. Molecul. Sci.*, **14**, 1870.
121 Kumar, J. P., Bhardwaj, N., Mandal, B.B. (2016) *RSC Adv.*, **6**, 105125–105136.
122 Siritientong, T., Bonani, W., Motta, A. et al. (2016) *Bioscience, Biotechnology, Biochem.*, **80**, 241–249.
123 Sangwong, G., Sumida, M., Sutthikhum, V. (2016) *Biocatalysis Agricultural Biotechnology*, **5**, 155–161.
124 Khyade, V.B. (2016) *World Scientific News*, **44**, 35–49.
125 Micheal, A.S., Subramanyam, M. (2014) *ISJ: Invertebrate Survival J.l*, **11**, 257–272.
126 Takechi, T., Wada, R., Fukuda, T. et al. (2014) *Biomedical Reports*, **2**, 364–369.
127 Sasaki, M., Kato, N., Watanabe, H., Yamada, H. (2000) *Oncol. Rep.*, **7**, 1049–1101.
128 Kaewkorn, W., Limpeanchob, N., Tiyaboonchai, W. et al. (2012) *Biological Research*, **45**, 45–50.

129 Zhaorigetu, S., Yanaka, N., Sasaki, M., et al. (2003). Athens, Greece, Spandidos.

130 Dash, R., Mandal, M., Ghosh, S. K. & Kundu, S. C. (2008). Silk sericin protein of tropical tasar silkworm inhibits uvb-induced apoptosis in human skin keratinocytes. *Molecular and Cellular Biochemistry*, **311**, 111–119.

131 Sano, M., Tamada, Y., Niwa, K. et al. (2009) *J. Biomaterials Sci., Polym. Edition*, **20**, 773–783.

132 Kato, N., Sato, S., Yamanaka, A., et al. (1998). Silk protein, sericin, inhibits lipid peroxidation and tyrosinase activity. *Bioscience, Biotechnology, and Biochemistry*, **62**, 145–147.

133 Aramwit, P., Damrongsakkul, S., Kanokpanont, S. & Srichana, T. (2010). Properties and antityrosinase activity of sericin from various extraction methods. *Biotechnology and Applied Biochemistry*, **55**, 91–98.

134 Chlapanidas, T., Faragò, S., Lucconi, G., et al. (2013). Sericins exhibit ros-scavenging, anti-tyrosinase, anti-elastase, and in vitro immunomodulatory activities. *International Journal of Biological Macromolecules*, **58**, 47–56.

135 Padamwar, M.N., Pawar, A.P., Daithankar, A.V., Mahadik, K R. (2005) *Cosmetic Dermatology*, **4**, 250–257.

136 Rajput, S.K., Singh, M.K. (2015) *IOSR J. Polym. Textile Engineering (IOSR-JPTE)*, **2**, 29–35.

137 Kim, H., Lim, Y.-J., Park, J.-H., Cho, Y. (2012) *Brit. J. Nutr.*, **108**, 1726–1735.

138 Adalı, T., Uncu, M. (2016) *J. Biological Macromol.*, **90**, 11–19.

139 Bhardwaj, N., Nguyen, Q.T., Chen, A.C. et al. (2011) *Biomaterials*, **32**, 5773–5781.

140 Yang, M., Shuai, Y., He, W. et al. (2012) *Int. J. Molecular Sciences*, **13**, 7762–7775.

141 Wang, J., Yang, Q., Cheng, N. et al. (2016) *Materials Sci. Engineering: C*, **61**, 705–711.

142 Asuncion, M.C.T., Goh, J. C.-H., Toh, S.-L. (2016) *Materials Sci. Engineering: C*, **67**, 646–656.

143 Ming, J., Li, M., Han, Y. et al. (2016) *Materials Sci. Engineering: C*, **59** 185–192.

144 Zhang, H., Liu, X., Yang, M., Zhu, L. (2015) *Materials Sci. Engineering: C*, **55**, 8–13.

145 Park, Y.R., Ju, H.W., Lee, J.M. et al. (2016) *Int. J. Biological Macromol.*, **93**, **Part B**, 1567–1574

146 Wangkulangkul, P., Jaipaew, J., Puttawibul, P., Meesane, J. (2016) *Materials, Design*, **106**, 428–435.

147 Bhattacharjee, P., Naskar, D., Maiti, T.K. et al. (2016) *Colloids and Surfaces B: Biointerfaces*, **143**, 431–439.

148 Shao, W., He, J., Sang, F., Ding, B. et al. (2016) *Materials Sci. and Engineering: C*, **58**, 342–351.

149 Rezvani, Z., Venugopal, J.R., Urbanska, A.M. et al. (2016) *Nanomedicine: Nanotechnology, Biology Medicine*, **12**, 2181–2200.

150 Matsumoto, A., Chen, J., Collette, A.L. et al. (2006) *J. Physical Chemistry B*, **110**, 21630–21638.

151 Nagarkar, S., Nicolai, T., Chassenieux, C., Lele, A. (2010) *Physical Chem., Chemical Physics*, **12**, 3834–3844.

152 Zhou, J., Zhang, B., Liu, X. et al. (2016) *Carbohyd. Polym.* **143**, 301–309.

153 Dyakonov, T., Yang, C.H., Bush, D. et al. (2012) *J. Drug Delivery*, **10**.

154 Zhang, W., Wang, X., Wang, S. et al. (2011) *Biomaterials*, **32**, 9415–9424.

155 Fernández-García, L., Marí-Buyé, N., Barios, J.A. et al. (2016) *Acta Biomater.*, **45**, 262–275.

156 Silva, R., Singh, R., Sarker, B. et al. (2016) *Int. J. Biological Macromol.*, **93**, Part B, 1420–1431.

157 Jin, Y., Kundu, B., Cai, Y. et al. (2015) *Colloids Surfaces B: Biointerfaces*, **134**, 339–345.

158 Lawrence, B.D., Omenetto, F., Chui, K., Kaplan, D.L. (2008) *J. Materials Sci.*, **43**, 6967.

159 Gil, E.S., Mandal, B.B., Park, S.-H. et al. (2010) *Biomaterials*, **31**, 8953–8963.

160 Wang, X., Yucel, T., Lu, Q. et al. (2010) *Biomaterials*, **31**, 1025–1035.

161 Hazra, S., Nandi, S., Naskar, D. et al. (2016) *Scientific Reports*, **6**, 21840.

162 Lawrence, B.D., Marchant, J.K., Pindrus, M. et al. (2009) *Biomaterials*, **30**, 1299–1308.

163 Wray, L.S., Rnjak-Kovacina, J., Mandal, B.B. et al. (2012) *Biomaterials*, **33**, 9214–9224.

164 Imsombut, T., Srisuwan, Y., Srihanam, P., Baimark, Y. (2010) *Powder Technology*, **203**, 603–608.

165 Wu, J., Zheng, Z., Li, G. et al. (2016) *Acta Biomaterialia*, **39**, 156–168.

166 Lozano-Pérez, A.A., Rivero, H.C., Pérez Hernández, M.D.C. et al. (2017) *Int. J. Pharm.*, **518**, 11–19.

167 Perteghella, S., Crivelli, B., Catenacci, L. et al. (2017) *Int. J. Pharmaceutics*, **520**, 86–97.

168 Zeng, D.-M., Pan, J.-J., Wang, Q. et al. (2015) *Materials Sci. Engineering: C*, **50**, 226–233.

169 Zhang, X., Fan, Z., Lu, Q. et al. (2013) *Acta Biomaterialia*, **9**, 6974–6980.

170 Khan, M.M.R., Tsukada, M. (2014) In *Silk Biomaterials for Tissue Engineering and Regenerative Medicine*, Kundu, S.C.(ed.), Woodhead Publishing, Cambridge.

171 Eslah, S., Tavanai, H., Morshed, M. (2016) *J. Textile Institute*, **107**, 949–957.

172 Sapru, S., Ghosh, A.K., Kundu, S.C. (2016) *Carbohyd. Polym.*

173 Shi, L., Yang, N., Zhang, H. et al. (2015) *Mater. Sci. Engineering: C*, **48**, 533–540.

174 Yun, H., Kim, M.K., Kwak, H.W. et al. (2016) *Int. J. Biological Macromol.*, **82**, 945–951.

175 Napavichayanun, S., Amornsudthiwat, P., Pienpinijtham, P., Aramwit, P. (2015) *Materials Sci. Engineering: C*, **55**, 95–104.

176 Aramwit, P., Palapinyo, S., Srichana, T., Chottanapund, S., Muangman, P. (2013a) *Arch. Dermatological Res.*, **305**, 585–594.
177 Siritientong, T., Angspatt, A., Ratanavaraporn, J., Aramwit, P. (2014) *Pharm. Res.*, **31**, 104–116.
178 Yang, M., Shuai, Y., Zhang, C. et al. (2014) *Biomacromolecules*, **15**, 1185–1193
179 Dinescu, S., Gălățeanu, B., Albu, M. et al. (2013) *BioMed Research Inter.*, 598056.
180 Song, D.W., Kim, S.H., Kim, H.H.,et al. (2016) *Acta Biomaterialia*, **39**, 146–155.
181 Kurland, N.E., Ragland, R.B., Zhang, A. et al. (2014) *Inter. J. Biological Macromol.*, **70**, 565–571.
182 Nayak, S., Dey, S., Kundu, S.C. (2014) *Inter. J. Biological Macromol.*, **65**, 258–266.
183 Kanoujia, J., Singh, M., Singh, P., Saraf, S.A. (2016) *Materials Sci. Engineering: C*, **69**, 967–976.
184 Suktham, K., Koobkokkruad, T., Saesoo, S. et al. (2016) *Colloids Surfaces B: Biointerfaces*, **148**, 487–495.
185 Yang, Y., Cai, Y., Sun, N. et al. (2017) *Colloids Surfaces B: Biointerfaces*, **151**, 102–111.
186 Napavichayanun, S., Yamdech, R., Aramwit, P. (2016) *Arch. Dermatological Res.*, **308**, 123–132.
187 Shao, W., He, J., Sang, F. et al. (2016) *Materials Sci. Engineering: C*, **62**, 823–834.

11

Polypeptides Synthesized by Ring-opening Polymerization of N-Carboxyanhydrides: Preparation, Assembly, and Applications

Yuan Yao[1], *Yongfeng Zhou*[2,**], *and Deyue Yan*[2,*]

[1] School of Materials Science and Engineering, East China University of Science and Technology, Shanghai, China
[2] School of Chemistry and Chemical Engineering, Shanghai Jiaotong University, Shanghai, China

11.1 Introduction

The successful synthesis of α-amino acid N-carboxyanhydride (NCA) in 1906 revealed a convenient procedure for preparing polypeptides with a high molecular weight through the ring-opening polymerization (ROP) of NCA [1]. Since Deming reported the first living initiating system for ROP of NCAs in 1997 [2, 3], a great number of synthetic polypeptides with controllable blocks and designable side-groups have been designed and prepared mainly for bio-related applications [4–6]. At present, polypeptides and their derivatives synthesized by ROP of NCAs still receive increasing attention not only because polypeptides have been important bio-compatible materials but also due to their adoptable multitude conformations in mimicking the functions of natural proteins.

The living polymerization of NCAs has been thoroughly studied and discussed in the last few decades. Many "active" transition metal and amine initiators have been developed to control the ROP of NCAs precisely [7, 8]. Based on these achievements, a large number of polypeptide-containing homo- and co-polymers, hyperbranched polypeptides, side-chain functionalized and side-chain grafted polypeptides are prepared with ingenious design for a variety of functions, leading to novel applications in different fields [7–12].

The self-assembly of amphiphilic polypeptides and polypeptide-containing polymers is focused extensively on the field of synthetic polypeptides.[7, 8] Similar to natural polypeptides, synthetic polypeptide blocks usually possess α-helix or β-sheet motifs. These secondary structures obviously influence the character of the self-assembly, and result in plenty of supermolecular structures for numerous bio-related applications.[5, 7–12] In addition, synthetic polypeptides can also act as templates to direct the formation of inorganic materials, as well as their natural counterparts, or be co-assembled with inorganic precursors to form inorganic-organic hybrids [13–15].

These precisely designed assembly structures of polypeptides can find applications in delivery vehicles, biomineralization, and tissue engineering [4–6]. Up to now, synthetic polypeptides have already been considered a type of important biomaterial. Based on the numerous papers and reviews that show the remarkable accomplishments from different aspects at present, this chapter comprehensively reviews the up-to-date advances in the preparation, self-assembly and the applications of polypeptide-based polymers.

11.2 Living Polymerization of NCAs

The discovery of the transition metal complex and the amine-based active initiators for the living polymerization of NCAs provided a powerful tool to manipulate the structure of polypeptides. In this section, we briefly introduce the two initiator families and highlight the recent advances in the living polymerization of NCAs.

11.2.1 Transition Metal Complexes

The ROP of NCAs can be initiated by many nucleophiles, in which primary amines and alkoxide anions are the most commonly adopted strategies (Figure 11.1). Some earlier works and reviews have presented the nucleophile initiators and their mechanisms in detail [1, 16, 17]. However, the initiators usually suffer board molecular weight distribution and many adverse reactions, and it is very difficult to prepare block copolypeptides [1].

In 1997, Timothy J. Deming applied zero valent transition metal nickel and cobalt complexes [i.e. bpyNi(COD)

* Email: dyyan@sjtu.edu.cn
** Email: yfzhou@sjtu.edu.cn

Bioinspired Materials Science and Engineering, First Edition. Edited by Guang Yang, Lin Xiao, and Lallepak Lamboni.

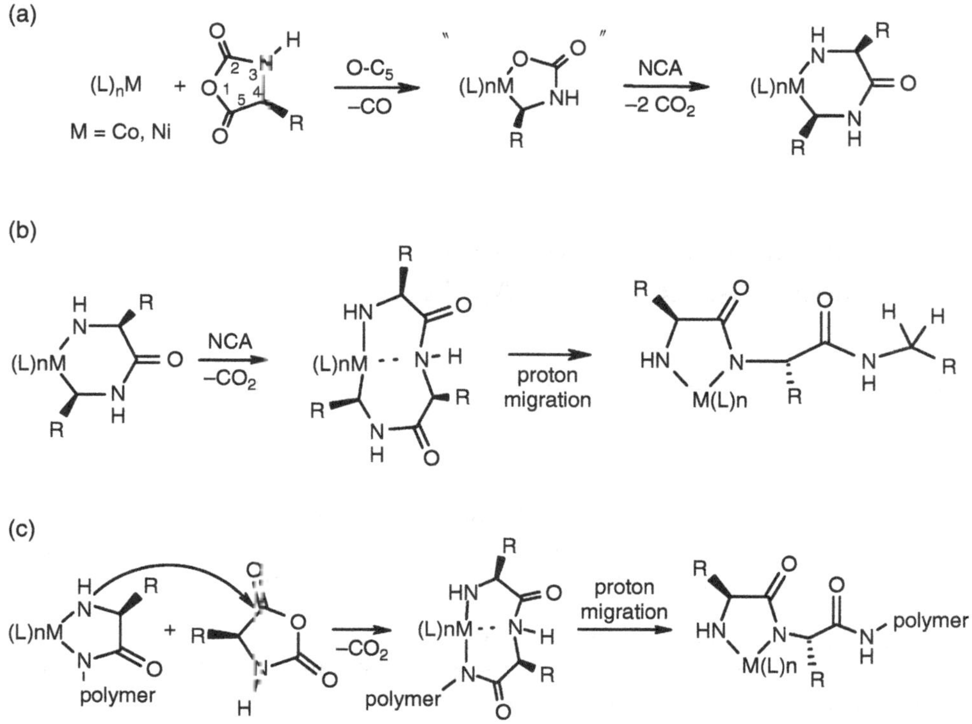

Figure 11.1 The mechanism of transition metal-initiated NCA polymerization. (a) oxidative addition of NCAs to zero valent cobalt and nickel complexes; (b) metallacycle ring contraction mediated by NCA addition; (c) chain propagation in transition metal-mediated NCA polymerization. *Source:* Reproduced with permission from [7].

and $(PMe_3)_4Co]$ as initiators to polymerize many NCAs into polypeptides with a high molecular weight and a narrow molecular weight distribution [2, 18, 19]. It was found that the complexes react with NCA monomers to give rise to metallacyclic complexes by oxidative addition to the anhydride bonds of NCA first, and then the complexes react with the second NCA monomer to form a 6-membered amido-alkyl metallacycle complex, which further reacts with another NCA, and simultaneously eliminates CO_2 to produce a 5-membered amido-amidate metallacycle complex via the migration of an amide proton to the metal-bound carbon. The amido-amidate metallacycle is proposed as the active intermediate of the living polymerization of NCAs. Next, the amido-amidate metallacycle complex attacks the nucleophilic amido group on the electrophilic C5 carbonyl of a NCA monomer, and regenerates the active species at the end of newly ring-open added NCA, upon elimination of CO_2 and after proton migration. In this way, the active metallacycle complex migrates along the growing polypeptide chain and finally results in polypeptides with a controllable molecular weight ($Mn = 500 \sim 500\,000$) and a narrow chain distribution length (polydispersion index, PDI, as $Mw/Mn \leq 1.2$) [2, 7, 8, 18–20].

The zero valent metal complex initiators provide a simple tool to produce block copolypeptides. However, because the zero valent metal complexes need to react with NCA to generate the active amido-amidate metallacycle complexes, they are not suitable for preparing other polymer/polypeptide hybrid blocks directly. Deming and his coworkers developed allyloxycarbonylaminoamides as universal precursors to amido-amidate nickelacycles. This amino acid derivative can form active NCA polymerization initiators with $Ni(COD)_2$ through tandem oxidative additions to nickel (0). This is a convenient approach to incorporate polypeptide and diverse polymers, and many other functional parts [21].

Interestingly, chiral donor ligand containing nickel initiators can initiate the enantioasymmetric polymerization of NCAs. Deming's group reported on a series of chiral nickel complexes initiators [22], in which the highest selectivity was recorded as enantiomer polymerization rate constants (kD/kL) of 5.2 with the nickel complex of (S)-4-tert-butyl-2-pyridinyl oxazoline. The result showed the possibility of synthesizing optically pure polypeptide with racemic monomers directly. Despite nickel and cobalt complexes, Deming's group also reported on ruthenium(II) amido-sulfonamide complex and the corresponding isoelectronic Cp*iridium(III) ($Cp^* = C_5Me_5$) complex initiating systems for the living NCA ROP [23]. The initiators showed much higher enantiomeric selectivity in the polymerizations of racemic NCAs compared to the nickel systems. Also, Lin prepared an amido-sulfonamide metallacycles initiator using platinum(II) for controlled polymerization of N^{ε}-carbobenzyloxy-L-lysine NCA (ZLys NCA) [24]. In addition, rare earth catalysts were also studied for NCA polymerization recently [25, 26]. In 2012, Ling et al. reported using rare earth metal complexes $Ln(BH_4)_3(THF)_3$ as the initiator for the living polymerization of NCAs, which is the newest extension of the metal complex strategy [26].

In general, transition metal complex initiators not only allow the preparation of well-defined polypeptides with high rate (ca. 30 ~ 60 min.), but also offer high enantiomeric selectivity by using special chiral ligands. The only minor drawback is the residue of metal in the polypeptide, and also that the initiators cannot be applied to N-substituted NCAs such as proline NCA, because the active species are formed via proton migration, which requires a proton at the *3*-N of the NCA [20].

11.2.2 Active Initiators Based on Amines

Amine initiators have the advantage of being readily available and do not need to be removed after polymerization. In the past few decades, several new approaches have been developed to achieve amine-initiated living polymerization of NCAs.

In 2003, the first breakthrough was reported by Schlaad et al. by using primary amine hydrochloride salts as the active initiators [27]. These salts are fast, the reversible deactivation of the reactive primary amines and the idea is similar to controlled radical polymerization strategies. In a dimethyl formamide (DMF) solution, a small amount of free amine appears by reversible dissociation of hydrochloride. As soon as the free primary amine reacts with an NCA, the resulting amine end group is immediately protonated and protected by hydrochloride, which effectively avoids the formation of NCA anions. The reaction continues and the amine end groups are still protected by hydrochloride until the NCA is exhausted. To increase the equilibrium concentration of free amine and the exchange rate between the amine and the amine hydrochloride, a higher reaction temperature (40–80 °C) was adopted in this work. Using amine hydrochloride end-capped polystyrene to polymerize ZLys NCA, the resulting polypeptide segments are well controlled with narrow PDI ($Mw/Mn \leq 1.03$). Also, PEG-amine hydrochloride macroinitiators [28] and poly(Bn-Glu)-block-poly(2-isopropyl-2-oxazoline) [29] were reported as an extension of this strategy.

Hadjichristidis reported on the primary amine-initiated NCA living polymerization under high vacuum conditions in 2004 [30]. By using high vacuum techniques, the impurities in the solvents and the primary amines were thoroughly removed and impurity-induced side reactions were reduced tremendously, which meant the characteristics of NCA ROP are in accordance with living polymerization. Furthermore, Hadjichristidis found that the purity of NCAs is also a key condition of living polymerization [30]. Only the NCAs recrystallized in a high vacuum are pure enough for primary amine-initiated living polymerization, whereas NCAs recrystallized in a dry box could not meet the requirements. In addition, a high vacuum eliminates the carbon dioxide produced during the propagation, which is beneficial to the life of active amine for complete polymerization. This technique is universal to the primary amine-initiated NCA ROP, including N-substituted NCAs.

Also in 2004, Giani et al. developed primary amine-initiated living NCA polymerization at a low temperature [31]. They studied n-hexylamine initiated ε-trifluoroacetyl-L-lysine NCA (TFA-Lys NCA) polymerization at different temperatures. At 20 °C, the polymer products consisted of 78% of dead chains (not amine-terminated) and only 22% living chains. In contrast, 99% of chains were living at 0 °C. The reason might be that primary amine-initiated chain propagation has a kinetic advantage over the side effects at 0 °C, although the propagation is much slower than that at room temperature. Following this discovery, a number of groups studied low temperature polymerization of NCAs. Shao's laboratory studied the synthesis of block copolypeptides of PBLG with poly-alanine, leucine, and phenylalanine segments at 0 °C, and found that more than 90% of PBLG segments were active for the second monomer addition [32]. Schouten e al. also reported the controllable polymerization of tert-butyl-L-glutamate NCA with other glutamate ester NCAs at 0 °C to prepare block copolypeptides [33]. Haise et al. studied ten different NCA monomers and found that on most occasions the NCA polymerization at 0 °C shows fewer side effects than at a higher temperature [34]. The other study combined a low temperature and a high vacuum and found the techniques are monomer-dependent [35]. The helicogenic NCA monomers (resulting in α-helix polypeptide segments) could be living polymerized at 20 °C if CO_2 was removed completely from the system, whereas non-helicogenic monomers (cannot form α-helix polypeptide segments) should be polymerized at 0 °C for living polymerization. A tetra-block copolypeptide of PBLG-PA-PZLL-PBLA was then prepared using this strategy.

Another innovative approach was developed by Cheng's group in 2007 [36–38]. They found that silylated amines could serve as initiators for the living polymerization of NCAs (Figure 11.2). Typically, Hexamethyldisilazane (HMDS) was applied for the polymerization of Z-Lys NCA and Bn-Glu NCA in DMF at an ambient temperature, and the resulting polypeptides showed well-controlled chain

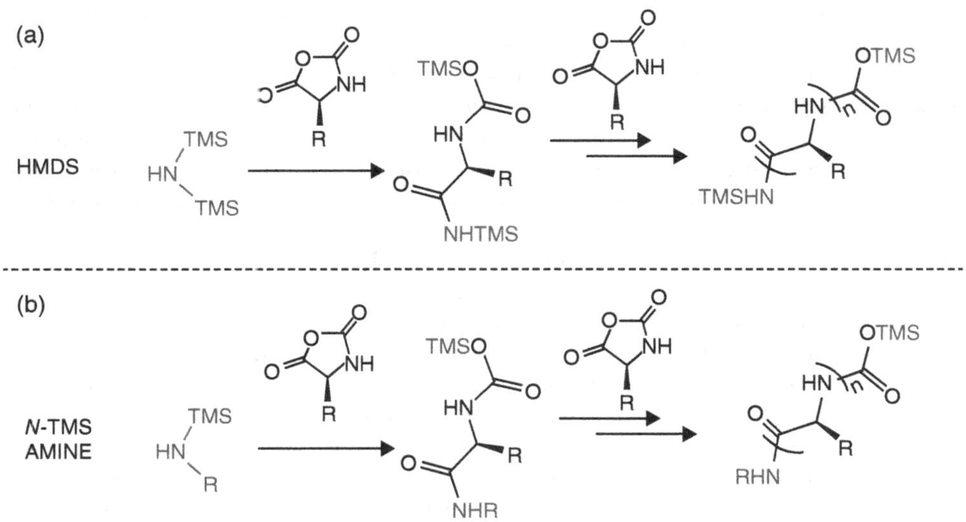

Figure 11.2 Organosilicon amines-mediated NCA polymerization. *Source:* Reproduced with permission from [36, 37].

length and narrow PDI [36]. The special property of the N-trimethylsilyl (TMS) group is the key factor of living polymerization. First, HDMS transfers a TMS group to the NCA, and the remaining silylamine is then added to the resulting intermediate to form a ring-opened molecular with a TMS-carbamate active end group, which can further initiate the NCA ROP and keep the chain propagation without traditional side effects. The reactions were completed in 12 ~ 24 hours. In addition, the technique could be extended to prepare other TMS amines, ranging from small molecules to polymers or containing a variety of functional groups (e.g. norbornene, alkyne, azide, PEG, etc.) [36–38]. The activity of these silylated amines is high enough to initiate some special NCAs ROP.

11.2.3 Recent Advances in Living NCA ROP Polymerization, 2013-2016

The living polymerization of NCA is still a hot issue recently in the area of "activated monomer mechanism" (AMM), except for only one report in 2013 by Wooly et al. on the "normal amine mechanism" (NAM) to obtain polypeptides with a controlled structure by using N2 flaws to get rid of the carbon dioxide generated during the NCA ROP [39].

In 2015, Hadjichristidis reported NCA living polymerization by using "allied amines" as the initiator [40, 41]. In these works, secondary amines or tertiary amines act cooperatively and in synergy with primary amine groups to trigger a fast and yet "living" polymerization of NCAs through the so-called accelerated amine mechanism through monomer activation. They further unveiled another NCA living polymerization that used aminoalcohols as initiators in the presence of N,N'-bis[3,5-bis(trifluoromethyl)phenyl]thiourea (TU-S) based on hydrogen-bonding organocatalysis [42]. The thiourea simultaneously activates the NCA monomers and reversibly deactivates the polymer chain-ends, silencing the tertiary amine through hydrogen bonding, which allows the polymerization to proceed under a well-controlled procedure.

In addition, Schlaad reported a primary ammonium/tertiary amine-mediated controlled NCA ROP in 2015 [43]. In this work, they showed a versatile way of controlling ROP of NCAs (BLG, L-Leu or L-Phe NCA) by using an initiator system composed of primary ammonium salt and tertiary amine. The polymerization rate could be tuned by varying the primary ammonium salt to the tertiary amine ratio because of an active (amine)-dormant (ammonium) equilibrium. The polymerization could be paused and resumed by adding HCl or TEA, respectively. Also, Barz et al. further developed the ammonium salt strategy by using the tetrafluoroborate salt of primary amine. The mechanism is similar to that of amine hydrochlorides [44, 45].

11.3 Synthesis of Traditional Copolypeptides and Hybrids

In the last 20 years, a large number of copolypeptides have been prepared through NCA ROP. The big family of "copolypeptides and hybrids" includes almost all kinds of polymers. The synthesis strategy of these copolypeptides and hybrids is introduced briefly in this section.

11.3.1 Random Copolypeptides

Commonly, random copolypeptides can be obtained readily through simultaneous copolymerization of two or more NCAs initiated by many compounds. Glatiramer Acetate, a very successful FDA-approved commercial drug to treat multiple sclerosis, is a random polypeptide prepared through NCA ROP [5]. Early works in this field were focused on the study of the reactivity ratios of NCA monomers and the amino acid sequence distribution because random copolypeptides are architecturally like the structures of natural polypeptides. Some groups, such as that of Kricheldorf and Kumar, made great contributions to this field. Among the structures prepared were those listed in a review article as the following [46]:

- random copolymers of γ-benzyl L-glutamate (BLG) and L-methionine, BLG and L-valine (Val);
- glycine and alanine;
- D,L-leucine and D,L-valine;
- ZLys and β-benzyl L-aspartate (BAsp);
- O-acetyl L-tyrosine with L-Val and glycine;
- L-alanine (Ala) and L-Val;
- L-Ala and sarcosine (Sar);
- ZLys and L-Val, and random terpolymers, such as of glycine, L-leucine (Leu), and L-Val

Other studies on L-Leu, L-Val, and BAsp terpolymer and L-Leu with BAsp, BAsp with L-Val, and L-Leu with L-Val bipolymers were also reported regarding the reactivity ratios.

Other works on random copolypeptides were focused on the functions and the applications rather than fundamental studies. The initiators applied in these systems included both ordinary nucleophilic compounds and active initiators. For example, ZLys NCA and O,O′-dicarbobenzoxy L-dihydroxyphenylalanine NCA were initiated by sodium tert-butoxide in THF, and further cross-linked to networks for moisture-resistant adhesives for a variety of substrates to mimic marine adhesives [47, 48]. Deming's group also prepared a series of amphiphilic random polypeptides composed of hydrophilic L-Lys and one of the hydrophobic amino acids L-Leu, L-PheAla, L-Isoleu, L-Val, or L-Ala. The polypeptides were applied to mimic many natural antimicrobial peptides [49]. Moreover, random co-polypeptides can be prepared through partial deprotection of homopolypeptides with protected side groups [50], but the strategy is rarely adopted.

11.3.2 Hybrid Block Polypeptides

The study of hybrid block polypeptide began in the mid-1970s [51]. Nowadays a wide variety of hybrid block polypeptide materials are prepared by different strategies including (1) a macromolecular initiator; (2) a dual functional initiator for step growth polymerization; and (3) a direct combination of blocks.

Hybrid block polypeptides can be obtained stepwise by using the macromolecular initiator approach. Most of this approach was applied in the preparation of a bio-compatible polymer block, such as polyethylene glycol (PEG), polylactic acid (PLA), and polycaprolactone (PCL) containing hybrid block polypeptides [6–8, 38]. The terminal groups of these blocks can be transformed to the amino group [38], or further modified to primary amine salt and N-TMS for the next step in controlled NCA ROP [27, 36]. Moreover, traditional living polymerizations, such as anionic polymerization, could also be terminated by special reagents that can be changed into amino end groups [52, 53]. Reversible addition-fragmentation chain-transfer (RAFT) and other living radical polymerization systems can also adapt the end-group transformation strategy to prepare the corresponding macromolecular initiators [54–57]. Further, the amido-amidate nickelacycle initiators could also be applied for the preparation of hybrid polypeptides. In this approach, N^{α}-allyloxycarbonyl-amino acid allyl amide terminated polymer blocks were prepared first, and then reacted with $Ni(COD)_2$ and bipyridyl to form an active coordination compound for initiating the next controlled NCA ROP. This approach was applied to prepare several di- and tri-hybrid block polypeptides [58, 59]. On the other hand, the whole polymerization could also be started from the polypeptide side. For instance, the amino group at the end of the polypeptide block could easily be transformed into active initiators for the next block through stable free-radical polymerization (SFRP), atom transfer radical polymerization (ATRP), RAFT, and other polymerizations [9].

Dual functional initiators are commonly used in the preparation of hybrid copolymers, including hybrid block polypeptides. One end of the dual functional initiator can initiate NCA ROP, and the other can be transferred for the next polymerization. For example, dual functional initiators containing potential $-NH_2$ (which means the $-NH_2$ groups were protected, or the group could be transformed to $-NH_2$), and the naked hydroxy group were applied to prepare hybrid block polypeptide with aliphatic polyester blocks such as poly(ε-caprolactone), PCL, and polylactide (PLA) [60–62]. Similarly, the compounds containing both $-NH_2$ and controlled radical polymerization active groups were reported to synthesize PBLG-block-Poly (N-isopropylacrylamide) (PNIPAM) [54] and poly(N-vinylpyrrolidone) (PNVP)-b-PBLG-b-PZLys [63].

Through "click" reaction, a polypeptide block could be combined with other blocks directly. As reported by Lecommandoux et al. [64], 1-azido-3-aminopropane was employed as a functional initiator for the polymerization of BLG-NCA, and the resulting azido group terminated PBLG. The block was further combined with excessive propargyl terminated dextran to prepare PBLG-block-dextran. The excess dextran was removed by dialysis against water. Another case of direct combination is the synthesis of poly(2-methyl-2-oxazoline)-b-poly[*O*-(tetra-*O*-acetyl-β-D-glucopyranosyl)-L-serine], which was reported by Tsutsumichi et al. [65]. The glucosylated-NCA was polymerized with n-hexylamine to obtain $-NH_2$ terminated polypeptide block, and the oxazoline was polymerized by using methyl trifluoromethanesulfonate as the initiator of cationic polymerization. The two blocks were then mixed equally to give the corresponding block copolymer.

11.3.3 Block Copolypeptides

Block copolypeptides can easily be obtained by stepwise adding different NCAs to an "active" initiator system. Transition metal mediated polymerization of NCAs was first adopted to prepare well-defined block copolypeptides with a controlled sequence and composition. The transition metal initiators offered enough high activity to polymerize NCAs in a relatively short time and had robust chain ends, allowing the preparation of multi-block liner copolypeptides [2, 6–8]. Furthermore, transition metal initiators also allow the monomers to be polymerized in any order, which is still a big challenge in most vinyl copolymerizations. However, it is difficult to prepare symmetric style block copolypeptides (ABA type, etc.) with transition metal initiators, whereas it is efficient to prepare asymmetric block copolypeptides with controlled molecular weights, narrow molecular weight distributions and high yields.

Although primary amines had been proved unsuitable for well-controlled NCA polymerization, there were still many reports on the preparation of block copolypeptides using these conventional initiators [4–8]. Unfortunately, most of these block copolypeptides did not properly make the structures predicted by monomer feed compositions and probably contained considerable amounts of homopolymers due to the side effects [1]. In contrast, low temperature and high vacuum techniques enable primary amines to be applied for the sequential polymerization of different NCAs to afford various well-defined block copolypeptides, and in addition, ABA-type block copolypeptides could be conveniently synthesized using this approach [30–35]. The application of N-TMS had also been reported for the preparation of block copolypeptides. This strategy could adapt the synthesis of ABA-type block copolymer with well-controlled molecular weights, and also could be applied to the synthesis of asymmetric block copolypeptides [36–38].

11.3.4 Non-linear Polypeptides and Copolypeptides

Despite traditional linear structured polypeptides and copolypeptides, star-shaped, hyperbranched, comb-like, and cyclic polypeptides have also received increasing consideration in the last 20 years. Compared with linear polypeptides, the special topological structures make these non-linear polypeptides suitable for various applications [11].

Star-like, dendritic, and hyperbranched polypeptides are usually applied as vehicles for drug and gene delivery [11, 46], In general, two different synthetic methodologies are adopted to generate star-shaped polypeptides, i.e., divergent and convergent. The commonly applied divergent, or "core first" approach allows simultaneous synthesis of polypeptide arms by using star-like initiators with several primary amine groups, including star-shaped small molecules and dendrimers. In addition, the multi-functional initiators could be asymmetrically designed to obtain hybrid star-like polypeptides with different arms. The convergent or "arm first" approach enables people to prepare star-shaped polypeptides by stitching linear polymer arms to a star-like core via coupling reactions. This strategy was also applied to prepare asymmetric star-shaped hybrid polypeptides. The up-to-date progress in the synthesis and applications of star-shaped polypeptides has recently been reviewed [11]. Moreover, recent advances in hydrogen bonding catalyzed living polymerization presented an easy way to prepare multi-armed polypeptides by using di-, tri-, or tetra-amino-alcohol initiators [42].

Hybrid dendritic/hyperbranched polypeptides polymers were usually prepared through amine-terminated dendritic/hyperbranched polymer cores-initiated NCA ROP [66, 67]. Commercially available hyperbranched polyethylenimine (PEI) might be the most commonly adopted core with initiator activity. For example, Tian et al. synthesized a series of hyperbranched PEI-*graft*-polypeptides by using hyperbranched PEI to initiate NCA ROP directly (Figure 11.3). The obtained PEI-*graft*-polypeptide further proved to be very effective in gene delivery [67–71]. On the other hand, hyperbranched polypeptides obtained through NCA ROP are seldom reported except some cases of hyperbranched PLys. The hyperbranched PLys could further serve as the core to conjugate with other polymer arms [72–75]. The discovery of "allied amines" mechanism would offer the chance of a new look at the reaction of multi amines (for instance, dendritic/hyperbranched PEI and PAMAM) and NCAs [40, 41].

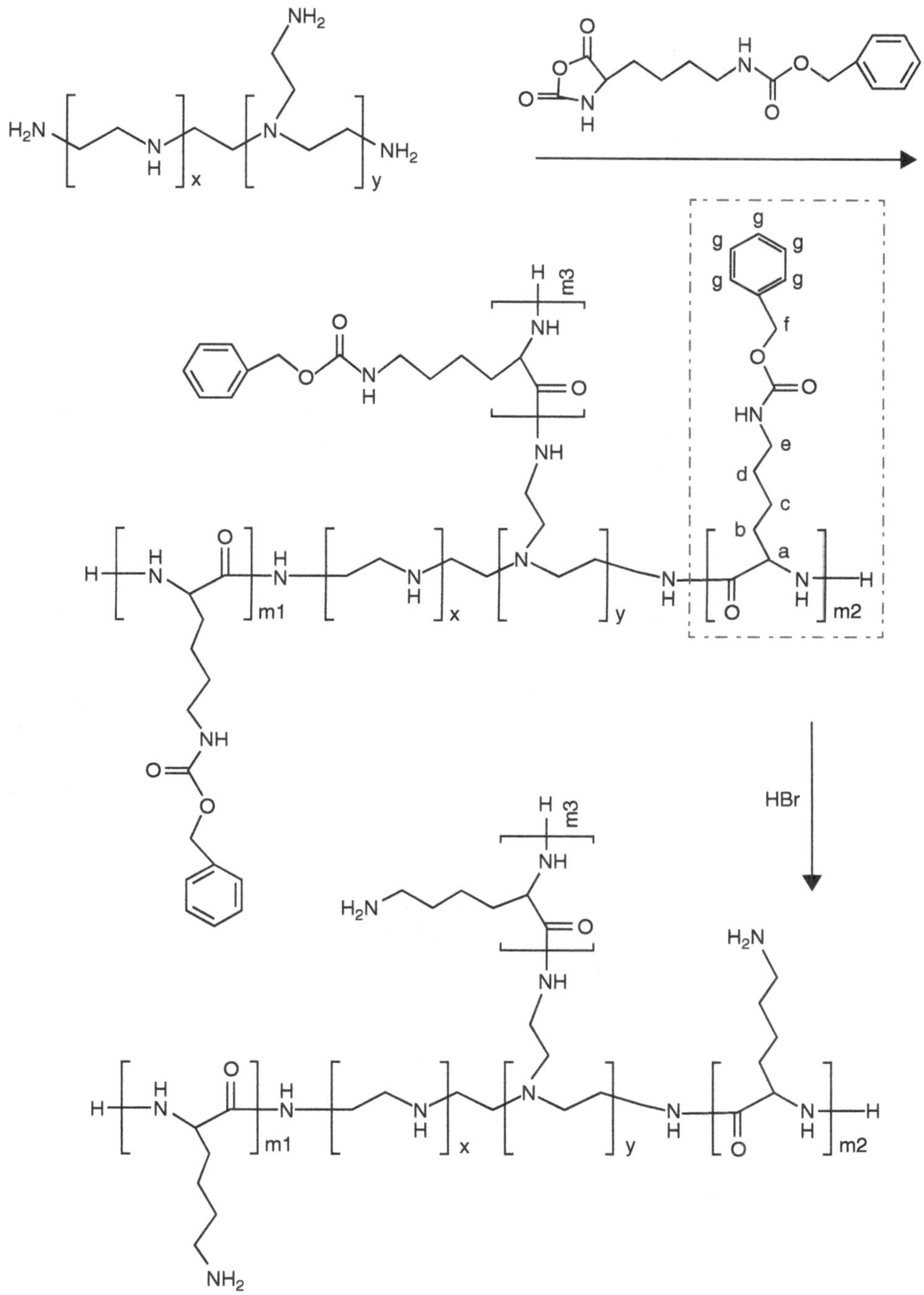

Figure 11.3 Synthesis of PEI-PLL copolymers. *Source:* Reproduced with permission from [71]

Comb or brush-like polypeptides are divided into three kinds: (1) polypeptide backbone/branch; (2) polypeptide backbone with other polymer branches; and (3) other polymer backbone with polypeptide branches. Three strategies have been developed to prepare comb-like copolypeptides as (1) the "grafting onto"; (2) the "grafting from"; and (3) the "grafting through" method. Although the strategy "graft onto" is seldom adopted since the combination of macromolecules is relatively difficult, there have been some interesting instances of this strategy as the polyelectrolyte-style copolypeptide-surfactant charge complex brushes. These complex brushes could form a helical conformation and further obtain hierarchical nanostructures in the solid state, or show interesting self-organized structures in solution [76–79]. Schmidt et al. compared the "grafting through" and the "grafting from" approach by preparing polymer brushes with PZLys and PBLG side chains [80]. For the "grafting through" approach, the macromonomers were prepared by conjugating the N-terminus of PZLys and PBLG with methacryloyl chloride and polymerized by free-radical polymerization. For the "grafting from" approach, poly[(2-aminoethyl)methacrylamide] with primary amino groups was synthesized as the initiator for NCAs. It was proved that the polymerization degree of the "grafting through" approach was fairly low. In contrast, almost each of the amino groups involved the initiation, and obtained peptide chains with an average length of approximately 10 units. According to the "graft from" strategy, a number of comb-like polypeptides were obtained. These interesting macromolecules took advantage of their topological structures for drug delivery, or assembly into special micro/nanostructures in solution and bulk [81–84].

Cyclic polypeptides were observed in the thermal polymerization of N-unsubstituted NCAs, but the products are always complex mixtures [85]. Secondary or tertiary amines may promote NCAs polymerizing to cyclic polypeptides. Sar-NCA, D,L-Leu-NCA, and D,L-Phe-NCA with imidazole initiator were reported by Kricheldorf et al., leading to mainly cyclic polypeptides in dioxane at 60 °C [85]. The formation of α-helix and β-sheet might hinder the cyclization due to the rapid precipitation, which occurred in the case of L-Ala-NCA polymerization [86]. Kricheldorf's group also discovered pyridine as an initiator for cyclic polypeptides preparation by using Sar-NCA, and it was further proved that a zwitterionic polymerization mechanism was involved in the polymerization [87]. Furthermore, sterically hindered trialkylamines were also found to initiate the cyclic polymerization of N-unsubstituted NCAs in dioxane [87]. Besides, a different approach was to synthesize linear telechelic polymers bearing amino end groups first, and then react with linking agents to form cyclic polypeptide [88]. At least a 50% yield was accomplished by using hydroxysuccinimide sebacate or sebasic acid bisimidazolide as the linking agent, but the PDI was broad due to the nature of linking reaction.

11.4 New Monomers and Side-Chain Functionalized Polypeptides

In the last few decades, there has been a strong motivation to synthesize side-chain functionalized polypeptides based on novel NCA monomers and/or facile link reactions. This mostly productive topic offers a different scope from the twenty amino acids composed from the natural proteins [12].

11.4.1 New NCA Monomers

Click reactions have been the most popular strategy to modify the side chain of polypeptides since the first report by Hammond in 2009 [89, 90]. In this procedure, clickable NCA monomers are prepared first, and then polymerized into polypeptide with clickable chain ending groups. After the click reaction with the corresponding reagents, the functionalized polypeptides are obtained finally. In general, clickable groups are alkyne, azide for 2 + 3 Husigen cycloaddition, or alkene for thiol-ene reactions [90]. Furthermore, alkyne-ending monomer and alkene-ending monomer can be added in sequence, and react with the corresponding reagents respectively to synthesize multi-functional block copolypeptides [90]. Another modification strategy is to introduce halogen (Br or Cl) groups at the end of the NCA side chains [91, 92]. After the NCA has polymerized, the halogen groups can be replaced by azide groups. This kind of monomer could also be applied to prepare brush-like polypeptides with polypeptide backbones while the end group can initiate living radical polymerization [91].

If the functional groups are inactive to the NCA rings, the corresponding NCA monomers can be prepared directly. The monomers are usually based on glutamate acid, lysine, and cysteine. This strategy has been successful in preparing some glycopolypeptides, oligo-PEGylated polypeptides, aromatic derivatives containing polypeptides and oligo-conducting polymer modified polypeptides in recent years [93–97]. It offers a 100% modification to the side chain of polypeptides, and usually results in a higher yield. However, this strategy is only suitable to introduce NCA inactive groups into the side chains.

11.4.2 Glycopolypeptides

Glycosylated polypeptides play diverse biological functions in nature. Therefore, synthesized glycopolypeptides received great attention not only for mimicking the natural functions but also for biomedical materials [12, 98]. As mentioned above, there are two fundamental routes to prepare glycopolypeptides based on NCA ROP: One is to synthesize glycosylated monomers directly, and the other is post-polymerization functionalization.

The post-polymerization functionalization strategy shows better adaptability that could incorporate more sugar functionality to polypeptides. In the early stage, the connection route relied traditionally on the formation of amide bonds to amine or carboxylic acid side-chain groups in lysine, glutamic acid, or aspartic acid residues

[99–101]. However, the glycosylation efficiency and loading ratio were usually high enough due to the limitations in reactivity and the solubility of the polypeptide residues, and steric crowding of the sugar units. In 2009, click reaction was introduced by Hammond et al. for side-chain functionalization [102], and soon afterward, Chen et al. adopted the strategy in the preparation of glycopolypeptides [103]. Three different azide functionalized monosaccharides were coupled with alkyne functional homopolypeptides synthesized from γ-propargyl-L-glutamate NCA. Zhang et al. then prepared poly(γ-3-chloropropyl-L-glutamate acid) [104]. After the conversion of chloro to azide groups, the polypeptides were coupled with alkyne functionalized D-mannose. This coupling chemistry was found to proceed with near complete conversion and a high glycan yield. The only disadvantage might be the hydrolytic instability of the glutamate ester linkages. To address this issue, some NCA monomers bearing clickable groups were prepared in 2010 [90]. For instance, Heise et al. reported the synthesis of poly(D/L-propargylglycine) through racemic propargylglycine amino acid [105]; Deming's group prepared azide-bearing polypeptide through N^{α}-Carboxybenzyl-L-azidonorvaline and Nα-carboxybenzyl-L-azidonorleucine NCA [106]; Schlaad synthesized poly(D/L-allyglycine) and incorporated monosaccharide through thiol-ene glycosylation [107, 108]. Poly(L-methionine) was functionalized with glycosyl triflate and iodoethyl glycoside as described by Deming and Kramer [109, 110]. These hydrolytically stable glycopolypeptides are suitable in biological environments.

Although the first glycosylated NCA, O-linked glycosylated-serine (glyco-Ser) NCA was prepared in 1966 [111], the glycosylated NCA direct polymerization received only a few reports in a long time because the purity of the NCA monomer could not meet the requirements of polymerization, and the glycols were too near to the backbone, which caused stronger steric and H-bonding interaction [98]. To overcome these difficulties, Deming's group developed a series of glycosylated L-lysine NCA (glycol-Lys NCA) monomers and a purification strategy [93, 112]. These NCAs are more stable because of their amide linkages, and provide long linker for a better steric environment. The corresponding glycopolypeptides can be obtained easily in less than 3 h with more than 150 residues. In 2011, a series of peracetylated monosaccharides (glucose, mannose, or galactose) were linked to carboxybenzyl-L-lysine via a thiourea linker by Gupta and Wenz [113, 114]. Gupta also reported several glycosylated lysine NCAs by other coupling modifications and the following conjugates of the corresponding glycopolylysines with other polymers [115, 116]. Furthermore, Deming and Kramer described a series glycosylated poly-L-cysteine [93, 117]. The glycosylated L-cysteine (glyco-Cys) NCA monomers were prepared by coupling allyl functionalized pre-protected sugars to L-cysteine using the thiolene click reaction, and then converted to corresponding NCAs via treatment with Cl_2CHOCH_3. These glyco-Cys NCAs can also be applied to prepare glycosylated block copolypeptides.

11.4.3 Water-soluble Polypeptides with Stable Helical Conformation

Water-soluble polypeptides with a stable helical conformation have received extensive interest because of their potential applications in drug and gene delivery and because they mimic the complex chiral behaviors of natural proteins. However, ordinary hydrophilic amino acid residues often limit the helicity of polypeptide chain due to side-chain charge repulsion. The first synthesis of water-soluble polypeptide with stable helical conformation through NCA ROP was reported by Deming et al. [118]. In this approach, short ethylene glycol (EG) segments protected lysine NCA monomer ($(EG)_2$-Lys NCA) was prepared and polymerized. CD spectra showed the polypeptide contained almost 100% α-helix in an aqueous solution. Similarly, a series of polylysine derivatives with glycosylated side chains were also found to be high α-helical in an aqueous media [93, 112–117]. In contrast, polyglutamate derivatives poly ((EG)x-Glu) were reported as thermal-responsive polypeptides by Li et al., and the α-helix content changes with the EG number [119, 120]. In addition, the helicity of poly (galactosyl-L-lysine) showed temperature dependence so that roughly 40% helicity was retained at 90 °C [12, 105].

Recently, Cheng et al. developed a series of hydrophilic ionic polypeptides with stable α-helix based on a PGlu skeleton (Figure 11.4) [121, 122]. Their strategy is to add inert spacers between the polypeptide skeleton and the ionic groups at the end of side chains. It was found that there should be at least 11 bonds between the skeleton and the end group to maintain the α-helix. Based on the result, a number of ionic polypeptides with stable conformations were developed through NCA ring-opening polymerization and further functionalization [5, 123].

Another subtle design of ionic polypeptide with helical conformation is a zwitterionic side chain. The interaction of two opposite ions weakens the whole deformation effect of the conformation [124]. In addition, poly(L-lysine-*random*-L-glutamic acid) shows a partly helical conformation while the ratio of lysine/glutamic acid residues is nearly 1:1 due to the same reason [125]. However, if the charges are near to the chain, the polypeptide would still show disordered conformation, which was proved by the study of poly(L-phosphorylcholine homoserine) [126].

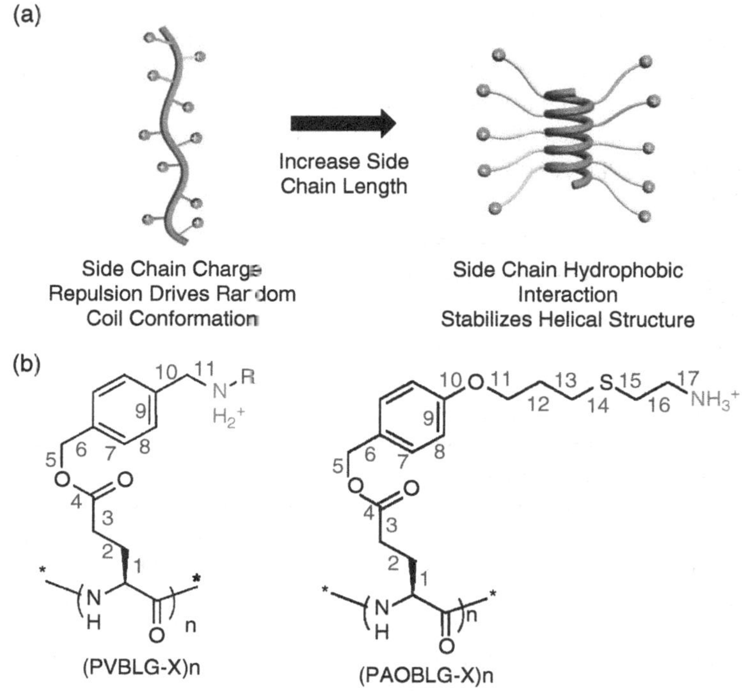

Figure 11.4 (a) Schematic illustration of the elongated hydrophobic side-chains leading to the formation of polypeptide helixes with unprecedented stability. (b) Structure of the polypeptides with 11 and 17 σ-bonds of side chain charge-backbone distances. *Source:* Reproduced with permission from [121, 122].

11.4.4 Stimuli-responsive Polypeptides

Stimuli-responsive polypeptides show great ability in responding to various internal/external triggers. Traditional pH-sensitive ionic polypeptides, such as PLys, PGlu and poly (aspartic acid) (PAsp) have been applied as pH-responsive blocks for several decades. In recent years, many stimuli-responsive polypeptides have been synthesized to match the developing bio-related applications [127–129].

The first thermo-responsive polypeptide (poly (L-EG_2Glu)) was accomplished by PEGylated-L-glutamate NCA ring-opening polymerization in 2011 (Figure 11.5) [119]. It was found that the secondary structure of the polypeptides is critical to the thermo-responsive property. The lower critical solution temperature (LCST) can be tuned with the length of oligo (ethylene glycol) (OEG). Similarly, side-chain PEGylated polycysteine also showed thermo-responsive property [130, 131]. These PEGylated polycysteines could be obtained through click reactions. Recently, side-chain PEGylated thermo-responsive polypeptides have been conjugated with other blocks to construct a thermo-responsive system [132, 133]. Some other polypeptide-contained block copolymers showed thermo-responsive properties in non-aqueous solvents. For example, PBLG-*block*-polydimethylsiloxane-*block*-PBLG (PBLG-*b*-PDMS-*b*-PBLG) was found to have thermo-induced gel-to-sol behavior in toluene due to its special rod (α-helix of PBLG)-coil (PDMS)-rod structure [134]. Some PEG-block-β-sheet polypeptides have also been found to have thermo-reversible gelation behavior in various solvents [132, 135–137]. The interesting polypeptide gels will be further introduced in Section 11.5.

The vast majority of redox-responsive polypeptides are based on thiol chemistry, such as disulfide bond and the thioether group. Deming's group reported the conformation of redox-responsible glycosylated polycystine [93, 138]. The polypeptide is water-soluble and shows α-helical conformation in solution. When the side-chain thioether linkages were oxidized to sulfone, the α-helical conformation was disrupted without loss of water-solubility. Redox-responsibility is more commonly achieved by controlling the formation/cleavage of disulfide bonds. For example, Zhang's group synthesized block copolymer poly(ethylene glycol)-*block*-poly(L-cysteine)-*block*-poly(L-phenylalanine) (PEG-*b*-PCys-*b*-PPhe) [139]. The thiol groups of PCys block can be reversibly cross-linked and broken under oxidative and reductive conditions. In addition, a disulfide bond could also be introduced as the linkage between blocks of the hybrid polypeptides to perform redox-sensitive cleavage under certain conditions [140, 141].

Light-responsive polypeptides are prepared mainly through the conjugation of side-chain light-sensitive functional groups or by introducing light-sensitive group contained initiators. Common light-responsive species, including azobenzene, spiropyran, coumarin, 2-nitrobenzyl and other functional groups, had been incorporated into various kinds of polypeptides for light-triggered reversible behaviors in earlier works [142–146]. When

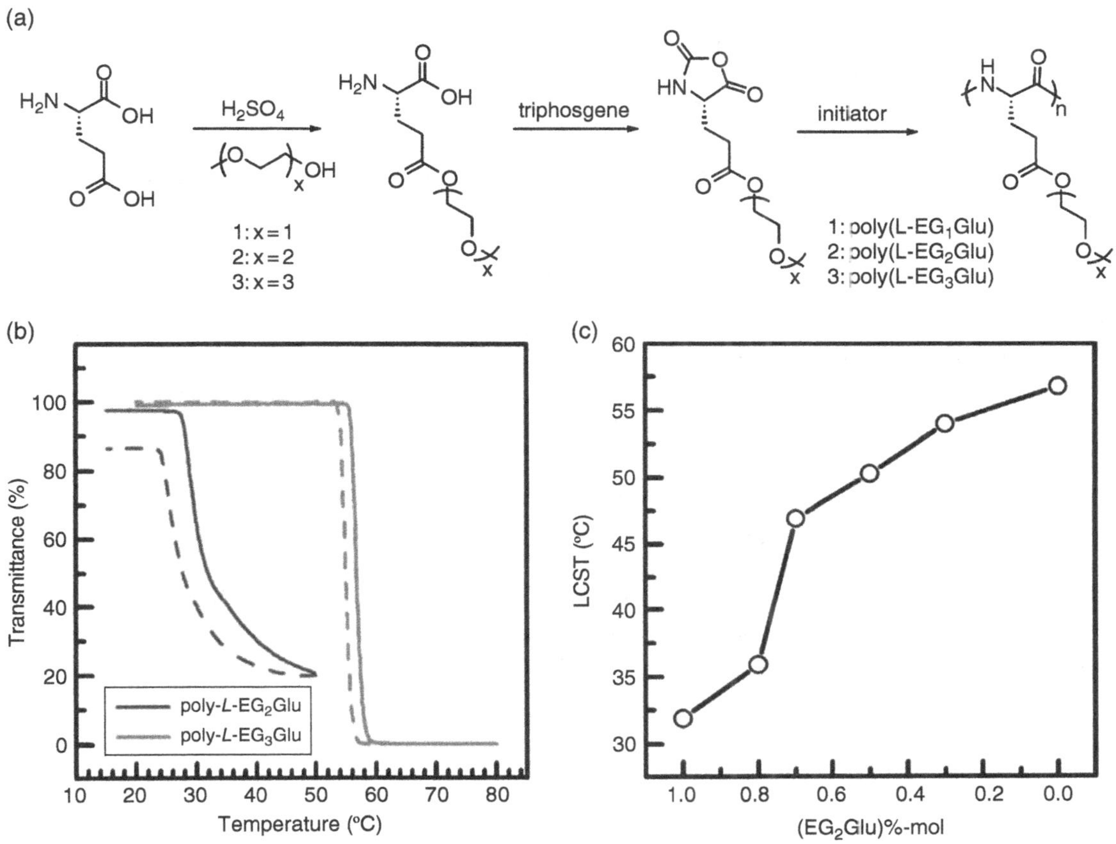

Figure 11.5 (a) Synthetic routes to poly(L-EGxGlu) homopolypeptides' (b) plots of transmittance as a function of temperature for aqueous solutions (2 mg mL^{-1}) of poly(L-EG_2Glu) and poly(L-EG_3Glu). Solid line: heating; dashed line: cooling. (c) LCST of poly(L-EG_2Glu-*co*-EG_3Glu) copolypeptides as a function of sample composition. *Source:* Reproduced with permission from [119].

exposed to light at a particular wavelength, the light-responsive polypeptides undergo some reversible physical/chemical property changes, such as the transition between α-helix and random coil, aggregation-disaggregation and the sol-gel process, viscosity, and solubility. Compared with other stimuli-responsive strategies, light-responsive polypeptides allow a remote light-controlled drug/gene release process in both a spatial and temporal manner, and might be particularly useful in bio-related therapy.

Recently, some novel pH-sensitive polypeptides prepared via a post-modification approach have been introduced as functional blocks for controlled release. For example, Hommand et al. reported a library of pH-responsive polypeptides through the click reaction between poly(γ-propargyl-L-glutamate) (PPLG) and amino azides [147]. A dual-responsive polypeptide with both temperature and pH responsibility was also prepared by grafting oligo(ethylene glycol) and tertiary amine groups to PPLG via click reaction [148, 149]. As well as the click reaction, other strategies were also adopted in this field. For instance, Kataoka's group reported the PAsp-based pH-responsive polypeptides by aminolysis reaction of poly(β-benzyl-L-aspartate) with polyamines [150].

Biomolecule-responsive polypeptides can be obtained by integrating the corresponding functional groups into the side chain of polypeptides. Glycopolypeptides are the most commonly studied biomolecule-responsive polypeptide because the sugar groups can be recognized by lectins, which are highly specific carbohydrate binding proteins [93, 98, 113, 115] Moreover, Chen's group reported a glucose-sensitive polypeptide by the incorporation of phenylboronic acid moieties into the polypeptide side chain. Phenylboronic acid can reversibly bind with glucose for an insulin delivery system [151]. In addition, some polypeptides designed to be sensitive to bio-catalyzed redox reactions could also be classified in this field. For instance, Deming et al. developed an enzyme-responsive copolypeptide by the incorporation of oxidable methionine units into hydrophobic copolypeptide, poly(L-methionine)$_{65}$-*b*-poly(L-leucine$_{0.5}$-*stat*-L-phenylalanine$_{0.5}$)$_{20}$ ($M_{65}(L_{0.5}/F_{0.5})_{20}$). Under a mild oxidation condition, the α-helix of poly(L-methionine) block underwent a transition to random coil accompanied by enhanced hydrophilicity. The obtained amphiphilic copolypeptide $M^{O}_{65}(L_{0.5}/F_{0.5})_{20}$ can self-assemble into vesicles in water (Figure 11.6). In the presence of dithiothreitol and methionine sulfoxide reductase enzyme, the vesicles can be disrupted due to the reduction of methionine sulfoxide to methionine and will release the encapsulated drug [152].

Besides, a number of functional groups were introduced into the polypeptide side chains to prepare light-, pH- and redox-responsive gels [153, 154]. For example, PLys and poly(L-ornithine) side chains were conjugated with coumarin moieties that could cross-link through photo-induced cyclodimerization, resulting in a sol-to-gel transition after light irradiation [155, 156]. Recently, Chen's group reported a hydrogel system based on tyramine and OEG grafted PGlu, horseradish peroxidase, and H_2O_2 [157]. The sol-to-gel transition was achieved through horseradish peroxidase catalyzed oxidative reaction between tyramine groups.

11.5 The Self-assembly of Polypeptides

The self-assembly of polypeptides is very interesting because polypeptide features both the natural protein and classic polymer properties. Polypeptides can form a variety of nano/micro structures under different conditions. In addition, polypeptides can also co-assemble with inorganic precursors or direct the formation of inorganic materials. Some reviews have given a comprehensive overview in the self-assembly of amphiphilic polypeptides [4–8]. In this section, some up-to-date advances are introduced and highlighted in this rapid developing field.

11.5.1 Chiral Self-assembly

Synthesized polypeptides are a promising chiral assembly candidate because natural polypeptides and proteins usually assemble in hierarchical chiral structures. In 2001, Boden et al. reported a statistical mechanical model for the self-assembly of rod-like, β-sheet-forming peptides into helical tapes [158, 159]. The helical tapes can further assemble into twisted ribbons, fibrils or

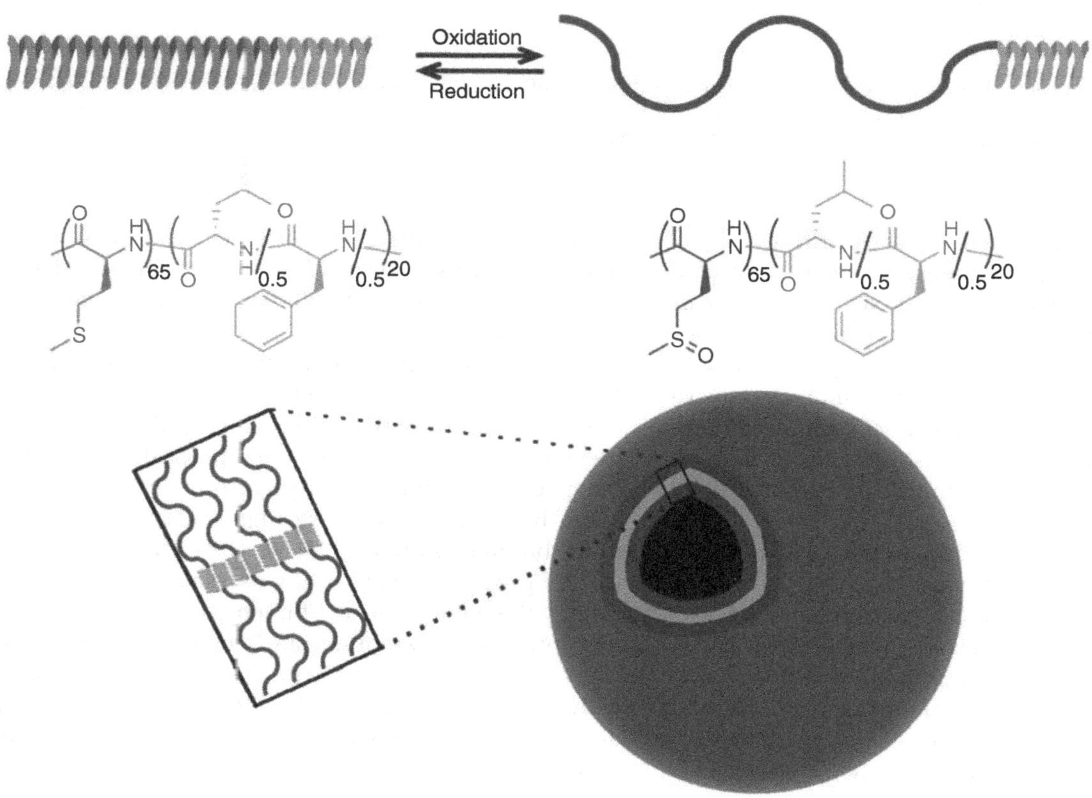

Figure 11.6 Schematic showing structure, redox properties, and proposed self-assembly of $M^{O}_{65}(L_{0.5}/F_{0.5})_{20}$ copolypeptides into vesicles. *Source:* Reproduced with permission from [152].

fibers at higher concentrations. Recently, the self-assembly behavior of hard helices has also been presented [160]. These results discovered polypeptides and their derivatives can assemble into micro/nano structures with chiral characters theologically.

Some helical polypeptide supermolecular structures have been observed in the last few decades [161]. Cheng and Lin reported the supermolecular polymerization of PGlu-grafted comb-like polymer into tubular superstructures (Figure 11.7a) [84]. The formation of these superstructures is derived from the association of the neighboring anti-parallel β-sheet of PGlu. In addition, the nucleation-controlled supramolecular polymerization of PGlu-grafted gold nanoparticles was also reported in 2013 (Figure 11.7b) [162]. Also, Cai and Lin reported a series of superhelical structures constructed by PBLG-b-PEG and PBLG [163, 164]. The rod-coil block copolymers can cooperatively self-assemble with rigid PBLG homopolymers into tobacco mosaic virus (TMV)-like core-shell superhelical structures with the block copolymers wrapping around the polypeptide homopolymer bundles. The chirality of the superhelical structures can be controlled by tuning the temperature and solvents [163].

11.5.2 Self-assembly with Inorganic Sources

Polypeptide-participated mineralization has also been a focus for almost 20 years because natural polypeptides and proteins play vital roles in biomineralization. In the early study of this area, anionic polypeptides and their block copolymers, such as PAsp, PGlu, PGlu-*b*-PEG, were introduced to control the formation of calcium carbonate as a mimic of natural biomineralization [13–15, 165–168]. PLys was then found to direct the formation of calcium carbonate nanocrystal clusters with a special morphology [169].

In 2000, Deming et al. reported the mineralization of silica directed by several polypeptides first [170]. Some silica bundles and beads were obtained in this study. In addition, by using PLys and its OEG derivatives, hexagonal silica platelets and helical silica platelets were prepared [171, 172]. It was confirmed that the α-helices play an important role in the formation of these platelets. With the help of the co-structure direction agent (CSDA), the α-helice of PLys/CSDA complex strands could self-assemble into very rare two-dimensional square symmetry through silica mineralization (Figure 11.8) [173]. Also, other silica

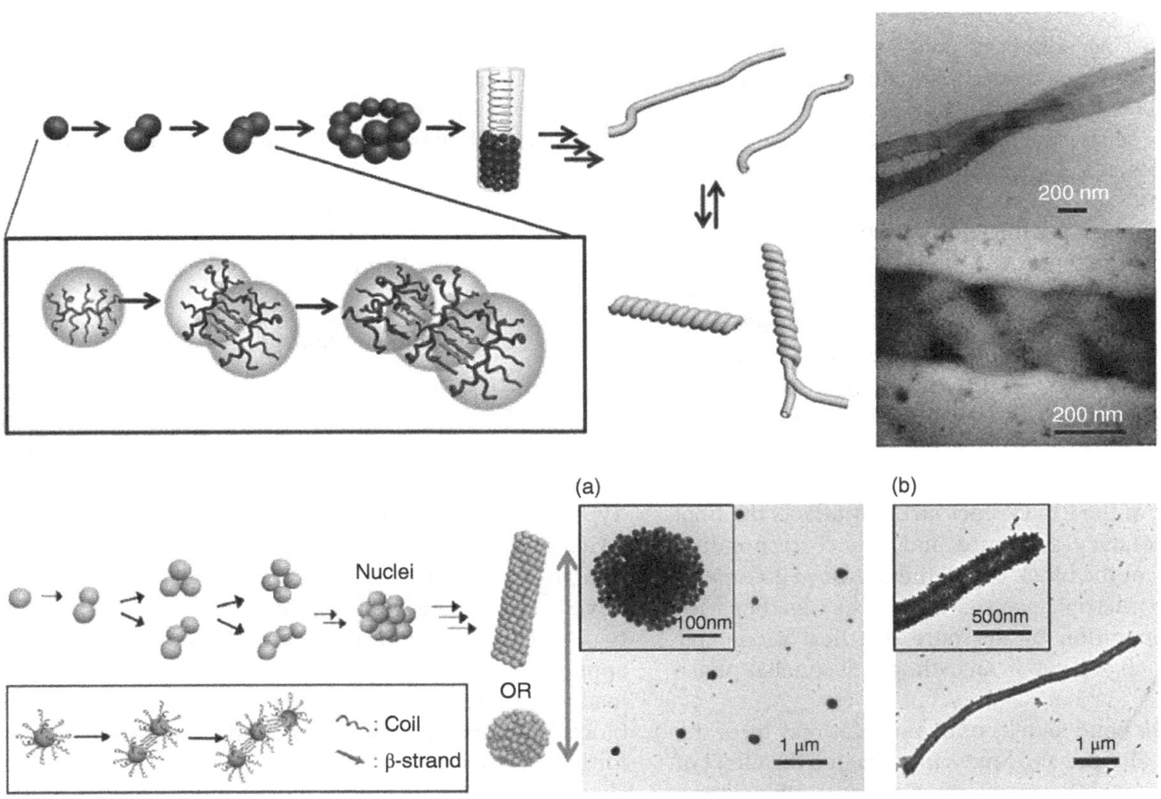

Figure 11.7 (a) Supramolecular polymerization of polypeptide-grafted comb-like polymers into tubular superstructures; (b) nucleation-controlled supramolecular polymerization of polypeptide-grafted gold nanoparticles. *Source:* Reproduced with permission from [84] and [162].

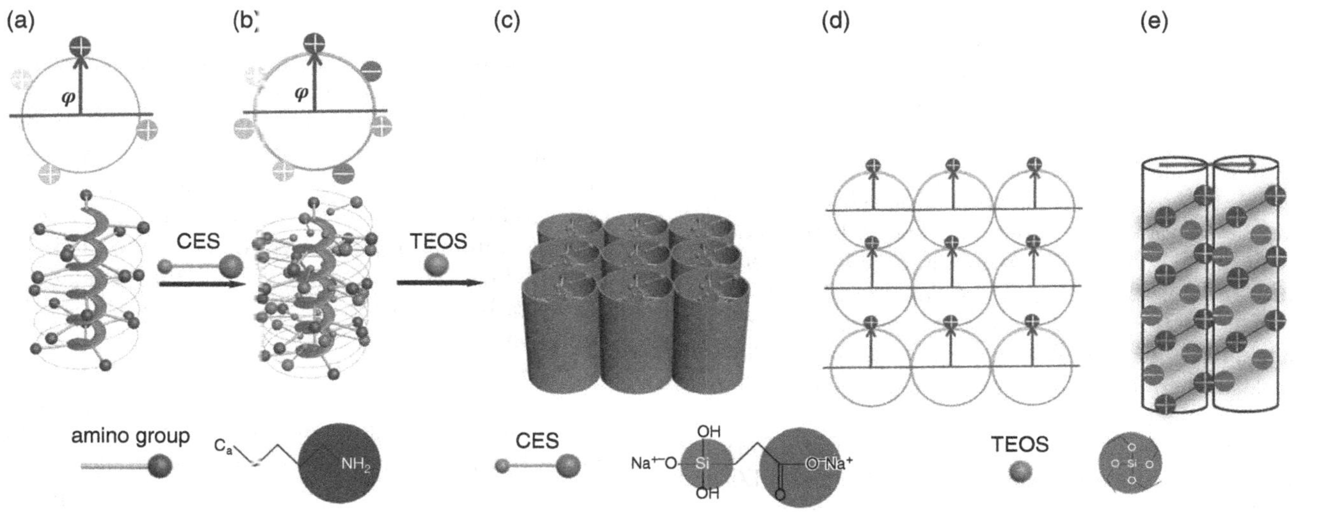

Figure 11.8 Illustration of the mechanism of formation for PLys-silica complex with a 2D square structure. *Source:* Reproduced with permission from [173].

morphologies, such as hollow spheres, nanoparticles, have also been prepared and summarized in former reviews [14, 174, 175].

11.5.3 Microphase Separation of Polypeptides

The microphase separation of polypeptide block copolymers is usually different from traditional block copolymers because the secondary structures of polypeptides leads to special phase separation behaviors. Unlike many other flexible synthetic polymers, polypeptides may possess either rigid rod-like α-helix or intermolecular hydrogen bonding (e.g. β-sheet forming polypeptides).

According to the reports of Klok, Lecommandoux, Lin and Mezzenga, the secondary structure of polypeptides offers opportunities to direct nanoscale structure formation, such as lamellar microphase separated structures or rods [176–178]. For example, Mezzenga studied the microphase separation diblock PBLG-*b*-poly(propylene oxide) (PPO), triblock PBLG-*b*-PPO-*b*-PBLG, and star-like PPO-*star*-PBLG [178]. He found that the polymerization degree of the PBLG block strongly affects the final α-helical secondary structure and the corresponding self-assembly of the block copolymers. Above 20 repeating units, the microphase separation goes together with the phase separation of the pure α-helical secondary structure, which ends up in an orthogonal lamellar phase or rods.

On the other hand, block copolypeptides may be of the rod-rod type. Floudas and Spiess investigated the effect of chain topology on the phase separation of diblock and star co-polypeptides of PBLG and PZLL [179]. It was found that linear diblock copolypeptides are nanophase separated with PBLG and PZLL domains comprising α-helices packed in a hexagonal lattice. In contrast, star-type copolypeptide showed only weak phase separation, and the helix persistence of PZLL and PBLG was reduced under the confinement. Furthermore, rod-rod type copolypeptides also allow other phase separation structures. Gkikas et al. studied the phase separation behavior of poly(γ-benzyl-L-glutamate)-*block*-poly(benzyl-L-hydroxyproline) (PBLG-*b*-PBLHyP) [180]. For shorter PBLHyP blocks, the hexagonal packing of PBLHyP helices was observed, while by increasing the length of the PBLHyP block and keeping at a similar PBLG block length, the packing is distorted. Zig-zag lamellar structures were obtained due to the mismatch in the packing periodicities of the PBLG and PBLHyP helices.

11.5.4 Self-assembly in Solution

During the last 20 years, a number of amphiphilic block polypeptides have been reported to self-assemble into various nanostructures, such as micelles, vesicles, fibers, tubes, ribbons, and other morphologies [4–8, 177]. Typically the block copolymers are designed as rod-coil style. The "rod" block usually contains the secondary structure as α-helix or β-sheet, the "coil" block is more flexible and provides the solubility [64]. In addition, the two blocks can be further functionalized for various applications. For example, the stimuli-responsive groups loaded polypeptide are usually introduced as the coil block for controlled release, and other functional groups for labeling, promotion of targeting, and cell uptake can also be incorporated into polypeptides through side-chain functionalization [181]. Amphiphilic polypeptides are also designed and one of the blocks is stimuli-responsible. These polypeptides have been widely applied for

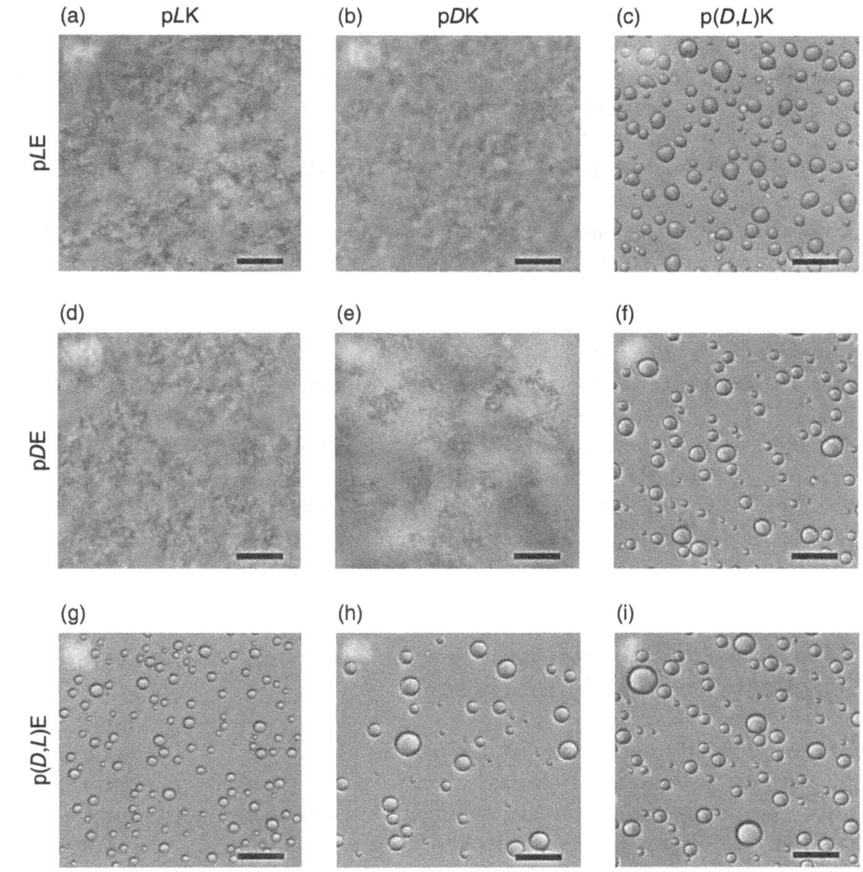

Figure 11.9 Bright-field optical micrographs showing the liquid coacervates or solid precipitates resulting from the stoichiometric electrostatic complexation of L, D, or racemic (D,L) poly(lysine) (K) with L, D or racemic (D,L) poly(glutamic acid) (E) at a total residue concentration of 6 and 100 mM NaCl. Complexes are formed from (a) pLK + pLE, (b) pDK + pLE, (c) p(D,L)K + pLE, (d) pLK + pDE, (e) pDK + pDE, (f) p(D,L)K + pDE, (g) pLK + p(D,L) E, (h) pDK + p(D,L)E, (i) p(D,L)K + p(D,L)E. Liquid coacervate droplets are only observed during complexation involving a racemic polymer. Scale bars, 25 μm. *Source:* Reproduced with permission from [184].

controlled drug, gene, and siRNA delivery [127–129]. For example, Cheng et al. prepared amphiphilic diblock rod-coil copolymer poly(ethylene glycol)-*block*-poly(γ-(4,5-dimethoxy-2-nitrobenzyl)-L-glutamate) (PEG-*b*-PL) for light-responsive drug delivery. The polypeptide can form vesicles with unprecedented densely packed multilayer membrane structures that benefited from the rigid helical hydrophobic block, and in response to ultraviolet light induced helix-to-coil transition and the eventual disassembly of vesicles [182].

The development of ionic polypeptide with a stabile helical conformation reveals a new view of the design of polypeptide assembled nanostructures because the stabile helical structure may provide considerable robust helical template for the coacervate-derived self-assembly [183]. A recent study on poly(γ-3-(4-(guanidinomethyl)-1*H*-1,2,3 triazol-1-yl)propyl-L-glutamate (PPLGPG) and diblock copolypeptide poly(ethylene glycol)-b-polypeptide (PEG-b-PPLGPG) coacervate with PGlu, PDGlu and PEG-b-PGlu showed the very interesting result that the stabile α-helices could prevent the formation of β-strands, which would drive the assembly into a solid state, thereby enabling coacervate formation between two chiral components, and then forming nanoscale coacervate-core micelles (Figure 11.9). On the other hand, chirality-selected phase behavior in ionic polypeptide complexes without a stable helical conformation has also been discussed [184]. It was found that the polypeptide chirality not only determines the physical state of the resulting polyelectrolyte complexes (liquid or solid), but also defines the strength of intermolecular interactions, and thus the material properties.

11.5.5 Polypeptide Gels

In 2002, Deming et al. developed block polypeptides composed of one charged block and the other hydrophobic segment containing rod-like α-helix or crystalline β-sheet for the preparation of robust hydrogels [185, 186]. The hydrophilic blocks are highly charged and dissolve readily in water. The hydrophobic domains feature regular conformations that aggregate and are insoluble in water. This special structure resulted in hydrogels that retain their mechanical strength up to 90 °C and recover rapidly after stress. Following this strategy, a library of block polypeptide were prepared. These hydrogels have been applied for tissue engineering both *in vitro* and *in vivo* [154].

Despite robust hydrogel systems, stimuli-responsive polypeptide gels also receive increasing focus [153, 154]. The sol-gel transition could be achieved in two ways. The first is the assembly/disassembly of the secondary conformation. This strategy is often adopted to prepare thermo-mechanical force-responsive polypeptide gels. One of the popular designs for thermo-responsive polypeptide hydrogelators is to introduce PEG/OEG segment into hybrid copolypeptides because the PEG block dehydrates and the polypeptide block shows more β-sheet content with the temperature increase [153]. Some β-sheet rich PEG-*b*-polypeptide systems also feature sonication-sensitive gel-to-sol property. The other style of stimuli-responsive polypeptide gels depends on the stimuli-sensitive side chain functional groups. The gelation process of this strategy often is accompanied with chemical reactions (for example, photo- or enzyme-induced cross-link) and was briefly mentioned in Section 11.4.4.

11.6 Novel Bio-related Applications of Polypeptides

The bio-related applications of polypeptides have been pursued for many years. In the last decade, polypeptides have been prepared for many bio-related applications such as carriers for DNAs, controlled drug release, and tissue engineering. Nowadays, the application of polypeptides has expanded to a number of new areas while it continues developing in "traditional" areas [4–8].

11.6.1 Drug Delivery

The most popular and important application of polypeptides is the carrier for controlled drug delivery. In this area, PEG-polypeptide, polyester-polypeptide, polyacrylate-polypeptide hybrid copolymer and polypeptide copolymer micelles have been studied systematically and many copolypeptides, such as PEG-b-PGlu/PAsp, PPO-b-PLys and poly(L-lactic acid) (PLLA)-PGlu/PAsp, have proved suitable for carrying a number of drugs, as shown in earlier works [4, 187, 188].

Stimuli-responsive polypeptides are promising carriers for adapting hydrophobic and cytotoxic chemotherapy drugs, and concisely releasing the drug molecules according to the microenvironment difference. For instance, tumor cells have a lower pH value, higher temperature and show over-expression of some enzymes. One common strategy to design stimuli-responsive polypeptide carriers for anti-cancer drugs is to introduce ionic amino acid residues or blocks, and manipulate the hydrophilic-hydrophobic triggering range to adapt the pH value of tumor tissues [125, 189, 190]. Recently, side-chain modification has been developed as an important strategy for this purpose. For example, a library of polypeptide with tertiary amines on the side-chain were prepared through the click reaction of PPLG and azide tertiary amines [147]. The response pH values of these polypeptides are in the range of 5.0–8.0, which offers the opportunity to be applied in the bloodstream (pH = 7.4), hypoxic tissue (pH = 6.5–7.0), and late endosomal to lysosomal systems (pH = 5.0–6.0).

Stimuli-responsive blocks can also be settled into polypeptides as spacers or cross-linkers to trigger the release of loaded drugs under certain condition. For example, redox-sensitive disulfide bonds has been introduced to connect two blocks of hybrid polypeptides, side chains, or NCA monomers [93, 138–141, 191]. These hybrid polypeptides could form micelles as the drug carriers, and, upon reduction, the cleavage of the disulfide-linker triggered the micellar rearrangement associated with the rapid release of the encapsulated drugs. Owing to self-cross-linking by oxidation of thiol groups in the air, cysteine derivatives have been used for the synthesis of redox-responsive block polypeptides by NCA ROP [108, 139]. These polypeptides can self-assemble into reversible shell-cross-linked micelles with the addition of a reducing agent, which can effectively prevent the loss of the loaded drugs and further release the drugs under reductive environment.

In addition, the development of a side-chain modification made it possible to prepare dual-responsive polypeptide carriers. Combining thermo- and pH-responsive blocks in one polypeptide chain [149] has been reported. The thermo-responsive block contains the OEG side group, and tertiary amine serves as the pH-responsive group. Furthermore, redox and pH dual-responsive hybrid polypeptide has been prepared based on lipoic acid (LA) and *cis*-1,2-cyclohexanedicarboxylic acid (CCA) decorated PEG-*b*-PLys (PEG-P(Lys-CCA/LA)) block copolymers [190].

11.6.2 Gene Delivery

Cationic polypeptides and hybrid polypeptides have been extensively studied as non-viral gene delivery vectors due to their biocompatibility and biodegradability [4–6]. The early studies on these gene vectors were focused on the reduction of toxicity, improvement of DNA binding and cell-penetrating capability, and facilitating the endosomal escape. Based on traditional cationic amino acid units, many efficient delivery vectors have been reported, such as oligo-histidine modified PLys [192–194]. poly(L-arginine) contained hybrids [195, 196], N-substituted poly(L-aspartamide) [197–199], PEG-conjugated polyserine derivatives [200, 201], and PEI-*graft*-polypeptides [67–71]. For instance, Tang et al. reported a randomly polymerized polypeptide as poly(ornithine-*co*-arginine-*co*-glycine-*co*-aspartic acid), Poly(O,R,G,D) through the NCAs polymerization followed by deprotection and guanidization [196]. This was a potential polymeric tumor-penetrating agent that can enhance tumor concentration and efficacy of nanomedicines in the treatment of solid tumors.

The development of helical ionic polypeptides has offered another promising way to develop novel non-viral gene delivery vectors in recent years. Gabrielson and Cheng reported cationic polypeptides named PVBLG-8 with a high molecular weight and a stable α-helix conformation [202, 203]. This novel polypeptide showed pretty good membrane activity such as cell-penetrating peptides (CPP) due to its helical conformation. PVBLG-8 features a higher molecular weight than CPP and a high charge density, which make it effective to condense and deliver DNA and siRNA to mammalian cells with higher efficient. The α-helix conformation of PVBLG-8 was demonstrated to be the key to effectively form pores on cell/endosomal membranes for cell penetration and gene transfection. Followed this study, trigger-responsive domains and supramolecular self-assembled nanoparticles were incorporated with PVBLG-8, and different molecular architectures of PVBLG-8 were designed to further improve the transfection efficiency [96, 204–207]. Recently, PVBLG-8, namely, PPABLG, has been incorporated into a siRNA containing hybrid nanoparticles (HNPs) system for high-efficiency delivery across the cell membrane. On the contrary, racemic PPABDLG with random coil conformation can also form similar HNPs system, but it is membrane inactive (Figure 11.10) [208]. Inspired by the strong DNA-binding and cell-penetrating capacities of poly-(L-arginine), side-chain guanidine group contained polypeptides with stable α-helix conformation were developed by Cheng's group [209]. These helical poly(arginine) mimics show superior cell membrane permeability up to two orders of magnitude higher than that of HIV-TAT peptide and excellent DNA and siRNA delivery efficiency. In addition, other cationic polypeptides with stable α-helix conformation have also been developed for cell penetration purposes, for instance, polypeptides bearing quaternary phosphonium side chains (PHLG-TEP) [124]. The helical polypeptide showed the desired cell penetration capability, and pretty low cytotoxicity.

11.6.3 Membrane Active and Antimicrobial Polypeptides

Cationic-charged membrane active peptides have been used as delivery carriers and antimicrobial drugs for many years. It was demonstrated the activities of these polypeptides are strongly related to their amphiphatic helical conformations [210–212]. Compared with some commonly used polymeric peptide analogues such as poly(β-peptides) and peptoids [213, 214], polypeptides received less attention in this area. Deming et al. first studied a series of polypeptides obtained by randomly co-polymerized ZLys NCA with a few hydrophobic amino acid NCAs [215]. The results showed polypeptides containing α-helical-favoring amino acids (e.g. Ala, Phe, and Leu) are generally much stronger membrane disruption agents than those containing amino acids favoring β-structures (e.g. Ile and Val). The later studies on poly(L-homoarginine)$_{60}$-b-poly(L-leucine)$_{20}$ ($R_{60}L_{20}$) showed it possesses superior cell-penetrating properties compared with the poly(L-lysine)$_{60}$-b-poly(L-leucine)$_{20}$ ($K_{60}L_{20}$) [216]. Hammond et al. grafted cationic polypeptides to chitosan, and these materials showed enough broad spectrum antimicrobial activities against a number of clinically significant bacteria and fungi [217].

Also benefiting by the discovery of helical ionic polypeptides, some novel antimicrobial peptides have been reported in recent years. The former mentioned polypeptides prepared through a click reaction between PPLG and azide amines showed non-hemolytic with moderate antimicrobial activities [147]. Recently, Cheng's group reported α-helical antimicrobial peptides (PHLG-BLMx) with radial amphiphilicity as antimicrobial peptides (AMP) [218]. The radially amphiphilic structure enables the polypeptide to strongly bind to the negatively charged bacterial surface and exhibits high antimicrobial activity against both gram-positive and gram-negative bacteria. In addition, the radially amphiphilic polypeptides can also be used as effective adjuvants for improving the permeation of commercial antibiotics in bacteria and enhanced antimicrobial activity by one to two orders of magnitude.

11.6.4 Tissue Engineering

Polypeptides are suitable for tissue engineering because of their bio-compatible property and diversity for different requirements of tissue engineering. In the early stage, polypeptide blocks, such as PLys and PGlu, were conjugated with other bio-compatible polymers including PLLA, PCL and PEG. Some corresponding tests revealed that these hybrid polypeptide block copolymers could promote cell adhesion and proliferation [219]. Polypeptide hybrids could be further incorporated with cell recognition motifs or growth factors to induce specific cellular responses [220, 221]. For instance, RGD tripeptide was grafted onto PEG-PLLA-PGlu to obtain PEG-PLLA-P(Glu-*g*-RGD) [221]. This hybrid polypeptide was complexed with PLLA to improve adhesion, spreading, and proliferation of human chondrocytes and 3T3 cells, and

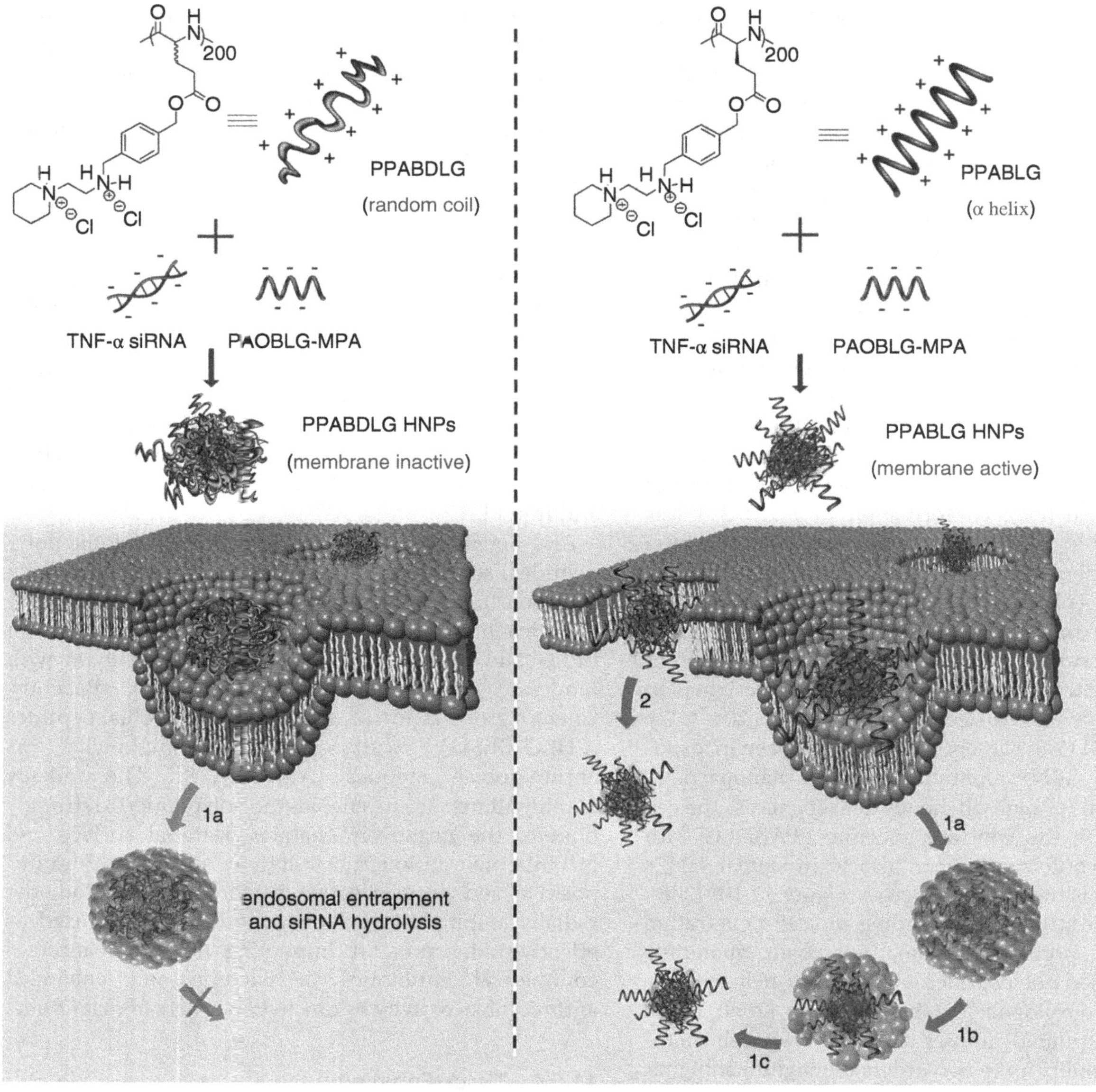

Figure 11.10 Intracellular kinetics of PPABLG/PAOBLG-MPA/siRNA and PPABDLG/PAOBLG-MPA/siRNA HNPs, highlighting the helicity-dependent membrane disruption capabilities of PPABLG toward effective cellular internalization as well as endosomal escape. *Source:* Reproduced with permission from [208].

combined with PLGA and hydroxyapatite (HA) to promote bone formation, better fusion interface, and enhanced bone growth. Also, polyelectrolyte complex scaffolds based on PGlu and chitosan were employed for delivery of adipose-derived stem cells to repair full thickness articular cartilage defects [222]. Polypeptide-modified nanostructures were even invested as tissue engineering scaffolds. For instance, PGlu-grafted silica nanoparticles were self-assembled through ice templating to form 3D macroporous materials. These materials showed potential as biocompatible 3D scaffolds in tissue engineering [223].

Polypeptide-based hydrogels are also a kind of promising tissue engineering material. After the discovery of diblock copolypeptide amphiphiles-based rapid recovering hydrogel scaffolds [185], a library of thermo- and salt-stable polypeptide hydrogels, such as KxLy, KxLyKz, and KxLyKzLyKx (herein, K is lysine and L is leucine) have been prepared and investigated for tissue engineering applications [185, 224, 225]. Although free PLys is known to be cytotoxic in solution, studies revealed that these ionic gels are suitable substrates for short-term cell culture. Some diblock polypeptide hydrogels have been applied successfully *in vivo* in mice central nervous system tissue and for the delivery of nerve growth factor [226]. Furthermore, polypeptide hydrogels are promising templates for organic-inorganic biocompatible porous materials where the porous gel scaffold can serve as the template for mineral growth [227, 228]. For example, $K_{170}L_{30}$ hydrogels were used as the templates for self-assembly of calcium phosphate nanocomposites. The composites were allowed to contain up to 50% of inorganic materials, close to the inorganic content of bone [227].

11.7 Conclusion

There have been marvelous advances in the NCA and polypeptide field during the last 20 years. The development of various new functional NCA monomers and the progress of precisely controlled ROP of NCAs provide convenient access to tailor-made polypeptide materials with well-defined topological structures and a certain performance. Throughout this brief review, remarkable milestones and recent breakthroughs in the synthesis, assembly, and application of polypeptide polymers have been highlighted. The blossoming of this field has opened a new view on synthetic biomaterials with great potential in increasing applications. Despite the promising prospects, the synthetic polypeptides are still a kind of polymer rather than a perfect mimic of natural proteins in terms of the complexity, activity, and functionality. There are several interesting and challenging research topics including the newly designed NCA monomers for special functions, moisture-resistant, and delicate conformations, integrating with other biological macromolecules, such as nucleic acids and polysaccharides to form hierarchical chiral micro/nanostructures with protein-like biological functions simply through self-assembly or phase separation. Hopefully, synthetic polypeptides will play a more important role in bio- and functional materials with the continuing developments in this field.

References

1 Kricheldorf, H. R. (2006) *Angew. Chem. Int. Ed.*, **45**, 5752–5784.
2 Deming, T.J. (1997) *Nature*, **390**, 386–389.
3 Deming, T.J. (1997) *J. Am. Chem. Soc.*, **119**, 2759–2760.
4 Deng, C., Wu, J., Cheng, R. et al. (2014) *Prog. Polym. Sci.*, **39**, 330–364.
5 Lu, H., Wang, J., Song, Z. et al. (2014) *Chem. Commun.*, **50**, 139–155.
6 Deming, T.J. (2007) *Prog. Polym. Sci.*, **32**, 858–875.
7 Cheng, J., Deming, T.J. (2012) *Top. Curr. Chem.*, **310**, 1–26.
8 Deming, T.J. (2013) *Adv. Polym. Sci.*, **262**, 1–38.
9 Menzel, H. (2013) *Adv. Polym. Sci.*, **253**, 1–36.
10 Wibowo, S.H., Sulistio, A., Wong, E.H.H. et al. (2014) *Chem. Commun.*, **50**, 4971–4988.
11 Byrne, M., Murphy, R., Kapetanakis, A. et al. (2015) *Macromol. Rapid Commun.*, **36**, 1862–1876.
12 Deming, T.J. (2016) *Chem. Rev.* **116**, 786–808.
13 Meldrum, F.C., Cölfen, H. (2008) *Chem. Rev.*, **108**, 4332–4432.
14 Liu, B., Cao, Y., Huang, Z. et al. (2015) *Adv. Mater.*, **27**, 479–497.
15 Patwardhan, S. V. (2011) *Chem. Commun.*, **47**, 7567–7582.
16 Huesmann, D., Birke, A., Klinker, K. et al. (2014) *Macromol.*, **47**, 928–936.
17 Zelzer, M., Heise, A. (2014) *J. Polym. Sci., Part A: Polym. Chem.*, **52**, 1228–1236.
18 Deming, T.J. (1998) *J. Am. Chem. Soc.*, **120**, 4240–4241.
19 Deming, T.J. (1999) *Macromol.*, **32**, 4500–4502.
20 Deming, T.J., Curtin, S.A. (2000) *J. Am. Chem. Soc.*, **122**, 5710–5717.
21 Curtin, S.A., Deming, T.J. (1999) *J. Am. Chem. Soc.*, **121**, 7427–7428.
22 Cheng, J., Deming, T.J. (1999), *J. Polym. Mat. Sci. Engr.*, **80**, 473–474.
23 Seidel, S.W., Deming, T.J. (2003) *Macromol.*, **36**, 969–972.
24 Peng, Y., Lai, S., Lin, C. (2008) *Macromol.*, **41**, 3455–3459.
25 Peng, H., Ling, J., Shen, Z. (2012) *J. Polym. Sci., Part A: Polym. Chem.*, **50**, 1076–1085.
26 Peng, H., Ling, J., Zhu, Y. et al. (2012) *J. Polym. Sci., Part A: Polym. Chem.*, **50**, 3016–3029.
27 Dimitrov, I., Schlaad, H. (2003) *Chem. Comm.*, 2944–2945.
28 Lutz, J.F., Schütt, D., Kubowicz, S. (2005) *Macromol. Rapid Comm.*, **26**, 23–28.
29 Meyer, M., Schlaad, H. (2006) *Macromol*, **39**, 3967–3970.
30 Aliferis, T., Iatrou, H., Hadjichristidis, N. (2004) *Biomacromol.*, **5**, 1653–1656.

31 Vayaboury, W., Giani, O., Cottet, H. et al. *Macromol. Rapid. Comm.*, **25**, 1221–1224.
32 Cao, H., Yao, J., Shao, Z. (2012) *Polym. Int.*, **61**, 774–779.
33 Nguyen, L.T.T., Vorenkamp, E.J., Daumont, C.J.M. et al. (2010) *Polymer*, **51**, 1042–1055.
34 Habraken, G.J.M, Wilsens, K.H.R.M., Koning, C.E., Heise, A. (2011) *Polym. Chem.*, **2**, 1322–1330.
35 Habraken, G.J.M., Peeters, M., Dietz, C.H.J.T. et al. (2010) *Polym. Chem.*, **1**, 514–524.
36 Lu, H., Cheng, J. (2007) *J. Am. Chem. Soc.*, **129**, 14114–14115.
37 Lu, H., Cheng, J. (2008) *J. Am. Chem. Soc.*, **130**, 12562–12563.
38 Lu, H., Wang, J., Lin, Y., Cheng, J. J. (2009) *J. Am. Chem. Soc.*, **131**, 13582–13583.
39 Zou, J., Fan, J., He, X. et al. (2013) *Macromol.*, **46**, 4223–4226.
40 Zhao, W., Gnanou, Y., Hadjichristidis, N. (2015) *Biomacromol.*, **16**, 1352–1357.
41 Zhao, W., Gnanou, Y., Hadjichristidis, N. (2015) *Chem. Comm.*, **51**, 3663–3666.
42 Zhao, W., Gnanou, Y., Hadjichristidis, N. (2015) *Polym. Chem.*, **6**, 6193–6201
43 Vacognea, C.D., Schlaad, H. (2015) *Chem. Comm.*, **51**, 15645–15648.
44 Conejos-Sánchez, I., Duro-Castano, A., Birke, A. et al. (2013) *Polym. Chem.*, **4**, 3182–3186.
45 Birke, A., Huesmann, D., Kelsch, A. et al. (2014) *Biomacromol.*, **15**, 548–557.
46 Hadjichristidis, N., Iatrou, H., Pitsikalis, M., Sakellariou, G. (2009) *Chem. Rev.*, **109**, 5528–5578.
47 Yu, M., Deming, T.J. (1998) *Macromol.*, **31**, 4739–4745.
48 Yu, M., Hwang, J., Deming, T.J. (1999) *J. Am. Chem. Soc.*, **121**, 5825–5826.
49 Wyrsta, M.D., Cogen, A.L., Deming, T.J. (2001) *J. Am. Chem. Soc.*, **123**, 12919–12920.
50 Higuchi, M., Takizawa, A., Kinoshita, T. et al. (1990) *Macromol.*, **23**, 361–365.
51 Yamashita, Y., Iwaya, Y., Ito, K. (1975) *Makromol. Chem.* **176**, 1207–1216.
52 Klok, H. A., Langenwalter, J.F., Lecommoux, S. (2000) *Macromol.*, **33**, 7819–7826.
53 Kim, K.T., Vermeulen, G.W.M., Winnik, M.A., Manners, I. (2005) *Macromol.*, **38**, 4958–4961.
54 Zhang, X., Li, J., Li, W., Zhang, A. (2007) *Biomacromol.*, **8**, 3557–3567.
55 Lai, J.T., Filla, D., Shea, R. (2002) *Macromol.*, **35**, 6754–6756.
56 Brzezinska, K.R., Deming, T.J. (2004) *Macromol. Biosci.*, **4**, 566–569
57 Wang, Y., Ling, J. (2015) *RSC Adv.*, **5**, 18546–18553.
58 Brzezinska, K.R., Deming, T.J. (2001) *Macromol.*, **34**, 4348–4354.
59 Brzezinska, K.R., Curtin, S.A., Deming, T.J. (2002) *Macromol.*, **35**, 2970–2976.
60 Degée, P., Dubois, P., Jérôme, R., Teyssié, P. (1993) *J. Polym. Sci., Part A: Polym. Chem.*, **31**, 275–278.
61 Gotsche, M., Keut, H., Höcker, H. (1995) *Macromol. Chem. Phys.*, **196**, 3891–3903.
62 Sun, J., Chen, X., Guo, J. et al. (2009) *Polymer*, **50**, 455–461.
63 Karatzas, A., Bilalis, P., Iatrou, H. et al. (2009) *React. Funct. Polym.* **69**, 435–440.
64 Schatz, C., Louguet, S., Le Meins, G.F., Lecommoux, S. (2009) *Angew. Chem., Inter. Ed.*, **48**, 2572–2575.
65 Tsutsumiuchi, K., Aoi, K., Okada, M. (1995) *Macromol. Rapid Commun.*, **16**, 749–755.
66 Shen, Y., Zhang, S., Wan, Y. et al. (2015) *Soft Matter*, **11**, 2945–2951.
67 Tian, H., Chen, X., Lin, H. et al. (2006) *Chem. Euro. J.*, **1b2**, 4305–4312.
68 Chen, J., Tian, H., Guo, Z. et al. (2009) *Macromol. Biosci.*, **9**, 1247–1253.
69 Guo, Z., Tian, H., Lin, L. et al. (2014) *Macromol. Biosci.*, **14**, 1406–1414.
70 Chen, J., Xia, J., Tian, H. et al. (2014) *Arch. Pharmacal Res.*, **37**, 142–148.
71 Tian, H., Lin, L., Jiao, Z. et al. (2013) *J. Controlled Release*, **172**, 410–418.
72 Egorova, A.A., Kiselev, A.V., Tarasenko, I.I. et al. (2009) *Russ. J. Bioorg. Chem.*, **35**, 437–445.
73 Wibowo, S.H., Wong, E.H.H., Sulistio, A. et al. (2013) *Adv. Mater.*, **25**, 4619–4624.
74 Chang, X., Dong, C. (2013) *Biomacromol.*, **14**, 3329–3337.
75 Kadlecova, Z., Rajendra, Y., Matasci, M. et al. (2013) *J. Controlled Release*, **169**, 276–288.
76 Gunari, N., Cong, Y., Zhang, B. et al. (2008) *Macromol. Rapid. Comm.*, **29**, 821–825.
77 Hanski, S., Houbenov, N., Ruokolainen, J. et al. (2006) *Biomacromol.*, **7**, 3379–3384.
78 Gkikas, M., Haataja, J.S., Ruokolainen, J. et al. (2015) *Biomacromol.*, **16**, 3686–3693.
79 Hammon, M.R., Klok, H.A., Mezzenga, R. (2007) *Macromol. Rapid Comm.*, **29**, 299–303.
80 Zhang, B., Fischer, K., Schmidt, M. (2005) *Macromol. Chem. Phys.*, **206**, 157–162.
81 Wu, X., Zhou, L., Su, Y., Dong, C. (2015) *Polym. Chem.*, **6**, 6857–6869.
82 Ding, J., He, C., Xiao, C. et al. (2012) *Macromol. Res.*, **20**, 292–301.
83 H. Lu, J. Wang, Y., Lin, J. Cheng, *J. Am. Chem. Soc.* 2009, **131**, 13582–13583.
84 Wang, J., Lu, H., Kamat, R. et al. (2011) *J. Am. Chem. Soc.* **133**, 12906–12909.
85 Kricheldorf, H.R., Von Lossow, K., Schwarz, G. (2005) *Macromol.*, **38**, 5513–5518.

86 Kricheldorf, H.R., Von Lossow, K., Schwarz, G. (2005) *J. Polym. Sci.,Part A: Polym. Chem.*, **43**, 5690–5698.

87 Kricheldorf, H.R., Von Lossow, K., Schwarz, G. (2006) *J. Polym. Sci. Chem. Ed*, **44**, 4680–4695.

88 Kricheldorf, H.R., Von Lossow, K., Schwarz, G., Fritsch, D. (2005) *Macromol. Chem. Phys.*, **206**, 1165–1170.

89 Engler, A.C., Lee, H.-I., Hammond, P.T. (2009) *Angew. Chem., Int. Ed.*, **48**, 9334–9338.

90 Quadir, M. A., Martin, M., Hammond, P.T. (2014) *Chem. Mater*, **26**, 461–447.

91 Liu, Y., Chen, P., Li, Z. (2012) *Macromol. Rapid Comm.*, **33**, 287–295.

92 Tang, H. Y., Zhang, D.H. (2010) *Biomacromol.*, **11**, 1585–1592.

93 Kramer, J.R., Deming, T.J. (2010) *J. Am. Chem. Soc.*, **132**, 15068–15071.

94 Kramer, J.R., Deming, T.J. (2012) *J. Am. Chem. Soc.*, **134**, 4112–4115.

95 Fu, X., Shen, Y., Fu, W., Li, Z. (2013) *Macromol.*, **46**, 3753–3760.

96 Yin, L., Tang, H., Kim, K.H. et al. (2013) *Angew. Chem., Int. Ed.*, **52**, 9182–9186.

97 Kumar, R.J., MacDonald, J.M., Singh, T.B. et al. (2011) *J. Am. Chem. Soc.*, **133**, 8564–8573.

98 Kramera, J.R., Deming, T.J. (2014) *Polym. Chem.*, **5**, 671–682.

99 Fiume, L., Di Stefano, G., Busi, C., Mattioli, A. (1994) *Biochem. Pharmacol.*, **47**, 643–650.

100 Di Stefano, G., Busi, C., Mattioli, A., Fiume, L. (1995) *Biochem. Pharmacol.*, **49**, 1769–1775.

101 Mahato, R.I., Takemura, S., Akumatsu, K. et al. (1997) *Biochem. Pharmacol.*, **53**, 887–895.

102 Engler, A., Lee, H.-I., Hammond, P. (2009) *Angew. Chem., Int. Ed.*, **48**, 9334–9338.

103 Xiao, C., Zhao, C., He, P. et al. (2010) *Macromol. Rapid Comm.*, **31**, 991–997

104 Tang, H., Zhang, D. (2010) *Biomacromol.*, **11**, 1585–1592.

105 Huang, J., Habraken, G., Audouin, F., Heise, A. (2010) *Macromol.*, **43**, 6050–6057.

106 Rhodes, A.J., Deming, T.J. (2013) *ACS Macro Lett.*, **2**, 351–354.

107 Sun, J., Schlaad, H. (2010) *Macromol.*, **43**, 4445–4448.

108 Krannig, K.S., Schlaad, H. (2016) *Macro-Glycolig, S. Methods, Protocols, Methods in Molecular Biology*, **1367**, 61–67.

109 Kramer, J.R., Deming, T.J. (2012) *Biomacromol.*, **13**, 1719–1723.

110 Kramer, J.R., Deming, T.J. (2013) *Chem. Comm.*, **49**, 5144–5146.

111 Rüde, E., Westphal, O., Hurwitz, E., Sela, M. (1966) *Immunochem.*, **3**, 137–151.

112 Kramer, J.R., Deming, T.J. (2010) *Biomacromol.*, **11**, 3668–3672.

113 Stöhr, T., Blaudszun, A. R., Steinfeld, U., Wenz, G. (2011) *Polym. Chem.*, **2**, 2239–2248.

114 Pati, D., Shaikh, A. Y., Hotha, S., Gupta, S.S. (2011) *Polym. Chem.*, **2**, 805–811.

115 Pati, D., Shaikh, A.Y., Das, S. et al. (2012) *Biomacromol.*, **13**, 1287–1295.

116 Pati, D., Kalva, N., Das, S. et al. (2012) *J. Am. Chem. Soc.*, **134**, 7796–7802.

117 Kramer, J.R., Rodriguez, A.R., Choe, U.-J. et al. (2013) *Soft Matter*, **9**, 3389–3395.

118 Yu, M., Nowak, A.P., Pochan, D.P., Deming, T.J. (1999) *J. Am. Chem. Soc.*, **121**, 12210–12211.

119 Chen, C., Wang, Z., Li, Z. (2011) *Biomacromol.*, **12**, 2859–2863.

120 Zhang, S., Chen, C., Li, Z. (2013) *Chin. J. Polym. Sci.*, **31**, 201–210.

121 Lu, H., Wang, J., Bai, Y. et al. (2011) *Nat. Comm.*, **2**, 206.

122 Zhang, Y., Lu, H., Lin, Y., Cheng, J. (2011) *Macromol.*, **44**, 6641–6644.

123 Shaikh, A.Y., Das, S., Pati, D. et al. (2014) *Biomacromol.*, **15**, 3679–3686.

124 Song, Z., Zheng, N., Ba, X. et al. (2014) *Biomacromol.*, **15**, 1491–1497.

125 Huang, Y., Tang, Z., Zhang, X. et al. (2013) *Biomacromol.*, **14**, 2023–2032.

126 Yakovlev, I., Deming, T.J. (2015) *J. Am. Chem. Soc.*, **137**, 4078–4081.

127 Huang, J., Heise, A. (2013) *Chem. Soc. Rev.*, **42**, 7373–7390.

128 He, C., Zhuang, X., Tang, Z. et al. (2012) *Adv. Healthcare Mater.*, **1**, 48–78.

129 Shen, Y., Fu, X., Fu, W., Li, Z. (2015) *Chem. Soc. Rev.*, **44**, 612–622.

130 Fu, X., Shen, Y., Fu, W., Li, Z. (2013) *Macromol.*, **46**, 3753–3760.

131 Ma, Y., Fu, X., Shen, Y. et al. (2014) *Macromol.*, **47**, 4684–4689.

132 Zhang, S., Fu, W., Li, Z. (2014) *Polym. Chem.*, **5**, 3346–3351.

133 Shen, J., Chen, C., Fu, W. et al. (2013) *Langmuir*, **29**, 6271–6278.

134 Kotharangannagari, V. K., Sánchez-Ferrer, A., Ruokolainen, J., Mezzenga, R. (2012) *Macromol.*, **45**, 1982–1990.

135 Huang, J., Hastings, C.L., Duffy, G. P. et al. (2013) *Biomacromol.*, **14**, 200–206.

136 Cheng, Y., He, C., Xiao, C. et al. (2013) *Biomacromol.*, **14**, 468–475.

137 He, X., Fan, J., Zhang, F. et al. (2014) *J. Mater. Chem., B*, **2**, 8123–8130.

138 Kramer, J.R., Deming, T.J. (2014) *J. Am. Chem. Soc.*, **136**, 5547–5550.

139 Wang, K., Luo, G., Liu, Y. et al. (2012) *Polym. Chem.*, **3**, 1084–1090.

140 Holm, R., Klinker, K., Weber, B., Barz, M. (2015) *Macromol. Rapid Commun.*, **36**, 2083–2091.
141 Takae, S., Miyata, K., Oba, M. et al. (2008) *J. Am. Chem. Soc.*, **130**, 6001–6009.
142 Sato, M., Kinoshita, T., Takizawa, A., Tsujita, Y. (1988) *Macromol.*, **21**, 1612–1616.
143 Ciardelli, F., Fabbri, D., Pieroni, O., Fissi, A. (1989) *J. Am. Chem. Soc.*, **111**, 3470–3472.
144 Kumar, S., Allard, J.-F., Morris, D. et al. (2012) *J. Mater. Chem.*, **22**, 7252–7257.
145 Liu, G., Dong, C. (2012) *Biomacromol.*, **13**, 1573–1583.
146 Kotharangannagari, V. K., Sanchez-Ferrer, A., Ruokolainen, J., Mezzenga, R. (2011) *Macromol.*, **44**, 4569–4573.
147 Engler, A.C., Bonner, D. K., Buss, H.G. et al. (2011) *Soft Matter*, **7**, 5627–5637.
148 Engler, A.C., Shukla, A., Puranam, S. et al. (2011) *Biomacromol.*, **12**, 1666–1674.
149 Chopko, C. M., Lowden, E.L., Engler, A.C. et al. (2012) *ACS Macro Lett.*, **1**, 727–731.
150 Nakanishi, M., Park, J.S., Jang, W. D. et al. (2007) *React. Funct. Polym.*, **67**, 1361–1372.
151 Zhao, L., Ding, J., Xiao, C. et al. (2012) *J. Mater. Chem.*, **22**, 12319–12328.
152 Rodriguez, A. R., Kramer, J.R., Deming, T.J. (2013) *Biomacromol.*, **14**, 3610–3614.
153 He, X., Fan, J., Wooley, K.L. (2016) *Chem. Asian J.*, **11**, 437–447.
154 Deming, T.J. (2014) *WIREs Nanomed. Nanobiotechnol.*, **6**, 283–297.
155 Ohkawa, K., Shoumura, K., Yamada, M. et al. (2001) *Macromol. Biosci.*, **1**, 149–156.
156 Yamamoto, H., Kitsuki, T., Nishida, A. et al. (1999) *Macromol.*, **32**, 1055–1061.
157 Ren, K., He, C., Cheng, Y. et al. (2014) *Polym. Chem.*, **5**, 5069–5076.
158 Aggeli, A., Nyrkova, I.A., Bell, M. et al. (2001) *Proc. Natl. Acad. Sci. U.S.A.*, **98**, 11857–11862.
159 Nyrkova, I.A., Semenov, A.N., Aggeli, A., Boden, N. (2000) *Eur. Phys. J B*, **17**, 481–497.
160 Kolli, H.B., Frezza, E., Cinacchi, G. et al. (2014) *Soft Matter*, **10**, 8171–8187.
161 Ho, R., Chiang, Y.W., Lin, S.C., Chen, C.K. (2011) *Prog. Polym. Sci.*, **39**, 376–453.
162 Wang, J., Xia, H., Zhang, Y. et al. (2013) *J. Am. Chem. Soc.*, **135**, 11417–11420.
163 Cai, C., Lin, J., Zhu, X. et al. (2016) *Macromol.*, **49**, 15–22.
164 Cai, C., Li, Y., Lin, J. et al. (2013) *Angew. Chem. Int. Ed.*, **52**, 7732–7736.
165 Cha, J., Bartl, M.H., Wong, M. et al. (2003) *Nano Lett.*, **3**, 907–911.
166 Cha, J., Birkedal, H., Euliss, L.E. et al. (2003) *J. Am. Chem. Soc.*, **125**, 8285–8289.
167 Wong, M., Cha, J., Choi, K.S. et al. (2002) *Nano Lett.*, **2**, 583–587.
168 Euliss, L.E., Trnka, T.M., Deming, T.J., Stucky, G.D. (2004) *Chem. Comm.*, **67**, 1736–1737.
169 Yao, Y., Dong, W., Zhu, S. et al. (2009) *Langmuir*, **25**, 13238–13243.
170 Cha, J., Stucky, G.D., Morse, D.E., Deming, T.J. (2000) *Nature*, **403**, 289–292.
171 Tomczak, M.M., Glawe, D.D., Drummy, L.F. et al. (2005) *J. Am. Chem. Soc.*, **127**, 12577–12582.
172 Bellomo, E.G., Deming, T.J. (2006) *J. Am. Chem. Soc.*, **128**, 2276–2279.
173 Yao, Y., Wang, D., Han, L., Che, S. (2013) *Chem. Eur. J.*, **19**, 15489–15492.
174 Jin, R., Yao, D., Levi, R. (2014) *Chem. Eur. J.*, **20**,7196–7214.
175 Patwardhan, S.V. (2011) *Chem. Commun.*, **47**, 7567–7582.
176 Klok, H.A., Lecommoux, S. (2006) *Adv. Polym. Sci.*, **202**, 75–111.
177 Cai, C., Wang, L., Lin, J. (2011) *Chem. Commun.*, **47**, 11189–11203.
178 Sánchez-Ferrer, A., Mezzenga, R. (2010) *Macromol.*, **43**, 1093–1100.
179 Gitsas, A., Floudas, G., Mondeshki, M. et al. (2008) *Biomacromol.*, **9**, 1959–1966.
180 Gkikas, M., Haataja, J.S., Seitsonen, J. et al. (2014) *Biomacromol.*, **15**, 3923–3930.
181 Holowka, E.P., Sun, V.Z., Kamei, D.T., Deming, T.J. (2007) *Nat. Mater.*, **6**, 52–57.
182 Song, Z., Kim, H., Ba, X. et al. (2015) *Soft Matter*, **11**, 4091–4098.
183 Priftis, D., Leon, L., Song, Z. et al. (2015) *Angew. Chem. Int. Ed.*, **54**, 11128–11132.
184 Perry, S.L., Leon, L., Hoffmann, K.Q. et al. (2015) *Nat. Comm.*, **6**, 6052.
185 Nowak, A. P., Breedveld, V., Pakstis, L. et al. (2002) *Nature*, **417**, 424–428.
186 Nowak, A. P., Breedveld, V., Pine, D.J., Deming, T.J. (2003) *J. Am. Chem. Soc.*, **125**, 15666–15670.
187 Kataoka, K., Harada, A., Nagasaki, Y. (2012) *Adv. Drug. Deliver. Rev.*, **64**, 37–48.
188 Hehir, S., Cameron, N.R. (2014) *Polym. Int.*, **63**, 943–954.
189 Yin, H., Lee, E.S., Kim, D. et al. (2008) *J. Controlled Release*, **126**, 130–138.
190 Wu, L., Zou, Y., Deng, C. et al. (2013) *Biomaterials*, **34**, 5262–5272.
191 Ding, J., Shi, F., Xiao, C. et al. (2011) *Polym. Chem.*, **2**, 2857–2864.
192 Thomas, J.J., Rekha, M.R., Sharma, C.P. (2012) *Mol. Pharm.*, **9**, 121–134.
193 Goncalves, C., Pichon, C., Guerin, B., Midoux, P. (2002) *J. Gene Med.*, **4**, 271–281.
194 Midoux, P., Monsigny, M. (1999) *Bioconjugate Chem.*, **10**, 406–411.

195 Futaki, S., Ohashi, W., Suzuki, T. et al. (2001) *Bioconjugate Chem.*, **12**, 1005–1011.
196 Yu, H., Tang, Z., Zhang, D. et al. (2015) *Macromol. Biosci.* **15**, 829–838.
197 Miyata, K., Oba, M., Nakanishi, M. et al. (2008) *J. Am. Chem. Soc.*, **130**, 16287–16294.
198 Suma, T., Miyata, K., Ishii, T. et al. (2012) *Biomaterials*, **33**, 2770–2779.
199 Uchida, H., Miyata, K., Oba, M. et al. (2011) *J. Am. Chem. Soc.*, **133**, 15524–15532.
200 Shim, M.S., Kwon, Y.J. (2010) *Biomaterials*, **31**, 3404–3413.
201 Sanjoh, M., Hiki, S., Lee, Y. et al. (2010) *Macromol. Rapid Commun.*, **31**, 1181–1186.
202 Gabrielson, N.P., Lu, H., Yin, L. et al. (2012) *Angew. Chem., Int. Ed.*, **51**, 1143–1147.
203 Zhang, R., Song, Z., Yin, L. et al. (2015) *WIREs Nanomed. Nanobiotechnol.*, **7**, 98–110.
204 Gabrielson, N. P., Lu, H., Yin, L. et al. (2012) *Mol. Ther.*, **20**, 1599–1609.
205 Yin, L., Song, Z., Kim, K.H. et al. (2013) *Biomaterials*, **34**, 2340–2349.
206 Yin, L., Song, Z., Kim, K.H. et al. (2013) *Adv. Mater.*, **25**, 3063–3070.
207 Yin, L., Song, Z., Qu, Q. et al. (2013) *Angew. Chem., Int. Ed.*, **52**, 5757–5761.
208 He, H., Zheng, N., Song, Z. et al. (2016) *ACS Nano.*, **10**, 1859–1870.
209 Tang, H., Yin, L., Kim, K.H., Cheng, J. (2013) *Chem. Sci.*, **4**, 3839–3844.
210 McMahon, H.T., Gallop, J.L. (2005) *Nature*, **438**, 590–596.
211 Gabriel, G.J., Tew, G.N. (2008) *Org. Biomol. Chem.*, **6**, 417–423.
212 Beevers, A.J., Dixon, A.M. (2010) *Chem. Soc. Rev.*, **39**, 2146–2157.
213 Mowery, B.P., Lindner, A. H., Weisblum, B. et al. (2009) *J. Am. Chem. Soc.*, **131**, 9735–9745.
214 Wu, C.W., Kirshenbaum, K., Sanborn, T.J. et al. (2003) *J. Am. Chem. Soc.*, **125**, 13525–13530.
215 Wyrsta, M. D., Cogen, A.L., Deming, T.J. (2001) *J. Am. Chem. Soc.*, **123**, 12919–12920.
216 Sun, V.Z., Li, Z., Deming, T.J., Kamei, D.T. (2011) *Biomacromol.*, **12**, 10–13.
217 Li, P., Zhou, C., Rayatpisheh, S. et al. (2012) *Adv. Mater.*, **24**, 4130–4137.
218 Xiong, M., Lee, M. W., Mansbach, R. A. et al. (2015) *Proc. Natl. Acad. Sci. U.S.A.*, **112**, 13155–13160.
219 Peng, H., Xiao, Y., Mao, X. et al. (2009) *Biomacromol.*, **10**, 95–104.
220 Deng, C., Tian, H., Zhang, P. et al. (2006) *Biomacromol.*, **7**, 590–596.
221 Zhang, P., Wu, H., Wu, H., et al. (2011) *Biomacromol.*, **12**, 2667–2680.
222 Zhang, K., Zhang, Y., Yan, S. et al. (2013) *Acta Biomater.*, **9**, 7276–7288.
223 Kar, M., Pauline, M., Sharma, K., et al. (2011) *Langmuir*, **27**, 12124–12133.
224 Li, Z., Deming, T.J. (2010) *Soft Matter*, **6**, 2546–2551.
225 Pakstis, L., Ozbas, B., Nowak, A. P. et al. (2004) *Biomacromol.*, **5**, 312–318.
226 Yang, C., Song, B., Ao, Y. et al. (2009) *Biomaterials*, **30**, 2881–2898.
227 Hu, Y., Yusufoglu, Y., Kanapathipillai, M. et al. (2009) *Soft Matter*, **5**, 4311–4320.
228 Xia, L., Liu, Y., Li, Z. (2010) *Macromol. Biosci.* **10**, 1566–1575.

12

Preparation of Gradient Polymeric Structures and Their Biological Applications

Tao Du[1], Feng Zhou[1], and Shutao Wang[2,]*

[1] *State Key Laboratory of Solid Lubrication, Lanzhou Institute of Chemical Physics, Chinese Academy of Sciences, Lanzhou, China*
[2] *CAS Key Laboratory of Bio-inspired Materials and Interfacial Science, CAS Center for Excellence in Nanoscience, Technical Institute of Physics and Chemistry, Chinese Academy of Sciences, Beijing, China*

12.1 Introduction

After long time of evolution and natural selection, gradients exist in many biological systems, such as bones and plant stems. For example, spatial variations in cell types present at the anterior cruciate ligament (ACL)-to-bone junction [1]. In the ligament and bone regions, the content of collagen is highest. Along the fibrocartilage interface, the relative collagen content decreases from ligament to bone (Figure 12.1a) [1]. In developmental biology, the morphogen forms a gradient of concentration to determine the fate of responding cells [2]. In plant stems, the fiber cap region exists a stiffness gradient (Figure 12.1b) [3, 4]. These gradients in living systems are often at levels that scientists can only envy. Thus, artificial gradient structures have aroused great interest because of their very wide range of applications not only in tissue engineering [5], but also in implantable materials [6], chemotactic gradients [7], cell migration [8], cell gradients [9, 10], and so forth.

Gradient structures mean that the chemical or physical properties of the materials themselves or some elements attached to the material surfaces gradually change along one or more given directions. Compared to the entrenched bulk properties of materials, the surface properties can easily be changed according to various surface modification methods. In the past few decades, a number of methods for surface modifications with a gradually varying chemical composition have been developed. The gradient surfaces are now receiving more attention, from basic research to practical application. In this chapter, we mainly highlight the recent advances in the preparation of gradient polymeric structures, including gradient hydrogels and polymer brushes, by using a wide range of polymerization methods, which allow exquisite control of the gradually varying chemical composition. Meanwhile, we show some biological applications of these gradient polymeric structures.

12.2 Gradient Polymeric Structures

12.2.1 Gradient Hydrogels

Hydrogels are networks of polymer chains that are hydrophilic and swell a lot in water, which are capable of converting chemical energy into mechanical energy [11]. Hydrogels can be made in a variety of structures and can mimic many of the physiological properties of tissues in the body. Some important biomedical applications use hydrogels as scaffolds for tissue engineering to repair tissues *in vivo* [12], and as the matrix for regulating cell behavior, such as adhesion, migration, and proliferation [13–15], because hydrogels can provide specific microenvironments for cells. Since hydrogels possess specific viscoelasticity and diffusive transport characteristics which are similar to the extracellular matrix (ECM) behavior [16], the formation of gradient hydrogels has been an effective method to facilitate multi-tissues transition (ligament, fibrocartilage, bone) [17].

The bulk properties of hydrogels such as stiffness, cross-linking density and porosity can be controlled spatially to make immobilized biological agents gradually change along one or more directions. Langer et al. reported a microfluidics/photopolymerization combined method for fabricating the poly(ethylene glycol) (PEG)-based hydrogels with tethered cell-adhesive ligands and cross-linking densities in a gradient distribution [18]. Two different monomer solutions (one including cell-adhesive ligands, Arg-Gly-Asp-Ser, RGDS) were injected into a PDMS microfluidic gradient maker consisting

* Email: stwang@iccas.ac.cn

Bioinspired Materials Science and Engineering, First Edition. Edited by Guang Yang, Lin Xiao, and Lallepak Lamboni.

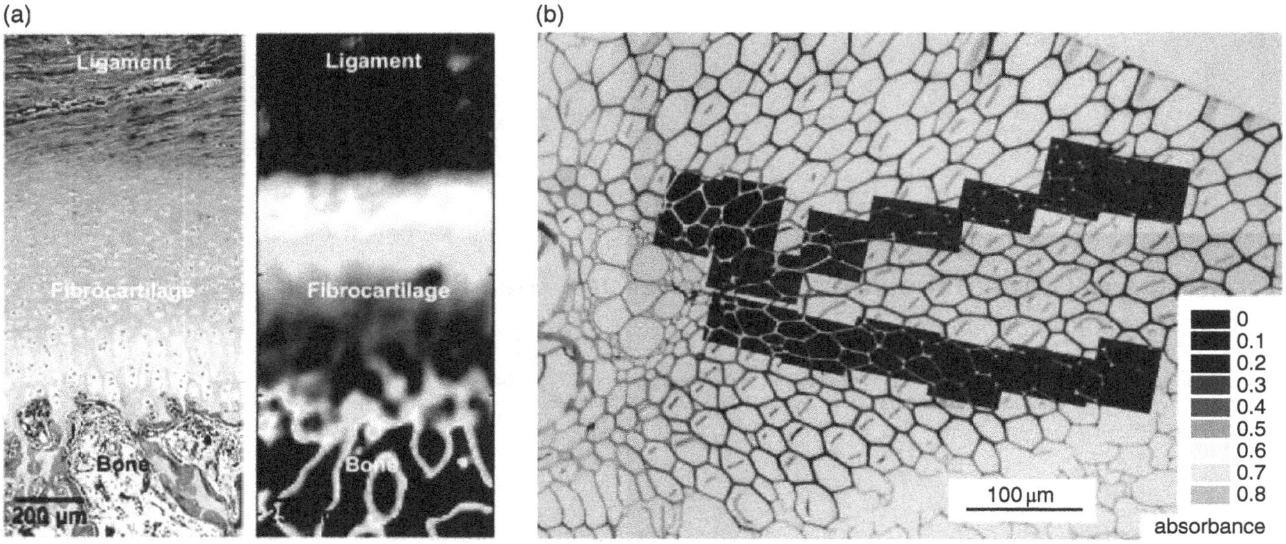

Figure 12.1 Gradient structures in biological systems. (a) The native ACL-bone interface (left) and Fourier Transform Infrared Spectroscopic (FTIR) image (right). (b) Composite image of a fiber cap cross-section. The 2D UV absorbance scans illustrate the degree of cell wall lignification. (red-to-blue color represents high to low levels of lignin). *Sources:* (a) Reproduced from [1] with permission, copyright 2008, Elsevier. (b) Reproduced from [3] with permission, copyright 2008, The Royal Society. (*See insert for color representation of the figure.*)

of microchannels, in which a gradient concentration was formed (Figure 12.2a). After being exposed to UV light through the PDMS mold, the solutions were photopolymerized into the gradient of tethered RGDS hydrogels. Human umbilical vein endothelial cells (HUVECs) were seeded onto the hydrogels. Compared with the homogeneous hydrogel without and with RGDS (A and B in Figure 12.2b), the attachment density of HUVECs varied from more to less across the hydrogel with the gradient cell-adhesive ligands (RGDS) (C in Figure 12.2b). This work showed that the gradient hydrogels could control the cell adhesion. Later, He et al. extended this strategy to fabricate centimeter-long gradient hydrogels [19]. One type of hydrogel precursor with low concentration was pre-filled in the microchannel. Another high concentration hydrogel precursor mixed with the chemicals of interest was injected from the inlet. The chemicals of interest solution flowed into the microchannel by a forward flow. The gradient was generated by the effect of molecular diffusion combined with evaporation-induced backward flow (Figure 12.2c). After photopolymerization, the thickness and porosity of the as-prepared hydrogel gradually varied along the chemical gradient direction from inlet to outlet (Figure 12.2d). The authors fabricated a gradient concentration of RGDS (which is the chemical of interest) along the PEG hydrogel to study the attachment behavior of HUVECs. Such spatial cell adhesive distribution can potentially be used in tissue engineering (e.g. cartilage tissue) [17].

Besides the above microfluidic methods, molecular diffusion is another strategy to fabricate gradient hydrogels. For this strategy, bioactive molecules are localized at one end, then diffuse through the hydrogel to generate a molecular concentration gradient. Peret et al. reported protein concentration gradients within PEG hydrogels [20]. The bovine serum albumin (BSA) or BSA-loaded microspheres were spatially localized within PEG hydrogels to generate a protein-releasing depot (Figure 12.3a). Similarly, Yi et al. reported hyaluronic acid (HA) hydrogels with the gradient of chemically immobilized bisphosphonate (BP) groups by diffusing the bisphosphonate acrylamide reagent into HA hydrogels (Figure 12.3b) [21]. The HA hydrogel matrix was first of all prepared by mixing HA-aldehyde and dually functional HA-hydrazide-thiol derivatives to give the hydrazone coupling. A solution of BP-acrylamide and photo-initiators diffused into the hydrogel matrix in a syringe to form a concentrated gradient of BP-acrylamide. After the thoil-ene addition of the matrix thiol groups and BP-acrylamide molecules, the BP groups were fixed to formed the HA-based hydrogel with gradient BP ligands (Figure 12.3b). This molecular diffusion method introduced bioactive molecules as localized sources, leading to spatial gradients that mimic a common paradigm in tissue development and regeneration [22].

Stiffness is an important influence on cell behaviors [23]. Many researchers used independent substrates with different stiffnesses [24], however, which suffered from limitations in the stiffness ranges, and were distorted from the tissue interface *in vivo* with the gradient stiffness. To overcome these shortcomings, using hydrogels with a successive stiffness gradient is a good strategy. Kim et al. reported a gradual freezing-thawing method

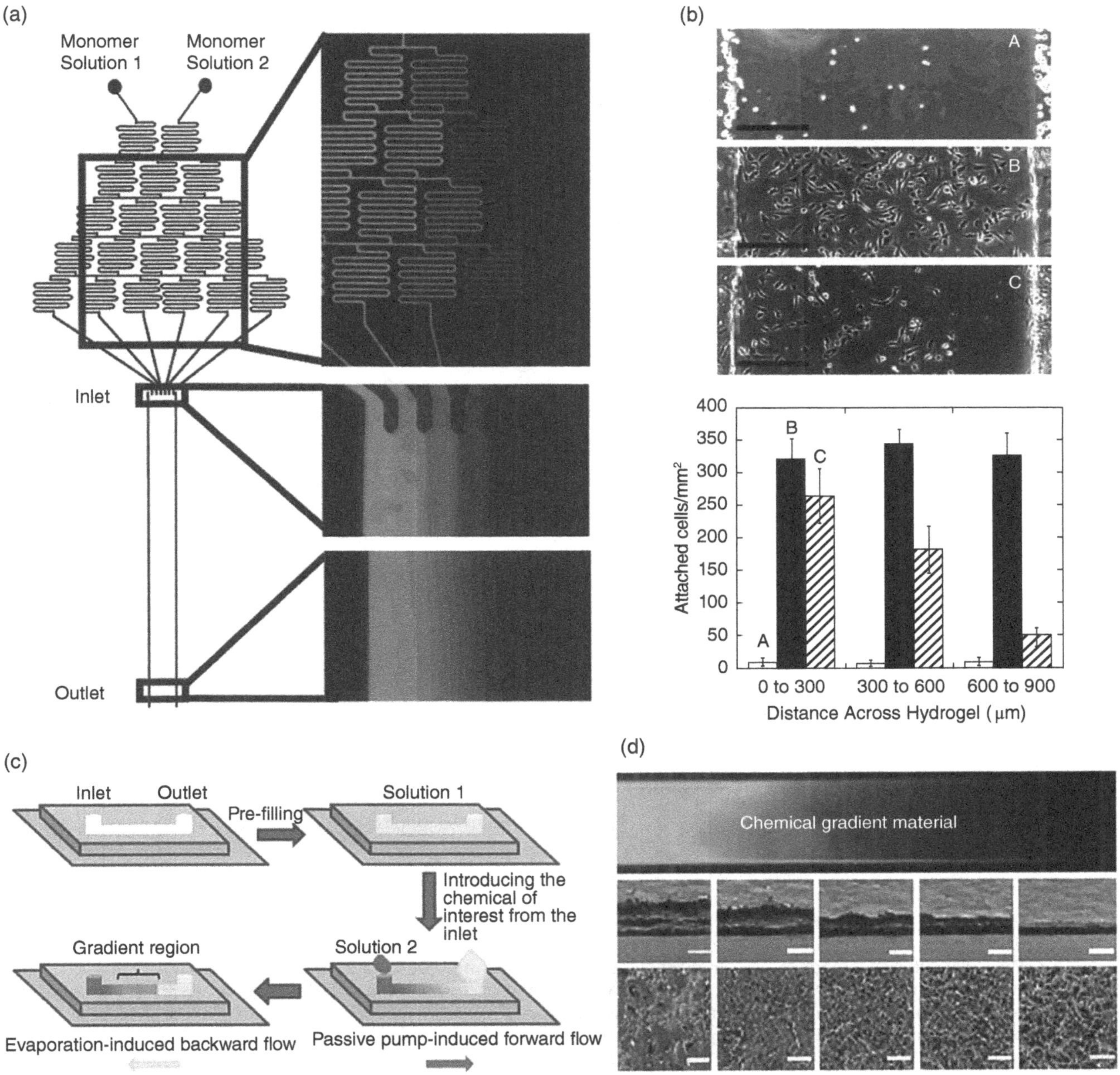

Figure 12.2 Examples of gradient hydrogels fabricated using the microfluidic methods. (a) Schematic of the microchannels used in the experiment. The monomer solution 1 contained rhodamine to make the fluorescent images illustrating the obvious gradient concentration at the inlet and outlet. (b) Light micrographs of HUVECs attached to the surface of PEG hydrogel without RGDS (A), with RGDS (B) and with a gradient of tethered RGDS (D). And quantification of HUVECs attachment on corresponding hydrogels (c) Schematic of gradient generation in a microfluidic channel. (d) The gradient PEG-DA hydrogels fabricated in (c) procedure. scale bar, 50 μm. *Sources:* (a) and (b) reproduced from [18] with permission, copyright 2004, American Chemical Society. (c) and (d) reproduced from [17] with permission, copyright 2011, Elsevier.

to prepare the stiffness gradient polyvinyl alcohol (PVA) hydrogel [25.] The PVA solution was put into a cylindrical mold, then made contact with the liquid nitrogen surface for 5 min to gradually freeze the PVA solution from the bottom (liquid nitrogen contacted side) to the top (the air side). During the PVA solution freezing process, ice generated in amorphous regions to induce the growth of polymer crystallinite points which were physical cross-linking points between the PVA chains. As the gradual freezing moved along the longitudinal direction, the

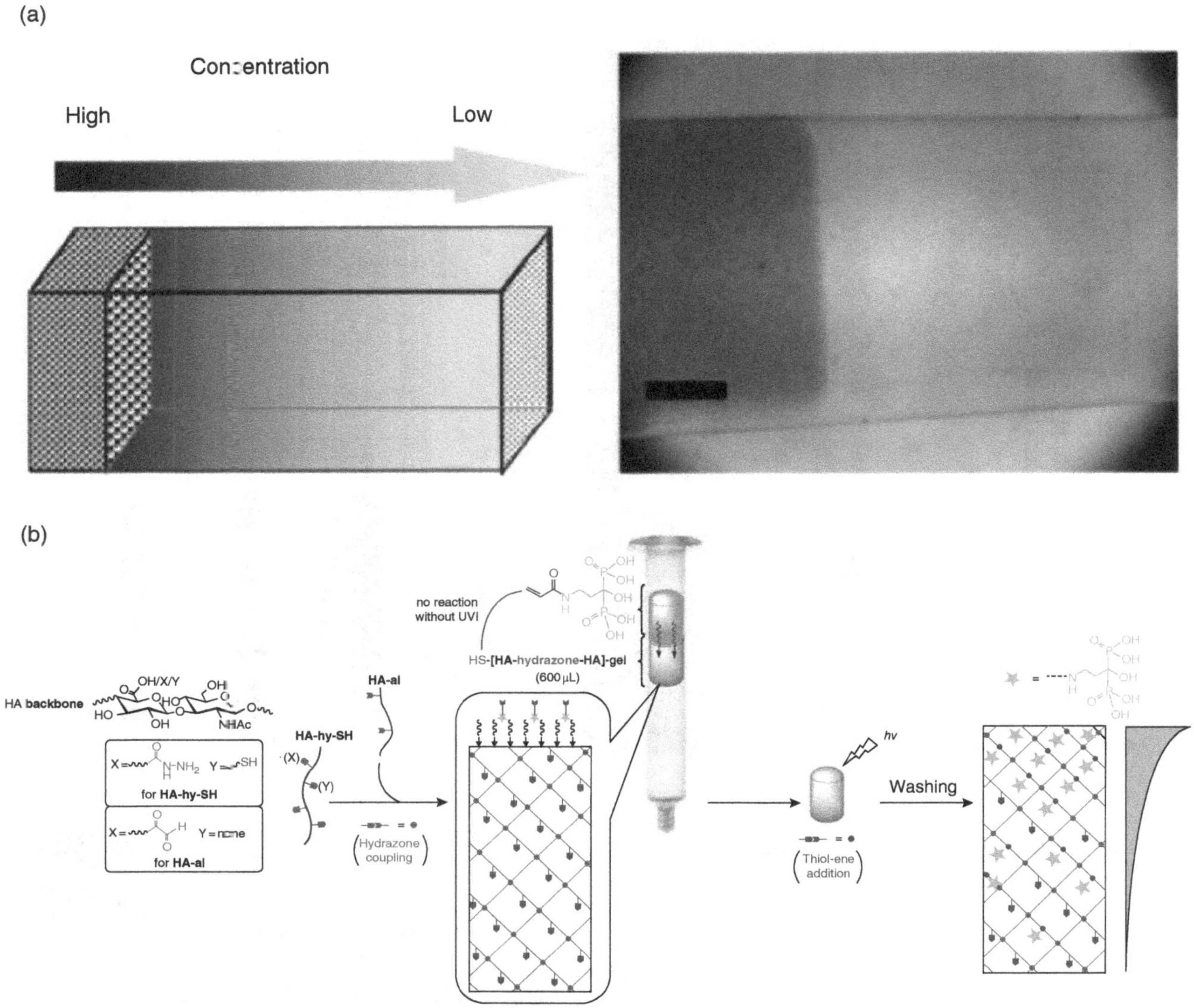

Figure 12.3 Molecular diffusion method to form gradient hydrogels. (a) Schematic representation of spatial localization of PLG microspheres within PEG hydrogels. The bright field image show distinct PLG microspheres-loaded depots, localized at left side of the PEG hydrogel (scale bar, 1 mm). (b) Schematic of preparation process of hyaluronic acid (HA) hydrogels with the gradient of chemically immobilized bisphosphonate (BP) groups involving two consecutive chemo-selective reactions. *Sources:* (a) Reproduced from [20] with permission, copyright 2008, John Wiley & Sons' (b) Reproduced from [21] with permission, copyright 2015, American Chemical Society.

cross-linking points at the bottom which are close to the freezing source were more than the far end of the top (Figure 12.4a). Then, the frozen PVA solution was thawed at room temperature. By repeating the above freezing-thawing process, the stiffness gradient PVA hydrogel was generated without any chemical cross-linkers. As expected, the as-prepared PVA hydrogel showed gradient mechanical strength and stiffness along the longitudinal direction. To investigate stem cell behavior, human bone marrow stem cells (hBMSCs) were seeded onto the stiffness gradient PVA hydrogel. The hBMSCs on stiffer hydrogels showed more cell spreading, while those on the softer hydrogels showed an elongated morphology (Figure 12.4b). As the result of cell differentiation behaviors, both stiff and soft PVA hydrogel sections promoted effective neurogenesis and osteogenesis of hBMSCs. Later, the same group reported a similar work to prepare a more wide-ranging stiffness gradient hydrogel by incorporating hyaluronic acid (HA) into the PVA solution [26]. In another work carried out by Rape et al., an HA-based hydrogel with an orthogonal combination of patterned stiffness and fibronectin density was prepared by free radical photopolymerization with a gradient photomask (Figure 12.4c) [27]. A photoconjugation reagent,

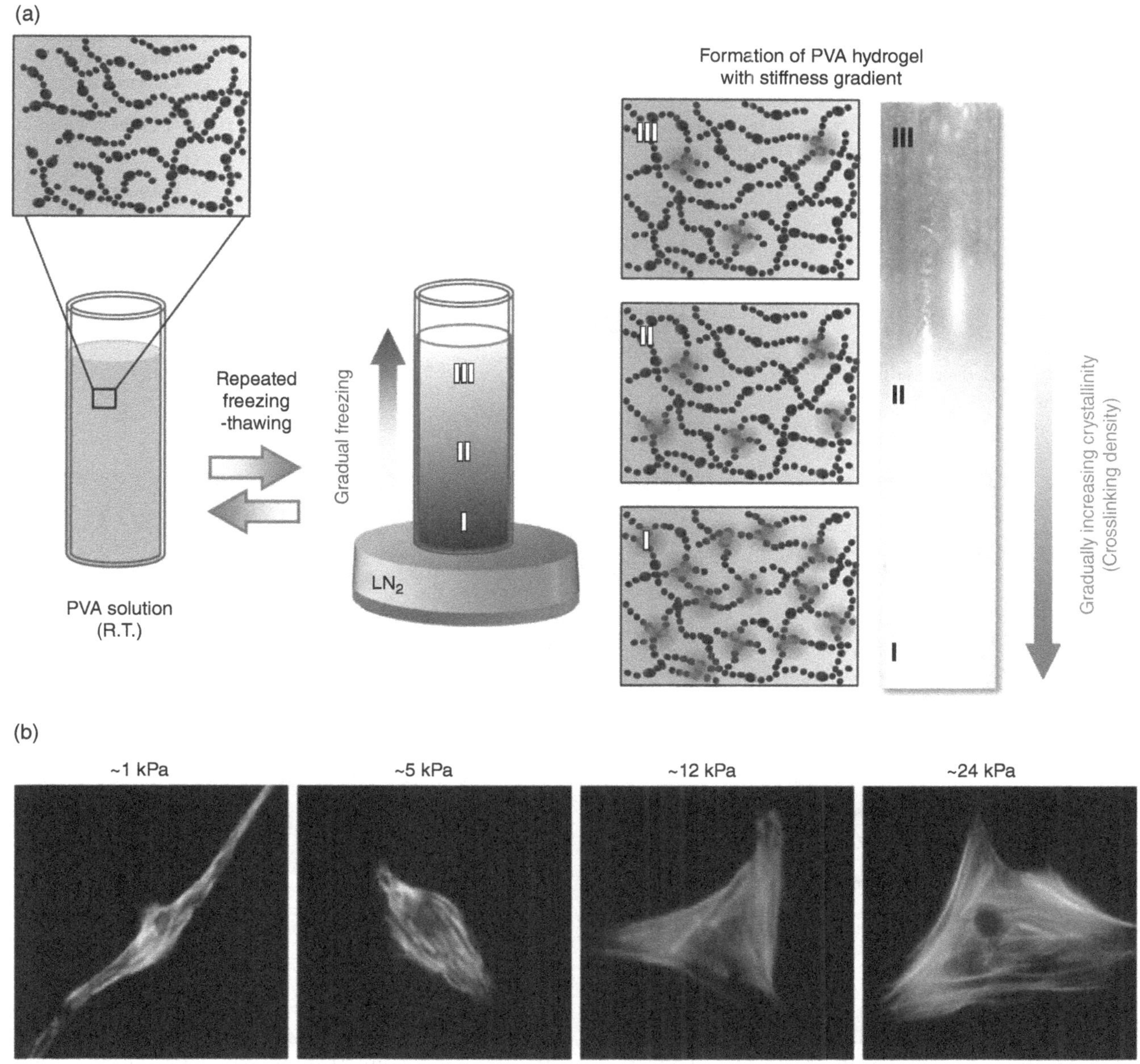

Figure 12.4 Gradient stiffness in hydrogels. (a) Schematic diagrams showing the stiffness gradient PVA hydrogel generated by gradual freezing-thawing method. The red dots represent cross-linking points in PVA hydrogel networks. (b) Immunofluorescent staining (Phalloidin) images of hBMSCs cultured on the stiffness gradient PVA hydrogel after 28 days. (c) Schematic diagrams of the process to prepare an HA-based hydrogel with orthogonal gradients. (d) The orthogonal gradients of both stiffness and ligand density. (e) U373-MG human GBM cells remained rounded and rarely exhibited lamellipodia in soft regions with low fibronectin. In contrast, cells spread extensively and developed large lamillipodia in high stiffness regions with high fibronectin. *Sources:* (a) and (b) reproduced from [25] with permission, copyright 2015, Elsevier. (c)–(e) reproduced from Rape (2015) [27] with permission, copyright 2015, Nature Publishing Group. (*See insert for color representation of the figure.*)

4,5-dimethoxy 2-nitrobenzyl-aminothiol (DMNBAT) was appended to HA-methacrylate for cross-linking with dithiothreitol (DTT) to form the HA-based hydrogels. Then, the hydrogel was incubated in the eosin Y-based photoinitiator. By covering with a gradient photomask to perform the free radical photopolymerization, the stiffness gradient was formed. By rotating the above gradient photomask at 90°, the gel was exposed to UV light to selectively conjugate the thiol groups with fibronectin on the gel through the cross-linker sulfo-SMCC. Finally, the orthogonal gradients of the stiffness and fibronectin density were generated in the single hydrogel (Figure 12.4d).

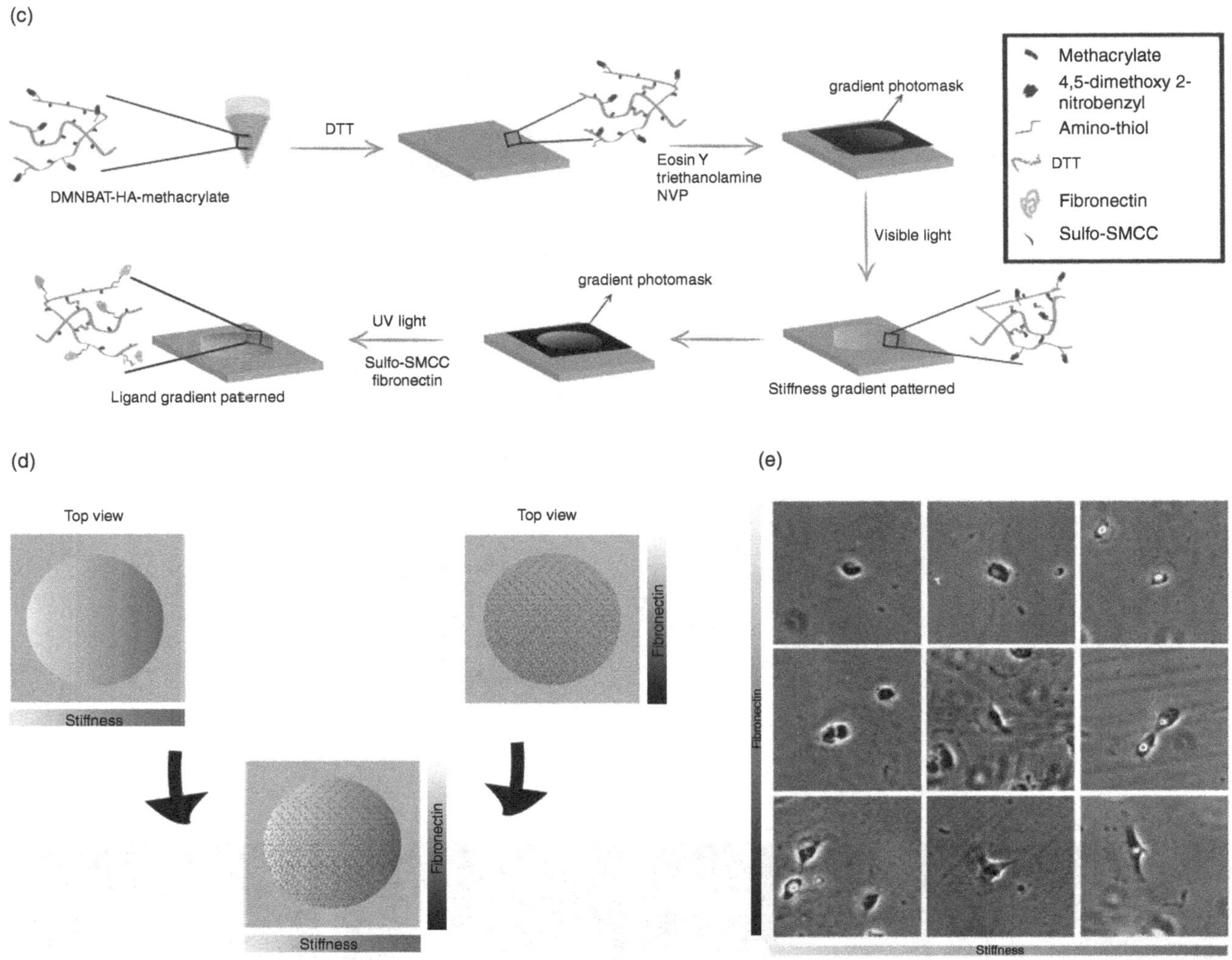

Figure 12.4 (Continued)

U373-MG human GBM cells were seeded onto as-prepared hydrogels to investigate the combined effect of stiffness and fibronectin density-regulated cell spreading. The results showed the cell spreading area increased both with fibronectin density for all stiffness values and with stiffness at all given fibronectin densities (Figure 12.4e). This method provides a new strategy to investigate the combinative variation of biochemical and biophysical parameters in a single platform.

Other methods for the preparation of gradient hydrogels, including electrophoresis [28], inkjet printing [29], photodegradation [30], and 3D printing [31], etc., have been developed to reveal comprehensive applications, most notably in the areas of tissue engineering and cell behaviors. However, the capability of hydrogel applications is often severely limited by their mechanical properties [32]. In the past few years, hydrogels with high strength and toughness, mimicking biological functionalities, such as cartilage and ligaments, have emerged as implantable materials [6].

12.2.2 Gradient Polymer Brushes

Polymer brushes refer to the ultrathin polymer coatings, by which chains of polymer molecules are tethered with some anchor points to a surface in such a way that the graft densities of the polymers are high enough to form a brush-type conformation [33]. Surface modification through grafting of polymer brushes has been a powerful tool to tailor the physical and chemical properties of interfaces and has made great achievements in surface and interface engineering [34]. Compared with the "graft to" approach [35], the "graft from" approach, in particular, by means of surface-initiated controlled radical polymerization (SI-CRP), has long been used as an essential tool to construct complex brush structures. Polymer brushes also have extensive applications in the biomedical field, because the treasure of chemical functional groups that are integrated into polymer brushes allows the achievement of various surface properties with bioactivity or bioinertness [36]. Crucially, architectural features play a significant role in modulating the behavior of polymer brushes [37]. The use of SI-CRP techniques accomplishes diverse surface-tethered polymer architectures, including homo, block, random, branched, binary, and gradient polymer brushes [34]. In this section, we will focus on the preparation of gradient polymer brushes via SI-CRP techniques.

Gradient polymer brushes allow the continuous variation of grafting densities or molecular weights of polymer chains along one or more directions [37]. Polymer brushes with the grafting densities gradient could be prepared by modifying the substrates with a concentration gradient of surface-anchored polymerization initiators. Wu et al. created the gradient of polymerization initiators on a silicon wafer substrate by vapor diffusion of a mixture of the trichlorosilane initiator and paraffin oil [38]. Then, the polyacrylamide brushes were prepared by surface-initiated atom transfer radical polymerization (SI-ATRP) to give the gradient architecture with "mushroom" to "brush" regime across the polyacrylamide layer. Subsequently, Genzer et al. extended this method to form a gradient of inactive OTS followed by backfilling with ATRP initiators. The poly(*tert*-butyl acrylate) (PtBA) brushes with the grafting densities gradient were grafted onto the substrate, and then converted into poly(acrylic acid) (PAA) brushes by chemically cleaving off the *tert*-butyl group (Figure 12.5a) [39]. The authors also undertook extensive research on the detailed structure of PAA brushes with the grafting densities gradient under different conditions [40]. Luzinov et al. used a gradient temperature stage to fabricate poly(glycidyl methacrylate) (PGMA) film with gradient thickness as a primary anchoring layer, which was then translated into the ATRP initiators gradient by reacting with 2-bromo-2-methylpropionic acid (BPA). Then, the polystyrene brushes with gradient grafting density were formed by SI-ATRP (Figure 12.5b) [41]. Coad et al. reported a new method for fabricating a gradient ATRP initiator surface by plasma copolymerizing 1,7-octadiene and allylamine. The gradient density of the surface-anchored amine groups was generated by gradually varying the feed composition of the two monomers. The ATRP initiators 2-bromo-2-methylpropionyl bromide (BiBB) were immobilized on the amine-functionalized substrate via acid-halide coupling to form the initiator gradient. Finally, poly(2-Hydroxyethyl methacrylate) (PHEMA) brushes with gradient grafting density were generated on the substrate (Figure 12.5c) [42]. Mei et al. prepared the PHEMA brushes with grafting density gradient on the silicon substrate. Fits of X-ray reflectivity profiles indicated that the structure of polymer chains changed from the "mushroom" to "brush" regime with the increasing graft densities. The bovine fibronectin was then adsorbed onto different regimes to test cell adhesion. The results showed the fibronectin adsorbed in the "mushroom" region resulted in cell adhesion and spreading, while it was repelled in the "brush" region with little cell adhesion (Figure 12.5d) [43]. Meanwhile, Genzer et al. prepared the gradient PHEMA brushes with varied grafting densities (σ) and polymer molecular weights (MW) in two orthogonal directions (Figure 12.5e) [44]. They also found PHEMA brushes played a role as the barrier for fibronectin adsorption, thus resulting in a concentration gradient of fibronectin in an opposite direction with respect to the grafting density of PHEMA brushes. The osteoblastic cells (MC3T3-E1) were seeded onto this MW-σ orthogonal PHEMA/fibronectin gradient substrate. The fluorescence images of MC3T3-E1 cells which were taken along the diagonal of the orthogonal substrate showed the cell density increased with the increasing amount of fibronectin on the surface, and the cell shape changed from a well-spread polygonal phenotype at high grafting density to an elongated spindle-shaped phenotype at low grafting density of fibronectin (Figure 12.5e). These studies indicated that gradient polymer brushes can regulate the protein adsorption to control cell behavior.

Besides the grafting density gradient, molecular weight gradient polymer brushes are another type of gradient architectures, which could be prepared by introducing certain gradient fields in polymerization processes. Genzer et al. reported a method for the preparation of surface-anchored poly(methyl methacrylate) (PMMA) brushes with gradient thickness by gradually draining the reaction solutions [45]. As shown in the schematic of the apparatus for this method (Figure 12.6a), the chamber was loaded with the ATRP solution which contained MMA monomer, metal/ligand, and the solvent. The chamber was then degassed with a stream of dry nitrogen. The ATRP initiator-functionalized silicon wafer was immersed into the reaction solution. During the reaction, the ATRP solution was removed gradually by a micropump, resulting in a stabile decrease along the three-phase contact line. The PMMA brushes with gradient thickness were obtained because of the different polymerization degree along the vertical direction. By decreasing the draining rate, the slope of the linear thickness was elevated but the thickness as a function of time

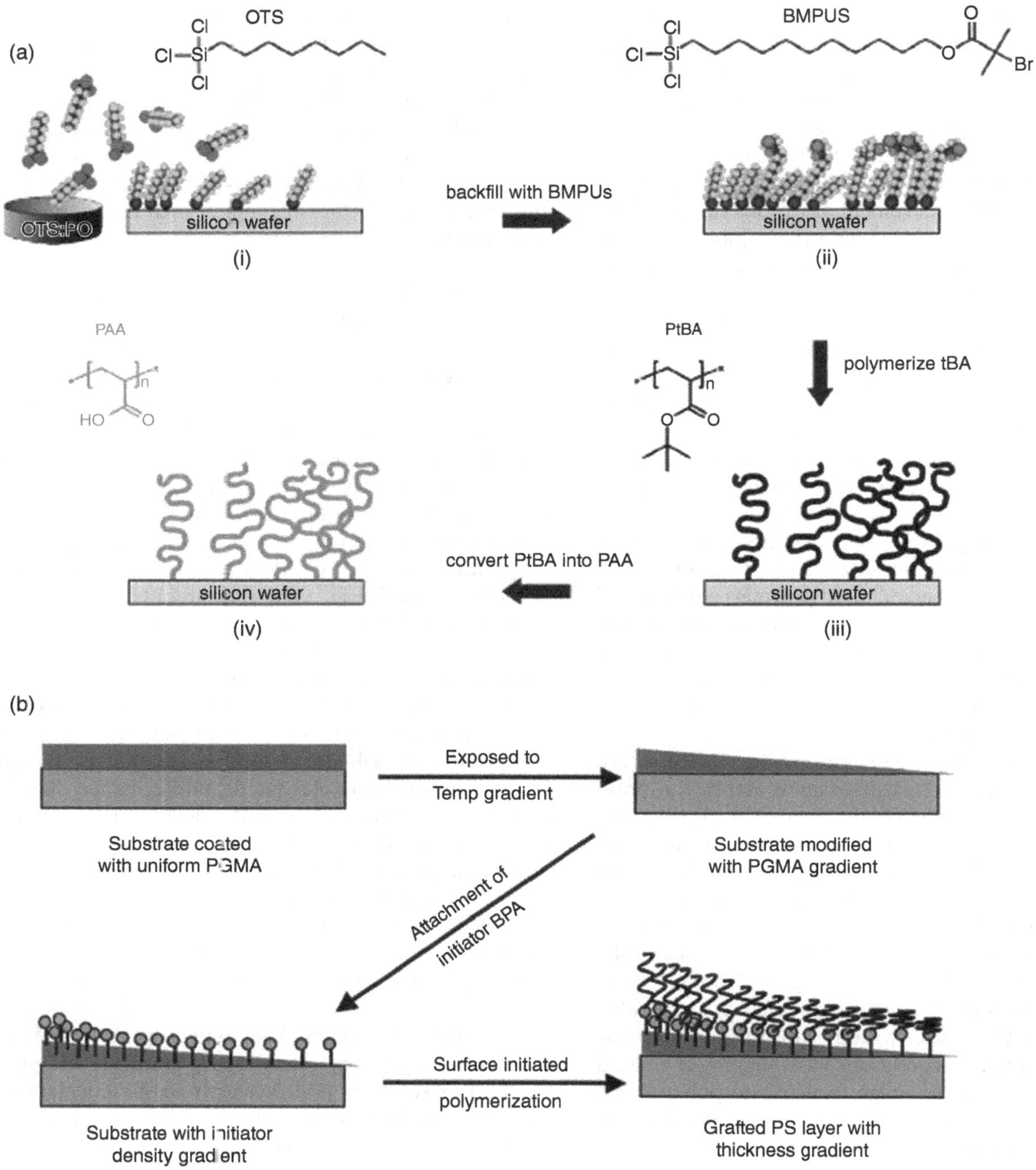

Figure 12.5 Grafting density gradient polymer brushes. (a) Schematic of the vapor diffusion method for preparation of grafting density gradient PAA brushes. (i) The OTS evaporated and diffused in the vapor phase to generate a concentration gradient along the substrate, then reacted with the -OH functional group bonded onto the substrate to form the gradient OTS SAM layer. (ii) The regions of the silica substrate which were not covered with the inactive OTS were backfilled with ATRP initiators. (iii)–(iv) PtBA brushes were grafted onto the substrates by the SI-ATRP technique, and subsequently converted into PAA brushes. (b) Schematic of the temperature gradient method for preparation of PS brushes with grafting density gradient. (c) Schematic of the plasma copolymerization method for preparation of PHEMA brushes with grafting density gradient. (d) Schematic of PHEMA brushes with gradient grafting density. The conformation of brushes changed from "mushroom" to "brush" regime. The adsorptive capacity of bovine fibronectin was inversely proportional to the grafting density of PHMEA brushes. (e) Contour plots of dry thicknesses of PHEMA brushes (left) and dry fibronectin thickness (right) in a MW-σ orthogonal gradient. The fluorescence images of MC3T3-E1 cells which are taken along the diagonal (marked with the numbers in the right contour plot) of the orthogonal substrate. *Sources:* (a) Reproduced from [39] with permission, copyright 2007, American Chemical Society. (b) Reproduced from [41] with permission, copyright 2005, American Chemical Society. (c) Reproduced from [42] with permission, copyright 2014, American Chemical Society. (d) Reproduced from [43] with permission, copyright 2005, American Chemical Society. (e) Reproduced from reference 44 with permission, copyright 2005, John Wiley & Sons.

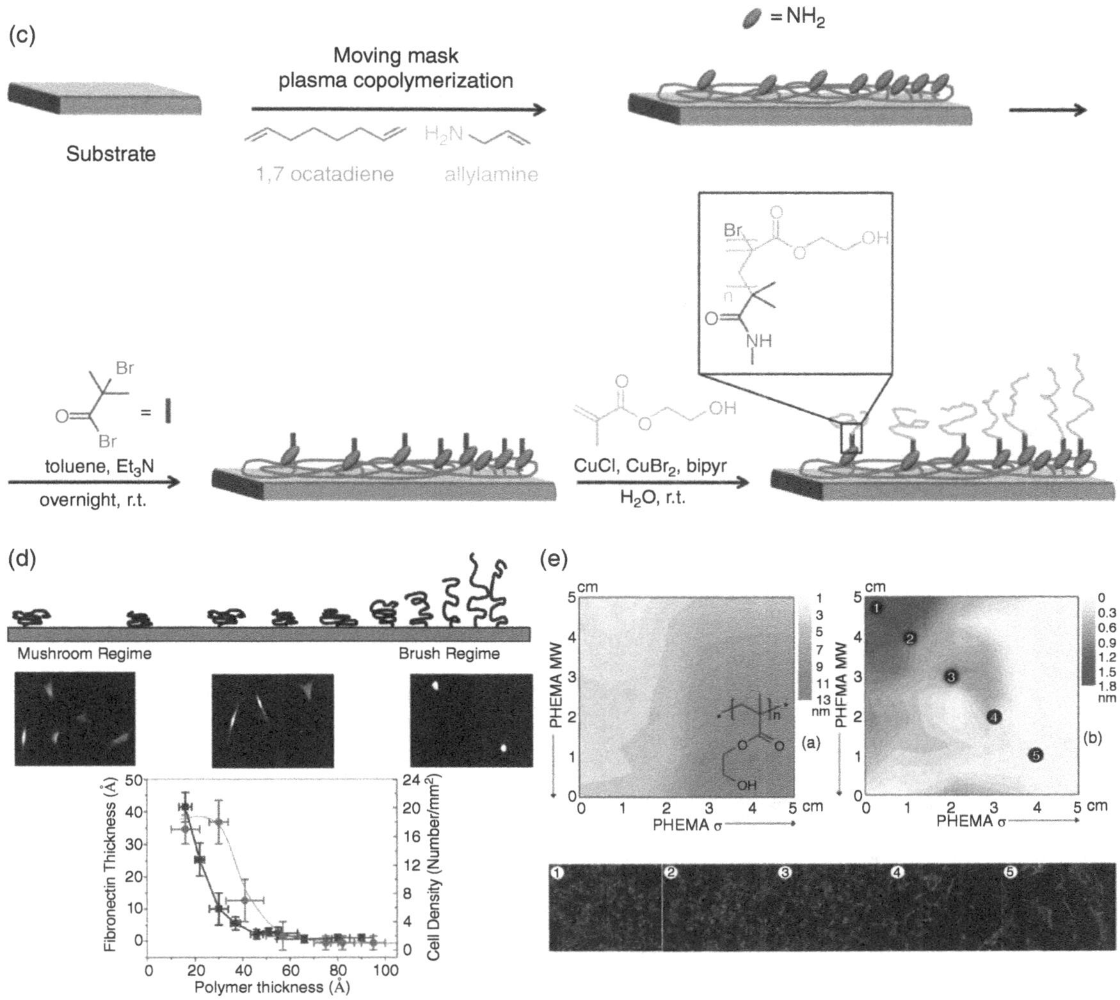

Figure 12.5 (Continued)

was unaffected (Figure 12.6b). This method can be used to study the polymerization kinetics of ATRP [46]. Xu et al. used the microchannel confined method to prepared PHEMA brushes with molecular weight gradient [47]. The microchannel was formed by placing a patterned PDMS stamp onto the ATRP initiator-functionalized silicon wafer. The ATRP solution was injected into the microchannel from the inlet at a controlled rate. After the solution reached the outlet, the patterned stamp was removed immediately. The silicon wafer was rinsed thoroughly with DMF. As the different regions on the surface were exposed to the ATRP solution with different times, the thickness of PHEMA brushes increased linearly with the distance from the outlet (Figure 12.6c) [47]. Metters et al. reported the synthesis of PMMA brushes with gradient thickness from photoiniferter-modified silicon wafer. By using a mobile photomask, the UV exposure time gradually varied along the initiator-modified substrate (Figure 12.6d) [48]. Further, by using this method, the authors reported the cell-adhesive RGD-ligand (Arg-Gly-Asp) density gradients controlled the 3T3 mouse fibroblast cell adhesion. Unlike the PHEMA brushes that play a role as the barrier for fibronectin adsorption described above, the poly(methacrylic acid) (PMAA) brushes contain carboxylic acid functionalities which can give the amidation reaction with primary amines of RGD-ligand. Thus, the concentration gradient of RGD-ligand was in the same direction with respect to the thickness of the PMAA brushes, resulting in the specific cell adhesion in a spatially defined manner across the substrate [49].

In recent years, external stimuli-regulated, controlled radical polymerizations have been successfully developed [50], such as electrochemically mediated ATRP (eATRP) [51], light-mediated living radical polymerization [52], sacrificial anode ATRP (saATRP) [53], and mechanically controlled ATRP [54]. These outstanding

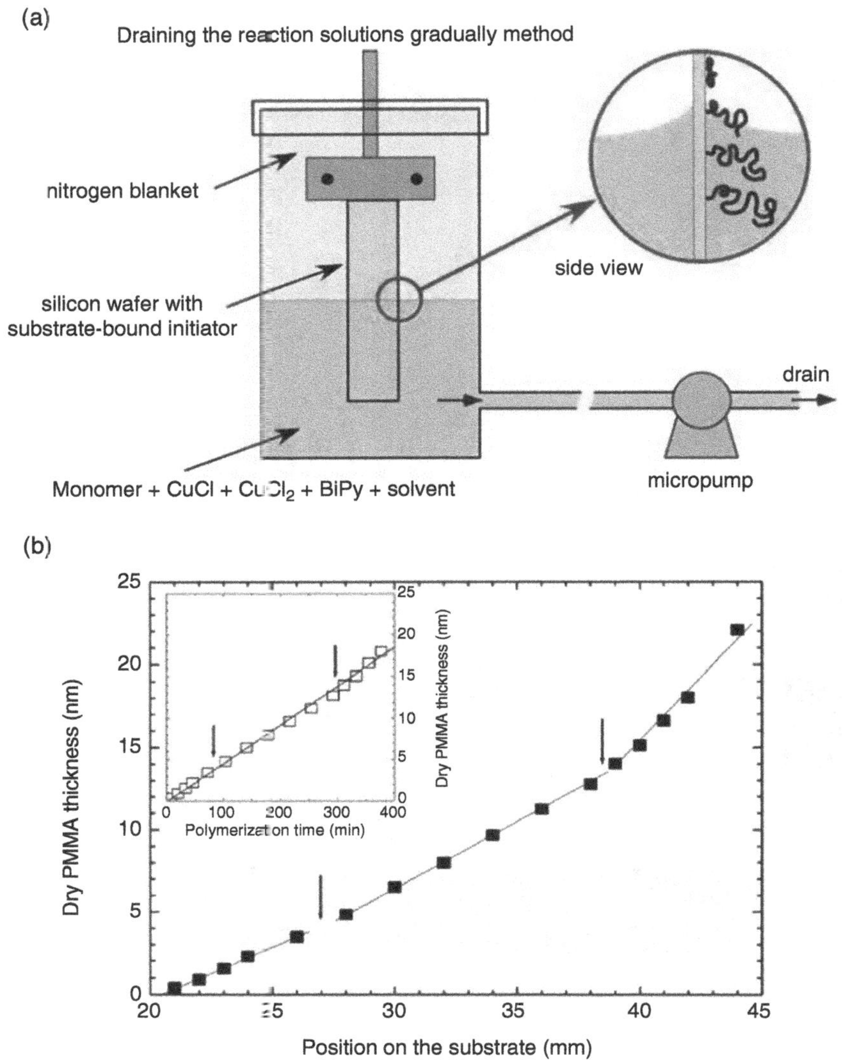

Figure 12.6 Molecular weight gradient polymer brushes. (a) Schematic of the apparatus for preparation of PMMA brushes with molecular weight gradient by gradually draining the reaction solutions. (b) The thickness of dry PMMA brushes vs position of the substrate. The inset is the thickness of dry PMMA brushes vs the polymerization time. (c) Schematic of the microchannel confined method for preparation of PHEMA brushes with molecular weight gradient. (d) Schematic of movable photomask method for preparation of PHEMA brushes with molecular weight gradient by photo-induced iniferter polymerization. (A), (B), and (C) show the procedure of gradient PHEMA brushes growth via the movable photomask. *Sources:* (a) and (b) reproduced from [45] with permission, copyright 2003, American Chemical Society. (c) Reproduced from [47] with permission, copyright 2005, American Chemical Society. (d) Reproduced from [48] with permission, copyright 2006, American Chemical Society.

methods greatly facilitate the surface modification by polymer brushes. The latest developments in gradient polymer brushes by the external stimuli-regulated CRP are presented here. Matyjaszewski et al. first reported the eATRP by using an electrochemical redox process at the electrode surface to precisely control the ratio of activator to deactivator catalyst [51]. Zhou et al. extended this method to surface modification by immobilizing the alkanethiol initiators on a gold substrate as the working electrode [55]. The applied potential reduced the Cu^{II}/L to Cu^{I}/L to generate a high local concentration of activating species near the initiator surface. The activating species reacted with alkanethiol initiators anchored on the gold surface to generate radicals which can trigger the polymerization (Figure 12.7a). 3-sulfopropyl methacrylate potassium salt (SPMA) and HEMA monomers in the solution polymerized to form the homogeneous films on the gold substrate. The SI-eATRP can be

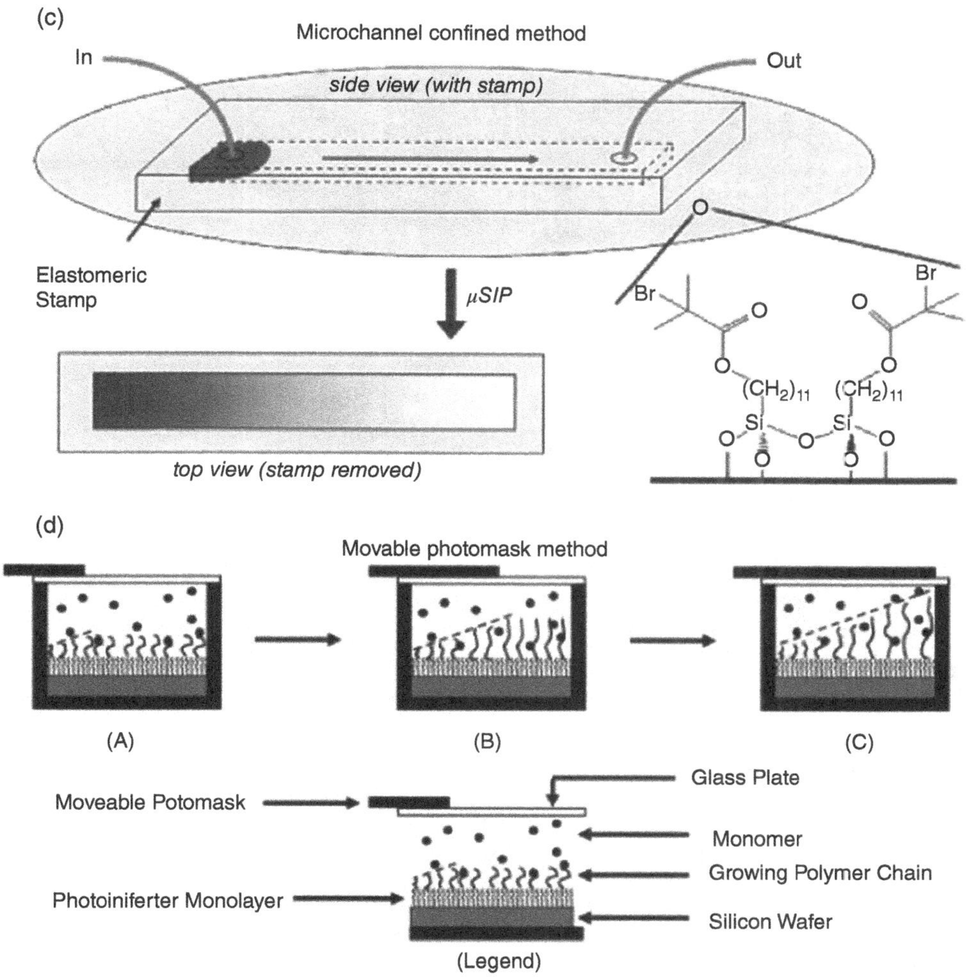

Figure 12.6 (Continued)

conducted in the presence of air and the polymerization solution can be reused several times. Furthermore, the authors used this method for non-conducting substrates [56]. First, the trichlorosilane initiator-modified non-conducting substrate (such as silicon wafer, PDMS) was fixed opposite to the working electrode, leaving a gap of a few hundred micrometers (*d* in Figure 12.7b) between the work electrode and substrate. A three-electrode set-up was inserted into the solution consisting of SPMA monomer, Cu^{II}/L, and the solvent. By a constant applied potential, which was determined by cyclic voltammograms, the Cu^{I}/L activator was generated at the surface of the working electrode, then diffused to the initiator-modified substrate to trigger the PSPMA brushes' growth (Figure 12.7b). The kinetics of the polymer brush growth (thickness) could be controlled by the distance of the gap between the working electrode and the initiator-modified substrate. More importantly, a stable concentration gradient of $[Cu^{II}/L]/[Cu^{I}/L]$ was generated in the gap by the diffusion effect. By simply tilting the initiator-modified substrate, the gradient polymer brushes could be prepared as a result of the different ratio of $[Cu^{II}/L]/[Cu^{I}/L]$ along the substrate interface, which led to different rates of polymerization. The thickness of as-prepared gradient polymer brushes was a linear continuous increase along one direction. By patterning the initiator via micro-contact printing (μCP) with a PDMS stamp, the "wedge" and "stair" patterned gradient PSPMA brushes also can be formed. Shortly thereafter, Shida et al. [57] used SI-eATRP combined with a bipolar electrode (BPE) [58] to fabricate the patterned and gradient polymer brush structure. The authors placed the initiator-modified glass substrate against the BPE, retaining the desired gap between them.

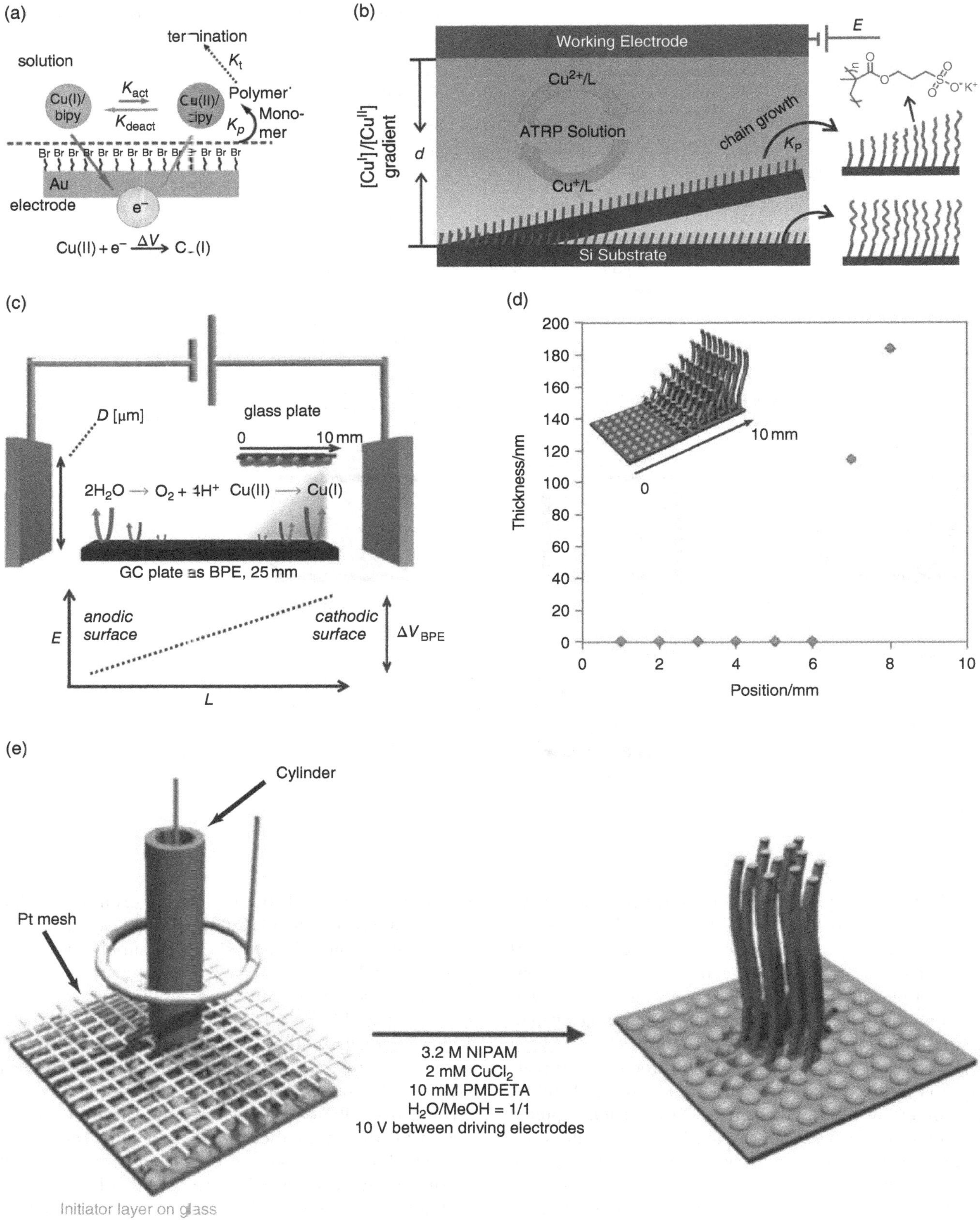

Figure 12.7 Gradient polymer brushes by surface-initiated electrochemically mediated ATRP (SI-eATRP) method. (a) Mechanism of SI-eATRP at a gold working electrode. (b) Schematic of SI-eATRP for non-conducting substrate through catalyst diffusion. (c) Schematic of the electrochemical apparatus used for SI-eATRP. The initiator-modified substrate with 10 mm length was placed against the BPE, leaving a gap with D μm. (d) The thickness of gradient PNIPAM brushes as a function of the position. (e) Schematic of the apparatus used for pattern polymer brushes. *Sources:* (a) Reproduced from [55] with permission, copyright 2012, John Wiley & Sons. (b) Reproduced from [56] with permission, copyright 2013 American Chemical Society. (c) to (e) reproduced from [57] with permission, copyright 2015, John Wiley & Sons.

By applying a constant voltage, a potential gradient was generated on the BPE. Thus, a concentration gradient of Cu^{I}/L catalyst was formed to make the gradient growth of poly(N-isopropylacrylamide) (PNIPAM) brushes (Figure 12.7c). The thickness of the obtained gradient polymer brushes at each position is shown in Figure 12.7d. In addition, the authors placed BPE under an insulating cylinder resulting in a circular potential distribution (Figure 12.7e). The patterned polymer brushes could be fabricated without pre-patterning the initiator on the substrate. These SI-eATRP methods opened a new avenue for the preparation of gradient polymer brushes.

Light is another external stimulus which can regulate controlled radical polymerizations and fabricate complex three-dimensional (3D) polymer brushes. Hawker et al. [52] reported a light-controlled living radical polymerization of methacrylates by using an Ir-based photoredox catalyst (Figure 12.8b) [59]. According to the proposed mechanism by the authors (Figure 12.8a), the Ir-based catalyst was excited by visible light with a photon to generate an Ir^{III*} species. The Ir^{III*} catalyst reacted with an alkyl bromide (Pn-Br) to form the radical (Pn•), which can trigger the polymerization. The Ir^{IV} then oxidized the radical chain-end back to the dormant (Pn-Br) as well as Ir^{IV} back to the initial Ir^{III} complex in the ground state. The whole procedure could be repeated under UV irradiation, resulting in a controlled/living radical polymerization process. More importantly, the most advantageous characteristic of this system is the ability to afford the temporal control over chain growth. Shortly thereafter, the same group fabricated the patterned and gradient polymer brushes via this method [60]. By using a photomask, patterned polymer brushes could be obtained because the chain growth was spatially confined to the exposed regions (Figure 12.8c). This method also could fabricate gradient PMMA brushes by adjusting the intensity of incident light using a grayscale photomask with a linear gradient (Figure 12.8d). From the optical micrograph (Figure 12.8f) and 3D atomic force microscopy (AFM) (Figure 12.8g), a nanoscale-inclined plane formed from the gradient PMMA brushes. A good linear correlation of the polymer brushes thickness as a function of position (Figure 12.8h) illustrated the superiority of this method in preparing the gradient and other complex 3D structural polymer brushes. The same group also prepared the gradient branched polymer brushes by this method [61]. Typically, MMA and HEMA were copolymerized on the initiator-modified silicon substrate to obtain the random copolymer P(MMA-co-HEMA) brushes. The propagating alkyl bromide chain ends were deactivated to prohibit the chain-end extension, resulting in the following polymer brushes only grafted from the backbone of brushes chain. Then, the HEMA units were functionalized with initiated groups to graft the arm brushes. By using the grayscale mask with a linear gradient, a gradient comb brush structure was prepared (Figure 12.8e). Although the light-mediated living radical polymerization has many advantages in fabricating the various 3D structural polymer brushes, this method also has some problems for more comprehensive application, e.g. the very high cost of the Ir-based catalyst, rigorous deoxygenation procedures, and the metal catalyst caused contamination. Thus, developing a metal-free CRP is an urgent challenge. In recent years, many groups have made great advances in designing the metal-free catalyst for light-controlled CRP [62–67]. These methods provide broad areas for fabricating the polymer brushes in an environmentally-friendly way.

The above methods often involve electrochemical apparatus or photocatalysts, and a large volume of ATRP solution. Thus, finding a simple and low-cost chemical substance as the external stimulus is a significant challenge for SI-CRP. Zhou et al. reported a method for the preparation of various 3D polymer brushes using a small amount of ATRP solution [68]. Figure 12.9a shows the experimental set-up of this method. An initiator-modified silicon substrate was fixed opposite a zinc (Zn) sheet. The distance of the gap (*D* in Figure 12.9b) could be adjusted by screws. When the assemblage was immersed into the ATRP solution consisting of Cu^{II}/L and monomer, a brown-red color appeared at the surface of Zn sheet as the evidence of Cu^{I}/L generating (Figure 12.9a). This is due to the redox reaction occurring at the Zn interface resulting in the reduction of Cu^{II}/L to Cu^{I}/L. Similar to the diffusion strategy of SI-eATRP, the Cu^{I}/L activator diffused toward the initiator-modified substrate, resulting in polymer brush growth (Figure 12.9b). Because the Zn played the role of the anode and itself participated in the redox reaction, the authors called this method a surface-initiated sacrificial-anode atom transfer radical polymerization (SI-saATRP). By tilting the Zn sheet in a certain angle (θ), a different ratio of $[Cu^{II}/L]/[Cu^{I}/L]$ along the y-direction resulted in different polymer-growth kinetics (Figure 12.9c). By changing the tilt angle, brush thickness gradients with different slopes were obtained (Figure 12.9c). The most prominent characteristic of SI-saATRP is using Zn with a variety of shapes to fabricate various 3D structural polymer brushes. As a proof-of-concept, the authors placed a Zn wire above the substrate, a ridge-like gradient PSPMA brush structure was formed (Figure 12.9d). Jordan et al. reported a similar work that used a copper (Cu) plate instead of Zn sheet without adding copper halide catalysts (Figure 12.9e) [69]. Gradient polymer brushes also could be prepared in the same way by tilting the copper plate (Figure 12.9f) [70]. It is worth noting that because the mechanism of Cu(0)-mediated CRP

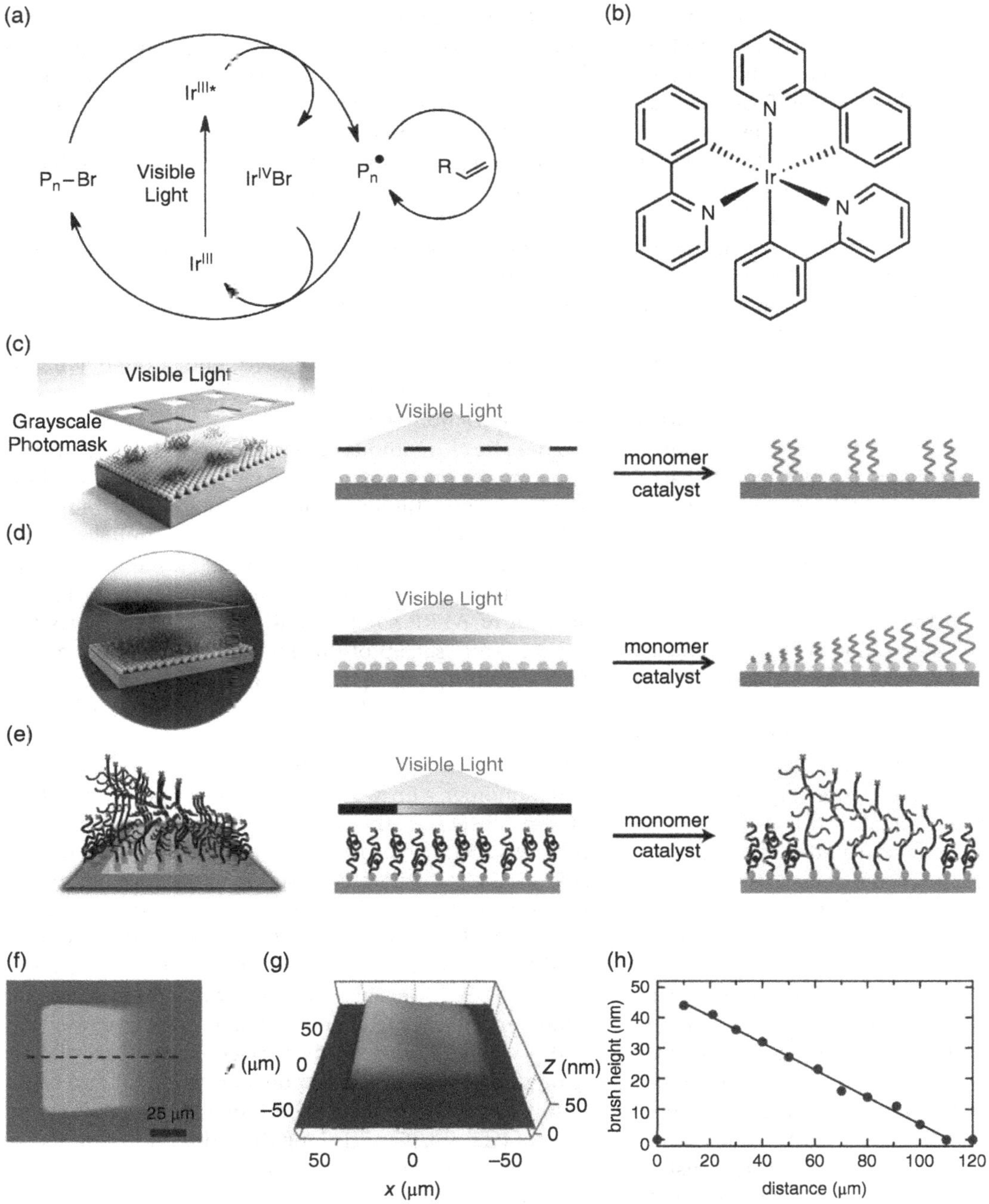

Figure 12.8 Gradient polymer brushes by light-mediated living radical polymerization. (a) Mechanism of the light-mediated living radical polymerization using a photoredox catalyst *fac*-[Ir(ppy)$_3$] (ppy = 2-pyridylphenyl) in (b). Schematic of the light-mediated CRP method for preparation of (c) patterned PMMA brushes by using a rectangular patterned photomask, (d) gradient PMMA brushes by using a photomask with a linear gradient neutral density filter and (e) gradient comb brush architecture. (f) Optical micrograph and (g) three-dimensional AFM image of nanoscale-inclined plane of gradient PMMA brushes. (h) Brushes' thickness as a function of position of gradient PMMA brushes. *Sources:* (a) and (b) reproduced from [52] with permission, copyright 2012, John Wiley & Sons. (c), (d), and (f–h) reproduced from [60] with permission, copyright 2013, John Wiley & Sons. (e) reproduced from [61] with permission, copyright 2016, John Wiley & Sons.

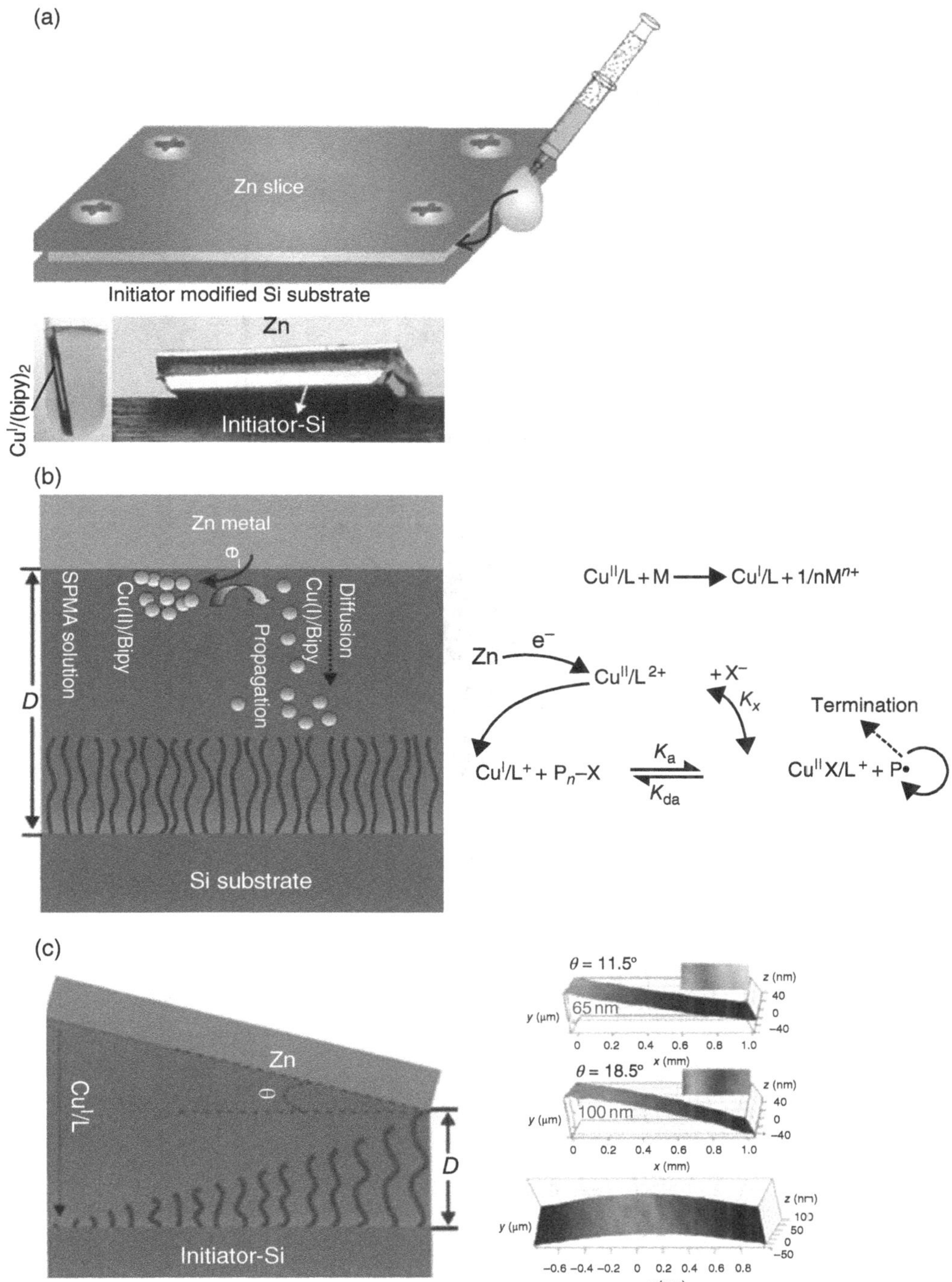

Figure 12.9 Gradient polymer brushes by surface-initiated sacrificial anode ATRP (SI-saATRP) and Cu(0)-mediated controlled radical polymerization (SI-CuCRP). (a) Experimental se-up and digital photo of the sandwiched architecture of SI-saATRP. A brown-red color appeared at the surface of Zn sheet. (b) The proposed mechanism of SI-saATRP. The "sacrificial anode" (Zn) reduced Cu^{II}/L to Cu^{I}/L, which diffuses to the surface of the initiator-modified silicon substrate. (c) Schematic of the preparation for gradient polymer brushes by tilting the Zn sheet in a certain angle (θ = 11.5° or 18.5°). Digital photos and 3D optical images of gradient PSPMA brushes in different tilting angles. (d) A ridge-like polymer brush structure was obtained by placing a Zn wire above the substrate. Plots of the brush thickness as a function of position with corresponding digital photo of ridge-like PSPMA brushes. (e) Schematic and experimental set-up of SI-CuCRP. The reaction solution consisted of monomer and ligand without copper halide catalysts. (f) Plots of the brush thickness vs position of gradient poly(methacryloyloxy)ethyl trimethylammonium chloride) (PMETAC) brushes. Inset is a digital photo of the substrate with macroscopic gradient PMETAC brushes. *Sources:* (a)–(d) reproduced from [68] with permission, copyright 2013, John Wiley & Sons. (e) Reproduced from Zhang (2015) [69] with permission, copyright 2015, The Royal Society of Chemistry. (f) Reproduced from Zhang (2015) [70] with permission, copyright 2015, The Royal Society of Chemistry.

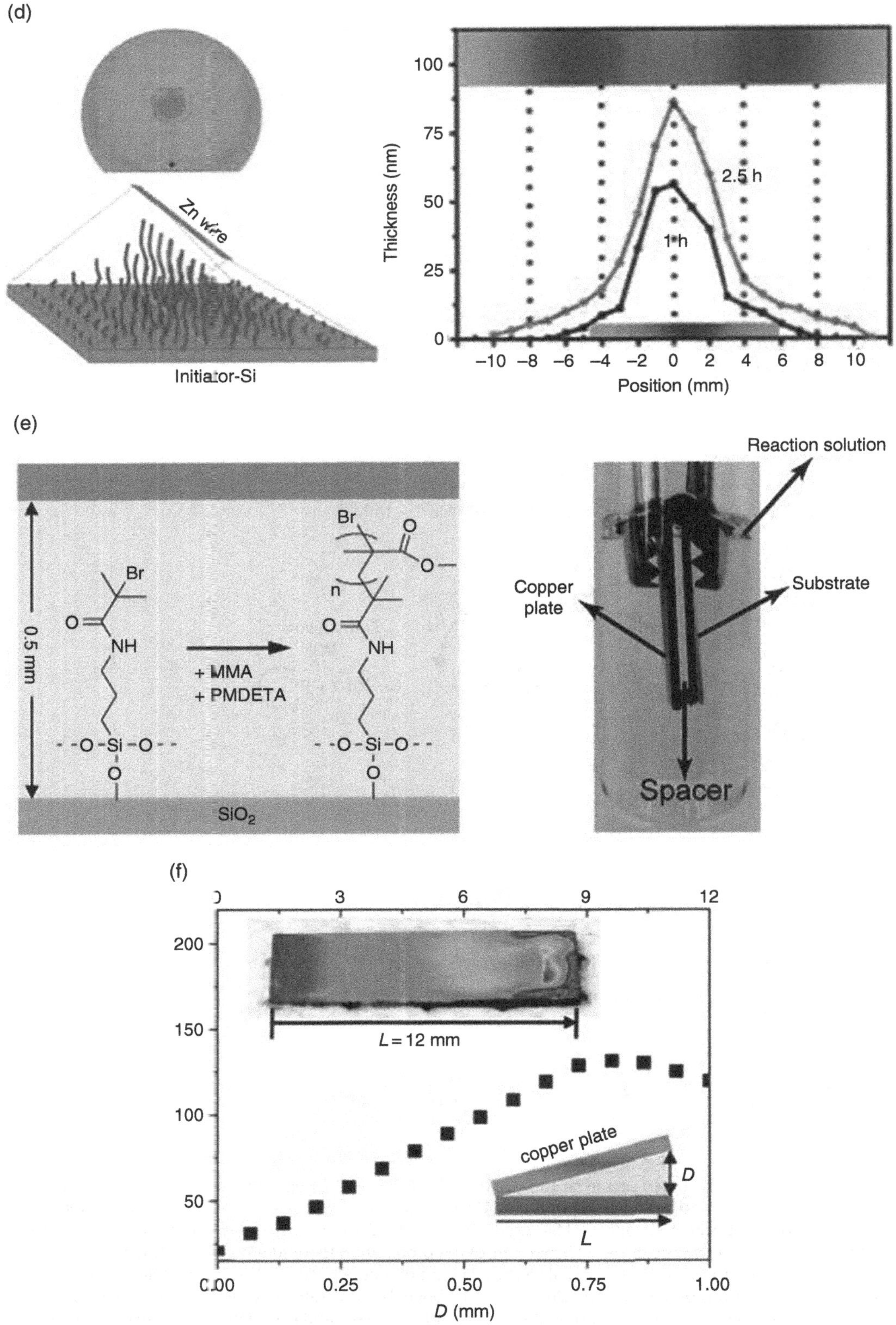

Figure 12.9 (Continued)

involves the single-electron transfer living radical polymerization (SER-LRP) [71, 72], the brush growth rate is faster than other reports. With the low cost of Zn or Cu and a small amount of reaction solution, these methods made the gradient polymer brushes accessible to a wider community and offered more comprehensive applications. For instance, Yoshida et al. designed a gradient self-oscillating polymer brush surface [73]. The initiator-modified substrate grafted with gradient poly (NIPAM-*r*-NAPMAM) by SI-saATRP. The ruthenium tris(2,2'-bipyridine) [$Ru(bpy)_3$], which is a catalyst of the Belousov-Zhabotinsky (BZ) reaction, was conjugated with the gradient binding sites of N-3-(aminopropyl)methacrylamide (NAPMAM), resulting in a spatial distribution of the immobilized catalyst on the polymer brush surface. When the gradient self-oscillating polymer brush substrate was immersed into a BZ reaction solution, the chemical waves which were spatiotemporal patterns created by the redox changes of the $Ru(bpy)_3$ catalyst in BZ reaction, moved from areas of low to high $Ru(bpy)_3$ concentration. This method can be used in spatiotemporally controlled biomolecular motor systems [74].

In the above-mentioned advances of SI-CRP, we summarized the cutting-edge techniques for fabricating gradient and 3D structural polymer brushes. Recently, Esser-Kahn et al. reported a mechanically controlled radical polymerization which initiated by ultrasound [54]. Hopefully, this work may offer some clues in developing new methods for surface modification with polymer brushes.

12.3 Gradient Polymeric Structures Regulated Cell Behavior

12.3.1 Gradient Cell Adhesion

Gradient polymeric structures have been used as advantageous platforms with applications in controlling cell behavior, such as cell adhesion, migration, and differentiation. Cell adhesion is a process in which one cell binds to another cell or to a surface. To accomplish this binding, cell adhesion molecules serve as intermediates that hold the cell onto another surface. One of the environmental parameters is the extracellular matrix (ECM) to which the cells adhere. The ECM comprises many proteins, including fibronectin, laminin, collagen, and proteoglycans. These proteins contain specific functional bonding sites that interact with integrin receptors which are located at the cell surface [75, 76]. For instance, an amino acid sequence Arg-Gly-Asp (RGD) is present in the cell-binding sites of fibronectin, laminin, and various other extracellular proteins [77]. RGD has been identified as bonding many different integrins and facilitates cell adhesion. As mentioned above, gradient polymer brushes could be functionalized with the RGD to gradually vary cell adhesion [43, 44, 49], In addition, click chemistry is also an effective strategy for the construction of biomaterials and biointerfaces [78, 79]. Guan et al. reported a multifunctional gradient biointerface which could guide the cell adhesion [80]. The authors first deposited the poly[(4-methylpropiolate-p-xylylene)-co-(p-xylylene)] onto the glass substrate using the chemical vapor deposition (CVD) technique. Then, the continuous gradient of biomolecules, such as the thiol-terminated PEG and the azide-modified RGD peptide, were immobilized on the substrate by the thiol-yne and alkyne-azide click reaction, respectively. The 3T3 fibroblast cells were seeded onto the PEG-modified substrate. As the cell repellent property of PEG [81], the gradient of cell density was reversed with the increasing PEG concentration (Figure 12.10a). The authors also fabricated the countercurrent gradients of PEG and RGD. In this way, the increasing cell-repellent property of PEG and the decreasing cell-adhesive property of RGD were distributed in parallel along the substrate. As expected, in the PEG increased direction, the gradient of cell density was also observed (Figure 12.10b), while with an enhanced influence on cell number (Figure 12.10c). These gradient biointerfaces may be applicable in the design of biomaterials. By using a convection and alternating flow approach, Khademhosseini et al. fabricated a poly(ethylene glycol-diacrylate) (PEG-DA) hydrogel with a thickness gradient [82]. The thickness of the gradient hydrogel gradually decreased from nearly 40 μm to less than 10 μm. This technique was also valid for the preparation of a thermally cross-linked collagen gradient with continuous changes in fibril density. More importantly, this gradient platform offered a simple method for preparing gradients of cell density.

Besides the above methods, there are some smart methods that can fabricate the gradients in cell density. Xia et al. reported a sedimentation method by inserting the fibronectin-modified glass slide into the homogeneous suspensions of cells (Figure 12.10d) [9]. Different gradients in cell density were generated by changing the tilt angle of the glass substrate. Few cells adhered at the starting point because the surface of the cell suspensions was at the same height with respect to the top of the glass slide. As shown in Figure 12.10e, different cell densities were observed at the 5 cm position, depending on the tilt angle. The larger tilt angle resulted in a higher cell density. Qin et al. used a high-throughput microdevice to monitor 3D cell invasion in real time, and generated opposing gradients for two types of cells by this sedimentation method [7]. Furthermore, Wang et al. reported a method for the rapid generation of cell gradients via nanotopographic interactions (Figure 12.10f) [10].

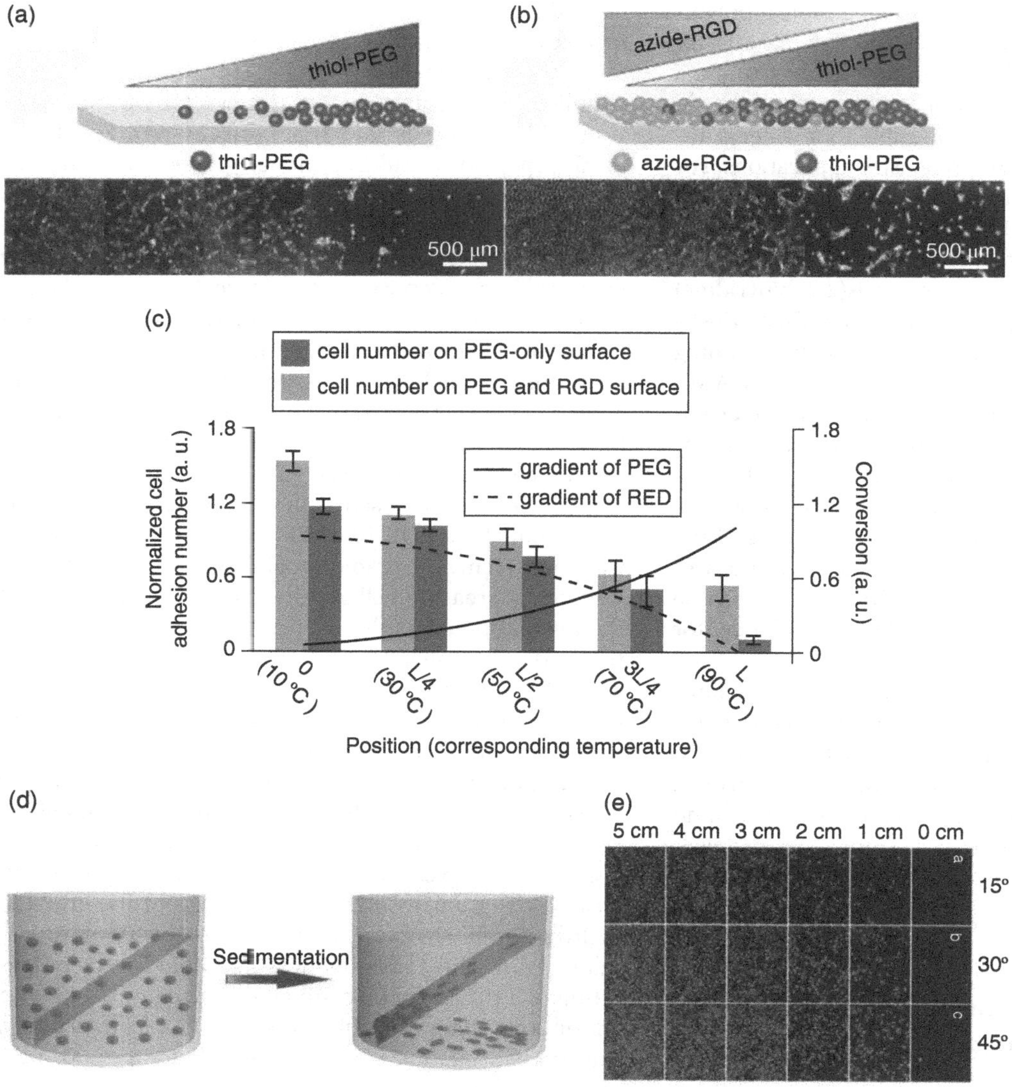

Figure 12.10 Gradient cell adhesion. (a) Schematic of the PEG concentration gradient prepared by click chemistry. 3T3 cells were cultured on this surface to show cell density gradient. (b) Schematic of the countercurrent gradients of PEG and RGD concentration. (c) Quantification of cell attachment on the substrate in (a) and (b). (d) Schematic of the sedimentation method for fabricating the cell density gradient. (e) Fluorescence micrographs showing cell density gradients generated on glass substrate at different tilt angles. (f) Schematic of the rapid generation of a cell gradient. Top view and side view SEM images showing the increased surface roughness and thickness of the nanodendritic gradient, respectively. The fluorescence micrographs showing the gradual increasing cell density along the direction with gradient nanodendrites. (g) The influence of chemical gradients on nanodendrites for generation of cell gradients. *Sources:* (a)–(c) reproduced from [80] with permission, copyright 2016, American Chemical Society. (d) and (e) reproduced from [9] with permission, copyright 2013, John Wiley & Sons. (f) and (g) reproduced from [10] with permission, copyright 2014, John Wiley & Sons. (*See insert for color representation of the figure.*)

The authors prepared the gradient nanodendritic silica by using candle soot as the template [83]. By holding the substrate above the flames of a candle at a tilting angle, a layer of carbon nanoparticle with a thickness gradient was deposited on the glass slide. Subsequently, a silica shell was coated on the carbon nanoparticles layer. After calcinating the hybrid carbon/silica composite, the gradient nanodendritic substrate was fabricated. Scanning electron microscope (SEM) images indicated the thickness and surface roughness of the nanodendrites increased along one direction (Figure 12.10f). By immersing the gradient nanodendritic substrates in a cell suspension and incubating for a short time, the number of attached cells increased along with the gradient nanodendrites (fluorescence images in Figure 12.10f). This is attributed to the enhanced topographic interactions

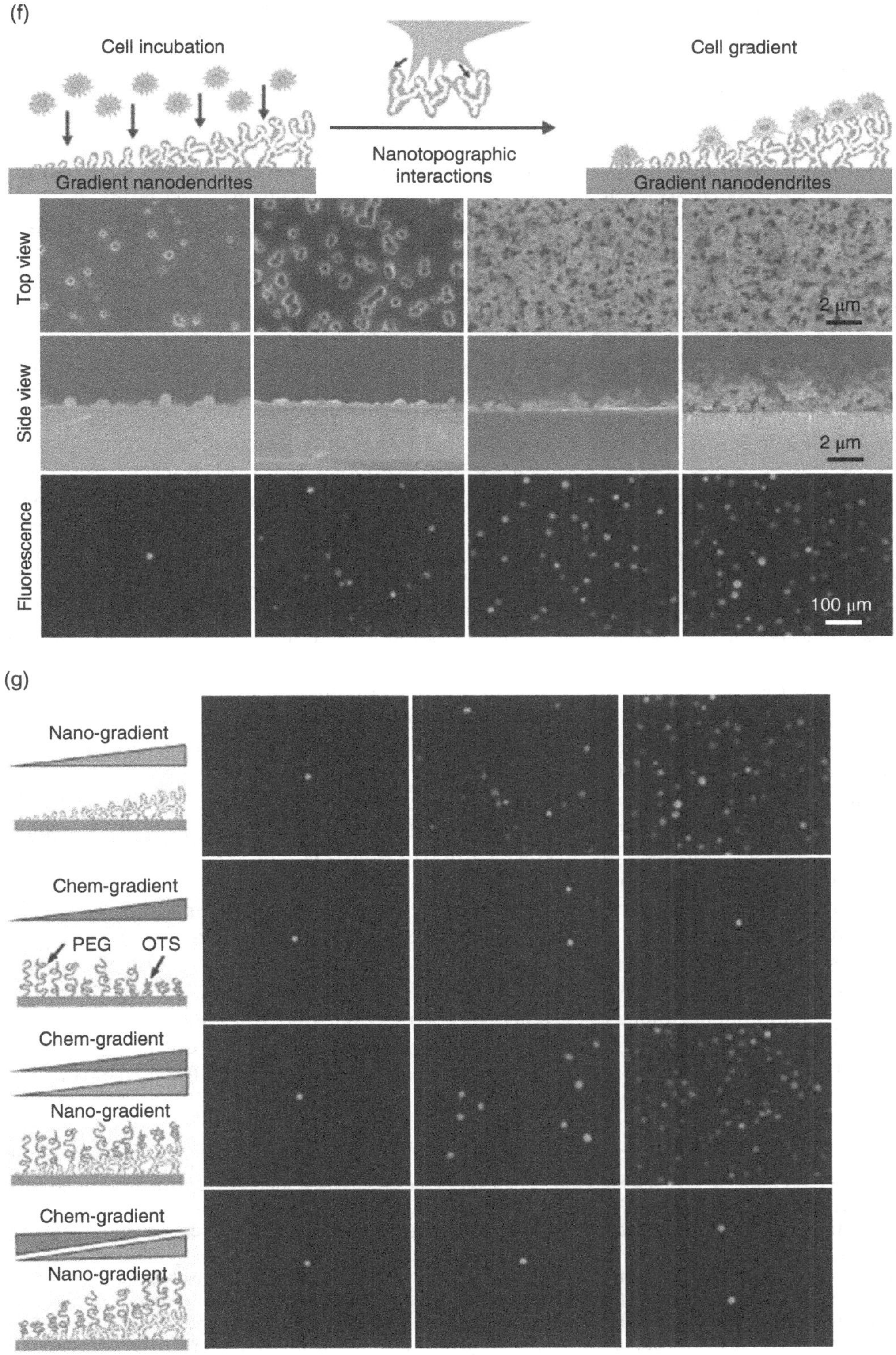

Figure 12.10 (Continued)

between the nanoscaled filopodia on the cell surface and the nanodendritic substrates [84–97]. In addition, the authors studied the effect of chemical gradients on nanodendritic substrates. A chemical gradient, on which are the typical chemical cues for repelling and promoting cell adhesion, was fabricated. There were almost no cells adhering on the PEG/OTS gradient surface (second line in Figure 12.10g), indicating that such a chemical gradient could not form the cell gradient. When the chemical gradient and the nanodendritic gradient were in the parallel direction, the gradient cell density was formed (third line in Figure 12.10g). When the chemical gradient and nanodendritic gradient are in the countercurrent, a few cells attached to the substrate (fourth line in Figure. X10g). These results implied that the chemical gradient offset the cell gradient induced by nanotopographic interactions.

12.3.2 Cell Migration

Cell migration is an essential process in the development and maintenance of multicellular organisms. Tissue formation during wound healing requires the arranged movements of cells [98]. In some important physiologic processes, e.g. cancer metastasis, the cells in random motion could acquire a directional motility phenotype and bias their motions in response to environmental cues [99]. Thus, regulation of cell migration can develop new therapeutic strategies to control tumor cells. Luo et al. fabricated a dynamic surface to spatiotemporally control the cell migration by the molecular gradients [8]. As shown in the process flow of Figure 12.11a, a hydrophobic hexadecanethiols pattern was fabricated on a gold substrate by micro-contact printing technique (the circular pattern in A of Figure 12.11b). The remaining unmodified regions were backfilled with tetra(ethylene glycol)-alkanethiol and nitroveratryloxycarbonyl (NOVC)-protected hydroquinone-terminated alkanethiols. The surface was inert to cell attachment because of the tetra(ethylene glycol). After illuminating under UV light through a gradient or geometrical patterned photomask, the NVOC group was deprotected, to reveal the hydroquinone group. The Swiss 3T3 fibroblast cells were seeded onto the surface and confined to the μCP hydrophobic regions (the third picture in A, Figure 12.11b). Upon an appropriate applied potential, the hydroquinone groups were converted to quinone groups. By adding the soluble RGD-oxyamine, the RGD peptide was rapidly installed on the quinone regions to form the oxime linkage (dashed line region in A of Figure 12.11b). The Swiss 3T3 fibroblast cells on the μCP regions began to sense the RGD peptides to migrate toward the patterned RGD regions (circular region to square region, fourth picture in A of Figure 12.11b). As another proof-of-concept, a more complex pattern was fabricated. The cells migrated down a gradient and traversed from one pattern to another (B in Figure 12.11b). The authors also fabricated a substrate in which were two RGD gradient patterns with different slopes (Figure 12.11c). The results indicated that cells on gradient 1 with a steeper slope responded faster than on gradient 2. This intelligent surface has great potential for investigating cell behavior.

The migration of neural stem cells (NSCs) from the central canal region toward the lesion site is very important in spinal cord injury. It has been reported that the stromal-cell-derived factor-1α (SDF1α) gradient secreted by the local astrocytes and endothelium can enhance the recruitment of endogenous NSCs to the lesion site via long-distance migration [100]. Dai et al. fabricated a continuous gradient of SDF1α on aligned collagen/poly(ε-caprolactone) (PCL) fibers using the electrospinning technique [101]. The fiber density gradually decreased along the radial direction toward the periphery. The gradient SDF1α along the radially aligned fibers was generated by the immobilization of the collagen-binding domain fused SDF1α (CBD-SDF1α). By varying the collection time, the gradient degree of the fiber density could be adjusted (Figure 12.11d). Neural stem cells were cultured around the periphery of a

Figure 12.11 Gradient polymeric structures-regulated cell migration. (a) Schematic of the process flow for generation of dynamic ligand surface gradients. (b) Fluorescent micrographs show cells at μCP regions migrated toward unveiled cell-adhesive photo-patterned regions. (c) Time-lapse micrographs showing the different migration rates of patterned cells up or down the RGD peptide gradients. (d) Different gradients of radially aligned fibers with varied collection time of electrospinning. Shallow (2 min), steep (8 min) and moderate (15 min) continuous gradient were generated with the increasing collection time. (e) Fluorescence micrographs showing the migration of NSCs on the CBD-SDF1α immobilized radially aligned fibers. (f) NSCs distribution on radially aligned fibers immobilized with CBD-SDF1α. NAT-SDF1α, PBS, and randomly oriented fibers immobilized with CBD-SDF1α were as the control groups. The dashed lines in all images indicate the border of NSCs seeding. (g) Schematic of the countercurrent density gradients of PHEMA brushes and YIGSR for selective guidance of ECs migration. (h) Migration traces of ECs and SMCs on countercurrent density gradient of PHEMA/YIGSR, single YIGSR gradient and PHEMA gradient. (i) Optical images of ECs and SMCs sheets cultured on PHEMA/YIGSR, YIGSR and PHEMA density gradient surfaces, respectively. *Sources:* (a)–(c) reproduced from [8] with permission, copyright 2011, American Chemical Society. (d)–(f) reproduced from [101] with permission, copyright 2016, John Wiley & Sons. (h) and (i) reproduced from [102] with permission, copyright 2014, American Chemical Society. (*See insert for color representation of the figure.*)

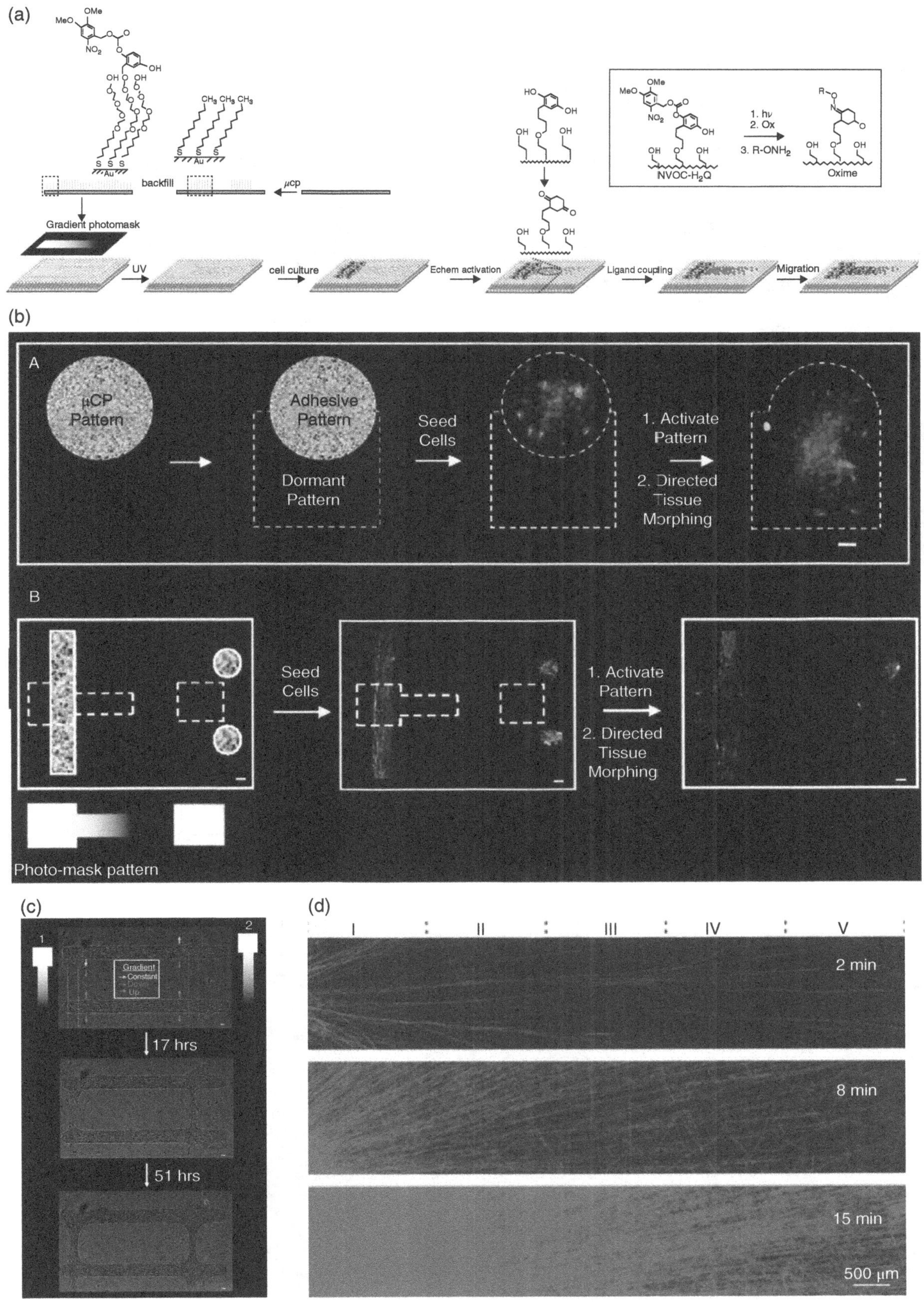
(a)
backfill
μcp
Gradient photomask
UV
cell culture
Echem activation
Ligand coupling
Migration
1. hν
2. Ox
3. R-ONH2
NVOC-H2Q
Oxime
(b)
A
μCP Pattern
Adhesive Pattern
Dormant Pattern
Seed Cells
1. Activate Pattern
2. Directed Tissue Morphing
B
Seed Cells
1. Activate Pattern
2. Directed Tissue Morphing
Photo-mask pattern
(c)
17 hrs
51 hrs
(d)
I
II
III
IV
V
2 min
8 min
15 min
500 μm

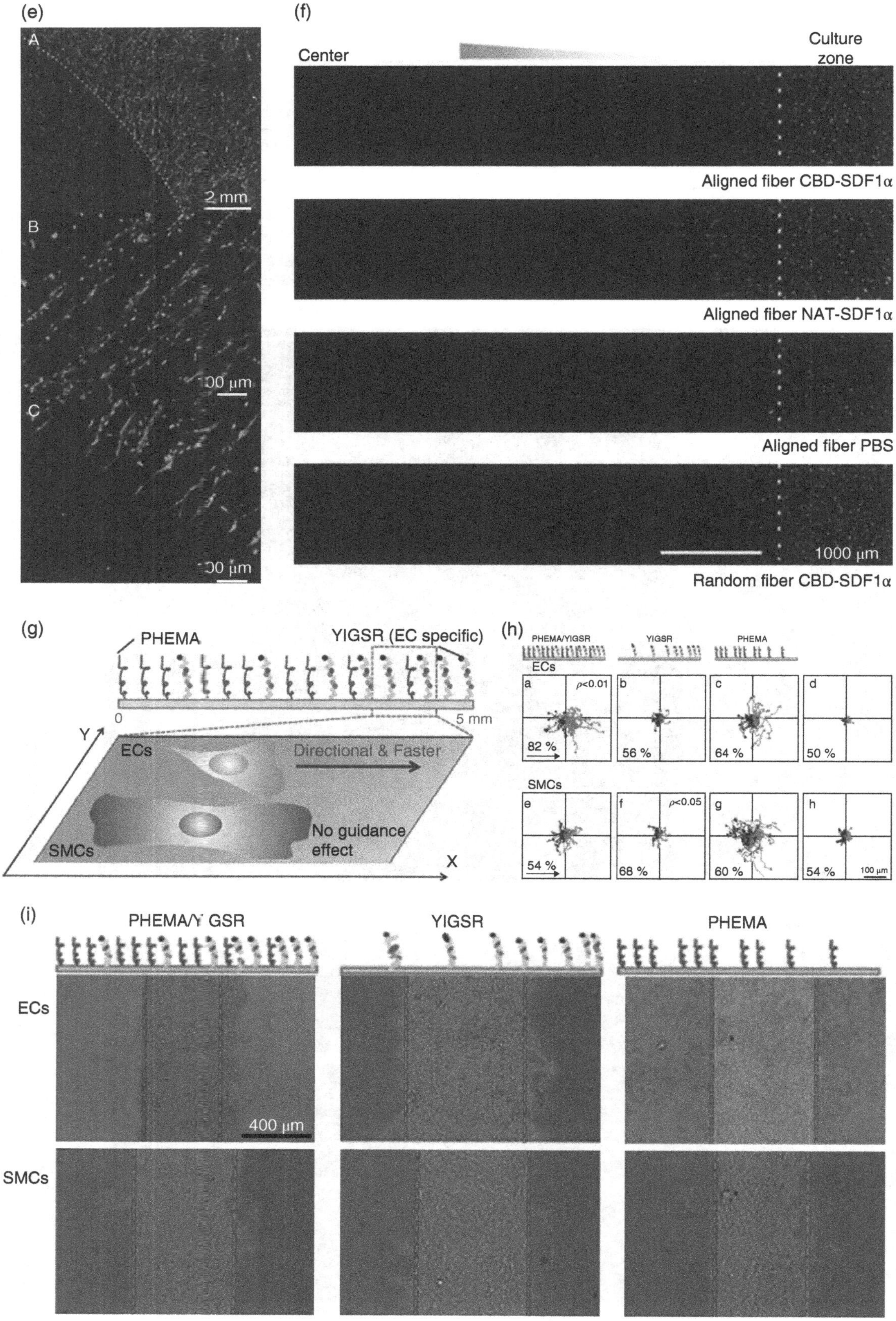

Figure 12.11 (Continued)

circular scaffold with gradient CBD-SDF1α immobilized fibers. The NSCs migrated toward the center of the scaffolds and exhibited an elongated shape (Figure 12.11e). The authors compared this gradient CBD-SDF1α immobilized fibers with other control groups. The results showed NSCs could effectively migrate toward the central region with higher SDF1α (Figure 12.11f). This study provided a new strategy for the design of gradient scaffolds for spinal cord injury. Gao et al. prepared the countercurrent density gradients of PHEMA brushes and endotheliocytes (ECs)-specific ligands Thy-Ile-Gly-Ser-Arg (YIGSR) on the same substrate for selective guidance of ECs migration (Figure 12.11g) [102]. As the decreased density of PHEMA brushes gradually enhanced the cell adhesive force, whereas the increased density of YIGSR peptides selectively enhanced the mobility of ECs, the ECs exhibited the directional migration on the gradient surface, whereas the mobility of the vein smooth muscle cells (SMCs) was not significantly influenced (Figure 12.11g). As shown in the migration traces in Figure 12.11h, 82% ECs moved to the X direction, suggesting almost unidirectional migration, whereas SMCs did not present obvious directional migration. For the control group of single gradient of YIGSR or PHEMA, the SMCs showed a relatively stronger mobility compared with ECs. These results indicated that the countercurrent density gradients of PHEMA/YIGSR were able to selectively guide the directional migration of ECs. The cell sheet assay showed the same trend, namely, the ECs sheets only migrated toward the X direction with nearly 100 μm, whereas the SMCs sheets did not show any preferential directionality (Figure 12.11i). Shortly thereafter, the same group fabricated a similar complementary density gradient with zwitterionic polymer brushes and neural cell adhesion molecule peptides to control the directional migration of Schwann cells [103]. These studies illustrated a new method for controlling the selective cell migration by a synergic effect of two signals.

12.4 Conclusion

In this chapter, we have discussed recent advances in the preparation of gradient polymeric materials, especially the gradient hydrogels and polymer brushes. We also described some biological applications for them. Gradient polymeric structures exhibit immense potential for generating a wide variety of properties on a single substrate. The emerging research reflects the broad scientific interest in the use of gradient polymeric structures to generate functional interfaces for biological applications. However, as cells contain nanoscale features whose sizes are compatible with the extracellular matrix structures [104], many studies require complex gradients on a nanoscale. Thus, the flourishing development of lithographical techniques compensates for the lack of the above-mentioned methods which can tailor elaborate confined structures of polymer brushes [105]. In addition, the combination of gradient hydrogels with 3D bioprinting techniques offers great potential for the production of complex tissues and organs [106]. It is foreseeable that the gradient polymeric materials will be used increasingly to expand our understanding of biointerfacial phenomena and to rapidly screen out optimal cellular conditions to regulate cell behavior, such as adhesion, migration, or differentiation. Although the gradient polymeric materials have made huge advances, many interdisciplinary applications need further study and exploration.

References

1 Moffat, K.L., Wang, I.N.E., Rodeo, S.A., Lu, H.H. (2009) *Clinics in Sports Medicine*, **28**, 157–176.
2 Gurdon, J.B., Bourillot, P. Y. (2001) *Nature*, **413**, 797–803.
3 Rüggeberg, M., Speck, T., Paris, O. et al. (2008) *Proc. Roy. Soc. B: Biological Sciences*, **275**, 2221–2229.
4 Speck, T., Burgert, I. (2011) *Ann. Rev. Mater. Res.*, **41**, 169–193.
5 Kim, M.S., Khang, G., Lee, H.B. (2008) *Progress in Polymer Science* (2008) **33**, 138–164.
6 Nonoyama, T., Wada, S., Kiyama, R. et al. (2016) *Adv. Mater.*, **28**, 6740–6745.
7 Zhang, Y., Zhou, L., Qin, L. (2014) *J. Amer. Chem. Soc*, **136**, 15257–15262.
8 Luo, W., Yousaf, M. N. (2011) *J. Amer. Chem. Soc.*, **133**, 10780–10783.
9 Liu, W., Zhang, Y., Thomopoulos, S., Xia, Y. (2013) *Angew. Chem. Int. Ed.*, **52**, 429–432.
10 Yang, G., Cao, Y., Fan, J., et al. (2014) *Angew. Chem. Int. Ed*, **53**, 2915–2918.
11 Buenger, D., Topuz, F., Groll, J. (2012) *Progress in Polymer Science*, **37**, 1678–1719.
12 Green, J. J., Elisseeff, J. H. (2016) *Nature*, **540**, 386–394.
13 Banerjee, A., Arha, M., Choudhary, S. et al. (2009) *Biomaterials*, **30**, 4695–4699.
14 Trappmann, B., Gautrot, J.E., Connelly, J.T. et al. (2012) *Nat Mater.*, **11**, 642–649.

15 Lee, T.T., Garcia, J.R., Paez, J.I., et al. (2015) *Nat Mater*, **14**, 352–360.
16 Slaughter, B.V., Khurshid, S.S., Fisher, O.Z. et al. (2009) *Adv. Mater.*, **21**, 3307–3329.
17 Seidi, A., Ramalingam, M., Elloumi-Hannachi, I., et al. (2011) *Acta Biomaterialia*, **7**, 1441–1451.
18 Burdick, J. A., Khademhosseini, A., Langer, R. (2004) *Langmuir*, **20**, 5153–5156.
19 He, J., Du, Y., Villa-Uribe, J.L., et al. (2010) *Adv. Funct. Mater.*, **20**, 131–137.
20 Peret, B.J., Murphy, W.L. (2008) *Adv. Funct. Mater.* **18**, 3410–3417.
21 Yi, Z., Zhang, Y., Kootala, S., et al. (2015) *ACS Appl. Mater. Interfaces*, **7**, 1194–1206.
22 Burdick, J.A., Murphy W.L. (2012) *Nature Comm.*, **3**, 1269.
23 Wells, R.G. (2008) *Hepatology*, **47**, 1394–1400.
24 Wen, J.H., Vincent, L.G., Fuhrmann, A. et al. (2014) *Nat Mater.*, **13**, 979–987.
25 Kim, T. H., An, D.B., Oh, S.H., et al. (2015) *Biomaterials*, **40**, 51–60.
26 Oh, S.H., An, D.B., Kim, T.H., Lee, J.H. (2016) *Acta Biomaterialia*, **35**, 23–31.
27 Rape, A.D., Zibinsky, M., Murthy, N., Kumar, S. (2015) *Nature Communications*, **6**, 8129.
28 Asoh, T.A., Matsusaki, M., Kaneko, T., Akashi, M. (2008) *Adv. Mater.* **20**, 2080–2083.
29 Ilkhanizadeh, S., Teixeira, A.I., Hermanson, O. (2007) *Biomaterials*, **28**, 3936–3943.
30 Kloxin, A.M., Benton, J.A., Anseth, K.S. (2010) *Biomaterials*, **31**, 1–8.
31 Cai, P., Layani, M., Leow, W. R. et al. (2016) *Adv. Mater.*, **28**, 3102–3110.
32 Sun, J.-Y., Zhao, X., Illeperuma, W.R.K. et al. (2012) *Nature*, **489**, 133–136.
33 Barbey, R., Lavanant, L., Paripovic, D., et al. (2009) *Chem. Rev.*, **109**, 5437–5527.
34 Zoppe, J.O., Ataman, N.C., Mocny, P. et al. (2017) *Chem. Rev.*, **117**, 1105–1318.
35 Minko, S., Patil, S., Datsyuk, V., et al. (2002) *Langmuir*, **18**, 289–296.
36 Krishnamoorthy, M., Hakobyan, S., Ramstedt, M., Gautrot, J.E. (2014) *Chemical Rev.*, **114**, 10976–11026.
37 Lin, X., He, Q., Li, J. (2012) *Chem. Soc. Rev.* **41**, 3584–3593.
38 Wu, T., Efimenko, K., Genzer, J. (2002) *J. Amer. Chem. Soc.*, **124**, 9394–9395.
39 Wu, T., Gong, P., Szleifer, I. et al. (2007) *Macromolecules*, **40**, 8756–8764.
40 Gong, P., Wu, T., Genzer, J., Szleifer, I. (2007) *Macromolecules*, **40**, 8765–8773.
41 Liu, Y., Klep, V., Zdyrko, B., Luzinov, I. (2005) *Langmuir*, **21**, 11806–11813.
42 Coad, B.R., Bilgic, T., Klok, H.-A. (2014) *Langmuir*, **30**, 8357–8365.
43 Mei, Y., Wu, T., Xu, C. et al. (2095) *Langmuir*, **21**, 12309–12314.
44 Bhat, R.R., Chaney, B.N., Rowley, J. et al. (2005) *Adv. Mater.*, **17**, 2802–2807.
45 Tomlinson, M.R., Genzer, J. (2003) *Macromolecules*, **36**, 3449–3451.
46 Tomlinson, M.R., Efimenko, K., Genzer, J. (2006) *Macromolecules*, **39**, 9049–9056.
47 Xu, C., Wu, T., Drain, C.M., et al. (2005) *Macromolecules*, **38**, 6–8.
48 Harris, B.P., Metters, A.T. (2006) *Macromolecules*, **9**, 2764–2772.
49 Harris, B.P., Kutty, J.K., Fritz, E.W. et al. (2006) *Langmuir*, **22**, 4467–4471.
50 Leibfarth, F.A., Mattson, K.M., Fors, B.P. et al. (2013) *Angew. Chem. Int. Ed.*, **52**, 199–210.
51 Magenau, A.J.D., Strandwitz, N.C., Gennaro, A., Matyjaszewski, K. (2011), *Science*, **332**, 81–84.
52 Fors, B.P., Hawker, C.J (2012) *Angew. Chem. Int. Ed.*, **51**, 8850–8853.
53 Park, S., Chmielarz, P., Gennaro, A., Matyjaszewski, K. (2015) *Angew. Chem. Int. Ed.*, **54**, 2388–2392.
54 Mohapatra, H., Kleiman, M., Esser-Kahn, A.P. (2017) *Nat Chem.*, **9**, 135–139.
55 Li, B., Yu, B., Huck, W.T.S. et al. (2012) *Angew. Chem. Int. Ed.*, **51**, 5092–5095.
56 Li, B., Yu, B., Huck, W.T.S. et al. (2013) *J. Amer. Chem. Soc.*, **135**, 1708–1710.
57 Shida, N., Koizumi, Y., Nishiyama, H. et al. (2015) *Angew. Chem. Int. Ed.*, **54**, 3922–3926.
58 Fosdick, S.E., Knust, K.N., Scida, K., Crooks, R.M. (2013) *Angew. Chem. Int. Ed.*, **52**, 10438–10456.
59 McNally, A., Prier, C.K., MacMillan, D.W.C. (2011) *Science*, **334**, 1114–1117.
60 Poelma, J.E., Fors, B.P., Meyers, G.F. et al. (2013) *Angew. Chem. Int. Ed.*, **52**, 6844–6848.
61 Narupai, B., Poelma, J.E., Pester, C.W. et al. (2016) *J. Polymer Science Part A: Polymer Chem.*, **54**, 2276–2284.
62 Treat, N.J., Sprafke, H., Kramer, J.W. et al. (2014) *J. Amer. Chem. Soc.*, **136**, 16096–16101.
63 Discekici, E.H., Pester, C.W., Treat, N.J. et al. (2016) *ACS Macr. Lett.*, **5**, 258–262.
64 Pan, X., Lamson, M., Yan, J., Matyjaszewski, K. (2015) *ACS Macr. Lett.*, **4**, 192–196.
65 Theriot, J.C., Lim, C.-H., Yang, H. et al. (2016) *Science*, **352**, 1082–1086.
66 Pearson, R.M., Lim, C.-H., McCarthy, B.G. et al. (2016) *J. Amer. Chem. Soc.*, **138**, 11399–11407.
67 Lim, C.-H., Ryan, M.D., McCarthy, B.G., et al. (2017) *J. Amer. Chem. Soc.*, **139**, 348–355.
68 Yan, J., Li, B., Yu, B. et al. (2013) *Angew. Chem. Int. Ed.*, **52**, 9125–9129.
69 Zhang, T., Du, Y., Muller, F. et al. (2015) *Polymer Chem.*, **6**, 2726–2733.

70 Zhang, T., Du, Y., Kalbacova, J. et al. (2015) *Polymer Chem.*
71 Percec, V., Guliashvili, T., Ladislaw, J.S. et al. (2006) *J. Amer. Chem. Soc.*, **128**, 14156–14165.
72 Whitfield, R., Anastasaki, A., Nikolaou, V. et al. (2016) *J. Amer. Chem. Soc.*, **139**, 1003–1010.
73 Masuda, T., Akimoto, A.M., Nagase, K. et al. (2016) *Science Adv.*, **2**, e1600902.
74 Kim, Y.S., Tamate, R., Akimoto, A.M., Yoshida, R. (2017) *Materials Horizons*, **4**, 38–54.
75 Buck, C.A., Horwitz, A.F. (1987) *Ann. Rev. Cell Biology*, **3**, 179–205.
76 Hynes, R.O. (1992) *Cell*, **69**, 11–25.
77 Ruoslahti, E., Pierschbacher, M. (1987) *Science*, **238**, 491–497.
78 Kolb, H.C., Finn, M.G., Sharpless, K.B. (2001) *Angew. Chem. Int. Ed.*, **40**, 2004–2021.
79 Lutz, J.-F., Zarafshani, Z. (2008) *Adv. Drug Deliv. Rev.*, **60**, 958–970.
80 Guan, Z.-Y., Wu, C.-Y., Wu, J.-T., et al. (2016) *ACS Appl. Mater. Interfaces*, **8**, 13812–13818.
81 Banerjee, I., Pangule, R.C., Kane, R.S. (2011) *Adv. Mater.*, **23**, 690–718.
82 Du, Y., Hancock, M. J., He, J. et al. (2010) *Biomaterials*, **31**, 2686–2694.
83 Deng, X., Mammen, L., Butt, H.-J., Vollmer, D. (2012) *Science*, **335**, 67–70.
84 Wang, S., Wang, H., Jiao, J., et al. (2009) *Angew. Chem. Int. Ed.*, **48**, 8970–8973.
85 Wang, S., Liu, K., Liu, J., et al. (2011) *Angew. Chem.Int. Ed.*, **50**, 3084–3088.
86 Liu, H., Li, Y., Sun, K., et al. (2013) *J. Amer. Chem. Soc.*, **135**, 7603–7609.
87 Liu, H., Liu, X., Meng, J., et al. (2013) *Adv. Mater.*, **25**, 922–927.
88 Liu, X., Chen, L., Liu, H. et al. (2013) *NPG Asia Mater.*, **5**, e63.
89 Zhang, P., Chen, L., Xu, T. et al. (2013) *Adv. Mater.*, **25**, 3566–3570.
90 Liu, X., Wang, S. (2014) *Chem. Soc. Rev.*, **43**, 2385–2401.
91 Liu, X., Zhang, F., Wang, Q. et al. (2014) *Small*, **10**, 4677–4683.
92 Yang, G., Liu, H., Liu, X. et al. (2014) *Adv. Healthcare Mater.*, **3**, 332–337.
93 Huang, C., Yang, G., Ha, Q. et al. (2015) *Adv. Mater.* **27**, 310–313.
94 Li, Y., Lu, Q., Liu, H. et al. (2015) *Adv. Mater.*, **27**, 6848–6854.
95 Ma, L., Yang, G., Wang, N. et al. (2015) *Adv. Healthcare Mater.*, **4**, 838–843.
96 Meng, J., Zhang, P., Zhang, F. et al. (2015) *ACS Nano*, **9**, 9284–9291.
97 Zhang, F., Jiang, Y., Liu, X.et al. (2016) *Nano Lett.*, **16**, 766–772.
98 Ridley, A.J., Schwartz, M.A., Burridge, K. et al. (2003) *Science*, **302**, 1704–1709.
99 Mahmud, G., Campbell, C.J., Bishop, K.J.M. et al. (2009) *Nat Phys.*, **5**, 606–612.
100 Imitola, J., Raddassi, K., Park, K.I. et al. (2004) *Proc. Nat. Acad. Sci.*, **101**, 18117–18122.
101 Li, X., Li, M., Sun, J. et al. (2016) *Small*, **12**, 5009–5018.
102 Ren, T., Yu, S. et al. (2014) *Biomacromolecules*, **15**, 2256–2264.
103 Ren, T., Yu, S., Mao, Z., Gao, C. (2015) *Biomaterials*, **56**, 58–67.
104 Kim, D.-H., Provenzano, P.P., Smith, C.L., Levchenko, A. (2012) *J. Cell Biology*, **197**, 351–360.
105 Chen, T., Amin, I., Jordan, R. (2012) *Chem. Soc. Rev.*, **41**, 3280–3296.
106 Pati, F., Gantelius, J., Svahn, H.A. (2016) *Angew. Chem. Int. Ed.*, **55**, 4650–4665.

Part III

Biomaterials

13

Bioinspired Materials and Structures: A Case Study Based on Selected Examples

Tom Masselter[1,3,4,], Georg Bold[1,3], Marc Thielen[1,2], Olga Speck[1,3,4], and Thomas Speck[1,2,3,4]*

[1] Botanic Garden of the University of Freiburg, Botany: Functional Morphology & Biomimetics, Faculty of Biology, Freiburg, Germany
[2] Freiburg Materials Research Center (FMF), Freiburg, Germany
[3] Freiburg Center for Interactive Materials and Bioinspired Technologies (FIT), Freiburg, Germany
[4] Networks of Competence Biomimetics and BIOKON e.V.

13.1 Introduction

The term "biomimetics" denotes the transfer of "ideas" derived from the study of living beings to technical applications, thereby using structures and functional principles of the organisms as biological concept generators to develop innovative and often "surprising" solutions to various technical challenges. These biological concept generators are the result of a long and complex evolutionary pathway during which the organisms (in our case, plants) evolved all the features which we focus on in this study. Among these are:

1) A light weight coupled with a high load-bearing capacity and excellent damping in unbranched beam-like plant stems (see Section 13.2.1 for an example of a linear beam called the "Technical Plant Stem" that was derived from extant plants).
2) A high load-bearing capacity as well as a benign fracture behavior in large stem-branch attachments in which the branches are not shed during ontogeny (Section 13.2.2).
3) Highly specialized functional features in large fruits and flowers of angiosperms such as the energy dissipation of large diaspores on impact, as e.g., the pomelo (Section 13.3), or the development of elastically deformable organs and structures to attract pollinators and ensure pollination, as is the case for *Strelitzia reginae* (Section 13.5).
4) A high potential to efficiently seal and finally heal wounds by fast and fail-safe self-repair processes (Section 13.4).

All these key features are enabled by the presence of complex structural hierarchies in the various plant organs and tissues. The understanding of the complex form-structure-function-relationships across several hierarchies is crucial for successful technical implementation in biomimetic products (see especially Section 13.3 for the pomelo).

Typically, the biological model organisms are smaller than the intended technical implementation for which they were chosen. However, it has been proven by experience that up-scaling often is quite possible: technical implementations based on abstractions of branchings in rather small (juvenile) dragon trees and columnar cacti (Section 13.2.2) or on the flower of *Strelitzia reginae* (Section 13.5.2) provide excellent examples for up-scaling. In *Strelitzia reginae*, the perch is a structure with a length roughly between 5 and 10 centimeters which has been transferred to the Flectofin® façade shading system that can be scaled up to 14 meters [1]. In a similar way, the analyses of small branchings of (juvenile) monocotyledons with a branch diameter between 5 and 10 millimeters have been transferred to structures with a diameter of 3–12 centimeters and will be further up-scaled (a diameter of 30 centimeters and beyond) for the implementation as load-bearing structures in architecture and construction engineering.

13.2 Fiber-reinforced Structures Inspired by Unbranched and Branched Plant Stems

Fiber-reinforced composites are used in many technical applications, such as aerospace and automotive engineering, bicycles, sports equipment, prostheses, etc. The development of biomimetic fiber-reinforced compound materials is facilitated by the comparability of the fiber-matrix structure in plants and in technical materials. On

* Email: tom.masselter@biologie.uni-freiburg.de

Bioinspired Materials Science and Engineering, First Edition. Edited by Guang Yang, Lin Xiao, and Lallepak Lamboni.

a larger scale, architecture and construction engineering are becoming important fields for applications of fiber-reinforced compound materials.

13.2.1 Technical Plant Stem

The "Technical Plant Stem" (Figure 13.1a) is a linear (i.e. unbranched) biomimetic fiber-reinforced product and was developed in close cooperation between the Plant Biomechanics Group Freiburg and the Institute of Textile Technology and Process Engineering (ITV) Denkendorf [2–5]. Various mechanical optimizations are integrated into the patented "Technical Plant Stem": good vibration damping, a light weight, a benign fracture behavior, good impact damping, and high flexural stiffness [6]. These properties are the result of a custom-made composite in which the fibers are aligned during the computer-controlled 3D-braiding pultrusion of the "Technical Plant Stem" in the direction of the main loads: with 0° (tensile/ compression/ bending loads), 30° (damping) and 45° (torsional loads).

While the development of such linear composites has become fairly standard in the last few years, the fabrication of custom-made composites becomes markedly more complicated when these structures are branched. Several concept generators led to an abstraction and the patent for an inner structure for a technical branching that can be used, for example, as an axle carrier [7].

13.2.2 Branched Fiber-reinforced Structures

Novel manufacturing methods had to be developed to meet the challenge of braiding biomimetic branched technical structures. A special braiding machine was developed by the Institute of Textile Technology and Process Engineering Denkendorf [8]. This was a major step toward producing biomimetic bifurcating structures (Figure 13.1b) in one continuous process. Other multiple-branched structures were produced at the Institute of Lightweight Engineering and Polymer Technology in Dresden with a variable braiding eye which is the main element in a novel patented manufacturing method [9, 10]. These developments were part of the priority program SPP 1420 that enabled the detailed study of natural concept generators for the development of novel biomimetic, branched, fiber-reinforced materials and structures [7, 11–18].

In another project that is part of the Collaborative Research Centre TRR 141 (www.trr141.de), larger branched structures are built for use as load-bearing pillars in architecture and construction engineering, for example, as roof-supporting struts (Figure 13.2a) This project is a collaboration between biologists of the Plant Biomechanics Group Freiburg and engineers of the Institute of Building Structures and Structural Design Stuttgart as well as of the Institute for Textile Engineering, Fiber-based Materials and Textile Machine Construction (ITFT) Stuttgart. The development of biomimetic branching joints for fiber composite truss structures in architecture and construction engineering is enabled by the investigation of selected permanent branchings of arborescent monocotyledonous plants with and without secondary growth; examples include: *Dracaena* [19], *Freycinetia*, *Nolina*, *Pandanus*, *Yucca* and *Semiarundinaria* (Figure 13.2b, c) and of dicotyledonous woody plants, e.g.,

(a) (b)

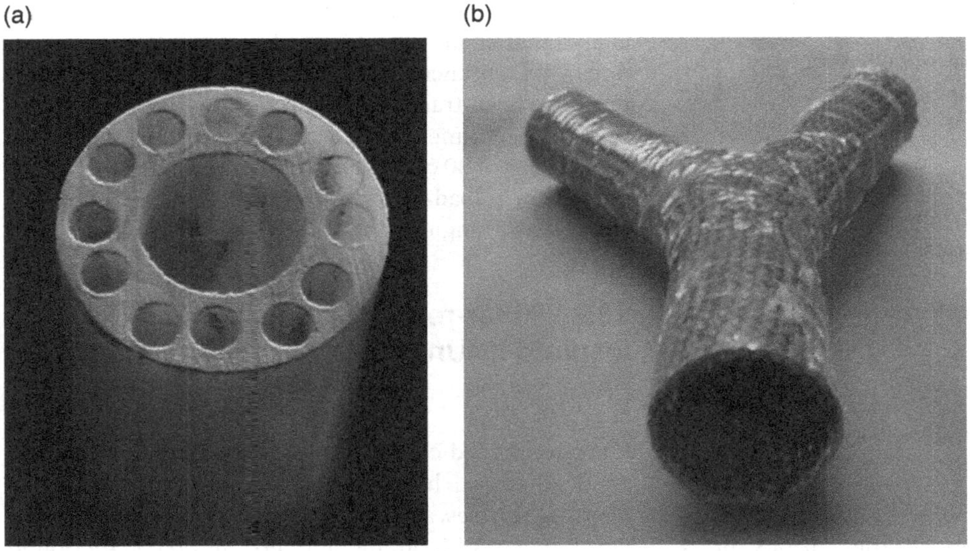

Figure 13.1 Biomimetic fiber-reinforced composites. (a) The linear (unbranched) 'Technical Plant Stem'. (b) Carbon fiber-reinforced polymer (CFRP) branching developed by the ITV Denkendorf that can be filled with concrete (Cc).

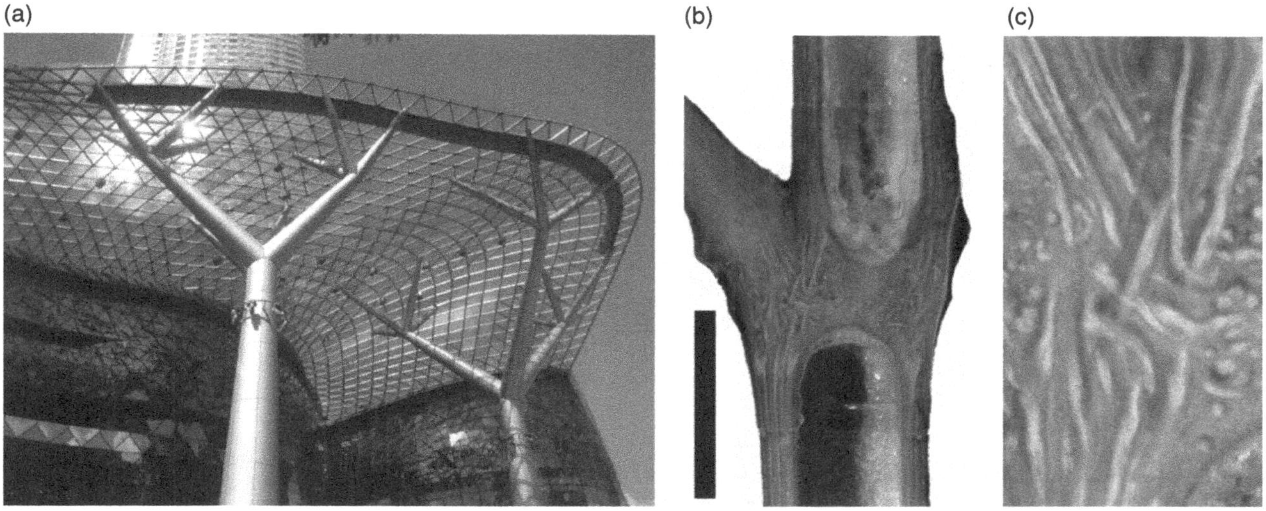

Figure 13.2 Technical and natural branchings. (a) Roof-bearing truss structure as example of technical implementation. (b) *Semiarundinaria* sp. as example of a biological concept generator, longitudinal section of stem-branch attachment. (c) Detail of longitudinal section in (b). Scale bar: 10mm.

Schefflera [20], *Angelica* and *Edgeworthia* as well as of the analyses of sedges and grasses such as *Carex* or *Secale*, in which a leaf-sheath encompasses the stem. Peripheral reinforcement structures as presented in herbaceous plants as, for example, in the mint family (Lamiaceae) will also be analyzed as to their suitability as role models for the development of large branched supporting structures in architecture.

In existing state-of-the art applications, linear (unbranched) fiber-reinforced polymer (FRP) hulls filled with concrete are used as load-bearing structures [21–25]. Results of experiments with these concrete-filled FRP tubes show the significant strengthening effect of the FRP jacket [26]. In the TRR141 project, branched biomimetic FRP jackets were produced by a combination of novel weaving and braiding techniques by the ITFT. In this method, the weaving technique lays the fibers in a hoop direction and a longitudinal direction while the new braiding technique enables the fibers to be laid in a double helix arrangement. A large demonstrator was built for the TRR 141 exhibition from Oct. 2017 to May 2018 in Stuttgart [27].

For the technical implementation in architecture, detailed analyses of the functional principles underlying the high load-bearing capacity of branchings in plants are indispensable. Novel methods like magnetic resonance imaging are adapted and used to compare the arrangement of fibers in loaded and unloaded biological and technical structures. This allows for the first time a 3D biomechanical analysis of a biological fiber-reinforced tissue to be performed. The biomechanical analyses are supported by Finite Element simulations of abstracted models of the biological concept generators.

First results for branched model structures indicate that critical notch stresses in the stem-branch-connection can be decreased by optimization of the outer shape and thereby distributing those stresses over a broader area (Figure 13.3). This finding corroborates results found for the branching regions of columnar cacti [18].

13.3 Pomelo Peel as Inspiration for Biomimetic Impact Protectors

Biological damping materials or structures normally can be found in or around fragile organs or tissues that need protection from excessive loading which otherwise would result in destructive accelerations and force peaks. Probably the most obvious of such structures are found in the animal kingdom, for example, the skull structures of woodpeckers and head-butting animals or the armor-plating skin of alligators and turtles [28–33]. But the plant kingdom also has examples of impact-protecting structures, such as fruit peels and nut shells, which protect the plant's seeds and embryos from being mechanically damaged [34, 35]. One of the most remarkable impact-protecting structures found in plants certainly is the thick foam-like pomelo (*Citrus maxima*) peel. It is able to dissipate considerable amounts of kinetic energy when the heavy fruit (1–2 kg) impacts on the ground after being shed, despite the fact that it mostly consists of comparatively "weak" parenchymatous, non-lignified cells. The amazing mechanical properties of the pomelo (peel) are mainly achieved by the hierarchical arrangement of its basic constituents – a construction principle shared by most biological structures – and thus constitutes an ideal role model for the

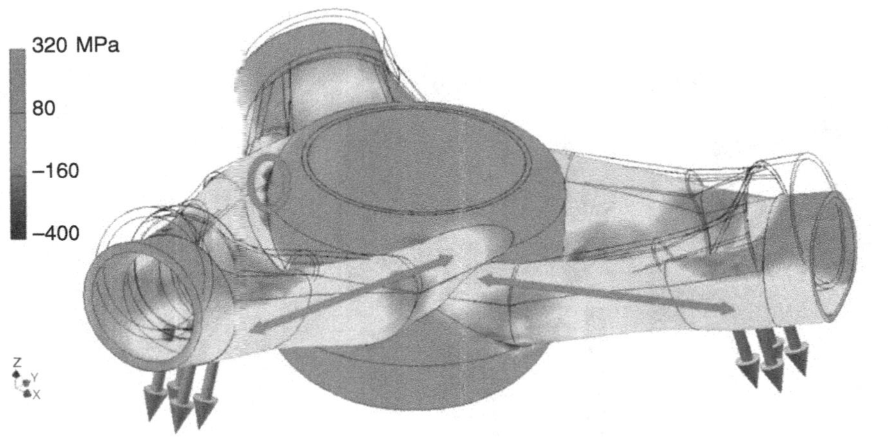

Figure 13.3 Stress distribution (first principal stress) in a simulation of a loaded model of a threefold branching. The thick multiple arrows indicate the direction and position of loading. The loose contours indicate the initial form of the branching. Tensile stresses are shown exemplarily by the double-headed arrows, and the compressive stresses by the ellipsis.

development of biomimetic impact-protective and energy-dissipating structures and devices [14, 36–40].

The structure-function-relationship of the pomelo peel was analyzed by the Plant Biomechanics Group Freiburg with the aim of developing – together with project partners from the foundry institute at the RWTH Aachen and from the chair of materials engineering at the TU Berlin – bioinspired cast metal foams for application as impact protectors. This research was part of the priority program SPP1420 "Biomimetic Materials Research: Functionality by Hierarchical Structuring of Materials" and will continue as part of the project "BISS: Bio Inspired Safety Systems" with the aim of transferring the underlying biological principles also to other material classes, for example, as polymer foams, and thus considerably expanding the field of application.

13.3.1 Hierarchical Structuring and its Influence on the Mechanical Properties

In the case of the pomelo, at least seven hierarchical levels can be distinguished (Figure 13.4). Whereas the fruit itself is defined as the first level, the second level distinguishes between the fruit pulp and the peel, which is the structure of most interest concerning the impact-protecting properties. It should, however, be noted that – depending on the extent of the impact –the pulp may also be involved in energy dissipation. The next smaller structural elements – hierarchical level 3 – are the relatively soft parenchymatic tissue consisting of a few outer cell layers called the flavedo, the albedo which constitutes the major part of the peel volume and the comparatively stiff three-dimensional network of vascular bundles traversing the whole peel. The higher stiffness of the vascular bundles arises from their lignification, which also renders them visible using a lignin-specific staining (in Figure 13.4 the lignified vascular bundles are stained with a solution of phloroglucinol and hydrochloric acid). Beside their physiological tasks (transport of water and assimilates), these regularly branched bundles also act as mechanical elements as they seem to help distribute local stresses and strains over large peel volumes [41]. From a technical point of view the albedo can be seen as a low density open-pored foam, whose struts correspond to the biological cells, and the flavedo can be seen as a compact closed-pored foam in which the biological cells represent foam cells. This means that the flavedo, due to its higher density, contributes to a uniform load transmission to an extensive area of the albedo, which for its part dissipates the major part of the kinetic energy while being compacted during impact.

During this compaction, kinetic energy is converted into deformation energy by performing work on the albedo cells (the struts of the open pored foam). At the next lower hierarchical level the albedo tissue appears to have a density gradient running perpendicular to the peel's surface, the middle part of the albedo being less dense, gradually becoming denser towards the flavedo, on the one side, and the fruit pulp, on the other side. The density was quantitatively assessed from peel thin-sections by counting the cell number per unit length because open-cell foams often are represented by idealized unit cells, e.g. Ashby cells (Figure 13.5a), consisting of solid struts surrounded by void space containing a gas or fluid [42, 43]. Because in the case of the albedo, the intercellulars represent the void space

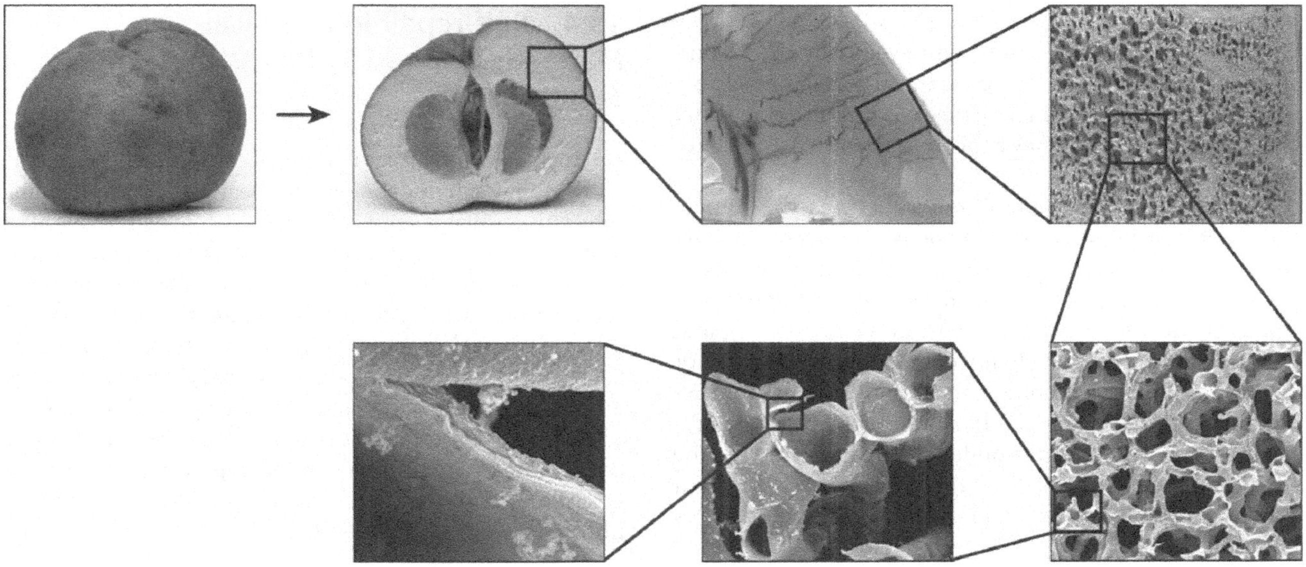

Figure 13.4 In case of the pomelo (*Citrus maxima*) at least seven hierarchical levels can be distinguished that range from the cm scale (whole fruit) to the μm scale (multilayered cell wall). The pomelo fruit – hierarchy level 1 – in this figure has a diameter of about 20 cm. The second hierarchy level distinguishes between the fruit pulp and the fruit peel. The latter is subdivided into flavedo, albedo, and vascular bundles (hierarchy level 3). The fourth hierarchical level is represented by the density gradient within the albedo, which – on the fifth hierarchy level – appears as an open-pored, foam-like structure consisting of struts (living biological cells) and void space (intercellulars). The afore-mentioned struts consist of the cell wall and the cell lumen (hierarchy level 6). The smallest-scaled hierarchical level considered in this approach is the cell wall itself, consisting of a fiber matrix compound.

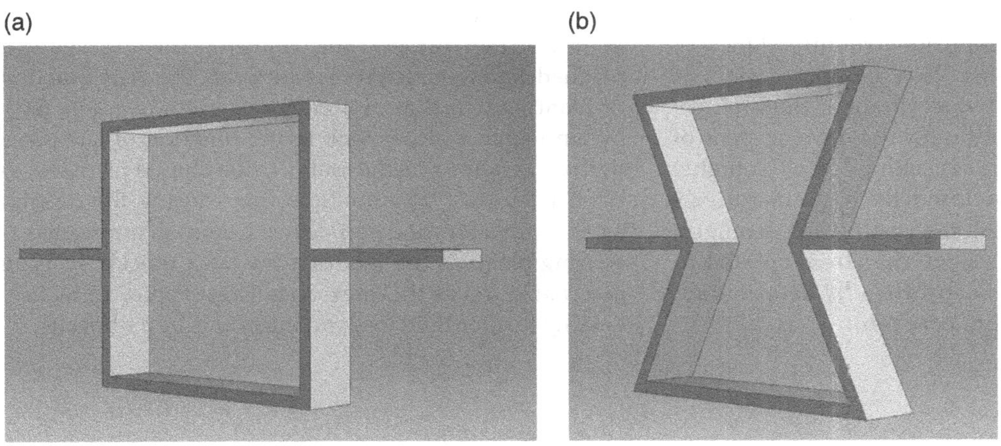

Figure 13.5 Typical idealized 2D-unit cells. (a) Open pored foam. (b) Re-entrant 2D-unit cell used to represent the mechanical principles on which auxetic foams are based.

of the foam-like structure, and the living biological cells constitute its struts, a higher number of cells per cross-sectional area implies smaller foam unit cells and thus a higher density of the tissue. This results in the effective cross-sectional area of supporting material (struts) being smaller in the less dense regions of the albedo leading to higher stresses in individual struts within the middle part of the albedo, because the same force is supported by fewer struts.

The observation that fresh pomelo peel samples collapse in a more or less uniform manner (in contrast to a progressive collapse) during impact experiments can be related to the fact that the "struts" constituting the albedo are living biological cells and thus have a hierarchical structure themselves [41]. In a simplified approach, biological plant cells can be seen as liquid-filled vessels (surrounded by a semi-permeable vacuolar membrane) which exert hydrostatic pressure (turgor) on an enclosing cell wall. When the albedo is compressed, the cells are bent leading first to cell wall tautening and stretching, and eventually the incompressible cell fluids may be forced through cell membranes and cell walls [41, 44]. Initially the cells thus become stiffer due to tautening, until they begin to leak and act as hydraulic dampers.

In contrast to most materials or structures, the pomelo peel contracts laterally when being compressed, a behavior termed as auxetic [45]. Auxetic (meta)materials have a negative Poisson ratio and can be seen as open-pored foams whose unit-cells are re-entrant (Figure 13.5b) and thus bend inwards when being compacted. This behavior results in a global stiffening and an increased energy dissipation of the foam at the location of impact as neighboring foam volumes are pulled into the vicinity of the impact. These types of (meta)material thus hold great potential for application in impact protectors and safety equipment.

13.3.2 Functional Principles for Biomimetic Impact Protectors

As the interaction between the diverse hierarchical levels is complex, it was not possible – but also neither necessary nor desirable – to simultaneously transfer all of them into cast metal prototypes. A prominent example of an abstraction process is the production of precision-cast bioinspired composite structures [46, 47]. The albedo "struts" having two damping regimes (structural and hydraulic) due to their hierarchical structure served as inspiration for the development of a composite metal structure consisting of a pure, highly ductile aluminum core enclosed by a comparatively stiff aluminum alloy (A356) shell with high strength. After fracture of the outer shell, the stress applied to such a structure can still be absorbed by deformation of the ductile pure aluminum core [46].

13.4 Self-repair in Technical Materials Inspired by Plants' Solutions

During Earth's history many different ways evolved to allow today's living beings to deal with external and internal damages and to ensure their survival after wounding. The capability for repair of structural defects on different length scales (from molecular to the entire organism) is the key to survival in all life forms. As well as for defects due to external as for internal influences, such as attacks by predators or internal growth processes, lesions in the micro- to macro-range play an important role. Among a multitude of repair processes in biology, in this chapter, two examples of self-repair in plants will be described and their (potential) transfer to technical products will be highlighted.

13.4.1 Plant Latex: Self-Sealing, Self-Healing and More

Plant latex can be found in more than 20 000 plant species from some 40 families [48–50]. Therefore, this milky plant sap can be assigned many different functions potentially varying from species to species. Among these mainly discussed in the literature were the functions as nutrition or waste storage system, as transport system, or as water reservoir [49, 50]. However, a consensus was reached among (most) researchers that the main function of plant latex is that of a defense system (e.g. [50, 51]). When taking a closer look at the function of latex as a defense system, a further subdivision should be made, as plant latex may take multiple roles during the defense against predators and pathogens. Common among latex-bearing plants is the fact that the latex quickly seals an external lesion as the latex coagulates (Figure 13.6), preventing further fluid loss and infections by pathogens.

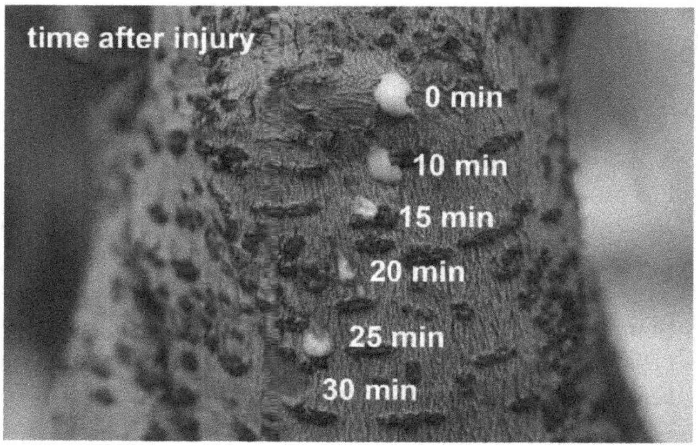

Figure 13.6 Latex of the Weeping Fig (*Ficus benjamina*) at various times after injury. When the cortex is freshly injured, the milky white latex oozes out quickly (0 min). As the latex coagulates, it turns more and more transparent (10–30 min).

Apart from quickly sealing a wound, two more functions of the plant latex can be mentioned here. On the one hand, the fresh and non-coagulated latex may (in many plants) easily wet various surfaces. This may lead to an agglutination of predator's mouth parts and thus prevent these predators from further feeding on the latex-bearing plant [51, 52]. Therefore, a latex which remains liquid over a long period of time is of great advantage as a coagulated (non-liquid) latex would not allow an efficient spreading on the predator's mouth parts. As in most latex solutions, poisonous ingredients are found, so a prolonged liquid phase additionally increases the poisoning effect of the latex against predators. On the other hand, quickly coagulating plant latex makes self-sealing more effective and may additionally contribute to self-healing of the plant cortex after injury by (at least partly) restoring its tensile strength. When injuring the bark of the Weeping Fig, after only 30 min (compare the time span for a complete coagulation of the latex, see Figure 13.6), a significant increase in the tensile strength is achieved (Figure 13.7). Subsequent experiments could prove that this fast self-healing effect can be attributed to the coagulation of the plant latex alone [53].

Thus, quick sealing of the lesion by spreading the latex can be interpreted as a first step in the plant's defense system, whereas longer-lasting subsequent steps such as the coagulation of the latex may even lead to a restoration of the mechanical strength of the plant. As recent studies show, the coagulation process may be either caused (primarily) by the physical process of evaporation (e.g. in tested *Euphorbia*-species) or may be significantly accelerated by additional chemical processes, as is the case of the Weeping Fig (*Ficus benjamina*) [53, 54] and the Para rubber tree (*Hevea brasiliensis*). The plant latex coagulation mechanism was studied in detail for the Para rubber tree (*Hevea brasiliensis*). In the intact plant – besides rubber particles and other substances – vacuolar structures containing the protein Hevein can be found within the liquid latex, which is stored under an overpressure of 8 bar and more in micro-tubes, the so-called laticifers. Upon injury, the latex oozes out of the lesion and the vacuolar structures burst due to the pressure drop between the pressure in the laticifer and the ambient pressure (1 bar) as well as due to shear stresses occurring in the effluent latex. Subsequently, the now released Hevein cross-links (with a contribution of Ca2+ ions) proteins on the surface of the rubber particles, resulting in the coagulation of the plant latex [55–57].

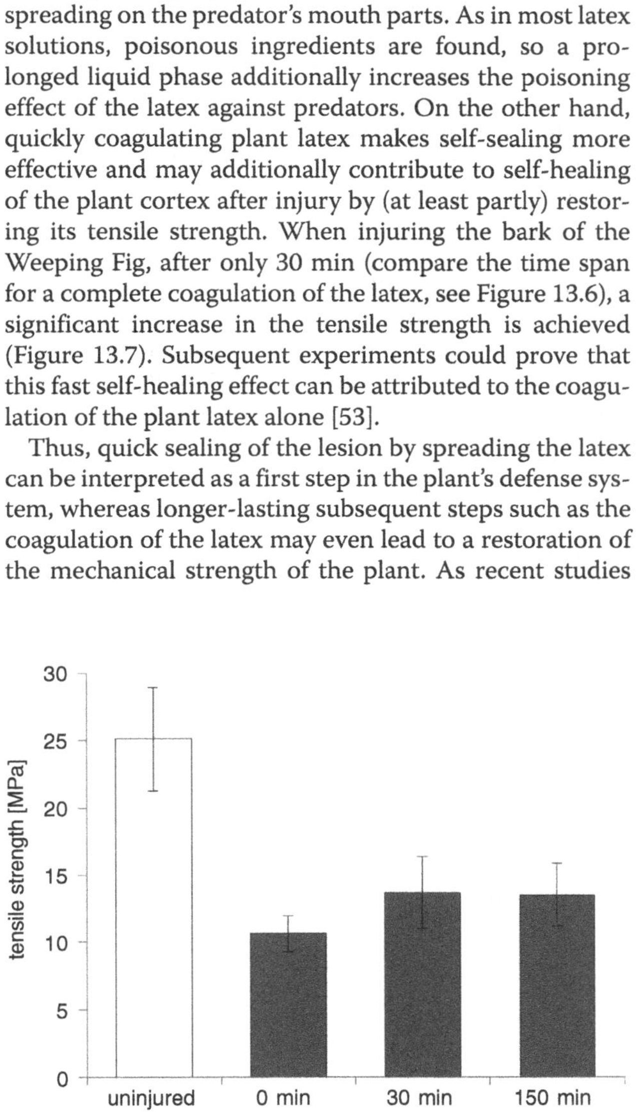

Figure 13.7 Restoration of tensile strength due to the coagulation of plant latex. When the bark of the Weeping Fig is injured (0 min), its tensile strength is reduced compared to an uninjured bark sample by more than 50%. A significant increase in tensile strength of the bark can be observed until only 30 min after injury due to the coagulation of the plant latex. A further increase in tensile strength cannot be observed after this time period (at least until several hours to few days after injury, when other mechanisms such as cell growth and cell proliferation cause an additional self-healing effect).

A transfer of this sophisticated self-sealing and self-healing system to technical materials was successfully applied for Nitrile Butadiene Rubber (NBR) blends in the framework of collaboration with the Institute for Macromolecular Chemistry of the University of Freiburg and the Freiburg Materials Research Center (FMF). Liquid polymers (hyperbranched polyethyleneimine, PEI) acting as self-sealing and self-healing agents were embedded in the NBR blends, forming micro-phase separated and pressurized domains due to the different thermal expansion of NBR and PEI. Severed NBR/PEI (with 12.5 phr of PEI with a molecular weight of 2000 g mol^{-1}) strips regained 44% of the initial tensile strength after rejoining and curing the samples for 12 h at 100 °C and then storing them for another 12 h at room temperature [58–61].

This procedure constitutes a biomimetic top-down process where, based on a technical challenge (equipping elastomers with self-sealing and self-healing functions), a solution was successfully transferred from a biological system to technology [62]. Another example of such a biomimetic top-down approach is explained in detail in the Section 13.4.2.

13.4.2 Wound Sealing in the Dutchmen's Pipe: Concept Generator for Self-Sealing Pneumatic Systems

An even more elaborate biological self-repair system than that due to the coagulation of plant latex can be found in the Dutchmen's pipe (*Aristolochia macrophylla*). In this plant, high radial and tangential stresses develop during secondary growth of the plant stem. This process results in internal (micro-)fissures of the outer stabilizing sclerenchymatous ring that are healed in several steps.

First, pressurized adjacent non-lignified parenchymatous cortex cells swell into these (micro-)fissures due to the disorder caused by the fissure in the stress-strain-field at the contact region between strengthening tissue and cortex. This relaxation process in the fissure is facilitated by the turgor pressure and the elastic/viscoelastic cell wall modifications in the repair cells. They thus provide for a quick sealing of the fissure. In a second step, the mechanical properties of the plant axis are (partly) restored by subsequent cell proliferation in a radial and tangential direction followed by synthesis and lignification of new cell wall material in the region of the fissures [58, 62–64].

Like all pneumatic structures also the membrane-based pneumatic light-weight Tensairity® technology suffers from a vulnerability of the membrane due to fissures. In order to prevent pneumatic systems from loss of functionality due to air leakage a rapid sealing of the fissure is necessary. Thus, the first step of the self-repair process in the Dutchmen's pipe (i.e. the quick sealing of fissures) has the potential to be successfully transferred to a self-repairing technical pneumatic structure. By coating the inner layer of the technical membrane with a closed cell polyurethane foam cured under slight overpressure of 1–2 bar, this coating passively seals fissures particularly rapidly, like the non-lignified plant cells seal the internal fissures in the Dutchmen's pipe. This process is markedly increased by the fact that due to the toroidal form of all pneumatic structures, the internal foam coating is under compression load, which supports the self-sealing process in terms of pressing together the damage edges and surfaces of the foam. By this biomimetic self-repairing foam coating, lesions in technical membranes are efficiently sealed. The self-sealing efficiency can reach up to 0.999 (a value of 1 corresponds to a complete seal, whereas a value of 0 corresponds to no seal) while allowing for only a slight increase in the weight of the technical system compared to alternative sealing solutions [64–68].

Thus, also this example constitutes a successful transfer of a biological self-sealing system into a technical solution via a biomimetic top-down approach [68]. Details of the top-down process of the self-sealing system of the Dutchmen's pipe into self-sealing technical pneumatic structures can be seen in Figure 13.8.

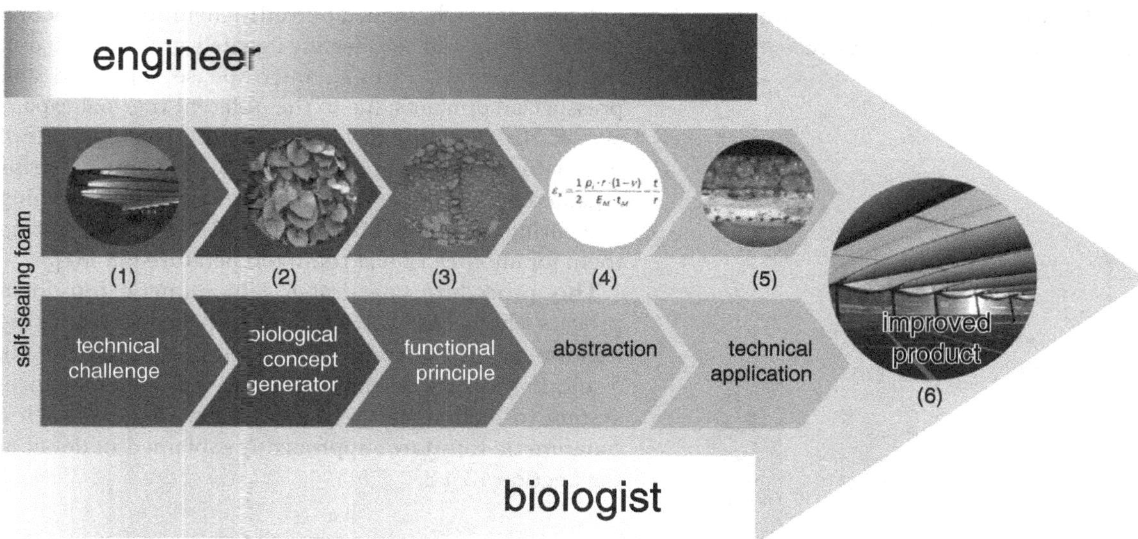

Figure 13.8 Top-down process in biomimetics, represented by the successful transfer of the self-sealing system of the Dutchmen's pipe to a biomimetic self-sealing foam coating for membranes of technical pneumatic structures. (1) After the question arises, whether biology might offer a solution for a technical challenge: can a technical pneumatic structure be equipped with an efficient light-weight self-sealing system?, a screening for suitable biological role models follows. (2) Self-sealing of internal fissures in the Dutchmen's pipe. (3) After a detailed investigation of the functional principle of the biological role model (rapid sealing of internal damage by swelling of pressurized, non-lignified parenchymatous cortex cells into the fissure), an abstraction of this operating principle is conducted. (4) The abstraction from pressurized, non-lignified biological cells to similar technical structures can be described by mathematical formulas expressing the interaction of material properties and geometric parameters. This functional principle can then be applied to the technical material. (5) Closed cell polyurethane foam coating under compression on the inside of the membrane, resulting in an enhanced product. (6) Self-sealing technical membrane which can be incorporated into a multitude of pneumatic systems.

13.5 Elastic Architecture: Lessons Learnt from Plant Movements

Adaptive building envelopes may render an important contribution to a more energy-efficient and sustainable architecture. Therefore, growing interest and urgent demand exist for adaptive building envelopes that can adjust themselves to changing external environmental conditions and/or to internal comfort requirements. In the present architecture, such systems typically are technically realized by either rigid components or soft textiles joined by highly strained hinges or bearings along a straight axis of translation and/or rotation. The main aim in bioinspired elastic architecture is to avoid such hinges or bearings, which need intensive maintenance and are prone to failure, due to the high localized mechanical loading. In particular for large-scale adaptive planar or curved architectural structures and due to the growing interest of architecture in non-standard geometries, technical solutions with an intrinsic adaptability to a wide range of spatial configurations are of increasing importance. In plants, movable organs are typically based on elastic deformation of the organ due to locally adapted stiffness of their components and do not show regions of highly concentrated strain by avoiding localized joint-like structures. The few examples among plants with small localized "joint-like" bending zones have no gliding elements but show a thin rod- or ribbon-shaped connection between the two (more) rigid elements that move relative to each other; examples include the pulvilli of mimosa (*Mimosa pudica*), the catapulting snap-tentacles of pimpernel sundew (*Drosera glanduligera*) and the staminal lever-arms of sages (*Salvia* spp.) [69–73]. For kinetic systems in building construction such as louvers or blinds, aspects such as "accuracy" or "velocity" are of less relevance whereas the criteria of "robustness" and "adaptability" are usually of major importance. Petals and leaves of plants show countless robust motion principles important for many biologically vital functions. They especially fulfil the criteria of "robustness" and "adaptability" demanded in architecture for louvers or blinds and therefore represent valuable and promising concept generators for an innovative elastic architecture.

13.5.1 Plant Movements: A Treasure Trove for Basic and Applied Research

Since Charles Darwin's classic work [74, 75], the movement of plants and especially of carnivorous plants has been an on-going field of interest for biological research [70, 76, 77]. With regard to movement and shape changes of plant surfaces, investigations have mainly focused on the processes of flower opening and closure [78], leaf growth and unfolding [79], and on the active movements of traps in carnivorous plants [70]. In the field of passive movements and shape changes attributable to external mechanical forces, special topics of interest are the mechanical deformation of flowers or flower parts during pollination [80, 81] and the passive reconfiguration of leaves under wind loads or of blades of kelp in the surf zone [82, 83]. Most studies in the field of flower opening and closure have focused on the influence of temperature and light, endogenous (circadian) rhythms, phytohormones, the ecological significance of flower movement, growth dynamics, and the development of complex flower shapes during blooming and the types of opening (diurnal/nocturnal, single/repetitive) [75]. Theoretical approaches have mainly dealt with the modeling of shapes and the shape development of flowers and leaves [84, 85].

Surprisingly, much less research has been performed with regard to the mechanics of opening and closure and to correlations and interactions between mechanics and changes in petal geometry during flower opening and closing [86]. Differential growth rates on the inner and outer side of petal lamina and/or petal midrib and/or reversible turgor changes have been hypothesized as the driving forces for petal movement in most species studied [78, 87]. New studies have established, however, that (at least in some species) differential growth rates of the edges and central part of the petal lamina drive the deployment of petals [88]. Studies of the movements in green leaves have focused on leaf unfolding during growth [79] and touch-sensitive and circadian leaf movements [70, 89, 90].

Most studies of leaf unfolding and form generation by differential growth rate are based on mathematical modeling. For simple angiosperm leaves, unfolding patterns can be numerically described by taking leaves as plane surfaces with straight parallel folds [79, 91]. For ribbon-shaped undulating monocot leaves and blades of kelp, the models again predict edge actuation attributable to differential growth rates of the leaf edge and center [92]. All active traps in carnivorous plants are interpreted as being modified leaves or leaf parts and as being developed from (originally) two-dimensional structural elements. Recent quantitative analyses of the active traps of Venus flytrap (*Dionaea muscipula*) [93], the waterwheel plant (*Aldrovanda vesiculosa*) [94–96], bladderworts (*Utricularia* spp.) [97], and pimpernel sundew (*Drosera glanduligera*) [69, 81] have focused on mechanical aspects of trap functioning and on the underlying structural features at the organ and tissue level.

Most of the above-mentioned studies rarely include considerations as to a potential transfer to bioinspired technical structures. Considerations of applications are either purely theoretical and/or include simple models on the lab bench scale [85, 88, 91, 98–100]. Even the most well-known compliant structures, namely the deployable solar panels or antennae used in satellites [101, 102], which are often discussed in relation to leaf and flower folding and unfolding patterns, are inspired by traditional Japanese folding techniques (Origami) more than by plants.

13.5.2 Flectofin®: a Biomimetic Façade-Shading System Inspired by the Deformation Principle of the "Perch" of the Bird of Paradise Flower

Compliant mechanisms are omnipresent in nature and, until now, have mostly been used for small technical devices, e.g. in the medical or packaging industries or in micro-electro-mechanical systems (MEMS) but are widely unknown on the larger scale of building construction [103]. During the last seven years simple demonstrators and prototypes of bioinspired elastic deployable building constructions based on planar and curved plant surfaces have been developed in close cooperation between biologists/biomimetic scientists of the Plant Biomechanics Group (PBG) at the University of Freiburg, civil engineers/architects of the Institute of Building Structures and Structural Design (ITKE) at the University of Stuttgart, and textile engineers/material scientists of the Institute of Textile Technology and Process Engineering ITV Denkendorf [1, 69, 104–107]. This R&D-project covers the flapping mechanism inspired by the deformation principle found in the "perch" of the bird of paradise flower (*Strelitzia reginae*) (Figure 13.9). In biomechanical tests it was established that the perch, which protects the stamen and ovary, can be released more than 3000 times with no evidence of structural or material fatigue, indicating the robustness of the biological system (Figure 13.10) The repetitive opening of the perch is caused by torsional bending movements which are well known and feared in civil engineering as an often catastrophic failure mode in open shell structures but has been "functionalized" in the biological concept generator.

The compliant flapping mechanism inspired by the bird of paradise flower has been transferred to a demonstrator with dimensions 2000 x 250 mm^2 and composed of hand-laminated glass-fiber-reinforced polymers

Figure 13.9 (a) Bird of paradise flower with perch which is bent down after landing of pollinating birds and served as concept generator for development of the biomimetic façade shading system Flectofin®. (b) Perch bent down by hand; due to torsional bending movement the perch opens and makes the previously protected stamen and ovary accessible for pollinators.

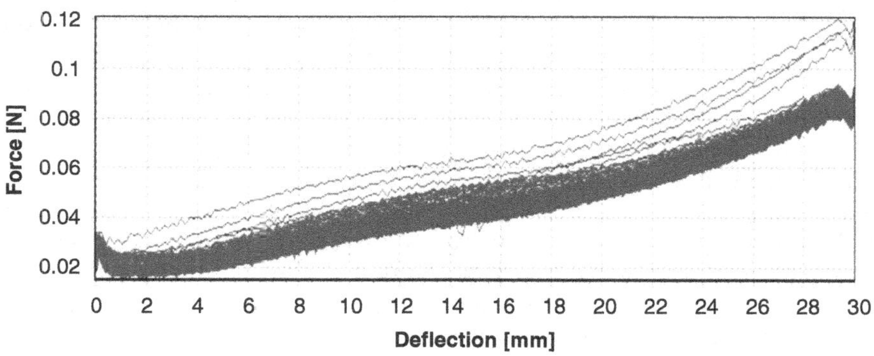

Figure 13.10 Force-deflection-curves of 3000 repetitive bending tests with a perch of a bird of paradise flower. The analysis proves that even after the perch was released 3000 times, no evidence of fatigue is visible. With the exception of the first 5–10 bending events, which are slightly higher than the subsequent ones, the force-deflection curves are very similar. *Source:* © Plant Biomechanics Group Freiburg.

(GFRP) (Figure 13.11) The structure was fine-tuned by local adjustment of the orientation and layering of the fibers and form optimization in the membrane-backbone transition region by using the tension-triangle-method [108, 109], and has been patented as registered trademark Flectofin® [107]. A small, slow and powerful displacement of one support by 25 mm is needed to activate a large, fast and wide deflection of the 2 m long fins [1, 81, 106, 110]. This external actuation can be replaced by an autonomous bending of the beam by using a bilayer fiber composite backbone with various thermal expansion coefficients and by including heating wires in the composite. Thus, a compliant mechanism (without any hinge) is possible that combines the kinematics of the petals of *Strelitzia reginae* with an actuation by a linear adaptive system.

Two other well-known examples of movements of curved two-dimensional plant structures are the active snap-traps found in the closely related Venus flytrap (*Dionaea muscipula*) and the waterwheel plant (*Aldrovanda vesiculosa*). The closing of the snap-trap leaf of a Venus flytrap involves snap-buckling of two lobes that abruptly switch their curvature from concave to convex [70, 93]. In contrast, the closing of the snap-trap leaf of the waterwheel plant involves a kinematic amplification by the bending deformation of the trap midrib, thereby changing its curvature during the trapping action, whereas the concave trap lobes do not change shape [70, 94–96, 110]. The actuation mechanism of *Aldrovanda vesiculosa* traps is based on two curved line folds [94, 95, 110]. By simulations and functional models at the lab scale level, it could be shown that both mechanisms stay operative, even when being scaled in size or distorted in shape, and therefore also represent valuable concept generators for bioinspired façade-shading systems [95, 96, 110]. Other examples, for which kinematics have been partially analyzed as to their potential as role models for bioinspired façade-shading systems include the garden lily (*Lilium* "Casa-Blanca") [110], flowers of *Ipomoea alba* [90, 110], and species of bladderwort (*Utricularia* spp.) [70, 97].

A successful example of the way that this bioinspired research can influence architecture is the Thematic Pavilion for the Expo 2012 in Yeosu, South Korea (Architects: soma Architecture, Vienna; Engineers: Knippers-Helbig, Stuttgart). The fins are made of GFRP plates that are slightly curved, only 9 mm thick and up to 14 m long [106, 111, 112].

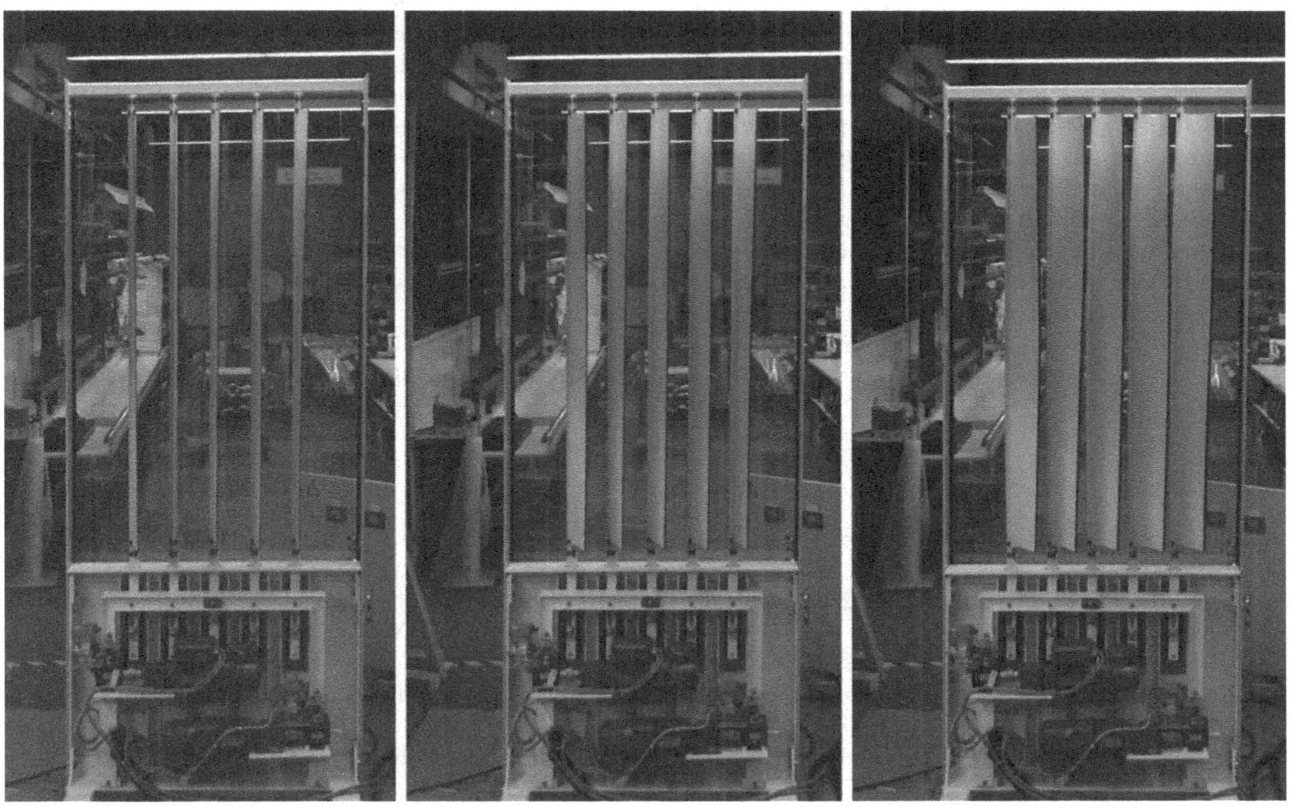

Figure 13.11 Demonstrator of the biomimetic façade shading system Flectofin® in different positions of closure caused by bending of the "backbone" due to a 25 millimeter displacement of hydraulic pistons in the basal support. *Source:* © Julian Lienhard, ITKE University of Stuttgart.

13.6 Conclusions

The present and past R&D-projects in the Plant Biomechanics Group Freiburg have proved that a detailed investigation of the biological concept generators is indispensable for a successful biomimetic abstraction and implementation in novel technical structures. The huge number of organisms with the specific structures and functions they have developed during evolution to adapt to differing environments represents the basis for all biomimetic R&D-projects.

The examples of biomimetic R&D projects as described above form the basis that allows the following conclusions:

1) A successful technical implementation is often based on not just one, but several biological concept generators, e.g. the "Technical Plant Stem" and the Flectofin®.
2) The technical product may have a (rather) distinct resemblance to biological concept generators, for example, the "Technical Plant Stem" to hollow plant axes as for example found in horsetails, or a (very) little resemblance as is the case for the bioinspired branched concrete filled structures as compared to branchings in nature as, for example, found in dragon trees.
3) Upscaling of the structure and the functional principles of biological organisms (in our case plants) is often possible in order to meet the mechanical demands of larger technical structures.
4) A deepened understanding of the form-structure-function-relationships of the biological role models is necessary for successful implementation in technical structures' development of biomimetic products. This requires detailed quantitative analyses of the plants by using a multitude of optical and mechanical devices in order to investigate the anatomy and morphology of the biological organisms on several hierarchical levels.
5) The potential areas of application of biomimetic developments are often not restricted to their primary goal as defined at the beginning of the project in top-down approach but are also suited for a wide range of technical solutions to differing specific challenges.

The results of present and past R&D projects in the Plant Biomechanics Group Freiburg demonstrate the high potential for a successful technical transfer and are leading to the development of prototypes that already incorporate "solutions inspired by nature." Biomimetics has a high innovation potential and offers the possibility for the development of sustainable technical products and production chains.

Acknowledgments

We thank our colleagues and their research groups for excellent cooperation within the projects presented above, especially Andreas Bührig-Polaczek (RWTH Aachen), Claudia Fleck (TU Berlin), Rolf Luchsinger (EMPA Dübendorf), Rolf Mülhaupt (University of Freiburg and FMF), Markus Milwich (ITV Denkendorf) and Jan Knippers (University of Stuttgart). For financial support we are grateful to: (1) the German Research Foundation (DFG) for funding within the priority program 1420 "Biomimetic Materials Research: Functionality by Hierarchical Structuring of Materials" and within the Transregional Collaborative Research Centre 141 "Biological Design and Integrative Structures"; (2) to the German Federal Ministry of Education and Research (BMBF) for funding within the scope of the program BIONA and within the scope of the funding program "Ideenwettbewerb: Bionik – Innovationen aus der Natur"; (3) to the EMPA Dübendorf for additional funding; and (4) to the German Federal Ministry of Education and Research (BMBF) "NanoMatTextil" for funding within the framework of "BISS: Bio Inspired Safety Systems."

References

1 Lienhard, J., Schleicher, S., Poppinga, S. et al. (2011) *J. Bioinspir. Biomim.* **6**, 1748.
2 Milwich, M., Speck, T., Speck, O. et al. (2006) *Am. J. Bot.* **93**, 1295.
3 Milwich, M., Speck, T., Speck, O., Planck, H. (2008) In *The Role of Plant Stems in Providing Biomimetic Solutions for Innovative Textiles in Composites*, Ellison, M.S., Abbot, A.G. (eds.), Blackwell, London, pp. 168–192.
4 Milwich, M., Speck, T. (2010) In *Learning from Nature: Lightweight Constructions Using the 'Technical Plant Stem'*, Pohl, G. (ed.), Woodhead Publishing in Textiles, Cambridge, pp. 290–309.
5 Rüggeberg, M., Burgert, I., Speck, T. (2010) *J. R. Soc. Interface*, **7**, 499.
6 Milwich, M., Linti, C., Stegmaier, T. et al. (2006) published patent WO 2006131344.
7 Dullenkopf, D., Groß, T., Kornprobst, W. et al. (2011) published patent DE 10 2009 043 *103, PCT/EP2010/005566.*
8 Müller, L., Milwich, M., Gruhl, A. et al. (2013) *Tech. Textil.* **56**, 231.
9 Hufenbach, W., Gruhl, A., Lepper, M., Renner, O. (2013) *Lightweight Design*, **6**, 44.
10 Hanke, U., Seliger, T., Gruhl, A. et al. (2014) published patent DE 102011006647 B4.

11 Haushahn, T., Schwager, H., Neinhuis, C. et al. (2012) *Bioinspir. Biomim. Nanobiomater.* **1**, 77.

12 Haushahn, T., Speck, T., Masselter, T. (2014) *Am. J. Bot.* **101**, 754.

13 Masselter, T., Haushahn, T., Schwager, H. et al. (2010) In *Biomimetic Fibre-Reinforced Composites Inspired by Branched Plant Stems*, Brebbia, C.A. (ed.), WIT Press, Southampton, Boston, USA, pp. 411–420.

14 Masselter, T., Barthlott, W., Bauer, G. et al. (2011) In *Biologically Inspired Products*, Bar-Cohen, Y. (ed.), CRC Press, Pasadena, USA, pp. 377–429.

15 Masselter, T., Eckert, S., Speck, T. (2011) *Beilstein J. Nanotechnol.* **2**, 173.

16 Masselter, T., Speck, T. (2011) In *Biomimetic Fibre-reinforced Compound Materials*, George, A. (ed.), Intech, Croatia, pp. 185–210.

17 Schwager, H., Haushahn, T., Neinhuis, C. et al. (2010) *Adv. Eng. Mater.* **12**, B695.

18 Schwager, H., Masselter, T., Speck, T., Neinhuis, C. (2013) *Proc. R. Soc. B* **280**, 20132244.

19 Hesse, L., Masselter, T., Leupold, J. et al. (2016) *Sci. Rep.* **6**, 32685.

20 Bunk, K., Fink, S., Speck, T., Masselter, T. (2017) *Trees*, **31**, 1793.

21 Fam, A., Rizkalla, S. (2002) *J. Compos. Constr.* **6**, 123.

22 Fam, A., Rizkalla, S. (2003) *Constr. Build. Mater.* **16**, 507.

23 Mirmiran, A., Shahawy, M. (1997) *J. Struct. Eng.* **123**, 583.

24 Pessiki, S., Harries, K.A., Kestner, J.T. (2001) *J. Compos. Constr.* **5**, 237.

25 Saafi, M., Toutanji, H.A., Li, Z. (1999) *ACI Mater. J.* **96**, 500.

26 Knippers, J., Cremers, J., Gabler, M., Lienhard, J. (2010) In *Construction Manual for Polymers + Membranes*, Edition Detail, Birkhäuser Verlag, Munich, Germany.

27 Bunk, K., Jonas, F.A., Born, L. et al. (2017) *Stuttg. Beitr. Naturkd. Serie C*, **82**, 112.

28 Hu, D.L., Sielert, K., Gordon, M. (2012) *J. Mech. Mater. Struct.* **6**, 1197.

29 Wang, L., Cheung, J.T.-M., Pu, F., et al. (2011) *PLoS ONE*, **6**, e26490.

30 Achrai, B., Wagner, H.D. (2015) *Mater. Sci. Eng. C*, **53**, 128.

31 Achrai, B., Wagner, H.D. (2013) *Acta Biomater.* **9**, 5890.

32 Farke, A.A. (2008) *J. Exp. Biol.* **211**, 3085.

33 Sun, C.-Y., Chen, P.-Y. (2013) *Acta Biomater.* **9**, 9049.

34 Jennings, J.S., Macmillan, N.H. (1986) *J. Mater. Sci.* **21**, 1517.

35 Staufenberg, G., Graupner, N., Müssig, J. (2015) *Bioinspir. Biomim.* **10**, 056006.

36 Wegst, U.G.K., Bai, H., Saiz, E., et al. (2015) *Nat. Mater.* **14**, 23.

37 Prüm, B., Seidel, R., Bohn, H.F., Speck, T. (2012) *Beilstein J. Nanotechnol.* **3**, 57.

38 Fratzl, P. (2007) *J. R. Soc. Interface* **4**, 637.

39 Dunlop, J.W.C., Fratzl, P. (2013) *Scr. Mater.* **68**, 8.

40 Fischer, S.F., Thielen, M., Loprang, R.R., et al. (2010) *Adv. Eng. Mat.* **12**, B658.

41 Thielen, M., Speck, T., Seidel, R. (2015) *R. Soc. Open Sci.* **2**, 140322.

42 Thielen, M., Schmitt, C.N.Z., Eckert, S. et al. (2013) *Bioinspir. Biomim.* **8**, 025001.

43 Ashby, M.F. (2006) *Phil. Trans. R. Soc. London A*, **364**, 15.

44 Thielen, M., Speck, T., Seidel, R. (2013) *J. Mater. Sci.* **48**, 3469.

45 Lakes, R. (1987) *Science*, **235**, 1038.

46 Fischer, S.F., Thielen, M., Weiß, P., et al. (2014) *J. Mater. Sci.* **49**, 43.

47 Bührig-Polaczek, A., Fleck, C., Speck, T. et al. (2016) *Bioinspir. Biomim.* **11**, 045002.

48 Lewinsohn, T.M. (1991) *Chemoecology*, **2**, 64.

49 Hunter, J. (1994) *Trees*, **9**, 1.

50 Agrawal, A.A., Konno, K. (2009) *Annu. Rev. Ecol. Syst.* **40**, 311.

51 Dussourd, D.E., Eisner, T. (1987) *Science*, **237**, 898.

52 Dussourd, D.E. (1995) *Ann. Entomol. Soc. Am.* **88**, 163.

53 Bauer, G., Speck, T. (2012) *Ann. Bot.* **109**, 807.

54 Bauer, G., Friedrich, C., Gillig, C. et al. (2014) *J. Roy. Soc. Interface*, **11**, 20130847.

55 D'Auzac, J., Prevot, J.C., Jacob, J.L. (1995) *Plant Physiol. Biochem.* **33**, 765.

56 Gidrol, X., Chrestin, H., Tan, H.L., Kush, A. (1994) *J. Biol. Chem.* **269**, 9278.

57 Wititsuwannakul, R., Pasitkul, P., Jewtragoon, P., Wititsuwannakul, D. (2008) *Phytochemistry*, **69**, 656.

58 Harrington, M.J., Speck, O., Speck, T. et al. (2015) In *Self-Healing Materials. Advances in Polymer Science*, Hager, M., van der Zwaag, S., Schubert, U. (eds.), Springer, Cham, Germany, pp. 307–344.

59 Schüssele, A.C., Nübling, F., Thomann, Y. et al. (2012) *Macromol. Mater. Eng.* **297**, 411.

60 Speck, T., Bauer, G., Flues, F. et al. (2013) In *Materials Design Inspired by Nature: Function through Inner Architecture*, Fratzl, P., Dunlop, J.W.C., Weinkamer, R. (eds.), Royal Society Chemistry, Cambridge, U. K. pp. 359–389.

61 Speck, T., Mülhaupt, R., Speck, O. (2013) In *Self-Healing Polymers*, Binder, W. (ed.), Wiley-VCH, Weinheim, Germany, pp. 61–89.

62 Speck, T., Speck, O. (2008) In *Design and Nature IV*, Brebbia, C.A. (ed.), WIT Press, Southampton, U. K. pp. 3–11.

63 Busch, S., Seidel, R., Speck, O., Speck, T. (2010) *Proc. R. Soc. B*, **277**, 2113.

64 Speck, O., Luchsinger, R., Rampf, M., Speck, T. (2014) *Konstruktion*, **9**, 72.

65 Rampf, M., Speck, O., Speck, T., Luchsinger, R. (2013) *Int. J. Eng. Sci.*, **63**, 61.

66 Rampf, M., Speck, O., Speck, T., Luchsinger, R.H. (2012) *J. Cell. Plast.* **48**, 49.

67 Rampf, M., Speck, O., Speck, T., Luchsinger, R.H. (2011) *J. Bionic Eng.* **8**, 242.
68 Speck, O., Speck, T. (2015) *Biol. Unserer Zeit*, **45**, 44.
69 Poppinga, S., Hartmeyer, S.R.H., Seidel, R. et al. (2012) *PLoS ONE*, **7**(9), e45735 DOI: 10.1371/journal.pone.0045735.
70 Poppinga, S., Masselter, T., Speck, T. (2013) *Bioessays*, **35**, 649.
71 Reith, M., Claßen-Bockhoff, R., Speck, T. (2006) In *Ecology and Biomechanics: A Mechanical Approach to the Ecology of Animals and Plants*, Herrel, A., Speck, T., Rowe, N.P. (eds.), Dekker Publishers/Taylor & Francis Group, Boca Raton, FL, USA, pp. 123–146.
72 Reith, M., Baumann, G., Claßen-Bockhoff, R., Speck, T. (2007) *Ann. Bot.* **100**, 393.
73 Volkov, A.G., Foster, J.C., Ashby, T.A. et al. (2010) *Plant Cell Environ.* **33**, 163.
74 Darwin, C. (1865) *Bot. J. Linn. Soc. London*, **9**, 1.
75 Darwin, C. (1875) *Insectivorous Plants*. John Murray, London, UK.
76 Skotheim, J.M., Mahadevan, L. (2005) *Science*, **308**, 1308.
77 Forterre, Y. (2013) *J. Exp. Bot.* **64**, 4745.
78 van Doorn, W.G., van Meeteren, U. (2003) *J. Exp. Bot.* **54**, 1801.
79 Kobayashi, H., Kresling, B., Vincent, J.F.V. (1998) *Proc. R. Soc. B*, **265**, 147.
80 Faegri, K., van der Pijl, L. (1979) *The Principles of Pollination Ecology*. Pergamon Press, Oxford, U. K.
81 Poppinga, S., Masselter, T., Lienhard, J. et al. (2010) In *Design and Nature V*. Brebbia, C.A., Carpi, A. (eds.), WIT Press, Southampton, UK, pp. 403–410.
82 de Langre, E. (2008) *Annu. Rev. Fluid Mech.* **40**, 141.
83 Niklas, K.J., Spatz, H.C. (2012) *Plant Physics*. University of Chicago Press, Chicago, USA.
84 Ben Amar, M., Müller, M.M., Trejo, M. (2012) *New J. Phys.* **14**, 085014.
85 Marder, M., Papanicolaou, N. (2006) *J. Stat. Phys.* **125**, 1069.
86 Kobayashi, H., Daimaruya, M., Fujita, H. (2003) *Solid Mech. Appl.* **106**, 207.
87 Yoshida, K., Miki, N., Momonoi, K. et al. (2009) *Proc. Jpn. Acad. Ser. B*, **85**, 187.
88 Liang, H., Mahadevan, L. (2011) *PNAS*, **108**, 5516.
89 Koller, D. (2011) *The Restless Plant*. Harvard University Press, Cambridge, MA, USA.
90 Schleicher, S., Lienhard, J., Poppinga, S. et al. (2010) In *Design and Nature V*, Brebbia, C.A. Carpi, A. (eds.), WIT Press, Southampton, U. K, pp. 479–490.
91 Kobayashi, H., Daimaruya, M., Vincent, J.F.V. (2000) In *Proceedings IUTAM–IASS Symposium on Deployable Structures, Theory and Applications*, Pellegrino, S., Guest, S.D. (eds.), Kluwer, Dordrecht, Netherlands, pp. 211–220.
92 Liang, H., Mahadevan, L. (2009) *PNAS*, **106**, 22049.
93 Forterre, Y., Skotheim, J. Dumais, J., Mahadevan, J. (2005) *Nature*, **433**, 421.
94 Poppinga, S., Joyeux, M. (2011) *Phys. Rev. E*, **84**, 041928.
95 Schleicher, S., Lienhard, J., Poppinga, S. et al. (2011) In *Proceedings of the IABSE-IASS Symposium 'Taller Longer Lighter'*, Nethercot, D., Pellegrino, S. et al. (eds.), London, U. K. 9 pp.
96 Schleicher, S., Lienhard, J., Poppinga, S. et al. (2011) In *Conference Papers of the International Adaptive Architecture Conference*, The Building Centre, London, U. K. 11pp.
97 Vincent, O., Weißkopf, C., Poppinga, S. et al. (2011) *Proc. R. Soc. B*, **278**, 2909.
98 de Focatiis, D.S.A., Guests, S.D. (2002) *Phil. Trans. R. Soc. B*, **360**, 227.
99 Jenkins, C.H. (2005) *Compliant Structures in Nature and Engineering*. WIT Press, Southampton, USA.
100 Kresling, B. (2000) In *Proceedings IUTAM–IASS Symposium on Deployable Structures, Theory and Applications*, Pellegrino, S., Guest, S.D. (eds.), Kluwer, Dordrecht, Netherlands. pp. 229–238.
101 Miura, K. (1993) *Int. J. Space Struct.* **8**, 3.
102 Furuya, H., Satou, Y., Inoue, Y., Masuoka, T. (2008) In *Proceedings of the 6th International Conference on Computation of Shell and Spatial Structures IASS-IACM 'Spanning Nano to Mega'*, Abel, J.F., Cooke, J.R. (eds.), Cornell University, Ithaca, NY, USA, 4 pp.
103 Howell, L.L. (2001) *Compliant Mechanisms*. John Wiley and Sons, New York, USA.
104 Knippers, J., Speck, T. (2012) *Bioinspir. Biomim.* **7**, DOI:10.1088/1748-3182/7/1/015002
105 Lienhard, J., Poppinga, S., Schleicher, S. et al. (2010) In *Design and Nature V*, Brebbia, C.A., Carpi, A. (eds.), WIT Press, Southampton, USA, pp. 469–477.
106 Speck, T., Knippers, J., Speck, O. (2015) *AD Architectural Design*, **85**, 34.
107 Knippers, J., Lienhard, J., Schleicher, S. et al. (2011) published patent EP 2 320 015 A2.
108 Mattheck, C. (2007) *Verborgene Gestaltgesetze der Natur*. Verlag Forschungszentrum Karlsruhe, Germany.
109 Mattheck, C. (2010) *Denkwerkzeuge nach der Natur*. Verlag Karlsruher Institut für Technologie, Karlsruhe, Germany.
110 Schleicher, S., Lienhard, J., Poppinga, S. et al. (2015) *Computer-Aided Design*, **60**, 105.
111 Knippers, J., Menges, A., Gabler, M. et al. (2012) In *Advances in Architectural Geometry*, Hesselgren, L., Sharma, S., Wallner, J. et al. (eds.), Springer, New York, USA, pp. 107–122.
112 Knippers, J., Scheible, F., Oppe, M., Jungjohann, H. (2012) *Bautechnik*, **90**, 341

14

Thermal- and Photo-deformable Liquid Crystal Polymers and Bioinspired Movements

*Yuyun Liu, Jiu-an Lv, and Yanlei Yu**

State Key Laboratory of Molecular Engineering of Polymers, Department of Materials Science, Fudan University, Shanghai, China

14.1 Introduction

Liquid crystals (LCs) are a typical organic material that show cooperative motion among molecules. Polymers are one of the most promising materials, because they have many attractive properties and characteristics, including being lightweight, inexpensive, fracture-tolerant, and pliable. The combination of LCs and polymers leads to LC polymers (LCPs) with fascinating properties which possess both the order of LCs and the good mechanical properties of polymers [1–3]. Especially in cross-linked LCPs (CLCPs), the orientation of the LCs is strongly coupled with the conformation of the polymer main chains. Taking advantage of thermo-induced change in the LCs' orientation and their cooperative motion, CLCPs undergo fast contraction above the phase transition temperature and revert to their original size by expanding as the temperature drops below the phase transition temperature [4–7]. Furthermore, light-responsive CLCPs performing such movements as contraction, bending, and twisting have also been obtained [8–10]. What is more, with the assistance of inspirational designs and advanced nanotechnologies, thermo- and photo-deformable CLCPs can potentially be used to mimic several features of living creatures based on the principles obtained from the natural materials.

Scientists take nature as a best source of inspiration and make it possible to develop diverse CLCP actuators, such as walking devices inspired by worms, oscillatory cantilevers inspired by hummingbirds and microrobots that behave like human hands and arms [11–15]. Since their functional response is similar to human or animals' muscles, CLCPs have gained the nickname "artificial muscles." Furthermore, CLCPs are able to be designed with various molecular directors and the resultant CLCP films generate complex shape deformations upon stimuli which mimic the mechanical behavior of plants, such as spring-like motions of tendrils [16].

Moreover, some creatures with micro/nanostructured surfaces, such as the superhydrophobic lotus leaf with ultralow water adhesion and rose petals which shows superhydrophobicity coexisting with strong adhesion force, exhibit fascinating surface wettabilities which have provided a model for the development of responsive CLCPs to extend and realize the applications as smart surfaces [17–19]. On one hand, the deformation induced a reversible roughness change of the microstructured CLCP surface; therefore, the contact angles of the microstructured surface changed accordingly [17, 18]. On the other hand, the water adhesive force of the superhydrophobic CLCP surfaces could be tuned between the water-repellent and the water-adhesive state, which provides a great many promising potential applications for smart no-loss microdroplet transfer [19]. Also, inspired by another type of biologic surface with elaborately periodical photonic nanostructured units, such as butterfly wings with a cornucopia of colors, several responsive CLCP photonic surfaces were developed to adjust the reflection spectra or structural colors, due to a change in the periodic micro/nano structures [20, 21].

This review briefly introduces the current state-of-the-art research on intriguing bioinspired movements of CLCPs, smart responsive micro/nanostructured CLCP surfaces, and their potential applications stimulated by temperature or light.

14.2 Thermal-responsive CLCPs

14.2.1 Thermal-responsive Deformation of CLCPs

De Gennes proposed the possibility of using CLCPs as artificial muscles, by taking advantage of their substantial contraction in the direction of the director axis and cooperative motion [22]. The basic principle behind the

* Email: ylyu@fudan.edu.cn

Bioinspired Materials Science and Engineering, First Edition. Edited by Guang Yang, Lin Xiao, and Lallepak Lamboni.

shape variation is the conformational change of the polymer backbone at LC-isotropic phase transition. For example, when the nematic CLCP films are heated above the nematic-isotropic phase transition temperature, the nematic order decreases and the films exhibit a general contraction along the alignment direction of the mesogens; the CLCP films revert to their original size by expanding as the temperature drops below the nematic-isotropic phase transition temperature (Figure 14.1).

There have been a number of studies on thermal-induced deformation of the CLCPs based on the LC-isotropic phase transition [23–39]. Finkelmann et al. prepared a variety of thermal-responsive polysiloxane CLCPs through the two-step method [24, 25]. First, well-defined weak networks are synthesized and then the mesogens of the film are aligned by a constant load to induce network anisotropy mechanical stretching after the first reaction. Second, cross-linking reactions fix the network anisotropy. Then the well-aligned elastomers are obtained (Figure 14.2). The two-step method has been widely used to fabricate monodomain polysiloxane CLCPs. Due to great anisotropy induced by mechanical stretching, reversible uniaxial thermocontraction of monodomain polysiloxane CLCP even reached 400%.

The alignment of LC mesogens significantly affects the bending behavior of the CLCP films. Urayama et al. investigated how the chiral arrangement of LC mesogens in twist-nematic polymer films induced the formation of helicoids and spiral ribbons [26]. Figure 14.3a shows the chiral mesogenic monoacrylate and the nonreactive chiral dopant. The twist nematic configuration was imprinted in the elastomer matrix by the cross-linking reaction with 1,6-hexanediol diacylate in the presence of a nonreactive chiral dopant. The director changing in orientation through the film's thickness leads to gradients in the induced strain. The orientation of mesogens changes smoothly by 90 ° from the bottom surface to the top surface (Figures 14.3b, c). As Figure 14.3d shows, the pitch of the formed ribbons can be tuned by temperature variation because of the coupling between the LC order and the elasticity. The shape of the ribbon was a helicoid with Gaussian saddle-like curvature at 330 K. Then it became almost flat at a certain temperature Tflat around 353 K. At the temperatures above Tflat, the spiral ribbons

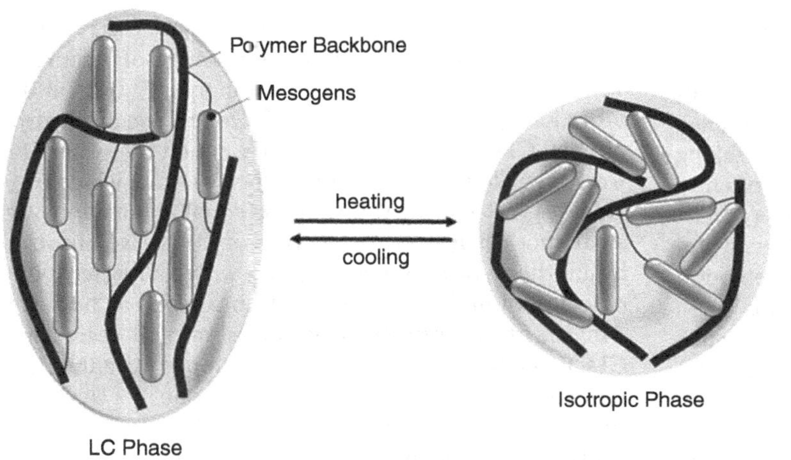

Figure 14.1 Schematic illustration of thermal-induced anisotropic contraction and expansion of the CLCPs.

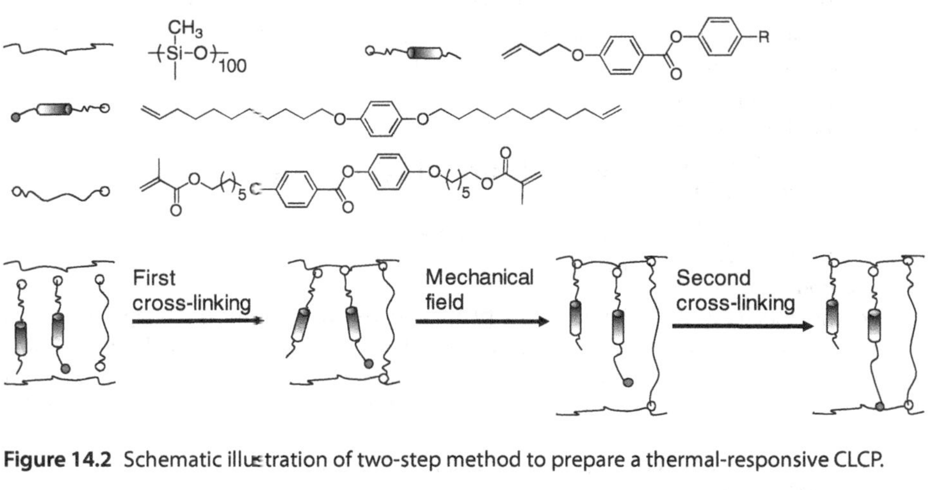

Figure 14.2 Schematic illustration of two-step method to prepare a thermal-responsive CLCP.

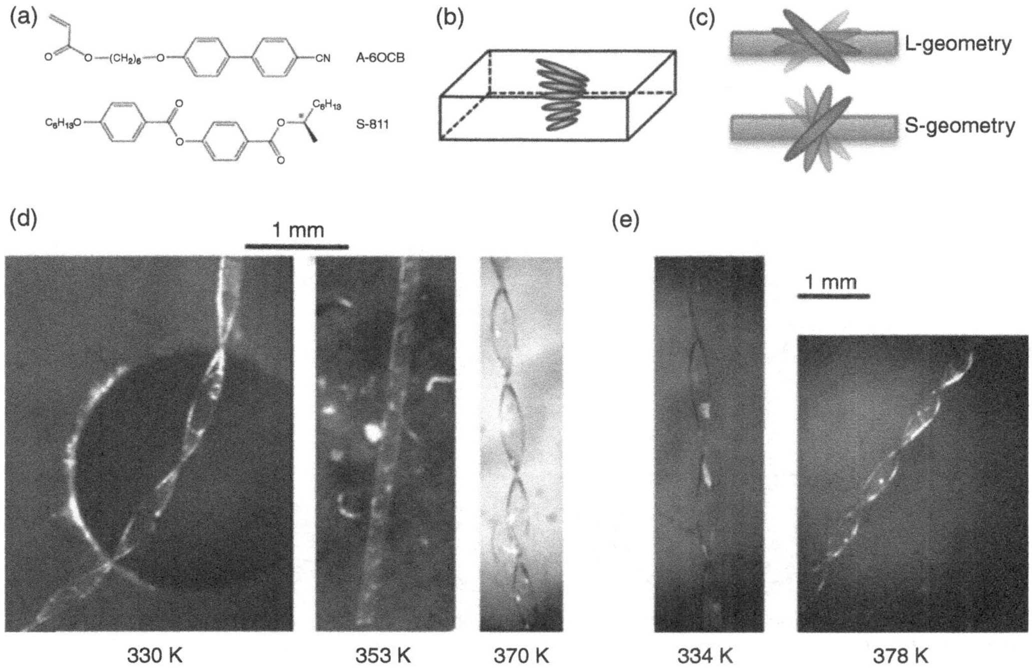

Figure 14.3 (a) Monoacrylate mesogenic monomer A-6OCB and chial dopant S-811. (b) Schematics of the director configuration of TNE ribbons. (c) Top view of the L- and S-geometry. Nematic director twists left-handedly by 90 ° between the top and bottom surfaces with the director at the midplane parallel to the long or short axis of the ribbon. (d) L-geometry helicoids formed by the narrow twist-nematic CLCP films: right-handed at 330 K; almost flat at 353 K; left-handed at 370 K. (e) Shape of the S-geometry ribbons: left-handed at 334 K; right-handed at 378 K. *Source:* Reprinted with permission from [26].

with L-geometry were right-handed, while those with S-geometry were left-handed (Figure 14.3e). What is more, the ribbons selected a helicoid or spiral ribbon depending on the width. The results of this study facilitate the understanding of physics for the shape formation of chiral materials and the designing of new chiral structures.

Real-world applications of the CLCPs are lacking because of the poor processibilities of the CLCPs. Yan Ji and Eugene Terentjev et al. showed that the processing bottleneck of the CLCPs can be overcome by introducing exchangeable links in place of permanent network cross-link [27]. As Figure 14.4a demonstrates, they used a transesterification reaction to generate an exchangeable CLCP. With the added catalyst, the fast breaking and reforming of ester bonds are above the activation barrier at a temperature Tv (Figure 14.4b), that is, the designed xCLCPs can be reshaped or reprocessed at T > Tv. At temperature below Tv, exchange reactions become slow exponentially, and the network acts as a classical covalently bonded thermoset. Reversible transesterification also allows an xCLCP film bonded with a non-LC epoxy elastomer from the reaction of the diglycidyl ether of bisphenol A and diacid above Tv (Figure 14.4c). As Figure 14.4d shows, as the temperature rises, the height of the dome-shaped CLCP actuator decreased because of the contraction induced by the thermal phase transition; on cooling back to the LC phase, the dome reappeared in its original shape. This configuration is very important for practical applications, especially in tactile (haptic) display technology, including dynamic Braille displays, when the actuation has to deliver a linear pushing force action.

In nature, the well-defined and precisely controlled structure of components on a molecular and mesoscopic scale is crucial for the complex processes found in living organisms [28]. Complex shape or surface deformation within soft materials may enable enhancement and miniaturization of devices in medicine, robotics, and aerospace [27, 28]. So far, various molecular directors of the CLCPs have been fabricated [29]. The CLCPs containing the twisted nematic phase and splayed configurations can bend or curl after applying the stimulus [30].

White and his colleagues reported the preparation of soft and ordered CLCPs, the director of which was written within small local volume elements (voxels) [31]. In monoliths with spatially patterned director, thermal stimuli transformed flat sheets into three-dimensional objects through controlled bending and stretching (Figure 14.5). The POM photograph shows the voxelated CLCP film contains nine defects in a square array (Figure 14.5b). Each defect is a point around which the director varies azimuthally by 360 °.

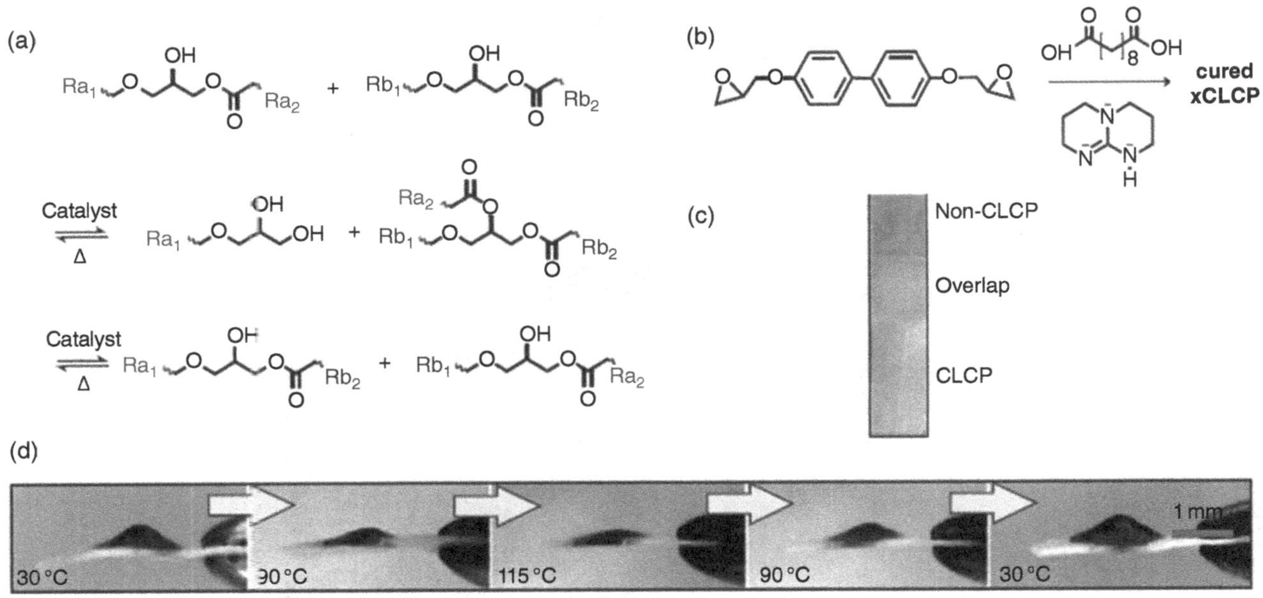

Figure 14.4 (a) A scheme of reversible transesterification. (b) The synthesis of xLCE, showing the mesogen and the spacer used at 1:1 ratio, and the triazabicyclodecene catalyst. (c) An xCLCP film bonded with a non-liquid-crystal epoxy elastomer could be joined together at a sufficiently high temperature $T > T_v$. (d) Reversible cycle of a dome-shaped molded CLCP actuator becoming completely flat on heating to the isotropic phase. *Source:* Reprinted with permission from [27].

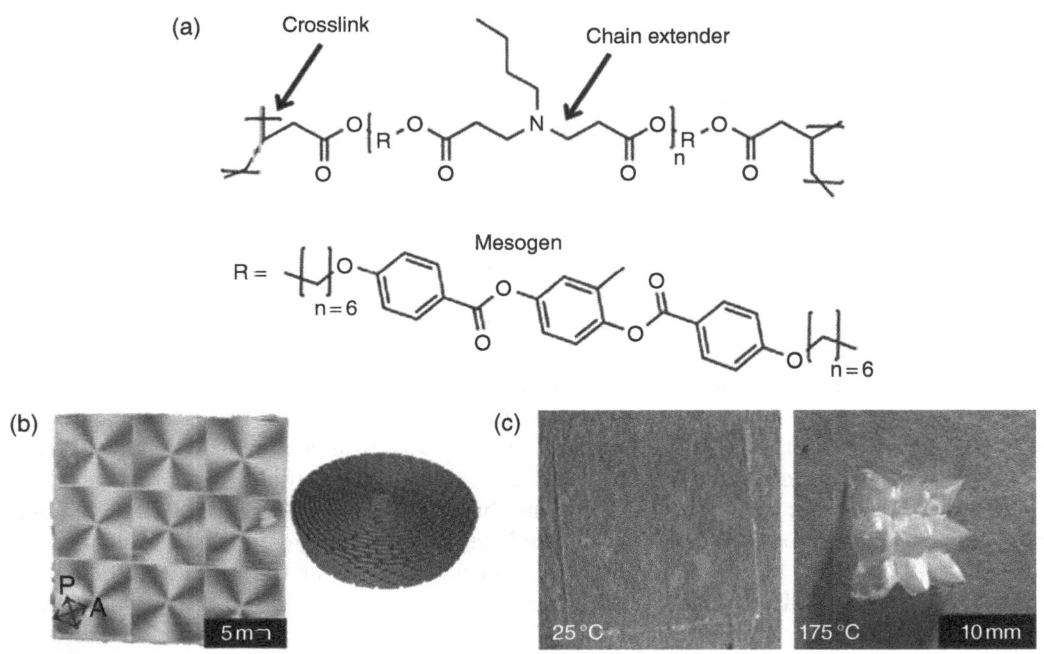

Figure 14.5 (a) Schematic of chemical structure of the main-chain CLCP that can be surface-aligned. (b) Representative photograph of CLCP film with nine +1 topological defects between crossed polarizers. As indicated in the inset schematic, the director orientation varies azimuthally around the defect. (c) Upon heating, nine cones arise from the CLCP film that reversibly flatten upon cooling. *Source:* Reprinted with permission from [31].

Macroscopic azimuthal contraction (along the director) and radial expansion around each defect center lead to the emergence of cones (Figure 14.5c). The programmable mechanical response of these materials could yield monolithic multifunctional devices or serve as reconfigurable substrates for flexible devices in aerospace, medicine, or consumer goods.

Inspired by the contraction-pulsation movement of a jellyfish, Zentel et al. fabricated one-piece micropumps from CLCPs core-shell particles via a microfluidic double-emulsion process, which provides a facile and continuous way to prepare core-shell particles with controlled size [32]. The microfluidic device comprises a triple co-flow channel (Figure 14.6a). The combination of concentric cylindrical and square cross-section geometries allows one fluid (the middle phase: the LC monomer) to flow in the voids between the circular and square capillaries, and another fluid (the inner phase: glycerol) to flow through the innermost circular channel. A schematic drawing of the micromanipulator is shown in Figure 14.6c. For better handling, the particle was heated to 90 °C, which was well above its glass-transition temperature, and pierced with the capillary tip in the rubbery state. The hot plate was then set to heat from 90 °C to 130 °C, and the core-shell particle was slowly lowered towards the hot plate leading to the expulsion of a considerable fraction of the glycerol through the shell hole and thus raising the liquid level in the capillary. Upon cooling, the particle reverted back to its original shape and the liquid was again sucked into the shell (Figure 14.6d). It is the first time actuating core-shell particles from CLCPs have been fabricated, which illustrates their potential application as micropumps and the feasibility of their integration in lab-on-chip systems.

The human eye possesses an enviable performance as a reference standard for optical ability, owning to the high flexibility and large dynamic range of the tunable iris. Zentel et al. implemented a circularly symmetric actuation of radially oriented CLCPs to realize a tunable aperture, which mimics the human iris constriction [33]. To keep the CLCPs-iris stable in shape (with no buckling) and to allow quick actuation in response to temperature, it is necessary to have the heating elements in a thicker CLCP film (about 300 μm thickness). Stretchable polyimide-based platinum heaters, as recently presented [34], were embedded in the CLCP before curing (Figure 14.7a). The heaters were directly inserted into the liquid CLCPs. The stretchable heater wire with a resistance of 3.3 ± 0.2 kΩ was very flexible in the actuation direction, so that the actuation was only slightly inhibited by the wires. The CLCP film was prepared from a monomer acrylate and a cross-linker, as Figure 14.7b shows. The orientation of the CLCP was determined using an electromagnet with a well-characterized field strength, which oriented large volumes of CLCPs.

Figure 14.7c demonstrates that a free standing iris could be held in different contraction states by maintaining a stable voltage at room temperature; the error bars in the graph indicate the standard deviation of 55 measurements. The required power levels for different actuation levels are also indicated in the graph. Figure 14.7d shows the different diameters of the radially oriented CLCP iris with integrated heaters. This entirely new form of tunable aperture may have significant potential in miniaturized photographic optics.

14.2.2 Bioinspired Thermal-responsive Nanostructure CLCP Surfaces

Surface properties, including composition, topography, and wettability, are of great importance in both fundamental research and practical applications. Inspired by the lotus leaf, scientists have developed many superhydrophobic surfaces, some of which show remarkable switching between a hydrophobic and a hydrophilic state under external stimuli [35]. Keller reported a microstructured responsive surface made of nematic CLCP and this surface had roughness-change-related switchable wetting properties. The microsized CLCP pillars prepared by a soft lithography technique underwent a contraction when heated upon phase transition temperature (TNI) [17]. During the heating-cooling cycles, the pillars showed a reversible contraction (Figure 14.8a). The contraction of the micropillars induced a reversible roughness change of the microstructured surface; therefore, the water contact angle (CA) of this microstructured surface changed with the temperature. As the temperature was gradually raised to above the transition temperature of the CLCP from nematic to isotropic phase (76 °C), for the first heating cycle, the CAs of the microstructured surface decreased from 127 ° at 40 °C to 86 ° at 81 °C (Figure 14.8b). When the responsive surface was cooled back to room temperature, CAs turned to 114 °, not fully back to 127 °. However, in the following thermal cycles, good reversibility was observed. CAs changed between 110 ° and 86 ° when the temperature of the structured surface alternated between 35 °C and 80 °C (Figure 14.8c). In this work, the surfaces exhibit only changes in the CAs as a result of temperature changes. However, tuning the adhesive force and mobility of pure water droplets on superhydrophobic surfaces still remains a great challenge, because it requires the precise coordination of surface chemistry response and surface roughness.

In the course of evolution, many organisms have developed unique strategies of light manipulation on surfaces

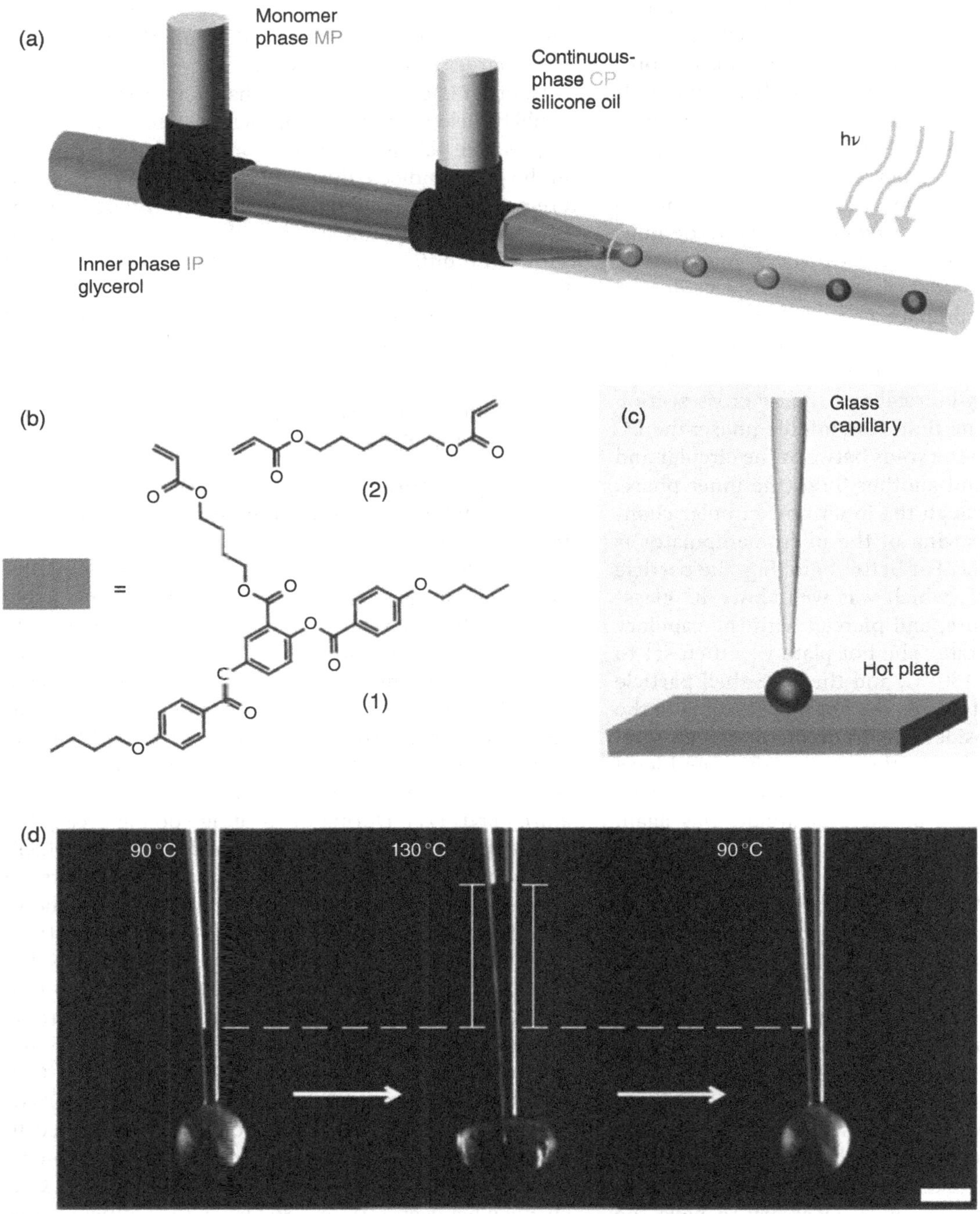

Figure 14.6 (a) The triple co-flowing geometry of the microfluidic set-up allows the fabrication of nematic shell particles filled with a glycerol core. Dispersed in the surrounding continuous-phase silicone oil, the monomer droplets are polymerized by ultraviolet light. (b) Chemical structures of the reactive mesogen (1) and cross-linker (2) in the monomer mixture. (c) Schematic drawing of the micromanipulator to penetrate the particle shell with a thin glass capillary. (d) When heated to the isotropic phase at 130 °C, the phase transition results in a deformation of the CLCP shell. The force exerted by the actuating shell pushes the liquid glycerol core into the capillary. Once the particle resumes its former shape in the nematic phase at 90 °C, the glycerol sinks to its original level (scale bar, 100 mm). *Source:* Reprinted with permission from [32].

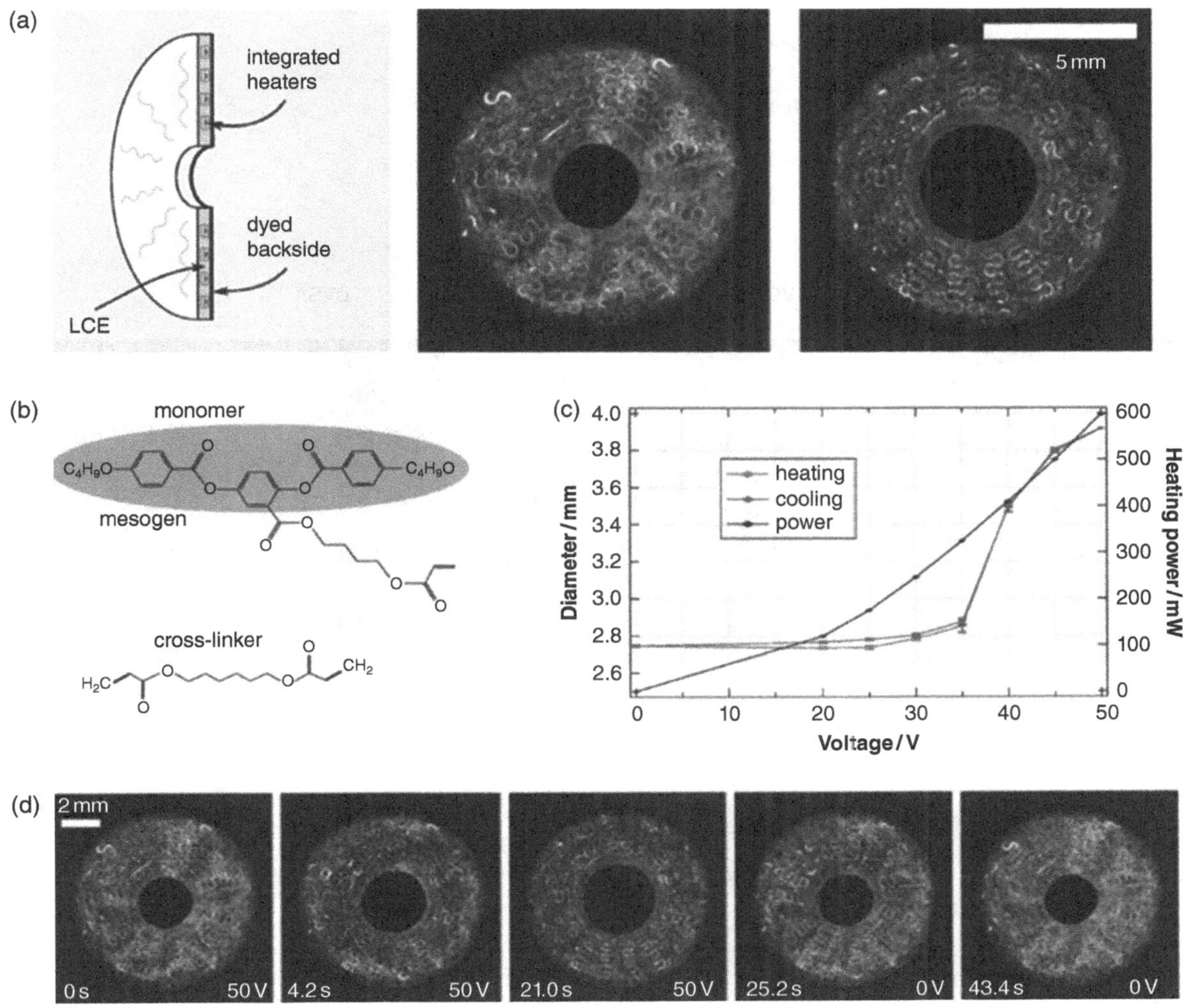

Figure 14.7 (a) Schematic diagram of the CLCP iris with integrated polyimide-based platinum heaters. (b) Chemical structures of the monomer and cross-linker. (c) Actuation of the iris at different voltages showing the controllability of discrete contraction states. The standard deviation of contraction (n = 55) is shown by the error bars. The power consumption of the iris in the different contraction states is indicated on the right axis. (d) A series of images showing different tuning states of the radially oriented CLCP iris with integrated heaters. *Source:* Reprinted with permission from [33].

that rely on intriguing combinations of a broad range of optical effects generated by materials with sophisticated, multiscale, hierarchical structural arrangements. By exploiting the optical principles underlying natural structural color, some researchers have developed different photonic crystals by using nanofabrication method. What is more, there is a strong demand for photonic materials with properties that can be tuned by external stimuli [36, 37]. CLCPs have also been introduced to generate responsive photonic crystals owing to their sharp shape change in response to external alteration. Li et al. fabricated thermal responsive photonic crystals using CLCPs and nematic LCs (5CB) as inverse opal materials. In the vicinity of the phase transition point of the mixture, the photonic band gaps of such inverse opal films exhibited a strong temperature dependence. As the molar ratio of CLCPs and nematic LCs changed, the character of their photonic band gaps also changed with increasing temperature, and therefore the wavelength of the resultant Bragg peak was tuned continuously [38]. However, the 5CB molecules were not linked to the CLCPs by covalent bonds. Later, they prepared a new type of electrothermally driven photonic crystal film based on highly cross-linked CLCP without 5CB [20]. When a 30 V electric field was applied on the CLCP inverse opal films to make them close to their TNI, the LC moieties became isotropic and the films started to contract along the nematic director axis. During this process, the holes in the films changed from round to ellipse, which led to a shift of the reflection peaks (Figure 14.9).

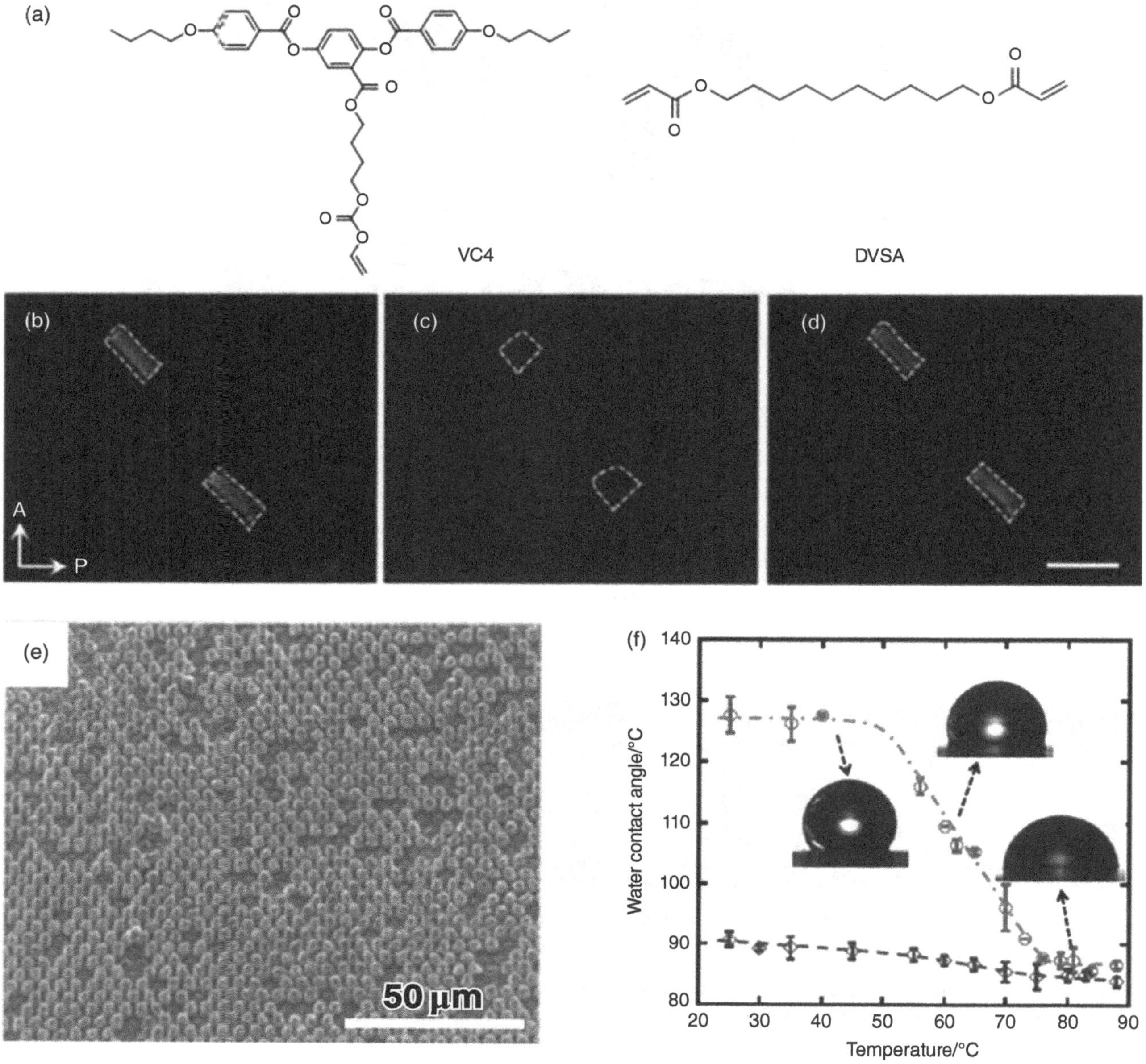

Figure 14.8 (a) Chemical structures of monomer VC4 and cross-linker DVSA used to prepare microsized CLCP pillars. (b–d) The shape of an isolated pillar with the diameter of 20 µm changes with temperature under polarizing optical microscope (from left to right, T = 30, 96, and 48 °C). Scale bar: 100 µm. A: Analyzer; P: Polarizer. (e) SEM images of structured surface with hexagonally-arranged micro-pillars. (f) Hydrophobicity changes for the responsive surface with micro-pillars at different temperatures. Red line (structured surface); blue line (flat surface). *Source:* Reprinted with permission from [17].

Subsequently, Guo et al. developed a bioinspired thermal-responsive micropatterned inverse opal film based on CLCPs [39]. The inverse opal films were fabricated by filling the silica opal glass cells with the LC monomer mixture. After UV photo-polymerization and removing the silica opal templates, the CLCP inverse opal films with bright structural color were obtained. Figure 14.10a displays the reflection spectra for the CLCP inverse opal films as a function of temperature. When the CLCP inverse opal film was heated from 30 °C to 140 °C, the reflection peak of the film had an obvious red-shift from 515 nm to 565 nm, leading to the change of the structural color from blue-green to yellow-green. As the film was cooled to 30 °C, the photonic band gap returned to its original wavelength as well as the structural color, which indicates that the thermal switching behavior of CLCP inverse opal films is reversible. Figure 14.10b illustrates the mechanism for the thermal-responsive behavior of the CLCP inverse. Such a

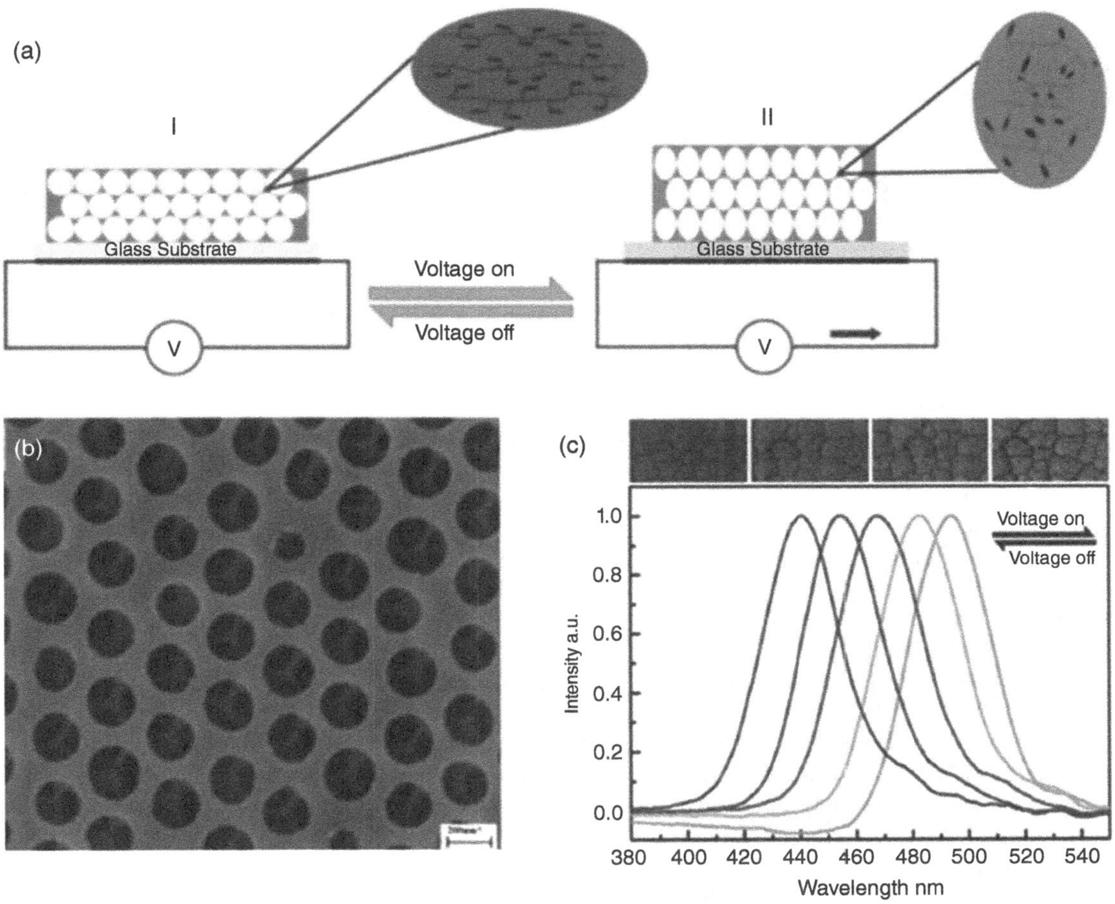

Figure 14.9 (a) Schematic illustration of the mechanism for the change of inverse opaline film based on CLCPs with different temperatures: (I) CLCP-based inverse opaline film in the nematic state of $T < T_{NI}$ (voltage off); (II) CLCP-based inverse opaline film in the isotropic state of $T > T_{NI}$ (voltage on). (b) SEM images of the inverse opaline polymer film. The bars in the images stand for 200 nm. (c) The photonic band-gap shifts and the induced structural colour changes reversible of the CLCP inverse opaline film driven by voltage. *Source:* Reprinted with permission from [20].

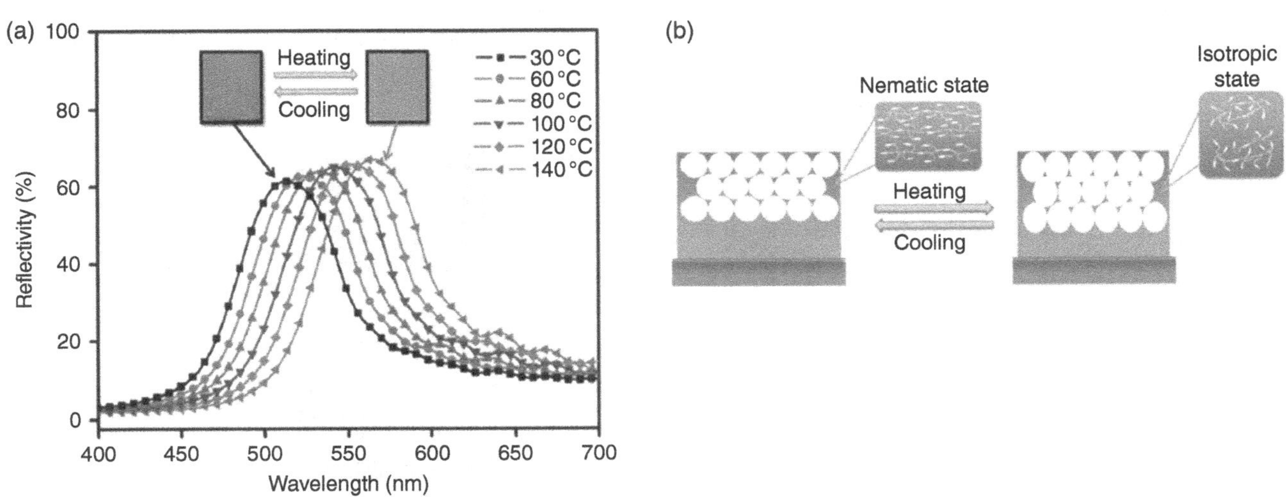

Figure 14.10 (a) Reflection spectra-reversible shifts and structural color-reversible changes of the inverse opal film as a function of temperature. The size of the CLCP inverse opal film in the image is about $12 \times 10\ mm^2$. (b) Schematic for the mechanism of the lattice space reversible change of CLCP inverse opal films induced by temperature variation. *Source:* Reprinted with permission from [39].

temperature-tuning effect in the photonic band gap should be of great interest in thermal switches and thermal sensors.

14.3 Photothermal-responsive CLCPs

Except for increasing the surrounding temperature directly, it is highly desirable for CLCPs-based actuators to activate in a highly controlled fashion without perturbing the environment, as in the example of infrared laser irradiation. Furthermore, promising methods have been developed to disperse multi-walled carbon nanotubes (CNTs) and single-walled CNTs (SWCNTs) in CLCP matrices such as polysiloxanes [40] and polyacrylates [41]. Irradiation of CNT/CLCP composites with visible to infrared light induces a deformation of the samples, which can either be two- or three-dimensional, depending on the initial alignment of the mesogens and the temperature distribution in the samples. For the photothermal-responsive CLCPs, the mechanism lies in that the incorporated nanophase materials absorb and convert photon energy into thermal energy and conduct the thermal energy over the matrices to heat the CLCPs. Then the CLCPs take advantage of the nematic-isotropic phase transition that occurs in the whole film upon heating. As a result, the bulk effect leads to a reversible and dramatic deformation.

Broer et al. used photoalignment technology to prepare circularly patterned CLCP films, which exhibited exotic shape deformations in response to infrared light [42]. As Figure 14.11a shows, a thin film of a cinnamate-based linearly photopolymerizable (LPP) polymer as an alignment layer (Staralign 2100) was spin-coated onto glass substrates. The director in the alignment layer was controlled by irradiation with linearly polarized UV light through a photomask. By slowly rotating the cell while irradiating through a photomask with a wedge-shaped opening, a series of different cells were prepared with a continuous change of the alignment direction throughout the LPP layer. Then by capillary action, the cells were filled with a mixture of polymerizable LCs doped with dyes that can convert near infrared (NIR) light into heat. After heating the LC samples to the nematic phase, photopolymerization is carried out to obtain a polymer film in which the alignment direction of the LCs has been frozen. As Figure 14.11 demonstrates, upon irradiation of infrared lamp, the free-standing film deformed into a conical shape within seconds, with the cone apex located at the center and pointing downward. When the lamp was switched off, the film returned to its original shape. In the case of a radial alignment pattern, the opposite deformations took place. By using different mask sizes and patterns, the possible range of dimensions and shapes can be greatly expanded.

In nature, some plants have leaves or flowers that move to point toward the sun, in a phenomenon called heliotropism. Inspired by this phenomenon, Jiang et al. have developed a reversible photo-thermomechanical SWNT-CLCP layer that has a strong actuation capability driven by natural sunlight [43]. The solar cells were installed on a platform that was connected to SWNT-CLCP actuators so that it can tilt under the actuation force (Figure 14.12a). SWNT absorbed and transformed the incoming sunlight into thermal energy, so that actuators changed to a contracted state toward the direction of the incident light, while other actuators not exposed to the sunlight remained in the relaxed state. Consequently, the platform holding the solar cells was driven by the contracted actuators, and self-adaptively tilted towards the sunlight, hence the increased photocurrent was output from the solar cells. Figure 14.12b shows the images of this device in one laboratory test; the platform with solar cells tilted toward the light source after the actuation (Figure 14.12c). As a result, a significant increase in the photocurrent output was observed from the solar cells in the artificial heliotropic devices, depending on the incident light direction.

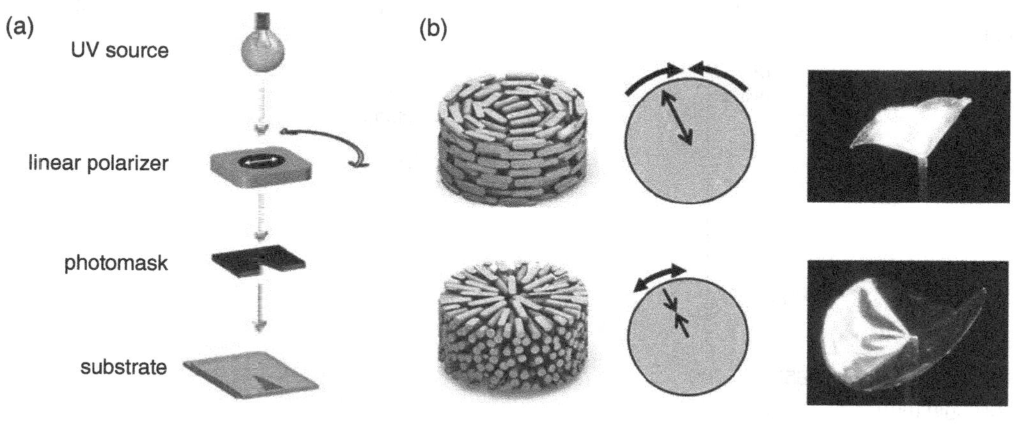

Figure 14.11 (a) Preparation of a patterned alignment cell. (b) Deformation of CLCP films with a complex order of mesogens: an azimuthal polymer film and a radial polymer film. The arrows along the radius and the azimuth indicate the direction of deformation. *Source:* Reprinted with permission from [42].

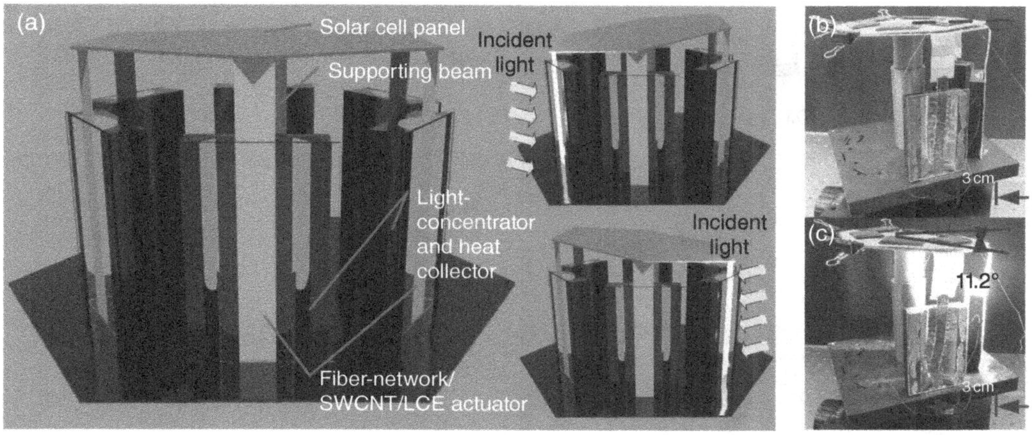

Figure 14.12 (a) Concept of artificial heliotropism. 3D schematic of the system (left) and 3D schematic of the heliotropic behavior (right). The actuator(s) facing the sun contracts, tilting the solar cell toward the sunlight. (b) Before irradiation. Intensity of the white light source is 100 mW/cm^2. (c) 30 s after irradiation was on. The solar cells tilt toward the sunlight leading to the increased photocurrent output. *Source:* Reprinted with permission from [43].

Different from the current man-made solar tracking systems which are generally realized through mechatronic designs, this system possesses the elegance and simplicity of natural heliotropism.

Recently, there has been an increasing interest in the field of bioinspired soft robotics, which can be used to carry soft, fragile payloads without causing damage and working in cluttered environments by employing large strain deformation, because of their soft materials-based structures and highly redundant degrees of freedom [44]. Taking advantage of the nematic-isotropic phase transition that occurs in the whole CLCP film upon exposure to heat, a reversible dramatic linear contraction/expansion has been developed (30–400%) [24, 45]. However, most thermal-responsive CLCPs just show contraction/expansion and the lack of a global bending motion significantly limits the soft robotic applications.

Chen et al. prepared wavelength-selective, infrared light-driven hinges composed of one active CLCP composite layer with infrared active fillers and one passive silicone layer. Infrared active fillers uniformly embedded in the CLCP matrix include SWNTs and NIR dyes, which can efficiently transform infrared light into thermal energy, thereby playing a role as nanoscale heaters [11]. Figure 14.13a shows a schematic illustration of the SWNT-CLCP composite/silicone bilayer film. The 0.1 wt% SWNT-CLCP/Silicone bilayer film started to bend when exposed to the infrared light (Figure 14.13b). Because the SWNT-CLCP layer was tightly bound to the silicone elastomer layer, the in-plane negative strain bent the underlying silicone elastomer layer. As the bilayer hinges can be used in conjunction with passive materials such as polycarbonate films, the various foldable structures are able to undergo reversible infrared-induced folding and unfolding. In particular, Figure 14.13c shows a foldable structure inspired by a Venus flytrap, which opened and closed repeatedly in response to the infrared light. The gripper consists of two curved polycarbonate films serving as arms that were connected by a number of bilayer hinges. The closing and opening of the arms of the gripper were controlled by turning on and off the infrared light. Also, an infrared light-driven inchworm walker was built by an asymmetric SWNT-LCE/silicone bilayer film and two polycarbonate films with different shapes (Figure 14.13d). In contrast to previous walking devices, it can climb up a hill at a 50° incline. The CLCP composite/silicone bilayer hinges not only show fast and reversible bending with a large strain owing to the bulk nematic-isotropic phase transition, but also exhibit infrared wavelength selectivity by using NIR dyes as fillers.

In addition, the photothermal mechanism is also applied to CLCP microparticles doped with gold nanocrystals [46]. Irradiation with a focused infrared laser beam on a submicrometer region within the microparticle induced the deformation into various shapes. Recently, Keller et al. also developed a novel experimental protocol of successfully employing polyacrylate based CLCPs incorporated with gold nanorods (AuNR) to fabricate CLCP/AuNR micropillars or microactuators [47]. With the inclusion of less than 1 wt% of AuNR, high maximum actuation strain (30%) and rapid response (a few seconds) were achieved in CLCP/AuNR micropillar actuators under 635 nm laser irradiation. The comparative experimental studies on the thermal and photothermal actuation behavior of the CLCP and CLCP/AuNR suggest that AuNR is an excellent candidate for developing high-performance CLCP actuators with photothermal actuation capability.

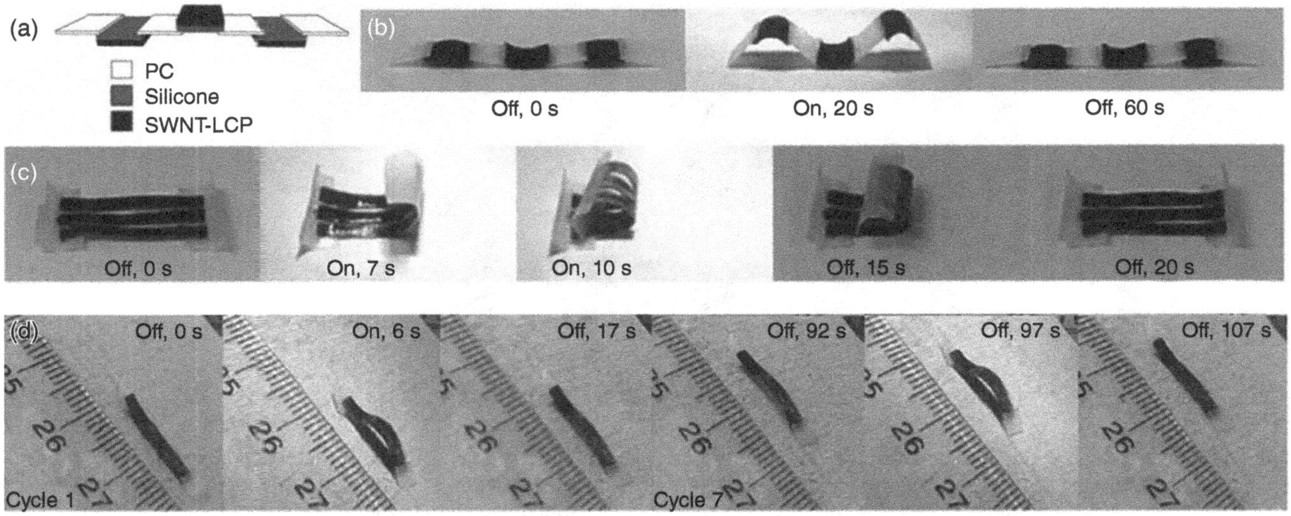

Figure 14.13 (a) Scheme of an origami structure of four polycarbonate (PC) films that are adhesively connected by three SWNT-LCP/silicone bilayer hinges. (b) Reversible folding and unfolding of the origami structure in response to continuous-wave (CW) NIR light (11.0 mW/mm^2). (c) Reversible closing and opening of a Venus flytrap-inspired gripper in response to CW NIR light (28.2 mW/mm^2). The structure has two curved PC films that are connected by three SWNT-LCP/silicone bilayer hinges. (d) The inchworm walker crawling up the wood substrate at a 50° incline in response to on and off cycles of CW NIR light (28.2 mW/mm^2). *Source:* Reprinted with permission from [11].

14.4 Light-responsive CLCPs

14.4.1 Light-responsive Deformation of CLCPs

Cooperative motion of molecules in LC phases may be most advantageous in changing the molecular alignment by external stimuli. The alignment of the majority of LC molecules will be changed if the alignment of a small portion of LC molecules is changed in response to an external stimulus. The energy needed to induce an alignment change of only 1 mol% LC molecules is enough to bring about the alignment change of the whole system. In other words, a huge amplification is possible in LC systems [48]. For example, when azobenzene moieties are cooperated into CLCPs and the resulting azo-CLCPs are irradiated to cause photochemical reactions of azobenzene moieties, an LC-isotropic phase transition of the mixtures can be induced isothermally [49]. Light is a good energy source that can be controlled remotely, instantly, and precisely. Light-driven soft CLCPs thus play an important role in realizing soft actuators, which can convert light energy into mechanical work directly by using a contactless laser beam.

The integration of photoactive mesogenic units into CLCPs has already given rise to remarkable results and is currently judged the most promising candidate for commercial applications [3]. Azobenzene has two isomers, the *trans* and *cis* forms, which can be reversibly switched to each other under photo-irradiation. The transform of the molecules further lead to the change of molecules size and polarity [50]. The rod-like *trans* form of the azobenzenes stabilizes the structure of LC phase, whereas the bent *cis* isomer tends to destabilize the ordered structure (Figure 14.14). This light-induced reduction of LC order has also been used to trigger a reversible deformation of CLCPs containing azobenzene just by changing the wavelength of actinic light. Finkelmann et al. reported the pioneering work on light-induced large contraction and elongation effects by using azobenzene as the sensitive chromophore in a CLCP film [51]. Since then, many attempts have been made to improve the performance of photomobile polymer materials [52–54].

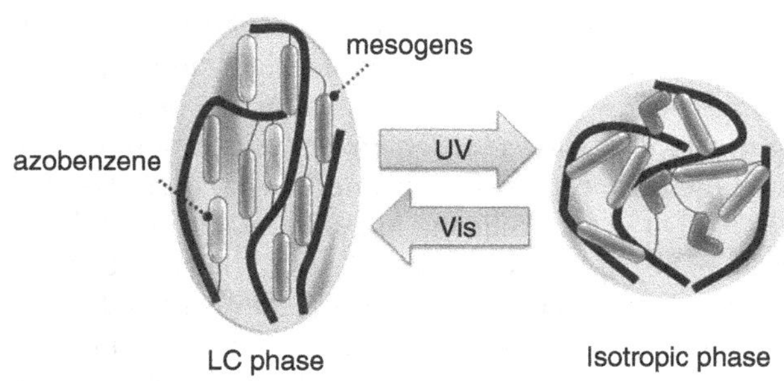

Figure 14.14 Schematic illustration of photoisomerization of the azobenzene chromophores and photochemical phase transition of the azobenzene-containing LCPs.

Another very interesting property of an azobenzene-based LCP film is its capability to perform three-dimensional movement, which could be advantageous in a variety of real manipulation applications. Ikeda et al. succeeded in demonstrating photo-induced bending and unbending of cross-linked polyacrylate films that bear azobenzene moieties [8, 55]. Because of the high concentration of the azobenzene moieties in the film and their high extinction coefficient of the azobenzene moieties at around 360 nm, most of the incident photons are absorbed by the film surface. Only the surface region facing the incident light undergoes the decrease in alignment order of LC and the phase transition. As a result, the contraction is generated only on the surface layer, causing the films to bend toward the irradiation source.

By means of the selective absorption of linearly polarized UV light in a macroscopically randomly aligned CLCP films, Ikeda et al. realized that the deformation direction of a polydomain CLCP film was chosen precisely along any direction, as Figure 14.15 shows [56]. The film bent toward the irradiation source in a direction parallel to the polarization of the light and completely reverted to its initial flat state upon exposure to visible light with a wavelength longer than 540 nm. The polydomain LC film consisted of many micro-sized domains of azobenzene moieties aligned in one direction in each domain. Although macroscopically the direction of alignment was random, upon exposure of the film to linearly polarized light, the selective absorption of light in a specific direction led to a contraction in specific domains where the azobenzene moieties were aligned along the direction of light polarization. This work represented a major milestone in the CLCPs deformation field.

In recent years, great efforts have been made to improve the mechanical and photoresponsive properties of photodeformable CLCP materials. Yu et al. found that the aligned nanostructure of CNTs sheet could effectively orient the CLCP mesogens along the length of CNTs without any other alignment layer. By using highly aligned CNT sheets, a new method to prepare photodeformable CLCP/CNT nanocomposite films was developed [57, 58]. The

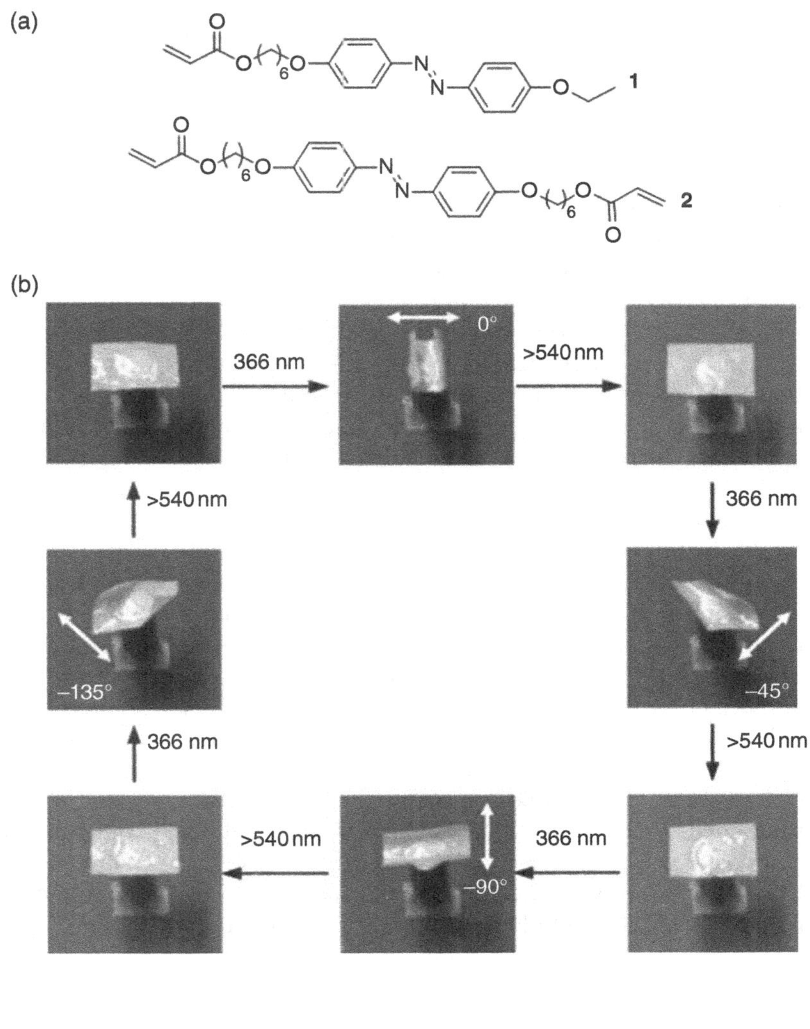

Figure 14.15 (a) Chemical structure of the LC monomer (1) and cross-linker (2) used for preparation of the polydomain CLCP film. (b) Images of the film bending in different directions in response to linearly polarized UV light with different polarization directions at $\lambda = 366$ nm. The film is subsequently flattened again by visible light at $\lambda = 540$ nm. Size of the film: 4.5 mm × 3 mm × 7 mm. *Source:* Reprinted with permission from [56].

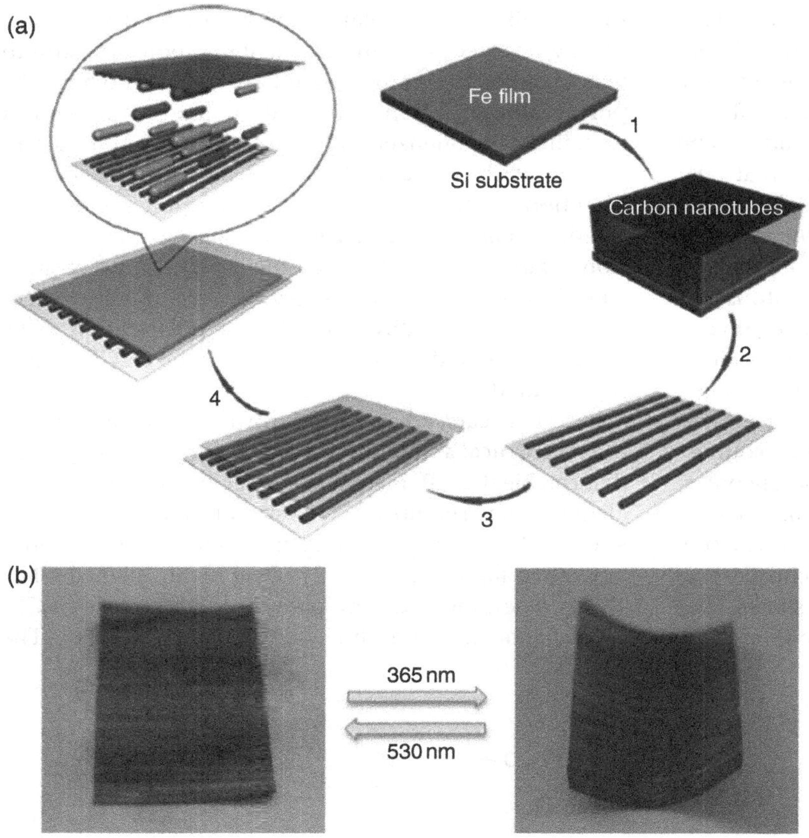

Figure 14.16 (a) Preparation of an oriented CLCP/CNT nanocomposite film in four steps: 1) growth of the CNT array; 2) formation and stabilization of the CNT sheet on a glass substrate; 3) preparation of the LC cell; 4) injection of a mixture containing monomers. (b) Bending behavior of a CLCP/CNT composite film upon exposure to UV (365 nm, 100 mW/cm^2, 50 s) and visible light (530 nm, 35 mW/cm^2, 140 s). *Source:* Reprinted with permission from [57].

CLCP/CNT composite film exhibited a rapid and reversible deformation under alternate irradiation of UV and visible light (Figure 14.16). This actuation is derived from the structure change in the composite film, which results from the photo-isomerization of the azobenzene moieties. Compared with the CLCPs prepared with the conventional mechanical rubbing method, the introduction of aligned CNTs remarkably improves the mechanical strength of the CLCP film and retains high electrical conductivity as well.

The photo-induced deformations mentioned above are interesting, however, UV light is not ideal for practical applications, due to considerations of safety, power consumption, and cost. For potential applications of light-driven actuators in biological systems, near-infrared (NIR) light, instead of UV or visible light, would be a more advantageous stimulus, because low-energy light penetrates deeper into tissues with less damage. Yu et al. incorporated up-conversion nanophosphors (UCNPs) into the azotolane-containing CLCP film and succeeded in generating fast bending of the resulting composite film exposed to CW NIR light at 980 nm. Under excitation with a CW 980 nm laser, the as-prepared UCNPs showed blue emission and the main up-conversion lamination (UCL) emission peaks at 450 nm and 475 nm overlapped the absorption band of the azotolane CLCP film (between 320 nm and 550 nm) perfectly; thus the UCL light triggered molecular-scale motion (*trans-cis* photo-isomerization of azotolane), which was converted into large macroscopic deformations (Figure 14.17) [59].

Recently, Yu et al. achieved red-light-controllable CLCP driven by low-power excited triplet-triplet annihilation-based UCL (TTA-UCL) [60]. A red-to-blue TTA-based up-conversion system was incorporated into rubbery polyurethane film and then assembled with an azotolane-containing CLCP film. The film bent toward the light source irradiated by 635 nm laser at low power of 200 mW/cm^2 (Figure 14.18). Furthermore, a biological experiment was conducted. A piece of pork with a thickness of 3 mm was put between the light source and the assembly film. After transmitting through the pork, the power density of 635 nm (power density = 500 mW/cm^2) was reduced to 86 mW/cm^2. The assembly film behind the pork bent toward the light source within 1 min. This demonstrates the possibility of using 635 nm red light as an excitation source to induce the photo-deformation of the CLCPs in biological systems because it can penetrate deeper in tissues and cause less harm to cells. This work was a

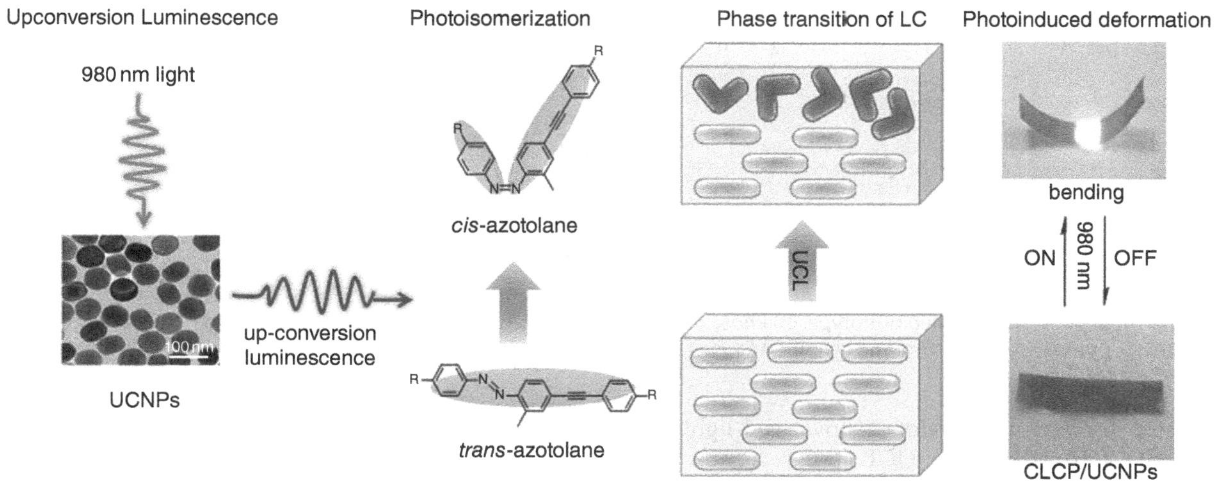

Figure 14.17 Schematic illustration of the mechanism of CW NIR-light-induced deformation of the azotolane CLCP/UCNP composite film, and photographic frames of the composite film bending in response to the NIR light at CW 980 nm and recovery after removing the light source. *Source:* Reprinted with permission from [59].

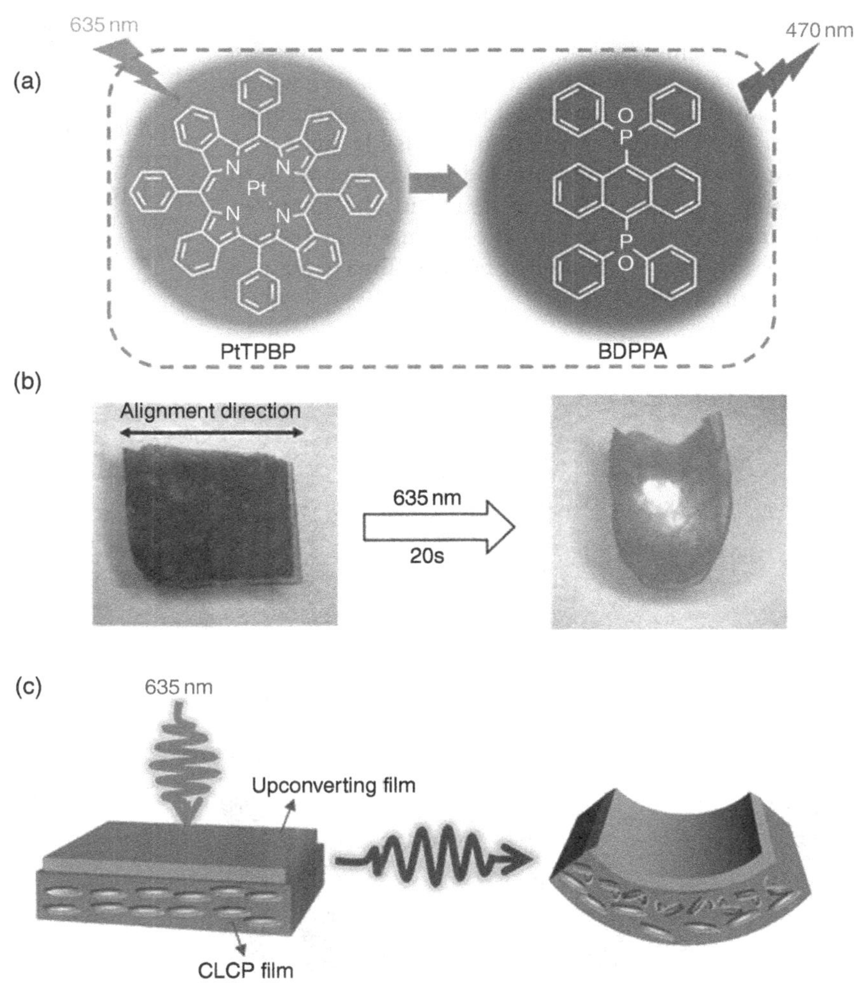

Figure 14.18 (a) Schematic illustration of TTA-UCL emission of PtTPBP (sensitizer) and BDPPA (annihilator) under excitation with 635 nm laser. (b) Red-light-induced deformation of the two layer assembly film. The power density of the 635 nm laser: 200 mW/cm^2. Thickness of upconverting film: 15 μm; thickness of of CLCP: 27 μm. (c) Mechanism for the photoinduced deformation of the as-prepared assembly films. The top layer represented the upconverting film; the bottom layer represented the CLCP film. *Source:* Reprinted with permission from [60].

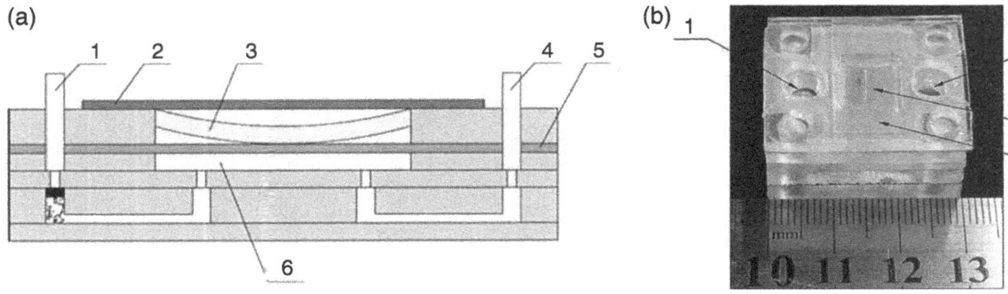

Figure 14.19 (a) The section of assembled prototype. (b) Photo of experimental prototype (1, inlet; 2, press plate; 3, photodeformable material; 4, outlet; 5, pump membrane; 6, pump chamber) *Source:* Reprinted with permission from [61].

significant step forward to develop a novel red-light controllable CLCP based on TTA-UCL.

Some interesting designs from the photo-responsive CLCPs present an opportunity to realize soft actuators in more applications. Yu et al. utilized photo-responsive CLCPs as a micropump [61]. Water was chosen as the pump medium. Upon UV light irradiation, downward bending of the film gave rise to the reduction of the pump chamber volume (Figure 14.19). The flow rate of the water varied in a stroke of the pump membrane, which means, the bending speed of the film decreased in this process. The smaller pressure would lead to a higher flow rate and a larger volume pumped in a stroke. On the irradiation of visible light, the upward unbending of the film led to the expansion of the pump chamber. Underpressure in the chamber drove the fluid to flow into the chamber. In further study, we utilized the bending of CLCP films to act as a valve membrane [62].

14.4.2 Bioinspired Soft Actuators

Bioinspired soft actuators are receiving great attention for use as novel devices, such as humanoid robots and micromachines. White and Bunning et al. reported oscillatory behavior of CLCPs with low azobenzene concentration like a hummingbird for the first time (Figure 14.20a) [12, 63]. Polarized irradiation with wavelengths in the 440–515 nm region of the visible spectrum, where both *trans* and *cis* isomers have comparable absorption, results in the realignment of molecules. Once *trans*-azobenzenes have become perpendicular to the polarization direction of the actinic light after repeating the *trans-cis-trans* cycles with visible light, they become inactive. Therefore, azobenzene molecules become perpendicularly aligned to the polarization direction of the incident light after a number of *trans-cis-trans* cycles (called the Weigert effect) [64]. The large laser intensity endowed a large enough moment of inertia for the azo-LCP cantilever above a threshold that

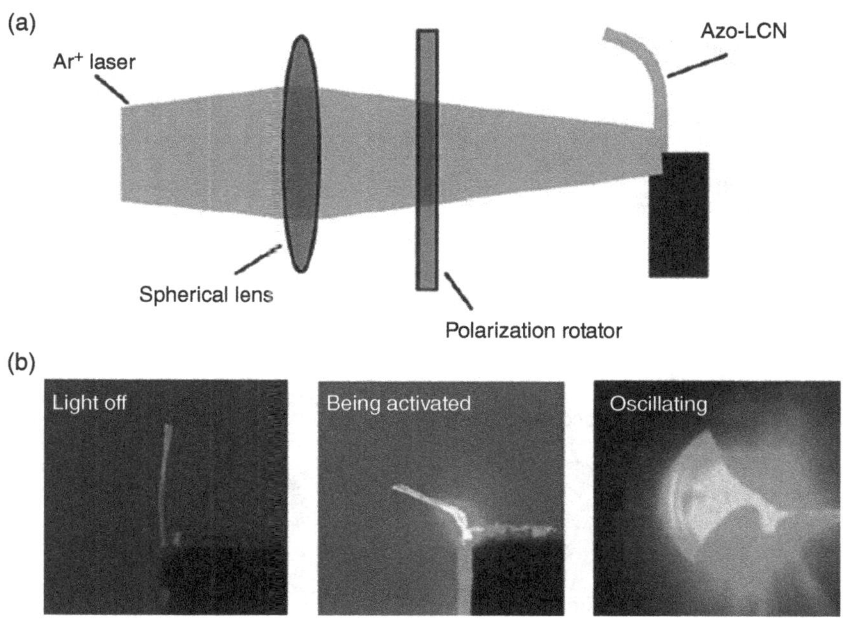

Figure 14.20 Oscillation of a CLCP cantilever induced by an argon ion laser. (a) Experimental set-up. (b) The photographs of the azo-CLCP cantilever oscillation "on" and "off." Size of the film: 2.7 mm × 0.8 mm × 50 mm. *Source:* Reprinted with permission from [63].

enabled the cantilever to deflect through the path of the laser beam, so that the back surface of the azo-CLCP was exposed to light. The azobenzene molecules at the back surface of the polymer subsequently underwent photo-isomerization. As the tip of the azo-CLCP again passed through the beam during the upstroke of the cycle, the front surface was re-exposed. This mechanism enabled fast oscillation up to 270 Hz (Figure 14.20b). Furthermore, the oscillation was driven by focused sunlight, showing the possibility of using the material for the conversion of energy from natural resources [65–68].

Ikeda et al. reported a full light-driven plastic motor with laminated films composed of a CLCP film and a polyethylene sheet [69]. A continuous plastic belt of the CLCP-laminated film was prepared by connecting both ends of the film, and then placed on the belt on a home-made pulley system as illustrated in Figure 14.21a. By irradiating the belt with UV light (366 nm, 240 mW/cm^2) from top right and visible light (>500 nm, 120 mW/cm^2) from top left simultaneously, the rotation of the belt was induced to drive the two pulleys in a counter-clockwise direction at room temperature. Furthermore, inspired by the movement of the inchworm, they demonstrated a unidirectional motion, an inchworm walk, of the CLCP laminated film with asymmetric end shapes (Figure 14.21b) [15].

Yu et al. first reported visible light-induced bending and unbending of azotolane-containing CLCPs, whose deformation even occurred upon irradiation with sunlight [70]. Then they prepared a visible-light-driven fully plastic microrobot [13]. The microrobot made of CLCP and polyethylene bilayer films consisted of several parts, including a hand, a wrist and an arm (Figure 14.22). Without the aid of any gears, bearings, or contact-based driving systems, the microrobot was manipulated to pick, lift, move, and place milligram-scale objects by irradiating different parts of the microrobot with visible light. More importantly, after lamination with polyethylene film, the mechanical strength was greatly enhanced. Compared with traditional electric robots, fully plastic microrobots have excellent processability and can be accurately controlled by remote and localized irradiation of light.

Some other fantastic movements of the CLCP films such as cilia-like motion have also been developed [14]. As Figure 14.23 shows, Oosten et al. used an azobenzene-containing film with a splay bent configuration which improved the bending strength of the material, and utilized inkjet printing technology to create cilia-shaped structures. Two parts of the CLCPs incorporated with different azobenzene chromophores allowed for a wavelength-selective control in the bending of the different segments. The strip was actuated in air using lights of different wavelengths, resulting in the steady-state responses. When the strip was illuminated with visible light, the top part with the molecule 1 showed a small bend. Then, the strip was illuminated with a combination of ultraviolet and visible light, resulting in rapid and strong bending over the whole length, assuming a final state completely bent into the light. With only ultraviolet illumination, the top part of the film relaxed and only the bottom part (with molecule 2) of the film was bent. With the light off, the film completely relaxed back to its original position. Thereby, the motion of natural cilia was mimicked.

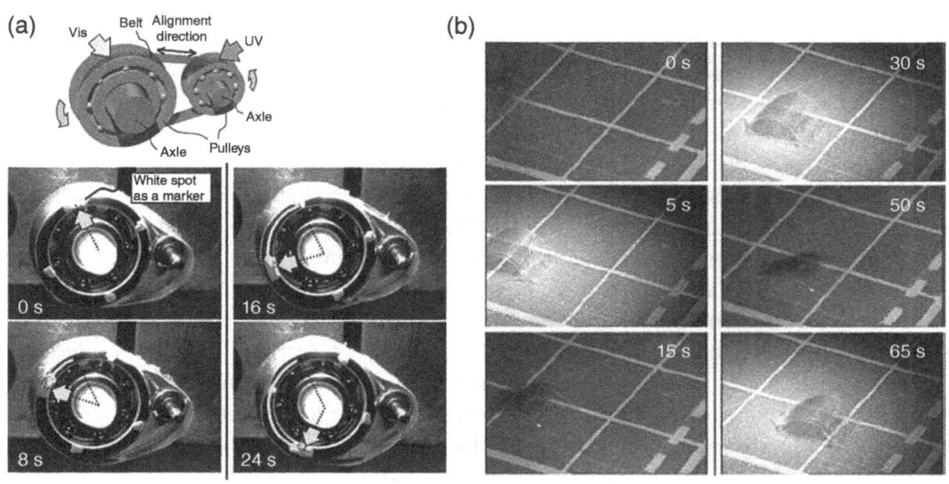

Figure 14.21 Three-dimensional motions of photomobile materials composed of bilayer structures of azobenzene CLCP and polyethylene layers. (a) Rotation of a light-driven plastic motor induced by simultaneous irradiation of UV and visible light. Size of the film: 36 mm × 5.5 mm. Thickness of the layers: PE, 50 mm; CLCP, 18 mm. (b) Series of photographs showing time profiles of the photoinduced inchworm walk of the CLCP-laminated film by alternate irradiation with UV (366 nm, 240mW/cm^2) and visible light (>540 nm, 120mW/cm^2) at room temperature. The film moved on the plate with 1 cm × 1 cm grid. *Source:* Reprinted with permission from [15, 69].

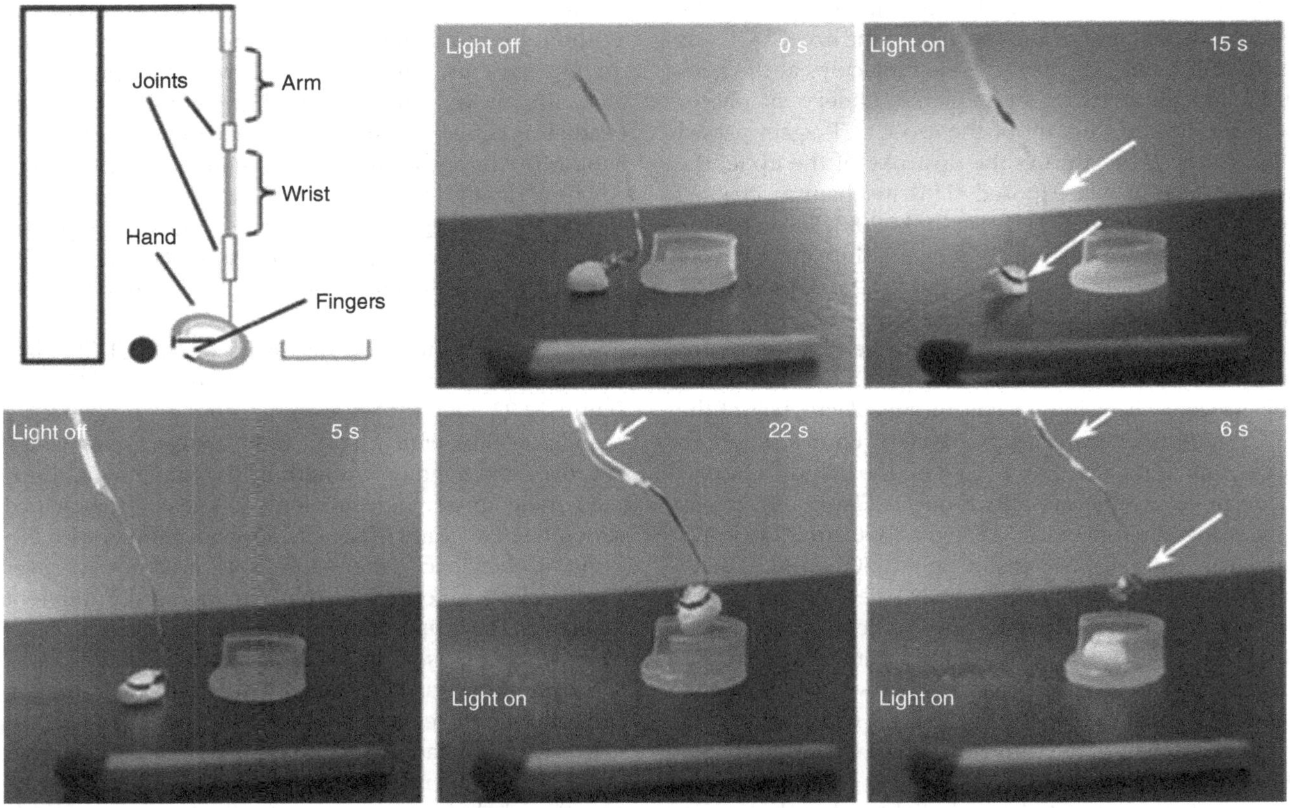

Figure 14.22 Schematic illustrations of the states of the microrobot during the process of manipulating the object. Photographs showing the microrobot picking, lifting, moving, and placing the object to a nearby container by turning on and off the light (470 nm, 30 mW/cm^2). Length of the match in the pictures: 30 mm. Thickness of PE and CLCP films: 12 mm. Object weight: 10 mg. *Source:* Reprinted with permission from [13].

Wiersma et al showed the walker motion of CLCP films [71, 72]. They employed a direct laser writing system to pattern the complex 3D CLCP structures with sub-micrometer resolution. Microrobots were therefore fabricated with the CLCPs acting as the walker's main body [72]. The walker's leg had a conical shape which was chosen to reduce the surface contact area, while 45° tilt of the leg created the adhesion asymmetry necessary for walking (Figures 14.24a, b). The artificial creature automatically performed various locomotion highly dependent on the interactions with the environment. The microscopic walker finished random or directional walking, rotation or jumping when placed on surfaces with different treating methods (Figure 14.24c).

Nature uses molecular-scale machines to drive every significant biological process and powers macroscopic mechanical motion in plants in highly complex processes. Examples of biological systems built on helical motion include powerful engines such as spasmoneme springs, seed pod opening, and tendril coiling. This study is based on the general concept that such plant-like helical deformations may similarly occur in artificial systems, which thus makes them capable of producing mechanical work.

Besides the bending behavior, the coiling movement of CLCPs in response to light was also reported. Based on a combination of a nematic LC and a small amount of chiral dopant, Fletcher et al. succeeded in a spring-like motion of the CLCP toward mimicking the mechanical behavior of plant tendrils (Figure 14.25) [16]. A small amount of chiral dopant S-811 was added into the mixture to induce a left-handed twist. As Figure 14.25a shows, the orientation of the LC molecule changes smoothly by 90° from the bottom surface to the top surface. The ribbons always deformed to accommodate the preferred distortion along the main axis of the ribbon, and this preferred distortion was determined by the orientation of the molecules, which was, in turn, determined by the cutting direction. With UV light irradiation, left-handed spiral ribbons B decreased in their macroscopic pitch, and the corresponding right-handed ribbons D worked just the opposite. The mixed-helicity springs comprising two opposite-handed helices displayed unwinding and winding motion simultaneously

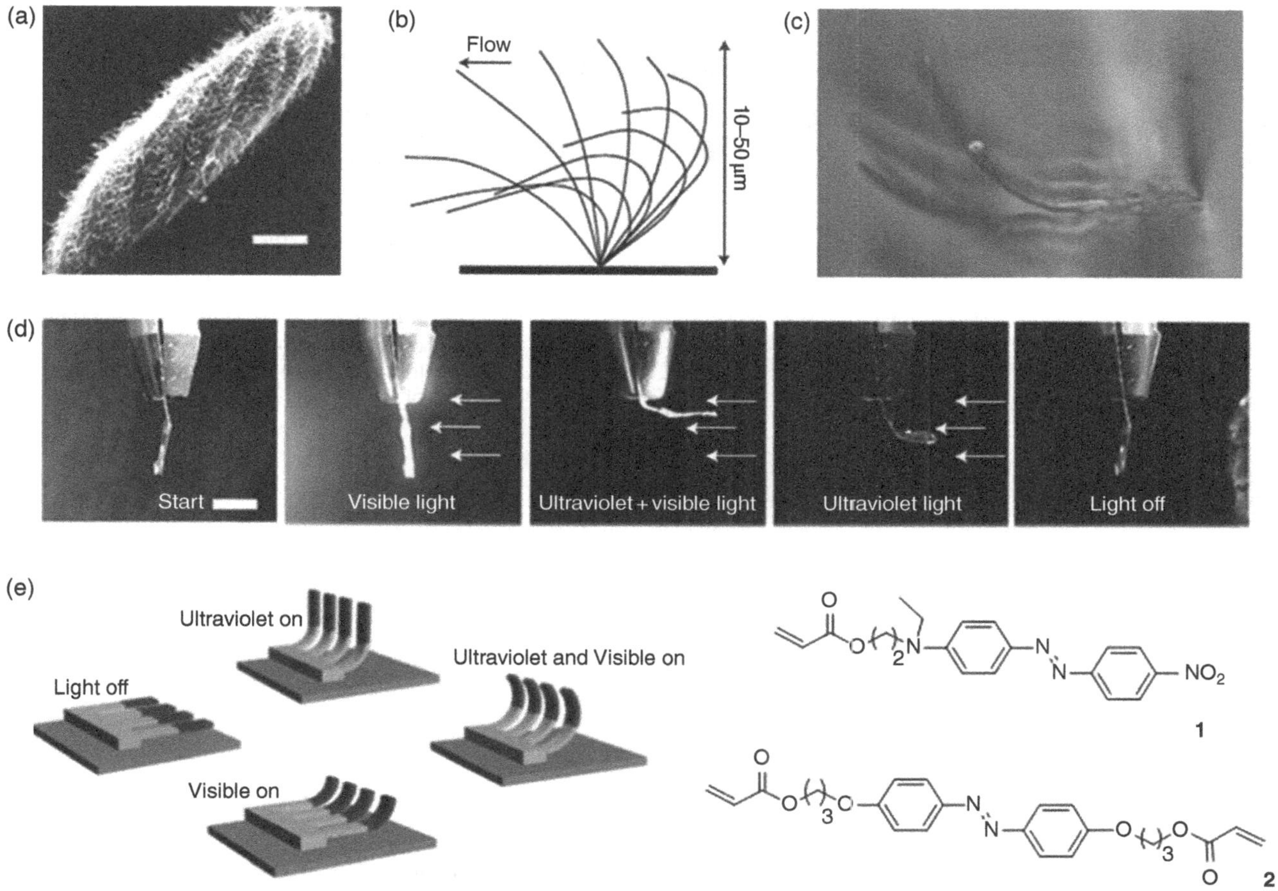

Figure 14.23 (a) An example of paramecia (scale bar 20 μm). (b) A paramecium uses the beating motion of the cilia, characterized by different forward and backward strokes, for self-propulsion. (c) Side view of the actuation of polymer cilia with ultraviolet light (1 W/cm^2) in water. (d) Response of a 10-μm-thick, 3-mm-wide, 10-mm-long polymer sample oriented in the splay-bend molecular organization through the thickness of the film to different colours of light (scale bar 5 mm). (e) Schematic illustration of an asymmetric motion of artificial cilia produced by controlling the wavelength of light. The sample is built with polymer-containing azobenzene 1 at the top (absorbing at < 390 nm) and polymer-containing azobenzene 2 at the bottom (absorbing in the 455–550 nm range). *Source:* Reprinted with permission from [14].

under UV irradiation, which successfully mimicked the movements of plant tendrils. (Figures 14.25b, 14.25c). These versatile actuators constitute a solid basis for designing novel functional materials that can provide an economic, clean and electrode-less alternative to conventional actuators.

14.4.3 Bioinspired Light-responsive Microstructured CLCP Surfaces

Control of the surface topology in the micrometer range by active switching between two states is of interest in a wide variety of applications [73–75]. Liu and Broer recently showed that artificial fingerprints can be switched between a flat state and a protruding state (Figure 14.26) [75]. The orientation of the helix axes is brought parallel to the substrate surface as schematically shown in Figure 14.26a. When the film was exposed to UV light the azobenzene underwent its *trans* to *cis* isomerization and decreased the order parameter of the LCs. At the points of planar anchoring, the surface formed protrusions while at the homeotropic positions it formed wells, as shown by confocal microscopy (Figure 14.26d). The resulting friction changes allowed use in a robot-like gripper. Mimicking robotic fingers they built a gripper that released its load by switching on the fingerprints. In addition to the use of dynamic friction in controlling grip and release, this could lead to versatile applications, such as a tracking light in a solar energy system, liquid mixing or flow regulating in microfluidic channels, and cell growth in biomedical applications.

Many biological systems are distinguished by their special superhydrophobic adhesion properties, such as

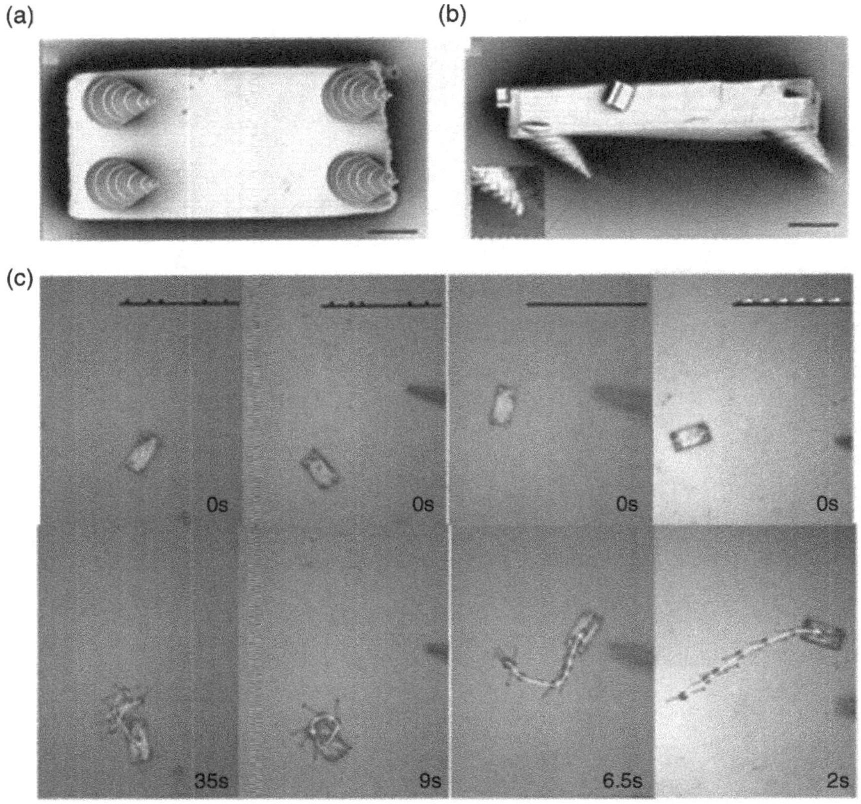

Figure 14.24 (a) SEM image of a microwalker lying upside down. Scale bar: 10 μm. (b) Side view of the microwalker with 500 nm leg tip shown in the inset. Scale bar: 10 μm. (c) Top row shows the initial state of microwalkers on different surfaces. Bottom row shows the microwalker randomly walking on the polyimidecoated glass surface, rotating with one leg stuck onto the polyimide-coated surface, walking with self-reorientation on the clean glass surface, walking in the direction determined by the grating groove pattern (vertical). Insets of the top row show the schematics of the surface. *Source:* Reprinted with permission from [72].

the superhydrophobic lotus leaf with ultralow water adhesion, rose petals that are superhydrophobic but stick water to them, rice leaves and butterfly wings with anisotropic water adhesion [76]. Surface wettability and adhesion are the paramount properties of solid surfaces, which are related closely to both surface chemical composition and morphological roughness. Under light irradiation, the chemical configuration of azobenzene changes between two states, with which the molecular polarity and surface free energy change accordingly, leading to a transition of surface wettability. During the light-induced transformation, various physical properties also change, such as surface microstructures and nanostructures, which are important in fabricating surfaces with unique wettability.

Therefore, the "rolling" and "pinning" of water droplets were achieved, giving rising to promising applications in microfluids. This is the first time that the photoresponsive CLCP materials were used to prepare superhydrophobic adhesion switchable surfaces, which is of great importance for no-loss microdroplet transfer.

Yu et al. designed and fabricated a light-responsive adhesion switch on a micro-arrayed azobenzene CLCP superhydrophobic surface, by which water droplets could be rapidly, precisely controlled [19]. Unlike contact angle (CA) switchable surfaces systems, our work put emphasis on the switching of sliding angle (SA) on the same surface, while the static CAs before and after switching were all in the superhydrophobic range. To introduce uniform and quantitatively controllable surface roughness to the azobenzene CLCP film, PDMS-soft-template-based secondary replication was utilized (Figure 14.27a). Benefiting from the wide and near-room-temperature nematic phase window of the mixed precursor used to prepare the azobenzene CLCP, the melt monomers were easily and sufficiently filled into the microgaps of the soft template and the azobenzene CLCP film were well replicated. Two sets of square-arrayed 10 mm^2 posts with space 5 mm (D5) and 15 mm (D15) were chosen (Figure 14.27b). For a surface with D5, although the water droplets stayed in a superhydrophobic state, the SAs in *trans* and *cis* states were both greater than 90 °, thus the adhesion change induced by photo-isomerization cannot be reflected.

It was found that a photo-controllable switch was successfully achieved with a 2 mL water droplet on surface with D15. After UV light (365 nm) irradiation for 6 s, the azobenzene mesogens at the surface of MA-F0208-D15 transferred to a *cis* state and the adhesion increases because of the growth of surface polarity. In this case, the surface showed a higher SA (a high water adhesion) and the 2 mL water droplet in the superhydrophobic state

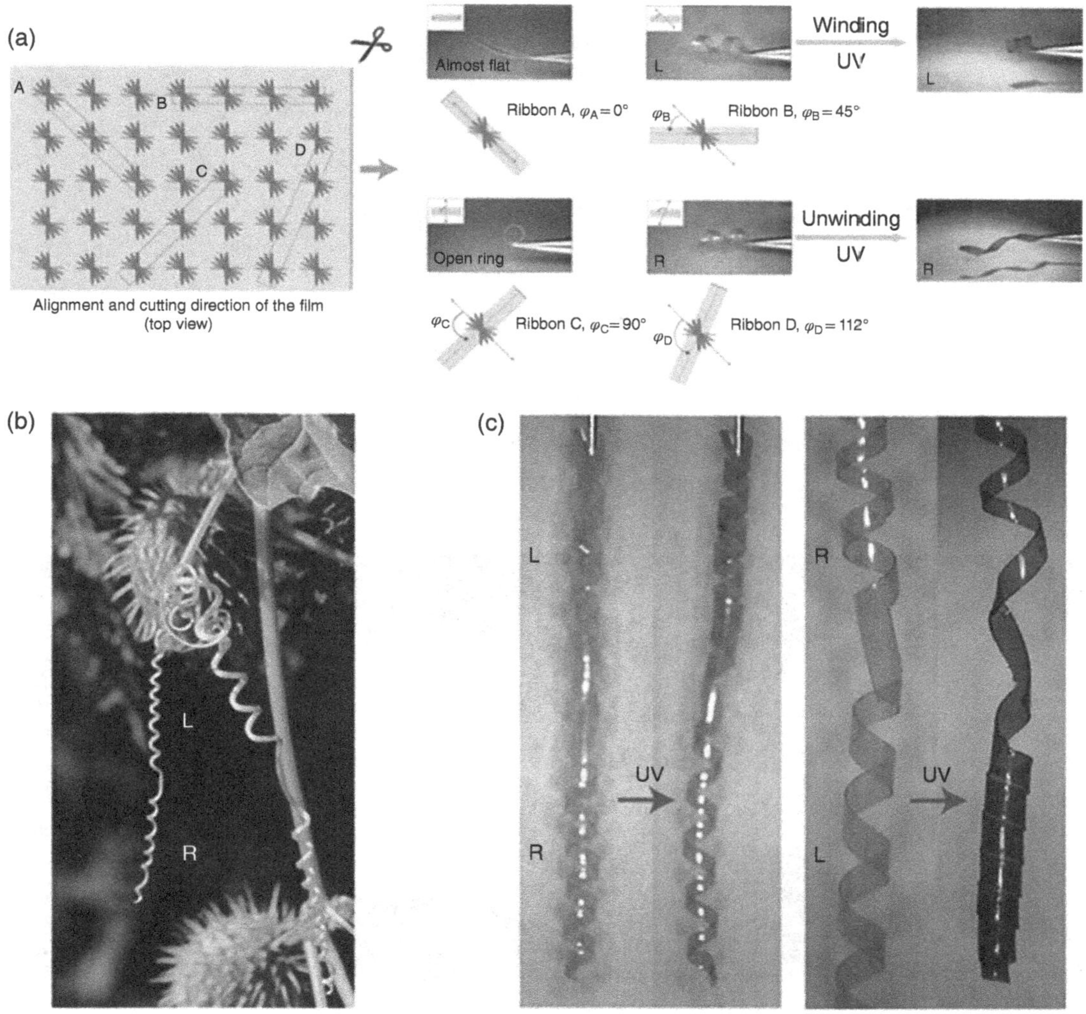

Figure 14.25 (a) The CLCP ribbons display a variety of shapes that depend on the direction in which they are cut. (Ribbon A: flat; Ribbon B: left-handed; Ribbon C: open ring; Ribbon D: right-handed). Ribbon B and Ribbon D irradiated for two minutes with UV light display winding and unwinding. (b) A coiled tendril of the wild cucumber plant. (c) A CLCP spring that displays a cucumber tendril-like shape, composed of two opposite-handed helices. On irradiation, the right-handed helix unwinds and the left-handed helix winds. *Source:* Reprinted with permission from [16].

was pinned to the surface (Figure 14.27c). Subsequently, after the visible light (530 nm) irradiation for 30 s, the azobenzene mesogens at the surface returned to a *trans* state and the surface reverted to the lower adhesive superhydrophobic state, thus the 2 mL water droplet could roll off the surface when the substrate was tilted more than 67.7 °.

Furthermore, Yu et al. demonstrated an *in situ* fully light-driven switching of superhydrophobic adhesion based on simply spin-coating a hydrophobic azo-polymer on an optimized micro-nanopost arrayed silicon substrate [77]. After UV light irradiation, a 2 μL water droplet can be pinned on the micro-nanopost array. Then, after visible light irradiation for about 10 s, the pinned droplet rolled off. Also, the micro/nanostructures play an important role in achieving the *in situ* fully light-driven switching of superhydrophobic adhesion. Switchable wettability on photoresponsive surfaces holds great promise in designing and developing bioanalytical methods, environmental clean-up, microfluidic devices, biochips, smart windows, controllable drug release, and intelligent devices.

The size of the arrays in CLCP films with different surface topographies (pillars and cones) was further decreased to nanoscale through a colloidal lithography technique by modulating different types of etching masks (Figure 14.28) [78]. Interestingly, these two species of films with the same chemical structure represented completely different wetting behavior of water adhesion and mimicked a rose petal and a lotus leaf,

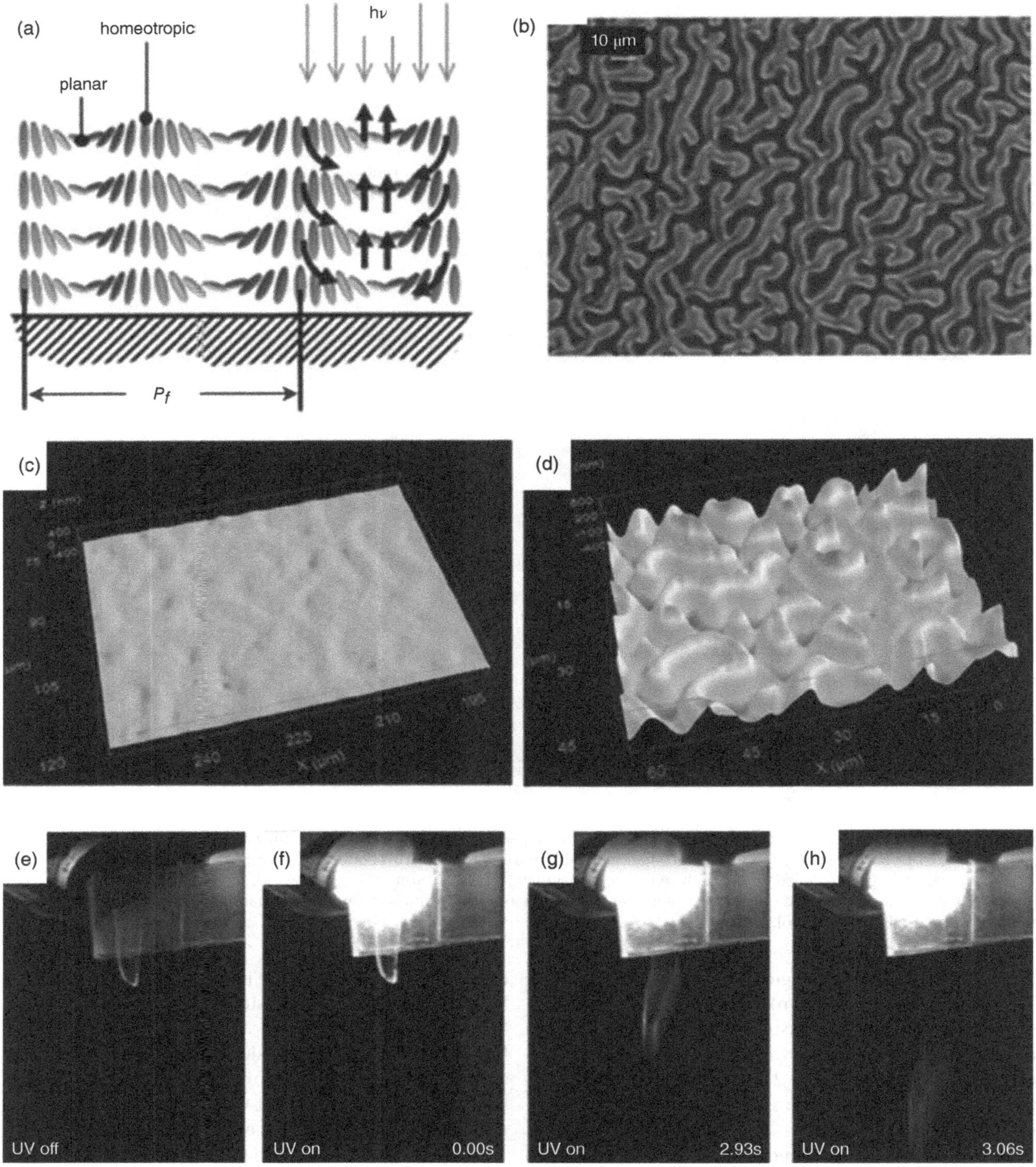

Figure 14.26 (a) Schematic representation of the dynamics of the fingerprints. (b) Polarized optical microscopy images of a fingerprint texture as observed between crossed polarizers. Bright regions correspond to planar and black areas to homeotropic orientation. (c) Confocal microscopic 3D image of the initial flat state of fingerprints. (d) 3D image of surface topographies under UV exposure. (e–h) Upon UV illumination, snapshots of a gripper that releases an object by changing friction of fingerprints. *Source:* Reprinted with permission from [75].

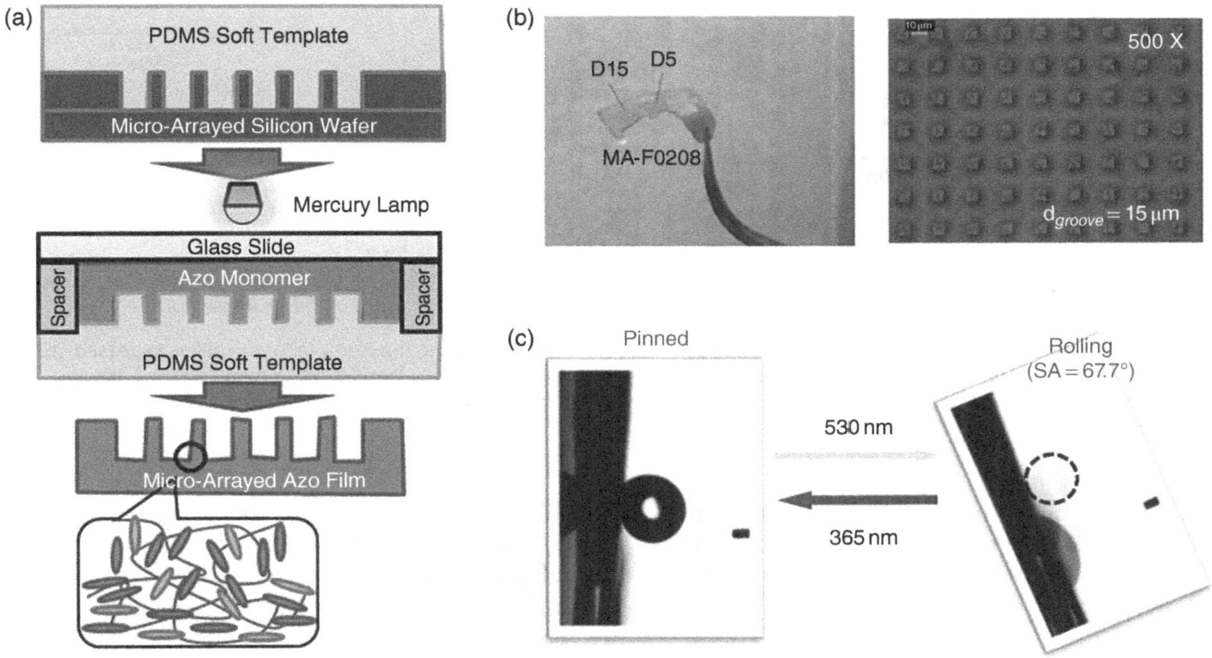

Figure 14.27 (a) Schematic illustration of the PDMS-soft-template-based secondary replication process to prepare the micro-arrayed azobenzene LCP film. (b) Optical photo of MA-F0208 with two patterned areas named D15 and D5. Large-area optical microscopic image and local amplified image (inset) of MA-F0208-D15. The patterns of D15 and D5 are all square-arrayed square posts with the post width of 10 mm. The spacings between two nearest posts for D15 and D5 are 15 mm and 5 mm, respectively. (c) Microscopic profiles of the rolling and pinned 2 mL water droplets. *Source:* Reprinted with permission from [19].

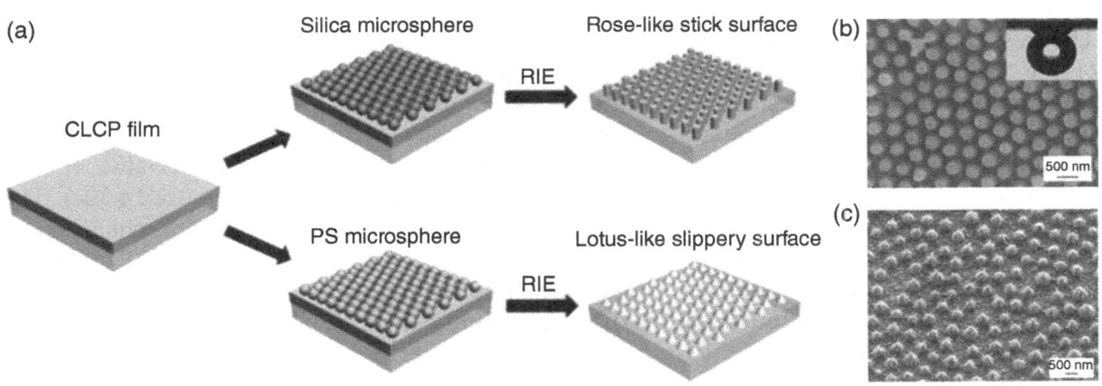

Figure 14.28 (a) Schematic procedure of the fabrication of microarrayed CLCP films. (b) SEM image of submicropillar arrayed CLCP film. (b, inset) The shape of a water droplet on the CLCP film when it is turned upside down, indicating its high water adhesion. (c) SEM image of the surface of submicrocone arrayed CLCP film. *Source:* Reprinted with permission from [78].

respectively. Both the submicropillar-arrayed film and the submicrocone-arrayed film exhibited superhyrophobicity, while the former demonstrated a very high SA greater than 90 ° which could pin the water droplet like a rose petal and the latter showed a small SA of only 3.1 ± 2.0 °. Compared to a replica molding technique and inkjet printing technology used to fabricate microstructured CLCPs, colloidal lithography technique is time-saving and can be modulated throughout the etching procedure, which finely regulates the structural parameters such as shapes and dimensions. Our work provides a new way to fabricate the CLCPs at the size of nanoscale.

In the biological world, numerous creatures such as butterflies, insects, and birds have exploited photonic structures to produce colorful reflections with important biofunctions in addition to unique brilliant structural coloration. These responsive photonic crystals have important applications in areas such as color displays,

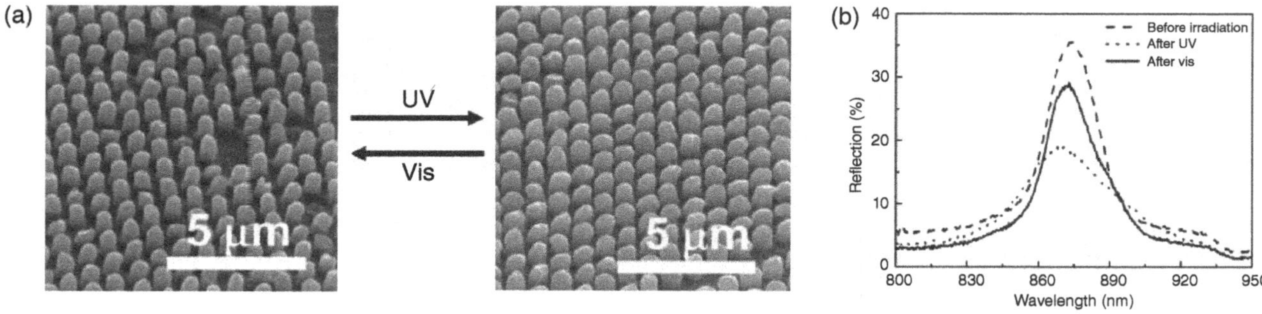

Figure 14.29 (a) SEM images of the azobenzene CLCP microarray. After irradiation of UV light, the diameter of the pillars increased. (b) Reflection spectra of the azobenzene CLCP microarray under the UV light irradiation (365 nm, 20 mW/cm^2, 15 min) and the following visible light irradiation (530 nm, 20 mW/cm^2, 5 min) with the angle of incidence of 60 °. *Source:* Reprinted with permission from [81].

biological and chemical sensors, inks and paints, and many optically active components [79]. For example, tunable photonic crystals may be used in the future as optical switches for the full automation of optical circuits [80].

Due to the light-responsive deformations, CLCPs have recently been considered a good candidate for future photonic crystals. Yu et al. fabricated a photo-responsive 2-D microarray with a period of about 1 μm using CLCP containing azobenzene groups by using the replica molding technique [81]. The CLCP microarray showed switchable behavior on the reflection spectra by alternate irradiation of UV and visible light, accompanied by the deformation of the CLCP pillars (Figure 14.29). Because the *trans-cis* photo-isomerization resulted in the alignment change of LC 14. especially in the side surface region of pillars, the pillars contracted along the long axis while they underwent expansion along the short axis, leading to the increase in the diameter of the pillars and variation in reflection spectra of the CLCP microarray. This was the first time CLCP had been used to fabricate the microarray with a period of about 1 μm and manipulate switchable behavior on the reflection spectra of the LC polymer microarray by light.

Compared with 2D photonic crystals, inverse opal materials have been demonstrated to exhibit excellent photonic crystal properties. Yu et al. demonstrated a light/thermal, dual responsive, inverse opal film fabricated from an azobenzene CLCP [21]. After infiltrating the ordered assembly of silica spheres with a precursor, the freestanding CLCP film was obtained by solidification and removing the SiO2 template (Figure 14.30a). The inverse opal films had good repeatability and different responsiveness. It was found that the reflection spectra of the inverse opal films showed switchable behavior when irradiated with UV light or heated. This phenomenon was ascribed to the change in the order of the periodic structure of the holes.

As shown in Figure 14.30b, a large amount of light was reflected at 677 nm wavelength and cannot pass through the CLCP inverse opal film in the initial state. When the film was exposed to UV light (365 nm, 50 mW/cm^2), the intensity of the reflection peak at 677 nm sharply declined. During this process, the maximum intensity of the reflection peak changed from 90% to 20%. This phenomenon in the CLCP inverse opal film was caused by the photo-induced deformation of the CLCP. The periodic holes in the inverse opal film underwent a change from regular to irregular in response to UV light. The intensity of the reflection peak at 677 nm sharply also declined with the increase of the temperature. When the film was heated to 90 °C, the reflection peak at 677 nm almost disappeared. Then the intensity of the reflection peak gradually increased and reached the maximum value of 70% when the temperature decreased (Figure 14.30c). The reflectivity of the inverse opal CLCP films changed to a greater extent compared to the 2D CLCP photonic crystals. Improving bioinspired photonic materials that cam mimic the reverse color-order diffraction found in the butterfly will increase our basic understanding of the optical interplay of hierarchical structures and provide a platform for the development of novel responsive photonic devices.

14.4 Conclusion

In the past decade, a great number of biomimetic devices have been developed by using CLCPs. The mechanical response of these CLCPs could yield various responsive soft actuators or serve as bionic substrates for multifunctional flexible devices in aerospace, lab-on-chip, or flexible electronic device. In order to better take advantage of CLCPs in specific applications, suitable designs and functions are required, for example, combining CLCPs with

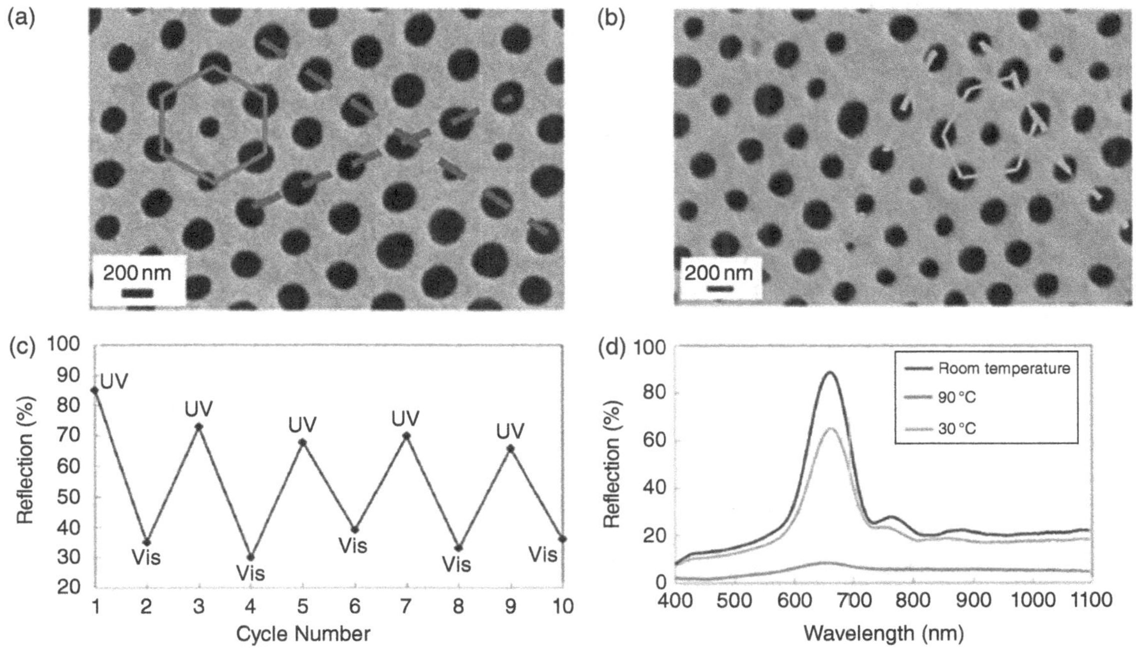

Figure 14.30 SEM images of the inverse opal film (a) before and (b) after UV light irradiation. (c) The maximum intensity of the reflection peak of the inverse opal film under alternating irradiation with UV and visible light at different times. (d) Reflection spectra of the inverse opal film as a function of temperature. *Source:* Reprinted with permission from [21]. (*See insert for color representation of the figure.*)

other polymers to improve the mechanical properties as well as reduce costs. In addition, new preparation techniques are also required to expand applications in micro/nano and complex structured devices, such as 3D printing and electrospinning. Researchers need to combine diverse CLCPs into micro-mechanical systems or microfluidic system. Only in this way will unexpected and yet unimagined functions be discovered. The developed functions would pave the way to applying mechanical response of CLCPs in practical applications which could help guide future research and bridge the gap between fundamental research and practical applications.

References

1 Ube, T., Ikeda, T. (2014) *Angew. Chem.* **126**, 10456; *Angew. Chem. Int. Ed.*, **53**, 10253.
2 Fleischmann, E.-K., Zentel, R. (2013) *Angew. Chem.* **125**, 8972; (2013) *Angew. Chem. Int. Ed.* **52**, 8810.
3 Ohm, C., Brehmer, M., Zentel, R. (2010) *Adv. Mater.*, **22**, 3366.
4 Yu, Y., Ikeda, T. (2006) *Angew. Chem.*, **118**, 5542; (2006) *Angew. Chem. Int. Ed.* **45**, 5416.
5 White, T., Broer, D. (2015) *Nat. Mater.*, **14**, 1087.
6 Ohm, C., Kapernaum, N., Nonnenmacher, D. et al. (2011) *Am. Chem. Soc.*, **133**, 5305.
7 Warner, M., Terentjev, E.M. (2003) *Liquid Crystal Elastomers*, Oxford University Press, Oxford.
8 Ikeda, T., Nakano, M., Yu, Y. et al. (2003) *Adv. Mater.*, **15**, 201.
9 Wei, J., Yu, Y. (2012) *Soft Matter*, **8**, 8050.
10 Liu, Y., Yu, Y. (2013) *Chin. J. Nat.*, **35**, 127.
11 Kohlmeyer, R.R., Chen, J. (2013) *Angew. Chem.*, **125**, 9404; (2013) *Angew. Chem. Int. Ed.*, **52**, 9234.
12 White, T., Tabiryan, N., Serak, S., Hrozhyk, U. (2008) *Soft Matter*, **4**, 1796.
13 F. Cheng, R. Yin, Y. Zhang, C.-C. Yen, Y. Yu, (2010) *Soft Matter*, **6**, 3447.
14 van Oosten, C.L., Bastiaansen, C.W.M., Broer, D. (2009) *Nat. Mater.*, **8**, 677.
15 Yamada, M., Kondo, M., Miyasato, R., et al. (2009) *J. Mater. Chem.*, **19**, 60.
16 Iamsaard, S., Asshoff, S.J., Matt, B., et al. (2014) *Nat. Chem.*, **6**, 229.
17 Wu, Z., Wei, R., Buguin, A. et al. (2013) *ACS Appl. Mater. Interfaces*, **5**, 7485.
18 Wu, Z., Buguin, A., Yang, H. et al. (2013) *Adv. Funct. Mater.*, **23**, 3070.
19 Li, C., Cheng, F., Lv, J. et al. (2012) *Soft Matter*, **8**, 3730.

20 Jiang, Y., Xu, D., Li, X. et al. (2012) *J. Mater. Chem.*, **22**, 11943.
21 Zhao, J., Liu, Y., Yu, Y. (2014) *J. Mater. Chem., C*, **2**, 10262.
22 de Gennes, P.G., Acad, C.R. (1975) *Sci. Paris Ser. B.*, **281**, 101.
23 Zhao, W., Lin, B., Zhang, X. et al. (2015) *Chin. J. Polym. Sci.*, **33**, 1431.
24 Wermter, H., Finkelmann, H. (2001) *e-Polymers*, **13**.
25 Schätzle, J., Kaufhold, W., Finkelmann, H. (1989) *Makromol. Chem.*, **190**, 3269.
26 Sawa, Y., Ye, F., Urayama, K. et al. (2011) *PNAS*, **108**, 6364.
27 Pei, Z., Yang, Y., Chen, Q. et al. (2014) *Nat. Mater.*, **13**, 36.
28 Forterre, Y., Skotheim, J.M., Dumais, J., Mahadevan, L. (2005) *Nature*, **433**, 421.
29 Kondo, M., Yu, Y., Ikeda, T. (2006) *Angew. Chem.*, **118**, 1406; (2006) *Angew. Chem., Int. Ed.*, **45**, 1378.
30 van Oosten, C., Harris, K., Bastiaansen, C., Broer, D. (2007) *Eur. Phys. J. E.*, **23**, 329.
31 Ware, T., McConney, M., Wie, J. et al. (2015) *Science*, **347**, 982.
32 Fleischmann, E., Zentel, R. (2012) *Nat. Comm.*, **3**, 1178.
33 Schuhladen, S., Preller, F., Rix, R. et al. (2014) *Adv. Mater.* **26**, 7247.
34 Petsch, S., Rix, R., Reith, P. et al. (2014) *in IEEE 27th Int. Conf. Micro Electro Mech. Syst,*. pp. 905–908.
35 Wen, L., Tian, Y., Jiang, L. (2015) *Angew. Chem.*, **127**, 3448; (2015) *Angew. Chem. Int. Ed.*, **54**, 3387.
36 Ge, J., Yin, Y. (2011) *Angew. Chem.*, **123**, 1530; (2011) *Angew. Chem. Int. Ed.*, **50**, 1492.
37 Zhao, Y., Xie, Z., Gu, H. et al. (2012) *Chem. Soc. Rev.* **41**, 3297.
38 Wu, G., Jiang, Y., Xu, D. et al. (2011) *Langmuir*, **27**, 1505.
39 H. Xing, J. Li, J. Guo, J. Wei, (2015) *J. Mater. Chem. C*, **3**, 4424.
40 Ji, Y., Huang, Y., Rungsawang, R., Terentjev, E.M. (2010) *Adv. Mater.* **22**, 3436
41 Yang, L., Setyowati, K., Li, A. et al. (2008) *Adv. Mater.* **20**, 2271.
42 de Haan, L., Sanchez-Somolinos, C., Bastiaansen, C. et al. (2012) *Angew. Chem.*, **124**, 12637; (2012) *Angew. Chem. Int. Ed.*, **51**, 12469.
43 Li, C., Liu, Y., Huang, X., Jiang, H. (2012) *Adv. Funct. Mater.* **22**, 5166.
44 Morin, S.A., Shepherd, R.F., Wai Kwok, S. et al. (2012) *Science*, **337**, 828.
45 Ahir, S., Tajbakhsh, A., Terentjev, E. (2006) *Adv. Funct. Mater.* **16**, 556.
46 Sun Y., Evans, J.S., Lee, T. et al. (2012) *Appl. Phys. Lett.* **100**, 241901.
47 Liu, X., Wei, R., Hoang, P.T. et al. (2015) *Adv. Funct. Mater.* **25**, 3022.
48 Yu, Y., Ikeda, T. (2009) *Smart Light-responsive Materials*, John Wiley & Sons, Hoboken, NJ.
49 Tazuke, S., Kurihara, S., Ikeda, T. (1987) *Chem. Lett.* **16**, 911.
50 Sung, J.-H., Hirano, S., Tsutsumi, O. et al. (2002) *Chem. Mater.*, **14**, 385.
51 Finkelmann, H., Nishikawa, E., Pereira, G.G., Warner, M. (2001) *Phys. Rev. Lett.*, **87**, 015501.
52 Wang, W., Wang, X., Cheng, F. et al. (2011) *Prog. Chem.*, **23**, 1165.
53 Zhan, Y., Liu, Y., Lv, J. et al. (2015) *Prog. Chem.*, **27**, 157.
54 Ikeda, T., Mamiya, J., Yu, Y. (2007) *Angew. Chem.*, **119**, 512; (2007) *Angew. Chem. Int. Ed.*, **46**, 506.
55 Yu, Y., Nakano, M., Shishido, A. et al. (2004) *Chem. Mater.*, **16**, 1637.
56 Yu, Y., Nakano, M., Ikeda, T. (2003) *Nature*, **425**, 145.
57 Wang, W., Sun, X., Wu, W. et al. (2012) *Angew. Chem.*, **124**, 4722; (2012) *Angew. Chem. Int. Ed.*, **51**, 4644.
58 Sun, X., Wang, W., Qiu, L. et al. (2012) *Angew. Chem.*, **124**, 8648; (2012) *Angew. Chem. Int. Ed.*, **51**, 8520.
59 Wu, W., Yao, L., Yang, T. et al. (2011) *J. Am. Chem. Soc.*, **133**, 15810.
60 Jiang, Z., Xu, M., Li, F., Yu, Y. (2013) *J. Am. Chem. Soc.*, **135**,16446.
61 Chen, M., Zhu, Y., Yu, Y. (2010) *Appl. Phys. A.* **100**, 39.
62 Chen, M., Zhu, Y., Yu, Y. (2011) *Appl. Phys. A.* **102**, 667.
63 Serak, S., Tabiryan, N., Vergara, R. et al. (2010) *Soft Matter*, **6**, 779.
64 Wu, Y., Ikeda, T., Zhang, Q. (1999) *Adv. Mater.* **11**, 4.
65 White, T., Serak, S., Tabiryan, N. et al. (2009) *J. Mater. Chem.*, **19**, 1080.
66 Lee, K., Smith, M.L., Koerner, H. et al. (2011) *Adv. Funct. Mater.* **21**, 2913.
67 Lee, K., Koerner, H., Vaia, R. et al. (2010) *Macromolecules*, **43**, 8185.
68 Lee, K., Tabiryan, N.V., Bunning, T., White, T. (2012) *J. Mater. Chem.*, **22**, 691.
69 Yamada, M., Kondo, M., Mamiya, J. et al. (2008) *Angew. Chem.*, **120**, 5064; (2008) *Angew. Chem. Int. Ed.*, **47**, 4986.
70 Cheng, F., Zhang, Y., Yin, R., Yu, Y. (2010) *J. Mater. Chem.*, **20**, 4888.
71 Zeng, H., Martella, D., Wasylczyk, P. et al. (2014) *Adv. Mater.* **26**, 2319.
72 Zeng, H., Wasylczyk, P., Parmeggiani, C. et al. (2015) *Adv. Mater.* **27**, 3883.
73 Liu, D., Bastiaansen, C., den Toonder, J., Broer, D. (2012) *Angew. Chem.*, **124**, 932; (2012) *Angew. Chem. Int. Ed.*, **51**, 892.
74 Liu, D., Bastiaansen, C., den Toonder, J., Broer, D. (2012) *Macromolecules*, **45**, 8005.
75 Liu, D., Broer, D. (2014) *Angew. Chem.*, **126**, 4630; (2014) *Angew. Chem. Int. Ed.*, **53**, 4542.

76 Wang, S., Song, Y., Jiang, L. (2007) *J. Photochem. Photobiol. C*, **8**, 18.

77 Li, C., Zhang, Y., Ju, J. et al. (2012) *Adv. Funct. Mater.* **22**, 760.

78 Zhan, Y., Zhao, J., Liu, W. et al. (2015) *ACS Appl. Mater. Interfaces*, **7**, 25522.

79 Hou, J., Zhang, H., Yang, Q. et al. (2014) *Angew. Chem.*, **126**, 5901; (2014) *Angew. Chem. Int. Ed.*, **53**, 5791.

80 S. Kubo, Gu, Z., Takahashi, K. et al. (2002) *J. Am. Chem. Soc.*, **124**, 10950.

81 Yan, Z., Ji, X., Wu, W. et al. (2012) *Macromol. Rapid Comm.*, **33**, 1362.

15

Tuning Mechanical Properties of Protein Hydrogels: Inspirations from Nature and Lessons from Synthetic Polymers

*Xiao-Wei Wang, Dong Liu, Guang-Zhong Yin, and Wen-Bin Zhang**

College of Chemistry and Molecular Engineering, Peking University, Beijing, China

15.1 Introduction

Hydrogels are three-dimensional polymeric networks containing large amounts of water [1]. While gelatins have been around in human life for thousands of years, they were not well understood until the emergence of the macromolecular hypothesis [2] introduced the idea of network formation in terms of "infinite molecular weight" and "percolation" [1]. It took several more decades for gels to evolve from simple viscoelastic solids into contact lenses, to superabsorbent materials, and further into "smart" hydrogels with phase transitions [3]. Since their compositions and physical states are similar to most biological tissues, hydrogels have been intensively studied as an excellent platform and matrix for engineered functions in the biological context [4]. For example, in living tissues, cells reside in a 3D extracellular matrix with heterogeneous compositions but well-defined spatiotemporal distribution of mechanical properties and biological cues [4]. Such highly sophisticated hierarchical structures enable the control of cell fate. Mimicking natural materials is being actively pursued [5]. With the development of the hydrogel theories and the increasingly powerful polymer chemistry, it is no longer uncommon to see synthetic hydrogels with excellent mechanical properties that rival their natural counterparts [6]. The diversity in terms of monomer choice, molecular weight control, topology variation, and post-polymerization functionalization offers a versatile platform for engineering hydrogel properties. Nevertheless, there remains a significant gap between synthetic hydrogel and natural extracellular matrices, especially in terms of bioactivity.

The rise of recombinant protein-based hydrogels in the 1990s was a major breakthrough [7–9]. Proteins are known as the workhorse of life [10]. They serve as the major structural component of the extracellular matrix; they are capable of performing diverse defined biological functions; and they interact with cells to control their behavior and direct their fate. Notably, protein production in cells is perhaps the most precise and programmable polymer chemistry process so far. The sequences, the length, the stereochemistry, and even folded structures can be precisely controlled and varied based on the genetic information, offering enormous opportunities for engineering their functions. In other words, proteins are perhaps the only material that is genetically encoded and programmable. Their structures and properties are, of course, still governed by the fundamental physical principles for polymers. Yet, their biological activities allow them to communicate with living systems, providing a unique access to biological mimicry of natural materials. Despite this advantage, the vulnerability of the protein structure (activity) and the usually low yield of protein production limit their practical application. This is in sharp contrast to synthetic materials' robust stability. Thus, synthetic polymer hydrogels and protein hydrogels are complementary to each other in many aspects. We need creative approaches to translate the knowledge about synthetic polymer hydrogels into protein hydrogels to achieve a synergy between the two systems for better control over its structure and function. The convergence of synthetic polymer and protein hydrogels is anticipated to create next-generation biomaterials. In this chapter, we draw a parallel comparison between these different systems and discuss ways to integrate them for designer biomaterials (Figure 15.1). Apparently, there are a lot of things that they can learn from each other and there is still plenty of room for improvement.

* Email: wenbin@pku.edu.cn

Bioinspired Materials Science and Engineering, First Edition. Edited by Guang Yang, Lin Xiao, and Lallepak Lamboni.

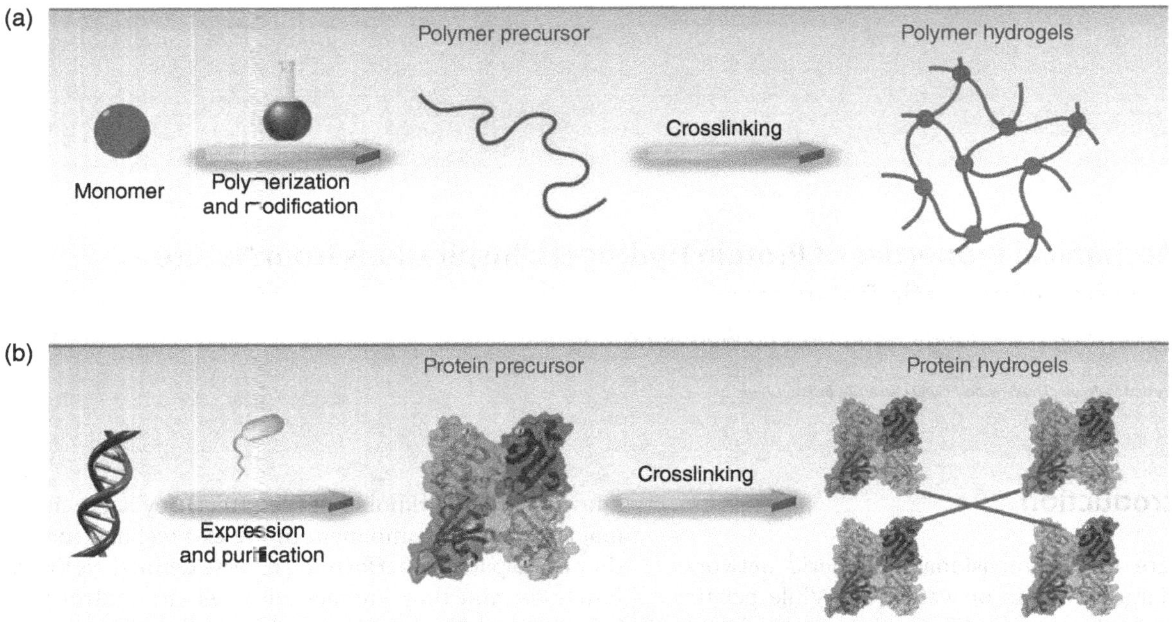

Figure 15.1 Schematic presentation of hydrogels based on synthetic polymers (a) and recombinant proteins (b).

15.2 What Are Different about Proteins?

Protein hydrogels are in principle one kind of polymer hydrogels. But there are many distinctive features that differentiate them from traditional synthetic polymer hydrogels. Here, we use the term "protein" to refer only to recombinant proteins with precise sequences and structures prepared by cellular synthesis. Proteins processed from natural resources are not included in this category. Polypeptides prepared by condensation polymerization or ring-opening polymerization are considered synthetic polymers since they usually are polydispersed. To understand the unique features of protein hydrogels, we would like first to compare synthetic polymers and recombinant proteins with the following general aspects.

15.2.1 Protein Structure and Function

Four grand challenges of polymer chemistry are found in the fine control of polymer length, sequence, stereochemistry, and topology [11]. Most synthetic polymers do not possess precise length and sequence, but rely on chain-length polydispersity and topology variation as effective means for tuning properties. Except in a crystalline state, synthetic polymers usually lack folded structures and are referred to as random coils since they are largely disordered. Their properties are thus more as a result of their collective interaction (such as molecular weight) but less as a result of their assembled structures. On the other hand, proteins exhibit a broad spectrum of order from intrinsically disordered proteins, to partially folded, to highly ordered and tightly folded structures (Figure 15.2a) [12]. Traditional wisdom has it that protein has to fold into certain hierarchical structures in order to be functional, as shown in globular proteins, such as enzymes, or in fibrous proteins with structural properties, such as collagens. Nowadays, it is increasingly recognized that the dynamic properties of protein structures also play an important functional role [13]. Different degrees of order may be favored by different purposes. Even for intrinsically disordered proteins (IDPs), there are actually many functional benefits that allow them to transit in between different functional states or interact with multiple ligands or to perform several tasks (moonlighting) [14]. The word "ordered" does not necessarily imply "function." The magic lies in the seemingly "random" placement of the amino acid building blocks with precise sequences. It is worth noting that while the function of proteins can diverge in a variety of directions, the structures of proteins are in fact strictly conserved during evolution. Most mutations only provide the fine tuning of their structure and substrate specificities [15, 16]. The folded structure can be preserved even though the whole amino acid sequence has changed up to 80% whereas sometimes only one single site mutation changes the entire folded structure. This interesting fact points to the importance of hierarchical physical structures and has significant implications for synthetic polymers. Although synthetic polymers can sometimes

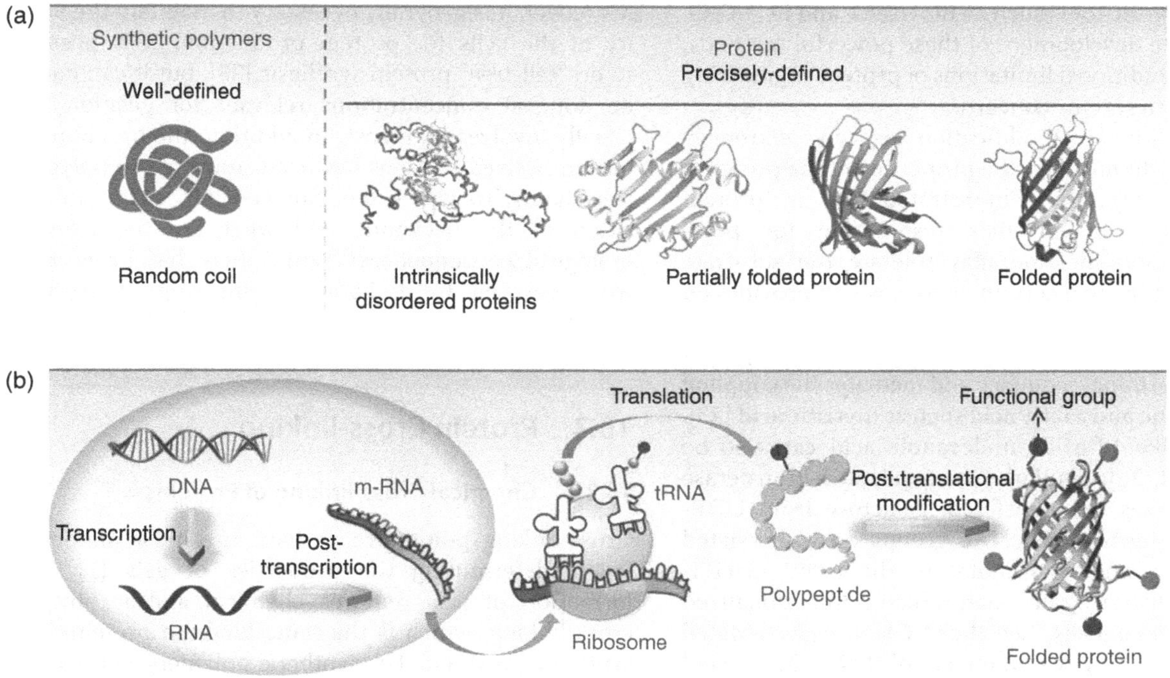

Figure 15.2 The unique features of protein (a) structure and (b) synthesis.

capture these features in simple models, as demonstrated in secondary structure formation and phase transitions, it is still not possible to design a meaningful monomer sequence and to prepare a synthetic functional polymer with precisely defined length and sequence using the state-of-the-art polymer chemistry. This problem can be circumvented by using genetic engineering and biosynthetic pathways of organisms to synthesize macromolecules with a predesigned structure and function. This method includes the biosynthesis of proteins but the principle can be extended to other genetically engineered macromolecules as well.

15.2.2 Protein Synthesis

For synthetic polymers, the formation of a polymer chain usually follows either a "step-growth" or a "chain-growth" mechanism [17]. Incorporation of multifunctional cross-linkers in both processes would change the final macromolecular architecture from linear to hyper-branched and further to a network. By contrast, the biosynthesis of proteins is a precise, consistent, and robust chain-growth process that is difficult to adapt and amend [10]. It uses the mRNA as the template and 20 natural amino acid building blocks to make the linear polypeptide which later folds into various structures (Figure 15.2b). To vary its composition and backbone topology, there are mainly two relatively well-established ways so far: (1) by incorporating noncanonical amino acids in either site- or residue-specific ways; or (2) by post-translational modification.

Site-specific introduction of unnatural amino acids uses stop (nonsense) codons or 4-base codons. Using a suppressor tRNA not recognized by the host aminoacyl-tRNA synthetases and a synthetase that can couple amino acid to this suppressor tRNA, unnatural amino acid analogues can be incorporated at a defined site of modification [18, 19]. The use of multiple mutually orthogonal aminoacyl-tRNA synthetases can further allow the incorporation of multiple unnatural amino acids at specific sites. This strategy is very useful for fine-tuning the functions of an enzyme or for interrogating the biological processes. Tirrell *et al.* have developed a method to globally substitute a particular amino acid with its structural analogues based on the promiscuity of the enzymes [20, 21]. The process would generally require a bacterial auxotroph and a media-shift procedure. Sometimes, it may require an over-expression of the tRNA synthetase to elevate its activity or even a directed evolution of tRNA synthetase to increase its tolerance for structural variants. A large variety of amino acid analogues have been successful incorporated. Due to the global substitution, it often leads to a distinct change in the overall physical properties, which is especially useful when these proteins are to be used as materials. This method has also been adapted to develop useful proteomtic tools such as BONCAT and FUNCAT [22]. With the development of these powerful methods, some of the traditional limitations of protein engineering are no longer a serious concern.

Post-translational modification is an extremely powerful way to diversify the properties of the proteins. Nature has elegantly demonstrated this in protein glycosylation. The enzymes responsible for post-translational modification may tolerate the substrate structural change to a certain degree, which provides an opportunity to introduce unnatural functional groups. For example, *N*-myristoyltransferase (NMT) recognizes a special *N*-terminal sequence and mediates the coupling between amine and a fatty acid such as myristic acid [23]. Surrogates like 12-azido-undecanoic acid can also be recruited [24, 25]. Similarly, β-1,4-galactosyltransferase (GalT) catalyzes the transfer of galactose from UDP-galactose to terminal GlcNAc groups on glycosylated proteins. By a single site mutation, the Y289L GalT is capable of transferring an unnatural ketone functionalized galactose to chemically "tag" the O-GlcNAc glycosylated proteins [26]. Using a mutant *E. coli* LplA (lipoic acid ligase), a foreign functional group can be covalently conjugated to the LplA acceptor peptide (LAP) genetically fused to the protein of interest, allowing site-specific incorporation of unnatural compositions at high specificity both *in vitro* and *in vivo*. This serves as the basis of the PRIME method [27]. Post-translational modification can also be used to change the backbone topology. The intein technology has been used to create cyclic proteins such as c-GFP [28]. Sortase-mediated ligation can also be used to prepare cyclic proteins [29–31]. More recently, the emergence of genetically encoded protein chemistry reactive pairs [32, 33], such as SpyTag and SpyCatcher [34], and SnoopTag and SnoopCatcher [35], offers an even more versatile platform for backbone topology engineering, giving rise to cyclic, tadpole, star, and even H-shape proteins with high efficiency [36]. With the aid of folding, the assembly-reaction synergy can further enable the cellular synthesis of protein catenanes [37]. In synthetic polymers, controlling the topology of the polymer backbone has proven useful in tuning the physical properties of the polymer. The ability to vary the protein backbone topology combines the advantage of both synthetic polymers and biopolymers, which is expected to be very powerful for tuning the composition and topology of proteins.

Despite such progress, protein synthesis remains an intricate design that hardly surrenders to any significant variation. In other words, we can use it, but we can hardly modify it. For example, it would be desirable to circumvent the start codon ATG and begin the peptide chain growth from a specific precursor (such as a peptide sequence), but that has not yet been demonstrated. Moreover, it is generally necessary to maintain the vitality of the cells for protein production. It is possible to do "cell-free" protein synthesis [38], but it can hardly be done at concentrations relevant for gelation and usually involves high cost. In addition, the backbone of biosynthesized proteins has been restricted to polypeptide chains. To what extent can we engineer the composition of the backbone and what are the relevant biological consequences? While there has been some work devoted to backbone engineering of artificial proteins [39], it remains to be accomplished *in vivo*.

15.3 Protein Cross-linking

15.3.1 Chemical Cross-linking of Proteins

Cross-linking points are perhaps the most important factor determining the properties of gels [1]. The formation of gels requires chemical and/or physical cross-linking, which is the same for both proteins and synthetic polymers. For synthetic polymers, gels can be conveniently made *in situ* from small molecule building blocks through chain growth polymerization along with multifunctional cross-linkers. As discussed previously, it is difficult to prepare proteins *in situ* at concentrations required for gelation or to incorporate a multifunctional amino acid cross-linker during expression. Hence, protein hydrogels are usually prepared with purified proteins following the step growth reactions (Figure 15.3a) [40]. This is also common for many synthetic polymers such as poly(ethylene glycol) (PEG), the gold standard for synthetic polymer hydrogels [41]. Most bioconjugation chemistry can be used to prepare these hydrogels [42]. However, most of them react with the functional groups already present on proteins (Figure 15.3b) [43]. Although it is convenient, it may not be very specific and could adversely affect the activity of proteins. It is highly desirable to use bio-orthogonal reactions that proceed rapidly and selectively under physiological conditions without interfering with the numerous functionalities present on proteins, which can be achieved either by using highly reactive functional groups typically not present in nature or by using unique amino acid sequences [44].

A straightforward way to achieve bio-orthogonality is to use highly reactive functional groups typically not present in nature. Prime examples include the condensation of ketones/aldehydes with amine (aminooxy or hydrazide compounds), the Staudinger ligation, the azide-alkyne cycloaddition, and other reactions involving alkenes [44]. The key to its application is the introduction of these unnatural functional groups into proteins, which is nontrivial and usually results in low protein yield, nonquantitative incorporation, and high cost. Typical

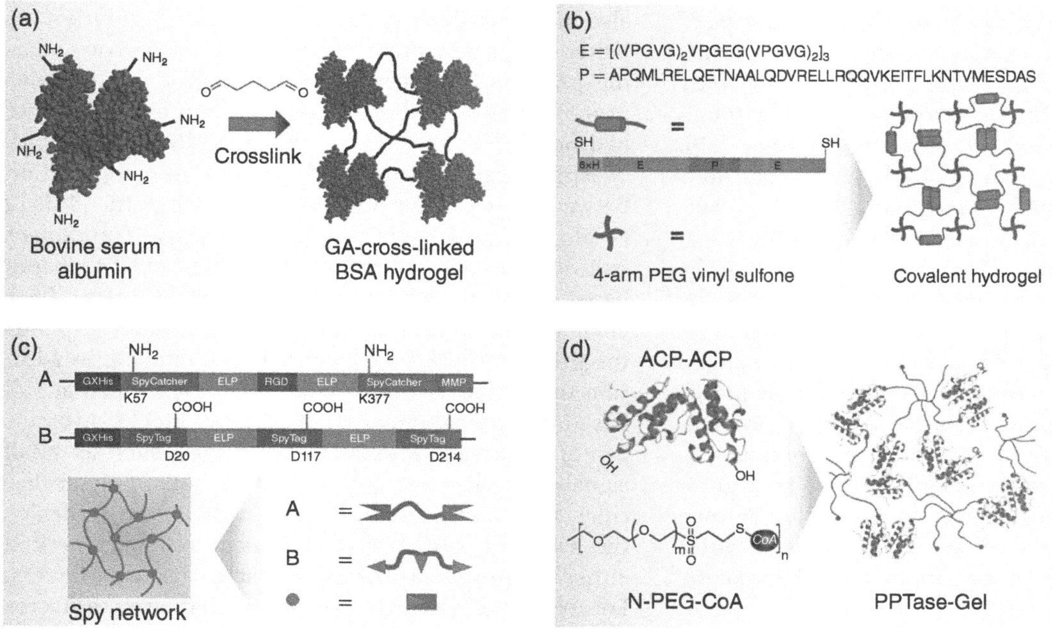

Figure 15.3 Chemical cross-linking of proteins. (a) The schematic of GA-cross-linked BSA hydrogel. (b) Artificial proteins EPE consist of terminal cysteine residues (-SH), elastin-like endblocks E, and the P midblock domain, and EPE requires covalent cross-linking with 4-arm PEG vinyl sulfone to form gels. (c) Spy network hydrogels. Top) Genetic constructs for the four protein precursors (A and B). Construct B contains an internal integrin binding RGD sequence to facilitate cell adhesion and an MMP-1 cleavage site at the C-terminus to allow matrix remodeling by encapsulated cells. Bottom) Schematic illustration of the covalently cross-linked gel formed by mixing protein precursors A and B. (d) Scheme of PPTase-mediated cross-linking of PEG hydrogels. *Sources:*(a) Redrawn from [40]. Copyright (2016) Nature Publishing Group. (b) Adapted with permission from [41]. Copyright (2016) Wiley-VCH. (c) Redrawn from [48]. Copyright (2014) National Academy of Sciences. (d) Adapted with permission from [50]. Copyright (2010) American Chemical Society.

methods were discussed in Section 15.2. Not only do they provide a bridge to achieve gelation with a synthetic component, these unnatural functional groups also offer the potential for new functions. For example, Tirrell et al. demonstrated that patterned protein surface could be prepared from the lithography of intrinsically photo-reactive proteins containing aryl azides [45].

How does cells perform chemistry with proteins within the biological reactivity space? The molecular recognition event serves as the key to ensure its high selectivity. Combined with natural amino acid side chains, a sequence of amino acids can be made ready for a unique and selective chemical/enzymatic modification with the additional benefit of being genetically encoded. Some of them can be applied to the preparation of all-protein-based hydrogels, such as intein-based technology [46, 47] and SpyTag-SpyCatcher chemistry (Figure 15.3c) [48, 49]. If the second component is not protein, many other reactions can also be used, providing a versatile platform for engineering their properties. For example, Lutolf et al. used phophopantetheinyl transferase (PPTase) to cross-link protein dimers bearing two phosphopantetheinyl moieties with CoA-functionalized PEG having serine residues (Figure 15.3d) [50]. In this system, gelation occurs only if PPTase is added with equal stoichiometry of the two functional groups. Similarly, other protein modification methods via enzymatic elaboration of peptide tags can also be adapted to achieve this purpose, including transglutaminase (TGase) technology [51], PRIME using lipoic acid ligase (LplA) and its mutants [27], the sortase A (SrtA) technology [52], the SNAP tag method and its recent adaption the "CLIP tag" method [53], as well as the HaloTag method [54]. An artificial ligase was developed by Howarth et al. by further splitting SpyCatcher into K-Tag and SpyLigase [55]. In the presence of a chemical chaperone trimethylamine *N*-oxide, the SpyLigase mediates the coupling of K-Tag and SpyTag. But the reaction is slow and the yield is not as satisfactory. Recently, a class of even more powerful Asn/Asp ligases called Butelase was discovered by Tam et al. in *Clitoria ternatea*, among which Butelase 1 was found to possess the fastest ligation rate to date [56–58]. These ligases have huge potential for applications on protein-containing bioactive hydrogels.

15.3.2 Physical Cross-linking of Proteins

Considered in many aspects to be complementary to chemical cross-linking, physical cross-linking has some unique features. First of all, the strength of cross-linking may vary across a broad range, depending on the type of secondary interactions and/or the number of weak bonds acting cooperatively. This feature has been elegantly used by Gong et al. in the development of a tough hydrogel-based solely on physical interactions [59]. Second, the dynamic nature of physical cross-links allows the development of injectable gels and/or self-healing gels, which are valuable for many practical applications [60]. Third, physical cross-links can be tuned on and off by various triggers, thus rendering the gel stimuli-responsive and even smart [61]. Therefore, supramolecular hydrogels have been intensively studied over the past decades and numerous structural motifs that possess well-defined multivalent secondary interactions have been developed, including the complementary quadruple hydrogen bonding motif, the host-guest complexation (e.g. cyclodextrin/cucurbit[n]uril/crown ether/pillar[n]arene and their corresponding guest molecules), the metal coordination, and electrostatic interactions.

Biomacromolecules provide a rich and versatile platform for tuning supramolecular assemblies. Even just for proteins, there is a vast library of protein-peptide, protein-ligand, and protein-protein interactions with varying strengths, distinct sizes, and high specificity. Hence, it is a natural fit to achieve protein gelation. Leucine zippers or other coiled-coil proteins forms well-defined oligomers via hydrophobic interactions between the interfaces. Tirrell was the first to use this as the cross-linking strategy for smart, stimuli responsive all-protein-based hydrogels (Figure 15.4a) [7]. This strategy also works in synthetic polymer-protein hybrid networks, as reported by Kopeček et al. [62]. Heilshorn et al. used the association between WW domain and the proline-rich peptide to achieve a mixing-induced, two-component hydrogel, which can be used to encapsulate murine adult neural stem cells and support their differentiation (Figure 15.4b) [63]. Strong protein-protein interactions are always favored for protein gelation. For example, the trimerization of CutA protein was found to be extremely stable and gelation can be easily achieved with the aid of the intein-technology [46, 47]. Antibody-antigen interactions are strong protein-protein interactions. Miyata et al. used rabbit IgG and goat anti-rabbit IgG as the antigen and antibody to produce a reversible responsive hydrogel [64]. Besides normally used anti-IgG/IgG, engineered single chain antibody fragment and peptide epitopes is also a good pair for high efficiency binding and recruitment, although only used in eukaryotic cells until now [65]. Other protein-peptide interactions, such as the binding between the 34 amino acid tetratricopeptide repeats (TPR) and the peptides DESVD [66] and the complexation between tax interactive protein 1 (TIP1) and PDZ domain peptide [67], have also been used to produce hydrogels (Figure 15.4c). Notably, a protein domain can be split into two associating partners (they can be protein/peptide or protein/protein), whose reconstitution can then be used as the hydrogel cross-links. For example, a physical "network of spies" can be prepared using the complexation between SpyCatcher and SpyTag-DA mutatnt [48]. Li et al. [68] also showed that both single-component hydrogel and two-component hydrogel can be prepared based on the protein fragments GN and GC from GL5, a GB1 mutant (Figure 15.4d). This significantly expands the protein toolbox for biomaterial scientists. In addition to the above-mentioned interactions which are all based on natural amino acid building blocks, proteins can also interact with other molecules, such as small signaling molecules, growth factors, ions, polysaccharides, and other ligands. This provides a way to bridge the synthetic system and biological system. One can either use protein as the matrix, or as the cross-linkers. For example, vascular endo-epithelial growth factor (VEGF) can bind with heparin, one kind of glycosaminoglycan. Kiick et al. constructed a four-arm PEG functionalized with modified heparin and produced the hydrogel by mixing the functionalized PEG with VEGF [69]. The genetically engineered bacterial gyrase subunit B (GyrB) with a hexahistidine tag can be grafted to polyacrylamide and subsequently dimerize in the presence of coumermycin, leading to hydrogel formation. This is reversible if Novobiocin is added to compete the binding [70].

It is apparent that biomolecular interactions are advantageous for achieving physical cross-linking. DNA hydrogels have shown extraordinary properties, including being injectable and shaping memory. For proteins, the situation may be more complicated because such interactions are sensitive to folded structure and the environment (concentration, ionic strength, temperature, etc.). But this also imparts the valuable stimuli-responsive, or smart hydrogels. In addition, since the size of the protein domains are usually relatively big, the hydrogels made of this are usually mechanically weak. In the following sections, we will discuss the strategies to make tough hydrogels and the realization of smart hydrogels.

15.4 Strategies for Mechanical Reinforcement

One of the major obstacles to the practical application of hydrogels is the poor mechanical properties: they tend to be brittle and weak. For synthetic polymer hydrogel, numerous strategies have been developed for mechanical enhancement. Gels can be made tougher, stretchier, and

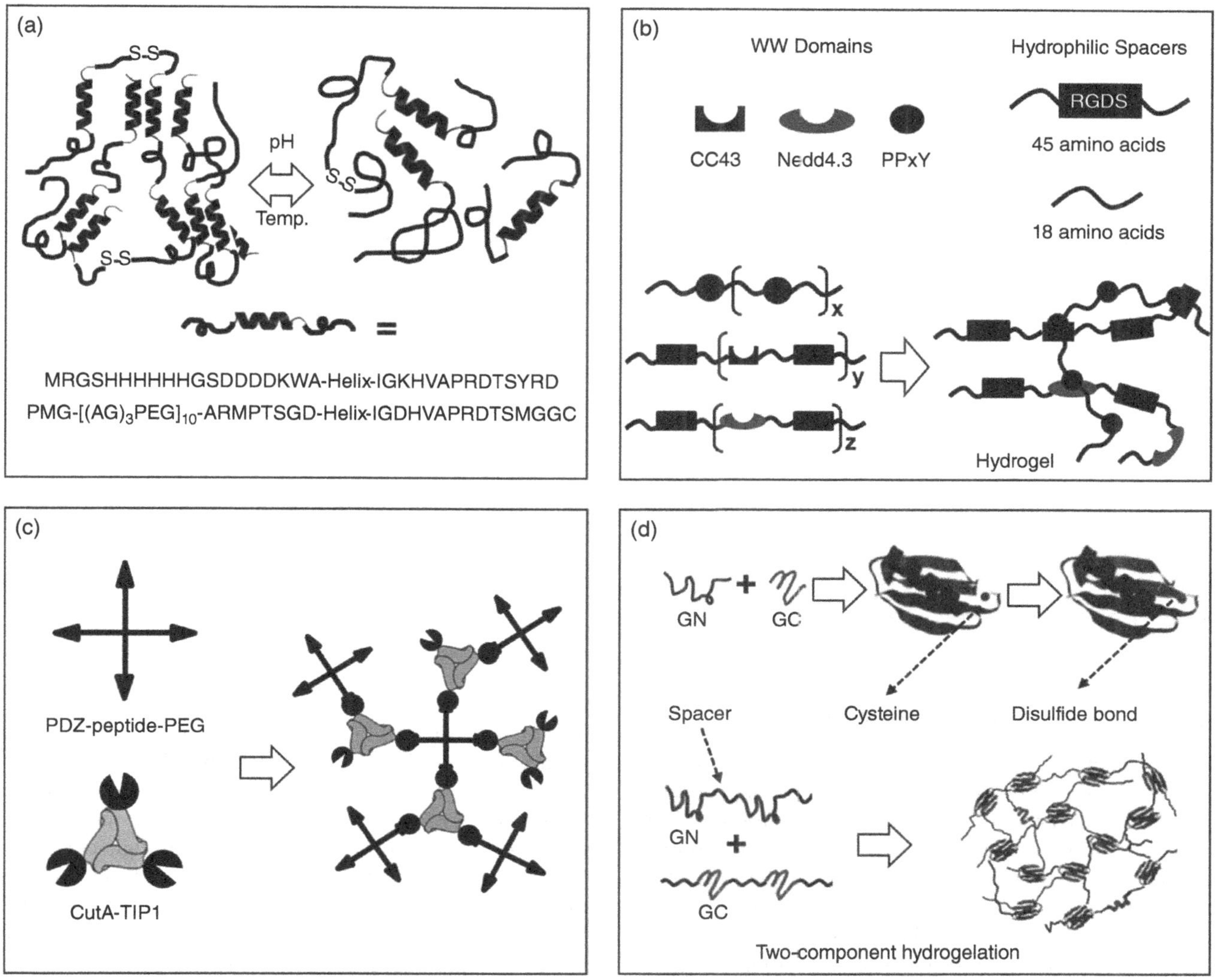

Figure 15.4 Physical cross-linking of proteins. (a) Proposed physical gelation of monodisperse copolymer. The chains are drawn as disulfide-linked dimers, joined through their COOH-terminal cysteine residues. (b) Schematic of the mixing-induced, two-component hydrogel. (Top left) Modular association domains assemble via molecular recognition. Two WW domains (CC43 and a Nedd4.3 variant) bind the same proline peptide (PPxY). (Top right) Hydrophilic spacers. (Bottom) Mixing components at constant physiological conditions results in hydrogel formation. (c) Schematic illustration of the hydrogel formed by the biospecific association between PDZ domain in TIP1 protein (CutA-TIP1) and PDZ domain recognizing peptide (PDZ-peptide-PEG). (d) Schematic showing hydrogelation based on protein fragment reconstitution. Top) Formation of reconstituted GL5 from GN and GC. Bottom) Schematic of fragment reconstitution induced two-component hydrogelation. The two tandem polyproteins, containing either GN or GC fragments, serve as multiple functional precursors for hydrogelation. *Sources:* (a) Redrawn from [7]. Copyright (1998) The American Association for the Advancement of Science. (b) Redrawn from [63]. Copyright (2009) National Academy of Sciences. (c) Redrawn from [67]. Copyright (2009) Elsevier Ltd. (d) Adapted with permission from [68]. Copyright (2015) Wiley-VCH.

more endurable. Actually, natural tissues can be regarded as hydrogels made of a hierarchically arranged complex mixture of cells, proteins, and polysaccharides with excellent mechanical properties. It is so far still difficult to develop synthetic hydrogels that can rival natural materials in all aspects. Mechanical properties are also known as an important signal for cells [71]. Combined with various growth factors, they direct the fate of the stem cells toward tissue formation [72]. In this section, we will discuss various strategies for mechanical enhancement (Figure 15.5). These strategies have been nicely demonstrated in synthetic polymer hydrogels, but have not yet been fully explored for their potential in protein-based hydrogels. We believe that protein-based hydrogels can also benefit from these strategies for better bio-mimetic materials.

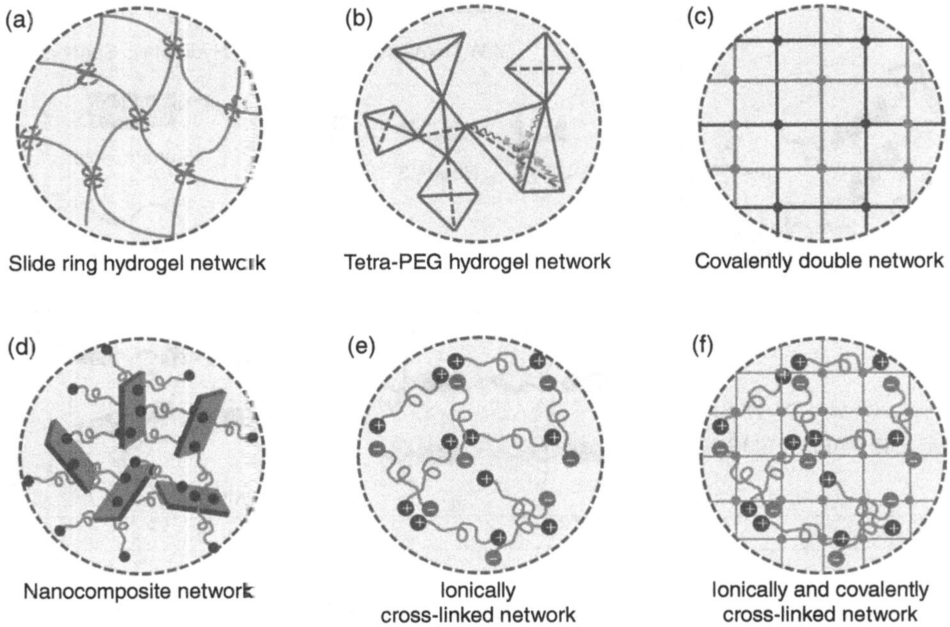

Figure 15.5 Strategies for mechanical reinforcement

15.4.1 Lessons from Synthetic Polymers

It is not exaggerating to say that cross-linking is the most critical factor that determines the mechanical properties of the gel. This has traditionally been understood through the classic network theory relating macroscopic modulus to microscopic cross-linking density and the properties of its constituent polymer in a quantitative scaling relationship. Generally speaking, the higher the cross-linking density is, the more brittle and stiffer the gel is. Covalent hydrogels usually have higher modulus while noncovalent physical gels are usually injectable or self-healing. However, since the stress is often concentrated at the tip of the break, catastrophic breaking of the chemical network is often observed. For practical applications, the gels need to be sufficiently robust to bear the mechanical loads while maintaining physical integrity when stressed over time. Therefore, the network should be able to dissipate energy effectively. The past decade has witnessed the rapid development of tough hydrogels. The principles of toughening mechanism are now relatively well understood: *the gel needs to be able to dissipate significant amounts of mechanical energy when deformed and retain the original configuration after deformation* [6, 73, 74]. Both the cross-linking point and the network structure can be engineered with this purpose in mind. Below, we will discuss these in the context related to protein-based hydrogels.

Energy dissipation can result from polymer chain fracture, reversible cross-linking, folding/unfolding events, and/or pull-out of fibers/fillers. Polymer chain fracture usually involves breaking covalent bonds, which is the most common mechanism for energy dissipation in hydrogels. However, the stresses are focused on the tips, crack forms, and propagations. Even though chemical bonds are usually strong, the energy cannot be effectively dissipated and thus most hydrogels are brittle and fragile. If fractured bonds can reform, as in dynamic covalent bonds and in physical networks, the energy can be effectively dissipated. The reversible cross-links constitute a dynamic network which is capable of "self-healing." Typical reversible covalent bonds include the imine formation, olefin metathesis, alkyne metathesis, and hindered urea bonds, etc., but this usually involves functional groups not present in biomacromolecules. Interestingly, most enzyme-mediated covalent bond formation is reversible, such as in sortase A mediated ligation. Recently, Giuseppone et al. have engineered reversible native chemical ligation in an attempt to develop dynamic covalent peptide library [75]. Despite great interest, all-protein-based hydrogels with dynamic covalent bonds have not yet been achieved. Physical cross-links are usually reversible and can have varying strengths. If multiple secondary interactions are working cooperatively and collectively, these physical bonds can be very strong with bond energy close to that of covalent bonds (e.g. the avidin-biotin interaction). It is often desirable to combine the advantages of both chemical and physical network by forming a chemical/physical network. The synergistic

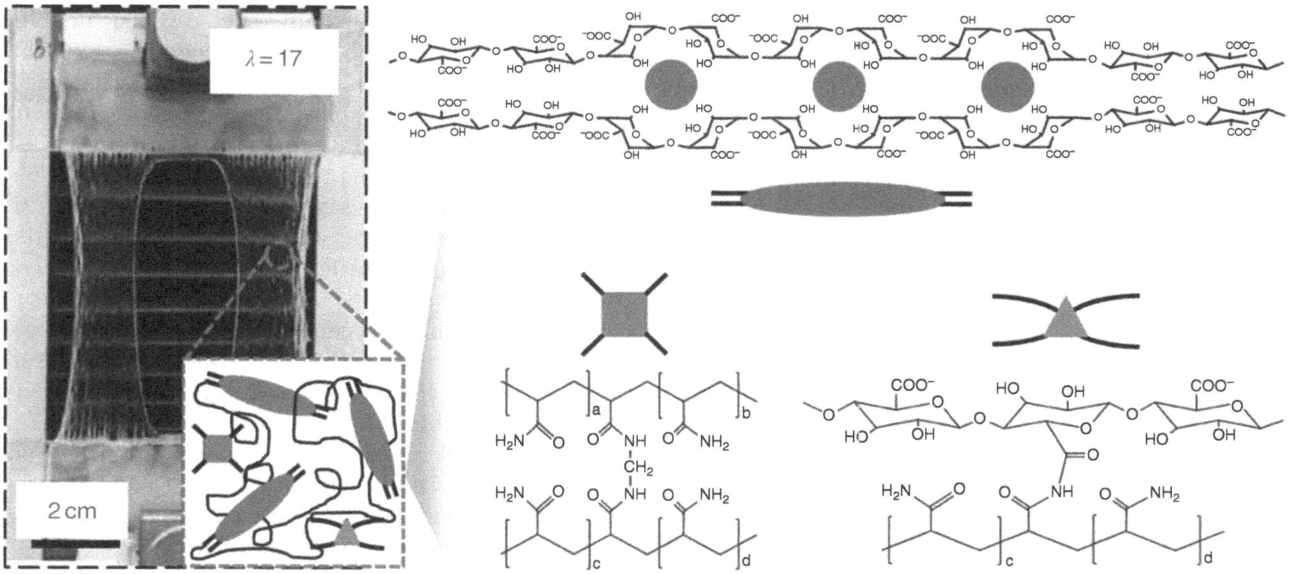

Figure 15.6 The gel containing the notch was stretched to 17 times its initial length, and the schematics of the hydrogel. *Source:* Redrawn from [62]. Copyright (2012) Nature Publishing Group.

effect has been observed in many cases. For example, Suo et al. used the covalent cross-links of polyacrylamide and the electrostatic crosslinks of Ca^{2+}-alginate interaction to develop a highly stretchable and tough hydrogel that can even tolerate the presence of a notch (Figure 15.6) [76]. Zheng et al. developed a simple one-pot method for the preparation of tough hydrogels by combining a polyacrylamide network with thermo-reversible agar gel [77]. More generally speaking, the combination of strong/weak interactions can all exhibit such improvements and may even be realized in a single polymer system. Gong et al. ingeniously utilized the random copolymerization of charged monomers to form a tough hydrogel *in situ* [59]. The random distribution of charges along the polymer chain leads to a wide distribution of aggregation domains and thus varying cross-linking strengths. The stronger ones hold the elasticity of the network and the weaker ones dissipate the energy through fracture and reformation.

Changing the connectivity/dimensionality of the cross-links is another effective way to tune mechanical properties. Dendrimers, molecular nanoparticles, quantum dots, nanorods, nanoplates, and nanosheets have all been used in synthetic polymer hydrogels for this purpose. Due to the extended connectivity, multiple (non) covalent bonds would require to be broken before the network fails. Hence, it could bear much higher loads. For example, a highly porous hydrogel with increased stiffness can be obtained using a fourth-generation poly(amido amine) dendrimer (G4 PAMAM) as the "island-like" cross-linker [78]. More drastic mechanical improvement can be achieved when fibers or nanosheets (or other fillers) are included, either as the cross-linkers or as composites, since it involves the additional pullout mechanism for energy dissipation. Aida et al. mixed clay and a dendritic molecular binder to prepare a high-water-content moldable hydrogel [79]. Compared to the one prepared from only triblock copolymers [80], the storage modulus is almost ten times higher [81]. Intriguing anisotropic properties can also arise from these nano-objects. For example, negatively charged unilamellar titannate nanosheets have been embedded into hydrogel with a cofacial arrangement [82]. The electrostatic repulsion between the nanosheets makes the hydrogel anisotropic and resistant toward orthogonally applied compressive forces, although it could still easily deform upon lateral shear. Gong et al. fixed lamellar micelle assemblies of a polymerizable surfactant into this system by photochemical gelation. The gel is photonic and can change color with applied stress [83, 84]. Although this is originally inspired by biological systems, no recombinant protein hydrogels have reproduced such properties. In fact, there are various low dimensional protein assemblies available, such as 1D CsgA proteins, the solenoid protein domains derived from antifreeze proteins, and 2D S-layered proteins. We anticipate that they will be promising candidates for making protein hydrogels nanocomposites.

Engineering the topology of cross-linking points could well be another valid strategy for mechanical reinforcement. The "slide-ring gels" are a great example of topological gels [85, 86]. The effect is in certain aspects similar to chain entanglements. It creates a variable and controllable mean mass distribution of the entanglement length. Hence, the gel is highly water-absorbent and stretchable. In nature bacteriophage HK97 forms multiple intersecting protein catenane structure in the capsid. This chain-mail structure imparts unusual stability to the capsid. This is reminiscent of the chain-mail armor for soldiers. It is highly desirable to apply this topological network structure engineering to hydrogels. Recently, Wang and Zhang reported a simple way to produce protein catenanes in *E. coli* (Figure 15.7) [37]. Combined with the versatility of protein structure and function, it is possible to prepare protein hydrogels with complex topologies and diverse functions. Although this direction remains less explored for protein hydrogels, it has huge potential.

Network structure engineering is very important for hydrogels. The heterogeneity of the network accounts in part for its weak mechanical properties. A straightforward way to make mechanical reinforcement is by making a uniform network with minimum amounts of defects. A classic example is the cross-linking of symmetrical tetrahedron-like PEGs. The networks thus formed are almost free from chain entanglements and network defects, making the gel really tough [87]. Tew et al. extended this strategy to make highly resilient hydrogels based on PDMS and PEO polymer system. The gel was formed in THF as a homogeneous system and toughening occurs after dialysis. The final water content was controlled by the PDMS composition [88]. One can also engineer the network structure by incorporating multiple networks. An interpenetrating network was first pursued but the effect was not impressive. In the early 2000s, Gong et al. found that it is important to interpenetrate two different types of network to achieve high toughness: a long-chain network for elasticity and a short-chain network for energy dissipation. The so-called "double" network later proved to be a huge success [73, 89]. The fracturing of the polymer chains in the short-chain network dissipates the mechanical energy. But since it cannot reform, the double network hydrogel suffers from fatigue. This is overcome by using the chemical/physical double network, as demonstrated in previous examples [76, 77]. For protein hydrogels, it is difficult to directly apply the double network principle for mechanical reinforcement since the conditions for two network formation are usually not compatible. Recently, a chemical/physical network has been constructed using thiol-functionalized telechelic elastin-like protein and a 4-arm PEG cross-linker [43]. The thiol-Michael reaction effectively makes the first network under denaturing conditions. Upon denaturant removal, the leucine zippers re-fold and re-associate to form the second physical network, improving the mechanical property of the hydrogel.

Finally, all of the above strategies can be combined in a single system and engineered across multiple length scales to achieve the best performance. The synergy can be even beyond one's expectations. For example, Glassman and Olsen successfully reinforced a shear-thinning protein hydrogel by thermo-induced aggregation of PNIPAM end-blocks (Figure 15.8) [90]. There are thus two independent physical networks in the final hydrogel: one from the associative protein zippers and the other from the aggregates. The elastic modulus is fivefold higher and the creep compliance is improved by orders of magnitude. More recently, an extremely stretchable and self-healable hydrogel was prepared using a hierarchical system of noncovalent and covalent cross-links: a micellar system with quadruple hydrogen bonding motif within the hydrophobic core and with covalent chains between the micelles [91]. The gel can be stretched up to 100 times its original length and heals within 30 s without external energy input, which is truly remarkable.

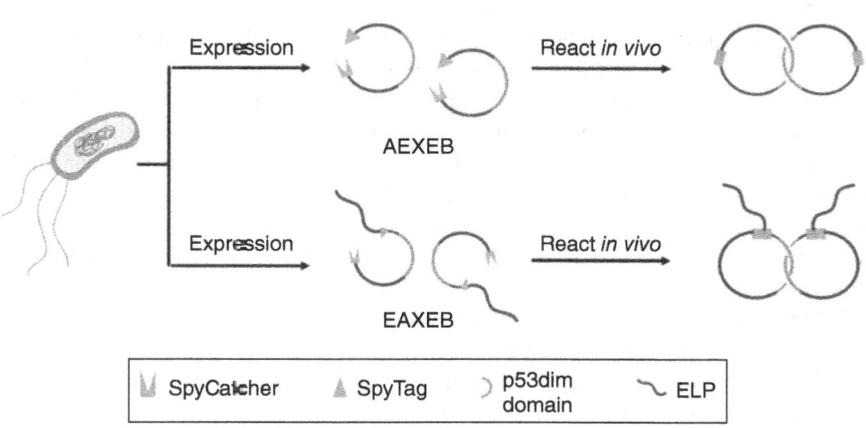

Figure 15.7 Biosynthesis of protein catenanes. *Source:* Adapted with permission from [37]. Copyright (2016) Wiley-VCH.

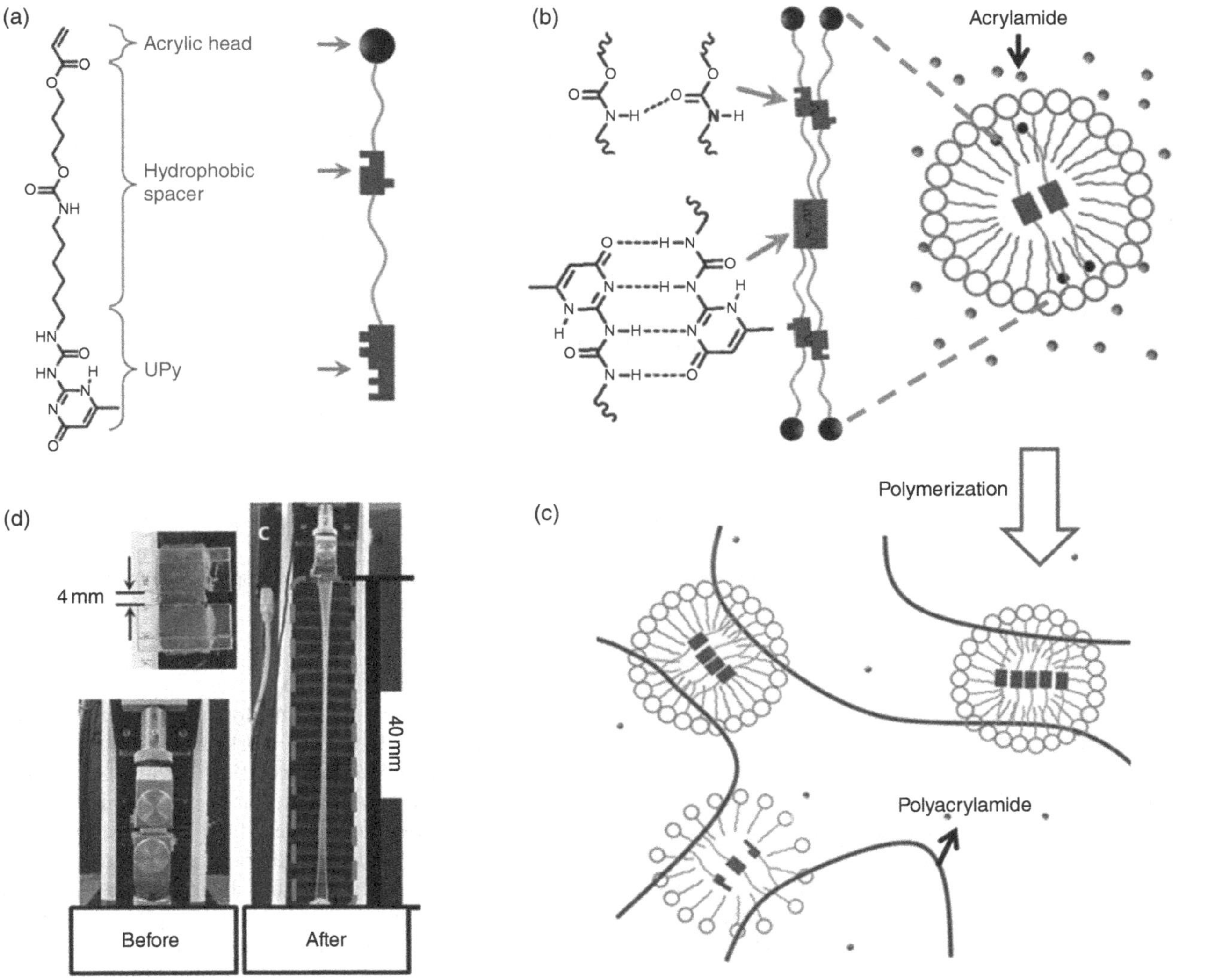

Figure 15.8 (a–c) Schematic of the micellar copolymerization of the UPyHCBA and acrylamide. (a) The structure of the UPyHCBA monomer; (b) SDS micelles loaded with UPyHCBA in an aqueous acrylamide solution; and (c) micellar copolymerization of acrylamide and UPyHCBA loaded in SDS micelles. (d) Image of the hydrogel materials before and after stretching. *Source:* Adapted with permission from [91]. Copyright (2016) Wiley-VCH.

15.4.2 Inspirations from Nature

For biological systems, while the principles are the same, nature adopts different approaches to ensure both mechanical properties and bioactive functions. The natural tissues, such as tendons and muscles, are remarkably strong. Such tissues are often constructed from a mixture of proteins, polysaccharides, and other compositions, forming a complex hierarchical structure [5]. Each component contributes to the final properties in a unique way. For example, the egg capsule wall of the whelk *Busycon canaliculum* can act as a shock absorber and withstand the impact of the sea wave for extended periods. This is partially due to the simple and reversible structural transition of egg capsule protein from α-helix to β-sheet during extension (Figure 15.9) [92, 93]. Single molecule force studies have also shown that the globular protein domains can be unfolded upon stretching, which also constitutes a unique mechanism to dissipate energy even if the protein itself is not a structural mechanical protein [94]. This has been nicely demonstrated by an immunoglobulin G-binding protein domain GB1 [95]. Nevertheless, it remains a challenge to relate the molecular events to the macroscopic gel properties in a quantitative way. It is difficult to calculate the force that is applied to a single protein domain during extension due to the heterogeneous network structure and the anisotropic extension. Inspired by this, Guan et al. show in a simple model system that the elastomer

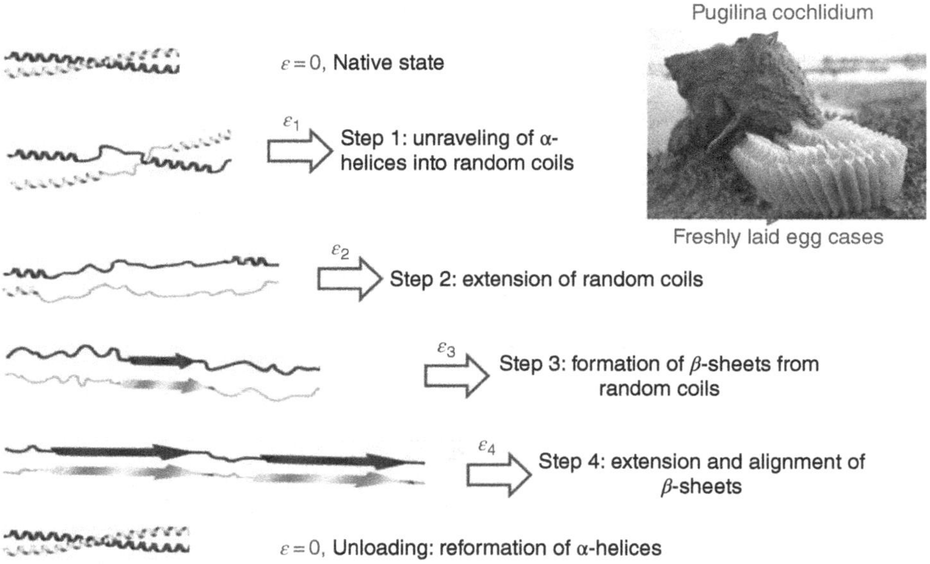

Figure 15.9 Schematic diagram of α-helix ↔ β-sheet transition during straining. *Source:* Redrawn from [92]. Copyright (2009) Nature Publishing Group.

containing cyclic molecules that can form intra-unit multiple hydrogen bonds is quite extensible and shows hysteresis behavior [96]. Moreover, the excellent functions of natural structural materials usually arise from a delicate hierarchal arrangement of hard and soft phases with a characteristic dimension across multiple length scales [5]. This is often difficult to mimic synthetically. One would expect that synthetic systems and biological polymers can learn from each other with regard to mechanisms for mechanical reinforcement. Last but not least, it is important to recognize that the mechanical properties of the hydrogel should be tailored to match a specific application. For each end application, a combination of different set of properties are desired while some of them may outweigh many others. Understanding the influence of properties on the molecular level is critical to be able to tailor the properties on demand.

15.5 Conclusion

In summary, the emergence of protein-based hydrogels offers a versatile platform for achieving bioactive materials. The convergence and synergy make protein hybrid hydrogels even more promising. It is well known that most organisms, such as cartilage, cornea, ECM, or muscles, can be regarded as hydrogels. The embedded cells and the matrix form a complex, integrated system that renews itself while keeping its function. and evolve itself. Ideal biomaterials should also be like this, or "living" materials. Being the workhorse of life, proteins offer great advantages in creating "living" materials. The state-of-the-art protein-based biomaterials have achieved considerable progress in this regard but their potential has not yet been fully realized. To achieve better integration with cells or devices, protein-based hydrogels can benefit from the development of more protein tools from millions of years of evolution in nature and from decades of protein engineering [93], as well as from the *de novo* design and rational computation of protein structure and functions [97–100]. Incorporation of unnatural amino acids would further expand the sequence space of proteins, offering great promise in diversifying the functional plasticity of proteins [101]. On the other hand, processing of protein materials is an important yet less pursued subject, requiring a high compositional heterogeneity and functional diversity with spatiotemporal precision without compromising the structural and functional integrity of proteins. Of course, the advantages of genetically designed materials can go well beyond proteins. The cellular machinery possesses unprecedented precision in synthesis over chemical synthesis. Taking advantage of this system, one can use synthetic biology to alter the natural pathway to create unnatural genetically designed products, often on a larger scale and at a lower cost, as in the example of the taxol precursor in *E. coli* [102].

Despite the many advantages of protein-based hydrogels, there are two factors limiting proteins' application as materials: immunogenicity and cost. First, there is often a misconception that proteins are completely "biocompatible." This is only partially true. While proteins are extremely useful, one cannot ignore the fact that proteins could elicit profound immunogenic responses and undesired reactions. But the origin and molecular foundation of immunogenicity have not yet been fully understood. Second, the cost of recombinant proteins is generally much higher than synthetically materials or natural biopolymers. To match the cost, it has to be able to accomplish something that traditional "biocompatible" materials like PDMS, chitosan, hyaluronic acids, etc. cannot do. The research should thus be oriented toward these niche applications and steered to address a specific set of problems combining principles of chemistry, physics, and biology. Only in this way can the potential of protein-based materials, especially those hybrids with synthetic components, be fully unleashed and the real impact can be generated. The future will witness the convergence of synthetic polymer hydrogels and protein hydrogels in synergy to address specific problems.

References

1 Demirci, U., Khademhosseini, A. (2016) Gels Handbook, World Scientific, Boston, MA.
2 Staudinger, H. (1920) *Ber. Dtsch. Chem. Ges.*, **53**, 1073–1085.
3 Dušek, K. (1993) *Responsive Gels: Volume Transitions*, Springer-Verlag, New York.
4 DeForest, C.A., Anseth, K.S. (2012) *Annu. Rev. Chem. Biomol. Eng.*, **3**, 421–444.
5 Wegst, U.G., Bai, H., Saiz, E. et al. (2015) *Nat. Mater.*, **14**, 23–36.
6 Zhao, X. (2014) *Soft Matt.*, **10**, 672–687.
7 Petka, W.A., Harden, J.L., McGrath, K.P. et al. (1998) *Science*, **281**, 389–392.
8 van Hest, J.C.M., Tirrell, D.A. (2001) *Chem. Commun.*, 1897–1904.
9 Wang, X.-W., Wang, X.-M., Ran, R.-H., Zhang, W.-B. (2014) *Polym. Bull.*, **8**, 44–55.
10 Whitford, D. (2005) *Proteins: Structure and Function*, John Wiley & Sons, Hoboken, NJ.
11 Müller, A.H.E., Matyjaszewski, K. (2009) *Controlled and Living Polymerizations: Methods and Materials*, Wiley-VCH, Weinheim.
12 Uversky, V.N. (2014) *Intrinsically Disordered Proteins*, Springer, New York.
13 Wright, P.E., Dyson, H.J. (1999) *J. Mol. Biol.*, **293**, 321–331.
14 Tompa, P. (2003) *J. Mol. Struc. Theochem.*, **666–667**, 361–371.
15 Arnold, F.H. (1998) *Acc. Chem. Res,.* **31**, 125–131.
16 Romero, P.A., Arnold, F.H. (2009) *Nat. Rev. Mol. Cell Biol.*, **10**, 866–876.
17 Odian, G.G. (2004) *Principles of Polymerization*, 4th edn., Wiley-Interscience, Hoboken, NJ.
18 Wang, L., Brock, A., Herberich, B., Schultz, P.G. (2001) *Science* **292**, 498–500.
19 Wang, L., Xie, J., Schultz, P.G. (2006) *Annu. Rev. Biophys. Biomol. Struct.*, **35**, 225–249.
20 van Hest, J. C. M., Kiick, K.L., Tirrell, D.A. (2000) *J. Am. Chem. Soc.*, **122**, 1282–1288.
21 Link, A.J., Mock, M. L., Tirrell, D.A. (2003) *Curr. Opin. Biotech.*, **14**, 603–609.
22 Ngo, J.T., Tirrell, D.A. (2011) *Acc. Chem. Res.*, **44**, 677–685.
23 Kahn, H.A.R.A. (2002) *Method Enzymol.*, **344**, 186–193.
24 Kulkarni, C., Kinzer-Ursem, T. L., Tirrell, D.A. (2013) *Chem. Bio. Chem.*, **14**, 1958–1962.
25 Heal, W.P., Wright, M.H., Thinon, E., Tate, E.W. (2011) *Nat. Protoc.*, **7**, 105–117.
26 Khidekel, N., Arndt, S., Lamarre-Vincent, N. et al. (2003) *J. Am. Chem. Soc.*, **125**, 16162–16163.
27 Uttamapinant, C., Sanchez, M.I., Liu, D.S. et al. (2013) *Nat. Protoc.*, **8**, 1620–1634.
28 Iwai, H., Lingel, A., Pluckthun, A. (2001) *J. Biol. Chem.*, **276**, 16548–16554.
29 Antos, J.M., Popp, M.W., Ernst, R. et al. (2009) *J. Biol. Chem.*, **284**, 16028–16036.
30 Wu, Z., Guo, X., Guo, Z. (2011) *Chem. Commun.*, **47**, 9218–9220.
31 van't Hof, W., Hansenova Manaskova, S., Veerman, E.C., Bolscher, J.G. (2015) *Biol. Chem.*, **396**, 283–293.
32 Zakeri, B., Howarth, M. (2010) *J. Am. Chem. Soc.*, **132**, 4526–4527;
33 Zakeri, B. (2015) *Chem. Bio. Chem.*, **16**, 2277–2282.
34 Zakeri, B., Fierer, J.O., Celik, E. et al. (2012) *Proc. Natl. Acad. Sci. USA*, **109**, E690-E697.
35 Veggiani, G., Nakamura, T., Brenner, M.D. et al. (2016) *Proc. Natl. Acad. Sci. USA*, **113**, 1202–1207.
36 Zhang, W.-B., Sun, F., Tirrell, D.A., Arnold, F.H. (2013) *J. Am. Chem. Soc.*, **135**, 13988–13997.
37 Wang, X.-W., Zhang, W.-B. (2016) *Angew. Chem. Int. Ed.*, **55**, 3442–3446.
38 Heyman, Y., Buxboim, A., Wolf, S.G. et al. (2012) *Nat. Nanotechnol.*, **7**, 374–378.
39 Deechongkit, S., Nguyen, H., Powers, E.T. et al. (2004) *Nature*, **430**, 101–105.
40 Ma, X., Sun, X., Hargrove, D. et al. (2016) *Sci. Rep.*, **6**, 19370.

41 Lin, C.C., Anseth, K.S. (2009) *Pharm. Res.*, **26**, 631–643.
42 Hermanson, G.T. (2013) *Bioconjugate Techniques*, 3rd edn., Elsevier: Waltham, MA.
43 Dooling, L.J., Buck, M.E., Zhang, W.B., Tirrell, D.A. (2016) *Adv. Mater.*, **28**, 4651–4657.
44 Sletten, E.M., Bertozzi, C.R. (2009) *Angew. Chem. Int. Ed.*, **48**, 6974–6998
45 Carrico, I.S., Maskarinec, S.A., Heilshorn, S.C. et al. (2007) *J. Am. Chem Soc.*, **129**, 4874–4875.
46 Ramirez, M., Guan, D., Ugaz, V., Chen, Z. (2013) *J. Am. Chem. Soc.*, **135**, 5290–5293.
47 Guan, D., Ramirez, M., Shao, L. et al. (2013) *Biomacromolecules* **14**, 2909–2916.
48 Sun, F., Zhang, W.-B., Mahdavi, A. et al. (2014) *Proc. Nat. Acad. Sci. U.S.A.*, **111**, 11269–11274.
49 Reddington, S. C., Howarth, M. (2015) *Curr. Opin. Chem. Biol.*, **29**, 94–99.
50 Mosiewicz, K.A., Johnsson, K., Lutolf, M.P. (2010) *J. Am. Chem. Soc.*, **132**, 5972–5974.
51 Heck, T., Faccio, G., Richter, M., Thony-Meyer, L. (2013) *App. Microbiol. Biotechnol.*, **97**, 461–475.
52 Tsukiji, S., Nagamune, T. (2009) *Chem. Bio. Chem.*, **10**, 787–798.
53 Gautier, A., Juillerat, A., Heinis, C. et al. (2008) *Chem. Biol.*, **15**, 128–136.
54 Los, G.V., Encell, L.P., McDougall, M.G. et al. (2008) *ACS Chem. Biol.*, **3**, 373–382.
55 Fierer, J.O., Veggiani, G., Howarth, M. (2014) *Proc. Nat. Acad. Sci. U.S.A.*, **111**, E1176–E1181.
56 Nguyen, G.K., Wang, S., Qiu, Y. et al. (2014) *Nat. Chem. Biol.*, **10**, 732–738.
57 Nguyen, G.K.T., Cao, Y., Wang, W. et al. (2015) *Angew. Chem. Int. Ed.*, **54**, 15694–15698.
58 Nguyen, G.K., Kam, A., Loo, S. et al. (2015) *J. Am. Chem. Soc.*, **137**, 15398–15401.
59 Sun, T.L., Kurokawa, T., Kuroda, S. et al. (2013) *Nat. Mater.*, **12**, 932–937.
60 Olsen, B.D., Kornfield, J.A., Tirrell, D.A. (2010) *Macromolecules*, **43**, 9094–9099.
61 Kopecek, J. (2007) *Biomaterials*, **28**, 5185–5192.
62 Wang, C., Stewart, R. J., Kopeček, J. (1999) *Nature*, **397**, 417–420.
63 Wong Po Foo, C.T., Lee, J.S., Mulyasasmita, W. et al. (2009) *Proc. Nat. Acad. Sci. U.S.A.*, **106**, 22067–22072.
64 Miyata, T., Asami, N., Uragami, T. (1999) *Nature*, **399**, 766–769.
65 Tanenbaum, M.E., Gilbert, L.A., Qi, L. S. et al. (2014) *Cell*, **159**, 635–646.
66 Grove, T.Z., Osuji, C.O., Forster, J.D. et al. (2010) *J. Am. Chem. Soc.*, **132**, 14024–14026.
67 Ito, F., Usui, K., Kawahara, D. et al. (2010) *Biomaterials*, **31**, 58–66.
68 Kong, N., Li, H. (2015) *Adv. Func. Mater.*, **25**, 5593–5601.
69 Yamaguchi, N., Zhang, L., Chae, B.-S. et al. (2007) *J. Am. Chem. Soc.*, **129**, 3040–3041.
70 Ehrbar, M., Schoenmakers, R., Christen, E.H. et al. (2008) *Nat. Mater.*, **7**, 800–804.
71 Engler, A.J., Sen, S., Sweeney, H.L., Discher, D.E. (2006) *Cell*, **126**, 677–689.
72 Even-Ram, S., Artym, V., Yamada, K.M. (2006) *Cell*, **126**, 645–647.
73 Gong, J.P. (2010) *Soft Matt.*, **6**, 2583.
74 Peak, C.W., Wilker, J.J., Schmidt, G., (2013) *Coll. Polym. Sci.*, **291**, 2031–2047.
75 Ruff, Y., Garavini, V., Giuseppone, N., (2014) *J. Am. Chem. Soc.*, **136**, 6333–6339.
76 Sun, J.Y., Zhao, X., Illeperuma, W.R. et al. (2012) *Nature*, **489**, 133–136.
77 Chen, Q., Zhu, L., Zhao, C., et al. (2013) *Adv. Mater.*, **25**, 4171–4176.
78 Wang, Y., Zhao, Q., Zhang, H. et al. (2014) *Adv. Mater.* **26**, 4163–4167.
79 Wang, Q., Mynar, J.L., Yoshida, M. et al. (2010) *Nature*, **463**, 339–343.
80 Hunt, J.N., Feldman, K.E., Lynd, N.A. et al. (2011) *Adv. Mater.*, **23**, 2327–2331.
81 Tamesue, S., Ohtani, M., Yamada, K. et al. (2013) *J. Am. Chem. Soc.*, **135**, 15650–15655.
82 Liu, M., Ishida, Y., Ebina, Y. et al. (2015) *Nature*, **517**, 68–72.
83 Yue, Y.F., Haque, M.A., Kurokawa, T. et al. (2013) *Adv. Mater.*, **25**, 3106–3110.
84 Yue, Y., Kurokawa, T., Haque, M.A. et al. (2014) *Nat. Commun.*, **5**, 4659.
85 Okumura, Y., Ito, K. (2001) *Adv. Mater.*, **13**, 485–487.
86 Ito, K. (2010) *Curr. Opin. Solid State & Mater. Sci.*, **14**, 28–34.
87 Sakai, T., Matusnaga, T., Yamamoto, Y. et al. (2008) *Macromolecules*, **41**, 5379–5384.
88 Cui, J., Lackey, M.A., Madkour, A.E. et al. (2012) *Biomacromolecules*, **13**, 584–588.
89 Gong, J.P., Katsuyama, Y., Kurokawa, T., Osada, Y. (2003) *Adv. Mater.*, **15**, 1155–1158.
90 Glassman, M.J., Chan, J., Olsen, B.D. (2013) *Adv. Funct. Mater.*, **23**, 1182–1193.
91 Jeon, I., Cui, J., Illeperuma, W.R. et al. (2016) *Adv. Mater.*, **28**, 4678–4683.
92 Miserez, A., Wasko, S.S., Carpenter, C.F., Waite, J.H. (2009) *Nat. Mater.*, **8**, 910–916.
93 Guerette, P.A., Hoon, S., Seow, Y. et al. (2013) *Nat. Biotechnol.*, **31**, 908–915.
94 Cao, Y., Li, H. (2007) *Nat. Mater.*, **6**, 109–114.

95 Lv, S., Dudek, D.M., Cao, Y. et al. (2010) *Nature*, **465**, 69–73.

96 Chung, J., Kushner, A.M., Weisman, A.C., Guan, Z., (2014) *Nat. Mater.*, **13**, 1055–1062.

97 Huang, P. S., Feldmeier, K., Parmeggiani, F. et al. (2016) *Nat. Chem. Biol.*, **12**, 29–34.

98 Brunette, T.J., Parmeggiani, F., Huang, P.S. et al. (2015) *Nature*, **528**, 580–584.

99 King, N.P., Bale, J.B., Sheffler, W. et al. (2014) *Nature*, **510**, 103–108.

100 Khoury, G. A., Liwo, A., Khatib, F. et al. (2014) *Proteins*, **82**, 1850–1868.

101 Yoo, T. H., Link, A.J., Tirrell, D.A. (2007) *Proc. Natl. Acad. Sci. U.S.A.*, **104**, 13887–13890.

102 Ajikumar, P.K., Xiao, W.-H., Tyo, K.E.J. et al. (2010) *Science*, **330**, 70–74.

16

Dendritic Polymer Micelles for Drug Delivery

Mosa Alsehli[1,2] *and Mario Gauthier*[1,*]

[1] *University of Waterloo, Waterloo, Ontario, Canada*
[2] *Taibah University, Madina, Saudi Arabia*

16.1 Introduction

Branched polymers have physical properties distinct from their linear analogues. For example, high-density polyethylene, a linear polymer without any branches, has a high degree of crystallinity, making it a tough material. On the other hand, the multiple short branches present along the backbone of low-density polyethylene lead to a lower degree of crystallinity and increased ductility [1, 2]. By controlling the degree of branching, it is thus possible to obtain polymers with properties tailored for specific applications.

Dendritic polymers are a class of branched macromolecules characterized by a highly branched tree-like architecture, incorporating multiple branching levels, resulting from the coupling reactions of either small molecule monomers (hyperbranched polymers and dendrimers; Figure 16.1a, b), or macromolecular building blocks (dendrigraft polymers; Figure 16.1c) [3–6]. These materials can be synthesized with good control over molecular size, branching functionality, and chemical functionality of the chain ends [7]. These molecules have a unique combination of features including a compact globular topology and diameters ranging from 1 to over 100 nm, the presence of internal cavities, and a large number of functional groups at their periphery. Due to these remarkable features, a broad spectrum of applications are being developed for these materials in areas including microencapsulation, drug delivery, nucleic acid therapy, diagnostic agents, light harvesting, and catalysis.

Dendrimers, the first family of dendritic polymers reported in the literature, ideally have a perfectly branched, defect-free structure. Their synthesis involves successive cycles of protection, condensation, and deprotection steps to obtain each branching level (generation). This stepwise synthesis leads to strictly controlled branching and a very narrow molecular weight distribution ($M_w/M_n < 1.01$) [6]. However, many synthetic steps are required to produce high molecular weight materials due to the use of low molecular weight monomers as building blocks. On the other hand, high molecular weight hyperbranched polymers can be obtained in one pot self-condensation reactions of AB_n-type monomers, but with poor control over their structure. The completely random branching process used in the synthesis of hyperbranched polymers provides the least-defined structures, and broad molecular weight distributions ($M_w/M_n > 2$) in most cases [8, 9]. The third family of dendritic polymers is dendrigraft polymers, typically obtained in a generation-based scheme analogous to dendrimers, but with cycles of polymerization and grafting reactions [10, 11]. The use of polymeric chain segments rather than small molecules as building blocks leads to very rapid growth, such that high molecular weights are obtained in only a few steps. In contrast to dendrimers, the random distribution of coupling sites over the substrate makes arborescent polymer growth less sensitive to side reactions, since the structural defects are also randomly distributed over the whole molecule. Under appropriate conditions very high molecular weights can be achieved quickly for these systems, while maintaining fairly narrow molecular weight distributions ($M_w/M_n < 1.1$).

The unique structure and interfacial properties of dendritic polymers make them suitable as nanocarriers for applications such as drug and gene delivery, and as diagnostic agents [12–15]. While polymeric micelles obtained by the self-assembly of amphiphilic block copolymers have been widely investigated for the encapsulation of drugs within their hydrophobic core, they can be very sensitive to variations in their environment. For example, self-assembled nanocarriers can become unstable and dissociate into free chains when subjected to pH or temperature variations, or upon dilution in the bloodstream. In contrast, dendritic polymers can be

* Email: gauthier@uwaterloo.ca

Bioinspired Materials Science and Engineering, First Edition. Edited by Guang Yang, Lin Xiao, and Lallepak Lamboni.

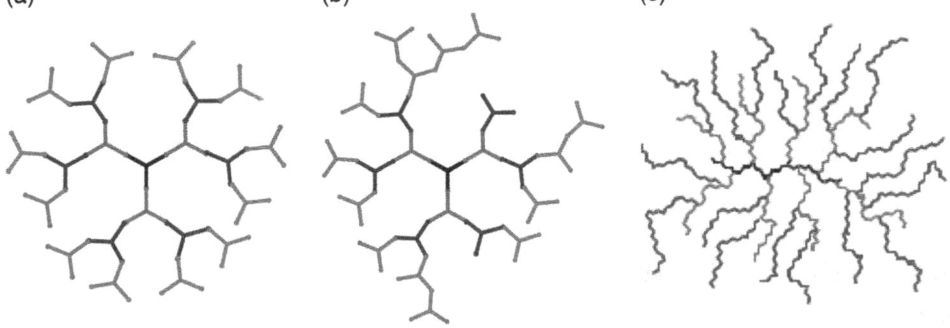

Figure 16.1 Schematic representation of dendritic polymers: (a) dendrimer; (b) hyperbranched polymer; and (c) dendrigraft polymer. *(See insert for color representation of the figure.)*

designed to behave like unimolecular micelles, namely, single molecules with an amphiphilic, covalently bound core-shell morphology. They are stable irrespective of their concentration or the solvency conditions, which makes them especially attractive as nanocarriers for drug delivery. The numerous advantages of dendritic polymers in drug delivery have been discussed in several papers [13, 14, 16–20]. Beyond providing a hydrophobic microenvironment that can greatly increase the solubility limit of the free drug in aqueous environments, these can provide a high loading capacity [13], are able to cross biological barriers [21–23], have tunable surface charges [22], and can improve the pharmaceutical and pharmacological properties of drugs, potentially without the need to alter the drug but rather through suitable design of the molecular architecture of the carrier [26, 27]. Their small size allows them to passively accumulate in tumors through the enhanced permeability and retention (EPR) effect [22, 25]. Active targeting can be achieved by introducing on the carrier ligands interacting with overexpressed cell receptors [26, 28], and the protection they afford minimizes drug degradation and loss [29, 30]. In these nanocarriers systems, the bioactive compound can either physically interact with the dendritic scaffold or be covalently conjugated with functional groups located on the branching units or the terminal units, and their release can be triggered and/or tuned in the biological environment through pH, enzymatic, redox, irradiation, or temperature-dependent processes [16, 18, 31].

In this chapter, we provide an overview of methods currently available for the synthesis, modification, and drug delivery application of dendritic polymers with various structures. The review is divided according to the three main subclasses of dendritic polymers, namely dendrimers, hyperbranched polymers, and dendrigraft polymers, respectively. For each dendritic polymer type, we first briefly describe well-established and more recent synthetic routes. Some of these approaches are illustrated using either seminal articles or examples which we believe illustrate well that particular strategy. We then discuss specific dendritic polymers showing promise as nanocarriers for drug delivery applications. While many systems have potential applications in drug delivery, only selected examples will be discussed here. Special attention is provided to the different modification methods proposed to achieve desirable nanocarrier properties. Through this approach, we hope to spark new ideas and to motivate continued investigations in this emerging research area.

16.2 Dendrimers

The most extensively investigated subset of dendritic polymers, dendrimers, were first synthesized in a so-called divergent (core-first) fashion and were termed "cascade molecules" by Vögtle in 1978 [32]. The word "dendrimer," coined by Tomalia in 1985 from the Greek words *dendri* (tree-branch-like) and *meros* (part), was first used in his paper on polyamidoamine (PAMAM) molecules [33]. Since this first report, many compositions and surface modifications have been reported for dendrimeric materials. The two most widely studied dendrimer families are the Tomalia-type PAMAM dendrimers and the Fréchet-type polyether compositions.

16.2.1 Dendrimer Synthesis: Divergent and Convergent Methods

Dendrimers have a core-shell structure with three distinct architectural components: (1) a central core; (2) an interior shell composed of repeating units; and (3) terminal functional groups forming the outer shell. The strictly controlled structure of dendrimers results from the layered assembly of repeating units surrounding the core, which is obtained through sequential reaction cycles. This can be achieved in two different ways, namely, by divergent (core-first) or convergent (arms-first) strategies as shown in Figure 16.2. These syntheses

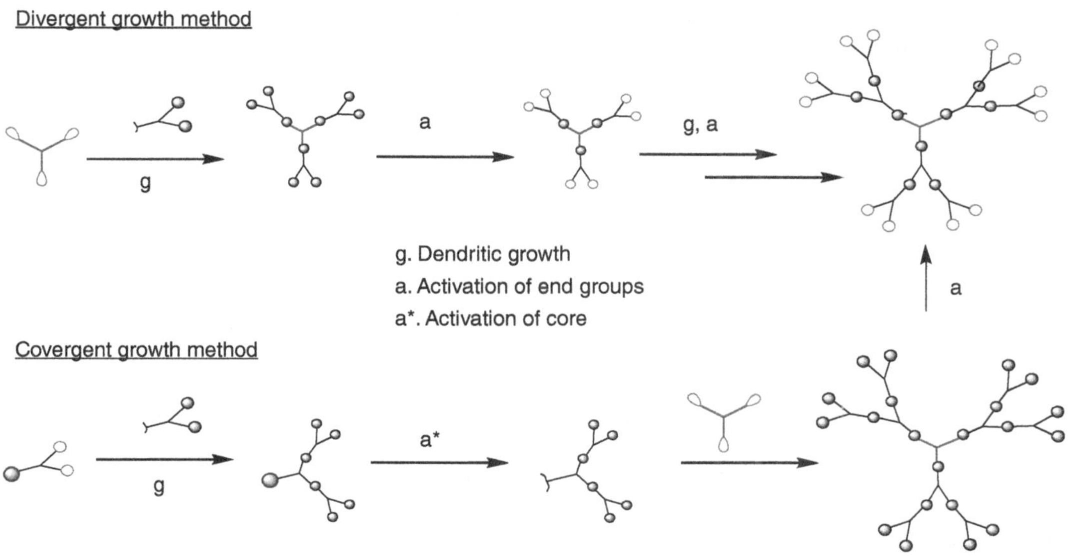

Figure 16.2 Divergent *vs.* convergent growth schemes for the synthesis of dendrimers.

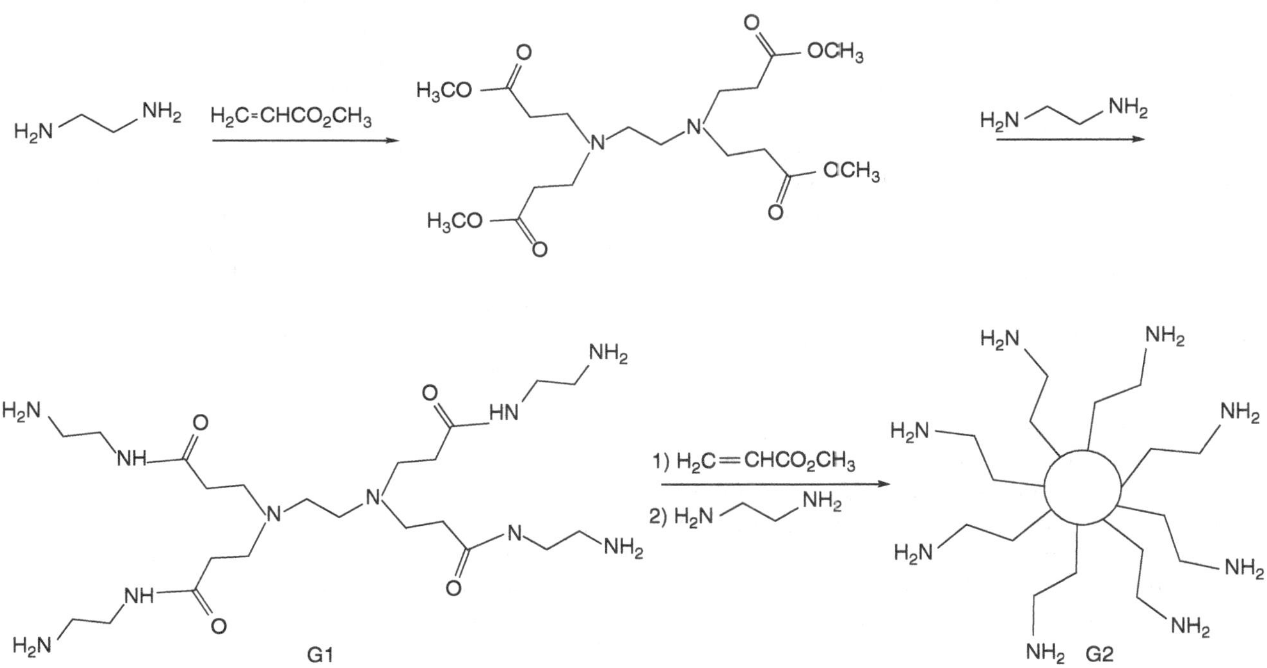

Figure 16.3 Divergent PAMAM dendrimer synthesis from an ethylenediamine (EDA) core.

involve repetitive stepwise growth and deprotection/activation protocols with careful purification steps for each generation.

In the divergent growth method a multifunctional core is coupled with an excess of reagent containing at least two protected branching sites, followed by purification and removal of the protecting groups. The synthesis of PAMAM dendrimers by the condensation of amines with acrylates is provided in Figure 16.3 to illustrate this strategy. An initiating core containing two or more amine functionalities is first reacted with an excess of methyl acrylate in a Michael addition, resulting in an alkyl ester branch for each amine proton. Amidation of the ester groups with a large excess of ethylenediamine yields the primary amine terminal groups of a "generation zero" (G0) dendrimer. Each amino group in this branched molecule is again reacted sequentially with methyl methacrylate and ethylenediamine to yield the first-generation (G1) PAMAM dendrimer, and so on. Each reaction cycle yields dendrimers of successive generation number, with increasing size and molecular weight.

Polyamidoamine or PAMAM dendrimers were the first dendritic materials commercialized, under the trade name Starburst®, and are probably the dendritic polymers most widely investigated to date. Dendrimers synthesized by the divergent approach require extensive purification at each reaction step, since a large excess of reagents is necessary to avoid defects in the structure due to incomplete reactions. This can be achieved by ultrafiltration or column chromatography. The divergent approach is currently the preferred strategy for the commercial production of dendrimers.

In the convergent growth method, individual dendrons or dendrimer wedges are first synthesized by protection/coupling/deprotection strategies analogous to the divergent approach described above, but starting from the periphery of the molecules. When the growing dendrons have reached the desired size (generation), they are coupled with a multifunctional core to give a complete dendrimer. In convergent growth a small number of reactive sites are involved at each step, which restricts the number of possible side reactions. Another advantage of this method over the divergent strategy is that the growth of the dendrons can be more easily monitored and controlled, since the dendrons are structurally simpler. Furthermore, there is a larger difference in molecular weight between the starting material and the product, which enables straightforward separation of the reactants and the product.

Newkome first reported a convergent dendron synthesis scheme in 1985 for the preparation of so-called arborol molecules, but the generation number and the molecular weight attained initially were very limited [34]. Hawker and Fréchet subsequently developed another convergent approach in 1990 [35], whereby they achieved the synthesis of polyether dendrimers by coupling benzyl bromide with 3,5-dihydroxybenzyl alcohol (DHBA) to create a dendrons (dendritic fragments), as shown in Figure 16.4. Conversion of the G1 dendritic benzyl alcohol to a benzylic bromide and further cycles of DHBA monomer coupling led to the following dendron generations. The dendritic fragments thus obtained, carrying a bromide functionality at their focal point, were finally coupled with a multifunctional core such as 1,1,1-tris(4′-hydroxyphenyl)ethane to obtain a symmetrical dendrimer. One of the most interesting advantages of the convergent method is the ability to obtain bifunctional dendrimers (e.g. Janus micelles) if dendrons containing functional groups of different types are coupled [36].

The main drawback of both the divergent and convergent dendrimer synthesis schemes described above is that they are tedious and time-consuming. Fortunately, several improvements have been suggested for these methodologies over the past years [37]. The need for deprotection or activation steps was thus eliminated by introducing the concept of orthogonality, namely, the use of functional groups reacting under different conditions [38]. For example, an orthogonal coupling strategy was applied to the synthesis of a G6 dendrimer from AB_2 monomer units containing two pairs of complementary coupling functionalities in just three steps [39]. The most commonly employed reaction for that purpose is the copper(I)-catalyzed Huisgen 1,3-dipolar cycloaddition (click reaction) between azides and primary acetylenes (CuAAC) as the coupling element, selectively forming 1,4-disubstituted 1,2,3-triazole rings as shown in Figure 16.5 [40, 41].

Unfortunately, complete removal of the copper (I) catalyst from the CuAAC reaction can be difficult, especially in the synthesis of dense (upper generation) dendrimers. Recently, Gonzaga et al. achieved the synthesis of high-generation dendrimers in high yield, and with only water-soluble salt byproducts [42]. The strategy was based on the metal-free cycloaddition of alkynes to azides with electron-deficient alkynes such as derivatives of acetylene dicarboxylate. The reaction between acetylenedicarboxylic acid and 2-bromopropanol yielded an activated monomer that coupled with the azides in a metal-free click reaction. More recently, click reactions were combined with orthogonal reactions to further improve dendrimer syntheses. For example, a combination of thiol–ene click addition (a reaction between a thiol and an alkene to form an alkyl sulfide) with esterification led to G5 dendrimers in only five steps [43]. Furthermore, a combination of thiol–ene click chemistry (radical-mediated addition of thiols across double bonds) with SN2 nucleophilic substitution reactions was employed to allow the synthesis of multifunctional dendrimers based on carbohydrate building blocks [44, 45]. In this approach (Figure 16.6), Roy and coworkers reacted a glucose-based AB_4 monomer (**8**), possessing four allyl groups and a N-(2-(glucosyloxy)ethyl)-2-chloroacetamide group, with thioacetylated **16** in the presence of a slight excess of 1 M MeONa–MeOH solution, to obtain dendrimer **17** (containing 24 allyl groups) in high yield. A subsequent thiol–ene reaction between **17** and thiogalactoside **18** provided glycodendrimer **19**, with 24 sugar residues. The uniformity of the structure was confirmed by gel permeation chromatography, which clearly showed a symmetrical Gaussian pattern with a low PDI of 1.064. Very recently, the same group reported the synthesis of G3 dendrimers with 108, 180, and 252 hydroxyl surface groups using AB3, AB5, and AB7 hypermonomers, respectively [46].

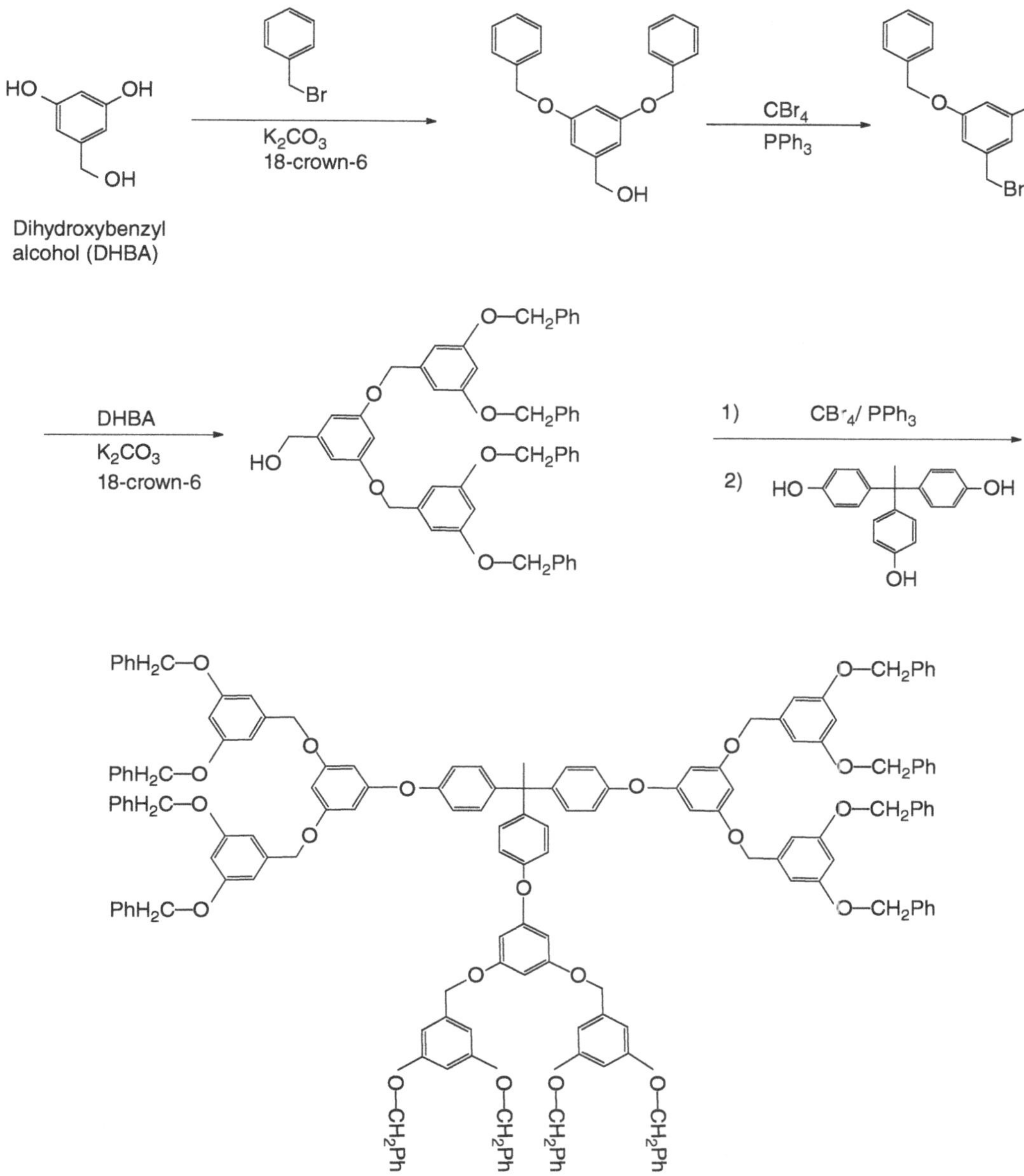

Figure 16.4 Synthesis of a Fréchet-type benzyl ether dendrimer by the convergent approach.

The dendrimers were constructed using a combination of microwave-assisted highly efficient CuAAC and thiol–ene reactions not requiring protection and deprotection steps. Low polydispersity indexes (PDI) of 1.03–1.08 confirmed the nearly monodisperse nature of the products.

Dendrimers possess several features making them attractive as nanocarriers for drug delivery applications, among which is the presence of open cavities allowing the encapsulation of poorly water-soluble drugs [47]. For example, dendrimers based on glycerol and succinic acid were used to encapsulate 10-hydroxycamptothecin (10HCPT), a poorly water-soluble anticancer drug [48]. The study showed that encapsulation not only increased the cellular uptake of the drug 16-fold as compared to the free drug, but it also improved the cellular retention of 10HCPT. The inability of nanocarriers acting by simple physical entrapment to precisely control the rate of drug release has led to a shift in attention toward the covalent conjugation approach. In this case, the release kinetics of the drug from the dendrimer are governed by the nature of the chemical linkage used to couple the drug with the dendrimer. A wide range of linking strategies have been reported for that approach. For example, drugs can be linked to the dendritic scaffold through amide, hydrazone, ester, imine, carbamate or disulfide groups, as well as enzymatically cleavable peptide

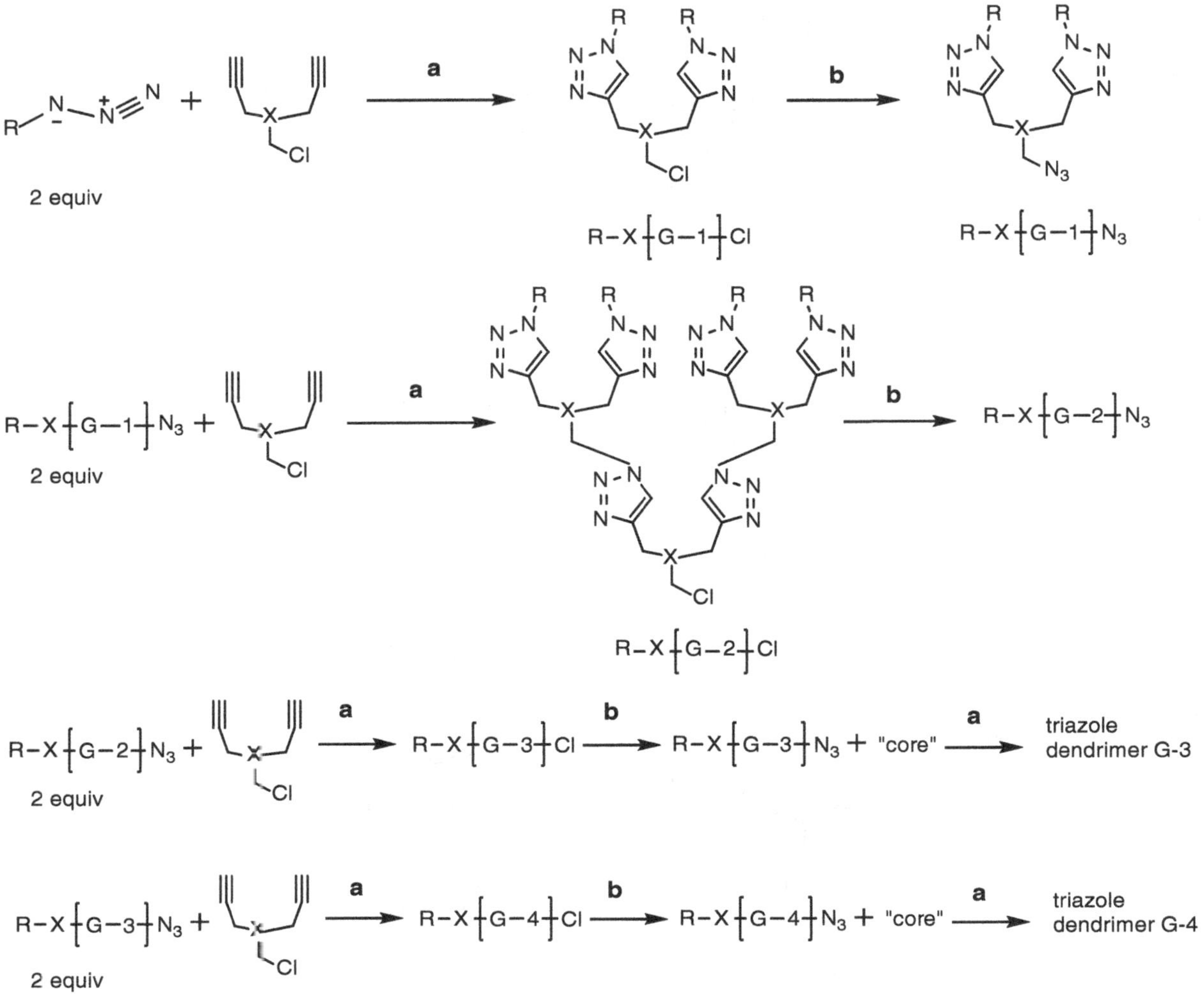

Figure 16.5 Convergent approach toward triazole dendrimers. (a) $CuSO_4$ (5 mol%), sodium ascorbate (10 mol%), H_2O/tBuOH (1:1); (b) 1.5 equiv NaN_3, CH_3COCH_3/H_2O (4:1).

sequences [31]. As a result, many researchers have attempted to develop new dendrimer systems taking advantage of their high density of functional groups for ease of drug conjugation. For instance, Kurtoglu et al. investigated the drug release properties of G4 polyamidoamine (PAMAM) dendrimer-ibuprofen conjugates with ester, amide, and peptide linkers. The amide-linked conjugates were relatively stable against hydrolysis, whereas the ester-linked conjugates displayed pH-dependent release behavior [20].

The development of targeted drug delivery is urgently required to further increase therapeutic efficacy and reduce toxicity. This was addressed, for example, using multifunctional dendrimers conjugated with drugs, targeting moieties, and imaging agents as drug delivery nanocarriers [49]. A PAMAM G5 dendrimer possessing 128 primary amino groups on its surface was thus used to covalently attach fluorescein isothiocyanate (FITC, an imaging agent), folic acid (FA) to target folate receptors over-expressed in cancer cells, and methotrexate (MTX), an anti-cancer drug conjugated through an ester bond. The results demonstrated that the PAMAM dendrimers were suitable for the targeted delivery of chemotherapeutic molecules capable of killing cancerous cells. Stimuli-responsive drug nanocarriers are another approach to promote therapeutic efficiency and reduce toxicity by controlling drug release at target sites. Different stimuli have been considered in the development of smart dendrimers for drug delivery, including specific enzymes, pH, temperature, and light. For instance, Zhang et al. prepared novel enzyme-sensitive dendrimer-doxorubicin conjugates for ovarian cancer therapy [50]. They obtained PEGylated peptide dendrimer–DOX conjugate nanocarriers through a two-step copper-catalyzed click reaction. The surface of the peptide dendrimer was modified with alkynyl groups twice.

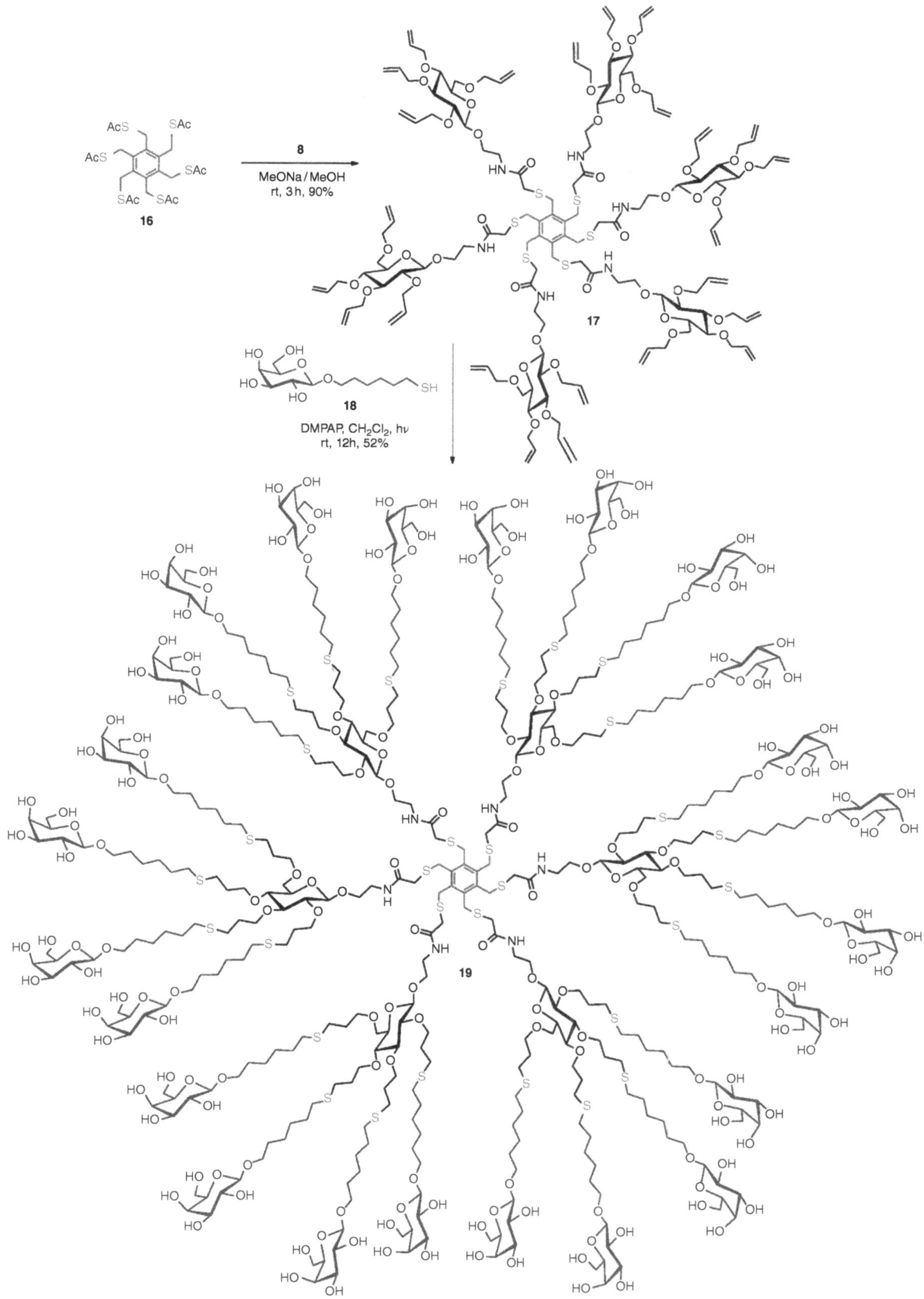

Figure 16.6 Synthesis of a dendrimer containing 24 galactoside groups. *Source:* Reproduced from [45], with permission from the Royal Society of Chemistry.

Azido-terminated PEG chains were first conjugated with a dendrimer via a click reaction, and then coupled with glycylphenylalanyl-leucylglycine tetra-peptide (Gly–Phe–Leu–Gly, GFLG)-modified DOX to the surface of the dendrimer through a click reaction as shown in Figure 16.7. The GFLG spacer served as an enzyme-responsive linker to connect the anti-cancer drug DOX to the PEGylated peptide dendrimer, to be cleaved by cathepsin B, over-expressed in tumor cells. The study showed a higher anti-tumor activity for the PEGylated peptide dendrimer–DOX conjugate as compared to free DOX.

Another study compared the anti-cancer activity of G4-paclitaxel dendrimers and PEG-paclitaxel conjugates. The data provided evidence that the G4-paclitaxel conjugate enhanced the activity of the drug

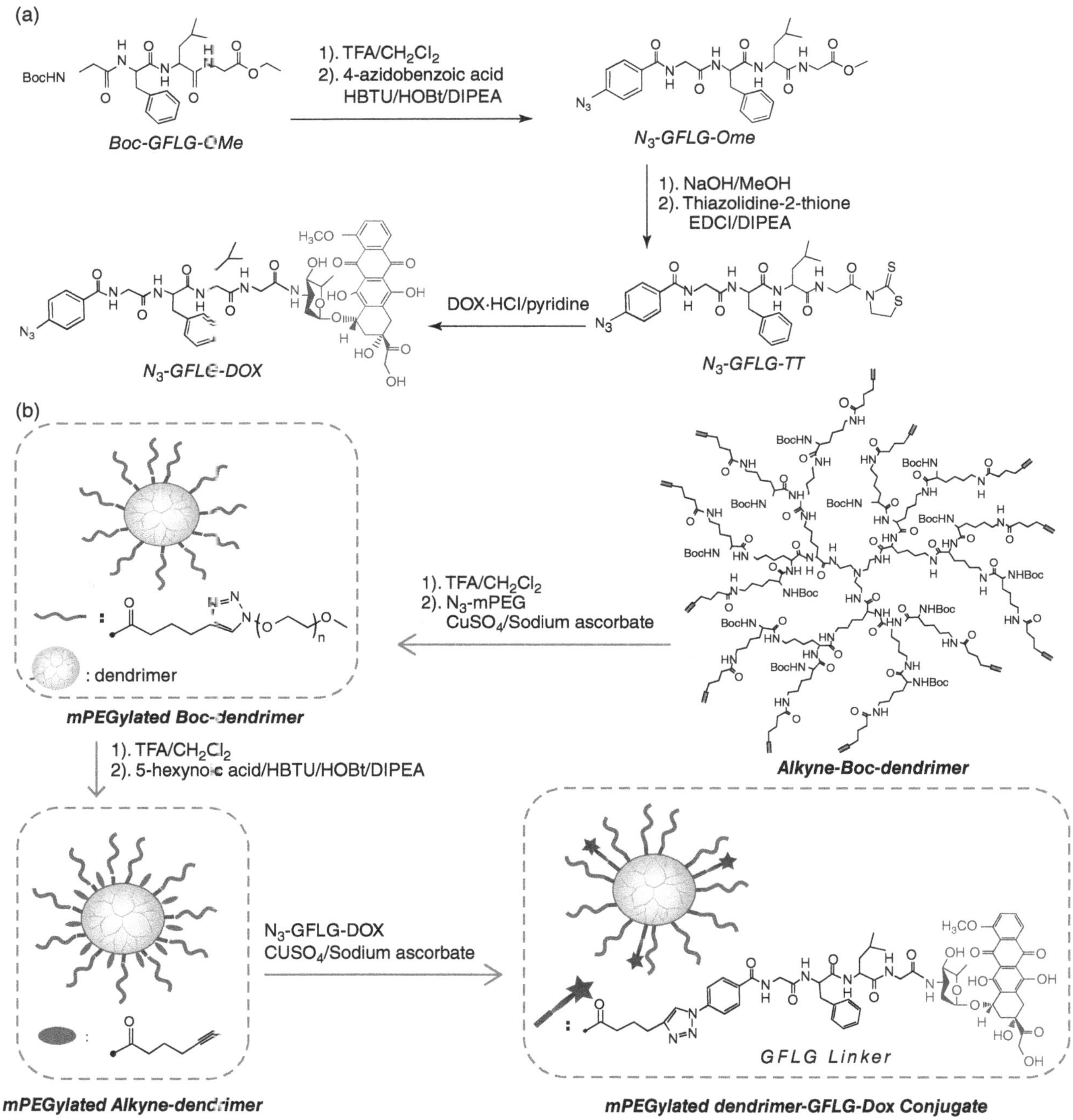

Figure 16.7 Synthesis of PEGylated dendrimer-GFLG-DOX conjugate. *Source:* Reproduced from [50], with permission from the Royal Society of Chemistry.

as compared to PEG-paclitaxel [51]. These results further demonstrate the great potential of dendritic polymers as drug carriers.

16.3 Hyperbranched Polymers

As early as 1952, Flory developed a theory concluding that highly branched polymers could be synthesized by the polycondensation of an AB_n-type monomer ($n \geq 2$), in which A and B functional groups can react with each other [52]. However, it was only in 1988 that the first experimental examples of hyperbranched polymers were reported by Gunatillake et al. [53]. Since then, hyperbranched polymers have also gained much attention in both academia and industry due to their unique properties and ease of preparation. They are obtained in a single reaction step (Figure 16.8), and hence are more readily available than other dendritic polymers, such as the dendrimers.

The synthetic strategies used for hyperbranched polymers fall into two main categories. In the single-monomer methodology (SMM), hyperbranched polymers are directly synthesized by the polymerization of an AB_n-type monomer, while in the double-monomer methodology (DMM), either two types of AB_n monomers or an $A_2 + B_3$ monomer pair serves to generate the molecules. The SMM methodology includes at least four specific approaches, depending on the reaction mechanism involved: The step-growth polycondensation of AB_n monomers, the self-condensing ring-opening polymerization (SCROP), the self-condensing vinyl polymerization (SCVP), and the proton transfer polymerization (PTP). The DDM methodology can be further divided into two main subclasses based on the monomer pair and reaction pathway selected, namely the "$A_2 + B_3$" approach and the couple-monomer methodology (CMM), which is a combination of the SMM and "$A_2 + B_3$" schemes [9]. Hyperbranched polymers are closer to conventional polymers than dendrimers in terms of molecular weight distribution, which can be even broader than for linear polymers. Furthermore, for hyperbranched polymers the addition of each monomer takes place randomly, and thus a large number of geometrical isomers can be formed, even for a given molecular weight [9].

The synthesis of hyperbranched polymers carried out as one-pot reactions offers limited control over the molecular weight of the products, and ultimately leads to gelation. Another problem is that intramolecular side reactions such as cyclization resulting from "backbiting" processes are common in these systems. One technique developed to minimize the occurrence of side reactions is a slow monomer addition protocol, whereby the monomer is added throughout the reaction. This method was termed "concurrent slow addition" [54, 55]. Polyfunctional initiators used in batch reactions have also shown promise in controlling the molecular weight and reducing the polydispersity index ($PDI = M_w/M_n$) of the product [54, 56.]

Kim and Webster reported the first example of a single-monomer polycondensation technique, using 3,5-dibromophenylboronic acid and aqueous sodium carbonate in the presence of a Pd catalyst [57]. A representation of this reaction and the resulting hyperbranched structure obtained are shown in Figure 16.9.

The most versatile approach to hyperbranched polymer synthesis is likely the double-monomer method (DMM), due to the wide range of monomers and chemical functionalities to which it can be applied. A hyperbranched analogue of the commercially available PAMAM dendrimers (HYPAM) was thus synthesized by a one-pot method as shown in Figure 16.10 [58]. This approach can be described as an $A_6 + B_6$ reaction of tris(2-aminoethyl)amine with tris(2-di(methyl acrylate)aminoethyl)amine.

In 1995, Fréchet et al. [59] first reported the radical polymerization of monomers by the so-called self-condensing vinyl polymerization (SCVP) technique to synthesize hyperbranched polymers with numerous reactive groups. This technique uses vinyl monomers also containing a pendant group that can be converted into an initiating moiety (represented as AB*), and thus can introduce branches in the polymer structure. Chain extension of the branched polymer with linear polymer chains can also be achieved by the atom

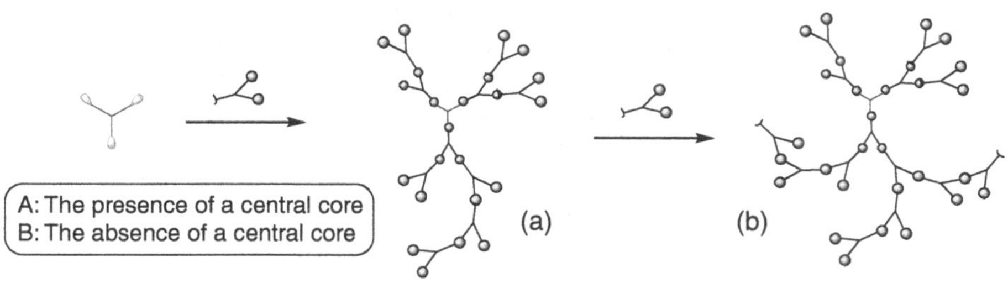

Figure 16.8 Core (a) and non-core (b) methods for the synthesis of hyperbranched polymers.

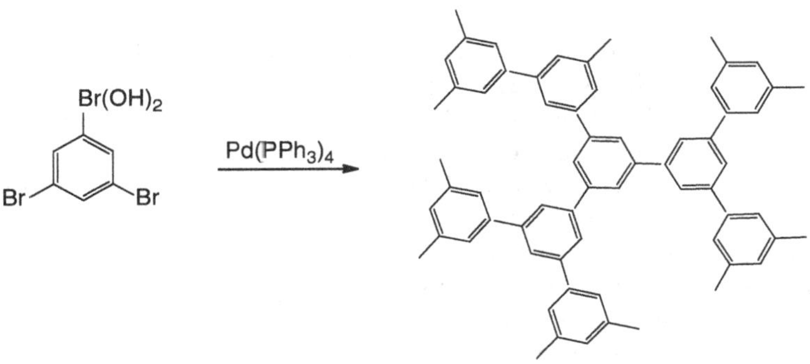

Figure 16.9 Hyperbranched polyphenylene synthesis by the single-monomer polycondensation method.

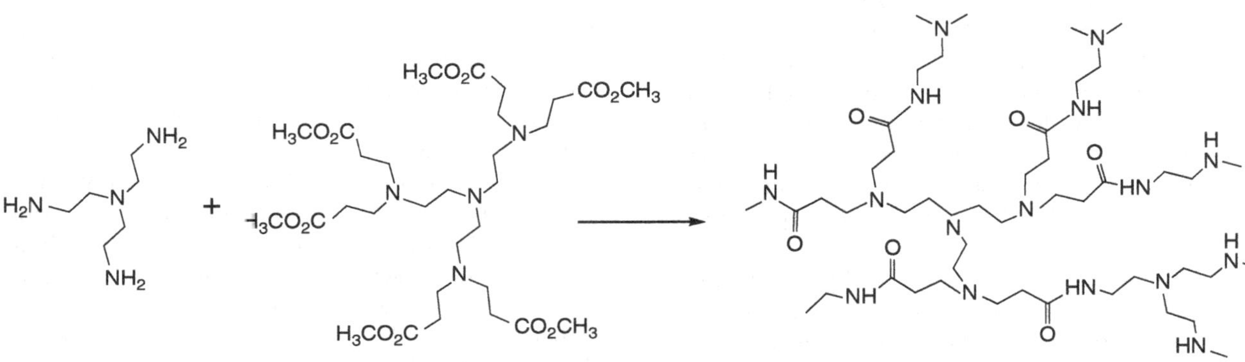

Figure 16.10 Synthesis of hyperbranched PAMAM (HYPAM).

transfer radical polymerization (ATRP) technique. The combination of these techniques can be used to produce hyperbranched polymers with various architectures and compositions, some of which may be promising as nanocarriers for drug delivery applications. For instance, Porsch et al. reported the synthesis of high molecular weight hyperbranched dendritic-linear polymer (HBDLP) hybrids in a two-step process via a combination of self-condensing vinyl (co)polymerization (SCV(C)P) and ATRP (Figure 16.11) [60]. A hydrophobic hyperbranched macroinitiator (HBMI) was first produced by self-condensing vinyl (co)polymerization (SCV(C)P) utilizing atom transfer radical polymerization (ATRP), using 1,1,1-tris(4-(2-bromoisobutyryloxy) phenyl)ethane (TBBPE) as a trifunctional ATRP initiator, and the inimer 2-(2-bromoisobutyryloxy)ethyl methacrylate (BBEMA) in combination with hexyl acrylate (HA). The active end groups were subsequently used to chain-extend the hyperbranched macroinitiators (HBMIs) with linear, hydrophilic poly(ethylene glycol) methacrylate (P(OEGMA)) segments via ATRP, to form a high molecular weight amphiphilic HBDLP. The chain-extension reactions proceeded in a controlled manner, resulting in polymers with reasonably uniform GPC traces and low polydispersity indices (PDI = 1.3–1.7).

The HBDLP was used to encapsulate doxorubicin to achieve sustained delivery. In addition, one of the HBMIs was chain-extended with a fluorine-containing copolymer, so that the unimolecular micelles would have potential as theranostic nanoagents.

The therapeutic efficacy of many anti-cancer drugs is limited by their poor penetration into tumors; targeted drug delivery is needed to address this challenge. Hyperbranched polymers are potentially useful as carriers of active drugs and, when coupled with targeting ligands, may further enhance the therapeutic efficacy and reduce the toxicity of the drug. For example, the RGD peptide is one of the promising ligands for targeting $\alpha v\beta 3$ integrin, over-expressed in various cancer cells. Seleci et al. [61] fabricated a tumor-targeting amphiphilic star-hyperbranched block copolymer derived from a hyperbranched substrate, namely, hyperbranched poly(methyl methacrylate)-*block*-poly(2-hydroxyethyl methacrylate) (PMMA-*b*-PHEMA), as a targeted drug delivery nanocarrier. The copolymer was prepared via sequential visible light-induced self-condensing vinyl polymerization (SCVP) and conventional vinyl polymerization. This involved a manganese-based visible light photo-initiator system ($Mn_2(CO)_{10}$) in the copolymerization of methyl methacrylate (MMA) with 2-bromoethyl methacrylate (BEMA), possessing both polymerizable and initiating sites within its structure. The subsequent visible light-induced polymerization of HEMA in the presence of the hyperbranched PMMA macroinitiator resulted in the formation of amphiphilic PMMA-b-PHEMA copolymers, possessing a hydrophobic hyperbranched PMMA core and hydrophilic PHEMA

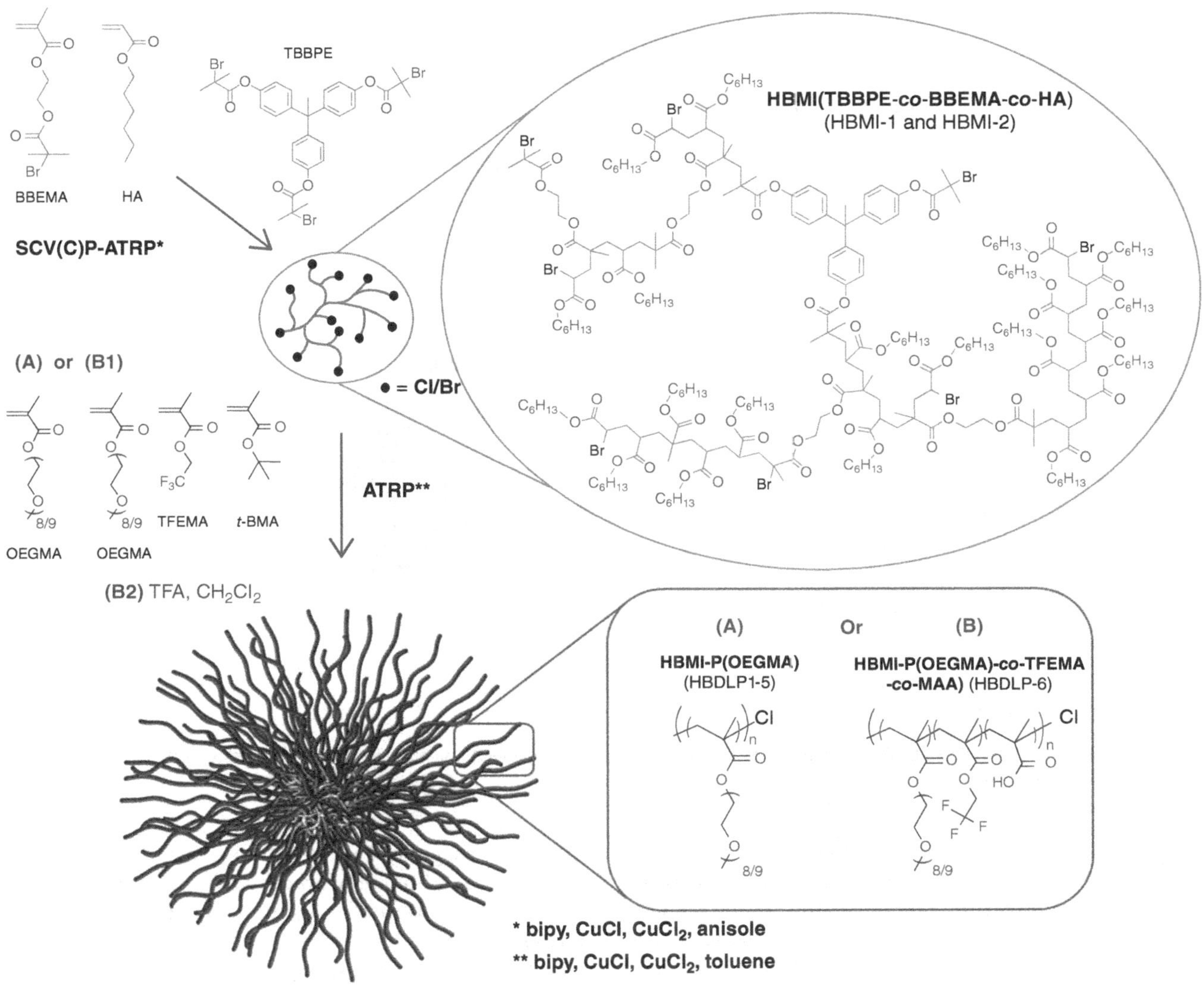

Figure 16.11 Synthesis of hyperbranched dendritic-linear polymers by sequential SCV(C)P-ATRP and ATRP techniques. *Source:* Reprinted with permission from [60]. Copyright 2014 American Chemical Society. (*See insert for color representation of the figure.*)

chains. DOX was successfully encapsulated within these copolymer micelles. It was concluded that the RGD-mediated amphiphilic star-hyperbranched block copolymers were more efficiently internalized by cells than star polymers without the targeting moiety on their surface.

While the physical entrapment of drugs by hyperbranched polymers has been successfully demonstrated, some limitations still exist for efficient drug encapsulation and controlled release by these systems. The major limiting issue is the relatively rapid and uncontrollable release of the drug from the hyperbranched polymer core. The preparation of covalently bonded hyperbranched polymer-drug conjugates stable in buffered solutions is the most obvious solution to this problem, and thus much research has focused on covalent drug attachment. Several types of covalent linker groups have been used to regulate drug release from HBP nanocarriers [31], as well as co-delivery systems based on HBP designed to promote therapeutic efficiency through synergistic effects. For example, Zhu et al. reported the covalent conjugation of a polyacylhydrazone (HPAH) hyperbranched polymer with DOX through acid-sensitive acylhydrazone bonds [62]. Subsequently, the autophagy inhibitor LY was loaded into the HPAH-DOX micelles core to yield an effective chemotherapeutic system for oral squamous cell carcinoma. The hyperbranched polymer HPAH was synthesized by the polycondensation of 2,3-butanedione (BD) and 1-(2-aminoethyl)piperazine tripropionylhydrazine (AEPNHNH$_2$), yielding a weight-average molecular weight (M_w) of 4.0×10^3 g/mol and a polydispersity index (PDI) of 1.6 (Figure 16.12) [63]. Then HPAH-DOX

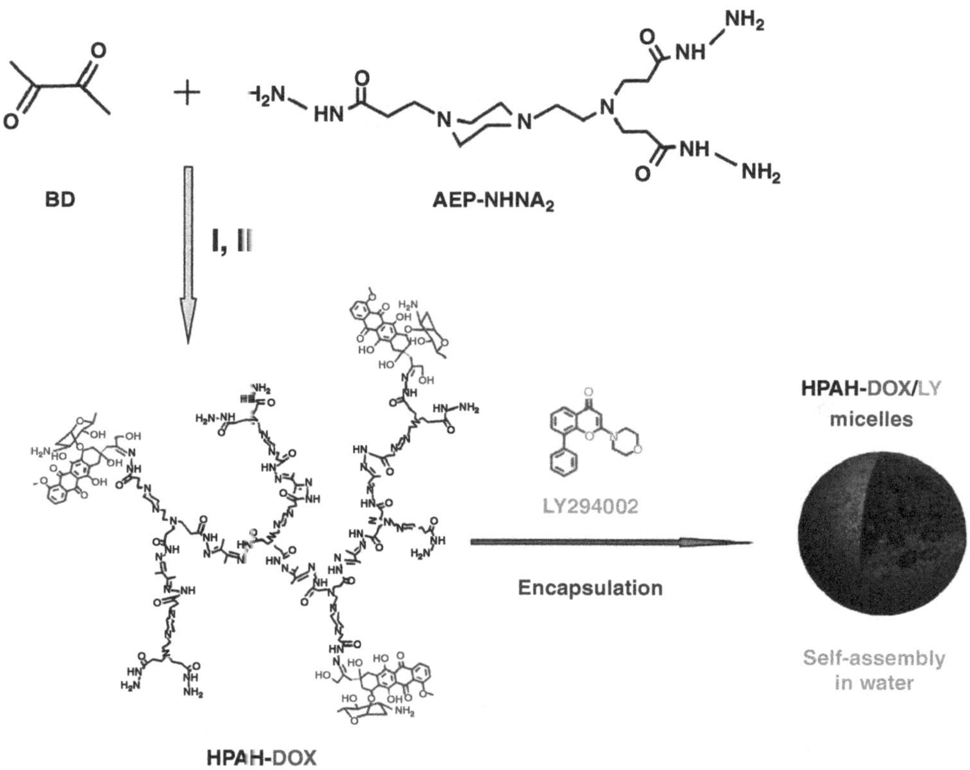

Figure 16.12 Synthesis of HPAH–DOX and autophagy inhibitor LY-loaded HPAH-DOX micelles. *Source:* Reprinted with permission from [63]. Copyright 2014 American Chemical Society.

was prepared by conjugation of DOX onto the surface of HPAH, through further condensation of its ketone functionality with the acylhydrazine groups. The study showed that DOX and LY were released in a controlled manner. Furthermore, the *in vitro* investigation revealed that, as compared to the HPAH-DOX micelles and a physical mixture of HPAH-DOX with LY, the LY-loaded HPAH-DOX micelles resulted in the synergistic inhibition of tumor cell proliferation. This HBP-based stimulus-responsive co-delivery system for chemotherapy drugs and autophagy inhibitors provides a promising platform for clinically applicable combination therapy.

Very recently, the same group synthesized redox-responsive hyperbranched polymer-drug conjugates by coupling a thiol-functionalized hyperbranched polymer with thiol-containing drugs [64]. The HBP poly((*S*-(4-vinyl) benzyl S′-propyltrithiocarbonate)-*co*-(poly(ethylene glycol) methacrylate)) or poly(VBPT-*co*-PEGMA) with multiple thiol groups was synthesized by a one-pot reaction via reversible addition-fragmentation chain transfer (RAFT) copolymerization (Figure 16.13). The branched poly(VBPT-*co*-PEGMA) substrate was obtained by RAFT copolymerization of a vinyl-containing trithiocarbonate RAFT agent and PEGMA with a weight-average molecular weight of 1.8×10^4 g/mol and a PDI around 2.0. The trithiocarbonate worked as both monomer and RAFT chain transfer agent. The trithiocarbonate terminal groups of poly(VBPT-*co*-PEGMA) were easily transformed into thiol groups using propylamine, to be coupled with the hydrophobic thiol-containing anticancer drug 6-mercaptopurine (MP) by a thiol-disulfide exchange reaction, resulting in the formation of the poly(VBPT-*co*-PEGMA)-S-S-MP conjugate.

Since the concentration of glutathione (GSH), a thiol-containing tripeptide capable of reducing disulfide bonds in biological environments, is much higher in tumors than in normal tissues, the redox-responsive HBP-drug conjugates selectively released the drug at tumor sites and thus increased the anticancer drug efficacy. This redox-responsive HBP-drug conjugate provides a promising platform for the delivery and controlled release of thiol-containing drugs or biological molecules.

Given the broad range of synthetic techniques and potential applications reported for hyperbranched polymers, the reader is referred to excellent reviews for more details on the synthetic strategies, characterization, properties, functionalization, and potential applications of hyperbranched polymers as therapeutic agents [22, 26, 65].

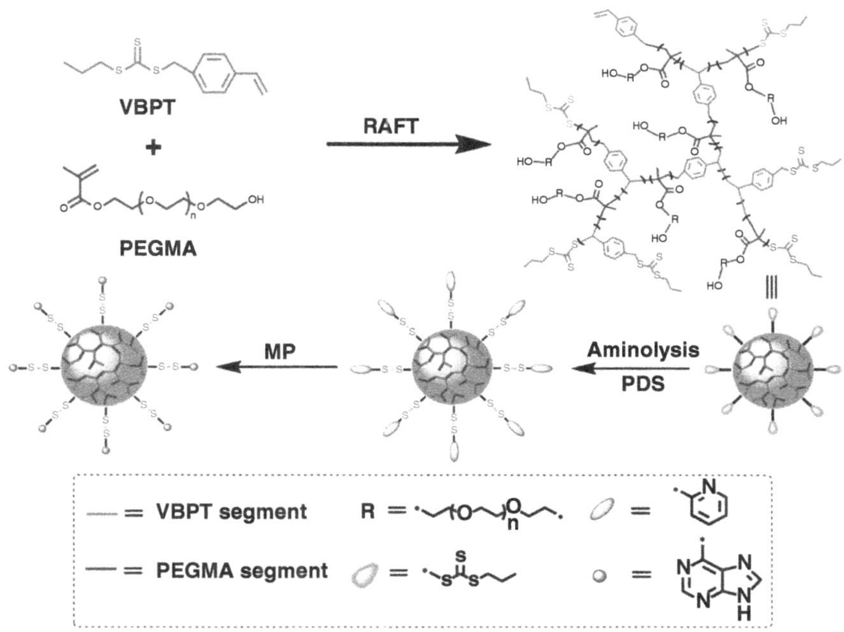

Figure 16.13 Synthetic route to poly(VBPT-*co*-PEGMA)-S-S-MP. *Source:* Reprinted with permission from [64]. Copyright 2014 American Chemical Society.

16.4 Dendrigraft Polymers

Dendrigraft polymers, referring collectively to arborescent and dendrimer-like star polymer structures, are the third and most recently reported subset of dendritic polymers [10, 11]. Whereas monomers are invariably employed in constructing dendrimers, reactive polymers serve as building blocks in successive grafting reactions for the synthesis of dendrigraft systems. Since a large number of coupling sites are randomly distributed on the grafting substrates, a rapid increase in molecular weight is observed per arborescent polymer generation and high molecular weight species are obtained in a few steps, while narrow molecular weight distributions ($M_w/M_n < 1.1$) are maintained in most cases. The closely related dendrimer-like star polymers, in contrast, have a terminally branched structure achieved through coupling sites strictly located at the chain ends of the previous generation [66]. As a result, a lower rate of molecular weight increase per generation is observed in these systems. Dendrigraft polymers have been synthesized by divergent (core first) approaches, including both *grafting onto* and *grafting from* methodologies, and by convergent (arms first) approaches, that are essentially *grafting through* methodologies. In the following sections we do not intend to present a comprehensive overview of all the synthetic strategies developed to date, but rather to provide examples illustrating the potential of these materials in drug delivery applications. For a more detailed overview of the synthesis and properties of dendrigraft polymers, the reader is directed to a review by Teertstra and Gauthier [5].

16.4.1 Divergent Grafting Onto Strategy

The divergent approach relies upon successive grafting reactions starting from a linear substrate (equivalent to the core in dendrimer syntheses) as illustrated in Figure 16.14. The first examples of dendrigraft polymer syntheses, using grafting onto strategies, were reported independently by Gauthier and Möller [10] and by Tomalia et al. [11] in 1991. Gauthier and Möller achieved the synthesis of branched polystyrenes, denominated arborescent polymers, by anionic grafting, while Tomalia et al. used cationic grafting to obtain branched polyethylenimines called Comb-burst® polymers. The synthesis of these materials starts by grafting linear side chains onto a linear polymer substrate functionalized with coupling sites to yield a comb-branched (generation zero or G0) polymer. The introduction of further coupling sites on the G0 polymer and grafting with side chains leads to a generation one (G1) polymer, and so on for subsequent generations.

The first example of the synthesis of arborescent polystyrene used chloromethyl coupling sites on the polystyrene substrates, and "living" polystyryllithium as side chains [10]. Repetition of the functionalization and grafting reactions as shown in Figure 16.15 led to arborescent polymers of generations up to G3. To achieve a high grafting yield, it was necessary to "cap" the polystyryl anions with a single 1,1-diphenylethylene unit to suppress the metal-halogen exchange reactions competing with coupling. To avoid the use of chloromethyl methyl ether, a potent carcinogen, a modified coupling method using acetyl coupling sites was subsequently developed by Li and Gauthier [67]. It was also

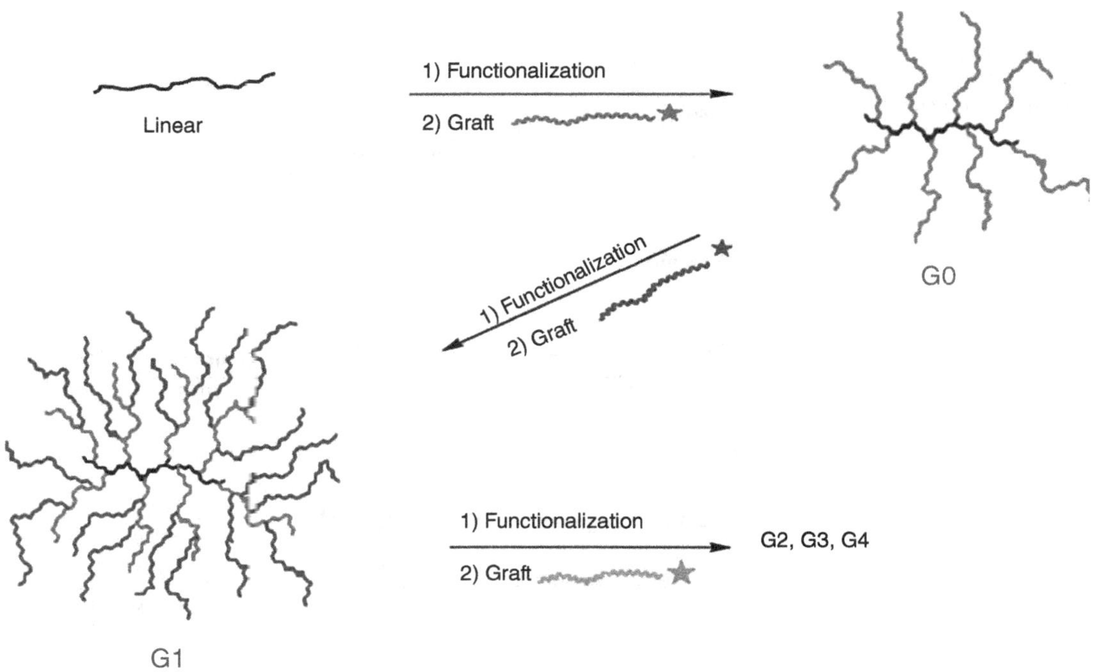

Figure 16.14 General grafting onto scheme for the synthesis of arborescent polymers.

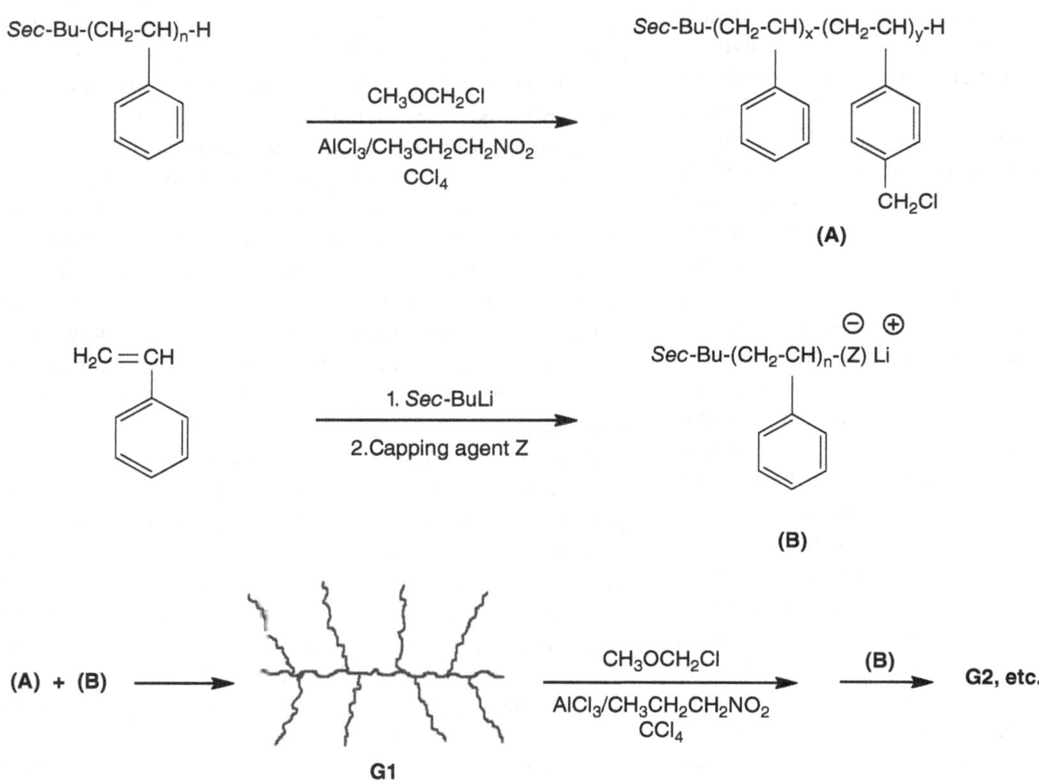

Figure 16.15 Arborescent polystyrene synthesis by grafting onto chloromethylated polystyrene substrates.

necessary to cap the polystyryl anions with a few 2-vinylpyridine units in that case, to minimize side reactions of the macroanions and increase the grafting yield. Repetition of acetylation and anionic grafting cycles led to generations of arborescent polymers with a low PDI (1.07–1.09).

The grafting onto strategy developed for arborescent polystyrene was subsequently extended to the synthesis of different arborescent homopolymers of polyisoprene [68] and polybutadiene [69], but also to graft copolymers potentially useful as drug nanocarriers such as arborescent polystyrene-*graft*-poly(2-vinylpyridine) [70] and polystyrene-*graft*-poly(methacrylic acid) [71]. Fully biocompatible arborescent polymers constructed from poly(γ-benzyl L-glutamate) (PBG) side chains were also synthesized recently via successive grafting reactions (Figure 16.16) [72]. In this case, linear PBG building blocks were obtained by ring-opening polymerization of γ-benzyl L-glutamic acid N-carboxyanhydride initiated by *n*-hexylamine. A fraction of the benzyl groups on a linear PBG substrate were deprotected and coupled with linear PBG using N,N'-diisopropylcarbodiimide and 1-hydroxybenzotriazole as coupling agents to give a generation zero (G0) arborescent PBG. Further partial deprotection and grafting cycles led to arborescent PBG reaching $M_n > 10^6$ g/mol and $M_w/M_n = 1.06$ for a G3 polymer. Amphiphilic PBG-based arborescent copolymers incorporating a hydrophilic shell of poly(ethylene oxide), polyglycidol or poly(glutamic acid), potentially very useful for drug release, have also been reported [73]. Very recently, doxorubicin hydrochloride (DOX·HCl) was loaded into hydrophilic arborescent poly(L-glutamic acid)-g-poly(ethylene oxide) copolymers via electrostatic interactions between the anionic polypeptide core and the cationic drug, to achieve a loading content of 14% w/w and a loading efficiency of 83%. Drug release from the loaded micelles was sustained over one month at 37 °C. The release rate was slow at physiological pH (7.4) but significantly increased at pH 5.5, mimicking the endosomal environment, suggesting pH-responsive release of DOX from the micelles [74].

The arborescent polystyrene-*graft*-poly(2-vinylpyridine) copolymers were also tested as model drug delivery systems. Njikang et al. thus investigated the encapsulation in aqueous solutions of various polycyclic aromatic hydrocarbon (PAH) probes within the hydrophobic core of arborescent polystyrene-*graft*-poly(2-vinylpyridine) (PS-*g*-P2VP) unimolecular micelles with different structures [75]. The study showed that the copolymers with a high polystyrene content were far better sorbents, albeit the capacity difference was less important for the less hydrophobic probes such as naphthalene. Variations in the structure of the unimolecular micelles also had a significant influence on the solubilization process: The highly hydrophobic probes were solubilized mostly in the hydrophobic core, while those of higher polarity remained in the ionic shell and the core-shell interfacial region. The solubilization capacity of copolymers of similar composition was also lower for the higher generation samples, presumably due to their increased structural rigidity. In a subsequent study, the release characteristics of these systems loaded with indomethacin and lidocaine as model drugs were investigated [76]. The *in vitro* release study of indomethacin and lidocaine in dilute HCl solutions displayed sustained release characteristics, with an initial burst followed by slower release. The arborescent PS-*g*-P2VP copolymers used in this investigation were obviously not biocompatible, and the *in vitro* test conditions different from those used in biological systems. These results were nevertheless useful in showing that the solubilization and release characteristics of arborescent copolymers could be controlled through variations in their structure and composition, and to understand the influence of branching on the release profile.

The efforts dedicated initially to the synthesis of dendrigraft polymers, namely, the arborescent and Comb-burst® polymers, have inspired many researchers to develop other strategies for the synthesis of related materials in recent years. A broad range of branched polymer architectures have been reported in the literature as a result of this, and the naming of these polymers likewise varied extensively. Selected examples of the different architectures reported, and a discussion of their potential as drug nanocarriers are provided below.

Dendrigraft polymers can be designed to have a core-shell morphology, to behave like covalently bonded unimolecular micelles. These can be designed to have therapeutic drug-carrying ability, to act as imaging agents, and may be functionalized with targeting ligands for a wide range of biomedical applications. For example, drug nanocarriers that also act as imaging agents can be easily monitored *in vivo* to determine their biodistribution by fluorescence microscopy [15]. Different strategies have been proposed to obtain fluorescent drug nanocarriers, including the encapsulation or conjugation of organic dyes, and the use of inorganic fluorescent nanocarriers or other metallic nanoparticles.[77] Yang et al. thus reported novel biodegradable photoluminescent polymers (BPLPs) [78] that were used by Gong et al. to construct fluorescent multi-arm star amphiphilic block copolymer drug nanocarriers [79]. This was achieved by first reacting citric acid with 1,8-octanediol, followed by condensation with L-cysteine via their carboxyl group to yield a hydrophobic fluorescent BPLP segment as shown in Figure 16.17. In a second step, an amphiphilic linear block copolymer BPLP-PEG-cRGD was synthesized by conjugating the hydrophilic PEG segment with cRGD peptide, and then with BPLP. Finally, fluorescent unimolecular micelles H40-BPLP-PEG-cRGD

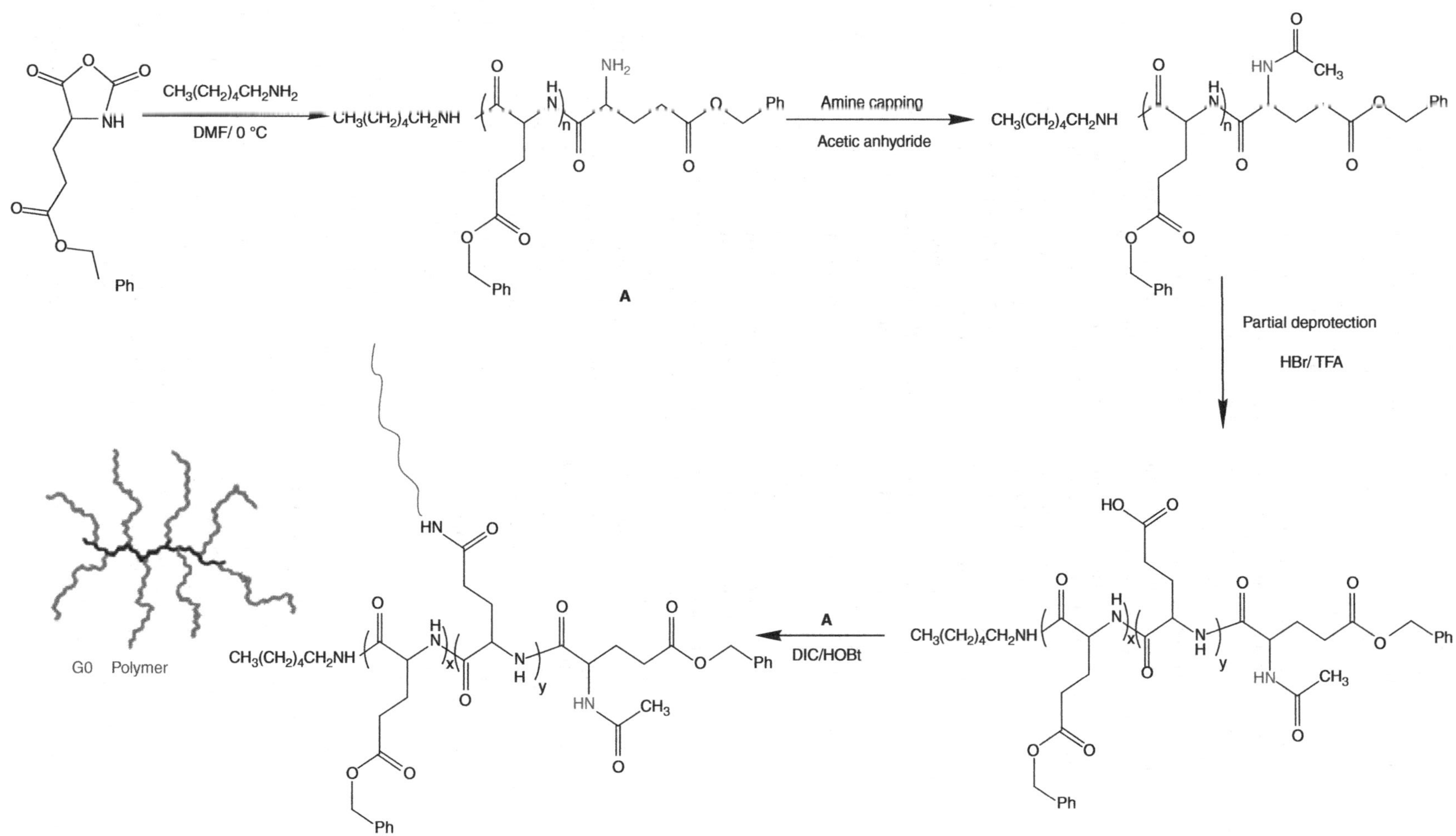

Figure 16.16 Synthesis of G0 arborescent PBG with a comb-branched structure.

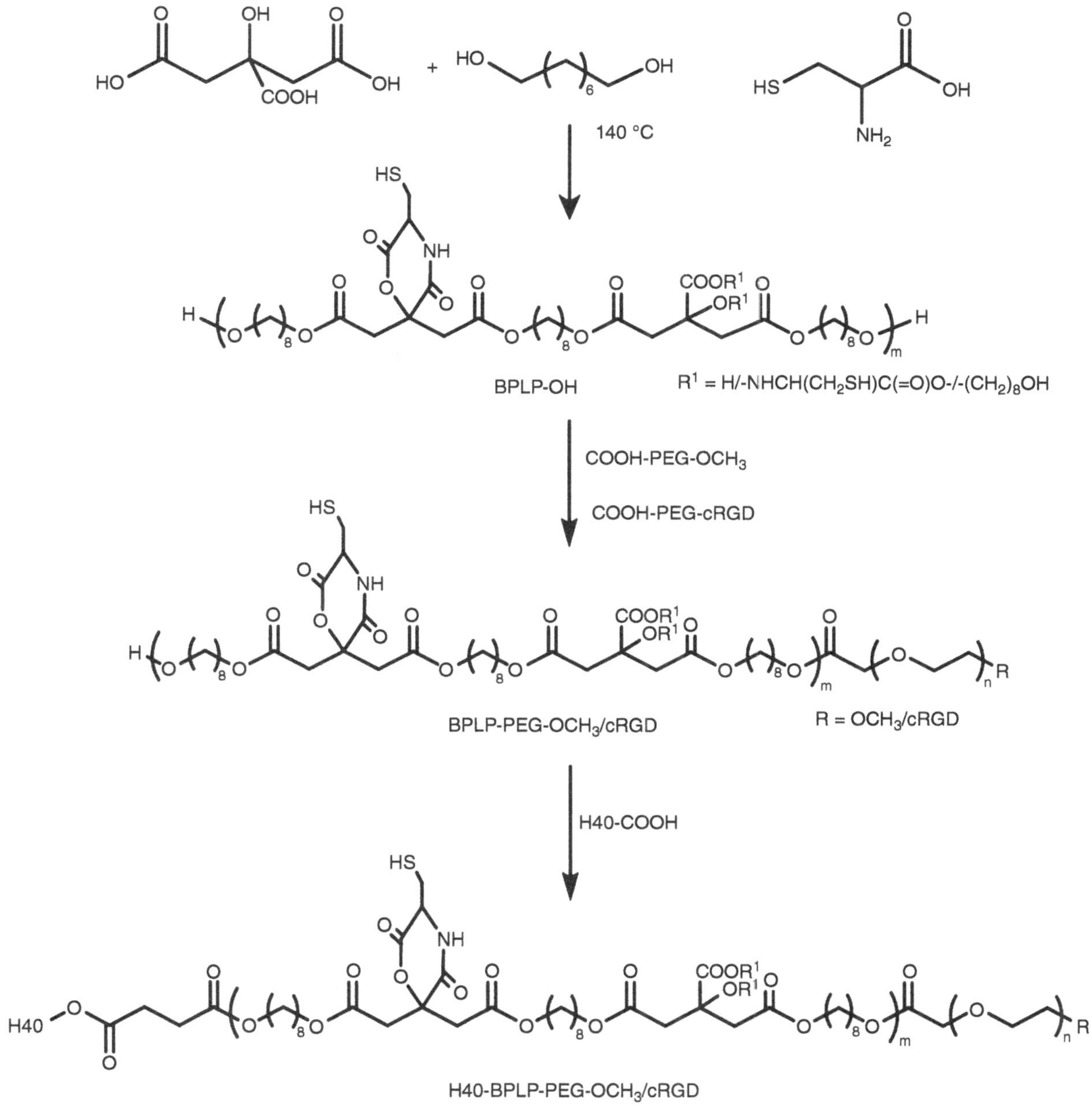

Figure 16.17 Synthesis of fluorescent multi-arm star amphiphilic block copolymer H40-BPLP-PEG-OCH_3/cRGD. *Source:* Reprinted from [79], with permission from John Wiley & Sons, Inc.

were obtained with a weight-average molecular weight of 1.6×10^5 g/mol and PDI of 1.43 by conjugating the BPLP-PEG-cRGD block copolymer with Boltron® H40 (a fourth-generation hyperbranched polymer) core through an esterification reaction. The successful encapsulation of DOX within the hydrophobic core was demonstrated by achieving a drug loading of 15.7 wt %. The multi-arm star amphiphilic block copolymer H40-BPLP-PEG-OCH_3/cRGD thus not only proved to be a promising candidate as a tumor-targeting drug nanocarrier, but also as a bioimaging probe.

One major limitation of the simple physical entrapment of a drug within the core of dendritic polymers is that it is difficult to precisely control the rate of drug release. This may lead to rapid drug release before the dendritic polymers reach their designated targets, and thus result in low therapeutic efficacy, apart from eliciting severe side effects [18]. As mentioned earlier, much attention has been paid to covalent drug conjugation to further improve drug delivery efficiency. However, the covalent conjugation chemistry must be carefully selected to allow bond cleavage and release of the drug

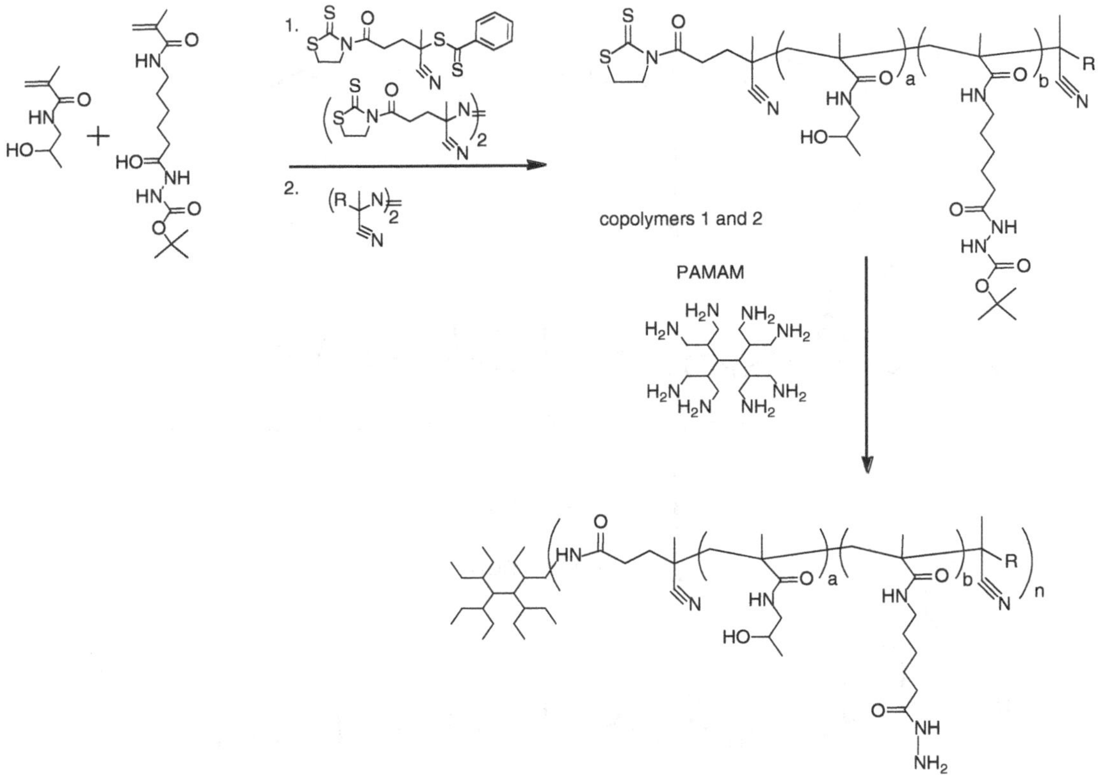

Figure 16.18 Synthesis of star copolymers prepared by RAFT polymerization.

molecules under specific biological conditions. For instance, water-soluble and pH-responsive star polymers analogous to the dendrimer-like systems, including imaging probes, were obtained for the treatment and/or visualization of solid tumors [80]. The star polymer nanocarriers were prepared by coupling semitelechelic or heterotelechelic *N*-(2-hydroxypropyl)methacrylamide copolymers containing thiazolidin-2-thione end groups with the terminal amino groups of PAMAM dendrimers. Controlled radical RAFT polymerization was used for the synthesis of a multivalent copolymer of *N*-(2-hydroxypropyl)methacrylamide with a Boc-protected hydrazide group-bearing monomer (copolymers 1 and 2 in Figure 16.18). The anticancer drug DOX was then attached via pH-sensitive hydrazine linkages. Finally, dibenzocyclooctyne (DBCO), a fluorescent dye, was conjugated with the star polymer by a copper-free click reaction. The study demonstrated that the nanocarriers were able to effectively deliver DOX to solid tumors in a controlled manner.

16.4.2 Divergent Grafting from Strategy

The divergent grafting from method starts with a substrate acting as a multifunctional initiator, from which polymer chains are grown. Unfortunately, in most cases it is difficult to characterize the structure of the graft polymers generated by this approach, and the molecular weight distribution of the products is generally broad; thus this method has received less attention than the grafting onto strategy. In spite of these limitations, this approach allowed the synthesis of novel dendritic polymers inaccessible by other methods. The first example of this methodology was reported in 1995 by Six and Gnanou for the synthesis of dendrimer-like star poly(ethylene oxide) through a strictly controlled terminal branching process [81]. This involved the synthesis of a star polymer, and conversion of the terminal groups to a difunctional initiator, to allow chain extension for the synthesis of the next generation. They also obtained amphiphilic and water-soluble dendrimer-like star polymers by similar strategies [82–84]. The preparation of a dendrimer-like star polymer incorporating three poly(ethylene oxide) and six poly(acrylic acid) side chains by a combination of living anionic polymerization of ethylene oxide (EO) and subsequent atom transfer radical polymerization (ATRP) of *tert*-butyl acrylate, followed by hydrolysis, is provided in Figure 16.19 to illustrate this strategy.

In comparison to traditional linear polymers, dendritic polymers have a large number of functional groups – on the surface of the dendritic core, along or at the end of

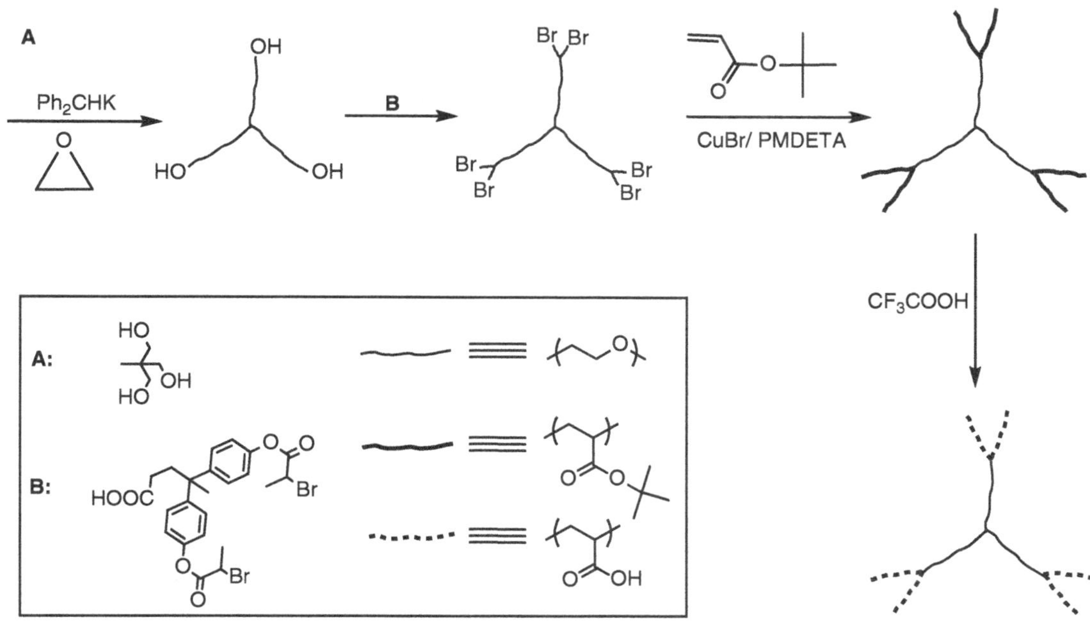

Figure 16.19 Synthesis of a dendrimer-like star copolymer with a poly(ethylene oxide) core and poly(acrylic acid) side chains.

the polymer-forming arms – which can be used for two, or even more, different tasks, such as delivering a drug, and serve as a targeting agent and a probe for imaging. Thus the construction of multifaced dendritic polymers – also called Janus micelles – with different types of peripheral groups to perform multiple functions is in high demand for biomedical applications. Recently, Gnanou et al. constructed Janus-type dendrimer-like poly(ethylene oxide) molecules (PEOs) of three generation-carrying terminal hydroxyl groups, on one side, and cleavable ketal groups, on the other, that were conjugated with folic acid (FA) to target folate receptors, and subsequently used to encapsulate camptothecin (CPT) as a therapeutic agent (Figure 16.20) [85]. The conjugation of both small molecules was accomplished by click chemistry based on 1,3-dipolar cycloaddition.

Structural variables such as the size and porosity of dendrigraft polymers can be designed to meet the specific requirements for drug delivery applications. For example, the loading and release properties of these materials can be tailored by adjusting the composition, the branching density, and the size of the side chains in the core and the corona. Very recently, Libera et al. evaluated the effects of macromolecular composition and structure on the encapsulation efficiency. They synthesized dendritic copolymers containing a hydrophobic core of poly(*tert*-butyl glycidyl ether) of controllable density, and a hydrophilic polyglycidol shell with a comparable molar mass and number of hydroxyl groups by anionic sequential polymerization [86]. The poorly water-soluble ruthenium complex $Ru(NH_3)_3Cl_3$, a potential anti-cancer drug, was either encapsulated within the hydrophobic core of the nanoparticles by entrapment from a suspension in water or water/methanol mixtures, or bound to the polymer shell by complexation with the hydroxyl groups. The solubilization capacity of two dendritic copolymers with different core densities did not differ significantly, but interestingly, the copolymer with the lower molar mass and smaller hydrodynamic radius core had a higher encapsulation efficiency in both encapsulation methods.

One advantage of dendritic nanocarriers is that they can be designed as delivery systems for combined drugs, to maximize their therapeutic effect on tumors. Thus, He et al. prepared a unimolecular micelle-like multilayered nanocarrier (PPDP) based on polyamidoamine (PAMAM) dendrimers as dual drug (paclitaxel and doxorubicin) containers, by combining a layer of hydrophobic oligo(3-caprolactone) (PCL) with hydrophilic PEGylated amidoamine dendrons within each molecule (Figure 16.21) [87]. Two nanocarrier variations were examined, either by encapsulating DOX and PTX in the different layers of the carrier (PPDP-enc-DOX + PTX), or by conjugating DOX through a pH-sensitive linkage on the exterior and encapsulating PTX in the core of the nanocarrier (PPDP-coj-DOX-enc-PTX) as shown in Figure 16.21. The study showed that for both systems, the release of PTX was achieved in a sustained manner over 48 h *in vitro*. On the other hand, the DOX release profile displayed a burst for the PPDP-enc-DOX + PTX system, while for the PPDP-coj-DOX-enc-PTX system the DOX release profile was sustained. Both systems displayed higher cytotoxicity

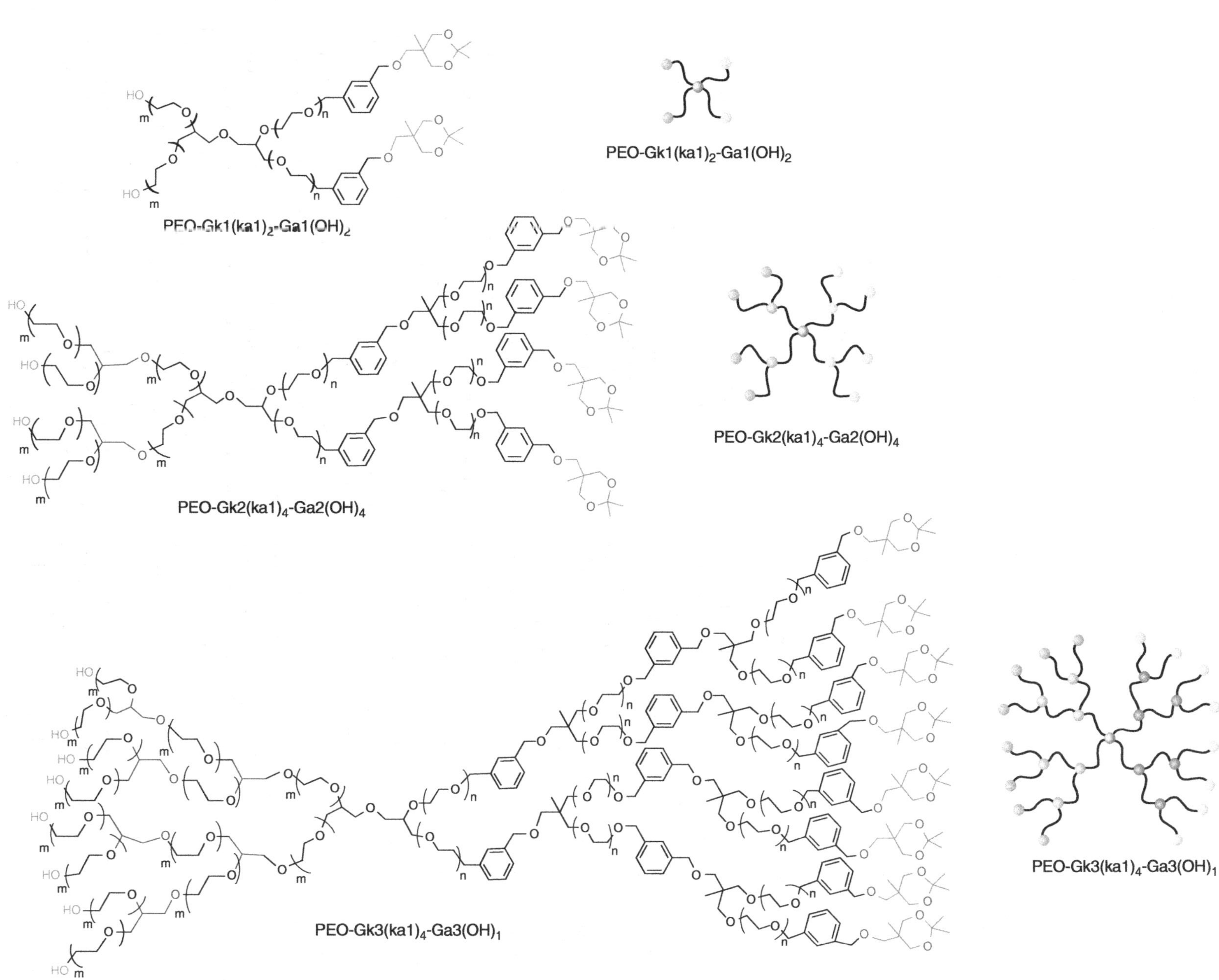

Figure 16.20 Structure of the three Janus-type dendrimer-like PEOs used for conjugation. Reprinted from [85], with permission from John Wiley & Sons, Inc.

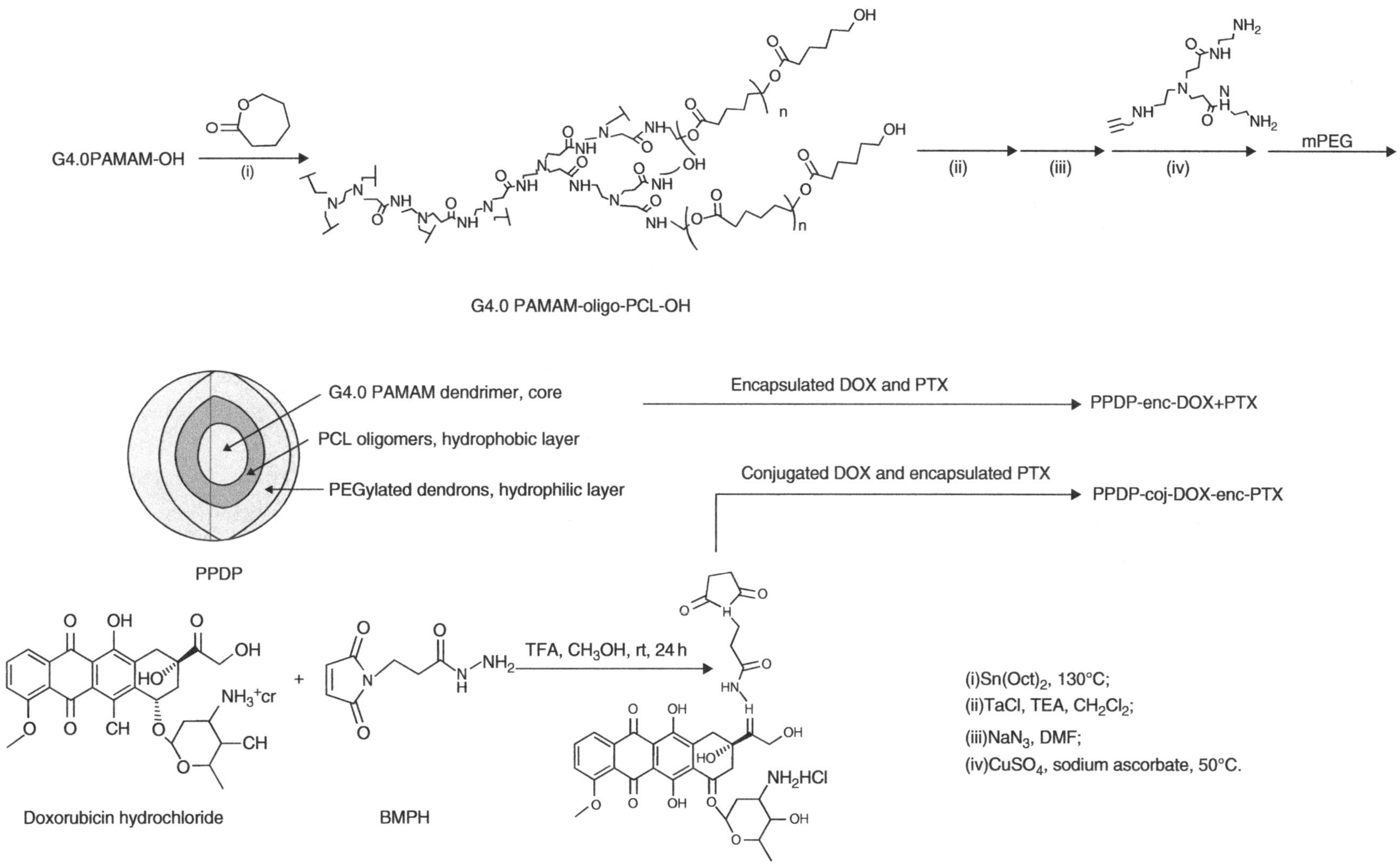

Figure 16.21 Synthesis of PAMAM dendrimer-based multilayered drug delivery systems. *Source:* Reproduced from [87], with permission from Royal Society of Chemistry.

toward MCF-7/ADR and MCF-7 cells than free DOX and PTX alone, due to the synergistic effect of the drugs.

Dendrimer-like star polymers are typically composed of a hydrophobic star polymer core and a shell of hydrophilic arms, which makes them very attractive for drug delivery applications. Their large number of surface functional groups not only can serve to conjugate drugs, but also ligand molecules for targeted delivery. For instance, a folate-functionalized degradable amphiphilic dendrimer-like star polymer (FA-DLSP), incorporating a well-defined poly(L-lactide) (PLLA) star polymer as hydrophobic core and six hydrophilic polyester dendrons based on 2,2-bis(hydroxymethyl)propionic acid forming a hydrophilic shell, was successfully synthesized by a combination of living ring-opening polymerization (ROP) of L-lactide and dendrimer synthesis [88]. To further enhance the water solubility of the unimolecular micelles, their surface was functionalized with carboxylic acid groups and coupled with PEG segments (Figure 16.22), and then with folate groups. The hydrophobic anti-cancer drug DOX was successfully encapsulated into the hydrophobic core of the micelles, with a drug loading of 4 wt % as determined by spectroscopic analysis. The *in vitro* release study showed that release of the encapsulated DOX was strongly pH-dependent: DOX was released more rapidly at pH 5.3 than at pH 7.4, because of the higher solubility of DOX in water at lower pH. Cellular uptake studies showed that the FA-DLSP micelles had greater cellular uptake than FA-deficient DLSP. The above results suggest that the FA-DLSP unimolecular micelles may be promising nanocarriers for targeted cancer therapy.

16.4.3 Convergent Grafting Through Strategy

Puskas et al. [89] developed the first convergent cationic polymerization strategy to synthesize arborescent isobutylene homopolymers in a single reaction pot. This was

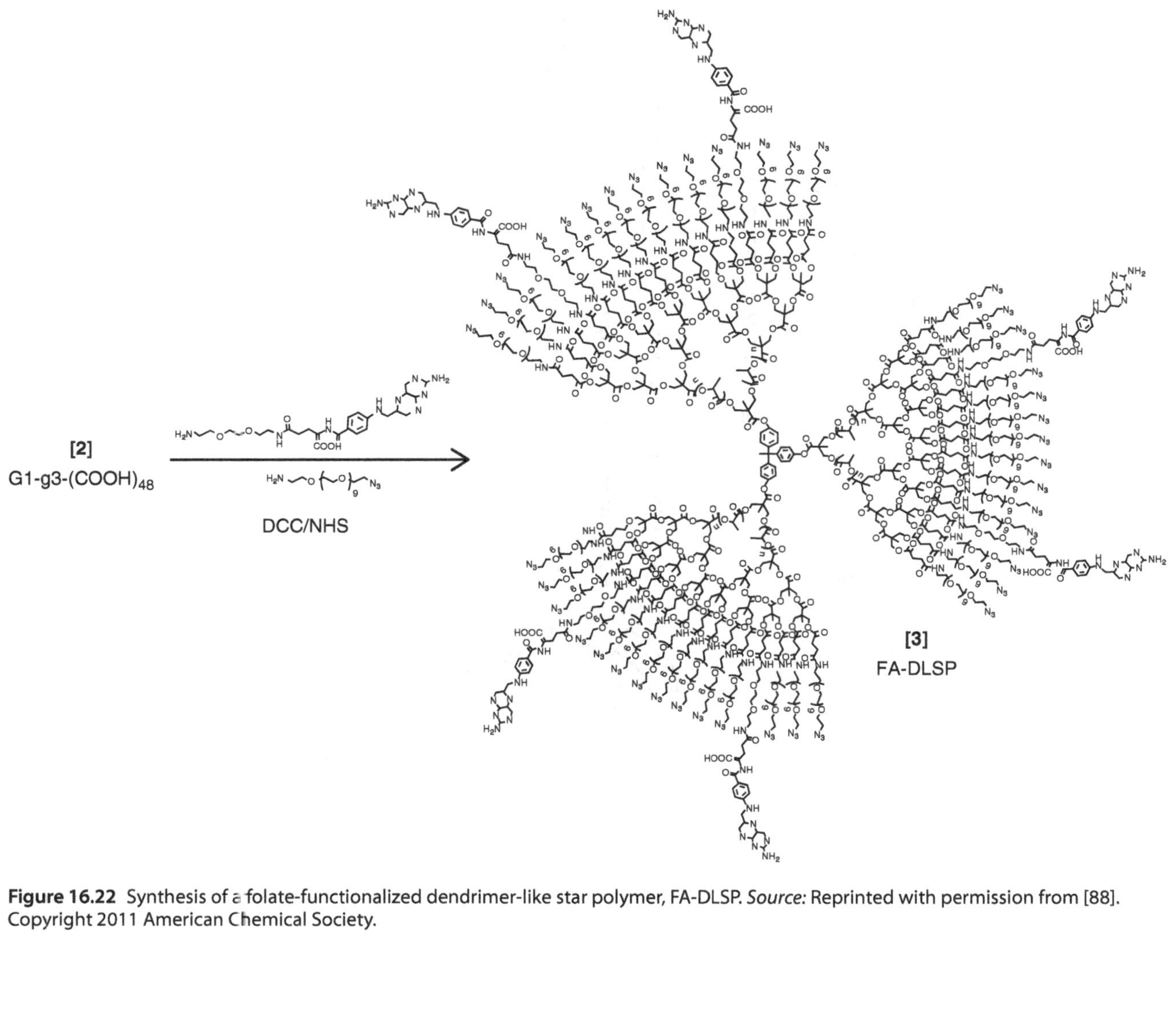

Figure 16.22 Synthesis of a folate-functionalized dendrimer-like star polymer, FA-DLSP. *Source:* Reprinted with permission from [88]. Copyright 2011 American Chemical Society.

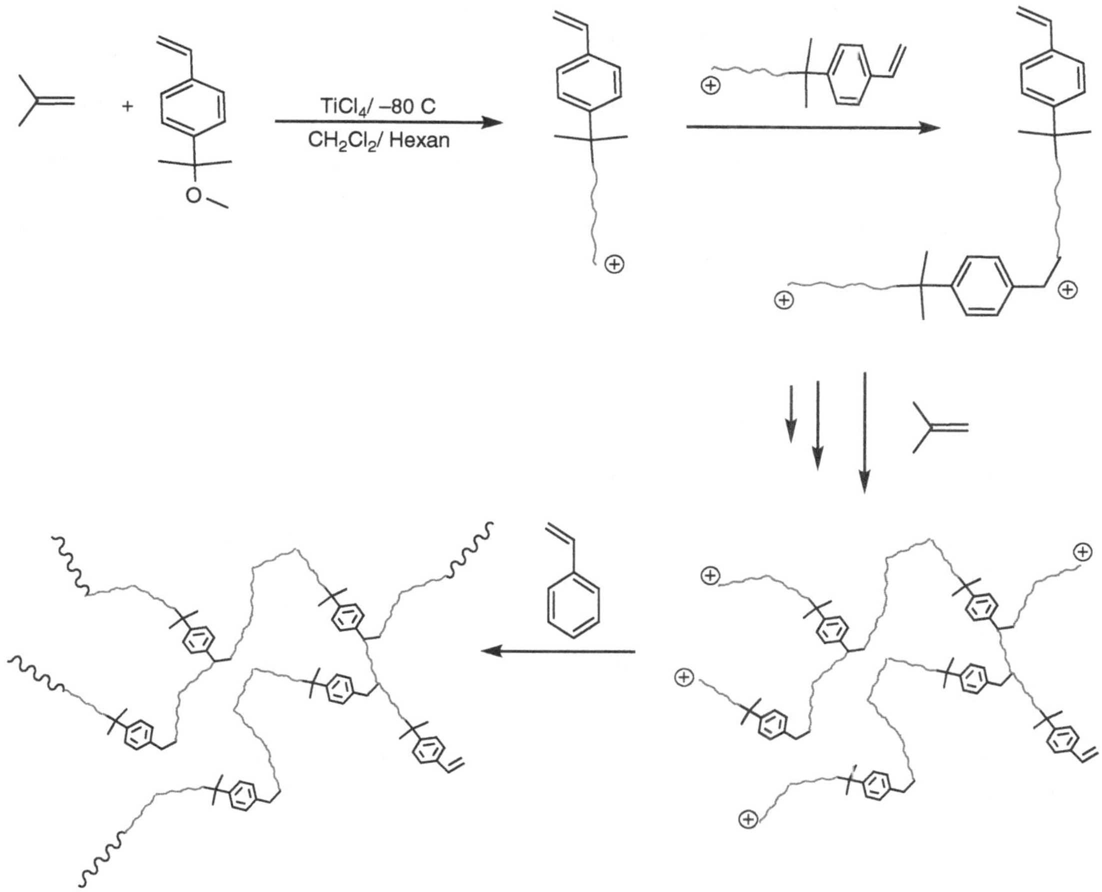

Figure 16.23 One-pot synthesis of hyperbranched polyisobutylene with polystyrene end-blocks.

achieved by the cationic copolymerization of isobutylene with a small amount of an inimer such as 4-(2-methoxyisopropyl)styrene to yield a "living" branched structure with multiple cationic active sites at the chain ends, which were subsequently used to initiate the polymerization of styrene as shown in Figure 16.23 [90].

The approval of poly(styrene-*b*-isobutylene-*b*-styrene) linear triblock copolymers by the Food and Drug Administration (FDA) as a coating material on the Taxus® coronary stent has opened a new door for polyisobutylene-based biomaterials. For instance, Puskas and Hoerr also studied drug release from arborescent polyisobutylene with polystyrene end-blocks. ElectroNanospray™, a proprietary technology for coating materials on the surface of medical devices, was used to coat coronary stents with arborescent polyisobutylene-*b*-polystyrene loaded with Dexamethasone, a model drug. The study showed that the drug release profiles were influenced both by the molecular weight of the dendritic polyisobutylene-*b*-polystyrene scaffold and by the spraying conditions used for the polymer-drug mixture [91].

Characteristics such as the size and porosity of amphiphilic dendrigraft copolymers can be designed to meet specific requirements for drug delivery applications by adjusting their composition, branching functionality, generation number, and the size of the side chains in their core and corona. Internal functionality can easily be introduced within the repeating units in many cases. Furthermore, as mentioned earlier, dendrigraft copolymers can be produced in one-pot reactions in some cases with very high molecular weights and relatively narrow molecular weight distributions. As a result, dendrigraft polymers seem to represent a good compromise solution between achieving a low dispersion comparable to dendrimers and maintaining a reasonably low production cost. We therefore believe that these materials have great potential in drug delivery applications.

16.5 Conclusion

Dendritic polymers are a unique and versatile class of branched polymers ideal for a number of high value applications, and particularly for drug delivery. Numerous methods have been reported to synthesize dendritic polymers while achieving control over the characteristics of the polymers, to achieve desirable properties or functionality. While the commercialization of dendrimers has been restricted by their tedious and costly synthesis, the one-pot syntheses reported for hyperbranched and dendrigraft polymers make these materials more viable for (large-scale) commercial production and applications. The field of dendritic drug delivery nanocarriers has rapidly evolved over the last decade, with an increasing number of candidates currently being under clinical evaluation. A wide range of dendritic polymer architectures and chemistries are currently being developed specifically for drug delivery applications. The collaboration between polymer chemists, medicinal and pharmaceutical scientists, and clinicians is critical for the development of clinical applications for practical uses of these nanocarriers.

References

1 Aggarwal, S.L., Sweeting, O.J. (1957) *Chem. Rev.*, **57**, 665–742.
2 White, J.L., Dharod, K.C., Clark, E.S. (1974) *J. Appl. Polym. Sci.*, **18**, 2539–2568.
3 Vögtle, F., Richardt, G., Werner, N. (2009) In *Dendrimer Chemistry*, Wiley-VCH Verlag GmbH & Co. KGaA, Weinheim, pp. 1–24.
4 Gauthier, M. (2007) *J. Polym. Sci., Part A: Polym. Chem.*, **45**, 3803–3810.
5 Teertstra, S.J., Gauthier, M. (2004) *Prog. Polym. Sci.*, **29**, 277–327.
6 Tomalia, D.A., Jørn B., Christensen, Boas., U. (2012) *Dendrimers, Dendrons, and Dendritic Polymers*, Cambridge University Press: Cambridge.
7 Gauthier, M., Tichagwa, L., Downey, J.S., Gao, S. (1996) *Macromolecules*, **29**, 519–527.
8 Jikei, M., Kakimoto, M. (2001) *Prog. Polym. Sci.*, **26**, 1233–1285.
9 Kim, Y.H. (1998) *J. Polym. Sci., Part A: Polym. Chem.*, **36**, 1685–1698.
10 Gauthier, M., Möller M. (1991) *Macromolecules*, **24**, 4548–4553.
11 Tomalia, D.A., Hedstrand, D.M., Ferritto, M.S. (1991) *Macromolecules*, **24**, 1435–1438.
12 Kaminskas, L.M., Porter, C.J.H. (2011) *Adv. Drug Deliv. Rev.*, **63**, 890–900.
13 Paolino, M., Ennen, F., Lamponi, S. et al. (2013) *Macromolecules*, **46**, 3215–3227.
14 Kannan, R.M., Nance, E., Kannan, S., Tomalia, D.A. (2014) *J. Internal Med.*, **276**, 579–617.
15 Menjoge, A.R., Kannan, R.M., Tomalia, D.A. (2010) *Drug Discovery Today*, **15**, 171–185.
16 Wu, W., Wang, W., Li, J. (2015) *Prog. Polym. Sci.*, **46**, 55–85.
17 Sowinska, M., Urbanczyk-Lipkowska, Z. (2014) *New J. Chem.*, **38**, 2168–2203.
18 Wu, L.-P., Ficker, M., Christensen, J.B. et al. (2015) *Bioconjugate Chem.*, **26**, 1198–1211.
19 Kurniasih, I.N., Keilitz, J., Haag, R. (2015) *Chem. Soc. Rev.*, **44**, 4145–4164.
20 Kurtoglu, Y.E., Mishra, M.K., Kannan, S., Kannan, R.M. (2010) *Int. J. Pharm.* **384**, 189–194.
21 Bhattacharya, P., Geitner, N.K., Sarupria, S., Ke, P.C. (2013) *PCCP*, **15**, 4477–4490.
22 Dong, R., Zhou, Y., Zhu, X. (2014) *Acc. Chem. Res.*, **47**, 2006–2016.
23 Lukowiak, M.C., Thota, B.N.S., Haag, R. (2015) *Biotechnology Adv.*, **33**, 1327–1341.
24 Zheng, Y., Li, S., Weng, Z., Gao, C. (2015) *Chem. Soc. Rev.*, **44**, 4091–4130.
25 Svenson, S., *Chem. Soc. Rev.*, **44**, 4131–4144.
26 Wang, D., Zhao, T., Zhu, X. et al. (2015) *Chem. Soc. Rev.*, **44**, 4023–4071.
27 Khandare, J., Calderon, M., Dagia, N.M., Haag, R. (2012) *Chem. Soc. Rev.*, **41**, 2824–2848.
28 Bugno, J., Hsu, H.-J., Hong, S. (2015) *Biomaterials Sci.*, **3**, 1025–1034.
29 Amir, R.J., Albertazzi, L., Willis, J. et al. (2011) *Angew. Chem. Int. Ed.*, **50**, 3425–3429.
30 Hu, Q.-Y., Berti, F., Adamo, R. (2016) *Chem. Soc. Rev*, **45**, 1691–1719.
31 Mura, S., Nicolas, J., Couvreur, P. (2013) *Nat. Mater.*, **12**, 991–1003.
32 Buhleier, E., Wehner, W., Vögtle, F. (1978) *Synthesis*, 155–158.
33 Tomalia, D.A., Baker, H., Dewald, J. et al. (1985) *Polym. J.*, **17**, 117–132.
34 Newkome, G.R., Yao, Z., Baker, G.R., Gupta, V.K. (1985) *J. Org. Chem.*, **50**, 2003–2004.
35 Hawker, C.J., Fréchet, J.M.J. (1990) *J. Am. Chem. Soc.*, **112**, 7638–7647.
36 Dengiz, C., Breiten, B., Gisselbrecht, J.-P. et al. (2015) *J. Org. Chem.*, **80**, 882–896.
37 Walter, M.V., Malkoch, M. (2012) *Chem. Soc. Rev.*, **41**, 4593–4609.
38 Wong, C.-H., Zimmerman, S.C. (2013) *Chem. Comm.*, **49**, 1679–1695.
39 Zeng, F., Zimmerman, S.C. (1996) *J. Am. Chem. Soc.*, **118**, 5326–5327.
40 Wu, P., Feldman, A.K., Nugent, A.K. et al. (2004) *Angew. Chem. Int. Ed.*, **43**, 3928–3932.
41 Kolb, H.C., Finn, M.G., Sharpless, K B. (2001) *Angew. Chem. Int. Ed.*, **40**, 2004–2021.

42 Gonzaga, F., Sadowski, L.P., Rambarran, T. et al. (2013) *J. Polym. Sci., Part A: Polym. Chem.*, **51**, 1272–1277.
43 Lowe, A.B. (2014) *Polym. Chem.*, **5**, 4820–4870.
44 Montañez, M.I., Campos, L.M., Antoni, P. et al. (2010) *Macromolecules*, **43**, 6004–6013.
45 Kottari, N., Chabre, Y.M., Shiao, T.C. et al. (2014) *Chem. Comm.*, **50**, 1983–1985.
46 Sharma, R., Zhang, I., Abbassi, L. et al. (2015) *Polym. Chem.*, **6**, 1436–1444.
47 Leiro, V., Garcia, J.P., Tomás, H., Pêgo, A.P. (2015) *Bioconjugate Chem.*, **26**, 1182–1197.
48 Morgan, M.T., Nakanishi, Y., Kroll, D.J. et al. (2006) *Cancer Res.*, **66**, 11913–11921.
49 Majoros, I.J., Williams, C.R., Becker, A., Baker, J.R. (2009) *Wiley Interdisciplinary Reviews: Nanomedicine and Nanobiotechnology*, **1**, 502–510.
50 Zhang, C., Pan, D., Luo, K., et al. (2014) *Polym. Chem.*, **5**, 5227–5235.
51 Khandare, J.J., Jayant, S., Singh, A. et al. (2006) *Bioconjugate Chem.*, **17**, 1464–1472.
52 Watson, K.D. (2001) in *Encyclopedia of Life Sciences*, John Wiley & Sons, Ltd, Chichester.
53 Gunatillake, P.A., Odian, G., Tomalia, D.A. (1998) *Macromolecules*, **21**, 1556–1562.
54 Hanselmann, R., Hölter, D., Frey, H. (1998) *Macromolecules*, **31**, 3790–3801.
55 Bharathi, P., Moore, J.S. (2000) *Macromolecules*, **33**, 3212–3218.
56 Schallausky, F., Erber, M., Komber, H., Lederer, A. (2008) *Macromol. Chem. Phys.*, **209**, 2331–2338.
57 Kim, Y.H., Webster, O.W. (1990) *J. Am. Chem. Soc.*, **112**, 4592–4593.
58 Pérignon, N., Mingotaud, A.-F., Marty, J.-D. et al. (2004) *Chem. Mater.*, **16**, 4856–4858.
59 Fréchet, J.M., Henmi, M., Gitsov, I., et al. (1995) *Science (New York, N.Y.)*, **269**, 1080–1083.
60 Porsch, C., Zhang, Y., Ducani, C., Vilaplana, F. et al. (2014) *Biomacromolecules*, **15**, 2235–2245.
61 Seleci, M., Seleci, D.A., Ciftci, M.et al. (2015) *Langmuir*, **31**, 4542–4551.
62 Zhu, L., Tu, C., Zhu, B. et al. (2011) *Polym. Chem.*, **2**, 1761–1768.
63 Saiyin, W., Wang, D., Li, L. et al. (2014) *Mol. Pharm.*, **11**, 1662–1675.
64 Zhuang, Y., Su, Y., Peng, Y. et al. (2014) *Biomacromolecules*, **15**, 1408–1418.
65 Gao, C., Yan, D. (2004) *Prog. Polym. Sci.*, **29**, 183–275.
66 Hirao, A., Sugiyama, K., Matsuo, A. et al. (2008) *Polym. Int.*, **57**, 554–570.
67 Li, J., Gauthier, M. (2001) *Macromolecules*, **34**, 8918–8924.
68 Yuan, Z., Gauthier, M. (2005) *Macromolecules*, **38**, 4124–4132.
69 Hempenius, M.A., Michelberger, W., Möller, M. (1997) *Macromolecules*, **30**, 5602–5605.
70 Kee, R.A., Gauthier, M. (2002) *Macromolecules*, **35**, 6526–6532.
71 Kee, R.A., Gauthier, M. (2008) *J. Polym. Sci., Part A: Polym. Chem.*, **46**, 2335–2346.
72 Whitton, G., Gauthier, M. (2013) *J. Polym. Sci. Part A: Polym. Chem.*, **51**, 5270–5279.
73 Whitton, G., Gauthier, M. (2016) *J. Polym. Sci., Part A: Polym. Chem.*, **54**, 1197–1209.
74 Alsehli, M., Gauthier, M. (2016) *Mater. Res. Soc. Symp. Proc.*, 1819, imrc2015s4d-o015, DOI: 10.1557/opl.2016.70.
75 Njikang, G., Gauthier, M., Li, J. (2008) *Polymer*, **49**, 1276–1284.
76 Njikang, G.N., Gauthier, M., Li, J. (2008) *Polymer*, **49**, 5474–5481.
77 Huang, C.K., Lo, C.L., Chen, H.H., Hsiue, G.H. (2007) *Adv. Funct. Mater.*, **17**, 2291–2297.
78 Gyawali, D., Zhou, S., Tran, R.T. et al. (2014) *Adv. Healthcare Materials*, **3**, 182–186.
79 Chen, G., Wang, L., Cordie, T. et al. (2015) *Biomaterials*, **47**, 41–50.
80 Chytil, P., Koziolová, E., Janoušková, O. et al. (2015) *Macromolecular Biosci.*, **15**, 839–850.
81 Six, J.-L., Gnanou, Y. (1995) *Macromolecular Symposia*, **95**, 137–150.
82 Feng, X., Taton, D., Ibarboure, E. et al. (2008) *J. Am. Chem. Soc.*, **130**, 11662–11676.
83 Feng, X.-S., Taton, D., Chaikof, E.L., Gnanou, Y. (2005) *J. Am. Chem. Soc.*, **127**, 10956–10966.
84 Gnanou, Y., Taton, D. (2001) *Macromolecular Symposia*, **174**, 333–341.
85 Feng, X., Pinaud, J., Chaikof, E.L. et al. (2011) *J. Polym. Sci., Part A: Polym. Chem.*, **49**, 2839–2849.
86 Libera, M., Formanek, P., Schellkopf, L. et al. (2014) *J. Polym. Sci., Part A: Polym. Chem.*, **52**, 3488–3497.
87 He, H., Wang, Y., Wen, H., Jia, X. (2014) *RSC Advances*, **4**, 3643–3652.
88 Cao, W., Zhou, J., Mann, A. et al. (2011) *Biomacromolecules*, **12**, 2697–2707.
89 Puskas, J.E., Grasmüller, M. (1998) *Macromolecular Symposia*, **132**, 117–126.
90 Puskas, J.E., Kwon, Y. (2006) *Polym. Adv. Technol.*, **17**, 615–620.
91 Puskas, J.E., Hoerr, R.A. (2010) *Macromolecular Symposia*, **291–292**, 326–329.

17

Bone-inspired Biomaterials

*Frank A. Müller**

Otto Schott Institute of Materials Research (OSIM), Friedrich Schiller University of Jena, Jena, Germany

17.1 Introduction

Most living beings mineralize at least a part of their bodies, usually for skeletal support (bone), but also for sexual attraction (antler), protection (armor, shells), or manducation (teeth). Biominerals like bone represent hybrid materials that combine organic (mainly collagen) and inorganic (carbonated hydroxy apatite (CHA)) phases to build up composites via a self-assembly process (Figure 17.1). These composites show outstanding mechanical properties because they are highly hierarchically structured at many different length scales and because they combine the failure mechanisms of different materials classes.

In bone tissue engineering and bone regeneration, highly porous scaffold materials have to provide a pathway for cell attachment, bone ingrowth, and vascularization. Typically, macroporous osteoconductive bioceramics (e.g. hydroxyapatite (HA), β-tricalcium phosphate (β-TCP)) are supplied commercially for bone grafting. However, approaches that mimic the biological model more closely should aim (1) exactly to recreate the composition of the naturally occurring phases; (2) to rebuild the porous structure of the biomineral; and (3) to adjust the interface and the interaction of the involved phases. To address the first requirement, modified simulated body fluids (SBF) were used to precipitate biomimetic apatite with a composition, morphology, and growth orientation equal to the mineral part of the vertebrate bone. To investigate chemical interactions in hybrid materials, collagen scaffolds and cellulose fibers were mineralized with apatite by using, for example, wet-chemical precipitation procedures. To rebuild the pore structure of biominerals, additive manufacturing technologies were used to fabricate cellular structures with bone-like mechanical properties and interconnected macro pores capable of inducing angiogenesis. Furthermore, damage-tolerant calcium phosphate scaffolds with an aligned and interconnecting porosity were prepared by ice templating and subsequent impregnation with a biodegradable polymer. Finally, future perspectives in the processing of bone-inspired materials for tissue regeneration will be highlighted.

17.2 Bone

The bone tissue of vertebrates represents a hierarchical composite material grown by a biomineralization process [1–3]. It consists of 60–70 mass % CHA, 20–30 mass % proteins (mainly collagen I) and water. At the lowest level of organization, the thin plate-shaped apatite crystals (50 × 25 × 3 nm in average dimension [1]) are embedded in and around the fibrous protein collagen (Figure 17.2). At the next level, these mineralized fibres lie bundled together and attached to each other. At the next level, these fibres form lamellae with a width of ~2 μm. The lamellae have various patterns. A very common one is a secondary osteon in which concentric lamellae form cylindrical structures, ~200 μm in diameter, surrounding a central blood vessel [4, 5]. At the highest level of organization, bone consists of two different structures, the cortical bone which is compact and builds up the outer bone shell and the cancellous bone which is highly porous and consists of many struts. These struts are not randomly arranged, but their orientation is related to the direction of loads on the bone [5, 6].

Bone apatite contains carbonate ions (3–9 mass- %) replacing the phosphate ions (B-type CHA) as well as the hydroxy ions (A-type CHA) of the HA crystal [7]. Figure 17.3 illustrates the ionic substitutions of biological apatite with a composition of $Ca_{10-x-y}Mg_y(HPO_4)_{x-z}(CO_3)_z(PO_4)_{6-x}(OH)_{2-x-w}(CO_3)_{w/2}$ ($0 \leq x,y \leq 1$; $z,w \leq x$) compared to stoichiometric HA. The [001]-direction (c-axis) represents the preferred growth orientation of the CHA crystals in bone. This direction is oriented parallel to the longitudinal axis of the

* Email: frank.mueller@uni-jena.de

Bioinspired Materials Science and Engineering, First Edition. Edited by Guang Yang, Lin Xiao, and Lallepak Lamboni.

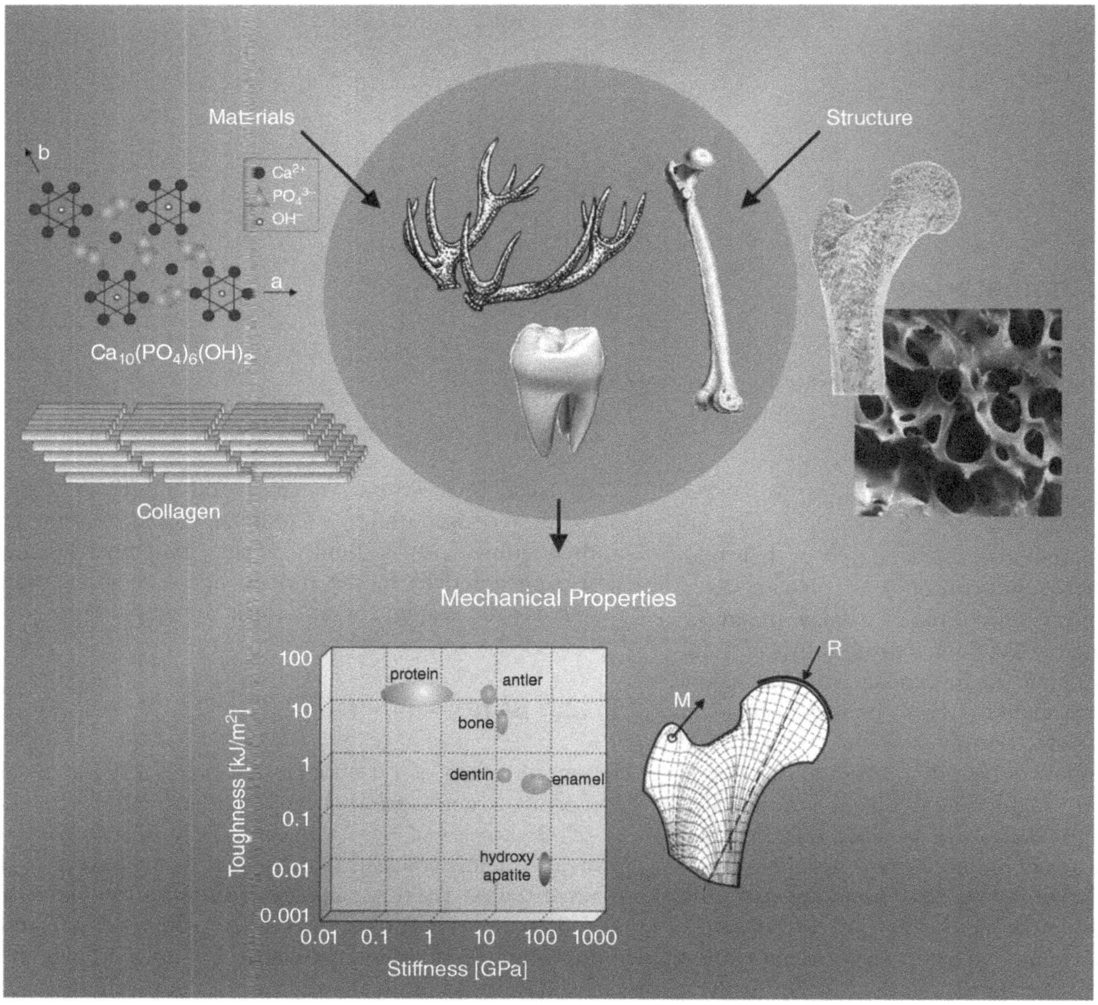

Figure 17.1 The bone family: structure, composition and properties.

collagen fibrils [8–10]. This hierarchical feature is an important characteristic of the biomineralization of calcium phosphates (CaP) [11].

The bones of vertebrates, as opposed to the shells of molluscs or human teeth, can be considered to be "living biominerals" since they contain cells which are under permanent activity. The bone formation process starts by the action of osteoblasts, special cells that synthesize and release the collagen matrix in the form of a jelly substance, the osteoid, which is subsequently mineralized by controlled deposition of apatite crystals. Another type of cells, the osteoclasts, catabolyze and resorb the bone. The main function of bone is related to the structural and mechanical support of the skeletal apparatus. Beyond that, bone constitutes a storage and hauling mechanism for the two essential elements, phosphorus and calcium. The dimensions on the nanoscale, the low crystallinity, and the non-stoichiometric composition are important features for the solubility and resorption capacity of biological apatite which significantly exceed those of the stoichiometric synthetic apatite [12, 13]. Consequently, under physiological conditions, bone is able to regenerate in the case of fracture and to respond to mechanical load [6] by a dynamic remodeling process of osteoblast-induced bone formation and osteoclast-driven bone resorption, respectively (Figure 17.4). The mechanical properties of bone depend on the interaction of all levels of organization and are dependent on the direction of load due to the pronounced anisotropy of the bone structure. Bone materials show outstanding mechanical properties due to their ability to combine the stiffness of inorganic materials with the toughness of polymers (Figure 17.1). Readers who are interested in learning more about the fracture mechanics of bone are referred to the book *Bones; Structure and Mechanics*,

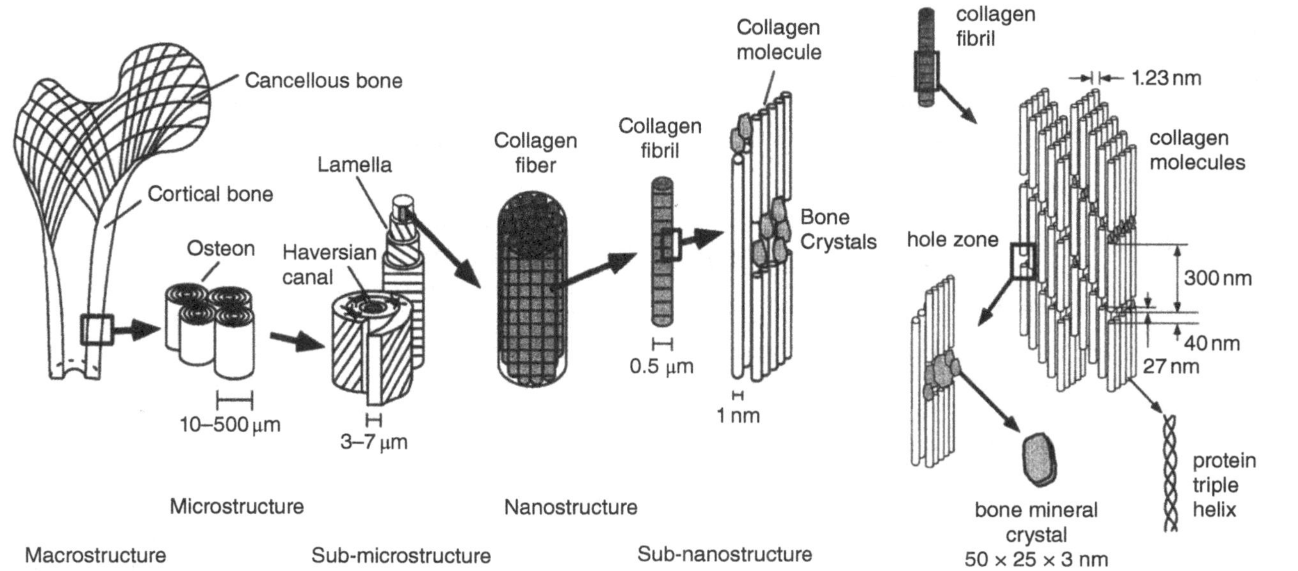

Figure 17.2 The hierarchical structure of bone [4].

Figure 17.3 Projection of the (001) plane of (a) stoichiometric hydroxy apatite and (b) ionic substitutions in bone apatite or biomimetic apatite. (*See insert for color representation of the figure.*)

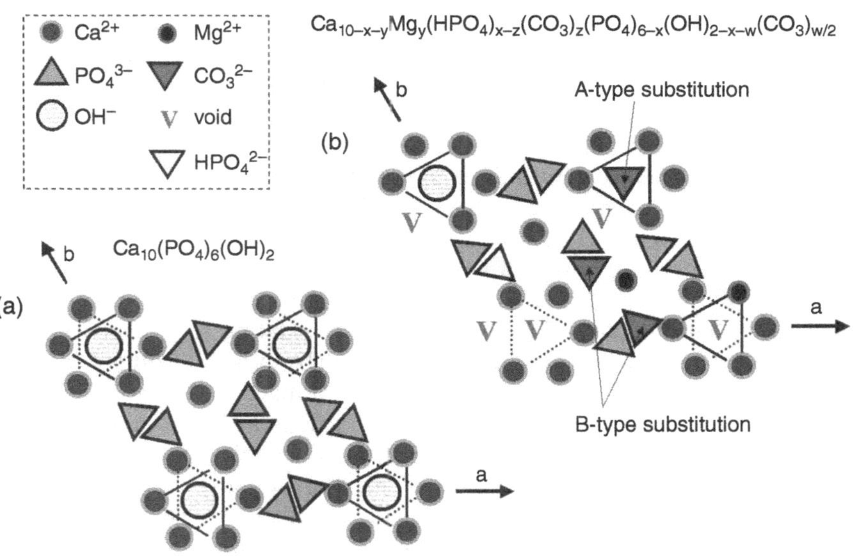

Figure 17.4 Bone remodeling and repair.

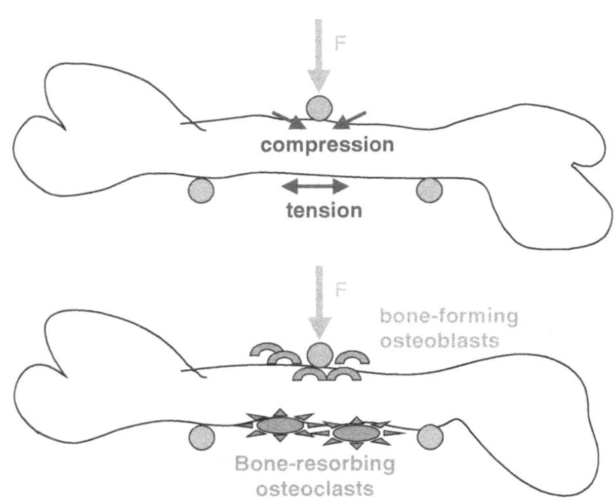

written by Professor John D. Currey [14], which gives the most comprehensive insight into bone structures and related mechanical properties.

17.3 Bone-like Materials

17.3.1 Biomimetic Apatite

Definition: Biomimetic apatite, precipitated under physiological conditions from a simulated body fluid (SBF), mimics the chemical, physical, and structural properties of biological bone apatite. After an implantation into a bony defect, it adopts the metabolic function of bone and allows the substantial integration of the implant.

Figure 17.3a shows the hexagonal unit cell of stoichiometric hydroxyapatite (HA, $Ca_{10}(PO_4)_6(OH)_2$). HA can be prepared very easily by solid state reactions at elevated temperatures or by a wet chemical precipitation, for example, from $Ca(OH)_2$ and H_3PO_4 and a subsequent calcination [11]. HA represents the most stable calcium orthophosphate with the lowest solubility at pH > 4.2 [15]. The bioactivity of stoichiometric HA is limited due to its chemical stability under physiological conditions. Therefore, HA only inadequately supports the dynamic remodeling processes of bone after an implantation. On the other hand, biogenic apatite that is present in bone is well replaced (Figure 17.3b). It contains carbonate ions replacing the phosphate ions (B-type CHA) as well as the hydroxy ions (A-type CHA) of the HA crystal, thereby increasing the solubility of bone apatite and facilitating the dynamic remodeling of bone tissue. The CHA crystals are plate-shaped with dimensions in the range of 25–50 nm and a thickness of 2–5 nm. The c-axis represents the preferred growth orientation of CHA crystals with (002) domain sizes around 20 nm. This direction is oriented parallel to the longitudinal axis of collagen fibrils. Therefore, a biomimetic apatite, which precipitated under physiological conditions and which mimics the structural and chemical properties of bone apatite represents an interesting alternative for bone regeneration materials due to its capacity to be actively integrated into the physiological bone remodeling processes.

In vivo studies have shown that bioactive materials form a CHA layer on their surface immediately after implantation into bone tissue [16]. It is generally accepted that the *in vitro* apatite formation from simulated body fluid (SBF), which resembles the inorganic part of human blood plasma, represents an indicator for the *in vivo* bioactivity of a material [17]. Beyond that, the use of SBF allows the materials' surfaces to be coated with a biomimetic apatite whose structure and composition are close to biological apatite.

To investigate the early stages of *in vitro* apatite formation, Langmuir monolayers of arachidic acid were exposed to SBF and characterized by Cryo-TEM investigations [18]. It was shown that the apatite formation starts with the aggregation of prenucleation clusters leading to the nucleation of an amorphous calcium phosphate (ACP) before the development of oriented apatite crystals occurs (Figure 17.5).

To investigate the composition and growth orientation of the apatite crystals, bioactive titanium was prepared by a two-step chemical treatment in HCl and NaOH and was exposed to SBF [19]. Figure 17.6 shows the formation of apatite depending on the time of exposure. After two days in SBF, EDX analyses reveal that Ca und P accumulate at the surface. After five days in SBF, small spheroids appear. They have a Ca:P ratio of 1.6 which is very close to the theoretical one of synthetic HA. After 10 days in SBF, a homogenous crystalline layer consisting of HA covers the whole titanium surface. From XRD analyses it is evident that the apatite crystals show a preferred growth orientation of their crystallographic c-axis perpendicular to the surface of the titanium substrate. Generally, the (211) peak is the most intense in HA powder, and the peak intensity ratio $I(002)/I(211)$ is 0.4. In the case of HA precipitated from SBF, it is 2.5. This is in the range of what is found in the longitudinal direction of long bones [20]. This preferred growth orientation was also confirmed by auto correlation analyses of high resolution TEM micrographs and SAED analyses (Figure 17.7) showing that the (002) diffraction ring is incomplete and forms an arc [21].

To investigate the influence of the hydrogen carbonate content in SBF on the degree of the carbonate substitution in precipitated apatite, SBF with different HCO_3 concentrations ranging from 5–27 mmol/l were prepared [22]. The first one is in the range of Kokuko's SBF [23], the last one represents the real concentration in human blood plasma [24]. Then the carbonate content in the precipitated apatite was determined by FTIR and Raman spectroscopy. For SBF5–20, a B-type apatite and for SBF27 an AB-type apatite was found. It was also found that the carbonate content increased from 3% for SBF5 to approximately 9% for SBF27. Looking at the chemical formula of B-type apatite, a maximum CO_3^{2-} substitution (z = 1 in Figure 17.8) leads to a maximum carbonate content of 6.4 mass-%. The carbonate content of apatite precipitated from SBF5, SBF10, SBF15 and SBF20 was below this level. Considering an AB-type apatite, the maximum carbonate substitution equals 10 mass-%. In the case of SBF27, which is the solution with the composition most similar to human blood plasma, the carbonate content of the precipitated apatite amounts to 9.3 mass-%. Hence, it can be assumed that the A-sites will be occupied after the B-sites are completely replaced. Moreover, this means that by using only SBF27 it is

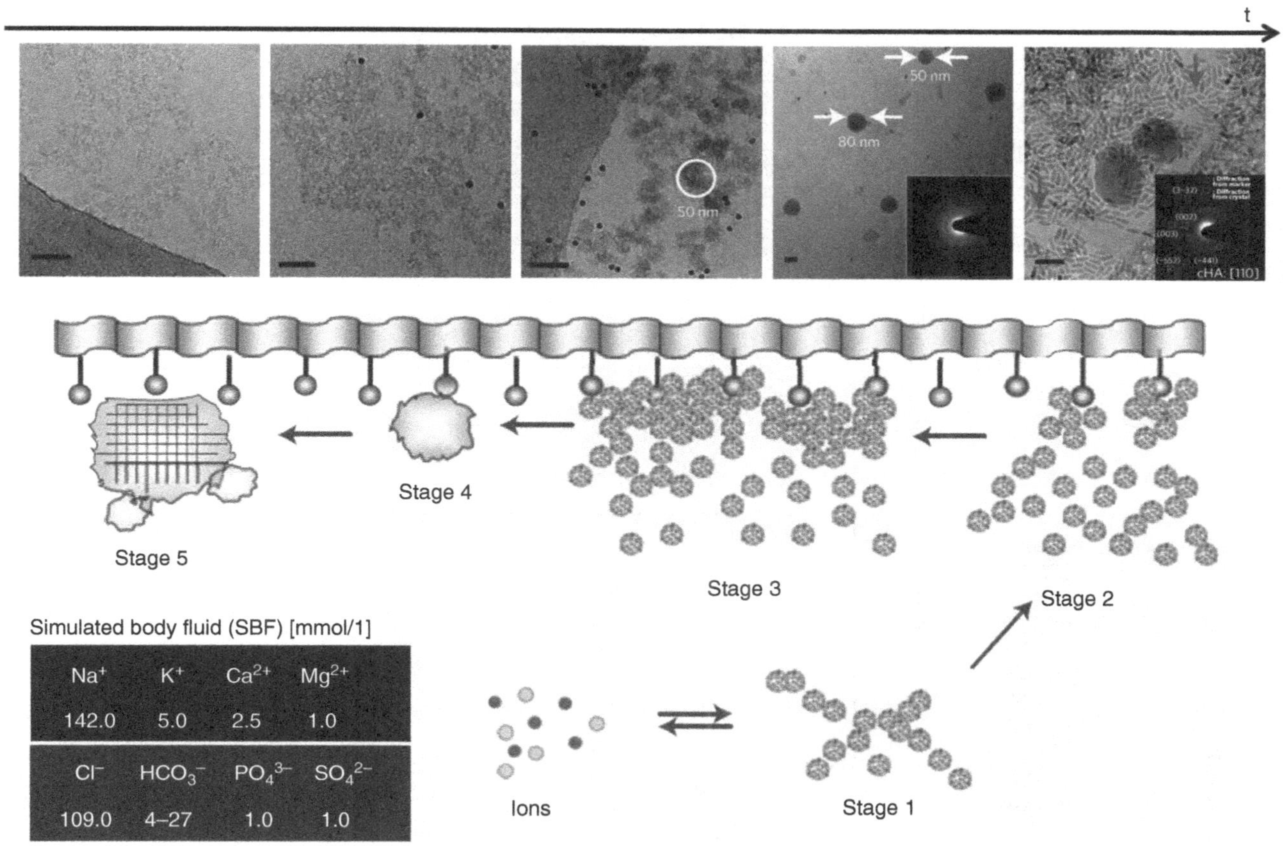

Figure 17.5 CryoTEM images and schematic illustration of the different mineralization stages during the surface-induced formation of apatite from SBF. *Source:* Adapted from [18].

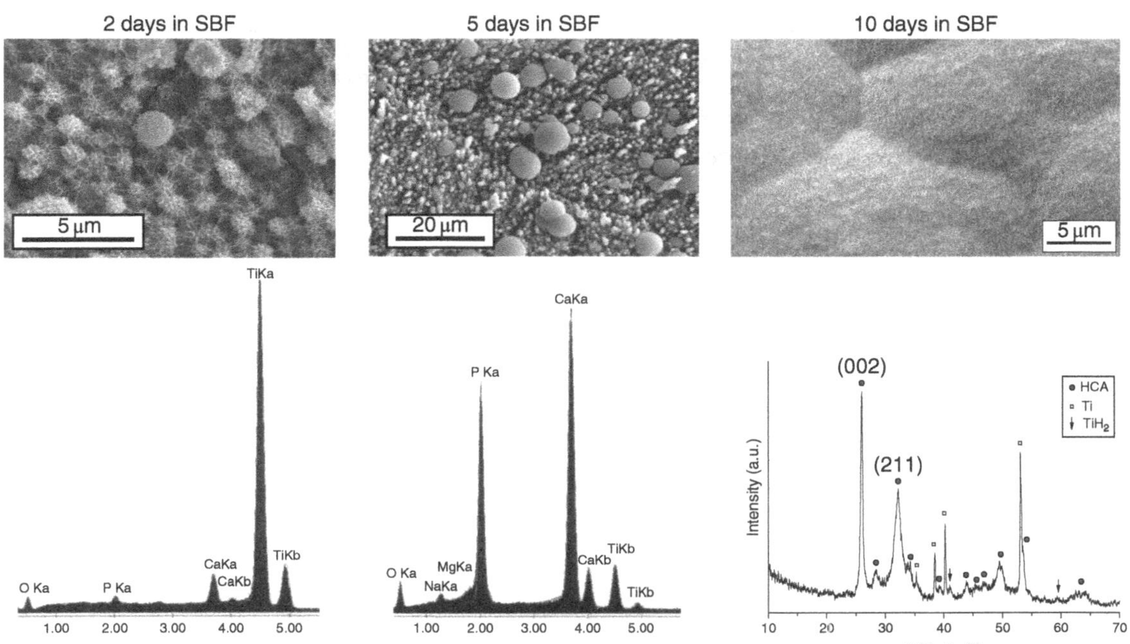

Figure 17.6 Biomimetic apatite formation on bioactive titanium surfaces. *Source:* Adapted from [19] and [21].

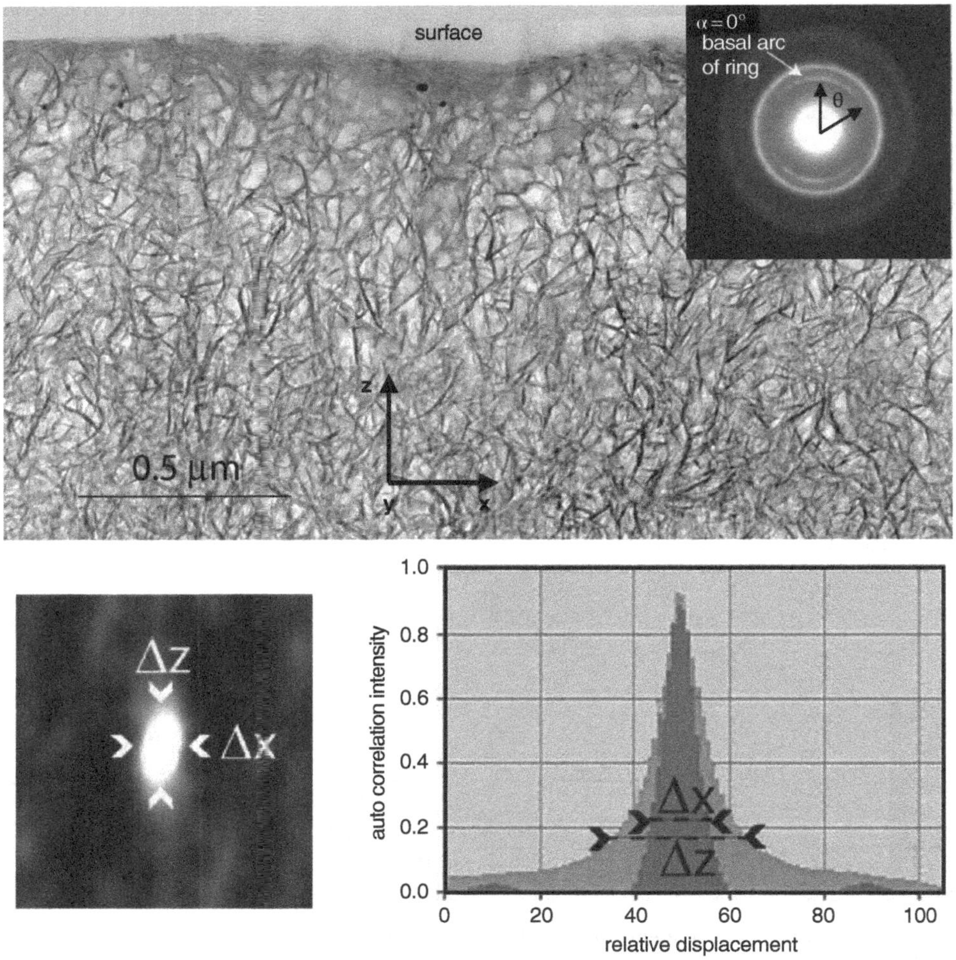

Figure 17.7 Preferred growth orientation of biomimetic apatite [21].

B-type HCA: $Ca_{10-x}(HPO_4)_{x-y}(CO_3)_y(PO_4)_{6-x}(OH)_{2-x}$ $\quad 0 < x,y < 1$

$(CO_3)^{2-} = 6.4$ wt–% $\quad x,y = 1$

AB-type HCA: $Ca_{10-x}(HPO_4)_{x-y}(CO_3)_y(PO_4)_{6-x}(OH)_{2-x-z}(CO_3)_{z/2}$ $\quad 0 < x,y,z < 1$

$(CO_3)^{2-} = 10$ wt–% $\quad x,y,z = 1$

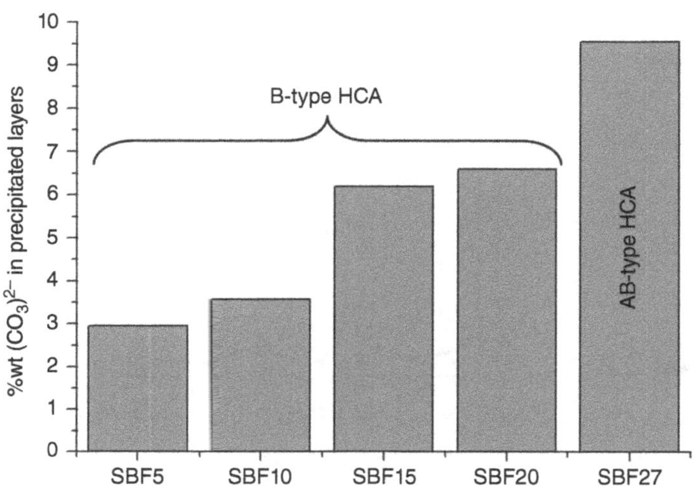

Figure 17.8 Carbonate substitution in biomimetic apatite, depending on the hydrogen carbonate content of SBF.

possible to achieve an AB-type substitution, which is also present in human bone tissue.

17.3.2 Bone-inspired Hybrids

The other important component of bone is the protein collagen type I. Collagen is responsible for bone's ductility, damage tolerance, and bending strength [14, 25]. Also, it serves as a scaffold for the mineralization of the apatite crystals. Although bone formation has been the focus of research for several decades, it is still controversially debated whether the exact mineralization mechanism is ion-based or induced by amorphous precursors [26]. Recently, bioinspired *in vitro* experiments have shown that calcium phosphate clusters and organic macromolecules present in a mineralization solution can form stable mineral droplets that enter the collagen fibrils and diffuse through the interior of the fibril, where they finally transform into oriented apatite crystals [27]. Numerous investigations were performed to mimic the mineralization of collagen [26] by using mineralization solutions that contain calcium and phosphate ions (e.g. SBF) [28], or by performing a direct nucleation of HA on self-assembled collagen fibers [29, 30]. Other approaches used coating technologies, such as the electrolytic deposition to prepare collagen/apatite composites by coating collagen fibers with hydroxyl apatite crystals [31].

Collagen products for biomedical applications are mainly of animal origin. Their processing is quite complex and comparatively expensive. Also, there might be some concerns regarding potential infection risks. Therefore, it is worth looking at alternatives that nature has to offer. Cellulose is the world's most abundant natural, renewable, and biodegradable polymer. Regenerated cellulose can be fabricated into a wide range of shapes with adjustable interconnecting porosity. Their good matching with the mechanical properties of bone, as well as their biocompatibility, have been demonstrated [32, 33]. Regenerated cellulose with a low degree of crystallinity is degradable, and the degree of absorbance by tissue can be controlled by modifying the structure of cellulose [34]. Nevertheless, a complete osseo-integration could not be observed after femoral implantation of regenerated cellulose hydrogels. Therefore, it would be beneficial to suggest a simple treatment that could induce apatite formation on the surface of cellulose structures when exposed to physiologic conditions. For this purpose, the biomimetic precipitation of CaP phases was studied on cellulose thin films prepared by the Langmuir–Blodgett technique using a trimethylsilyl cellulose derivative [35]. It was shown that the chemical pretreatment of cellulose with supersaturated $Ca(OH)_2$ can induce CaP formation after soaking in 1.5*SBF. The procedure was then transformed to regenerated cellulose fabrics (Lyocell®). The cellulose fibers were homogeneously covered with a 30-μm-thick bone-like CHA layer after soaking in 1.5*SBF for 21 days (Figure 17.9a) [36]. The biocompatibility of activated as well as coated cellulose fabrics thus was demonstrated [37]. Thus, chemically pretreated cellulose fabrics with adjustable porosity are considered suitable to build up novel scaffold architectures for tissue engineering.

To accelerate the biomimetic formation of bone-like apatite on cellulose templates, a SBF solution with a high degree of supersaturation (5*SBF) was used [38]. After generating CaP nuclei on the cellulose fibers in 5*SBF with high Mg^{2+} and HCO^{3-} concentrations, the cellulose templates were immersed in a modified 5*M-SBF which was optimized with respect to crystal growth kinetics by reduced Mg^{2+} and HCO^{3-} concentrations. After 48 hours a CHA layer with a thickness of 20 μm was obtained

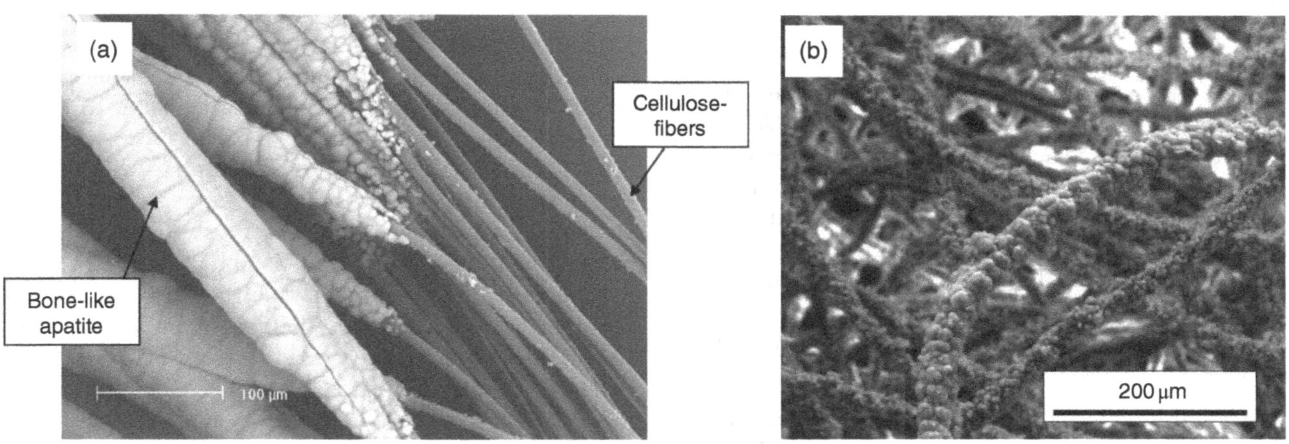

Figure 17.9 Apatite formation on (a) chemically pre-treated cellulose fibers after exposure to 1.5*SBF; and (b) untreated cellulose fibers after exposure to 5*SBF solutions.

(Figure 17.9b). The carbonate content of the CHA layer increased from 4 mass-% after 4 h of exposure in 5*M-SBF to 8 mass-% after 72 h [38].

17.4 Bone-like Scaffolds

In bone tissue engineering and bone regeneration, highly porous scaffold materials provide a pathway for cell attachment, bone in-growth, and vascularization. To improve osteoconduction and integration within the living bone, material features like porosity, pore size, pore geometry, and pore connectivity have to be controlled in a suitable range. Micropores (1–10 μm) have to promote the rapid hydration of the implant. Pore sizes should be in the range of 100 μm for the in-growth of living cells and in the range of 500 μm for vascularization. Typically, macroporous osteoconductive bioceramics (e.g. HA, β-TCP) are supplied commercially for bone grafting. The pore geometry of such scaffolds is usually derived from porogens burnt out during sintering [39, 40], by a slurry foaming process [41, 42], or by the conversion of natural materials, such as coral or bovine cancellous bone [43–45]. However, problems that often occur are mainly related to the inadequate mechanical properties of the resulting scaffolds that do not match the intended site for implantation and handling, or to an insufficient interconnectivity of the porous network that does not favor tissue integration and vascularization [46]. Also, it is still a challenge to prepare complex porous structures with different hierarchical levels of porosity by these conventional methods.

17.4.1 Additive Manufacturing

Additive manufacturing technologies provide a promising alternative to overcome the above-mentioned problems. They offer the possibility of building up complex geometries by a bottom-up approach that uses layer-by-layer methods. The way in which the layers are produced and the materials in which parts can be built vary significantly between the different RP processes [47]. The starting points for all methods, however, are either 3D CAD models or data from medical applications, such as an MRI scan. The latter is of specific interest to prepare patient-specific bone grafts and implants for cranio- and maxillofacial reconstructions [48–50]. The original data is then reformatted into an STL file and sliced horizontally to generate each layer. The manufacturing technologies to build up each layer are based on five significantly different processing principles.

1) *Stereolithography*: A photo-curable monomer is exposed to a selective UV light source (laser beam or lamp/mask) in order to harden the polymer (only suitable for specific polymers).
2) *Laminated object manufacturing*: Sheets of a material are cut to shape and stacked on top of each other (mainly applied for paper but generally suitable for all classes of materials).
3) *Fused deposition modeling*: A polymer melt is extruded and deposited as a hot filament (for polymer-based materials).
4) *Selective sintering/selective melting*: A powdered material is sintered/melted together using a locally controllable heat source (laser beam or electron beam) (all classes of materials with restrictions in the case of ceramics).
5) *3D printing*: A liquid binder is deposited onto a powdered material to glue the powder together (all classes of materials).

An overview of bone tissue engineering using additive manufacturing technologies has recently been provided by Bose et al. [51].

Figure 17.10 illustrates the processing principle of powder-based methods using the example of the selective electron beam melting (SEBM) and 3D printing (3DP). Both processes start with the homogeneous application of a powder layer on a process platform using a doctor blade or a roller. Then, an electron beam, in the case of SEBM or a binder in the case of 3DP, scans the powder layer and creates a cross-section of the part by fusing the powder particles. Subsequently, the process platform is lowered by the thickness of one layer, a new powder layer is applied and the process is repeated until the whole part has been built.

SEBM was used to prepare cellular structures of Ti alloys – a diamond-like structure with a porosity around 80% and a hatched structure with a porosity around 60% (Figure 17.11) [52]. Comparing the mechanical properties of these structures with those of human bone, the diamond structure exhibits properties comparable with trabecular bone, whereas the hatched structure's properties are closer to those of cortical bone. The comparison also shows that the elastic moduli of bone and of cellular Ti–6Al–4V processed by SEBM are in a similar range, which should minimize stress-shielding effects of the biomaterial after implantation into bony defects.

The 3DP of CaP cements allows the preparation of complex geometries with defined open porosities by using the setting reaction of a reactive CaP powder and a H_3PO_4 binder (Figure 17.12). The resolution of this process is around 100 μm in the x, y, and z-direction [53]. In addition to the printed porosity, the cements show an

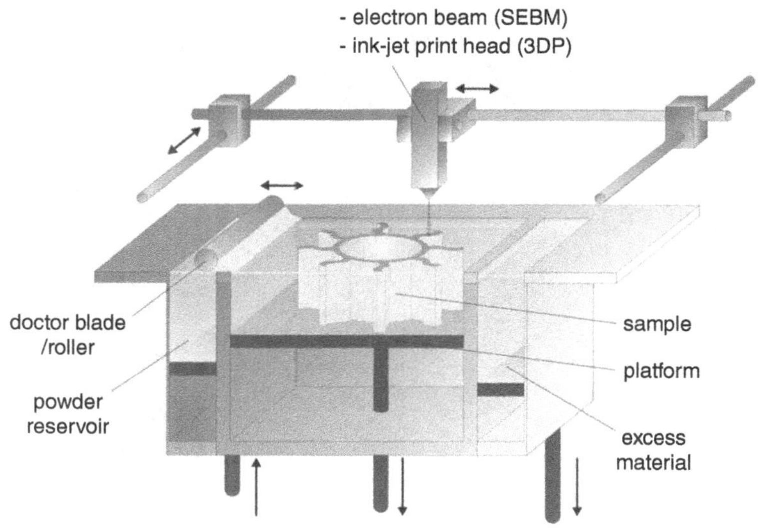

Figure 17.10 Schematic illustration of the SEBM and 3D printing process.

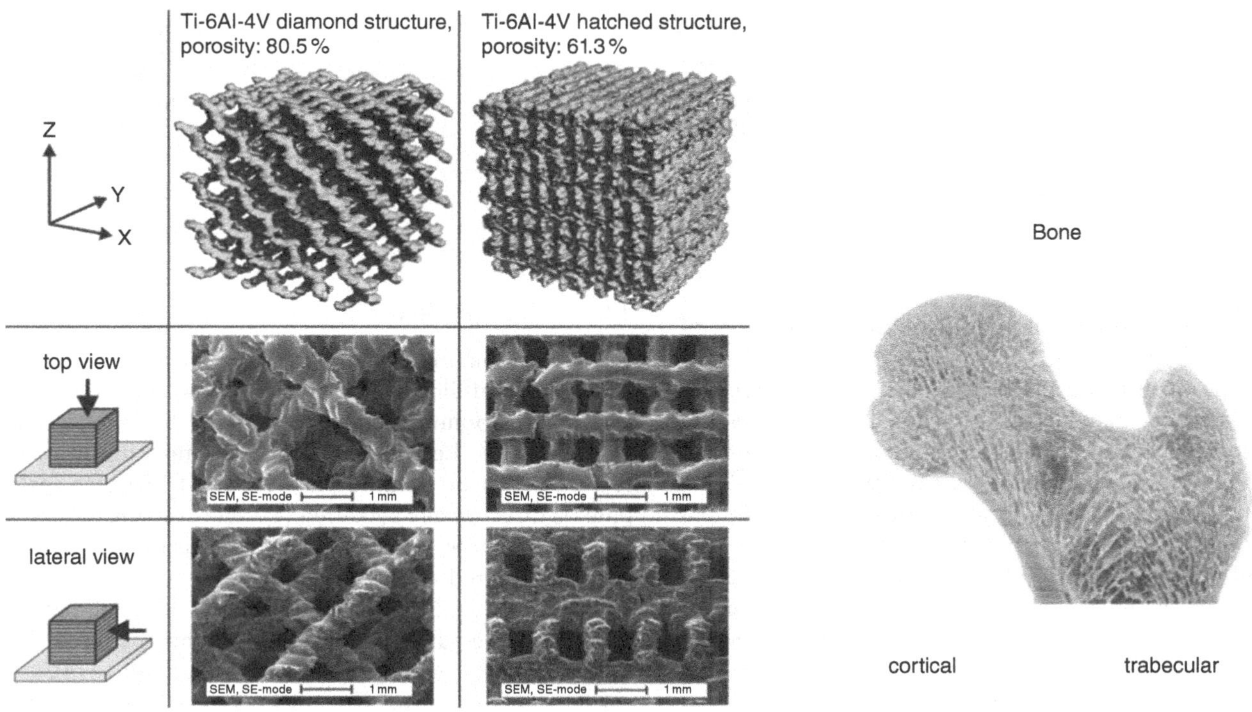

	Ti-6Al-4V diamond structure	Ti-6Al-4V hatched structure	cortical	trabecular
E ∥	1.6 GPa	12.9 GPa	18.2 GPa	1.1 GPa
E ⊥	0.9 GPa	3.9 GPa	11.7 GPa	
σ ∥	29.3 MPa	148.4 MPa	205 MPa	25.0 MPa
σ ⊥	21.0 MPa	127.1 MPa	137 MPa	

Figure 17.11 Mechanical properties of bone in comparison to titanium scaffolds prepared by SEBM. *Source:* Adapted from [52].

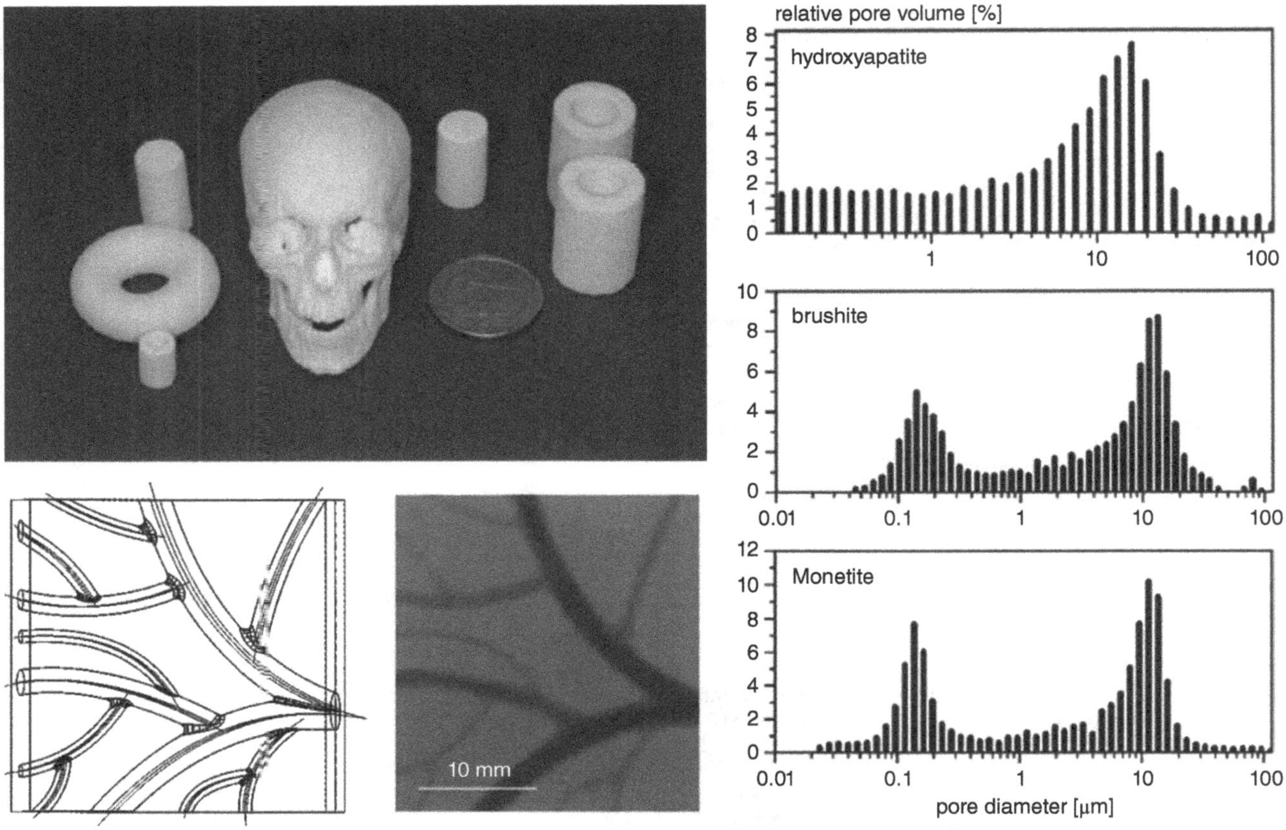

Figure 17.12 3D printed CaP cements. *Source:* Adapted from [53, 54].

intrinsic bimodal porosity with maxima in the range of 0.1 μm and 10 μm [54].

An important and critical issue in bone regeneration is the vascularization capacity of implants exceeding a size of around 1 cm^3. In order to produce model implants to investigate the spatially localized release of organic (VEGF) and inorganic (Cu) angiogenic factors, a Y-shaped pore channel with one closed and two open ends was printed [55]. Growth factors were placed at the end of the closed pore. Vascularization was examined after implantation into mice for 15 days. Control samples without angiogenic growth factors had only tissue ingrowth of 2 mm. On the other hand, vascular tissue was observed in the main opening and closed pore channels for brushite implants that had been loaded with VEGF or copper sulfate. Intramuscular implantation of brushite and monetite implants with integrated closed pore channels resulted in bone formation within the channels, indicating osteoinductivity of these resorbable materials [56].

17.4.2 Ice Templating

The ice-templating process provides another possibility to prepare scaffolds featuring an open and aligned porosity similar to bone structures [57]. Figure 17.13 schematically illustrates the principle of the ice-templating process. A suspension of ceramic particles (red) in water (pale blue) is directionally solidified (1). The growing ice crystal lamellae (gray) repel and enrich the particles by forming a structured frozen body with aligned ice crystals surrounded by particle-rich lamellae (2). Ice sublimation at reduced pressure leads to a green ceramic scaffold (3) that is densified by sintering at high temperatures (4).

Ice-templated structures show similarities with the hierarchical hybrid structure of several biominerals like nacre, trabecular bone, and antler (Figure 17.14). Therefore, ice templating seems to be a key to the design of bioinspired hierarchical materials. Ice-templated scaffolds consisting of HA have demonstrated their suitability to serve as a bone replacement material because they exhibit unusually high compressive strengths up to 65 MPa for 56% open porosity [58, 59].

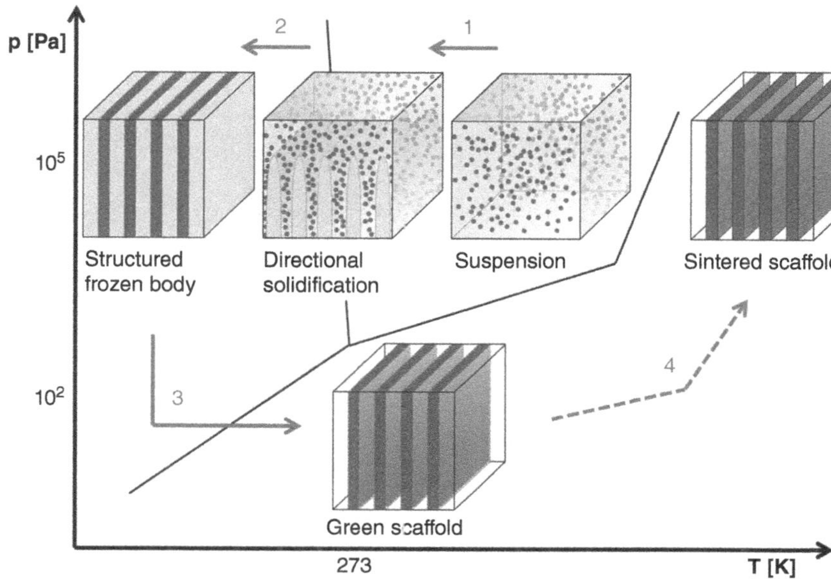

Figure 17.13 Scheme of the ice-templating process.

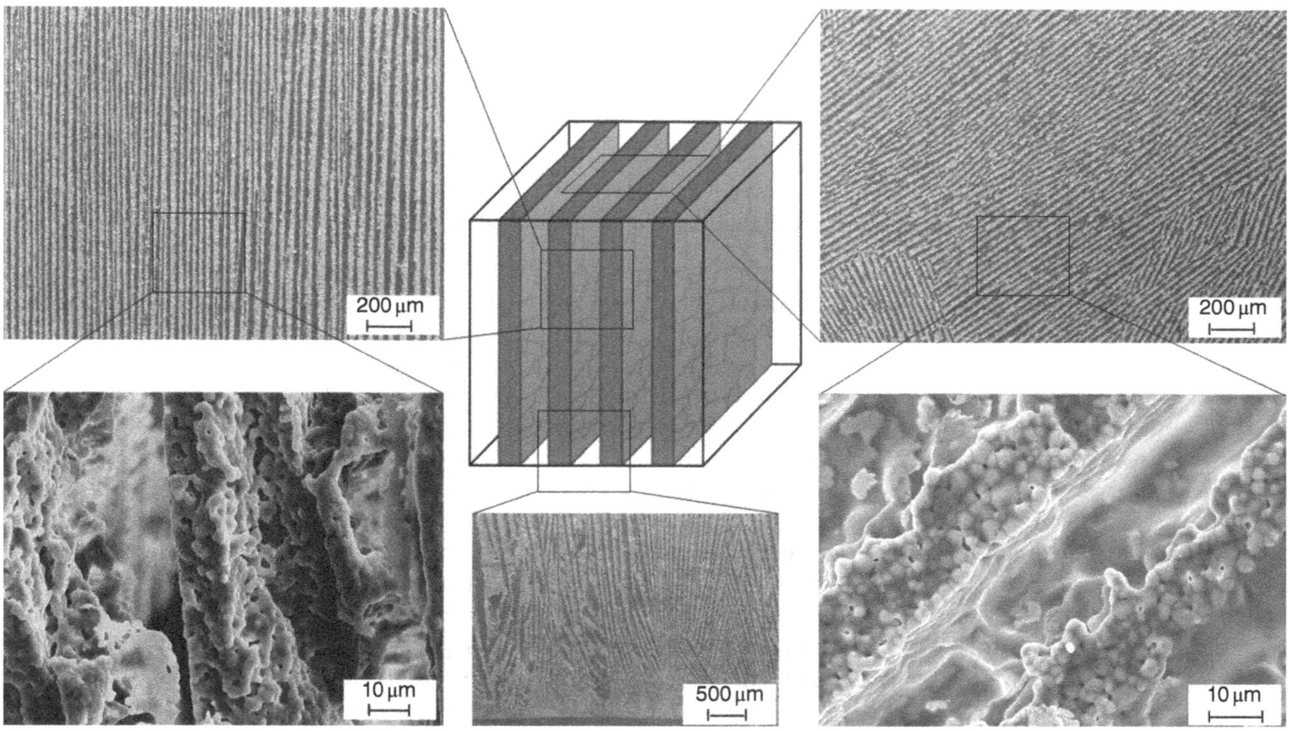

Figure 17.14 SEM micrographs of the morphology of an ice-templated β-TCP scaffold prepared with a 20 vol-% initial solid content and a 20 µm s^{-1} ice front velocity [60].

A general analytical solution of the heat conduction equation was proposed to control structural sizes over the whole sample height during the ice templating of β-TCP suspensions [60]. With this novel approach it is possible to obtain constant solidification front velocities and hence constant structural sizes at high solidification velocities and/or long processing times [61]. β-TCP is a bioresorbable and osteoconductive bioceramic that represents an interesting material to generate porous scaffolds for bone regeneration. It was found that the structural sizes and mechanical properties of sintered β-TCP scaffolds depend on the solid content of the suspensions and on the applied

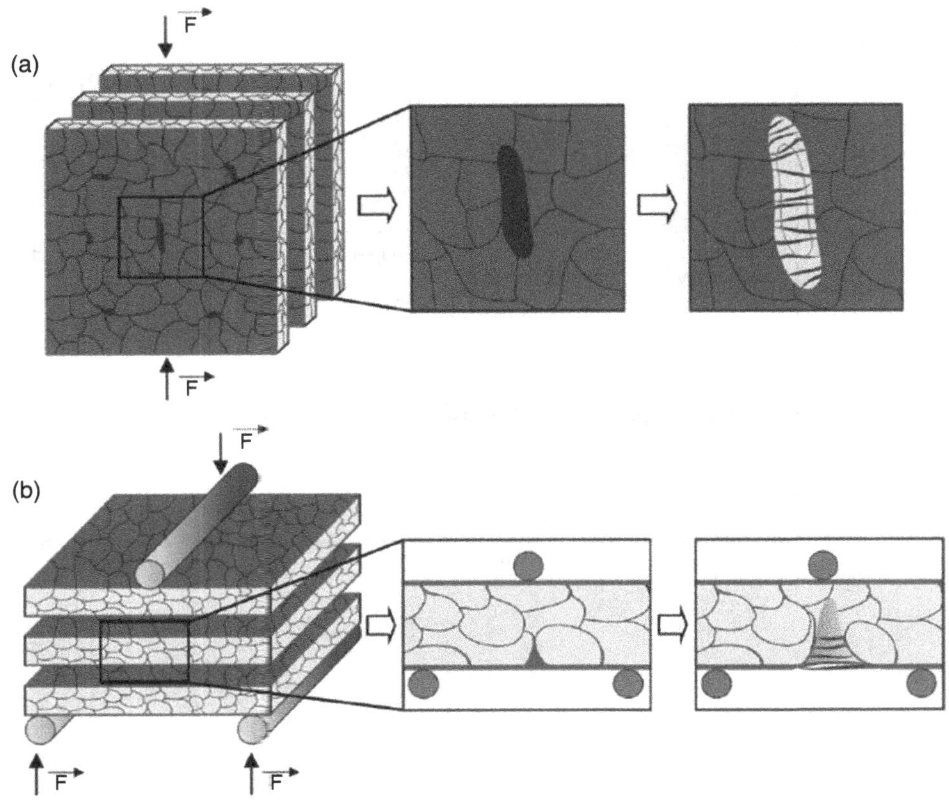

Figure 17.15 Scheme of the failure mechanism of a composite scaffolds under (a) bending mode and (b) compression mode [62].

ice front velocity [60]. With increasing solid content, the porosity of β-TCP scaffolds decreases from 82% for 10 vol-% suspensions to 49% for 30 vol-% suspensions. This porosity is independent of the applied ice front velocity. For a given porosity, the onset ice front velocity influences the pore widths and ceramic lamella thicknesses. With increasing velocity, the structural sizes decrease. Simultaneously, the compression strength of scaffolds increases by more than double due to the reduced amount of critical size defects. The highest strength of 40 MPa was obtained for scaffolds with a porosity of 49% at a high solidification velocity of 30 μm s^{-1}. Pore widths of up to 40 μm were obtained for scaffolds with an 82% porosity prepared at a low solidification velocity of 10 μm s^{-1} [61].

Sintered β-TCP scaffolds were impregnated with the biodegradable biopolymer poly(ε-caprolactone) (PCL) to prepare organic/inorganic hybrids and to investigate the influence of the impregnation on the mechanical properties of the porous composites [62]. Only a small amount of polymer was used to solely infiltrate the micro pores and to coat the ceramic lamellae with a thin polymer film. The macroporosity of the scaffolds is mostly preserved during the solution-mediated PCL impregnation The infiltrated samples show a damage-tolerant non-catastrophic failure mechanism which can be explained by the formation of cold drawn PCL fibers bridging micro- and macrocracks (Figure 17.15). PCL (blue) present on the surface and inside surface flaws and micropores of the β-TCP lamellae (yellow) forms fibers during crack extension. The PCL is cold-drawn to fibers during the crack growth and crack extension. The cold-drawn fibers with increased strength and modulus of elasticity bridge the crack and transfer load across the crack. Fibrils also glue the scaffold together even after the failure of the ceramic lamellae. Moreover, the compressive and flexural strength of the impregnated samples with high porosity increases by approximately 100% due to the strain hardening of the PCL fibers [62]. The PCL impregnation of ice-templated β-TCP scaffolds seems to be a promising pathway to produce bioresorbable bone replacement materials with adjustable porosities, pore sizes, and mechanical properties. Because of their damage tolerance, these materials could be of particular interest to the design of novel bioinspired implant architectures for load-bearing applications.

The ice-templating process is not only limited to suspension systems but can also be applied to solutions of polymers such as agar [63], alginate [64], polyvinylalcohol [65], or chitosan/gelantine [66]. The ice-templating process was successfully applied to water-based cellulose solutions.

Porous scaffolds consisting of regenerated cellulose have been prepared [67]. The open pores show a columnar morphology and a preferred alignment in the direction of the heat flow. However, in contrast to ceramic suspensions, the pore sizes of cellulose scaffolds can only be slightly adjusted by the processing conditions.

17.5 Conclusion

It can be predicted that the use of additive manufacturing technologies to prepare bone-inspired biomaterials will further increase in the coming years, due to their ability to fabricate customized medical devices that can be tailored to patient-specific and defect-specific needs. Future challenges are mainly related to the combination of different materials in one single process to build up functional hybrid materials mimicking the composition of bone. Also, the generation of smart hierarchical structures is still challenging because most additive manufacturing technologies are not suitable for generating sub-micrometer structures that are required to mimic the features of the natural extracellular matrix (ECM). Here, the combination of additive manufacturing with conventional methods, such as electrospinning, ice templating, or biomineralization, opens new possibilities to generate bioinspired 3D structures with unrivalled properties and multifunctionality [68].

Another future trend is the biofabrication of scaffolds that directly integrate living cells and growth factors to print tissues and organs [69]. Currently, the materials being used are selected mainly because of their biocompatibility or processability. This limits the range of usable materials and could be overcome by new molecular design strategies to synthesize 3D printable hydrogels [70]. Beyond just mimicking surface functional groups to optimize cell attachment, biomechanics (stiffness) and cell in-growth (porosity) of hydrogels have to be adjusted to control the proliferation and differentiation of cells in the scaffold. For this purpose viscoelastic hydrogels with a tunable stress relaxation behavior similar to ECM might be the key to designing a new generation of bioinspired materials that promote bone regeneration [71].

References

1 Weiner, S., Wagner, H.D. (1998) *Ann. Rev. Mater. Sci.*, **28**, 271.
2 Weiner, S., Nudelman, F., Sone, E. et al. (2006) *Biointerphases*, **1**, P12.
3 Fratzl, P., Weinkamer, R. (2007) *Prog. Mater. Sci.*, **52**, 1263.
4 Rho, J. Y., Kuhn-Spearing, L., Zioupos, P. (1998) *Med. Eng. Phys.*, **20**, 92.
5 Currey, J.D. (2005) *Science*, **309**, 253.
6 Wolff, J. (2010) *Das Gesetz der Transformation der Knochen – 1892*, Reprint: Pro Business, Berlin, Germany.
7 Wopenka, B., Pasteris, J.D. (2005) *Mater. Sci. Eng., C*, **25**, 131.
8 Schmidt, W.J. (1936) *Naturwissenschaften*, **24**, 361.
9 Nightingale, J.P., Lewis, P. (1971) *Nature*, **232**, 334.
10 White, S.W., Hulmes, D.J.S. Miller, A., Timmins, P.A. (1977) *Nature*, **266**, 421.
11 Dorozhkin, S.V., Epple, M. (2002) *Angew. Chem. Int. Ed.*, **41**, 3130.
12 Vallet-Regí, M., González-Calbet, J.M. (2004) *Prog. Sol. State Chem.*, **32**, 1.
13 Landi, E., Tampieri, A., Celotti, G. (2005) *Biomaterials*, **26**, 2835.
14 Currey, J.D. (2002) *Bones: Structure and Mechanics*, Princeton University Press, Princeton, NJ.
15 Elliot, J.C. (1994) *Structure and Chemistry of the Apatites and Other Calcium Orthophosphates*, Elsevier Press, Amsterdam.
16 Hench, L.L. (1998) *J. Am. Ceram. Soc.*, **81**, 1705.
17 Kokubo, T. (1991) *Biomaterials*, **12**, 155.
18 Dey, A., Bomans, P.H.H., Müller, F.A. (2010) *Nat. Mater.*, **9**, 1010.
19 Jonášová, L., Müller, F.A., Helebrant, A. (2004) *Biomaterials*, **25**, 1187.
20 Nakano, T., Kaibara, K., Tabata, Y. (2002) *Bone*, **31**, 479.
21 Müller, F.A., Müller, L., Caillard, D., Conforto, E. (2007) *J. Cryst. Growth*, **304**, 464.
22 Müller, L., Müller, F.A., (2006) *Acta Biomater.*, **2**, 181.
23 Kokubo, T., Takadama, H. (2006) *Biomaterials*, **27**, 2907.
24 Jalota, S., Bhaduri, S.B., Tas, A C. (2006) *J. Mater. Sci.: Mater. Med.*, **17**, 697.
25 Gao, H., Ji, B., Jäger, I.L. et al. (2003) *Proc. Nat. Acad. Sci.*, **100**, 5597.
26 Olszta, M.J., Cheng, X., Jee, S.S. et al. (2007) *Mater. Sci. Eng. R.*, **58**, 77.
27 Nudelman, F., Pieterse, K., George, A. et al. (2010) *Nat. Mater.*, **9**, 1004.
28 Scharnweber, D., Born, R., Flade, K. (2004) *Biomaterials*, **25**, 2371.
29 Rovieri, N., Falini, G., Sidoti, M.C. et al. (2003) *Mater. Sci. Eng. C*, **23**, 441.
30 Tampieri, A., Sandri, M., Landi, E. et al. (2008) *Adv. Appl. Ceram.*, **107**, 298.
31 Fan, Y., Duan, K., Wang, R. (2005) *Biomaterials*, **26**, 1623.
32 Chauveaux, D., Barbie, C., Barthe, X. et al. (1990) *Clin. Mater.* **5**, 215.

33 Poustis, J., Baquey, C., Chauveaux, D. (1994) *Clin. Mater.*, **16**, 119.
34 Miyamoto, T., Takahashi, S., Ito, H. et al. (1989) *J. Biomed. Mater. Res.*, **23**, 125.
35 Cromme, P., Zollfrank, C., Müller, L. et al. (2007) *Mater. Sci. Eng. C*, **27**, 1.
36 Müller, F.A., Jonášová, L., Cromme, P. et al. (2004) *Key Eng. Mater.*, **254–256**, 1111.
37 Müller, F.A., Müller, L., Hofmann, I. et al. (2006) *Biomaterials* **27**, 3955.
38 Hofmann, I.F., Müller, L., Greil, P., Müller, F. (2006) *Surf. Coat. Technol.*, **201**, 2392.
39 Kim, H.W., Lee, S.Y., Bae, C.J. et al. (2003) *Biomaterials*, **24**, 3277.
40 Descamps, M., Duhoo, T., Monchau, F. et al. (2008) *J. Eur. Ceram. Soc.*, **28**, 149.
41 Sepulveda, P. (1997) *Am. Ceram. Soc. Bull.*, **76**, 61.
42 Studart, A.R., Gonzenbach, U.T., Tervoort, E., Gauckler, L.J. (2006) *J. Am. Ceram. Soc.*, **89**, 1771.
43 Weber, J.N., White, E.W., Lebiedzik, J. (1971) *Nature*, **233**, 337.
44 Roy, D.M., Linnehan, S.K. (1974) *Nature*, **247**, 220.
45 Poumarat, G., Squire, P. (1993) *Biomaterials*, **14**, 337.
46 Hutmacher, D.W. (2001) *J. Biomater. Sci.-Polym. Ed.*, **12**, 107.
47 Upcraft, S., Fletcher, R. (2003) *Assembly Autom.*, **23**, 318.
48 Leong, K.F., Cheah, C.M., Chua, C.K. (2003) *Biomaterials*, **24**, 2363.
49 Hutmacher, D.W., Sittinger, M., Risbud, M.V. (2004) *Trends Biotechnol.*, **22**, 354.
50 Lowmunkong, R., Sohmura, T., Suzuki, Y. et al. (2009) *J. Biomed. Mater. Res. B*, **90**, 531.
51 Bose, S., Vahabzadeh, S., Bandyopadhyay, A. (2013) *Mater. Today*, **16**, 496.
52 Heinl, P., Müller, L., Körner, C. et al. (2008) *Acta Biomater.*, **4**, 1536.
53 Gbureck, U., Hölzel, T., Klammert, U. et al. (2007) *Adv. Funct. Mater.*, **17**, 3940.
54 Gbureck, U., Vorndran, E., Müller, F.A., Barralet, J.E. (2007) *J. Control. Release*, **122**, 173.
55 Gbureck, U., Hölzel, T., Doillon, C. et al. (2007) *Adv. Mater.*, **19**, 795.
56 Habibovic, P., Gbureck, U., Doillon, C.J. et al. (2008) *Biomaterials*, **29**, 944.
57 Deville, S. (2008) *Adv. Eng Mater.*, **10**, 155.
58 Deville, S., Saiz, E., Nalla, R.K., Tomsia, A.P. (2006) *Science*, **311**, 515.
59 Deville, S., Saiz, E., Tomsia, A.P. (2006) *Biomaterials*, **27**, 5480.
60 Flauder, S., Gbureck, U., Müller, F.A. (2014) *Acta Biomater.*, **10**, 5148.
61 Stolze, C., Janoschka, T., Schubert, U.S. et al. (2016) *Adv. Eng Mater.*, **18**, 111.
62 Flauder, S., Sajzew, R., Müller, F.A. (2015) *ACS Appl. Mater. Interfac.*, **7**, 845.
63 Tong, H.M., Noda, I., Gryte, C.C. (1984) *Colloid Polym. Sci.*, **262**, 589.
64 Zmora, S., Glicklis, R., Cohen, S. (2002) *Biomaterials*, **23**, 4087.
65 Zhang, H., Hussain, I., Brust, M. et al. (2005) *Nat. Mater.*, **4**, 787.
66 Meghri, N.W., Donius, A.E., Riblett, B.W. et al. (2010) *Biol. Biomed. Mater.*, **92**, 71.
67 Flauder, S., Heinze, T., Müller, F.A. (2014) *Cellulose*, **21**, 97.
68 Giannitelli, S.M., Mozetic, P., Trombetta, M., Rainer, A. (2015) *Acta Biomater.*, **24**, 1.
69 Murphy, S.V., Atala, A. (2014) *Nat. Biotechnol.*, **32**, 773.
70 Jungst, T., Smolan, W., Schacht, K. et al. (2016) *Chem. Rev.*, DOI 10.1021/acs.chemrev5b00303.
71 Chaudhuri, O., Gu, L., Klumpers, D. et al. (2016) *Nat. Mater.*, DOI 10.1038/NMAT4489.

18

Research Progress in Biomimetic Materials for Human Dental Caries Restoration

*Yazi Wang[1], Fengwei Liu[1], Eric Habib[2], Ruili Wang[1], Xiaoze Jiang[1], X.X. Zhu[2,**], and Meifang Zhu[1,*]*

[1] *State Key Laboratory for Modification of Chemical Fibers and Polymer Materials, College of Materials Science and Engineering, Donghua University, Shanghai, China*
[2] *Département de Chimie, Université de Montréal, Montréal, Québec, Canada*

18.1 Introduction

Tooth decay, also known as dental caries, is the final outcome of many factors such as acidogenic bacteria, saliva, and fermentable carbohydrates. It is the leading cause of oral pain and tooth loss [1]. Among the common restorative materials for dental caries, resin composites possess advantageous esthetics, acceptable biocompatibility, and convenient clinical manipulation. Hence, they have been extensively studied and used [2]. However, composite restorations also have their limitations. Recent reports have shown that secondary caries and bulk fracture were the main causes for the failure of composite restorations [3, 4]. As dentistry and materials science advance, researchers have gained a deeper understanding of the composition and structure of natural teeth, and they have attempted to solve these problems mentioned above using a biomimetic approach. Hydroxyapatite ($Ca_{10}(PO_4)_6(OH)_2$, HA) with a Ca/P ratio of 1.67 is the main component of dental hard tissues [5]. Scientists have tried to use HA as a reinforcing filler to develop biomimetic dental resin composites with high biocompatibility and bioactivity. Considering that the excellent performance of natural dental hard tissues largely depends not only on the high content of HA, but also on the unique hierarchical microstructure, other researchers have focused on the search for novel biomimetic materials, or methods to construct enamel-like microstructure. In this chapter, we review the progress of HA-filled dental resin composites and the biomimetic construction of tooth enamel.

18.2 Tooth Structure

The human tooth consists of four distinct structures: enamel, dentin, pulp, and cementum [6] (Figure 18.1). Enamel is the outermost part of tooth that envelops the crown. It consists of approximately 96% HA, 3% water, and 1% organic material by weight [7]. Natural enamel has extraordinary mechanical and anti-abrasive properties, and it has been proven that these properties are largely determined by its unique hierarchical microstructure [8]. The typical structural units of enamel are prisms, which are the bundles of aligned nanorod-like hydroxyapatite crystals (Figure 18.2). The interfacial regions between prisms are termed interprisms. The main difference between the prisms and the interprisms is the crystal orientation. Crystals in the prisms tend to align, whereas the minerals in the interprisms are less ordered [9–11]. Dentin, the main body of the human tooth, is located below the enamel, consisting of approximately 70% HA, 20% organic material, and 10% water by weight [12]. It plays a supporting role, and absorbs stresses from the enamel, preventing its fracture [6]. The mechanical properties of dentin are related to the microstructure and composition of peritubular and intertubular dentin [13, 14]. Pulp is the innermost section that is covered by the dentin, containing tooth innervation and vascularization [13, 15]. Finally, cementum is also a mineralized hard tissue, which serves to join the root of the tooth to the alveolar bone through the periodontal ligament [16].

* Email: zhumf@dhu.edu.cn
** Email: julian.zhu@umontreal.ca

Bioinspired Materials Science and Engineering, First Edition. Edited by Guang Yang, Lin Xiao, and Lallepak Lamboni.

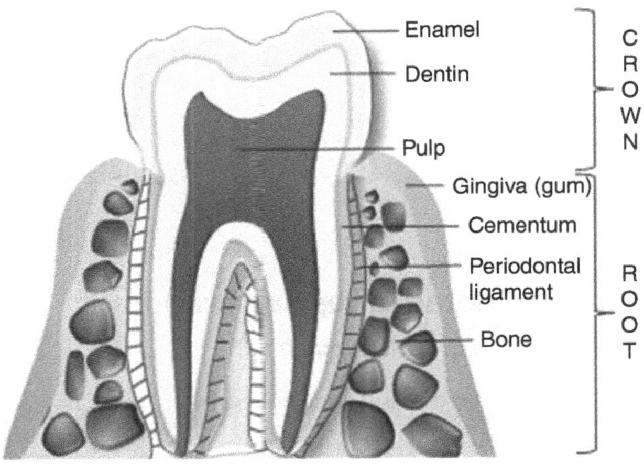

Figure 18.1 The structure diagram of human tooth [17].

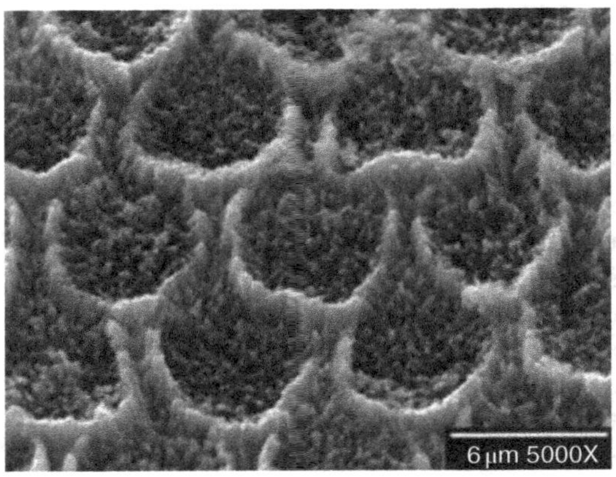

Figure 18.2 SEM micrograph of a demineralized enamel specimen indicates that the prisms and interprisms are comprised of nanorod-like apatites [18].

18.3 The Formation Mechanism of Dental Caries

Dental caries can be subdivided into many categories according to different principles and criteria, but the basic formation mechanism is the same for all types [19]. The caries process begins on the tooth surface, covered by a biofilm (dental plaque) where the bacteria can ferment dietary carbohydrates and produce organic acids [1]. The acids diffuse through the biofilm into exposed dental hard tissues, and then dissociate to generate hydrogen cations. These protons can dissolve the minerals, and cause a loss of minerals from the hard tissue by freeing calcium and phosphate ions into the saliva [20]. The process mentioned above is known as demineralization. The opposite process is termed remineralization, a redeposition process where calcium and phosphate ions are deposited back onto the tooth from saliva. The dissolution and redeposition of HA crystals are associated with the value of pH and the concentration of calcium and phosphate ions in saliva, which could be simply described by Eq. 18.1 [7]. When the demineralization continues and can no longer be reversed by remineralization, a dental cavity eventually forms.

$$Ca_{10}(PO_4)_6(OH)_{2(s)} + 8H^+_{(aq)} \rightleftharpoons 10Ca^{2+}_{(aq)} + 6HPO_4^{2-}{}_{(aq)} + 2H_2O \quad (18.1)$$

18.4 HA-filled Biomimetic Resin Composites

An ideal artificial replacement for tooth tissues would be a material with similar physico-chemical and mechanical characteristics to those of the natural substance. From this perspective, synthesized HA is a better choice as a filler to construct dental restorative resin composites, since it is similar in nature to HA in enamel and dentin. Researchers have thus been trying to fabricate HA-filled biomimetic resin composites with superior mechanical characters. These HA-filled bioactive resin composites show promising results for self-healing through HA-induced remineralization on the surface of restorative resin composites. Herein, we introduce recent progress in HA-filled dental restorative resin composites.

18.4.1 Particulate HA as Filler in Dental Restorative Resin Composites

Particulate HA has been studied in dental restorative resin composites. Based on the knowledge of tooth composition, Labella et al. [21] prepared novel HA-filled dental resin composites in 1994. They proved that silanized HA significantly improved flexural strength, diametral tensile strength, and Vickers hardness of Bis-GMA or UDMA-based dental composites, when compared with untreated HA, showing great potential to design biomimetic biomaterials for treating caries.

Santos' investigation [22, 23] showed that flexural modulus of particulate HA-filled resin composites were significantly improved through increasing HA loading, but the mechanical properties were not significantly affected by the surface treatment on HA. The increase of HA loading and silanization contributed to decreasing water absorption and equilibrium uptake. The aggregation and porosity of the HA filler would have adverse effects on water absorption as shown in Figure 18.3. In addition, a dense and continuous calcium phosphate layer was formed on the surface of the

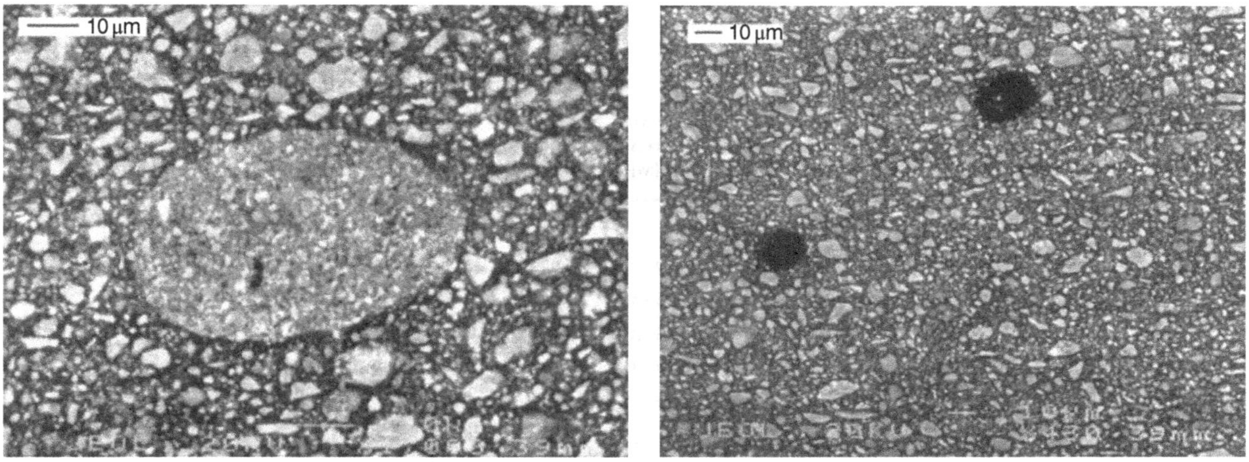

Figure 18.3 SEM micrograph of the cross-section of the dental resin composites incorporating unsilanized HA (left) and silanized HA (right) [22].

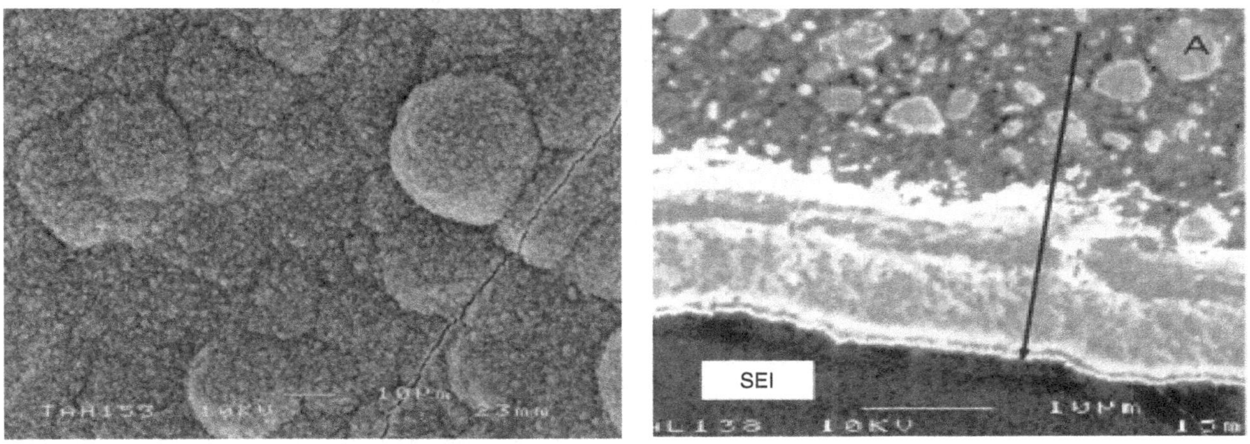

Figure 18.4 SEM images of the surface (left) and secondary electron image of the cross-section (right) of the composites with silanized HA after soaking in SBF for 4 weeks [23].

composite incorporating silanized HA after 4 weeks immersion in simulated body fluid (SBF), which showed good *in vitro* bioactivity and potential self-healing characteristics (Figure 18.4). The remineralization of dental resin composites not only would improve marginal adaptation of teeth restorations, but also would delay bacterial accumulation and penetration, halting a potentially recurring caries-active process [24].

Arcís et al. [25] studied the effect of microscopic and nanoscopic HA particles on the mechanical properties of resin composites. The strength, modulus, and hardness were favored when microscopic HA was used as reinforcing filler. It was indicated that more work would be needed to further improve the dispersion of nanoscopic HA particles and the interaction between particles and the matrix. Similar work was carried out by Domingo et al. [26, 27]. The composites using only HA nanoparticles were unsuitable for clinical use due to their extremely high solubility in water and poor mechanical properties as shown in Table 18.1. However, HA nanoparticles would be expected to fill the voids left by the packing of the HA microparticles, increasing the filler loading, and thus the mechanical properties. The author also pointed out that the higher refractive index of the HA filler was the main disadvantage for photoinitiated polymer, leading to more light scattering. The optical, physical, and mechanical properties could be further improved by changing the crystal size and shape characteristics of the inorganic filler, and the compositions of organic matrix.

This early work investigated the viability of particulate HA fillers as biomimetic dental fillers. In general, dental resin composites with HA particles showed mechanical properties inferior to conventional dental resin composites,

Table 18.1 Evaluation of dental resin composites with microscopic and nanoscopic HA particles [26].

Mat.	HA wt %	Coupling Agent	Young's modulus E (GPa)	Flexural strength σ (MPa)	Vickers hardness No.	Surface roughness Before Ra (μm)	Surface roughness After Ra (μm)	Elution-in-water (wt %)	Water-uptake (wt %)
Series 1: Bis-GMA:TEGDMA(60:40 wt %)									
A	Nano. 50	Sodium citrate	3.9 (0.8)*	26 (6)	38 (6)	3.9 (1.4)	1.7 (0.9)	5.2 (2.1)	7.4 (1.6)
B	Micro. 60	—	4.5 (0.6)	38 (6)	34 (9)	1.5 (0.8)	0.9 (0.5)	1.4 (0.8)	2.1 (1.0)
C	Micro. 60	Hydrosuccinic a.	5.2 (0.8)	25 (3)	55 (10)	1.6 (1.0)	1.5 (0.8)	>10	6.3 (2.1)
D	Micro. 60	Citric a.	5.9 (0.7)	48 (4)	56 (10)	1.0 (0.6)	0.7 (0.4)	0.7 (0.1)	0.9 (1.1)
E	Micro. 60	Acrylic a.	7.3 (1.1)	67 (7)	48 (6)	1.0 (0.5)	0.8 (0.4)	0.3 (0.1)	2.8 (1.5)
F	Micro. 60	Methacrylic a.	5.9 (0.6)	46 (8)	53 (6)	1.8 (0.8)	0.7 (0.3)	0.5 (0.4)	0.9 (0.6)
Series 2: Bis-GMA:HEMA (60:40 wt %)									
A′	Nano. 50	Sodium citrate	2.8 (0.5)	17 (3)	41 (6)	2.0 (0.9)	1.2 (0.4)	>10	>10
B′	Micro. 60	—	4.3 (0.5)	40 (3)	43 (7)	0.7 (0.5)	1.3 (0.7)	2.1 (0.6)	3.1 (1.2)
C′	Micro. 60	Hydrosuccinic a.	3.9 (0.2)	24 (2)	41 (5)	1.1 (0.5)	6.3 (3.1)	>10	7.2 (1.3)
D′	Micro. 60	Citric a.	6.9 (1.2)	52 (7)	39 (8)	0.9 (0.5)	2.4 (0.8)	2.5 (1.1)	2.5 (1.0)
E′	Micro. 60	Acrylic a.	6.2 (0.8)	56 (9)	46 (7)	0.7 (0.4)	3.9 (1.6)	1.7 (1.2)	5.0 (0.3)
F′	Micro. 60	Methacrylic a.	6.2 (1.0)	52 (8)	49 (12)	0.7 (0.2)	0.9 (0.5)	0.6 (0.2)	1.1 (0.7)
Herculite XRV^{MR}: Bis-GMA-based composite									
Filler: 0.6 μm parts., 59% volume			7.9 (1.0)	80 (17)	87 (16)	1.1 (0.6)	1.5 (0.5)	1.6 (0.7)	0.3 (0.1)

Note: *mean value (standard deviations), a. = acid.

severely limiting their development and advancement for clinical applications.

18.4.2 Novel Shapes of HA as Fillers in Dental Restorative Resin Composites

Recently, novel shapes of HA have been developed and utilized, with the aim of enhancing mechanical properties and improving the bioactivity and biocompatibility of dental composites. HA whiskers are typical inorganic whiskers with small diameter and very high strength and modulus due to their perfect crystalline structure. HA nanofibers have a larger aspect ratio than HA whiskers. Compared to particulate HA, HA whiskers and nanofibers have superior reinforcing and toughening effects, and would be very useful to enhance dental resin composites.

Zhang et al. [28] evaluated failure modes under Hertzian indentation, and the behavior in water of Bis-GMA-based resin composites with HA whisker, or nanoscopic HA powder. As the filler loading increased, the decrease of failure load for whisker was smaller than that of the HA particle. For whisker-filled composites, crack propagation was inhibited through whisker's usual toughening mechanisms, indicating more energy was absorbed during failure. HA nanoparticles tended to aggregate in the composite, leaving voids inside and around the aggregates, which further facilitated water infiltration. Comparatively, HA whisker-filled materials had lower water-uptake and solubility than HA nanoparticle-filled materials.

Zhang et al. [29, 30] also investigated the effect of silanization and filler loading of HA whiskers and HA nanoparticles on the mechanical properties of Bis-GMA-based dental resin composites. The reinforcing effect was significantly influenced by filler morphology. HA whiskers showed typical toughening mechanisms, such as crack deflection, crack bridging, pinning, and whisker pullout, and provided better mechanical properties in composites compared with HA nanoparticles. However, the silanization did not have an obvious effect on the strength or modulus of the resin composites. The bioactivity and biocompatibility of the HA whiskers and HA nanoparticle-filled dental resin composites were also studied. The morphology did not cause significant variations in the proliferation and viability of L929 cells, or the apatite-forming ability of the composites in SBF [31]. Such HA whisker-filled resin composites may be promising for load-bearing bioactive restorations.

Chen et al. [32] studied the effect of HA nanofibers on the mechanical properties of resin composites with or without microscopic silica. When the resin composites contained 3 wt% HA nanofibers with 57% microscopic silica or only 10 wt% HA nanofibers, the flexural strength was significantly increased (over 120 MPa). However, HA nanofibers tended to form bundles when increased to an higher filler loading, leading to more mechanical defects, and a decline in strength. A good dispersion of HA nanofibers, and good HA-matrix interfacial properties were critical to exert the reinforcing effect. The HA nanofibers, which have similar chemical composition to tooth minerals, might be a potential biomimetic filler for dental composites, but more investigations are needed on their mechanical properties in simulated oral environment, volume shrinkage, water sorption, biocompatibility, and bioactivity.

In order to improve HA whisker-matrix interfacial properties, Liu et al. [33] synthesized poly(Bis-GMA)-grafted silanized hydroxyapatite whiskers (PGSHW), and investigated the effects of graft polymerization and graft ratio of PGSHW on the physical and mechanical properties of these Bis-GMA-based dental resin composites. It was shown that small amounts of poly(Bis-GMA) grafted onto HW effectively enhanced the HW-matrix interfacial compatibility and bonding, and improved the dispersion of HW in the resin matrix. These in turn increased the mechanical strength and decreased the volume shrinkage and water sorption without affecting *in vitro* bioactivity. Further, Liu et al. [34] used hybrid fillers of PGSHW (surface graft polymerization on HW for 1 h) and nanoscopic silica to solve the remaining problems of resin composites with only PGSHW as a filler, such as lower packing density, higher refractive index of HA, and the resulting lower monomer conversion and mechanical properties. Multimodal hybrid fillers, which combined the advantages of both PGSHW and nanoscopic silica, enhanced the interfacial properties, and increased the filler packing density. PGSHW/silica (mass ratio 2:4) hybrid fillers dramatically improved the flexural strength, flexural modulus, and compressive strength of the resin composite, which were 39.1%, 61.1%, and 50.1% higher than those of the resin composite with only PGSHW as a filler, respectively. Meanwhile, the water absorption of the resulting composites was effectively decreased. Compared to resin composites containing only silica filler, resin composites containing PGSHW had superior apatite-forming ability in SBF as shown in Figure 18.5. These new PGSHW/silica-filled resin composites possessed good mechanical properties, and desirable bioactivity, a combination not available in current HA-filled dental resin composites. They may be promising in reducing the occurrence of both restoration fracture and secondary caries.

Liu et al. [35] fabricated novel urchin-like hydroxyapatite (UHA)-filled dental resin composites, and investigated the reinforcing effect of UHA on the mechanical properties of the resin composites. UHA has a radical structure, combining advantages of both whiskers and spheres (Figure 18.6): it could be well wetted by monomers, leading to a better dispersion in the matrix compared with HA whiskers and HA nanoparticles. The unique claw-like structure was embedded in the resin matrix, increasing the UHA-matrix interfacial contact area and forming a stronger combination. The silanized UHA substantially improved the mechanical properties of the dental resin, and the higher UHA loading (30 wt%) continuously improved the flexural modulus and microhardness without compromising the flexural strength. Besides, impregnation of silanized UHA into dental composites (with silica nanoparticles), the flexural strength, flexural modulus, and compressive strength were further increased by 50.3%, 40.0%, and 13.1%, respectively, in comparison with the composite filled with silica nanoparticles alone.

Overall, these HA fillers with novel morphologies efficiently reinforced the dental resin composite and may be potential candidates as fillers for constructing biomimetic dental resin composites, as opposed to particulate HA. The obtained resin composites possessed significantly enhanced mechanical properties (flexural strength 120–150 MPa, flexural modulus 7–10 GPa, compressive strength 300–400 MPa) and desirable *in vitro* bioactivity. These HA-filled dental resin composites are very promising as strong and bioactive restoration materials for curing caries.

18.4.3 Challenges and Future Developments

Though HA-filled resin composites have been studied for many years, some problems still remain. HA nanoparticles tend to aggregate, and the HA-matrix interfacial interaction and bonding were poor, causing unfavorable influence on physical and mechanical properties. The higher refractive index of HA is another disadvantage for photo-initiated dental resin composites due to the light scattering, leading to a lower photopolymerization efficiency, and a decreased monomer conversion. To resolve these problems, possible approaches would be surface modification on the HA filler, hybrid filling techniques, and more effective mixing methods. To date, most of the work concerning dental resin composites containing HA fillers has focused on the basic mechanical properties, stability (water sorption and solubility) in water, and *in vitro* bioactivity. More work is

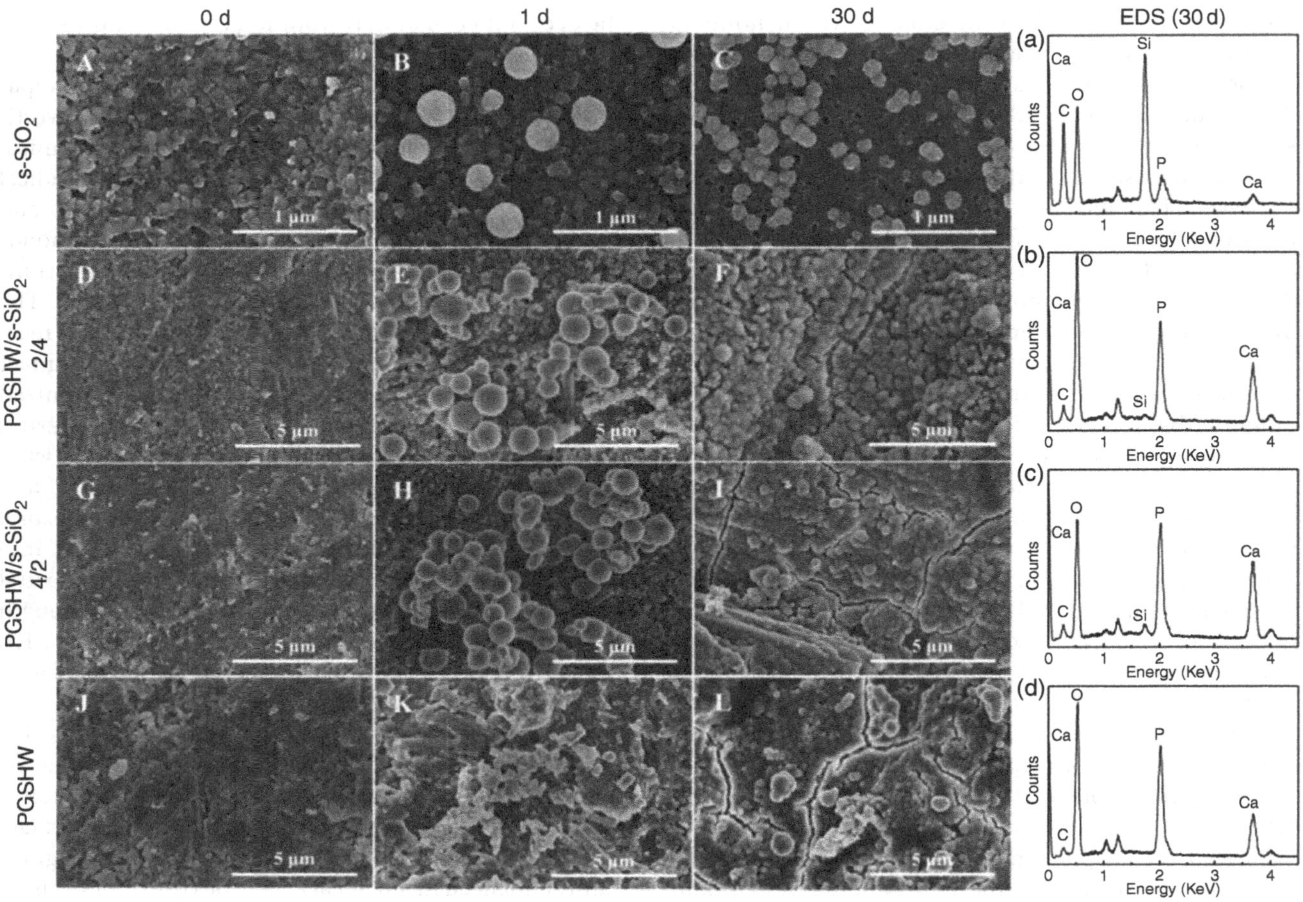

Figure 18.5 SEM images and EDS profiles of the surface of the resin composites with silanized silica (s-SiO_2), PGSHW/s-SiO_2 (2:4, 4:2), and PGSHW after soaking in SBF for 0, 1, and 30 days [34].

needed to systematically evaluate the performance and viability of HA-filled bioactive dental resin composites, such as polymerization shrinkage, shrinkage stress analysis, the remineralization mechanism and its self-healing effect, physical and mechanical properties under simulated oral environment, and *in vivo* bioactivity and biocompatibility.

18.5 Biomimetic Synthesis of Enamel Microstructure

As the outermost layer of the tooth, the demineralization of the enamel is facilitated under the acidic conditions caused by bacterial metabolism and food residue, and its regeneration is extremely difficult once the damage has been done [36]. Current restorative materials for enamel defects cannot recreate the hierarchical microstructure of the enamel, but previous studies have shown that the typical prism structure of enamel contributed greatly to its mechanical properties. Hence, increasing attention has been paid recently to the biomimetic construction of enamel-like microstructures. Some natural or synthetic substances such as amelogenins, peptides, biopolymer gels, dendrimers, and surfactants/chelators can induce the biomimetic mineralization of HA, which are summarized in this section.

18.5.1 Amelogenins-containing Systems

Amelogenins make up more than 90% of the extracellular enamel matrix with molecular masses typically in the range of 20–25 kDa [37]. The sequence of amelogenins comprises three prominent amino acid domains: (1) a hydrophobic N-terminal domain rich in tyrosine; (2) a hydrophobic central region rich in proline; and (3) a hydrophilic C-terminal domain with negatively charged acidic residues [6, 38, 39]. Amelogenins can self-assemble to form nanosphere aggregates due to their characteristic structure: an overall hydrophobic character, but a strikingly hydrophilic C-terminal domain [40].

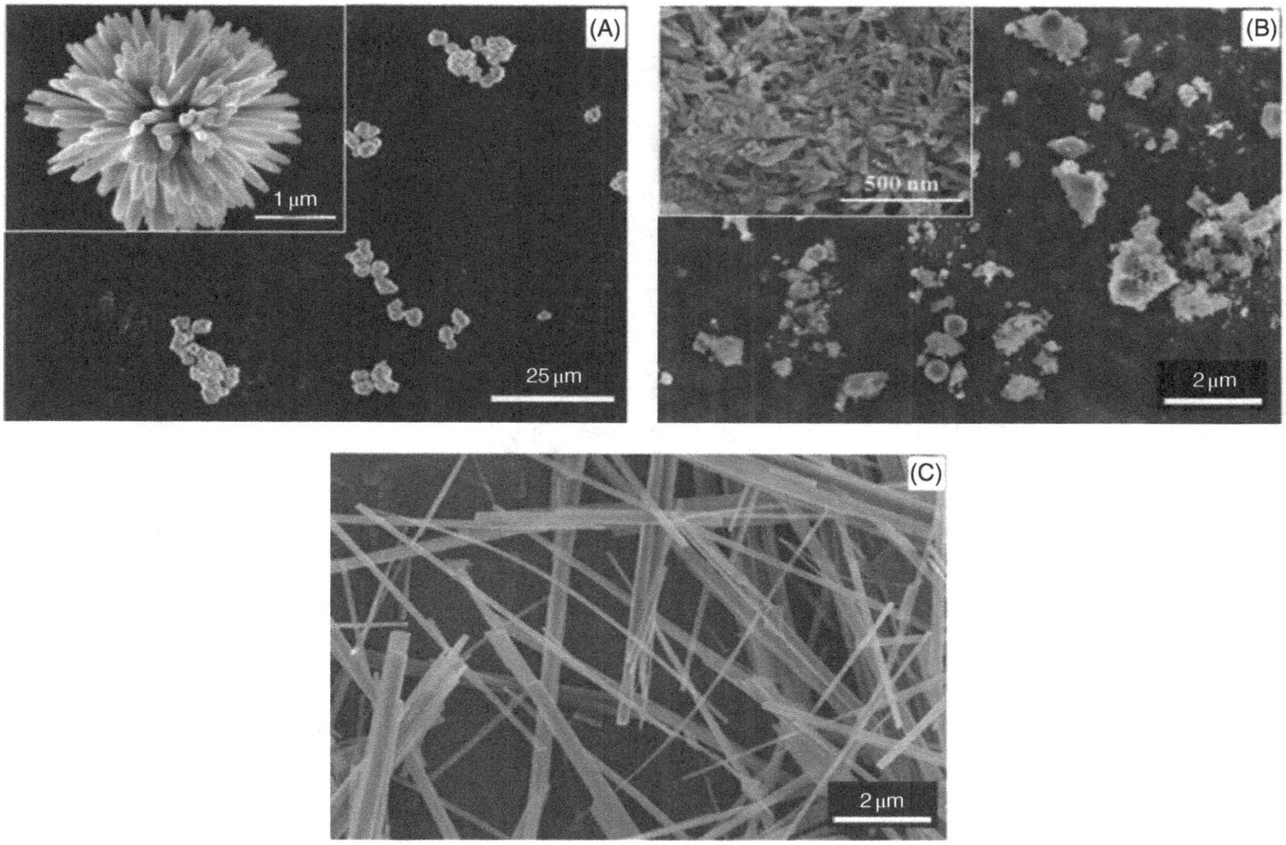

Figure 18.6 Representative SEM images of (A) UHA, (B) irregular particulate hydroxyapatite(IPHA), and (C) HW [35].

They are considered to play a substantial role in controlling the orientation and elongation of enamel crystals *in vivo* [41, 42].The hypothetical functions of amelogenins involved in the formation of enamel mainly include: (1) the prevention of premature intercrystalline fusion; and (2) the control of crystal morphology and subsequent elongation [43, 44]. The early enamel crystals are only 1–2 nm thick and have high surface energy, resulting in a tendency to interact with adjacent crystals [45]. To inhibit premature crystals fusion, the amelogenin nanospheres bind to the initial crystals and space them out without destroying the crystal growth [46].

Having a good understanding of the structure and function of amelogenins can help us to prepare biomaterials with an enamel-like microstructure. Fan et al. [47] fabricated an enamel-mimicking composite coating through an electrolytic deposition (ELD) technique. Parallel bundles of nanocrystals grown on a substrate surface were observed in the presence of recombinant full-length amelogenin rP172, whereas only nano-sized spherical aggregates were observed in the presence of truncated amelogenin rP148 lacking the hydrophilic C-terminal domain. In another modified biomimetic mineralization experiment, Fan et al. [48] confirmed that the concentration of rP172 also had an important effect on the mineral microstructure. Oriented bundles of needle-like fluoridated hydroxyapatite could be formed at a concentration of 33 μg/mL rP172 in the presence of 1 mg/L NaF, whereas no similar growth was observed at a concentration of below 33 μg/mL rP172. Compared with previous amelogenin-containing systems, Ruan et al. [49] created an amelogenin-chitosan system to prepare an organized, enamel-like mineralized layer anchored to natural enamel surface tightly (Figure 18.7). The introduction of the chitosan hydrogel perhaps inhibited the occurrence of secondary caries because of its pH-responsive and antimicrobial properties. However, the hardnesses and elastic modulus of the reconstructed enamel-like layer repaired by the amelogenin-chitosan system were less than the half of those of healthy enamel, which need to be further improved for practical clinic applications.

18.5.2 Peptides-containing Systems

The extraction and purification of amelogenin are relatively difficult [50]. Short peptides could act as effective, low-cost alternative to proteins, which can be produced easily

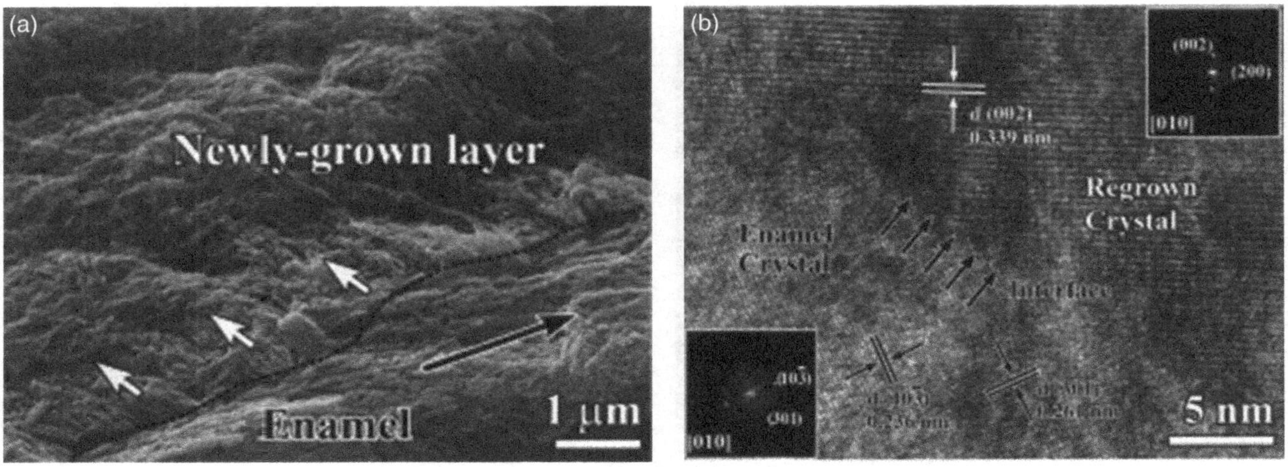

Figure 18.7 (a) Cross-section SEM image of the repaired layer after remineralization in amelogenin–chitosan system for 3 days fused to the surface of the natural enamel. The white and black arrows indicate the crystallographic orientations of the crystals in the newly grown layer and natural enamel, respectively. The dotted line shows the boundary of the natural enamel and the newly grown layer. (b) HR-TEM image of the interface between the enamel and regrown crystal, showing seamless growth of the repaired crystal on the enamel. The black arrows indicate the interface between regrown and enamel crystals. (Inset) FFT images corresponding to enamel and regrown crystals [49].

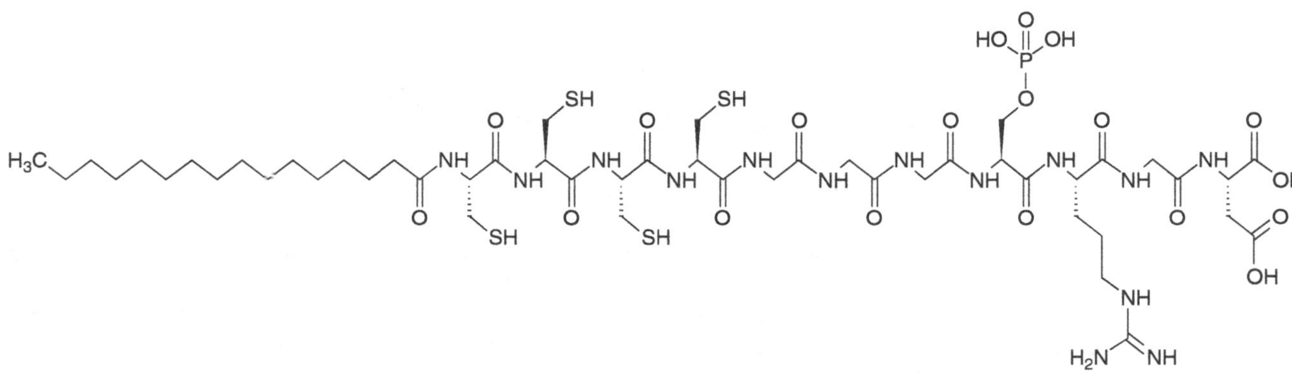

Figure 18.8 Chemical structure of the peptide amphiphiles [54].

by standard solid-phase synthesis [51]. Over the past decade, peptide amphiphiles (PAs) have attracted extensive attention in many fields [52]. PAs are a class of molecules that combine a hydrophobic block, commonly an alkyl chain, with a hydrophilic peptide segment (Figure 18.8). They have a similar structure to amphiphilic surfactants and can self-assemble into a variety of nanostructures. The driving forces contributing to the self-assembly of PAs in water mainly include: hydrophobic interactions from alkyl tails, hydrogen bonding among the peptide segments, and electrostatic repulsions of the charged amino acids. The first two are attractive forces, which could promote the aggregation of PA molecules, whereas electrostatic repulsions tend to separate the PA molecules. These assemblies with specific size and morphology are finally formed when the three forces reach a balance [51]. Hartgerink et al. [53] synthesized PA molecules which could self-assemble into cylindrical micelles in an aqueous environment. The incorporation of the phosphoserine residue into the peptide region could make the self-assembled micelles a highly phosphorylated surface, interacting strongly with calcium ions and leading to local supersaturation, followed by the oriented nucleation of the crystals. The mineralization experiments showed that the PA micelles could nucleate and grow hydroxyapatite on their surfaces, and the c axes of the HA crystals are in parallel with the long axes of the micelles.

Peptides can be designed to respond spontaneously to external stimuli such as pH, ionic strength, temperature, or light. This will be a significant property for practical applications, because these peptides could infiltrate the irregular cavities of the tooth in a monomeric fluid state, then could self-assemble and solidify in the presence of chemical or physical stimuli [55]. Kirkham et al. [56]

developed a type of β-sheet-forming peptide, which could spontaneously form three-dimensional biomimetic scaffolds at pH below 8.0 or in the presence of salt. These assembled peptides were capable of promoting enamel remineralization.

18.5.3 Biopolymer Gel Systems

Biopolymers such as gelatin and agarose in gel form have been widely used in the mineralization process due to their biocompatibility, biodegradability, and bioactivity [57]. Busch [58] used a glycerine-enriched gelatin gel to form an enamel-like fluoroapatite layer on human enamel slices. However, glycerine is not biocompatible enough to apply in the oral environment. The regeneration of enamel prism-like structure without the presence of glycerine was reported by Cao et al. [59] through an agarose hydrogel biomimetic mineralization model. The morphology of the precipitates on the etched enamel surface was characterized after the model was incubated for 2, 4, and 6 days, respectively. With the extension of incubation time, new rod-like crystals on the enamel slice gradually became ordered and closely packed along the c axis. The amount of hydrogel matrix in the spaces between the rod-like crystals also decreased. After 6 days of incubation, the adjacent crystals fused together side by side to form the bundles of prism-like structure due to the strong attraction between them and there was hardly any hydrogel matrix found (Figure 18.9). The c axes of the regenerated crystals were perpendicular to the underlying slice surface. The mean elastic modulus and mean nanohardness of the regenerated enamel after 6 days of incubation were 89 and 3 GPa, respectively, which approximated the values of 90 and 4 GPa of the untreated enamel. Recently, Wu et al. [60] used the agarose hydrogel mineralization system to precipitate hydroxyapatites onto demineralized dentin with the aid of electrophoresis. The hypothesized function of electrophoresis is to promote the diffusion of calcium and phosphate ions into the interior of the

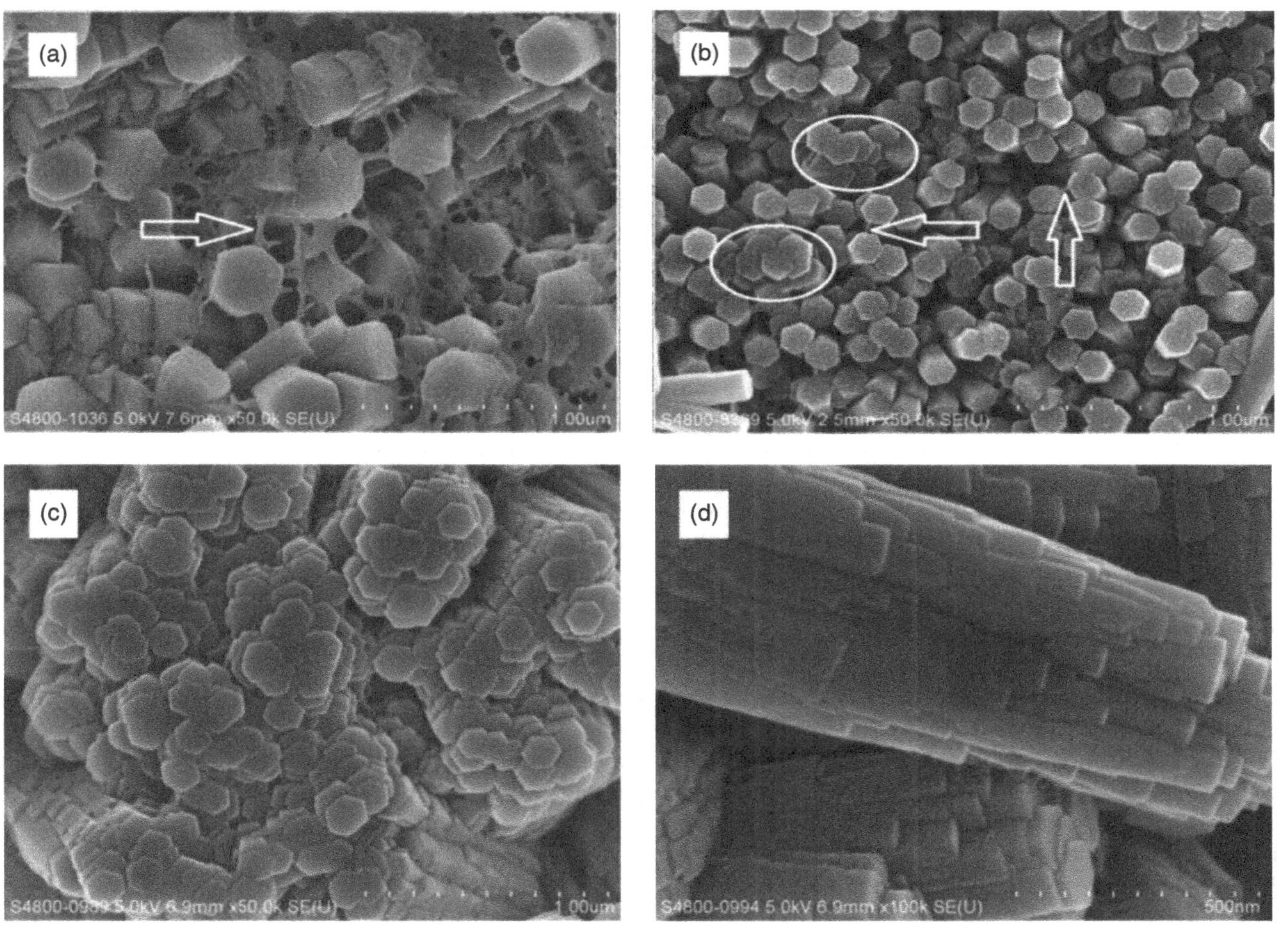

Figure 18.9 SEM images of the regenerated mineralized tissue after 2 days (a), 4 days (b), and 6 days (c, d) of incubation. The arrow indicates hydrogel matrix and the oval indicates the rudiment of the enamel prism-like bundles [59].

demineralized collagen matrix and then accelerate mineral deposition. A simultaneous regeneration of both enamel-like and dentin-like tissues was obtained in this experiment, which showed a promising application in the treatment of dentin defects.

18.5.4 Dendrimers-containing Systems

Dendrimers, some of which were referred to as "artificial proteins," possess monodispersity, well-defined size, and easily modifiable surface groups. Their typical structure has many branches radiating from a central core [61]. As the first synthetic dendrimer, poly(amidoamine) (PAMAM) dendrimer was studied to prepare HA crystals [62]. The size and morphology of HA could be controlled by the PAMAM dendrimer with different surface groups. Zhou et al. [63] synthesized HA under hydrothermal conditions, assisted with carboxylic terminated PAMAM (PAMAM-COOH), and polyhydroxy terminated PAMAM (PAMAM–$C(CH_2OH)_3$). The results revealed that the needle-like HA was obtained in the absence of PAMAM, while the rod-like HA was obtained in the presence of PAMAM. Their aspect ratios followed the order: control > PAMAM-COOH > PAMAM–$C(CH_2OH)_3$. The generations and concentrations of PAMAM dendrimer also affect the growth of crystals. Yan et al. [64] used a hydrothermal technique to synthesize the HA crystals with controlled size and shape, in the presence of an amido terminated PAMAM dendrimer with different generations and concentrations. With the increase of generations of the PAMAM dendrimer from G1.0 to G4.0, the particle sizes of HA decreased from 82 to 38 nm, and the shapes of HA also changed from rod-like to ellipsoid-like. The particle sizes of ellipsoid-like HA were reduced from 55 to 25 nm when the concentration of G4.0 PAMAM dendrimer increased from 5.0 to 40 g/ L. In the oral environment with flowing fluids, the free dendrimer unanchored on the enamel surface could not induce the crystal growth of HA effectively. To increase the binding capability of PAMAM dendrimer onto the substrate surface, Wu et al. [65] designed an alendronate (ALN) conjugated PAMAM-COOH dendrimer (ALN-PAMAM-COOH), of which the ALN moiety acted as an anchor for the dental hard tissue and the -COOH moiety induced *in situ* remineralization of HA. Compared with PAMAM-COOH, ALN-PAMAM-COOH had a stronger binding strength to enamel surface without affecting the oriented growth of HA. On the basis of the above research, Chen et al. [62] further developed a phosphate-terminated PAMAM dendrimer (PAMAM-PO_3H_2) with a similar function to ALN-PAMAM-COOH to control the remineralization of HA, the schematic demonstration is shown in Figure 18.10. Compared with ALN-PAMAM-COOH, the synthesis of PAMAM-PO_3H_2 was relatively simple. The -PO_3H_2 group not only provided strong HA-binding strength, but also had a stronger affinity for calcium ions than the -COOH. A newly regenerated HA layer with prism-like structure was finally formed under the regulation of PAMAM-PO_3H_2.

18.5.5 Surfactants/Chelators-containing Systems

With respect to constructing the enamel prism-like structure through the introduction of surfactants or chelators, Chen et al. [66] synthesized HA nanorods and then modified them with surfactant docusate sodium salt which could allow the nanorods to self-assemble at a water/air interface. A prism-like structure was finally acquired with a length of 400 nm and a cross-section of 100 nm. However, the typical human enamel prism structure is approximately 5 μm in cross-section and 1–2 mm in length. To acquire the size and morphology similar to that of human enamel prism, Chen et al. [67] designed a hydrothermal technology containing chelator ethylenediaminetetraacetic acid calcium disodium salt (EDTA-Ca-Na_2) to directly grow the prism-like fluorapatite layers on metal plates. The prepared EDTA-Ca-Na_2/$NaH_2PO_4 \cdot H_2O$/NaF mixture with various substrates (iron, titanium, mica, glass, silicon plates) was transferred into an autoclave and heated at 121 °C at a pressure of approximately 2 atm for 10 h. The function of EDTA in the above formulation composition was to control the release of Ca^{2+}, and influence the conversion and growth of crystals. The results showed that only the iron plate substrate could form the prism-like structure (5–10 μm in cross-section) consisting of individual crystals (100–300 nm in cross-section), which approximated the size of human enamel prisms. Chen et al. [68] further developed a novel method containing chelator N-(2-hydroxyethyl)ethylene-diamine-N,N',N'-triacetic acid (HEDTA). Compared with the above method, the differences mainly included (1) constructing the prism-like structure under near-physiological conditions (pH 6.0, 37 °C, 1 atm), and (2) growing the fluoridated hydroxyapatite directly on the natural enamel surface, not on the metal plates. The regenerated enamel layer had similar mechanical properties and microstructure to the natural one, and was still firmly bonded to the surface of enamel after ultrasonic treatment (40 kHz, 240 W) for 1 h (Figure 18.11).

18.5.6 Challenges and Future Developments

Much work regarding the biomineralization of HA has been carried out to acquire the enamel-like microstructure. However, the current methods to develop

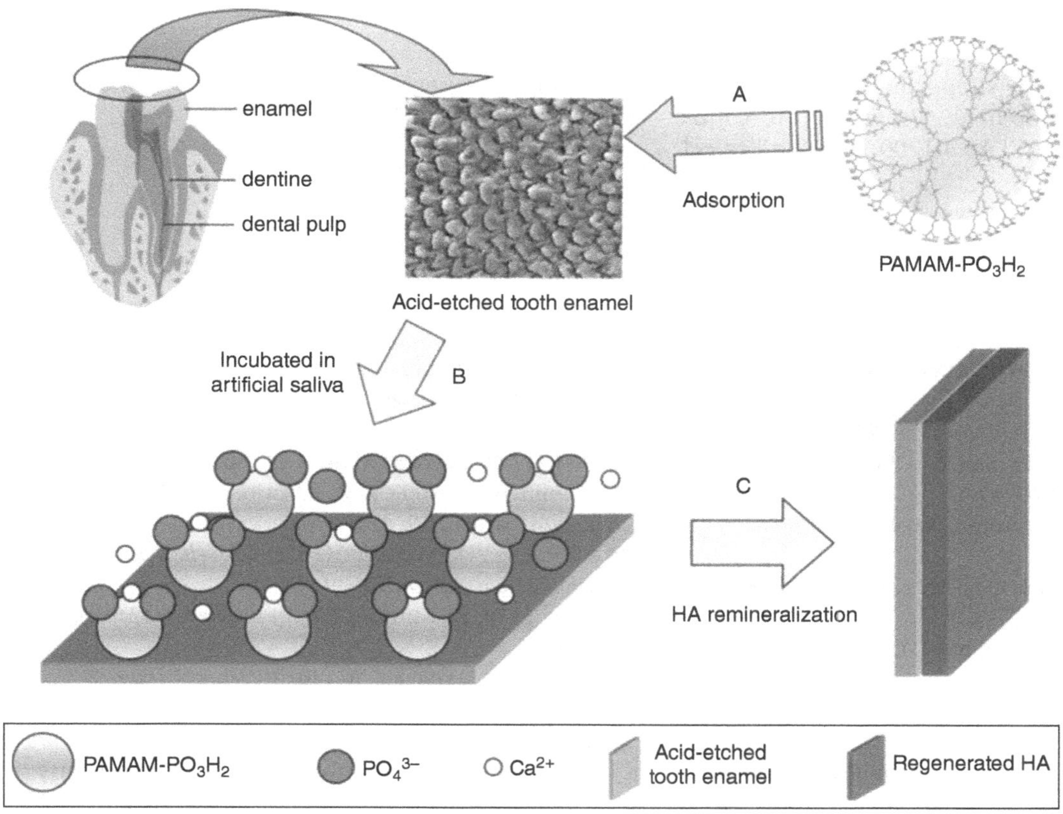

Figure 18.10 Schematic demonstration of remineralization of HA induced by PAMAM-PO_3H_2 [62].

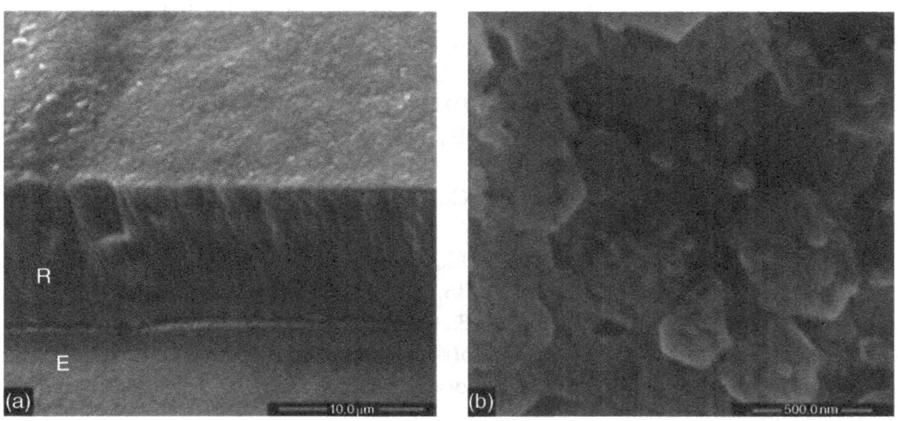

Figure 18.11 SEM images of regenerated layer after ultrasonic treatment. (a) The regenerated layer is firmly bonded to the surface of natural enamel. E stands for enamel and R stands for regenerated layer. Scale bar: 10.0 μm. (b) The regenerated layer is composed of individual crystals. Scale bar: 500.0 nm [68].

well-organized prism patterns were usually complex and required long periods of time, which were not suitable for clinical applications. Much attention was paid to the enamel-like morphology, but the systematic evaluation of performances, particularly the overall mechanical properties, was still insufficient. In addition, more experiments should be done *in vivo*. The future biomimetic materials for curing tooth defects should meet these clinical requirements as much as possible, allowing easy operation, fast restoration, improved mechanical properties, and good biosafety.

Acknowledgments

This work was financially supported by National Key Research and Development Program of China (2016YFA0201702/2016YFA0201700) and the Project of Shanghai International Science and Technology Cooperation Fund (14520710200).

References

1 Selwitz, R.H., Ismail, A.I., Pitts, N.B. (2007) *The Lancet*, **369**, 51–59.
2 Chan, K.H., Mai, Y., Kim, H. et al. (2010) *Materials*. **3**, 1228.
3 Ferracane, J.L. (2011) *Dental Mater.*, **27**, 29–38.
4 Xu, H.H.K., Moreau, J.L., Sun, L., Chow, L.C. (2011) *Dental Mater.*, **27**, 762–769.
5 Kantharia, N., Naik, S., Apte, S. et al. (2014) *Bone*, **34**, 1.71.
6 Palmer, L.C., Newcomb, C.J., Kaltz, S.R. et al. (2008) *Chem Rev.*, **108**, 4754–4783.
7 Li, X., Wang, J., Joiner, A., Chang, J. (2014) *J. Dent.*, **42**, Suppl. 1, S12–S20.
8 Si, Q., An, X., Wang, Z. et al. (2013) *Adv. Mater. Res.*, **658**, 67–72.
9 Cui, F-Z., Ge, J. (2007) *J. Tissue Engin. Regen. Med.*, **1**, 185–191.
10 Moradian-Oldak, J. (2012) *Frontiers Bioscience, Landmark*, **17**, 1996–2023.
11 Tang, R., Wang, L., Orme, C.A. et al. (2004). *Chem.*, **116**, 2751–2755.
12 Chun, K., Choi, H., Lee, J. (2014) *J. Dental Biomech.*, **5**, 1758736014520809-.
13 Angker, L., Swain, M.V. (2006) *J. Mater.Res.*, **21**, 1893–1905.
14 Kinney, J.H., Pople, J.A., Marshall, G.W., Marshall, S.J. (2001) *Calcif Tissue Int.*, **69**, 31–37.
15 Busch, S., Schwarz, U., Kniep, R. (2001) *Chem. Mater.*, **13**, 3260–3271.
16 Ho, S.P., Goodis, H., Balooch, M. et al. (2004) *Biomaterials*, **25**, 4847–4857.
17 Sowmya, S., Bumgardener, J.D., Chennazhi, K.P. et al. (2013) *Prog. Polym. Sci.*, **38**, 1748–1772.
18 Wang, L., Tang, R., Bonstein, T. et al. (2005) *J. Phys. Chem. B*, **109**, 999–1005.
19 Featherstone, J.D.B. (2004) *J. Dental Res.*, **83** Spec No C, C39–C42.
20 Featherstone, J.D.B. (2000) *J. Amer. Dental Assoc.*, **131**, 887–899.
21 Labella, R., Braden, M., Deb, S. (1994) *Biomaterials*, **15**, 1197–1200.
22 Santos, C., Clarke, R.L., Braden, M. et al. (2002) *Biomaterials*, **23**, 1897–1904.
23 Santos, C., Luklinska, Z.B., Clarke, R.L., Davy, K.W.M. (2001) *J. Mater. Sci., Mater. Med.*, **12**, 565–573.
24 Peters, M.C., Bresciani, E., Barata, T.J.E., et al. (2010) *J. Dental Res.*, **89**, 286–291.
25 Arcís, R.W., López-Macipe, A., Toledano, M. et al. (2002) *Dental Mater.*, **18**, 49–57.
26 Domingo, C., Arcís, R.W., López-Macipe, A. et al. (2001) *J. Biomedical Mater. Res.*, **56**, 297–305.
27 Domingo, C., Arcís, R.W., Osorio, E. et al. (2003) *Dental Mater.*, **19**, 478–486.
28 Zhang, H., Darvell, B.W. (2012) *J. Mech. Behav. Biomedical Mater.*, **10**, 39–47.
29 Hongquan, Z., Ming, Z. (2010) *Biomedical Mater.*, **5**, 054106.
30 Zhang, H., Darvell, B.W. (2012) *Dental Mater.*, **28**, 824–830.
31 Zhang, H., Zhang, M., Shen, Y. et al. (2010) Paper presented at 2010 3rd International Conference on Biomedical Engineering and Informatics (BMEI 2010)., Vols 1–7. 1640–1644.
32 Chen, L., Yu, Q., Wang, Y., Li, H. (2011) *Dental Mater.*, **27**, 1187–1195.
33 Liu, F., Wang, R., Cheng, Y. et al. (2013) *Mater. Sci. Engineering, C*, **33**, 4994–5000.
34 Liu, F., Jiang, X., Zhang, Q., Zhu, M. (2014) *Composites Sci. Techn.*, **101**, 86–93.
35 Liu, F., Sun, B., Jiang, X. et al. (2014) *Dental Mater.* **30**, 1358–1368.
36 Li, L., Mao, C., Wang, J. et al. (2011) *Adv. Mater.*, **23**, 4695.
37 Fincham, A.G., MoradianOldak, J. (1996) *Connective Tissue Res.*, **35**, 151–156.
38 Lokappa, S.B., Chandrababu, K.B., Dutta, K. et al. (2015) *Biopolymers*, **103**, 96–108.
39 Ruan, Q., Moradian-Oldak, J. (2015) *J. Mater. Chem. B*. **3**, 3112–3129.
40 Wen, H.B., Fincham, A.G., Moradian-Oldak, J. (2001) *Matrix Biol.*, **20**, 387–395.
41 Kwak, S-Y., Green, S., Wiedemann-Bidlack, F.B. et al. (2011) *Eur. J. Oral Sci.*, **119**, 103–111.
42 Ping-An, F., Conway, J.F., Margolis, H.C. et al. (2011) *Proc. Nat. Acad. Sci. USA*, **108**, 14097–14102.
43 Fincham, A.G., Moradian-Oldak, J., Diekwisch, T.G.H. et al. (1995) *J. Structural Biol.*, **115**, 50–59.
44 Xiudong, Y., Lyun, W., Yueling, Q. et al. (2010) *J.Physical Chem. B*, **114**, 2293–2300.
45 Moradian-Oldak, J. (2001) *Matrix Biol.*, **20**, 293–305.

46 Moradian-Oldak, J., Tan, J., Fincham, A.G. (1998) *Biopolymers*, **46**, 225–238.
47 Fan, Y., Sun, Z., Wang, R. (2007) *Biomaterials*, **28**, 3034–3042.
48 Fan, Y., Sun, Z., Moradian-Oldak, J. (2009) *Biomaterials*, **30**, 478–483.
49 Ruan, Q., Zhang, Y., Yang, X. et al. (2013) *Acta Biomaterialia*, **9**, 7289–7297.
50 Li, Q-L., Ning, T-Y., Cao, Y. et al. (2014) *BMC Biotechnology*, **14**.
51 Cui, H., Webber, M.J., Stupp, S.I. (2010) *Peptide Sci.*, **94**, 1–18.
52 Cavalli, S., Albericio, F., Kros, A. (2010) *Chem. Soc. Rev.*, **39**, 241–263.
53 Hartgerink, J.D., Beniash, E., Stupp, S.I. (2001) *Sci.*, **294**, 1684–1688.
54 Hartgerink, J.D., Zubarev, E.R., Stupp, S.I. (2001) *Curr. Opin. Solid State Mater. Sci.*, **5**, 355–361.
55 Firth, A., Aggeli, A., Burke, J.L. et al. (2006) *Nanomedicine*, **1**, 189–199.
56 Kirkham, J., Firth, A., Vernals, D. et al. (2007) *J. Dental Res.*, **86**, 426–430.
57 Silverman, L., Boskey, A.L. (2004) *Calcif Tissue Int.*, **75**, 494–501.
58 Busch, S. (2004) *Angew. Chem. Int. Ed.*, **43**, 1428–1431.
59 Cao, Y., Mei, M.L., Li, Q-L. et al. (2014) *ACS Appl. Mater. Interf.*, **6**, 410–420.
60 Wu, X-T., Cao, Y., Mei, M.L. et al. (2014) *Crystal Growth Design.*, **14**, 5537–5548.
61 Svenson, S., Tomalia, D.A. (2012) *Adv. Drug Del. Rev.*, **64**, 102–115.
62 Chen, M., Yang, J., Li, J. et al. (2014) *Acta Biomaterialia*, **10**, 4437–4446.
63 Zhou, Z-H., Zhou, P-L., Yang, S-P., et al. (2007) *Mater. Res. Bull.*, **42**, 1611–1618.
64 Yan, S-J., Zhou, Z-H., Zhang, F. et al. (2006) *Mater. Chem. Phys.*, **99**, 164–169.
65 Wu, D., Yang, J., Li, J. et al. (2013) *Biomaterials*, **34**, 5036–5047.
66 Chen, H., Clarkson, B.H., Sun, K., Mansfield, J.F. (2005) *J. Coll. Interf. Sci.*, **288**, 97–103.
67 Chen, H., Tang, Z., Liu, J. et al. (2006) *Adv. Mater.*, **18**, 1846-+.
68 Yin, Y., Yun, S., Fang, J., Chen, H. (2009) *Chemical Comm.*, 5892–5894.

Index

Bioinspired Materials Science and Engineering, First Edition. Edited by Guang Yang, Lin Xiao, and Lallepak Lamboni.

o

p